T0145417

Emergence, Complexity and Computation

Volume 33

The Emergence, Complexity and Computation (ECC) series publishes new developments, advancements and selected topics in the fields of complexity, computation and emergence. The series focuses on all aspects of reality-based computation approaches from an interdisciplinary point of view especially from applied sciences, biology, physics, or chemistry. It presents new ideas and interdisciplinary insight on the mutual intersection of subareas of computation, complexity and emergence and its impact and limits to any computing based on physical limits (thermodynamic and quantum limits, Bremermann's limit, Seth Lloyd limits…) as well as algorithmic limits (Gödel's proof and its impact on calculation, algorithmic complexity, the Chaitin's Omega number and Kolmogorov complexity, non-traditional calculations like Turing machine process and its consequences,…) and limitations arising in artificial intelligence field. The topics are (but not limited to) membrane computing, DNA computing, immune computing, quantum computing, swarm computing, analogic computing, chaos computing and computing on the edge of chaos, computational aspects of dynamics of complex systems (systems with self-organization, multiagent systems, cellular automata, artificial life,…), emergence of complex systems and its computational aspects, and agent based computation. The main aim of this series it to discuss the above mentioned topics from an interdisciplinary point of view and present new ideas coming from mutual intersection of classical as well as modern methods of computation. Within the scope of the series are monographs, lecture notes, selected contributions from specialized conferences and workshops, special contribution from international experts.

More information about this series at http://www.springer.com/series/10624

Andrew Schumann

Behaviourism in Studying Swarms: Logical Models of Sensing and Motoring

Andrew Schumann
Department of Cognitive Science
University of Information Technology
and Management in Rzeszow
Rzeszow
Poland

ISSN 2194-7287 ISSN 2194-7295 (electronic)
Emergence, Complexity and Computation
ISBN 978-3-030-08271-0 ISBN 978-3-319-91542-5 (eBook)
https://doi.org/10.1007/978-3-319-91542-5

Softcover re-print of the Hardcover 1st edition 2019

Printed on acid-free paper

This Springer imprint is published by the registered company Springer International Publishing AG part of Springer Nature
The registered company address is: Gewerbestrasse 11, 6330 Cham, Switzerland

Contents

Chapter 1
Introduction

The notion of *swarm intelligence* [1–3] was first introduced in [4] to denote the collective behaviour of decentralized and self-organized systems. Now, this notion is used in robotics to design a population of robots interacting locally among themselves and reacting locally to their environment with an emergent effect when all the local reactions of them are being cumulated into the one collective reaction. There are many natural examples of swarm intelligence: ant colonies [5, 6], bee colonies [7–9], fish schooling [10, 11], bird flocking and horse herding [12], bacterial colonies with a kind of social behaviour [13–18], multinucleated giant amoebae *Physarum polycephalum* [19], etc. The main feature of all these systems is that their individual agents behave locally without any centralized control, but their interactions lead to the emergence of global behaviour of the whole group that cannot be reduced to subsystems additively.

The swarm intelligence is actively studied now, because swarms (ants, bees, some social bacteria, *Physarum polycephalum*, etc.) can solve logistic and transport problems very effectively [20]. For instance, there is a collective navigation of bacterial swarms [21, 22] and there is an effective path finding by amoebae and a possibility of traffic optimization by them [23–27]. Swarms can easily solve some complex (NP-hard) logistic problems: (i) the Travelling Salesman Problem can be solved by ants [28] and by amoebae [29]; (ii) the Steiner Tree Problem can be solved by amoebae [30]; (iii) the Generalized Assignment Problem can be solved by bees [31]; (iv) mazes can be solved by amoebae [32, 33], etc. As we see, even unicellular organisms can solve logistic problems effectively. Also, they can be involved in constructing algorithms for simulating the crowd evacuation [34] and for simulating transport systems such as the route systems in China [35] and in the United States of America [36].

The main characteristics of any swarm consists in a possibility to optimising the own traffic in reactions to attractants and repellents. *Attractants* are things or sites in the environment, such as food pieces and sex pheromones, which attract individuals of swarm. *Repellents* are things or sites in the environment, such as predators, which

A. Schumann, *Behaviourism in Studying Swarms: Logical Models of Sensing and Motoring*, Emergence, Complexity and Computation 33,
https://doi.org/10.1007/978-3-319-91542-5_1

repel individuals of swarm. For example, ants and bees exploit the food sources, they find out, very effectively so that they can solve the Travelling Salesman Problem and the Generalized Assignment Problem. Fish schools and horse herds avoid predators very effectively, too. The same swarm behaviour is observed even among human beings in the following two cases: (i) an addictive behaviour such as the behaviour of alcoholics [37] or gamers—in this case the role of human attractants causing addiction increases strongly; (ii) an escape panic [38]—in this case the role of human repellents like a terrorist act increases strongly, also.

Hence, by placing attractants and repellents at different sites we can manage and program the swarm behaviour. This opportunity allows us to design a *biological computer*—an abstract machine (i) with inputs presented by stimuli coming from attractants and repellents and (ii) with outputs presented by the swarm reactions to appropriate stimuli. This computer can be realized on different swarms differently. The point is that different matters are attractants and repellents for different animals. They differ a lot even for microorganisms. Nevertheless, their logic and mathematics are the same.

In the biological computer we have two main stages: (i) *sensing*—when a swarm detects neighbour attractants and repellents; and (ii) *motoring*—when this swarm reacts to founded attractants and repellents, e.g. to exploit attractants and to avoid repellents.

1.1 Designing a Biological Computer

1.1.1 From Controlling Reflexes to Biological Computer

Designing the biological computer is based on *behaviourism* as theoretical frameworks explaining the animal behaviour by reflexes produced by responses to external stimuli with a possibility to track individual's history of reinforcement and punishment implied by those stimuli. Each *reflex* is a direct response of animal to a stimulus, connecting the stimulus to a behaviour within two basic modes: either attracting (e.g. reinforcement) or repelling (e.g. punishment) coming from this stimulus. If we know what attracting and repelling matters are for the animal in fact, we can manage a control of its reactions by placing these matters. It means, simply put that *controlling stimuli causes controlling behaviour* according to behaviourism. So, we assume, the environment determines each animal behaviour.

In the theory of reflexes there are distinguished the following two types of responses/reactions/reflexes: (i) the *unconditioned response* to a stimulus (then an appropriate stimulus is called *unconditioned*); (ii) the *conditioning response* to a stimulus (then an appropriate stimulus is called *conditioning*). The unconditioned stimuli are represented by biologically active matters: either *attractants* (e.g. food or sex pheromones attracting an animal) or *repellents* (e.g. places evaluated automatically as dangerous by an animal). These stimuli are strongly connected to some direct responses from the very beginning, i.e. they are given a priori, just because of chemical processes of organisms. In the meanwhile, at the beginning, unconditioned stimuli are biologically neutral and assume no direct reactions of organism, i.e. they are rather ignored under standard conditions. Nevertheless, they can be associated with some unconditioned responses, too. Then they become non-neutral, but attractive (if they were associated with attractants) or negative (if they were associated with repellents). For the first time, Ivan Pavlov (1849–1936) demonstrated this learning process through experiments with dogs, e.g. he showed that a neutral stimulus such as a bell can be paired with a food by a dog, if this dog was previously feeded paralleled by this bell many times. After this association, the organism will exhibit a conditioned response to the conditioned stimulus when the conditioned stimulus will be presented alone. Thus, conditioning responses are not given a priori, they are a posteriori, i.e. they are learnt and after that they play the roles of attractants or repellents, although they are not biologically active matters in fact.

At the time of Pavlov, one thought that a possibility of conditioning responses is one of the basic features of brain and nervous system. But it is known now through experiments that it is a fundamental feature of any adaptive behaviour including the adaptive behaviour of unicellular organisms, see [39, 40]. Conditioning responses mean a memory, when an organism was taught that a neutral stimulus is a context of appearing an attractant or repellent. The knowledge of this context allows the organism to behave more adaptively in the feature. Thus, the possibility of conditioning responses is connected to a possibility of life to have a memory and then to be more adaptive to the environment.

Let us consider some examples of memory of unicellular organisms. From [41], we know that the amoeba *Physarum polycephalum* can learn the patterns of shocks at regular intervals, and then it changes its behaviour in anticipation of the next shock to come. In [42], this experiment is performed in the following manner. Unfavorable conditions for *Physarum polycephalum* are presented as three consecutive pulses at constant intervals. Under these unfavorable conditions, the amoebae reduce their locomotive speed in response to each episode. When the amoebae move under the favorable conditions, they spontaneously reduce their locomotive speed at the time when the next unfavorable episode is expected to occur. This fact shows that we deal with the anticipation of impending environmental change. *For the amoeba the regular interval of shocks became a conditioning stimulus.*

Another experiment with the amoeba *Physarum polycephalum* was performed in [39] to show that the temperature fluxes can become a conditioning stimulus for the amoebae. So, we know that the amoebae avoid cold temperatures and under these conditions they become slow. In the experiment there was shown that when the temperature changes had stopped, the amoebae slowed down, anticipating a cold flux. It is an evidence of *memory of temperature fluxes*.

All these experiments demonstrating a kind of memory of unicellular organisms can be explained by *chemotaxis*, the unconditioned response to some chemical cues (such as pheromones), that is different for static gradients and dynamic cues, see [43], where this difference is shown for the migration of *Dictyostelium* cells in response to the source of traveling waves of chemoattractant during aggregation.

Hence, some primitive forms of conditioning reflexes (that is a primitive memory) can be detected even at the level of cellular proteins (including actin filaments) responding to the changes of the environment, such as shock intervals or cold fluxes. Not only neurons can cumulate conditioning responses, as Pavlov thought first, but also cellular proteins. This basic feature of chemotaxis to have a short memory is involved into the particle swarm optimization [3, 44] and in designing *organic memristors*—organic devices with a memory. For example, in [40] for designing circuit models there was applied an ability of slime mould of *Physarum polycephalum* to both memorize the period of temperature and humidity variations and anticipate the next variations to come. Some other organic memristors are proposed in [45–49].

Thus, in designing the biological computer we can remain at the level of cellular proteins (first of all, actin filaments) responsible for chemotaxis and their reactions. One of the possible realizations of biological computer has been obtained by us in the project *Physarum Chip: Growing Computers from Slime Mould* [50, 51] supported by the Seventh Framework Programme (FP7-ICT-2011-8). In the project, we have focused on theoretical and experimental laboratory studies on sensing and motoring of plasmodia of *Physarum polycephalum*. My part has consisted in developing logics for programming the plasmodium behaviour. The project has been resulted in the design and prototyping of a *Physarum Chip*. The main objectives we have realised are to design a distributed biomorphic computing device built and operated by the slime mould of *Physarum polycephalum*. Let us recall that *Physarum polycephalum* belongs to the species of order *Physarales*, subclass *Myxogastromycetidae*, class *Myxomycetes*, division *Myxostelida*. Plasmodium is a vegetative phase of *Physarum polycephalum*, having a form of single cell with a myriad of diploid nuclei. It behaves and moves as a giant amoeba. Typically, the plasmodium forms a network of protoplasmic tubes connecting the masses of protoplasm at the food sources which has been shown to be efficient in terms of network length and resilience.

The *Physarum* Chip is a network of processing elements made of the slime mould's protoplasmic tubes acting as a non-linear transducer of information, while templates of tubes coated with conductor act as fast information channels. This chip has parallel inputs (optical, chemo- and electro-based) and parallel outputs (electrical and optical) and it can solve a wide range of computation tasks, including optimisation on graphs, computational geometry, decentralized robot control, logic and arithmetical computing [33, 34, 51–65], etc.

The idea of biological computer satisfies the presupposition of *radical behaviourism*: attractants, repellents and conditioning stimuli (connected to attractants or repellents) as environmental variables can completely control observable behaviours of animals [66–68]. This presupposition is used also in functional analysis of behavioural psychology to define relationships between stimuli and responses on the basis of explicating the following four elements: motivating operations, trigger of behaviour, behaviour itself, consequence of behaviour [69]. The radical behaviourism is applied in the *animal training*, too. Some basic principles of this training: (i) using attractants for a *positive reinforcement* (an expected behaviour is resulted as a conditioning response connected to an attractant); (ii) using repellents for a *negative reinforcement* (an expected behaviour is followed by the removal of repellent, i.e.

there was a repellent and a conditioning response was associated with it, then the removal of repellent causes appearing an opposite conditioning response); (iii) using repellents for a *positive punishment* (an expected behaviour is resulted as a conditioning response connected to a repellent); (iv) using attractants for a *negative punishment* (an expected behaviour is followed by the removal of attractant, i.e. there was an attractant and a conditioning response was associated with it, then the removal of attractant causes appearing an opposite conditioning response).

Thus, *the biological computer is a logic for radical behaviourism*. In other words, it defines how different Boolean combinations of attractants and repellents determine the patterns of animal behaviour, first of all their swarm behaviour.

1.1.2 Bounded Rationality and Cognitive Biases

In *logical positivism* (called *logical empiricism* or *physicalism* also), there was a fundamental presupposition that, on the one hand, any complex of facts is a combination of elementary facts and, on the other hand, each elementary fact is described by an atomic proposition and the complex of facts is described by an appropriate Boolean complex of true atomic propositions and by all logical consequences drawn from this Boolean complex:

> Every statement about complexes can be analysed into a statement about their constituent parts, and into those propositions which completely describe the complexes. [...] The totality of true propositions is the total natural science (or the totality of the natural sciences). [...] The specification of all true elementary propositions describes the world completely. The world is completely described by the specification of all elementary propositions plus the specification, which of them are true and which false (Ludwig Wittgenstein, *Tractatus Logico-Philosophicus*).

This presupposition of logical positivists would mean for radical behaviourism that (i) the complex of swarm behavioural patterns can be reduced to a composition of some elementary swarm patterns, (ii) if we know an appropriate attractant or repellent for each elementary pattern, then from a complex of attractants and repellents we can deduce a complex of patterns. Hence, following logical positivism we can suppose that the animal behaviour is *rational*—it is the same under the same conditions with the same complex of attractants and repellents. Each attractant is more or less preferable and each repellent is more or less avoidable. Then this rationality would mean that at each step if a swarm faces several attractants, then the most preferable attractant is always contained among chosen attractants; and if it faces several repellents, the most avoidable repellent is always contained among rejected repellents. In other words, the rational animal tries to maximize its satisfaction and, at the same time, to minimize its frustration. If preferences affect decisions indeed, then a Boolean complex of stimuli based on these preferences determine a complex of behavioural patterns. Is it so? Can we design the biological computer as a complex of Boolean compositions?

The assumption of rationality of agents is fundamental for game theory and decision theory. According to these theories, each rational agent always follows his or her preferences in his or her choices of items or strategies. So, the rationality of human beings in microeconomics is understood in the same way of utility-maximizing consumers and profit-maximizing firms. Each rational agent of the market follows the same decision model to maximize their own profits: I choose items to maximize my satisfaction as a consumer and I offer something to maximize my profits as a firm. I offer a thing that is preferable for A to obtain by exchange of a thing of A that is preferable for me (e.g. I can offer commodities to obtain moneys to invest them elsewhere). Hence, the free market is efficient only because it offers the best opportunity for individuals to maximize their commodities and satisfaction to create an economy that could insure maximum utility for everybody. We face a zero reflexion of rational humans there. Everyone is transparent for others, e.g. satisfies their expectations, if all them are rational. This transparency in decisions means that rational agents appeal only to facts which are the same for everyone, therefore these agents always can agree (see the Aumann's agreement theorem [70, 71]). Preferences are instances of facts that should be seen before meeting. Furthermore, due to rationality and zero reflexion of individuals we can always agree about long-term joint actions. From these actions socioeconomic institutions have resulted, therefore the latter yield outcomes deemed preferable for the majority. This point of view is called constructivism [72] according to which "all worthwhile social institutions were and should be created by conscious deductive processes of human reason" [73].

In other words, according to constructivism (e.g. ideas of rationality as profit-maximizing), we deal just with either facts which are the same for everyone or conventions (e.g. agreements) which appear only after rationally motivated discussions and hence are evident for the majority. Thus, according to constructivism, the propositions being considered make claims about the world and these claims are true just in the case the world is as it is claimed to be, or they are conventions that are understandable for everybody who wants to understand. For example, if I can see preferences of other people as such indeed, then I always can offer them something for my profit. We suppose there that facts are correlated with propositions in accordance with logical positivism, i.e. there is a reference (correspondence) relation that maps facts into (onto) propositions if we assume that each fact can be described by one and only one (some) proposition(s) or there is a reference relation that maps propositions into (onto) facts if we assume that each proposition describes one and only one (some) fact(s). This approach to semantics was inspired by Aristotle and hammered out by Bertrand Russell (1872–1970) and it is called *Russellian semantics* [74, 75]. Rational agents can always agree, because they can verbally appeal to the same facts or to the same conventions, i.e. they use the Russellian semantics.

Nevertheless, there are many empirical evidences which contradict the assumption that, on the one hand, consumer behaviour is reasonably characterized as the maximization of expected lifetime utility subject to a budget constraint and conditional on the available information and, on the other hand, firm behaviour is characterized as the maximization of expected profit subject to investments and conditional on the correct programming of consumer behaviour. For example, Thaler [76, 77] shows the *bounded rationality* and impatience of consumers and, as a result, he proposes the behavioural life-cycle theory emphasizing self-control, mental accounting, and framing. According to his hypothesis, consumers maintain mental accounts that lead them to treat various components of their wealth as non-fungible, e.g. instead of the wealth-maximizing choice consumers can postpone the receipt of income in order to control spending. In other words, very often consumers refer not to facts, but to *mental accounting* which give no weight to the economically relevant cost of their decision. Evidently, for mental accounting there are no Russellian semantics, since for different people mental accounts can be different as well, i.e. they are not transparent for everyone a priori. Psychologists Amos Tversky and Daniel Kahneman established a research programme of studying *cognitive heuristics and biases*, i.e. the programme of studying contexts and patterns which influence on human decisions beyond rationality [78] (i.e. beyond logical positivism and classical game theory and decision theory).

Propositions whose meanings depend upon contexts of utterances are called *performative*. They are studied within the *Austian semantics* [74, 75] based on the ideas of John Langshaw Austin (1911–1960). So, meanings of performative propositions depend on contexts and can change in time. For example, investors in their performative propositions (e.g. forecasts) might be optimistic, while at other times, they might be pessimistic. The point is that with the lapse of time investors can build different families of intensions (intersected concepts) as the meaning of phenomenon. For instance, under conditions of investor exuberance entrepreneurs prefer to create closed-end funds, because they forecast that they can sell fund shares for more than they are worth. These conditions are contexts for one family of meanings of performative propositions connected to closed-end funds. Nevertheless, there can be other families as well, e.g. conditions, where a closed-end fund is being liquidated. Then investors forecast that at liquidation, the fund price will equal net asset value. In finance there exists the notion of reference points at which investors make their decisions. Reference points can be regarded as different families of intersected concepts (different meanings of the performative proposition ‘I sell asset X’) also. Let us assume that an investor invests \$100,000 (purchase price) and hopes to sell for \$150,000 (aspiration price). If the selling price is reduced to be \$120,000, this is a gain compared to the purchase price as a reference point but a loss compared to the reference point of the aspiration price.

Thus, we can point out that contexts, which determine the changing meanings of performative propositions, are framed by behavioural scenarios that can vary in different situations. For example, usually there is a phenomenon of managerial optimism [79], i.e. managers overestimate the probability that the future performance of their firm will be positive. Behavioural scenarios used in finance or everyday life may be more or less rational. Complete irrational scenarios are called *cognitive biases*. Let us enumerate some key examples of cognitive biases: anchoring (prior estimations); mental accounting (individuals categorize their money for separate accounts based on factors such as the source of money and intended purpose of the money); framing (people may frame the risky outcome differently, either considering the worst case or the best case); confirmation and hindsight bias (paying more attention to information that supports our opinions); gambler's fallacy (ignoring pre-existing events); herd behaviour (copying actions of a larger group due to the social pressure of conformity, e.g. some investors may follow others to buy the same stocks); confidence (optimistic or pessimistic assessment of one's ability to perform, e.g. in a period of economic recession, people are less confident and more pessimistic about future economic conditions and as a consequence they avoid risky decisions); overreaction and availability bias (overreacting to dramatic and unexpected news occurrences, e.g. bad news including falling stock prices), see [80–89]. Cognitive biases in finance are studied within behavioural accounting (or human resource accounting) to consider and integrate human behaviour into accounting decisions in an organization [90].

In contrast to cognitive biases, *cognitive heuristics* is grounded on efficient, but (as well) non-logical rules which people often apply to make effective decisions beyond logical reasoning—e.g. decisions in risk situations, prompt decisions, creative decisions and so on [78].

As we see, in behavioural economics and experimental psychology it was made evident through experiments that classical game theory and decision theory assuming the rationality of agents are too abstract, because real agents are too far from rational. The problem is that preference relations are not so strongly linked to decisions. As a consequence, a Boolean composition of preferences does not give a complex preference for a complex decision. Mathematically, it means that complexes of preferences as well as complexes of behavioural patterns are not additive.

Let us return to our task of designing the biological computer. From these examples taken from behavioural economics, we can assume that logical positivism is unapplied in the logical modelling of the swarm behaviour, as well—because we deal with a non-additivity of all natural behaviours. As a result, a Boolean composition of attractants and repellents cannot give an expected complex of behavioural patterns in the standard way of constructing inductive sets.

Hence, due to recent results in behavioural economics we know that even in human cognitions (which evaluated as the most rational among all animals) we cannot avoid cognitive biases and cognitive heuristics which are a substantial part in our decision making not only in everyday situations, but also in firms. Therefore, cognitive reflexion and cognitive biases are evaluated now as a natural mechanism of cognitions [91, 92].

Some primitive forms of cognitive biases and cognitive heuristics of unicellular organisms were found out by us within the project *Physarum Chip: Growing Computers from Slime Mould* [50]. Even unicellular organisms behave differently in stress or under favourable conditions. For instance, under the favourable condition (meeting attractants) bacteria behave more predictable in choosing a trajectory of motion—so, if a concentration of attractant increases, bacteria tumble less frequently. In the stress (meeting repellents) bacteria try to change a trajectory stochastically—so, if a concentration of repellent increases, bacteria tumble more frequently. As we see, we face a duality in basic reactions of bacteria to their environment. It is a kind of the most primitive "cognitive heuristics" detected at the bacterium level—to be more or less predictable in a favourable or stress situation, respectively.

Usually, bacteria react only to one stimulus: either to one attractant or to one repellent. Nevertheless, they can build up swarms, also—*biofilms* even with a common sensing quorum for collective reactions to many stimuli simultaneously [13]. There are different forms of biofilms—from groups combining individuals mechanically (such as *Escherichia coli* bacteria) to true swarms (such as *Paenibacillus vortex* bacteria).

Many animals are able to form swarms, but even swarm animals can behave individually, too. For example, cercariae—parasites of the genus *Trichobilharzia* Skrjabin and Zakharov, 1920 (*Schistosomatidae* Stiles and Hassall, 1898)—behave as free-swimming individuals in a favourable situation that is represented for them by a normal natural environment where hosts can be ever found. However, under some additional conditions, cercariae behave as one large group. For example, cercariae of *Bilharziella polonica* (Trematoda: Schistosomatidae) can attach themselves to the surface-film of the water and throw out a secretion from the mouth, which forms a kind of web and instantly clings to the feathers of any duck with which it comes in contact [93]. Another case of swarming of cercariae is presented by *Rat king cercaria* (or Rattenkönig), see an example with *Clypeomorus batillariaeformis* Habe and Kosuge in [94]. These cercariae bind to each other by the lower half of their tail to aggregate into a swarm. Within this swarm they synchronise their movement to move as one giant organism. When a fish eats this mass, the cercariae begin digging into the fish's mouth.

Amoebae of *Physarum polycephalum* can behave both individually and collectively. Therefore, they can be considered a swarm, although they are unicellular organisms in fact, but with myriad nuclei. We have discovered that the main feature in perceiving external signals even by one cell due to actin filament networks is that the same signal can be perceived differently—it depends on the cell shape and many other internal circumstances. As a result, one unicellular organism bears more outputs than inputs in possible reactions to external signals. So, even one cell can have "cognitive biases" and behave quite unpredictably and irrationally under the same conditions. It means that amoebae of *Physarum polycephalum* and *Amoeba proteus* can modify their own elementary actions—they possess a kind of "free will".

Let us remember that there are the following two complementary mechanisms in perceiving signals: *lateral inhibition* (the *stress condition* which increases the contrast of one signal to react more directly and efficiently and to be better focused) and *lateral activation* (the *comfortable condition* which decreases the contrast of signals to perform concurrent actions). Both mechanisms like many others in perceiving stimuli are scale-invariant. In particular, lateral inhibition and lateral activation are ubiquitous events that occur over many scales including within the cell during assembling and disassembling of actin filament waves, between groups of neurons within the visual cortex to process visual cues, and between active zones of swarms to react to their environments. In this book, different logical systems for modeling effects of lateral inhibition and lateral activation are considered. An intelligent behaviour of animals is called swarm in the book, because, on the one hand, the patterns of the swarm behaviour is most intelligent among animals (at the same time they are well visible and studied well experimentally) and, on the other hand, these patterns are scale-invariant (they can be detected even at the level of some unicellular organisms, such as *Physarum polycephalum* or *Amoeba proteus*). The swarm behaviours should be distinguished from social behaviours. So, in social behaviours, we deal with symbolic values and symbolic interactions which are out of pure behaviourism in the strict sense. Thus, this book offers some logical systems for studying the swarm behaviours.

1.1.3 Experiments

My theoretical models of swarm behaviour were verified on experimental data with *Physarum polycephalum*, *Schistosomatidae*, and people addicted to alcohol.

So, my research of *Physarum polycephalum* is fully based on the experimental results obtained within the joint project *Physarum Chip: Growing Computers from Slime Mould* [50] mainly at the following three institutions: (i) Unconventional Computing Centre, University of the West of England (Bristol, UK); (ii) CNR-IMEM, National Research Council (Consiglio Nazionale delle Ricerche) (Parma, Italy); (iii) Institute of Plant Sciences, University of Graz (Graz, Austria). The detailed presentation of results can be found in [95]. The majority of experiments were conducted by Andrew Adamatzky (Bristol).

Experiments with *Physarum polycephalum* were usually performed by culturing plasmodia on wet paper towels, by feeding them with oat flakes (attractants), and by moistening paper towels regularly. These paper towels are a nutrient-poor substrate. In this condition, the plasmodium behaves in respect to attractants more predictably. Each experiment does not exceed 5–7 days usually. There are two types of environment for plasmodia: (i) 2% agar gel or paper towels as nutrient-poor substrate, and (ii) 2% oatmeal agar as a nutrient-rich substrate.

Another set of experimental results was granted for me by Ludmila Akimova. She conducted some experiments with parasites of *Schistosomatidae* (Trematoda: Digenea) at the State Scientific and Production Amalgamation "The Scientific and Practical Center of the National Academy of Sciences of Belarus for Bioresources" (Minsk, Belarus). She studied their group behaviours.

The third set of empirical data was obtained jointly with Vadim Fris, the Chief Psychiatrist of the rehabilitation centre "Iscelenie" (Minsk, Belarus). We have carried out a statistical research of group behaviour of 107 people addicted to alcohol who have been actually treated in this centre.

I am grateful to Andrew Adamatzky, Ludmila Akimova, Vadim Fris and all other collaborators within the project *Physarum Chip: Growing Computers from Slime Mould*.

1.2 Methodology

1.2.1 Tools for Studying Non-additive Behavioural Patterns

Experimental results of behavioural economics show that behaviours cannot be defined by additive measures, e.g. we cannot apply classical game theory and classical decision theory. Nevertheless, the non-additivity of behavioural phenomena and everyday decisions does not mean that they cannot be studied mathematically. So, there is a *p-adic probability theory* [96–102] (probability theory, where probability measures are defined not by real numbers, but by p-adic numbers), which is based on non-additive measures. These measures are used also in p-adic quantum mechanics developed since the publications of Igor Volovich [103]. Notice that the p-adic numbers, where p is prime, are an alternative, to the real numbers, extension of the field of rational numbers by the construction of Cauchy sequences [104].

The main feature of p-adic probabilitics (or more generally, non-Archimedean probabilities or probabilities on infinite streams [101, 102, 105–107]) is that they do not satisfy additivity. For instance, we cannot obtain a partition for any set into disjoint subsets whose sum gives the whole set. Using p-adic (non-Archimedean) probabilities we can disprove the *Aumann's Agreement Theorem* [107, 108] and develop new mathematical tools for game theory, in particular we can define context-based games by means of coalgebras or cellular automata [109, 110]. In these context-based games we can appeal solely to non-Archimedean probabilities. These games can describe and formalize complex reflexive processes of behavioural finances (such as short selling or long buying). As we see, avoiding additivity in probability theory can have many important results of an epistemic nature in game theory.

Thus, my research is devoted to the analysis of some new logical methods, such as p-adic or non-Archimedean many-valued logic and stream-valued [111–113] probability theory, which can be applied in behavioural sciences from the viewpoint of non-additivity of behavioural patterns. Such methods could be used in decision support systems as well.

Usually, in business intelligence the following mathematical methods are used: classification, regression, time series analysis, association rules, or clustering. All these methods have developed within mathematical statistics and sciences, such as conventional probability theory or econometrics. However, the possibilities for using the newest logical methods based on stream values are not yet conclusively clarified. Among these methods the most significant for my research are as follows: coalgebra [114], non-well-founded set theory [115, 116], concurrent [117] and massive-parallel computing [118, 119]. The logical theory of sensing and motoring in the swarm behaviour is studied in my work within these methods. For example, the Aumann's agreement theorem uses inductive sets as the main construction. These sets are postulated upon the axiom of foundation [115]. However, there is a mathematics without this axiom (the so-called non-well-founded mathematics). Within the limits of this mathematics the Aumann's agreement theorem is impossible [107, 108]. Hence, the main goal of my research is to consider unconventional logical tools for analyzing the swarm sensing and motoring as non-well-founded data, i.e. as data without logical atoms (these data cannot be measured by additive measures).

For the logical studies of sensing and motoring of swarms I use a special metalogic as a methodology of logical researches that combines many-valued logic, epistemic logic, illocutionary logic, unconventional computing, process calculi, cellular automata, etc. This metalogic deals with a metalanguage, which is rich enough for the identification of boundaries of behavioural patterns. This metalanguage regulates and contains formal conceptual equipment, languages which can be involved in cognitive and behavioural sciences.

1.2.2 Tools of Unconventional Computing

My work in modelling the swam behaviour is carried out within *unconventional computing* [54, 118–122], a new subject of scientific researches combining different disciplines such as natural sciences, computation theory, applied mathematics, logic, etc. In this new subject many significant results are obtained and I apply them in the logical modelling of swarm sensing and motoring as well.

The most basic notion in the conventional approach to algorithms implicitly used in conventional data mining is the least fixed point presupposing the induction principle (*Noetherian induction*), e.g. this notion is fundamental for recursion (in the meaning of computable functions). However, the conventional treatment of an algorithm is broken in unconventional computing models. The main originality of unconventional computing is that just as conventional models of computation make a distinction between the structural part of a computer, which is fixed, and the data on which the computer operates, which are variable, so unconventional models assume that both structural parts and computing data are variable [54, 118]. Therefore the essential point of conventional computation is that the physics is segregated once and for all within the logic primitives. For instance, reaction-diffusion computing [118] is one of the unconventional models built on spatially extended chemical systems, where both the data and the results of the computation are encoded as concentration profiles of the reagents, i.e. both change permanently. As a result, they cannot have a single centralized source of control. Therefore the conventional definition of algorithm is unapplied for unconventional computing models, because this definition is grounded on controlling sequences of all steps.

Instead of recursion, in unconventional computing one can apply *corecursion* as a type of operation that is dual to recursion [113]. Corecursion is typically used to generate infinite data structures. The rule for primitive corecursion on codata is dual to that for primitive recursion on data. Instead of descending on the argument, we ascend on the result. Notice that corecursion creates potentially infinite codata, whereas ordinary recursion analyzes necessarily finite data. By induction and recursion we use the notion of least fixed points, whereas by coinduction and corecursion we use the notion of greatest fixed points (for more details concerning coinduction and corecursion see [111–113]). Within the concurrent and massive-parallel computing paradigm instead of induction and recursion only coinduction and corecursion can be used [117, 123]. There are many studies in this field. The point is that induction and recursion determine enumerable collections of finite structures, while coinduction and corecursion determine non-enumerable collections of infinite structures, including all possible modifications of behavioural patterns. As a result, conventional logic cannot model interactive and massive-parallel processes of behaviours which are non-enumerable and circular by definition.

Thus, in my research I appeal to interactive-computing/concurrency paradigm. The latter assumes coinductive and corecursive methods to be used in computations. Although in such paradigms, computation models may be abstract in the same measure as in Turing machines, they assume a combination of real contexts (physics) and logic in the computation, as well as that found in cellular automata [119] or in other unconventional computing models. In conventional computing, the computation is performed in a closed-box fashion, transforming a finite input, determined by the start of the computation, to a finite output, available at the end of the computation, in a finite amount of time. Therefore physical implementation does not play a role in such a computation which may be considered a process in a black box. As opposed to the conventional approach, in interactive models inputs and outputs may be infinite and computation in each point of the lattice may proceed simultaneously and independently. In the biological computer we cannot use an aprioristic approach with induction and recursion [109, 124]. However, if we want to apply logical and mathematical tools to the biological computer we can refer to an interactive-computing/concurrency paradigm [117].

The interactive-computing paradigm can describe concurrent/parallel computations whose configuration may change during the computation. Within this paradigm, there were proposed concurrency calculi, also known as process algebras [113, 117, 125, 126]. Process algebras are used in formalizing business processes including decision making processes.

1.2.3 Anti-platonism

From the standpoint of philosophy, I avoid Platonism in setting up the new approach to analysing behavioural patterns. The presuppositions of Platonism allowed us to generate the axiomatic method in mathematics. According to Platonism, there exist entities with a conclusive absolute status of existence (primary objects of mathematics, e.g. notions of point or line in geometry) and there are propositions with the highest persuasiveness and evidence (axioms). Actually, Platonism is a corner stone of classical mathematics. The axiomatic method aims to obtain a maximum of information from a minimum of premises. David Hilbert (1862–1943) in his program of foundations of mathematics considered the axiomatic method (respectively, Platonic assumptions) as a paramount factor in mathematical thinking. In the main points of his program, Hilbert postulated a formal theory that should have included all classical mathematics with appropriate proofs of inconsistency, completeness and decidability of this theory by finite methods. Hilbert's program in its first meaning became unsolvable, but it has caused the mathematics development. The formal theory obtained by Hilbert should be constructed on the basis of axioms of two types: (i) logical axioms which set reasoning rules and (ii) non-logical axioms by which a primary object of mathematics is defined. In the decision approach proposed in my work, I appeal to massively parallel proof theory, where there are no axioms [109, 124]. Thus, anti-Platonic logic and mathematics is understood as satisfying the following basic presuppositions:

- there are no propositions that are true on any class $\mathfrak{K}$ of models (i.e. there are no logical axioms);
- there are no non-logical axioms, we draw conclusions from premises having equal status, all formulas are inferred from hypotheses;
- finally, any object has only a relative existence.

Within these assumptions, derivations without using axioms are built up, therefore there is no sense in distinguishing logic and theory (i.e. logical and non-logical axioms), derivable and provable formulas, etc. In such a formal theory, the conclusion is understood as massive-parallel computing, i.e. the deduction is considered a transition of an automaton. This formal proof theory is reduced to an appropriate cellular automaton which is called a proof-theoretic cellular automaton, whose cell states are regarded as well-formed formulas of a logical language. As a result, in deduction we do not obtain derivation trees and instead of the latter we find derivation traces, i.e. a linear evolution of each singular premise. Thereby some derivation traces may be circular, i.e. some premises could be derivable from themselves, and hence some derivation traces may be infinite. Propositions from circular derivations are said to be analytic a posteriori. Other propositions are called synthetic a posteriori. Proof-theoretic cellular automata can be used for aposterioristic decision making [109, 110]. Aposterioristic logics can be used in formulating context-based decision rules in games [127, 128]. Usually, for representing databases of games, payoff matrices are involved. However, in the case of coinductive databases we cannot appeal to payoff matrices. For example, we cannot appeal to them if we deal with games limited by some unstable contexts or with infinite games. I have shown that some kinds of coinductive databases for making decisions could be presented by payoff cellular automata, a kind of proof-theoretic cellular automata [109]. Notice that such efforts to build a formal theory of massively parallel and competing decision making are relatively new.

1.2.4 Object-Oriented Programming Language

On the basis of process-algebraic formalization of swarm propagations within the interactive-computing/concurrency paradigm, in the project *Physarum Chip: Growing Computers from Slime Mould*, Krysztof Pancerz and me have developed a new *object-oriented programming language* OPL-Ph for controlling the swarm behaviour by placing attractants and repellents. This language has been used by us for controlling plasmodia of *Physarum polycephalum*.

The plasmodium behaves differently under comfortable and stress conditions. The comfortable conditions are determined by a *nutrient-rich substrate* of plasmodium feeding and the stress conditions are determined by a nutrient-poor substrate. So, if the plasmodium feeds under the condition of nutrient-poor substrate, then it can distinguish all attractants and repellents, and, as a result, those attractants and repellents involved in the stimulation of plasmodium gives a topology which can be defined as a *Voronoi diagram* [129, 130]. Within one Voronoi cell a reagent has a full power to attract or repel the plasmodium. In other words, within this cell the reagent determines the plasmodium behaviour completely. On the board of Voronoi cells, the plasmodium cannot choose its one further direction and splits. Within the same Voronoi cell two active zones will fuse. Now, suppose that the plasmodium feeds under the condition of nutrient-rich substrate. This means that for the plasmodium there is an information noise and the plasmodium cannot define where precisely attractants are located indeed. Hence, in the case of nutrient-poor substrate with well distinguished localizations of attractants and repellents we can fully manage the plasmodium behaviour and propose a biological version of storage modification machines. These machines are well defined in our programming language OPL-Ph [131–137]. Notice that under the conditions of nutrient-rich substrate, storage modification machines on plasmodia cannot be conventional. The point is that under these conditions we cannot approximate elementary (atomic) acts, i.e. we deal with a massive-parallel behaviour of plasmodia (they are to be expanded in all possible directions). Hence, we must extend our OPL-Ph by some unconventional tools to make OPL-Ph applicable for the case of nutrient-rich substrate, also. Rough sets on transitions have become these new tools for us [62, 138–141].

Within our programming language OPL-Ph, we can check possibilities of practical implementations of storage modification machines on a swarm behaviour and their applications to behavioural science such as behavioural economics and game theory [109, 127, 128, 133, 138, 140, 142–144]. The proposed OPL-Ph can be used for developing programs for swarms by the spatial configuration of stationary nodes. Geometrical distribution of stimuli can be identified with a low-level programming language for controlling the swarm behaviour.

Any swarm propagation is stimulated by attractants and repellents and on the basis of the ladder diagram language we can use *Programmable Logic Controllers* for programming swarms. Rungs of the ladder can consist of serial or parallel connected paths of swarm propagation. A kind of connection depends on the arrangement of regions of influences of individual stimuli. If both stimuli influence a swarm, we obtain alternative paths for its propagation. It corresponds to a parallel connection (i.e. the *OR gate*). If the stimuli influence the swarm sequentially, at the beginning only the first one, then the second one, we obtain a serial connection (i.e. the *AND gate*). The *NOT gate* is imitated by the repellent avoiding the swarm propagation. In the proposed approach, we have assumed that each attractant (repellent) is characterized by its region of influence in the form of a circle surrounding the location point of the attractant (repellent), i.e. its center point. The intensity determining the force of attracting (repelling) decreases as the distance from it increases. A radius of the circle can be set assuming some threshold value of the force. The swarm members must occur in a proper region to be influenced by a given stimulus. This region is determined by the radius depending on the intensity of the stimulus. Controlling the swarm propagation is realised by activating/deactivating stimuli. Logic values for inputs have the following meaning in terms of states of stimuli: 0 means that attractant/repellent is deactivated, 1 means that attractant/repellent is activated. Logic values for outputs have the following meaning in terms of states of stimuli: 0 is an absence of swarm members at the attractant, 1 is a presence of swarm members at the attractant.

Also, we can adopt more abstract models than distribution of stimuli to program swarms which can be identified with programming in the high-level language, such as Petri nets. We have shown how to build *Petri net models on swarm behaviours* [136, 137]. In our approach, we use Petri nets with inhibitor arcs. Inhibitor arcs are used to disable transitions. Inhibitor arcs test the absence of tokens in a place. A transition can only be if all its places connected through inhibitor arcs are empty. This ability of Petri nets with inhibitor arcs is used to model behaviour of repellents. In the proposed Petri net models, we can distinguish three kinds of places:

- Places representing a swarm. For such a place, the presence of the token means that a member of swarm is present in the origin point. Otherwise, the absence of the token means that there is no swarm members.
- Places representing input attractants or repellents. For such a place, the presence of the token means that the attractant/repellent is activated. Otherwise, the absence of the token means that attractant/repellent is deactivated.
- Places representing output attractants. For such a place, the presence of the token denotes the present of swarm members at the attractant (the swarm members occupy the attractant). Otherwise, the absence of the token denotes the absence of swarm members at the attractant.

In the AND gate, the transition T represents the flow (propagation) of swarm from the origin place to the output attractant. T is enabled to fire if both attractants are activated. In the OR gate, the transitions T_1 and T_2 represent the alternative flows of plasmodium from the origin place to the output attractant. T_1 is enabled to fire if the first attractant is activated. T_2 is enabled to fire if the second attractant is activated. In the NOT gate, the transition T represents the flow (propagation) of plasmodium from the origin place to the output attractant. T is enabled to fire if the repellent is deactivated.

In our OPL-Ph we have analysed biological mechanisms for *Physarum polycephalum* decision making and reconstructed the so-called concurrent games of plasmodia [109, 127, 128, 133, 138, 140, 142–144]. So, attractants may be regarded as payoffs for *Physarum polycephalum*. Plasmodia occupy attractants step by step. By different localizations of attractants we can affect on *Physarum polycephalum* motions differently. We can interpret the stimuli for plasmodium motions as Boolean functions on payoffs. Boolean functions are designed by logical gates mentioned above. In this way we have designed the *zero-sum game* between plasmodia of

Physarum polycephalum and *Badhamia utricularis*, the so-called PHY game. To simulate *Physarum polycephalum* games, we have created a specialized software tool. The tool works under the client-server paradigm. The server window contains:

- a text area with information about actions undertaken;
- a combox for selecting one of two defined situations;
- start and stop server buttons.

A communication between clients and the server is realized through text messages containing statements of OPL-Ph. The locations of attractants and repellents are determined by the players during the game. At the beginning, origin points of *Physarum polycephalum* and *Badhamia utricularis* are scattered randomly on the plane. During the game, players can place stimuli. New veins of plasmodia are created. The server sends to the clients information about the current localization of origin points of *Physarum polycephalum* and *Badhamia utricularis*, localization of stimuli as well as a list of edges, corresponding to veins of plasmodia, between active points.

So, we have proposed a game-theoretic visualization of morphological dynamics with nonsymbolic interfaces between living objects and humans. These non-symbolic interfaces are more general than just sonification and have a game-theoretic form. The user interface for this game is designed on the basis of the following game steps: first, the system of OPL-Ph generates locations of attractants and repellents; second, we can chose n plasmodia/agents of *Physarum polycephalum* and m plasmodia/agents of *Badhamia utricularis*; third, we obtain the task, for example to reach as many as possible attractants or to construct the longest path consisting of occupied attractants, etc.; fourth, we can choose initial points for *Physarum polycephalum* transitions and initial points for *Badhamia utricularis* transitions; fifth, we start to move step by step; sixth, we define who wins, either *Physarum polycephalum* or *Badhamia utricularis*. Strategies in games between *Physarum polycephalum* and *Badhamia utricularis* are described by rough sets defined on transition systems [62, 138–141]. We claim that bio-inspired games might wake new interests in designing new games and new game platforms. These games can simulate zero-sum games between different swarms.

To sum up, in the project *Physarum Chip: Growing Computers from Slime Mould*, Krysztof Pancerz and me have developed *the object-oriented programming language OPL-Ph that can be used as a basic programming language for the biological computer.* I am grateful to Krysztof Pancerz for his collaboration.

My book contains fundamental theoretical results for designing object-oriented programming languages for controlling swarms.

In Chap. 2, I show that some unicellular organisms, such as *Physarum polycephalum* or *Amoeba proteus*, have basic features of swarms in their behaviours due to actin filament networks. Notice that *Amoeba proteus* behaves as a swarm at short ranges and *Physarum polycephalum* behaves as a swarm on long distance. It means that in order to design the biological computer it is enough to design an artificial cell with actin filament networks.

In Chap. 3, I concentrate on chemical dynamics explaining sensing and motoring of swarms. The formalization of this dynamics is grounded on reaction-diffusion computations [53, 118, 129, 145–147] explaining the phenomenon of auto-waves of chemical self-organized systems. One of the best examples how chemical reactions can transmit signals and be 'intelligent' is the Belousov-Zhabotinsky reaction. The difference of living systems from the Belousov-Zhabotinsky reaction is that the living systems are instable in the meaning that they can change their decisions and they can behave differently under the same outer conditions if they are in stress or not. I propose a process-algebraic formalization of *Physarum polycephalum* behaviour. Within this formalization we can formulate some performative propositions of plasmodia, such as 'I would like to eat Ψ', 'I fear Ψ', and 'I am almost satisfied by Ψ'. If the plasmodium is under stress and its behaviour can be completely controlled, then we can design some conventional logic gates on it, such as the Flip-Flop circuit, the FEYNMAN gate, the TOFFOLI gate, the FREDKIN gate, the CNOT gate, the Full Adder circuit, etc. If the plasmodium is under favourable conditions, then we can design some unconventional logic gates based on non-linear permutation groups.

In Chap. 4, I show that each swarm behaviour can be formalized in the way of process algebra, described in Chap. 3. On swarms we can implement Kolmogorov-Uspensky machines [52], Schönhage's storage modification machines, and random-access machines, but these implementations would have a low accuracy. The problem is that any swarm can follow some emergent patterns which are fully eliminated in conventional automata such as Kolmogorov-Uspensky machines. It is worth noting that conventional automata are suitable for simulating swarms just at their discovering stage—when they are looking for attractants without an information noise. At the logistic stage they are constructing networks, which change permanently. So, in this case some emergent patterns appear and conventional automata become unapplied. In this chapter, I propose a logic of strings imitating the swarm paths and having logical values on p-adic integers. This logic can simulate the swarm behaviour under different conditions, including stress and information noise, and at different stages, including the discovering stage and the logistic stage.

In Chap. 5, I construct the following non-Archimedean logics for simulating swarm paths: (i) the logic $\text{Ł}\Pi\forall_\infty$ with truth values on the set of hyperrational or hyperreal numbers of [0, 1] (this logic is an extension of fuzzy logic $\text{Ł}\Pi\forall$ defined in [148–151]); (ii) the logic $BL\forall_\infty$ with truth values on the set of p-adic integers (it is an extension of fuzzy logic $BL\forall$ defined in [148–151]). Both logics are called non-Archimedean, because their truth values assume infinitesimals (infinitely small numbers) or infinitely large integers and, therefore, for them the Archimedean axiom is not valid. Some versions of standard fuzzy logic $BL\forall$ are as follows: Łukasiewicz's logic, Gödel's logic, and Product logic. I propose a non-Archimedean extension of them. Then I formalize $BL\forall_\infty$ with values on hyperrational or hyperreal numbers of [0, 1] and with values on p-adic integers. Also, I define non-Archimedean probabilities (probabilities on hyperrational or hyperreal numbers of [0, 1] and probabilities on p-adic numbers) and I formalize the non-Archimedean probability logic. By using p-adic valued probabilities, I define knowledge operators on some propositions describing experiments with swarms, such as 'Attractants, which can be occupied by agent i', 'Attractants accessible for the attractant N_i by a road system', etc. For these knowledge operators the Aumann's agreement theorem is not valid (as I show in Chap. 10). Within the p-adic valued logic I define p-adic valued adder and p-adic valued subtracter on swarm behaviours and possible p-adic valued natural deductions within the swarm paths.

In Chap. 6, it is shown that the swarm behaviour cannot be measured by additive measures because of the phenomenon of individual-collective duality, when we cannot approximate atomic individual acts (the matter is that a swarm can behave individually, as one agent, or collectively, as a group of agents with concurrent actions). So, the double-slit experiment with *Physarum polycephalum* as well as with any other swarm demonstrates that we cannot approximate its single actions. But we can use p-adic fuzzy and probability measures which are not additive to simulate the swarm behaviour. For describing experiments with swarms, I introduce a p-adic fuzzy syllogistic. On the basis of p-adic valued logic, it is possible to design p-adic valued fuzzy logic controllers of swarm propagations.

In Chap. 7, there is defined a syllogistic system for atomic propositions with two quantifiers: 'all neighbour attractants/repellents of an attractant/repellent' and 'some neighbour attractants/repellents of an attractant/repellent'. This syllogistic for swarms is verified in the p-adic valued universe of stimuli for controlling the swarm behaviour. In this chapter, I examine a syllogistic under stress (Aristotelian) and a syllogistic under favourable conditions (non-Aristotelian). The non-Aristotelian syllogistic can be used, for example, for simulating the propagation of cercariae—parasites of the genus *Trichobilharzia* Skrjabin and Zakharov, 1920 (*Schistosomatidae* Stiles and Hassall, 1898).

In Chap. 8, the behaviour of swarms is represented as a bio-inspired game, for which all basic definitions are verified in the experiments with swarms, such as *Physarum polycephalum*. This game depends upon localizations of attractants and repellents in the p-adic universe, i.e. it depends on the number $p - 1$ how many attractants and repellents can be detected by a swarm at each time step of its propagation. If there are not more than 4 attractants or repellents at each step, we deal with the 5-adic universe. In this 5-adic universe we can define a Go game as a zero-sum game between two species of the slime mould: *Physarum polycephalum* and *Badhamia utricularis*. The bio-inspired game defined in this chapter is especially interesting taking into account the fact that we cannot use classical game theory built on additive measures for simulating the swarm behaviour. The bio-inspired game theory proposed by me is based on context-based (or hybrid) actions assuming an unlimited variety of possible modifications of elementary acts.

In Chap. 9, there is introduced a basic logic for preference relations of swarm. We know that each action of swarm can have many possible modifications. As a consequence, preference relations of swarm are being modified, also. In this logic I focus on the following two ways of modifications in sensing and motoring: (i) the stress condition with increasing the proto-symbolic value of item; and (ii) the favourable condition with decreasing the proto-symbolic value of concrete items. Then I define an illocutionary logic for preference relations under stress and favourable conditions. This logic is verified on the group behaviour of alcoholics, too. Their behaviour can be managed by localization of places for meeting to drink.

In Chap. 10, I propose a generalization of bio-inspired games in the form of reflexive games. In these games we try to cheat other players. These games are based on non-Archimedean probabilities, as well, and by using these probabilities we can reject the Aumann's agreement theorem. Instead of the Aumann's statement, we can prove that the reflexion disagreement theorem is valid. According to this theorem, I can always cheat.

In Chap. 11, I show that bio-inspired and reflexive games can be represented in the form of especial cellular automata, where the more modified a cellular-automatic transition rule is at each time step, the more reflexive our game is. So, the Belousov-Zhabotinsky reaction can be described as a cellular automaton, for which the transition rule does not change. It means that it is a game of zero reflexion. If a transition rule is being modified at each time step, we deal with a reflexion of infinite level. Human beings are able to some higher reflexive levels. The infinite level is just an abstraction.

In Chap. 12, the rule from the Talmud *qal wa-ḥomer* is analysed. It is shown that this rule can simulate a spatial dynamics of swarm propagation under the favourable conditions, e.g. it was exemplified by the ant networking. Then I show that some analogy reasoning similar to *qal wa-ḥomer* are applied by mathematicians in cognitions of proof threes of several theorems simultaneously.

References

1. Bonabeau, E., Dorigo, M., Theraulaz, G.: Swarm Intelligence: From Natural to Artificial Systems. Oxford University Press (1999)
2. Kennedy, J., Eberhart, R.: Swarm Intelligence. Morgan Kaufmann Publishers, Inc. (2001)
3. Zelinka, I., Chen, G. (eds.): Evolutionary Algorithms, Swarm Dynamics and Complex Networks: Methodology, Perspectives and Implementation, vol. 26. Springer (2017)
4. Beni, G., Wang, J.: Swarm intelligence in cellular robotic systems. In: Dario, P., Sandini, G., Aebischer, P. (eds.) Robots and Biological Systems: Towards a New Bionics?, pp. 703–712. Springer, Berlin (1993)
5. Dorigo, M., Stutzle, T.: Ant Colony Optimization. MIT Press (2004)
6. John, A., Schadschneider, A., Chowdhury, D., Nishinari, K.: Characteristics of ant-inspired traffic flow. Swarm Intell. **2**(1), 25–41 (2008). https://doi.org/10.1007/s11721-008-0010-8
7. Karaboga, D.: An idea based on honey bee swarm for numerical optimization. Technical report-tr06, Engineering Faculty, Computer Engineering Department, Erciyes University (2005)
8. Karaboga, D., Akay, B.: A comparative study of artificial bee colony algorithm. Appl. Math. Comput. **214**(1), 108–132 (2009)
9. Michener, C.: Comparative social behavior of bees. Annu. Rev. Entomol. **14**, 299–342 (1969). https://doi.org/10.1146/annurev.en.14.010169.001503
10. Abrahams, M., Colgan, P.: Risk of predation, hydrodynamic efficiency, and their influence on school structure. Environ. Biol. Fishes **13**(3), 195–202 (1985)
11. Viscido, S., Parrish, J., Grunbaum, D.: Individual behavior and emergent properties of fish schools: a comparison of observation and theory. Marine Ecol. Prog. Series **273**, 239–249 (2004)
12. Reynolds, C.W.: Flocks, herds, and schools: a distributed behavioral model. Comput. Graph. **21**, 25–34 (1987)

13. Ben-Jacob, E.: Social behavior of bacteria: from physics to complex organization. Eur. Phys. J. B **65**(3), 315–322 (2008)
14. Ingham, C.J., Ben-Jacob, E.: Swarming and complex pattern formation in paenibacillus vortex studied by imaging and tracking cells. BMC Microbiol. **36**, 8 (2008)
15. Ingham, C.J., Kalisman, O., Finkelshtein, A., Ben-Jacob, E.: Mutually facilitated dispersal between the nonmotile fungus aspergillus fumigatus and the swarming bacterium paeni bacillus vortex. Proc. Nat. Acad. Sci. U.S.A. **108**(49), 19731–19736 (2011)
16. Ivanitsky, G.R., Kunisky, A.S., Tzyganov, M.A.: Study of 'target patterns' in a phage-bacterium system. In: Krinsky, V. (ed.) Self-organization: Autowaves and Structures Far From Equilibrium, pp. 214–217. Springer, Heidelberg (1984)
17. Margenstern, M.: Bacteria inspired patterns grown with hyperbolic cellular automata. In: HPCS, pp. 757–763 (2011)
18. Passino, K.M.: Biomimicry of bacterial foraging for distributed optimization and control. Control Syst. **22**(3), 52–67 (2002)
19. Tsuda, S., Aono, M., Gunji, Y.P.: Robust and emergent Physarum-computing. BioSystems **73**, 45–55 (2004)
20. Kassabalidis, I., El-Sharkawi, M.A., Marks, R.J., Arabshahi, P., Gray, A.A.: Swarm intelligence for routing in communication networks. In: 2001 Global Telecommunications Conference, GLOBECOM '01, vol. 6, pp. 3613–3617. IEEE (2001)
21. Ariel, G., Shklarsh, A., Kalisman, O., Ingham, C., Ben-Jacob, E.: From organized internal traffic to collective navigation of bacterial swarms. New J. Phys. **15**, 12501 (2013)
22. Shklarsh, A., Finkelshtein, A., Ariel, G., Kalisman, O., Ingham, C., Ben-Jacob, E.: Collective navigation of cargo-carrying swarms. Interface Focus **2**, 689–692 (2012)
23. Nakagaki, T., Iima, M., Ueda, T., Nishiura, Y., Saigusa, T., Tero, A., Kobayashi, R., Showalter, K.: Minimum-risk path finding by an adaptive amoeba network. Phys. Rev. Lett. **99**, 68–104 (2007)
24. Nakagaki, T., Yamada, H., Tothm, A.: Path finding by tube morphogenesis in an amoeboid organism. Biophys. Chem. **92**, 47–52 (2001)
25. Shirakawa, T., Yokoyama, K., Yamachiyo, M., Gunji, Y.-P., Miyake, Y.: Multi-scaled adaptability in motility and pattern formation of the Physarum plasmodium. Int. J. Bio-Inspir. Comput. **4**, 131–138 (2012)
26. Watanabe, S., Tero, A., Takamatsu, A., Nakagaki, T.: Traffic optimization in railroad networks using an algorithm mimicking an amoeba-like organism. Physarum plasmodium. Biosystems **105**(3), 225–232 (2011)
27. Whiting, J.G.H., de Lacy Costello, B., Adamatzky, A.: Transfer function of protoplasmic tubes of Physarum polycephalum. Biosystems **128**, 48–51 (2015)
28. Dorigo, M., Gambardella, L.M.: Ant colonies for the travelling salesman problem. Biosystems **43**(2), 73–81 (1997). https://doi.org/10.1016/S0303-2647(97)01708-5
29. Zhu, L., Aono, M., Kim, S.J., Hara, M.: Amoeba-based computing for traveling salesman problem: long-term correlations between spatially separated individual cells of Physarum polycephalum. Biosystems **112**(1), 1–10 (2013). https://doi.org/10.1016/j.biosystems.2013.01.008
30. Tero, A., Nakagaki, T., Toyabe, K., Yumiki, K., Kobayashi, R.: A method inspired by Physarum for solving the Steiner problem. Int. J. Unconv. Comput. **6**(2), 109–123 (2010)
31. Ozbakir, L., Baykasoglu, A., Tapkan, P.: Bees algorithm for generalized assignment problem. Appl. Math. Comput. **215**(11), 3782–3795 (2010). https://doi.org/10.1016/j.amc.2009.11.018
32. Nakagaki, T., Yamada, H., Toth, A.: Maze-solving by an amoeboid organism. Nature **407**, 470–470 (2000)
33. Ntinas, V.G., Vourkas, I., Sirakoulis, G.C., Adamatzky, A.: Oscillation-based slime mould electronic circuit model for maze-solving computations. IEEE Trans. Circuits Syst. **64-I**(6), 1552–1563 (2017)
34. Kalogeiton, V.S., Papadopoulos, D.P., Georgilas, I., Sirakoulis, G.C., Adamatzky, A.: Cellular automaton model of crowd evacuation inspired by slime mould. Int. J. Gen. Syst. **44**(3), 354–391 (2015). https://doi.org/10.1080/03081079.2014.997527

35. Adamatzky, A., Yang, X., Zhao, Y.: Slime mould imitates transport networks in China. Int. J. Intell. Comput. Cybern. **6**(3), 232–251 (2013). https://doi.org/10.1108/IJICC-02-2013-0005
36. Adamatzky, A., Ilachinski, A.: Slime mold imitates the united states interstate system. Complex Syst. **21**(1) (2012)
37. Schumann, A., Fris, V.: Swarm intelligence among humans—the case of alcoholics. In: Proceedings of the 10th International Joint Conference on Biomedical Engineering Systems and Technologies—Volume 4: BIOSIGNALS, (BIOSTEC 2017), pp. 17–25. ScitePress (2017)
38. Helbing, D., Farkas, I., Vicsek, T.: Simulating dynamical features of escape panic. Nature **407**(6803), 487–490 (2000). https://doi.org/10.1038/35035023
39. Shirakawa, T., Gunji, Y.-P., Miyake, Y.: An associative learning experiment using the plasmodium of Physarum polycephalum. Nano Commun. Netw. **2**, 99–105 (2011)
40. Traversa, F.L., Pershin, Y.V., Di Ventra, M.: Memory models of adaptive behavior. IEEE Trans. Neural Netw. Learn. Syst. **24**(9), 1437–1448
41. Ball, P.: Slime mould displays remarkable rhythmic recall. Nature **451**, 385 (2008). https://doi.org/10.1038/451385a
42. Saigusa, T., Tero, A., Nakagak, T., Kuramoto, Y.: Amoebae anticipate periodic events. Phys. Rev. Lett. **100**(1), 018101 (2008)
43. Skoge, M., Yue, H., Erickstad, M., Bae, A., Levine, H., Groisman, A., Loomis, W.F., Rappel, W.J.: Cellular memory in eukaryotic chemotaxis. Proc. Nat. Acad. Sci. U.S.A. **111**(40), 14448–14453 (2014). https://doi.org/10.1073/pnas.1412197111
44. Wang, Y., Li, B., Weise, T., Wang, J., Yuan, B., Tian, Q.: Self-adaptive learning based particle swarm optimization. Inf. Sci. **181**(20), 4515–4538 (2011)
45. Dimonte, A., Berzina, T., Pavesi, M., Erokhin, V.: Hysteresis loop and cross-talk of organic memristive devices. Microelectron. J. **45**(11), 1396–1400 (2014). https://doi.org/10.1016/j.mejo.2014.09.009
46. Erokhin, V.: On the learning of stochastic networks of organic memristive devices. Int. J. Unconv. Comput. **9**(3–4), 303–310 (2013)
47. Erokhin, V., Howard, G.D., Adamatzky, A.: Organic memristor devices for logic elements with memory. Int. J. Bifurc. Chaos **22**(11) (2012). https://doi.org/10.1142/S0218127412502835
48. Pershin, Y.V., La Fontaine, S., Di Ventra, M.: Memristive model of amoeba learning. Phys. Rev. E **80**(2), 021926 (2009)
49. Pershin, Y.V., Di Ventra, M.: Memristive and memcapacitive models of Physarum learning. In: Advances in Physarum Machines, pp. 413–422. Springer (2016)
50. Adamatzky, A., Erokhin, V., Grube, M., Schubert, T., Schumann, A.: Physarum chip project: growing computers from slime mould. Int. J. Unconv. Comput. **8**(4), 319–323 (2012)
51. Schumann, A., Pancerz, K.: Logics for physarum chips. Studia Humana **5**(1), 16–30 (2016). https://doi.org/10.1515/sh-2016-0002
52. Adamatzky, A.: Physarum machine: implementation of a Kolmogorov-Uspensky machine on a biological substrate. Parallel Process. Lett. **17**(4), 455–467 (2007)
53. Adamatzky, A.: Physarum machines: encapsulating reaction-diffusion to compute spanning tree. Naturwisseschaften **94**, 975–980 (2007)
54. Adamatzky, A.: Physarum Machines: Computers from Slime Mould. Series on Nonlinear Science. Series A. World Scientific (2010)
55. Adamatzky, A.: Slime mould logical gates: exploring ballistic approach. In: Applications, Tools and Techniques on the Road to Exascale Computing, vol. 1, pp. 41–56 (2010)
56. Adamatzky, A.: Slime mould computing. Int. J. Gen. Syst. **44**(3), 277–278 (2015)
57. Adamatzky, A., Schubert, T.: Slime mold microfluidic logical gates. Mater. Today **17**(2), 86–91 (2014)
58. Berzina, T., Dimonte, A., Cifarelli, A., Erokhin, V.: Hybrid slime mould-based system for unconventional computing. Int. J. Gen. Syst. **44**(3), 341–353 (2015). https://doi.org/10.1080/03081079.2014.997523
59. Jones, J., Mayne, R., Adamatzky, A.: Representation of shape mediated by environmental stimuli in Physarum polycephalum and a multi-agent model. JPEDS **32**(2), 166–184 (2017)

60. Jones, J.D., Adamatzky, A.: Towards Physarum binary adders. Biosystems **101**(1), 51–58 (2010)
61. Mayne, R., Adamatzky, A.: Slime mould foraging behaviour as optically coupled logical operations. Int. J. Gen. Syst. **44**(3), 305–313 (2015). https://doi.org/10.1080/03081079.2014.997528
62. Pancerz, K., Schumann, A.: Rough set models of Physarum machines. Int. J. Gen. Syst. **44**(3), 314–325 (2015)
63. Schumann, A.: *p*-adic valued logical calculi in simulations of the slime mould behaviour. J. Appl. Non-Class. Log. **25**(2), 125–139 (2015). https://doi.org/10.1080/11663081.2015.1049099
64. Schumann, A.: Towards slime mould based computer. New Math. Nat. Comput. **12**(2), 97–111 (2016). https://doi.org/10.1142/S1793005716500083
65. Whiting, J.G., de Lacy Costello, B.P., Adamatzky, A.: Slime mould logic gates based on frequency changes of electrical potential oscillation. Biosystems **124**, 21–25 (2014)
66. Baum, W.: Understanding Behaviorism: Behavior, Culture, and Evolution. Blackwell Pub (2005)
67. Cheney, C., Ferster, C.: Schedules of Reinforcement. Copley Publishing Group (1997)
68. Skinner, B.F.: About Behaviorism. Random House Inc, New York (1976)
69. Bandura, A.: Self-efficacy mechanism in human agency. Am. Psychol. **37**, 122–147 (1982)
70. Aumann, R.J.: Agreeing to disagree. Ann. Stat. **4**(6), 1236–1239 (1976)
71. Aumann, R.J.: Notes on Interactive Epistemology. Mimeo, Hebrew University of Jerusalem, Jerusalem (1989)
72. Berger, P., Luckmann, T.: The Social Construction of Reality. Anchor Books (1971)
73. Smith, V.L.: Constructivist and ecological rationality in economics. Am. Econ. Rev. **93**(3), 465–508 (2003)
74. Barwise, J., Etchemendy, J.: The Liar. Oxford University Press, New York (1987)
75. Barwise, J., Moss, L.: Vicious Circles. Stanford (1996)
76. Thaler, R.: Saving, fungibility, and mental accounts. J. Econ. Perspect. **4**(1), 193–205 (1990)
77. Thaler, R.: Psychology and saving policies. Am. Econ. Rev. **84**(2), 186–192 (1994)
78. Tversky, A., Kahneman, D.: Judgments under uncertainty: Heuristics and biases. Science **185**(4157), 1124–1131 (1974)
79. Heaton, J.B.: Managerial optimism and corporate finance. Financ. Manag. **31**(2), 33–45 (2002)
80. Arnold, V., Sutton, S.: Behavioural Accounting Research: Foundations and Frontiers. American Accounting Association (1997)
81. Barberis, N., Huang, M.: Mental accounting, loss aversion, and individual stock returns. J. Finance **56**(4), 1247–1292 (2001). https://doi.org/10.1111/0022-1082.00367
82. Barberis, N., Huang, M., Santos, T.: Prospect theory and asset prices. Working Paper 7220, National Bureau of Economic Research (1999). https://doi.org/10.3386/w7220
83. Barberis, N., Shleifer, A., Vishny, R.W.: A model of investor sentiment. Working Paper 5926, National Bureau of Economic Research (1997). https://doi.org/10.3386/w5926
84. Chen, J., Hong, H., Stein, J.C.: Forecasting crashes: trading volume, past returns, and conditional skewness in stock prices. J. Financ. Econ. **61**(3), 345–381 (2001). https://doi.org/10.1016/S0304-405X(01)00066-6
85. Gilovich, T., Griffin, D., Kahneman, D. (eds.): Heuristics and Biases: The Psychology of Intuitive Judgment. Cambridge University Press (2002)
86. Goldberg, J., Nitzch, R.V.: Behavioural Finance. Wiley (2001)
87. Kahneman, D., Tversky, A.: Prospect theory: an analysis of decision under risk. Econometrica **47**(2), 263–291 (1979)
88. Kahneman, D., Tversky, A. (eds.): Choices, Values and Frames. Cambridge University Press (2000)
89. Shleifer, A.: Inefficient Markets: An Introduction to Behavioural Finance. Oxford University Press (2000)
90. Hofstedt, T.R., Kinard, J.C.: A strategy for behavioral accounting research. Account. Rev. **45**(1), 38–54 (1970)

91. Frederick, S.: Cognitive reflection and decision making. J. Econ. Perspect. **19**(4), 24–42 (2005)
92. Gigerenzer, G., Brighton, H.H.H.: Why biased minds make better inferences. Top. Cogn. Sci. **1**(1), 107–143 (2009)
93. Taylor, E., Baylis, H.: Observations and experiments on a dermatitis-producing cercaria and on another cercaria from limna stagnalis in great britain. Trans. Royal Soc. Trop. Med. Hyg. **24**(2), 219–244 (1930). https://doi.org/10.1016/S0035-9203(30)92001-9
94. Beuret, J., Pearson, J.C.: Description of a new zygocercous cercaria (opisthorchioidea: Heterophyidae) from prosobranch gastropods collected at heron island (great barrier reef, australia) and a review of zygocercariae. Syst. Parasitol. **27**(2), 105–125 (1994). https://doi.org/10.1007/BF00012269
95. Adamatzky, A. (ed.): Advances in Physarum Machines: Sensing and Computing with Slime Mould, vol. 21. Springer (2016)
96. Khrennikov, A.: p-adic probability interpretation of Bell's inequality. Phys. Lett. A **200**(3–4), 219–223 (1995)
97. Khrennikov, A.: Interpretations of probability and their p-adic extensions. Theory Probab. Appl. **46**(2), 256–273 (2001)
98. Khrennikov, A.: Toward theory of p-adic valued probabilities. Stud. Logic, Grammar Rhetor. **14**(27), 137–154 (2008)
99. Khrennikov, A., Yamada, S., van Rooij, A.: The measure-theoretical approach to p-adic probability theory. Ann. Math. Blaise Pascal **6**(1), 21–32 (1999)
100. Khrennikov, A.Y., Schumann, A.: Logical approach to p-adic probabilities. Bull. Sect. Log. **35**(1), 49–57 (2006)
101. Schumann, A.: Non-archimedean valued predicate logic. Bull. Sect. Log. **36**(1–2), 67–78 (2007)
102. Schumann, A.: Non-archimedean fuzzy and probability logic. J. Appl. Non-Class. Log. **18**(1), 29–48 (2008)
103. Volovich, I.V.: Number theory as the ultimate theory. Technical report, CERN preprint, CERN-TH.4791/87 (1987)
104. Koblitz, N.: p-adic Numbers, p-adic Analysis and Zeta Functions, 2nd ed. Springer-Verlag (1984)
105. Schumann, A.: DSm models and non-archimedean reasoning. In: Smarandache, F., Dezert, J. (eds.) Advances and Applications of DSmT (Collected works), vol. 2, pp. 183–204. American Research Press, Rehoboth (2006)
106. Schumann, A.: Non-archimedean valued sequent logic. In: Eighth International Symposium on Symbolic and Numeric Algorithms for Scientific Computing (SYNASC'06), pp. 89–92. IEEE Press (2006)
107. Schumann, A.: Reflexive games and non-archimedean probabilities. P-adic Numbers, Ultrametr. Anal. Appl. **6**(1), 66–79 (2014)
108. Schumann, A.: Probabilities on streams and reflexive games. Oper. Res. Decis. **24**(1), 71–96 (2014). https://doi.org/10.5277/ord140105
109. Schumann, A.: Payoff cellular automata and reflexive games. J. Cell. Autom. **9**(4), 287–313 (2014)
110. Schumann, A.: Towards context-based concurrent formal theories. Parallel Process. Lett. **25**, 1540,008 (2015)
111. Pavlović, D., Escardó, M.H.: Calculus in coinductive form. In: Proceedings of the 13th Annual IEEE Symposium on Logic in Computer Science, pp. 408–417 (1998)
112. Rutten, J.J.M.M.: Behavioral differential equations: a coinductive calculus of streams, automata, and power series. Theor. Comput. Sci. **308**, 1–53 (2003)
113. Rutten, J.J.M.M.: A coinductive calculus of streams. Math. Struct. Comput. Sci. **15**(1), 93–147 (2005)
114. Rutten, J.J.M.M.: Universal coalgebra: a theory of systems. Theor. Comput. Sci **249**(1), 3–80 (2000)
115. Aczel, A.: Non-Well-Founded Sets. Stanford University Press (1988)
116. Barwise, J., Moss, L.: Hypersets. Springer Verlag, New York (1992)

117. Milner, R.: Communicating and Mobile Systems: The π-calculus. Cambridge University Press, Cambridge (1999)
118. Adamatzky, A., De Lacy Costello, B., Asai, T.: Reaction-Diffusion Computers. Elsevier (2005)
119. Wolfram, S.: Universality and complexity in cellular automata. Physica D **10**, 1–35 (1984)
120. Khrennikov, A., Schumann, A.: p-adic physics, non-well-founded reality and unconventional computing. P-adic Numbers, Ultrametr. Anal. Appl. **1**(4), 297–306 (2009)
121. Schumann, A.: Non-well-founded probabilities within unconventional computing. In: Durham, U.K. (ed.) 6th International Symposium on Imprecise Probability: Theories and Applications (2009)
122. Schumann, A.: Proof-theoretic cellular automata as logic of unconventional computing. Int. J. Unconv. Comput. **8**(3), 263–280 (2012)
123. Hennessy, M., Milner, R.: Algebraic laws for nondeterminism and concurrency. JACM **32**(1), 137–161 (1985)
124. Schumann, A.: Towards theory of massive-parallel proofs. cellular automata approach. Bull. Sect. Log. **39**(3/4), 133–145 (2010)
125. Dam, M.: Proof systems for pi-calculus logics. In: de Queiroz, R. (ed.) Logic for Concurrency and Synchronisation, pp. 145–212. Kluwer (2003)
126. Rutten, J.J.M.M.: Processes as terms: non-well-founded models for bisimulation. Math. Struct. Comp. Sci. **2**(3), 257–275 (1992)
127. Schumann, A., Pancerz, K., Adamatzky, A., Grube, M.: Bio-inspired game theory: the case of Physarum polycephalum. In: Suzuki, J., Nakano, T. (eds.) Proceedings of the 8th International Conference on Bio-inspired Information and Communications Technologies (BICT'2014), pp. 9–16. Boston, Massachusetts, USA (2014)
128. Schumann, A., Pancerz, K., Adamatzky, A., Grube, M.: Context-based games and Physarum polycephalum as simulation model. In: Proceedings of the Workshop on Unconventional Computation in Europe. London, UK (2014)
129. Adamatzky, A.: Reaction-diffusion algorithm for constructing discrete generalized voronoi diagram. Neural Netw. World **6**, 635–643 (1994)
130. de Lacy Costello, : B.P.J., Adamatzky, A., Ratcliffe, N.M., Zanin, A., Purwins, H.G., Liehr, A.: The formation of voronoi diagrams in chemical and physical systems: experimental findings and theoretical models. Int. J. Bifurc Chaos **14**, 2187–2210 (2004)
131. Pancerz, K., Schumann, A.: Principles of an object-oriented programming language for Physarum polycephalum computing. In: Proceedings of the 10th International Conference on Digital Technologies (DT'2014), pp. 273–280. Zilina, Slovak Republic (2014)
132. Pancerz, K., Schumann, A.: Some issues on an object-oriented programming language for Physarum machines. In: Bris, R., Majernik, J., Pancerz, K., Zaitseva, E. (eds.) Applications of Computational Intelligence in Biomedical Technology, Studies in Computational Intelligence, vol. 606, pp. 185–199. Springer International Publishing, Switzerland (2016)
133. Schumann, A., Pancerz: Physarumsoft—a software tool for programming Physarum machines and simulating Physarum games. In: Ganzha, M., Maciaszek, L., Paprzycki, M. (eds.) Proceedings of the 2015 Federated Conference on Computer Science and Information Systems (FedCSIS'2015), pp. 607–614. Lodz, Poland (2015)
134. Schumann, A., Pancerz, K.: Towards an object-oriented programming language for Physarum polycephalum computing. In: Szczuka, M., Czaja, L., Kacprzak, M. (eds.) Proceedings of the Workshop on Concurrency, Specification and Programming (CS&P'2013), pp. 389–397. Warsaw, Poland (2013)
135. Schumann, A., Pancerz, K.: Timed transition system models for programming Physarum machines: extended abstract. In: Popova-Zeugmann, L. (ed.) Proceedings of the Workshop on Concurrency, Specification and Programming (CS&P'2014), pp. 180–183. Chemnitz, Germany (2014)
136. Schumann, A., Pancerz, K.: Towards an object-oriented programming language for Physarum polycephalum computing: A Petri net model approach. Fundam. Informaticae **133**(2–3), 271–285 (2014)

137. Schumann, A., Pancerz, K.: Petri net models of simple rule-based systems for programming Physarum machines: extended abstract. In: Suraj, Z., Czaja, L. (eds.) Proceedings of the 24th International Workshop on Concurrency, Specification and Programming (CS&P'2015), vol. 2, pp. 155–160. Rzeszow, Poland (2015)
138. Pancerz, K., Schumann, A.: Rough set description of strategy games on Physarum machines. In: Adamatzky, A. (ed.) Advances in Unconventional Computing, Volume 2: Prototypes, Models and Algorithms, Emergence, Complexity and Computation, vol. 23, pp. 615–636. Springer International Publishing (2017)
139. Schumann, A., Pancerz: Roughness in timed transition systems modeling propagation of plasmodium. In: Ciucci, D., Wang, G., Mitra, S., Wu, W.Z. (eds.) Rough Sets and Knowledge Technology. Lecture Notes in Artificial Intelligence, vol. 9436. Springer International Publishing (2015)
140. Schumann, A., Pancerz, K.: A rough set version of the go game on Physarum machines. In: Suzuki, J., Nakano, T., Hess, H. (eds.) Proceedings of the 9th International Conference on Bio-inspired Information and Communications Technologies (BICT'2015), pp. 446–452. New York City, New York, USA (2015)
141. Schumann, A., Pancerz, K.: Physarumsoft: an update based on rough set theory. AIP Conf. Proc. **1863**(1), 360005 (2017). https://doi.org/10.1063/1.4992534
142. Schumann, A.: Go games on plasmodia of Physarum polycephalum. In: 2015 Federated Conference on Computer Science and Information Systems, FedCSIS 2015, Lódz, Poland, 13–16 Sept 2015, pp. 615–626 (2015). https://doi.org/10.15439/2015F236
143. Schumann, A.: Syllogistic versions of go games on Physarum. In: Adamatzky, A. (ed.) Advances in Physarum Machines: Sensing and Computing with Slime Mould, pp. 651–685. Springer International Publishing, Cham (2016)
144. Schumann, A., Pancerz, K.: Interfaces in a game-theoretic setting for controlling the plasmodium motions. In: Proceedings of the 8th International Conference on Bio-inspired Systems and Signal Processing (BIOSIGNALS'2015), pp. 338–343. Lisbon, Portugal (2015)
145. Adamatzky, A., De Lacy Costello, B.: Experimental logical gates in a reaction-diffusion medium: the XOR gate and beyond. Phys. Rev. E **66**, 46–112 (2002)
146. Adamatzky, A., Wuensche, A.: Computing in spiral rule reaction-diffusion hexagonal cellular automaton. Complex Syst. **16**, 4 (2007)
147. Schumann, A., Adamatzky, A.: Towards semantical model of reaction-diffusion computing. Kybernetes **38**(9), 1518–1531 (2009)
148. Běhounek, L., Cintula, P.: Fuzzy class theory. Fuzzy Sets Syst. **154**(1), 34–55 (2005)
149. Cintula, P.: The $l\pi$ and $l\pi\frac{1}{2}$ propositional and predicate logics. Fuzzy Sets Syst. **124**(3), 21–34 (2001)
150. Cintula, P.: Advances in the $l\pi$ and $l\pi\frac{1}{2}$ logics. Arch. Math. Log. **42**(5), 449–468 (2003)
151. Hájek, P.: Metamathematics of Fuzzy Logic. Springer, Netherlands, Dordrecht (1998)

Chapter 2
Actin Filament Networks

The plasmodium of *Physarum polycephalum* is very sensitive to its environment and reacts to stimuli by its appropriate motions. The sensitive stage as well as the motor stage of these reactions are explained by hydrodynamic processes, based on fluid dynamics, with participating of actin filament networks. This chapter is devoted to *actin filament networks* as a computation medium, see [1–8]. The point is that actin filaments with a participating of many other proteins like myosin are sensitive to outer cellular stimuli (attractants as well as repellents) and they appear and disappear at different places of the cell to change the cell structure, e.g. its shape. Due to assembling and disassembling actin filaments, some unicellular organisms like *Amoeba proteus* can move in responses to different stimuli. As a result, these organisms can be considered a simple reversible logic gate, where outer cellular signals are its inputs and the motions are its outputs. In this way, we can implement different logic gate on the amoeboid behaviours. These networks can embody arithmetic functions within p-adic valued logic. Furthermore, within these networks we can define the so-called diagonalization for deducing undecidable arithmetic functions.

In physics we analyze the world by applying some mathematical equations. So, the main question is that whether the mathematical tools are enough for understanding physical systems. In other words, whether physical systems can be explicated by means of solvable or decidable computational functions in fact. Let us remember that a function is *decidable* if there is an algorithm (a finite procedure) in a Turing machine saying how to compute this function by finite steps. Otherwise the function is *undecidable* or *unsolvable*. In theory of computation there are different classifications of computational functions according to their inherent difficulty from functions simulated by deterministic Turing machines (decidable arithmetic functions) to unsolvable computational problems (undecidable arithmetic functions). And in applied physics there is the same trouble—we should know whether mathematical equations of physics can be computed in general in fact.

A. Schumann, *Behaviourism in Studying Swarms: Logical Models of Sensing and Motoring*, Emergence, Complexity and Computation 33,
https://doi.org/10.1007/978-3-319-91542-5_2

In unconventional computing any physical system is regarded as a computer. *Physarum polycephalum* is one of the unicellular organisms best studied from the standpoint of computation theory [9–29]. In the project *Physarum Chip Project: Growing Computers From Slime Mould* [30], we have designed some processors on the basis of the *Physarum polycephalum* motions. There are some other models to formalize the plasmodium behaviour: a memory model [31, 32], an auto-oscillatory model [33, 34] and many others. In the memory model, there are used some memory circuit elements, such as resistors, capacitors and inductors with memory whose state depends on the history of signals applied [31, 32]. In the auto-oscillatory model, autooscillations and autowaves are explicated on the basis of properties of the actomyosin filament system [34]. The plasmodium of *Physarum polycephalum* consists of different proteins, among them actin filament networks responsible for the plasmodial cytoskeleton participate in the intelligent behaviour of *Physarum polycephalum* [35, 36].

In this chapter, I try to consider which decidable and undecidable arithmetic functions can be implemented on the medium of actin filament networks. In this way I show that any cell can be considered a computer due to its actin filament networks, in particular *Amoeba proteus* can be regarded as a logic gate. In actin filament networks we can define decidable and undecidable arithmetic functions.

Hence, I support the idea of *orchestrated objective reduction*, i.e. of a hypothesis that our consciousness originates from processes inside neurons, rather than from connections between neurons [37] and actin filament networks participate in intelligent reactions of each cell (as well as of neuron) to its environment [38]. So, in this chapter I am going to show the role of actin filaments (as an intercellular mechanism) playing in the biological intelligence. The main result of this research is in coding amoeboid motions by arithmetic functions and in expressing undecidable arithmetic functions within actin filament networks. The point is that each cell due to actin filament waves has more outputs than inputs. Therefore, some arithmetic functions implemented in this behaviour can be undecidable. It means that actin filament networks implement a stronger intelligence than any Turing machine.

This chapter is focused on computational aspects of actin filament waves and introduces some idealizations in considering the amoeboid movement. Thus, I exemplify actin filament networks by behaviours of *Amoeba proteus*, rather than by *Physarum polycephalum*, because the actin filament mechanism such as shear deformation, bending deformation, and orientation deformation in responding to outside signals is more visible on amoebas. Activated and deactivated zones of actin filament waves are used for coding the amoeboid motions by arithmetic functions. Hence, the physics of this mechanism can be different, e.g. we have different motions of *Amoeba proteus* and *Physarum polycephalum*, but in any case the mathematics of these motions with activated and deactivated zones of actin filament waves remains the same. This research is concentrated just on mathematics—possibilities to implement decidable an undecidable arithmetic functions within actin filament networks. See [3–5].

2.1 Reversible Logic Gates on Actin Filaments

2.1.1 Unconventional Computing on Actin Filaments

In machine learning there have been used some biologically inspired networks such as *artificial neural networks*, where we have a fixed number of processors involved into computations. So, in neural networks we deal with a system $\langle N, V, w \rangle$, where N is the set of processors called "neurons", V is a set of tuples $\{(i,j): i,j \in N\}$ whose elements are connections between neuron i and neuron j, and w is a function from V to $\mathbf{R}$ such that $w((i,j))$, for short $w_{i,j}$, is called the weight of the connection between neuron i and neuron j. Usually, the weights of a neural network are defined as a *Hinton diagram* presented as a square weight matrix W with the row number indicating, where the connection begins, and the column number indicating, which neuron is the target. Meanwhile, the weight 0 designates a non-existing connection.

However, elementary computational units understood as processors in real biological networks are never fixed. They are being built and rebuilt permanently in responses to different stresses as external conditions. One of the best examples of these units in biological networks is presented by *actin filaments*. Their networks are most important in remodelling cell configurations and in the cell motility. The point is that actin filament networks are engaged in changing the cell shape, for example in the division of one cell into two daughter cells and in the protrusion of parts of the cells, e.g. in the cell deformation by means of growing pseudopodia during phagocytosis. Meanwhile, the actin filaments are being assembled and disassembled during the time. As a consequence, we face a permanent assembly and disassembly of actin filament networks. These changes in networking are cell responses to different stimuli. For instance, actin filament networks by the own reconstruction can transmit internal stresses and govern the spatial organization of the cytoskeleton. So, these networks can provide signal transduction pathways and make a mechanical equilibrium of the cell and its environment.

The actin filament networks react to external stresses. These stresses are inputs for the networks. There are three main types of the actin filament networks: *unstable bunches* (parallel unbranched filaments), *trees* (branched filaments), *stable bunches* (cross-linked filaments), see: [39–41]. Outputs are different for different network types. For unstable bunches and trees, the outputs are represented by chemotaxis, cell spreading, nerve growth-cone movement, etc. For stable bunches, the outputs have the form of mechanic stress transduction such as (i) a tensional stress in the distortion to the network; (ii) a curvature stress in the deformation of the network; (iii) an orientational stress in the deformation of the network.

Due to actin filaments, the amoeboid cell motility is possible. In robototechnics, one species of the Mycetozoan group of the Amoebozoa, which is characterized by this kind of motility, is best studied. It is *Physarum polycephalum*. Each *plasmodium* of *Physarum polycephalum* has both an external stationary ectoplasm and an internal liquid endoplasm that moves to pseudopodia. The amoeboid motility is classified by the following three stages: (i) the stage of growing pseudopodia; (ii) the stage of attaching the pseudopodium to a substrate; (iii) the stage of dragging up the rest of the cell, see [42–44]. Such a movement requires an oscillatory mode of contractility system where an actin filament network is being assembled and disassembled for an equilibrium between ectoplasm and endoplasm [45], to change the cell shape [46]. Hence, the *Physarum polycephalum* plasmodium motility that is so intelligent and can be programmed by using chemotaxis so that we can design computers on their media is based on some actin filament network properties.

Actin filament networks are a universal mechanism in the reception and further transmission of external stimuli/stresses in any biological organism. By chemicals it is possible to govern an actin polymerisation and depolymerisation, i.e. the filament assembly and disassembly. For instance, on the one hand, cytochalasin B, the cell-permeable mycotoxin, strongly inhibits the formation of actin filament networks. On the other hand, the mycotoxin of *Amanita phalloides* strongly activates the aggregation of all cell G-actin into filaments which become Ph-actin and cannot be depolymerised and thus this taxin avoids any dynamics of cytoskeleton.

Actin filament networks are studied from the point of view of biochemistry in the following papers: [38, 47–56]. From the point of view of computer science, actin filament networks started to be studied just recently: [1, 2, 6–8].

Hence, theoretically we can assume that it is possible to synthesize and then govern/program an *artificial actin filament network*. This network is fundamental in any cell reaction to stimuli. As a result, this artificial network could be considered a biological computer/chip as such. Its computational processes would be primary for organic intelligence in principle.

For programming the artificial actin filament network, it is better to appeal to an *object-oriented programming* that models objects rather than actions and simulates aposterioristic data rather than aprioristic logic. In this programming for simulating the motions of actin filament waves we are based on process-algebraic formalizations of the artificial actin filament network. So, we define some instructions in the terms of *process algebra* like as follows: add node, remove node, add edge, remove edge. Adding and removing nodes can be implemented through activation and deactivation of attractants, respectively. Adding and removing edges can be implemented by means of repellents put in proper places in the space. An activated repellent can avoid a transition between attractants. Adding and removing edges can change dynamically over time. To model such a behaviour, we propose a high-level model, based on timed transition systems. In this model we define the following four basic forms of transitions (motions): *direct* (direction, a movement from one point, where the artificial cell or metacell is located, towards another point, where there is a neighbouring attractant), *fuse* (fusion of two actin filament fronts at the point, where they meet the same attractant), *split* (splitting one actin filament front into two active fronts, where two neighbouring attractants with a similar power of intensity are located), and *repel* (repelling of actin filament front or inaction).

Some computational features of programming the artificial actin filament network are as follows:

- We deal with a system $\langle N', V', w' \rangle$, where (i) N' is a non-well-founded set of processors called "filaments"; this set is non-well-founded, because it is impossible to divide N' into atoms or even just into excluded subsets n_j which form a partition of $N' = \bigsqcup_j n_j$; in other words, processors are being redesigned permanently and they can appear and disappear and ever change own features; (ii) V' is a set of tuples $\{(i_t, j_t): i_t, j_t \in N'\}$ whose elements are connections between filament i_t and filament j_t at time step t; hence, the set V' is non-well-founded, too, as its cardinality can change during the time t; (iii) w' is a function from V' to $^*\mathbf{R}$, where $^*\mathbf{R}$ is a set of hyperreal numbers such that $w'((^*i, ^*j))$, where $^*i = i_0, i_1, i_2, i_3, \ldots$ and $^*j = j_0, j_1, j_2, j_3, \ldots$, for short $w_{^*i,^*j}$, is called the weight of the connection between filament $^*i$ and filament $^*j$ at each time step $t = 0, 1, 2, 3, \ldots$; notice that a filament $^*i$ can be hidden (not present) at actual time.
- Each filament behaves as an *artificial organism* (e.g. as an artificial cell): (i) it can grow up; (ii) its behaviour can be attracted/directed by chemotaxis. In cells there are ever filaments, some of them 'die' and some others 'born'.
- Each unstable actin filament network behaves as an *artificial swarm*: (i) it can grow up; (ii) its behaviour can be attracted/directed by chemotaxis [57]; (iii) there can be a fusion of two swarms (two actin filament networks) into one swarm (one network); (iv) there can be a splitting of one swarm (one network) into two swarms (two networks). In cells there are ever some actin filament networks.
- Each cross-linked actin filament network behaves as a *metaswarm*: it reacts to mechanical external stimuli and it can be partly reorganized.

Thus, in constructing the actin filament network $\langle N', V', w' \rangle$ we have the following three levels of computations: (i) an artificial organism (filament); (ii) an artificial swarm (unstable bunch or tree); (iii) a metaswarm (stable bunch). On all three levels we face an instability of computation substratum. So, we assume that an *actin filament chip* changes its configurations during the computation processes. In designing this chip we can appeal to swarm computing, e.g. to slime mould computing. In the same measure as the neural networks, the actin filament networks can be used in pattern recognitions. So, due to the actin filament networks any slime mould or amoeba can recognize its dynamic environment and occupy the pieces of food/attractants.

2.1.2 Actin Filaments as "Swarms"

In any cell there is a huge number of actin monomers or *globular actin* (the so-called G-actin) denoted by A_i. Monomers are involved into computations only within actin filaments, minimally consisting of three actin monomers. G-actin A_i can assemble into double helical filaments of 7–8 nm in diameter and of several microns in length.

Filaments are (re)designed by the processes of nucleation, polymerization, and depolymerization:

- *Nucleation*. For nucleation that allows monomers to be assembled into filaments it is enough to bound three G-actin monomers at first. Then for enlargement an actin filament must be polarised with a barbed end that is plus at which monomer addition is faster than at the pointed end that is minus. So, actin filaments are morphologically asymmetric with different kinetic characteristics at two ends, and this fact helps to polymerize filamentous actin (the so-called F-actin) from the head (plus end) to the tail (minus end), see please [58–63].

- *Polymerization.* It is an association of new filaments mediated by actin cross-linking proteins: α-actinin, fascin, fimbrin, and filamin. This assembly is carried out on the basis of adding new monomers at the barbed end. Meanwhile, at the steady state, when a filament finishes to grow, the net rate of disassembly matches the rate of assembly at the plus end [64–68].
- *Depolymerization.* It is a dissociation of filaments which takes place when the critical concentration for actin polymerization is less than the dissociation constants at the two filament ends [69].

Thus, an actin filament consists of two filament strands in the helical form. For each actin monomer A_i in both strands there are three possible states: *bound on the left*, lA_i; *bound on the right*, rA_i; and *bound on both sides*, bA_i. Out of any filament, the monomer A_i is considered *free*, fA_i, or turned off. Monomers in the first strand are distinguished by $sA_i^\bullet$, where $s \in \{l, r, b\}$. Monomers in the second strand are distinguished by sA_j°, where $s \in \{l, r, b\}$.

Definition 2.1 Polymerization is a process calculus defined as follows:

$$names \ := fA_i \,|,\ lA_i^\bullet \mid rA_i^\bullet \mid bA_i^\bullet \mid lA_j^\circ \mid rA_j^\circ \mid bA_j^\circ;$$

$$\&, bounding \ := \textbf{if } fA_j^\star \& rA_i^\star, \textbf{ then}$$

$$fA_j^\star \textbf{ is } rA_{i+1}^\star \textbf{ and } rA_i^\star \textbf{ is } bA_i^\star \mid$$

$$\textbf{if } fA_j^\star \& lA_i^\star, \textbf{ then}$$

$$fA_j^\star \textbf{ is } lA_{i-1}^\star \textbf{ and } lA_i^\star \textbf{ is } bA_i^\star \mid$$

$$\textbf{if } lA_j^\star \& rA_i^\star, \textbf{ then}$$

$$lA_j^\star \textbf{ is } bA_{i+1}^\star \textbf{ and } rA_i^\star \textbf{ is } bA_i^\star \mid$$

$$\textbf{if } fA_i^\star \& fA_j^\star, \textbf{ then}$$

$$fA_i^\star \textbf{ is } rA_{j+1}^\star \textbf{ and } fA_j^\star \textbf{ is } lA_j^\star \textbf{ or}$$

$$fA_i^\star \textbf{ is } lA_i^\star \textbf{ and } fA_j^\star \textbf{ is } rA_{i+1}^\star,$$

where $\star \in \{\circ, \bullet\}$;

$$P, processes \ := \& \mid P_1 | P_2.$$

In this definition, we have defined filaments as growing from the right side. So, a free monomer $fA_i^\star$ can interact with another free monomer $fA_j^\star$ and, as a result, we can have a strand $rA_{j+1}^\star lA_j^\star$ or a strand $rA_{i+1}^\star lA_i^\star$. A free monomer $fA_i^\star$ can also interact with a monomer $lA_j^\star$, and, as a result, we obtain a strand $\ldots bA_j^\star lA_{j-1}^\star$. A free monomer $fA_i^\star$ can also interact with a monomer $rA_j^\star$, and, as a result, we obtain a strand $rA_{j+1}^\star bA_j^\star \ldots$. A monomer $rA_i^\star$ can associate with a monomer $lA_j^\star$ to give a strand $\ldots bA_{i+1}^\star bA_i^\star \ldots$.

Definition 2.2 Depolymerization is a process calculus defined as follows:

$$names \ := fA_i \,|,\ lA_i^\bullet \mid rA_i^\bullet \mid bA_i^\bullet \mid lA_j^\circ \mid rA_j^\circ \mid bA_j^\circ;$$

$$\overline{\&}, unbounding \ := \textbf{if } bA_{i+1}^\star \overline{\&} bA_i^\star, \textbf{ then}$$

$$bA_{i+1}^\star \textbf{ is } lA_j^\star \textbf{ and } bA_i^\star \textbf{ is } rA_i^\star \mid$$

$$\textbf{if } bA_{i+1}^\star \overline{\&} lA_i^\star, \textbf{ then}$$

$$bA_{i+1}^\star \textbf{ is } lA_{i+1}^\star \textbf{ and } lA_i^\star \textbf{ is } fA_i^\star \mid$$

$$\textbf{if } bA_j^\star \overline{\&} rA_{j+1}^\star, \textbf{ then}$$

$$bA_j^\star \textbf{ is } rA_j^\star \textbf{ and } rA_{j+1}^\star \textbf{ is } fA_{j+1}^\star \mid$$

$$\textbf{if } rA_{i+i}^\star \overline{\&} lA_i^\star, \textbf{ then}$$

$$rA_{i+1}^\star \textbf{ is } fA_{i+1}^\star \textbf{ and } lA_i^\star \textbf{ is } fA_i^\star;$$

$$P, processes \ := \overline{\&} \mid P_1 | P_2.$$

Hence, a monomer $lA_i^\star$ can dissociate from a strand $\ldots bA_{i+1}^\star lA_i^\star$ or a strand $rA_{i+1}^\star lA_i^\star$ and to become a free monomer $fA_i^\star$. A monomer $rA_{i+1}^\star$ can dissociate from a strand $rA_{i+1}^\star lA_i^\star$ or a strand $rA_{i+1}^\star bA_i^\star \ldots$ and to become a free monomer $fA_{i+1}^\star$. A strand $\ldots bA_{i+1}^\star bA_i^\star \ldots$ can be divided into two strands $\ldots lA_{i+1}^\star$ and $rA_i^\star \ldots$.

Let $[sA_i^\star]$, where $s \in \{l, r, b\}$, mean a value of $sA_i^\star$: (i) $[sA_i^\star] = 1$ iff $sA_i^\star$ is exited; and (ii) $[sA_i^\star] = 0$ iff $sA_i^\star$ is not excited. A signal is transmitted in filaments $\ldots bA_{i+1}^\star lA_i^\star$ form the left to the right hand and the signal transmission defined as follows:

$$[bA_i^\circ] = \begin{cases} 1, \text{ if either } [bA_{i+1}^\circ] \wedge [bA_{i+1}^\bullet] = 1 \text{ or} \\ \quad [bA_i^\circ] \wedge [bA_i^\bullet] = 1; \\ 0, \text{ otherwise.} \end{cases}$$

$$[bA_i^\bullet] = \begin{cases} 1, \text{ if either } [bA_{i+1}^\bullet] \wedge [bA_{i+1}^\circ] = 1 \text{ or} \\ \quad [bA_i^\bullet] \wedge [bA_i^\circ] = 1; \\ 0, \text{ otherwise.} \end{cases}$$

$$[lA_i^\circ] = \begin{cases} 1, \text{ if } [bA_{i+1}^\circ] \wedge [bA_{i+1}^\bullet] = 1; \\ 0, \text{ otherwise.} \end{cases}$$

$$[lA_i^\bullet] = \begin{cases} 1, \text{ if } [bA_{i+1}^\bullet] \wedge [bA_{i+1}^\circ] = 1; \\ 0, \text{ otherwise.} \end{cases}$$

Each filament is instable because of the faster net loss of G-actin at the pointed end than at the barbed end and the faster net addition at the barbed end than at the pointed end. If there is an equilibrium of two rates of association and dissociation, it gives rise to a *treadmilling* when there is a net flow of actin subunits through the filament [70]. The rate of treadmilling may be altered by the inhibition of disassembly at the pointed end, e.g. by phosphate [69]. Increasing the rate of monomer dissociation at the pointed end, e.g. by cofilin can accelerate treadmilling, also. Proteins that control treadmilling of actin filaments and their adhesions are classified as follows:

$$C, control \ := \ \text{SeqP} \mid \text{CrossP} \mid \text{SevP} \mid \text{NucP},$$

where

- SeqP are *sequestering proteins*: (i) β-thymosins which sequester G-actin to prevent spontaneous nucleation; (ii) profilin which interacts with actin monomers to enhance nucleotide exchange [71, 72];
- CrossP are *cross-linking proteins*: (i) α-actinin which cross-links the actin filaments; (ii) vinculin, talin, and zyxin which link the cortex to the plasma membrane [73–75];
- SevP are *severing proteins*: (i) cofilin (actin depolymerization factor) which sever F-actin to generate more filament ends for assembly or disassembly and enhances subunit dissociation; (ii) gelsolin and the Arp2/3 complex (containing the actin-related proteins, Arp2 and Arp3) which cap filament ends to regulate addition or loss of actin subunits, see: [76, 77];
- NucP are *nucleating proteins*: (i) formin, the Arp2/3 complex which nucleate filament growth; (ii) integrins which nucleate the formation of assemblies of structural and signaling proteins for filament adhesions [65, 78–81].

Actin filaments can be organized as bunches in response to the activation of signalling pathways by external stimuli. The basic processes in filament bunching are as follows:

- *Anchoring*. As response to external stimuli the actin filaments anchor to membranes. Meanwhile, filaments are attached at their plus ends so that the filament elongation occurs at anchored ends which can cause a membrane deformation in the form of growing pseudopodia [82, 83].
- *Parallel orientation*. Usually, actin filaments are short, randomly oriented, and not bundled. External stimuli organize actin filaments in linear patterns with orientation of filament heads towards stimuli.
- *Branching*. The Arp2/3 complex anchors the pointed end of the future daughter filament to the existing mother filament. Then the daughter filament grows up at

its barbed end. As a result, a branch appears. Continuing in the same way, a tree is assembled. More often the Arp2/3 complex nucleates the tree assembly close to the cell membrane at the point of external stimulus [54].

- *Cross-linking*. The actin filaments are cross-linked and bound to the cytoskeleton due to the actin-bounding proteins: profilin, the Arp2/3 complex, filamin, spectrin, and α-actinin [84]. These proteins transform unbundled actin filaments with small adhesions into the bundled actin filaments with larger adhesions. Notice that the actin cytoskeleton built up by cross-linkers is used then for mechanotransduction and signal transmission [64, 85–88].
- *Adhesion*. Integrins and myosin II are mainly responsible for adhesion and actin organization as bunches [89]. For adhesion stability talin activates integrins and links them to actin. Adhesions disassemble at the back edge of the lamellipodium, which is a region of active actin depolymerization [74]. About adhesion mechanisms please see [78].

Each actin filament is attracted by a stimulus so that the filament grows up towards this stimulus. In case we have many actin filaments in one bunch, they behave as a swarm in front of stimuli attracting each individual of that bunch. This grouping behaviour can be represented as the following process calculus:

Definition 2.3 Let $P_{bunch} = \{n_1, n_2, \ldots, n_k\}$ be initial states of transitions for k filament bunches, $A_{bunch} = \{a_1, a_2, \ldots, a_j\}$ be a set of external stimuli (stresses) localized at different places, $V_{bunch} = \{r_1, r_2, \ldots, r_i\}$ be a set of filament trees. Then the bunch transition system, $TS_{bunch} = (S_{bunch}, E_{bunch}, T_{bunch}, I_{bunch})$, is defined in the following manner:

- $\sigma : P_{bunch} \cup A_{bunch} \to S_{bunch}$ assigning a state to each original point of bunches as well as to each external stimulus;
- $\tau : V_{bunch} \to T_{bunch}$ assigning a transition to each filament tree attracted by one stimulus;
- $\iota : P_{bunch} \to I_{bunch}$ assigning an initial state to each bunch localization.

Each event of the set of events E_{bunch} is assigned to bunch transitions in accordance with the following process types:

- *direction* (the filament tree grows from one state/localization/initial point to another state/stimulus),
- *fusion* (the filament tree grows from different states/localizations to the same one state/stimulus),
- *splitting* (the filament tree grows from one state/localization/initial point to different states/stimuli),
- *repelling* (the filament tree can dissociate).

Thus, we have a hierarchic architecture in our object-oriented programming: (i) a polymerization and depolymerization of each actin filament for transmitting signals (Definitions 2.1 and 2.2); (ii) grouping of actin filaments into filament waves; (iii) a collective behaviour of actin filament bunches/waves (Definition 2.3).

Table 2.1 The CNOT gate in the matrix form

	00	01	10	11
00	1	0	0	0
01	0	1	0	0
10	0	0	0	1
11	0	0	1	0

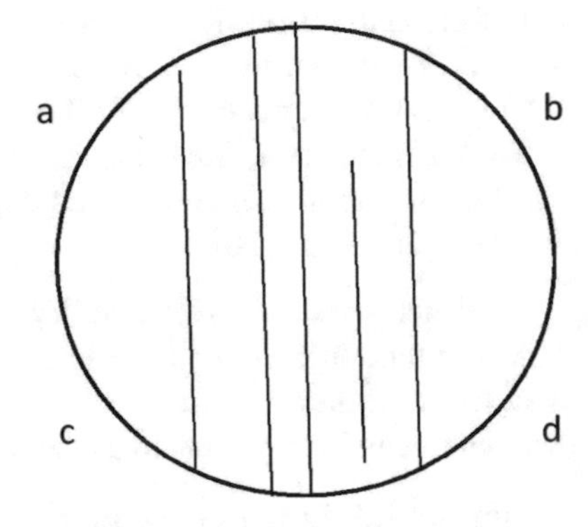

Fig. 2.1 The first possible state of reacting to two stimuli (one from *ab* and one from *cd*): mapping 00 into 00, i.e. the string [*ab*] is mapped into the string [*cd*], see [3]

The system TS_{bunch} just defined behaves like the slime mould computer [11] and it can solve the same tasks: transporting net optimization, pattern recognition, etc. Hence, in TS_{bunch} we can design reversible logic gates. Let us remember that in these gates the number of inputs and outputs is the same. Let us show how we can implement the *CNOT gate* (the 2-bit controlled-NOT gate) into filament trees. In the CNOT gate, the four possible input bit strings are 00, 01, 10, 11 and these are mapped into 00, 01, 11, and 10 respectively (see Table 2.1).

In the unexcited state filaments are chaotically oriented. Let us consider an artificial plane cell with the four zones on the cell surface: a, b, c, d, see Fig. 2.1. Assume that two external stimuli are transmitted from both fronts: *ab* and *cd*. At the same time, suppose that the protein pool of the cell activates the polymerization and branching of filaments only in the four possible ways pictured in Figs. 2.1, 2.2, 2.3 and 2.4. Then the cell can be regarded as the CNOT gate.

2.1.3 Actin Filaments as "Metaswarm"

Highly cross-linked filaments are used in cells for transferring the mechanical stimulus resulted by mechanical forces applied to cell surfaces. These forces generate elastic stress waves which rapidly propagate through actin stress fibers. Such external mechanical stresses can imply a diffusion of actin filament networks if filaments were uncross-linked semi-dilute [90, 91]. So, the rotation around their axis cause

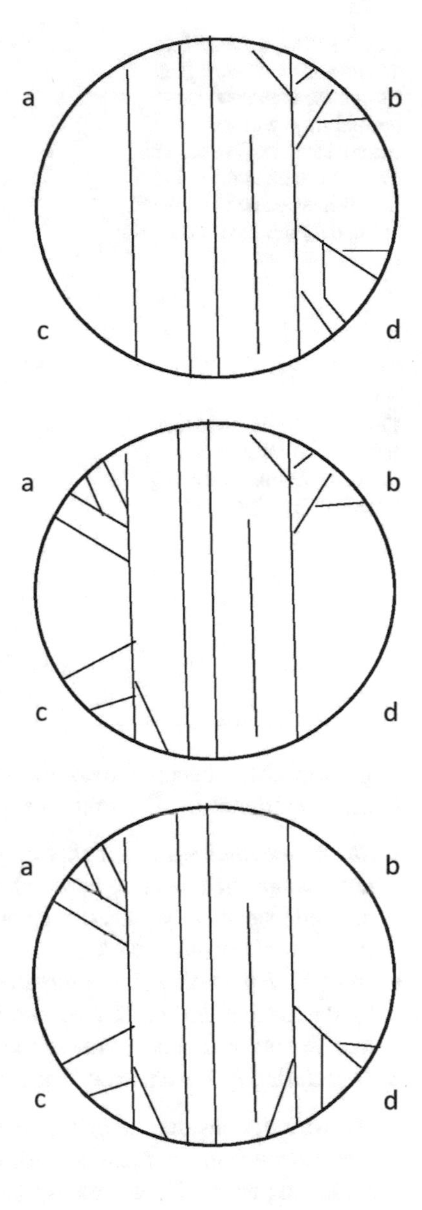

Fig. 2.2 The second possible state of reacting to two stimuli: mapping 01 into 01, see [3]

Fig. 2.3 The third possible state of reacting to two stimuli: mapping 11 into 10, see [3]

Fig. 2.4 The fourth possible state of reacting to two stimuli: mapping 10 into 11, see [3]

colliding with other filaments for uncross-linked filament networks, but in the case of cytoskeleton the mechanical stimulus is being transmitted by filament strands. As a consequence, in response to the physical force the stress fiber displacement is activated in the cytoskeleton [59, 92–95]. Some forms of that displacement which transmit the external stress: fiber inertia, fiber viscoelasticity, and cytosolic damping. Hence, if highly cross-linked actin filament networks are organized in parallel arrays

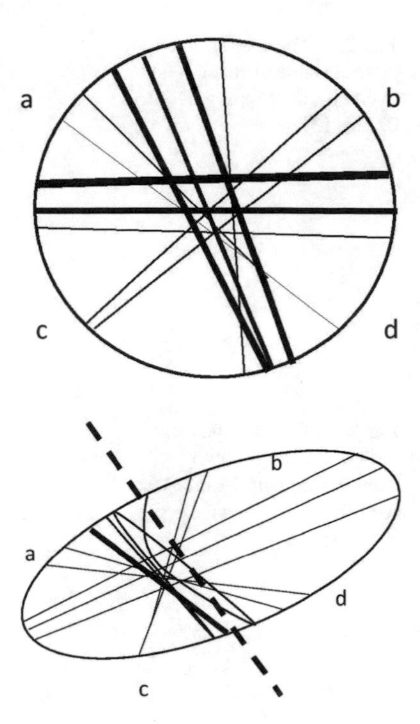

Fig. 2.5 The artificial cell considered the CNOT gate. Its cytoskeleton is at the normal stage without external stresses. We assume that after stresses the string $[ab]$ is transformed into the string $[cd]$ and vice versa, see [3]

Fig. 2.6 The first possible state of the cell after one mechanical stress: mapping 00 into 00, see [3]

of filaments, they become stress fibers with the following kinds of deformation which strongly influence on transmitting signals [35, 81, 96–99]:

- *Shear deformation*. In a highly cross-linked actin filament network both sides do not change their lengths under shear, but the diagonals are stretched and compressed respectively. This stretch or compression causes a large tensional stress in the actin filament network.
- *Bending deformation*. It is a curvature stress that is a result of force that is generated by the local extension of the material on the convex side and compression of the material on the concave side of the bend.
- *Orientation deformation*. It is an mechanical stress, when the orientation changes.

Thus, we can consider the cell a *reversible logic gate* which has the same number of inputs (mechanical stresses) and outputs (stress transmissions). Let us examine an artificial plane cell pictured in Fig. 2.5. Then let us concentrate on the four zones on the cell surface: a, b, c, d. Each letter runs over the two values: 1 or 0. It has the value 1 if a mechanical stress goes through an appropriate zone. Notice that the stress transmission depends on the cytoskeleton structure and there may be zones, where the mechanical stress cannot go through at all. Let us suppose that the cell pictured in Fig. 2.5 can have only four kinds of deformation in transmitting mechanical signals: from Figs. 2.6, 2.7, 2.8 and 2.9. Then this cell can be regarded as the CNOT gate.

Each filament in a reversible gate can be numbered from 1 to n, where n is the number of all filaments engaged into this gate. Then we create an $n \times n$ matrix, G,

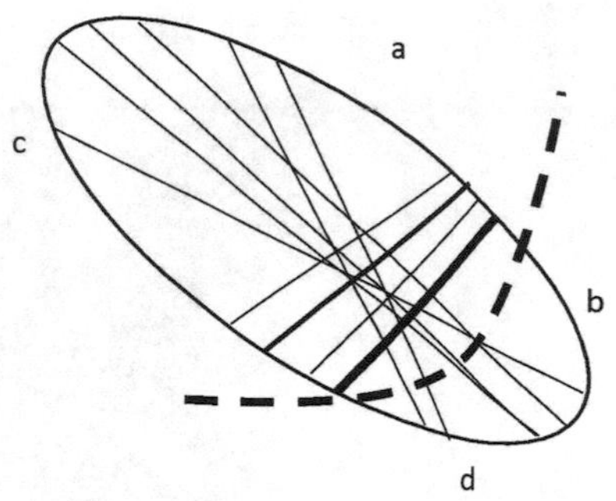

Fig. 2.7 The second possible state of the cell after one mechanical stress: mapping 01 into 01, see [3]

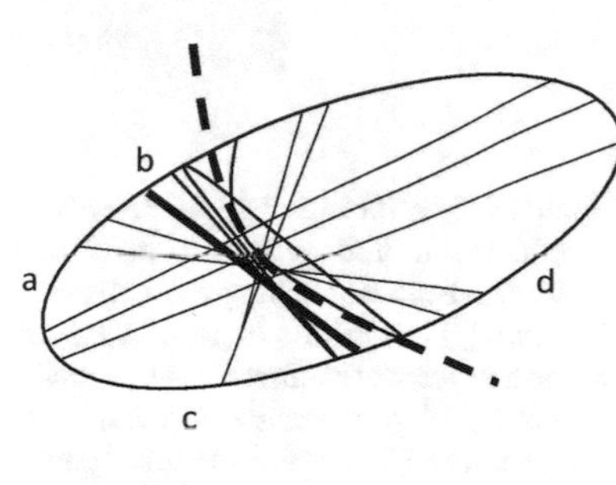

Fig. 2.8 The third possible state of the cell after one mechanical stress: mapping 11 into 10, see [3]

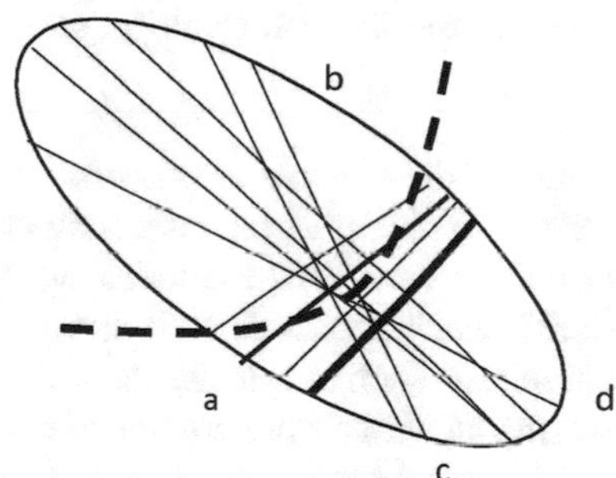

Fig. 2.9 The fourth possible state of the cell after one mechanical stress: mapping 10 into 11, see [3]

such that all elements along its diagonal and below are zero and the value of an element above the diagonal is either one or zero depending on whether the i-th and j-th filaments are cross-linked.

2.1.4 Motility of Amoeba Proteus *and Actin Filaments*

Let us consider now some real examples.

Reactions of actin filament networks to external stimuli are better visible in the amoeba behaviour than in the plasmodium motions. *Amoeba proteus* (see Fig. 2.10) is very sensitive to the environment and reacts directly to external stimuli by the motility of its shape. This shape changes due to the cytoplasmic streaming that extends

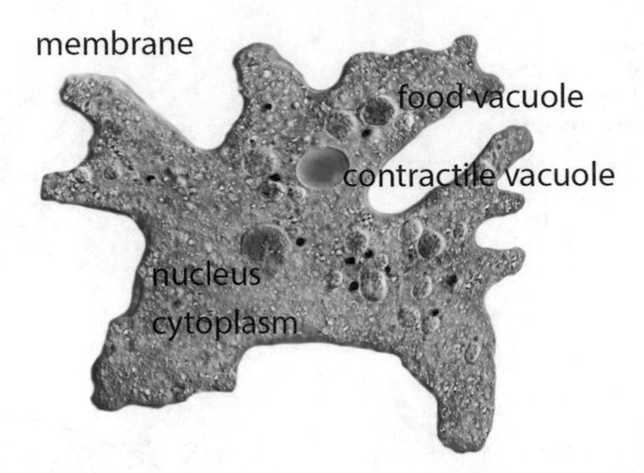

Fig. 2.10 The morphology of *Amoeba proteus*: (1) *contractile vacuole* (a water bubble within the endoplasm of *A. proteus* to regulate the water content of the cell); (2) *nucleus* (a membrane bound organelle containing the cells genetic information and responsible for the actions of the amoeba); (3) *cytoplasm* (a gel-like substance that allows *A. proteus* to form its pseudopodia and preform its respective functions, it contains actin filaments which are responsible for sensitivity of *A. proteus*); (4) *food vacuole* (a vacuole with a digestive function, containing a food for *A. proteus*); (5) *membrane* (it contains the inner part of the cell such as organelles and cytoplasm, it has good regenerative abilities and elasticity), see [5]

pseudopodia towards *attractants* (food). So, the amoeboid locomotion is committed forward if the amoeba detects an attractant. Meanwhile, for *A. proteus* there exist *repellents*, as well: the amoeba avoids strong light (and it moves towards the weaker light), also it avoids dark (it moves towards light) and many other conditions: some chemicals (such as salt), obstacles, anode (it moves towards cathode), cold (it prefers warm), and hot (it prefers not so hot), etc.

The amoeboid reactions to attractants and repellents are studied well and explained by *actin filaments* or *F-actin* (see Fig. 2.11), i.e. the protein which is organized into higher-order structures, forming linear bundles, two-dimensional networks, and three-dimensional semisolid gels. Actin monomers polymerize to form thin, flexible fibers (actin filaments) 5–9 nm in diameter and up to several micrometers in length. Actin filaments are connected to the plasma membrane, where they form an *actin cortex* that provides mechanical support (see Fig. 2.13). If there is an attractant before the cell, actin filaments form a *wave* to change the cell shape to allow the movement of the cell surface to build a *pseudopodium* by cross-linked filaments (see Fig. 2.12).

Actin filaments under different external conditions can be assembled and disassembled and these reactions are regulated by actin-binding proteins, see [100]. For instance, on the one hand, *cofilin* remains bound to actin monomers following filament disassembly and sequesters them in the ADP-bound form. On the other hand, *profilin* stimulate the incorporation of actin monomers into filaments. Also, there are *actin-binding proteins* connecting two different actin filaments into bundles or even into networks which can crosslink perpendicular filaments.

Fig. 2.11 The *actin filament* or *F-actin* is a linear polymer of globular actin monomers (G-actin). F-actin is flexible and has a helical repeat every 37 nm. It ranges from 5–9 nm in diameter. It has a rotation of 166.15° around the axis. Each G-actin has tight binding sites that mediate head-to-tail interactions with two other actin monomers, in this way actin monomers are oriented in the same direction and their polymerization gives a distinct polarity at the ends of the actin filament: the plus and minus ends. At these ends there are different rates of the actin filament grow so that we have the plus end to which monomers are added five to ten times faster than to the slow-growing minus end, see [4]

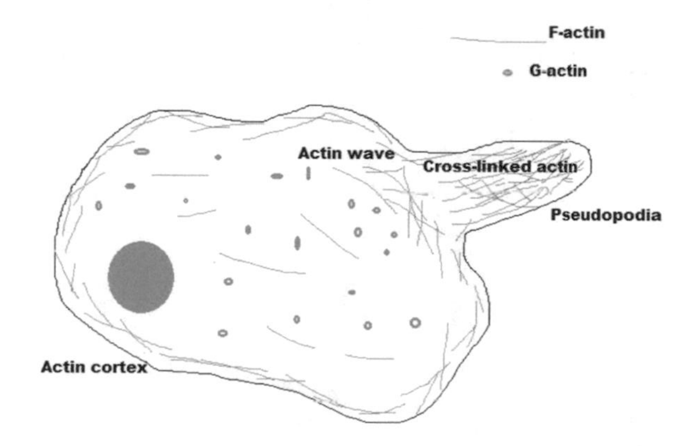

Fig. 2.12 Actin filaments responsible for the motility of *Amoeba proteus*, see [4]

First of all, actin filaments form the *cell cortex*, which lies adjacent to the plasma membrane to support it (Fig. 2.13a). This cytoskeleton is dynamic and sensitive to the cell surroundings. Each external force (each taxis) acting on the actin cortex are transmitted by signaling pathways to directly react to the external environment. If actin filaments are assembled in parallel with the same polarity direction, they propagate some projections, called *microvilli*, by adding new monomers at the plus ends adjacent to the plasma membrane (Fig. 2.13b). For the cell migration actin filaments are crosslinked to propagate membrane protrusions in the form of *filopodia* (Fig. 2.13c) or *lamellipodia* (Fig. 2.13d). They are being formed also to probe the cell microenvironment. The more stable bundles of actin filaments are represented by *stress fibers* (Fig. 2.13e) which allow the cell to form a track system for cargo transport. In the cell they build up networks which change their topology by reactions to the external forces.

Hence, actin filaments are instable, they can assemble and disassemble rapidly by polymerization and depolymerization respectively. For more details see [38, 48–56, 101].

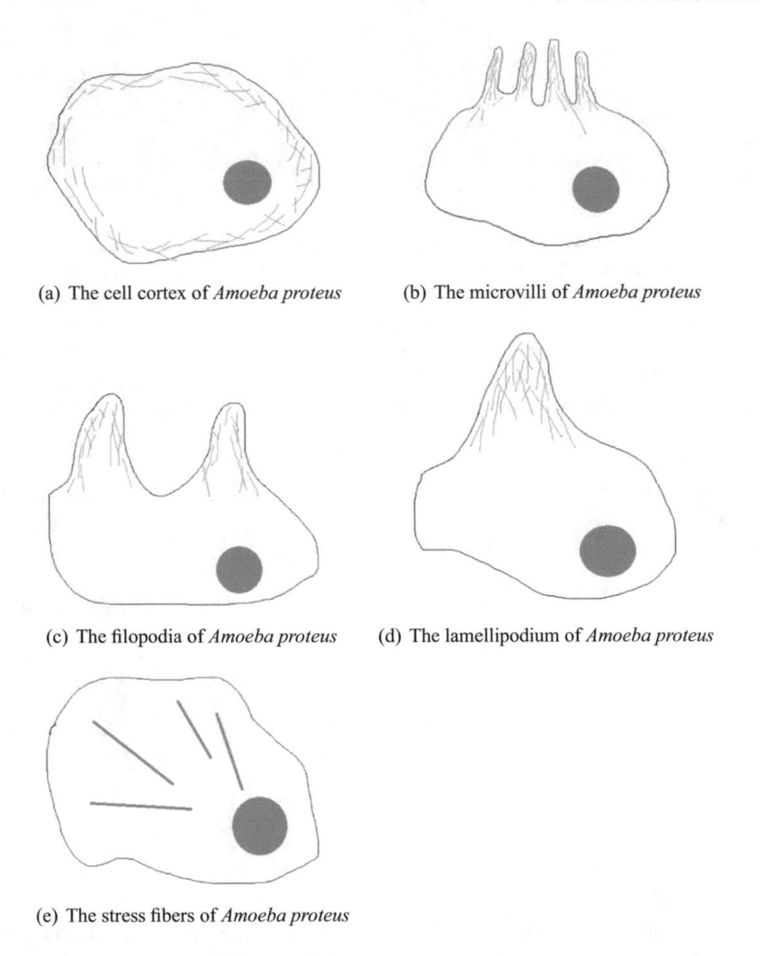

(a) The cell cortex of *Amoeba proteus* (b) The microvilli of *Amoeba proteus*

(c) The filopodia of *Amoeba proteus* (d) The lamellipodium of *Amoeba proteus*

(e) The stress fibers of *Amoeba proteus*

Fig. 2.13 The role of actin filaments in changing the shape of *Amoeba proteus*, see [4]

2.1.5 Actin Filament Zones as Fuzzy Logic Gates

Any unicellular organism like the amoeba of *Amoeba proteus* or the plasmodium of *Physarum polycephalum* uses actin filament networks to react to external signals well so that we can consider these organisms as logic gates [11–13, 16, 20, 21, 26, 27, 29, 30, 102]. Let us consider the situation of one signal represented just by one attractant, see Fig. 2.14. Hence, we deal here with a logic gate with one input. How many outputs does this gate have? We see at Fig. 2.14 that the number of outputs is infinite and represented by a continuous interval, e.g. for the amoeba the outputs run over the closed interval $[Outp_1, Outp_2]$ at least. So, we have one discrete input and one interval as a continuous-valued output. This interval is understood as an active zone of actin filament polymerization.

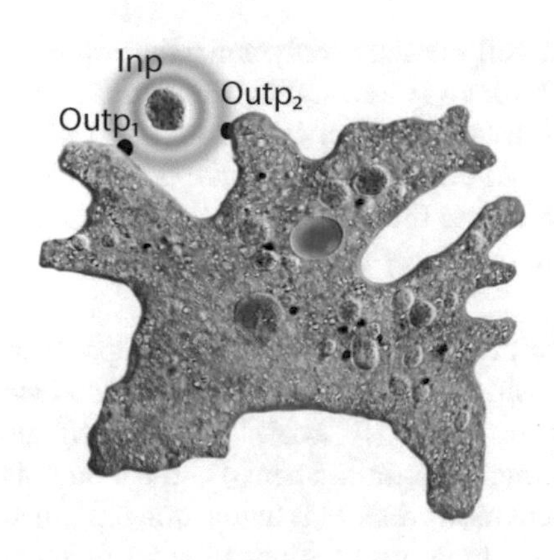

(a) The amoeba of *Amoeba proteus* as a logic gate with one input

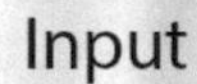

(b) The plasmodium of *Physarum polycephalum* as a logic gate with one input

Fig. 2.14 **a** The amoeba of *Amoeba proteus* feels that there is an attractant (food particle) above denoted as *Inp*. This attractant is first detected at the two points of the amoeba body: $Outp_1$ and $Outp_2$. Then it is detected by a large upper zone of the body. **b** The plasmodium of *Physarum polycephalum* feels that there is an attractant (food particle) above. At the end it is also detected by a large zone of the plasmodium body, see [5]

However, the situation is much worse, because we deal not with a well-bounded interval $[Outp_1, Outp_2]$, but with a *fuzzy set* A_{Inp} such that $[Outp_1, Outp_2]$ is covered by A_{Inp}. This A_{Inp} is defined as follows. Let μ be a membership function for each actin filament x from the cell U such that $\mu\colon U \mapsto [0, 1]$. Each outside signal *Inp* gives a set A_{Inp} of active zone, where actin filaments are polymerized in the direction of *Inp*. This set is determined by the membership function μ: $A_{Inp} = \{x\colon \mu(x) > 0\}$.

At the momentum of signal *Inp*, the shape of the amoeba of *Amoeba proteus* or the plasmodium of *Physarum polycephalum* can be different. Therefore, the same *Inp* can give different fuzzy sets A_{Inp}, see Fig. 2.14.

Let us assume that we have two outside signals Inp_1 and Inp_2. Then we can define their fuzzy compositions [103]:

$$A_{Inp_1} \vee A_{Inp_2} = \{x\colon \mu_{A_{Inp_1} \vee A_{Inp_2}} = \max(\mu_{A_{Inp_1}}, \mu_{A_{Inp_2}})\};$$

$$A_{Inp_1} \wedge A_{Inp_2} = \{x\colon \mu_{A_{Inp_1} \wedge A_{Inp_2}} = \min(\mu_{A_{Inp_1}}, \mu_{A_{Inp_2}})\};$$

$$\neg A_{Inp_1} = \{x\colon \mu_{\neg A_{Inp_1}} = 1 - \mu_{A_{Inp_1}}\};$$

$$\neg A_{Inp_2} = \{x\colon \mu_{\neg A_{Inp_2}} = 1 - \mu_{A_{Inp_2}}\}.$$

Continuing in the same way, we can define fuzzy compositions for many signals $Inp_1, Inp_2, \ldots, Inp_n, \ldots$

Each outside signal *Inp* perceived by a cell causes a polymerization of actin filaments in the direction of *Inp*. As a result, there is assembling an actin filament network to change the cell shape. This cell transformation can be regarded as an automaton with a discrete time $t = 0, 1, \ldots$ At each time step t, the cell shape is different which causes the difference in reactions to the same outside signal *Inp*, i.e. this shape transformation is a transition from one fuzzy set A^t_{Inp} at time t to another fuzzy set A^{t+1}_{Inp} at time $t + 1$.

In the beginning, let us define transitions from one shape states to other shape states just for crisp sets of active zones. These transitions can be defined as an *Euclidean cellular automaton* [104] over a parameter space $P = ([0, \infty)^3)^n$ presented by the inputs $(i_1, i_2, \ldots, i_n)$ (i.e. external forces acting on actin filaments) and the outputs $(o_1, o_2, \ldots, o_n)$ (i.e. reactions of actin filament networks). This automaton is defined as a 4-tuple $\langle \mathscr{P}, I, F, T \rangle$ where $\mathscr{P} \subset 2^P$ is a finite set of states of actin filament networks given as subsets of P; $I \subset \mathscr{P}$ is the set of initial states; $F \subset \mathscr{P}$ is the set of accepting states; and $T : P \times \mathscr{P} \to \mathscr{P}$ is the transition function that assigns for each parameter setting $\mathbf{v} = (i_1, i_2, \ldots, i_n) \in P$ and each state $\mathbf{s} \in \mathscr{P}$ a next state $\mathbf{t} = T(\mathbf{v}, \mathbf{s})$. The parameter $\mathbf{v} \in P$ is defined as a neighbourhood for $(o_1, o_2, \ldots, o_n)$ with a 3-dimensional radius ε: $\mathbf{v} = \{(i_1, i_2, \ldots, i_n) : |i_k - o_k| \leq \varepsilon,\ k = \overline{1, n}\}$. In this automaton we deal with a continuous domain and with a finite set of states, i.e. with subsets P_i of P indexed from a finite index set S. If $P_i \cap P_j = \emptyset$ for all $i, j \in S$, we call the Euclidian cellular automaton deterministic; and if $\bigcup_{i \in S} P_i = P$, we call it complete.

The input $(i_1, i_2, \ldots, i_n)$ is treated as n different signals from one side of the plasmodium or amoeba body. Each i_k for $k = \overline{1, n}$ is a 3-dimensional coordinates of the signal source. If there is no signal, its coordinates are $(0, 0, 0)$. The output $(o_1, o_2, \ldots, o_n)$ as n active zones of actin filament polymerization. Each i_k for $k = \overline{1, n}$ is a 3-dimensional coordinates of the center of polymerization wave. If the active zone is deactivated, its coordinates are $(0, 0, 0)$. Let us assume that in a resting state the active zone is designated by a sphere with a diameter equal to $2a_0$, see Fig. 2.15. In an excited state this sphere is deformed to an ellipsoid so that an appropriate square $ACBD$ is transformed into a parallelogram $A_1C_1B_1D_1$. Let $2a_1$ and $2b_1$ be conjugate diameters connecting the points of ellipsoid tangency with the parallelogram (i.e. the larger and smaller side of the parallelogram, respectively) and δ be an angle between them. Then the two main axes (i.e. the largest and smallest diameters of the ellipsoid, respectively) are defined as follows:

$$2a = \sqrt{2(a_1^2 + b_1^2 + \sqrt{(a_1^2 + b_1^2)^2 - 4a_1^2 b_1^2 \sin^2 \delta}}),$$

$$2b = \sqrt{2(a_1^2 + b_1^2 - \sqrt{(a_1^2 + b_1^2)^2 - 4a_1^2 b_1^2 \sin^2 \delta}}),$$

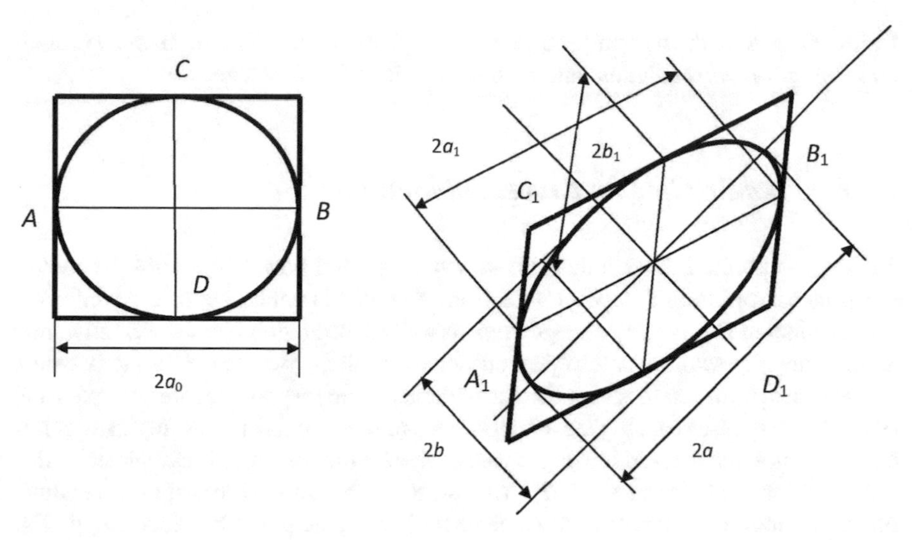

Fig. 2.15 The deformation of the sphere in the square *ACBD* into the ellipsoid in the parallelogram $A_1C_1B_1D_1$, see [5]

and the angle α between the largest diameter of the ellipsoid and the largest side of the parallelogram as follows:

$$tg\ \alpha = \frac{1}{2}((1 - \frac{b^2}{a^2})tg\,\delta - \sqrt{(1 - \frac{b^2}{a^2})tg^2\,\delta - 4\frac{b^2}{a^2}})$$

The largest diameter of the ellipsoid, $2a$, shows the direction of propagation of active zone of actin filament polymcrization.

Thus, for the case of crisp sets, we have examined a zone A_{Inp} of actin filament polymerization as a sphere with a diameter equal to 2_{a_0}, see Fig. 2.15. Nevertheless, in fact this A_{Inp} is not a crisp sphere, but a *fuzzy sphere* (e.g. fuzzy circles are considered in [105]): instead of points (a, b, c) we take fuzzy points $\widetilde{P}(a, b, c)$ and instead of diameter $d = 2_{a_0}$ we take a fuzzy number $\widetilde{d}$. Continuing in the same way, we can define a *fuzzy ellipsoid* defined by fuzzy points $\widetilde{P}(a, b, c)$, the largest fuzzy diameter $\widetilde{d_{max}}$, the smallest fuzzy diameter $\widetilde{d_{min}}$, and by a fuzzy angle $\widetilde{\alpha}$ showing a direction of polymerization. For more details about fuzzy figures see [105].

Thus, the *Euclidean cellular automaton* defined above can simulate the amoeboid motions if we concentrate on transitions as deformations from one fuzzy ellipsoids denoting active zones of polymerization at time t to other fuzzy ellipsoids denoting active zones at time $t + 1$. This automaton having n inputs and n outputs is an implementation of an appropriate reversible logic gates. In the FREDKIN gate, the parameter space $P = ([0, \infty)^3)^3$. The input (i_1, i_2, i_3) of this Euclidean cellular automaton shows the localizations of attractants and repellents in the FREDKIN motions of the amoeba or the plasmodium. The output (o_1, o_2, o_3) shows the localization of active zones of actin filament polymerization in accordance with the FREDKIN motions.

Using this gate in the way of [106], we can implement the adder on *Amoeba proteus* and *Physarum polycephalum* and then some arithmetic functions.

2.1.6 Fredkin Gate on Amoeboid Motions

Taking into account the fact that it is known in general how it is possible to control the polymerization and depolymerization of actin filaments, we can consider the plasmodium of *Physarum polycephalum* as well as the amoeba (more correctly, their actin filament networks) as a logical automaton with the two values: 1 and 0, where (i) the external forces, to which the actin filaments are responding, are its inputs and (ii) all the responses causing the amoeboid movement are its outputs. In this way it is easier to implement reversible logic gates, where a unique input is associated with a unique output and vice versa [107]. In these gates, the automaton maps each distinct bit string input of the length n into a distinct bit string output of the same length. For example, the *FREDKIN gate* is reversible, for the input $[Inp_1 Inp_2 Inp_3] \in \{0, 1\}^3$ it gives the output $[Outp_1 Outp_2 Outp_3] \in \{0, 1\}^3$ by the following rule: $Outp_1 = Inp_1$, $Outp_2 = \mathrm{OR}(\mathrm{AND}(\mathrm{NOT}Inp_1, Inp_2), \mathrm{AND}(Inp_1, Inp_3))$, and $Outp_3 = \mathrm{OR}(\mathrm{AND}(Inp_1, Inp_2), \mathrm{AND}(\mathrm{NOT}Inp_1, Inp_3))$, i.e. we deal with the three inputs Inp_1, Inp_2, Inp_3, and the three outputs $Outp_1$, $Outp_2$, $Outp_3$. See Table 2.2.

The FREDKIN gate can be implemented by the motility as follows. Let us assume that we have a configuration of the three active zones of actin filament polymerization depicted as ellipses A, B, C in Fig. 2.16c. And this configuration is the same for the amoeba of Fig. 2.16a and for the plasmodium of Fig. 2.16b. Then let us take the input string $[Inp_1 Inp_2 Inp_3]$ with the following meaning: (i) $Inp_1 = 0$ if an appropriate zone of the microenvironment contains an obstacle (barrier) consisting of repellents and $Inp_1 = 1$ otherwise; (ii) $Inp_2 = 1$ if an appropriate zone of the microenvironment

Table 2.2 The FREDKIN gate in the permutation matrix form. The input [000] is mapped to the output [000], [001] is mapped to [001], etc.

	000	001	010	011	100	101	110	111
000	1	0	0	0	0	0	0	0
001	0	1	0	0	0	0	0	0
010	0	0	1	0	0	0	0	0
011	0	0	0	1	0	0	0	0
100	0	0	0	0	1	0	0	0
101	0	0	0	0	0	0	1	0
110	0	0	0	0	0	1	0	0
111	0	0	0	0	0	0	0	1

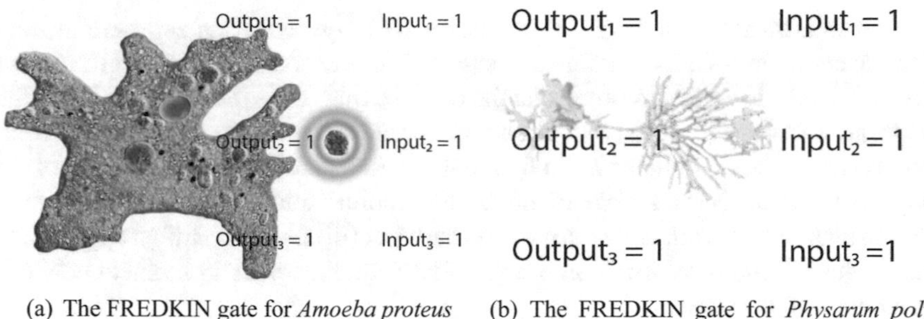

(a) The FREDKIN gate for *Amoeba proteus*

(b) The FREDKIN gate for *Physarum polycephalum*

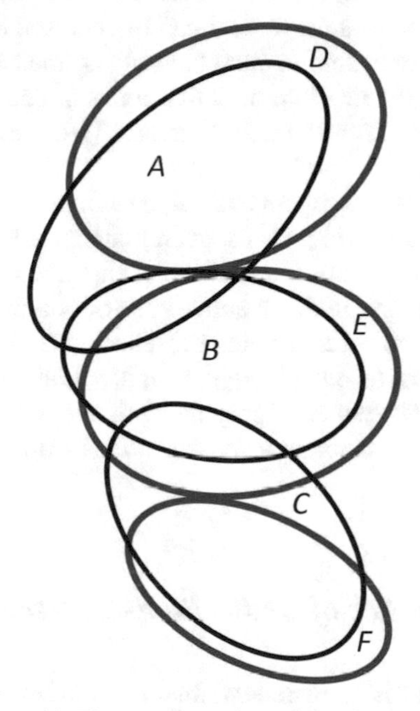

(c) The transformation of the three actin filament zones

Fig. 2.16 **a** The FREDKIN gate for *Amoeba proteus* with the three inputs: $Inp_1 = 1$, $Inp_2 = 1$, $Inp_3 = 1$, and the three outputs: $Output_1 = 1$ (the cell zone $Output_1$ moves right up), $Output_3 = 1$ (the cell zone $Output_3$ moves right down), $Output_2 = 1$ (the cell zone $Output_2$ moves right). **b** The FREDKIN gate for *Physarum polycephalum* with the three inputs: $Inp_1 = 1$, $Inp_2 = 1$, $Inp_3 = 1$, and the three outputs: $Output_1 = 1$ (the protoplasmic tubes $Output_1$ move right up), $Output_3 = 1$ (the protoplasmic tubes $Output_3$ moves right down), $Output_2 = 1$ (the protoplasmic tubes $Output_2$ moves right). **c** Fuzzy ellipsoids A, B, C designating the three active zones of actin filament polymerization are transformed into fuzzy ellipsoids D, E, F, causing the movement of the amoeba [in (**a**)] as well as of the plasmodium [in (**b**)], see [5]

contains an attractant and $Inp_2 = 0$ otherwise; (iii) $Inp_3 = 0$ if an appropriate zone of the microenvironment contains an obstacle (barrier) consisting of repellents and $Inp_3 = 1$ otherwise. Let us assume that the output string $[Outp_1 Outp_2 Outp_3]$ with the following meaning: (i) $Outp_1 = 1$ if an appropriate zone of the cell has a deformation because of assembling an actin filament network and $Outp_1 = 0$ otherwise; (ii) $Outp_2 = 1$ if an appropriate zone of the cell has a deformation because of assembling an actin filament network and $Outp_2 = 0$ otherwise; (iii) $Outp_3 = 1$ if an appropriate zone of the cell has a deformation because of assembling an actin filament network and $Outp_3 = 0$ otherwise.

Thus, we can implement some reversible logic gates on the amoeboid motions based on actin filament networks, see Figs. 2.16, 2.17, 2.18, 2.19, 2.20, 2.21, 2.22 and 2.23. Hence, the amoeboid movement can be represented as a reversible logic gate if and only if we involve among attractants also some bariers/repellents. These reversible logic gates on the medium of active zones of actin filament polymerization can implement decidable arithmetic functions [106], especially p-adic valued arithmetic functions [102].

The amoeboid motion is intelligent and can be simulated and controlled. Attractants and repellents are outside signals for actin filament networks and these signals cause the polymerization of actin filaments to change the cell shape which causes an appropriate amoeboid motion. Each active zone of polymerization is regarded as a fuzzy ellipsoid directed towards the signal. Hence, we have reversible logic gates with inputs corresponding to outside signals and outputs corresponding to fuzzy ellipsoids designating active zones of polymerization. I have considered the FREDKIN gate as an example of automaton on transitions from one fuzzy ellipsoids to other fuzzy ellipsoids.

2.1.7 Neural Properties of Actin Filament Networks

The *actin filament network* is more general than *artificial neural networks*. The main difference is that in the latter the processors ('neurons') do not disappear, because they are fixed, but in the actin filament networks the processors ('filaments') appear and disappear permanently. These filaments are combined into a wave front that represents an active zone of actin filament polymerization.

It is worth noting that the actin filaments are responsible for remodeling neurons in many-cellular organisms possessing the nervous system, also. In this system the actin filaments change the shape and structure of dendritic spines in the same way as they do it for the amoeba motility. G-actin is distributed throughout the whole axon and the whole dendrite and it can be polymerized into F-actin to form new spines as well as to stabilize the spine volume. As a result, the actin filaments form new synapses to increase the cell communication. The filament polymerization promotes *long-term potentiation* increasing the spine volume and the cell communication. The filament depolymerization leads to a *long-term depression* decreasing the spine volume and the cell communication.

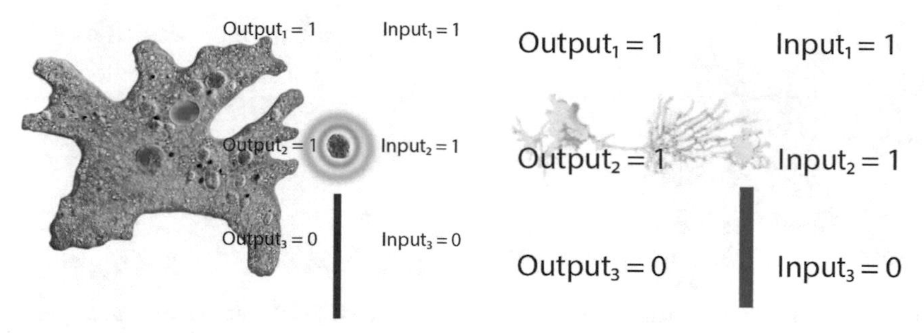

(a) The FREDKIN gate for *Amoeba proteus*

(b) The FREDKIN gate for *Physarum polycephalum*

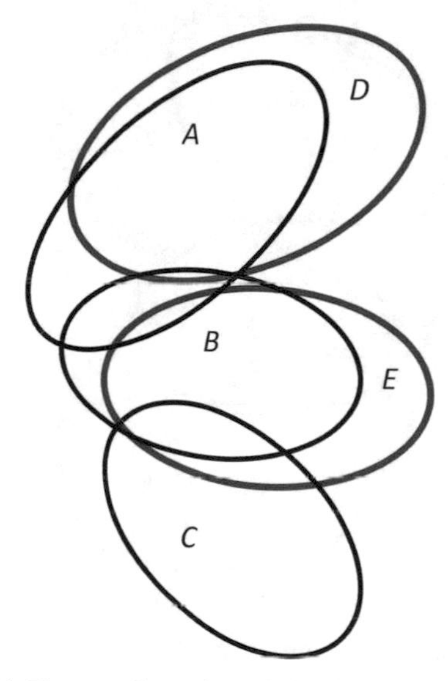

(c) The transformation of the three actin filament zones

Fig. 2.17 **a** The FREDKIN gate for *Amoeba proteus* with the three inputs: $Inp_1 = 1$, $Inp_2 = 1$, $Inp_3 = 0$, and the three outputs: $Output_1 = 1$ (the cell zone $Output_1$ moves right up), $Output_3 = 0$ (the cell zone $Output_3$ does not move), $Output_2 = 1$ (the cell zone $Output_2$ moves right). **b** The FREDKIN gate for *Physarum polycephalum* with the three inputs: $Inp_1 = 1$, $Inp_2 = 1$, $Inp_3 = 0$, and the three outputs: $Output_1 = 1$ (the protoplasmic tubes $Output_1$ move right up), $Output_3 = 0$ (the protoplasmic tubes $Output_3$ do not move), $Output_2 = 1$ (the protoplasmic tubes $Output_2$ moves right). **c** Fuzzy ellipsoids A, B designating the two active zones of actin filament polymerization are transformed into fuzzy ellipsoids D, E, causing the movement of the amoeba [in (**a**)] as well as of the plasmodium [in (**b**)]. Fuzzy ellipsoid C designating one active zone of actin filament polymerization did not change and was deactivated, see [5]

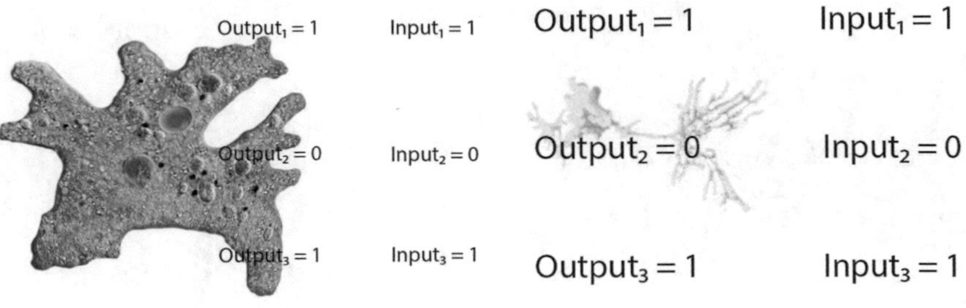

(a) The FREDKIN gate for *Amoeba proteus*

(b) The FREDKIN gate for *Physarum polycephalum*

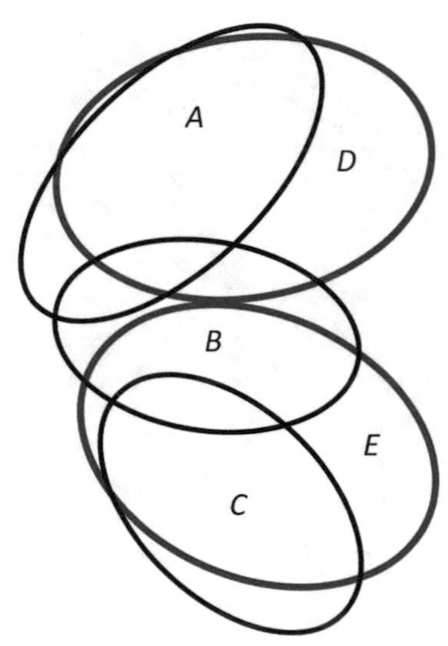

(c) The transformation of the three actin filament zones

Fig. 2.18 **a** The FREDKIN gate for *Amoeba proteus* with the three inputs: $Inp_1 = 1$, $Inp_2 = 0$, $Inp_3 = 1$, and the three outputs: $Output_1 = 1$ (the cell zone $Output_1$ moves right up), $Output_3 = 1$ (the cell zone $Output_3$ moves right down), $Output_2 = 0$ (the cell zone $Output_2$ does not move). **b** The FREDKIN gate for *Physarum polycephalum* with the three inputs: $Inp_1 = 1$, $Inp_2 = 0$, $Inp_3 = 1$, and the three outputs: $Output_1 = 1$ (the protoplasmic tubes $Output_1$ move right up), $Output_3 = 1$ (the protoplasmic tubes $Output_3$ move right down), $Output_2 = 0$ (the protoplasmic tubes $Output_2$ do not move). **c** Fuzzy ellipsoids A, C designating the two active zones of actin filament polymerization are transformed into fuzzy ellipsoids D, E, causing the movement of the amoeba [in (**a**)] as well as of the plasmodium [in (**b**)]. Fuzzy ellipsoid B was deactivated and then divided and merged with D and E, see [5]

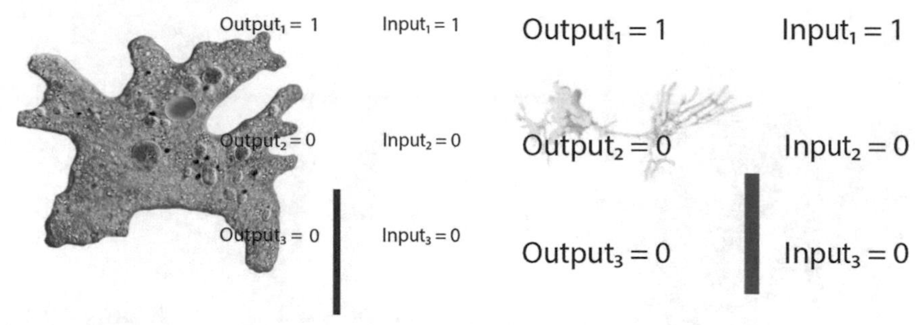

(a) The FREDKIN gate for *Amoeba proteus*

(b) The FREDKIN gate for *Physarum polycephalum*

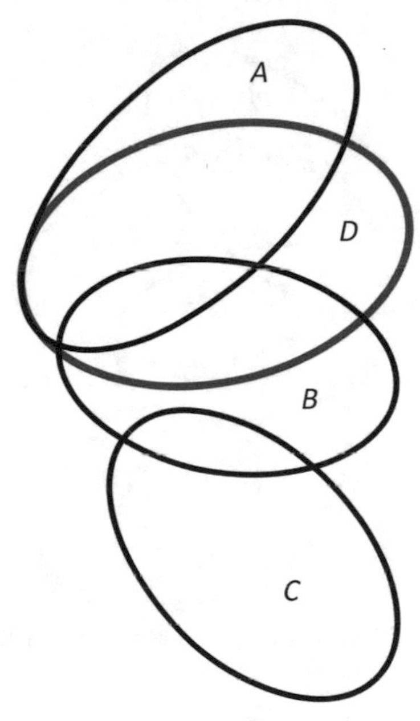

(c) The transformation of the three actin filament zones

Fig. 2.19 **a** The FREDKIN gate for *Amoeba proteus* with the three inputs: $Inp_1 = 1$, $Inp_2 = 0$, $Inp_3 = 0$, and the three outputs: $Output_1 = 1$ (the cell zone $Output_1$ moves right up), $Output_3 = 0$ (the cell zone $Output_3$ does not move), $Output_2 = 0$ (the cell zone $Output_2$ does not move). **b** The FREDKIN gate for *Physarum polycephalum* with the three inputs: $Inp_1 = 1$, $Inp_2 = 0$, $Inp_3 = 0$, and the three outputs: $Output_1 = 1$ (the protoplasmic tubes $Output_1$ move right up), $Output_3 = 0$ (the protoplasmic tubes $Output_3$ do not move), $Output_2 = 0$ (the protoplasmic tubes $Output_2$ do not move). **c** Fuzzy ellipsoid A designating one active zone of actin filament polymerization is transformed into fuzzy ellipsoid D, causing the movement of the amoeba [in (**a**)] as well as of the plasmodium [in (**b**)]. Fuzzy ellipsoids B, C were deactivated and fuzzy ellipsoid B was merged with fuzzy ellipsoid D, see [5]

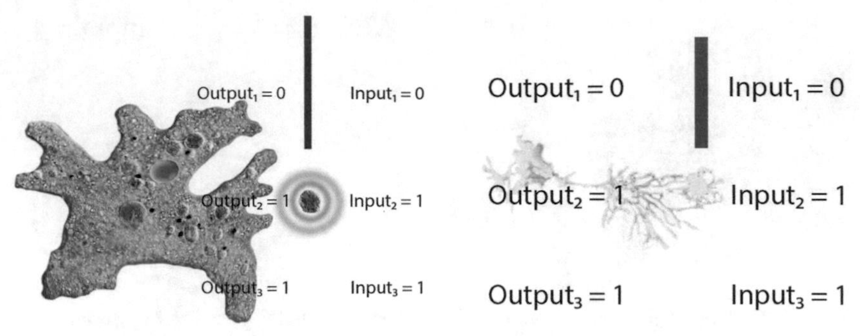

(a) The FREDKIN gate for *Amoeba proteus* (b) The FREDKIN gate for *Physarum polycephalum*

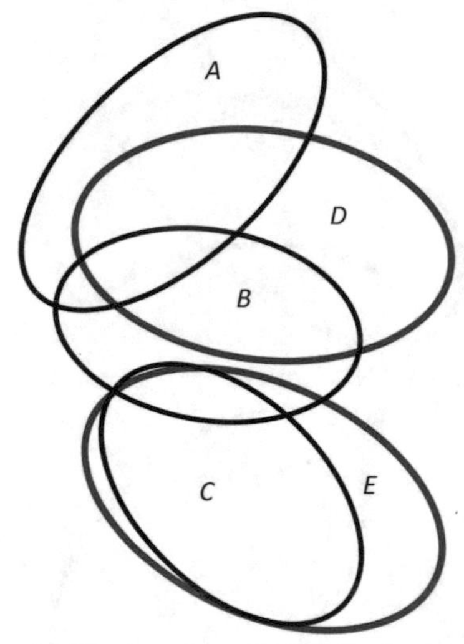

(c) The transformation of the three actin filament zones

Fig. 2.20 **a** The FREDKIN gate for *Amoeba proteus* with the three inputs: $Inp_1 = 0$, $Inp_2 = 1$, $Inp_3 = 1$, and the three outputs: $Output_1 = 0$ (the cell zone $Output_1$ does not move), $Output_3 = 1$ (the cell zone $Output_3$ moves right down), $Output_2 = 1$ (the cell zone $Output_2$ moves right). **b** The FREDKIN gate for *Physarum polycephalum* with the three inputs: $Inp_1 = 0$, $Inp_2 = 1$, $Inp_3 = 1$, and the three outputs: $Output_1 = 0$ (the protoplasmic tubes $Output_1$ do not move), $Output_3 = 1$ (the protoplasmic tubes $Output_3$ move right down), $Output_2 = 1$ (the protoplasmic tubes $Output_2$ move right). **c** Fuzzy ellipsoids B, C designating the two active zones of actin filament polymerization are transformed into fuzzy ellipsoids D, E, causing the movement of the amoeba [in (**a**)] as well as of the plasmodium [in (**b**)]. Fuzzy ellipsoid A was deactivated and merged with fuzzy ellipsoid D, see [5]

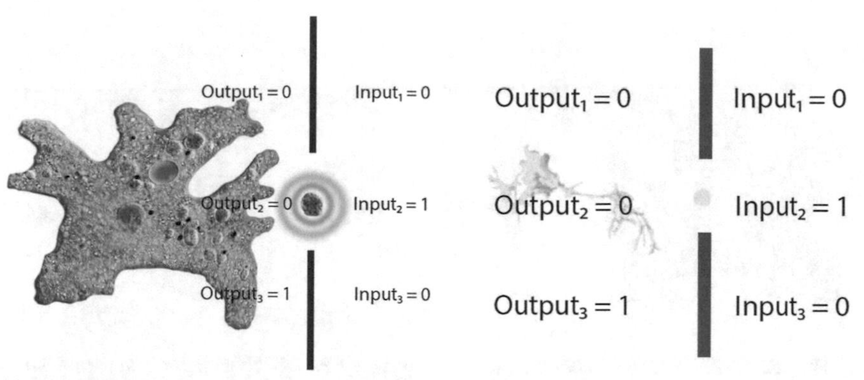

(a) The FREDKIN gate for *Amoeba proteus* (b) The FREDKIN gate for *Physarum polycephalum*

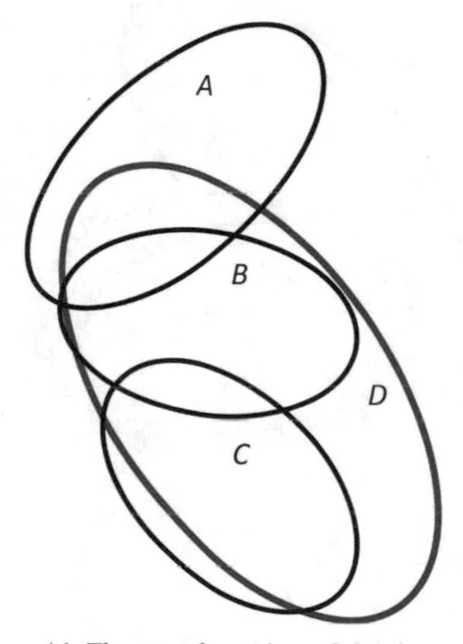

(c) The transformation of the three actin filament zones

Fig. 2.21 **a** The FREDKIN gate for *Amoeba proteus* with the three inputs: $Inp_1 = 0$, $Inp_2 = 1$, $Inp_3 = 0$, and the three outputs: $Output_1 = 0$ (the cell zone $Output_1$ does not move), $Output_3 = 1$ (the cell zone $Output_3$ moves right down), $Output_2 = 0$ (the cell zone $Output_2$ does not move). **b** The FREDKIN gate for *Physarum polycephalum* with the three inputs: $Inp_1 = 0$, $Inp_2 = 1$, $Inp_3 = 0$, and the three outputs: $Output_1 = 0$ (the protoplasmic tubes $Output_1$ do not move), $Output_3 = 1$ (the protoplasmic tubes $Output_3$ move right down), $Output_2 = 0$ (the protoplasmic tubes $Output_2$ do not move). **c** Fuzzy ellipsoid *C* designating one active zone of actin filament polymerization is transformed into fuzzy ellipsoid *D*, causing the movement of the amoeba [in (**a**)] as well as of the plasmodium [in (**b**)]. Fuzzy ellipsoid *A* was deactivated totally. Fuzzy ellipsoid *B* was merged with fuzzy ellipsoid *D*, see [5]

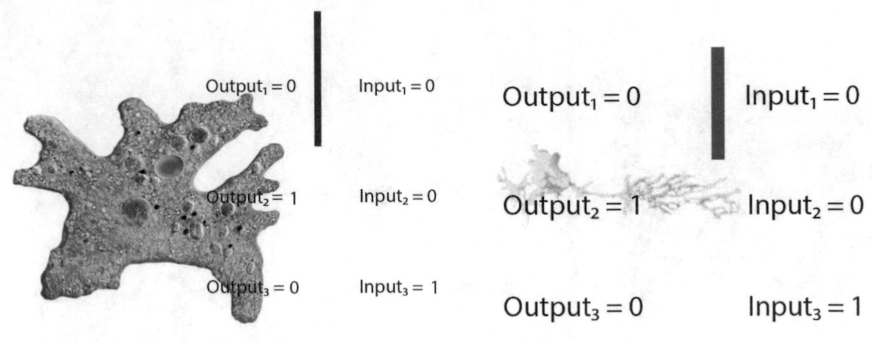

(a) The FREDKIN gate for *Amoeba proteus* (b) The FREDKIN gate for *Physarum polycephalum*

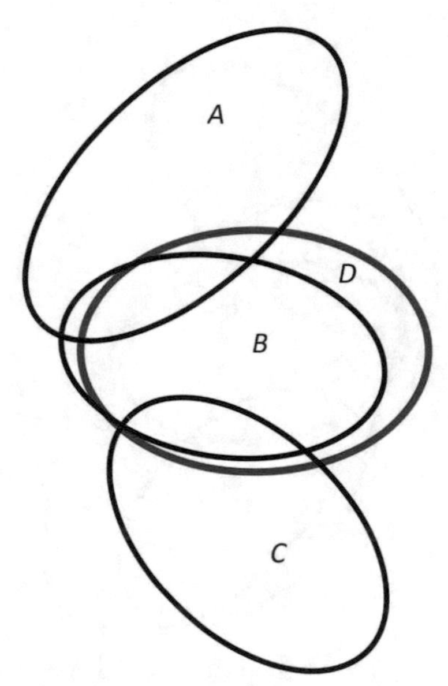

(c) The transformation of the three actin filament zones

Fig. 2.22 **a** The FREDKIN gate for *Amoeba proteus* with the three inputs: $Inp_1 = 0$, $Inp_2 = 0$, $Inp_3 = 1$, and the three outputs: $Output_1 = 0$ (the cell zone $Output_1$ does not move), $Output_3 = 0$ (the cell zone $Output_3$ does not move), $Output_2 = 1$ (the cell zone $Output_2$ moves right). **b** The FREDKIN gate for *Physarum polycephalum* with the three inputs: $Inp_1 = 0$, $Inp_2 = 0$, $Inp_3 = 1$, and the three outputs: $Output_1 = 0$ (the protoplasmic tubes $Output_1$ do not move), $Output_3 = 0$ (the protoplasmic tubes $Output_3$ do not move), $Output_2 = 1$ (the protoplasmic tubes $Output_2$ move right). **c** Fuzzy ellipsoid *B* designating one active zone of actin filament polymerization is transformed into fuzzy ellipsoid *D*, causing the movement of the amoeba [in (**a**)] as well as of the plasmodium [in (**b**)]. Fuzzy ellipsoids *A*, *C* were deactivated, see [5]

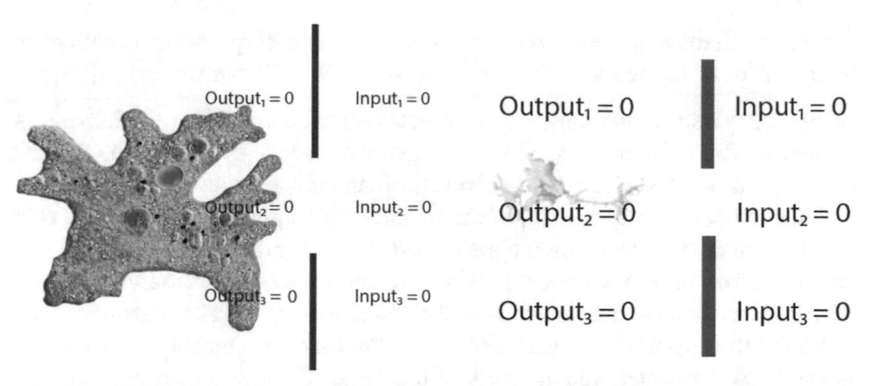

(a) The FREDKIN gate for *Amoeba proteus* (b) The FREDKIN gate for *Physarum polycephalum*

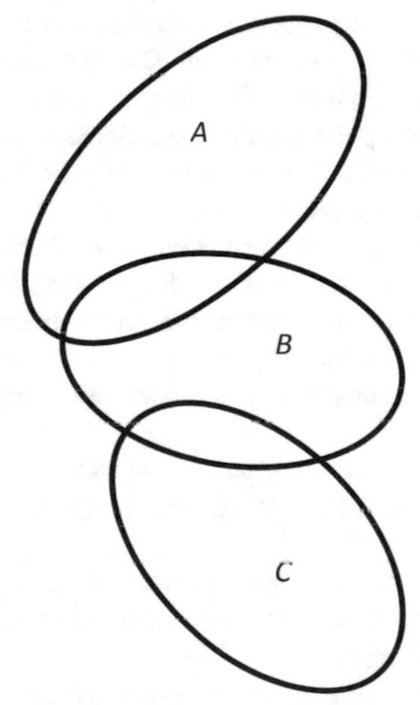

(c) The transformation of the three actin filament zones

Fig. 2.23 **a** The FREDKIN gate for *Amoeba proteus* with the three inputs: $Inp_1 = 0$, $Inp_2 = 0$, $Inp_3 = 0$, and the three outputs: $Output_1 = 0$ (the cell zone $Output_1$ does not move), $Output_3 = 0$ (the cell zone $Output_3$ does not move), $Output_2 = 0$ (the cell zone $Output_2$ does not move). **b** The FREDKIN gate for *Physarum polycephalum* with the three inputs: $Inp_1 = 0$, $Inp_2 = 0$, $Inp_3 = 0$, and the three outputs: $Output_1 = 0$ (the protoplasmic tubes $Output_1$ do not move), $Output_3 = 0$ (the protoplasmic tubes $Output_3$ do not move), $Output_2 = 0$ (the protoplasmic tubes $Output_2$ do not move). **c** Fuzzy ellipsoids A, B, C designating the three active zones of actin filament polymerization were deactivated, causing the rest of the amoeba [in (**a**)] as well as of the plasmodium [in (**b**)], see [5]

In the actin filament networks we find out some analogous phenomena with some basic properties of the neural networks (see [108, 109] and compare to [14]):

- *Lateral inhibition*. In neurons, a presynaptic cell excites inhibitory interneurons and they inhibit neighbouring cells in the neural network. As a result, the contrast of the signal is made more visible. In actin filaments, neighbouring bundles are inhibited to increase the intensity of the signal. As a consequence, one active zone of actin filament polymerization appears instead of several possible zones.
- *Lateral activation*. In neurons, a presynaptic cell excites activation interneurons and they activate neighbouring cells in the neural network. As a consequence, the contrast of the signal is made less visible. In actin filaments, neighbouring bundles are activated to decrease the intensity of the signal. Several active zones of actin filament polymerization appear towards this signal.
- *Feedback/recurrent inhibition*. In neurons, a presynaptic cell transmits the signal to a postsynaptic cell, and the postsynaptic cell in turn transmits it to an interneuron, which then inhibits the presynaptic cell. Due to this circuit there is a limitation for the excitation and the rhythmic changing in the transmission of the signal is possible. The similar takes place for the actin filament bundles causing the generating of rhythmic behaviours. One active zone of actin filament polymerization is rhythmically transformed into another active zone.
- *Feedback/recurrent excitation*. A presynaptic cell excites a postsynaptic neuron and the postsynaptic neuron excites in turn the presynaptic cell. It is used for learning and memory processes. In actin filaments, recurrent excitation accumulates the external stimuli as a positive feedback to continue the same pattern of behaviour. One active zone of actin filament polymerization is continuously transformed into another active zone.
- *Feedforward inhibition*. A presynaptic neuron excites an inhibitory interneuron that inhibits the next neuron. The actin filaments ignore some signals if they 'see' neighbour repellents.
- *Feedforward excitation*. A presynaptic neuron excites a postsynaptic neuron. In actin filaments, we have a direct action in changing the actin filaments caused by one external attracting stimulus.
- *Convergence/divergence*. A postsynaptic neuron receives a convergent input from a number of different presynaptic neurons and this postsynaptic neuron makes further divergent connections to other postsynaptic neurons. Convergence allows a cell to receive a signal from many cells and divergence allows a cell to transmit the signal further. Secreting a cyst wall of the amoeba is an example of this effect for the actin filament bundles in *Amoeba proteus*.

As we see, the actin filament networks are more complex than neural networks and the basic neuronal properties are analogous to appropriate properties of actin filaments. Nevertheless, the actin filament networks are not studied well recently from the point of view of mathematics. About memory of unicellular amoeboid organisms, please see [31, 32, 110–115].

2.2 Arithmetics in Actin Filament Networks

2.2.1 p-Adic Valued Arithmetic Functions in Actin Filament Networks

Let us consider the discrete time $t = 0, 1, 2, \ldots$ assuming that at each time step t the actin filaments of the amoeba face not more than n attractants or repellents and react to n stimuli. Then the amoeba motion can be examined as an arithmetic function $f_{2^n}(x) = y$, where $x, y \in \{0, 1, \ldots, 2^n - 1\}$, e.g. in the FREDKIN gate (Figs. 2.16, 2.17, 2.18, 2.19, 2.20, 2.21, 2.22 and 2.23) we deal with the arithmetic function f_{2^3}, where the inputs and the outputs of Table 2.2 are rewritten as natural numbers: $x_0x_1x_2 = \sum_{i=0}^{2} x_i \cdot 2^i$. For example, $000 = 0$ and $111 = \sum_{i=0}^{2} 1 \cdot 2^i = 7$. Hence, if we have n signals at the time step t, then *Amoeba proteus* or plasmodium of *Physarum polycephalum* calculates an arithmetic function f_{2^n} at this t. What f_{2^n} is in fact, depends on the topology of n stimuli (their intensity, localization, combination, etc.). For instance, the combination of inputs $[Inp_1 Inp_2 Inp_3]$, where Inp_2 is an attractant or its absence and Inp_1, Inp_3 are barriers consisting of repellents or their absence, gives us the FREDKIN gate if we have a configuration of the three active zones of actin filament polymerization depicted as ellipses A, B, C in Fig. 2.16c.

Thus, if the amoeba meets not more than n stimuli at $t = 0, 1, 2, \ldots$, then we obtain a sequence of functions:

$$f_{2^n}^{t=0} f_{2^n}^{t=1} f_{2^n}^{t=2} \ldots, \tag{2.1}$$

where at each $t = i$ the arithmetic function $f_{2^n}^{t=i}$ can be different. Let us denote this sequence by f. It can be considered a *p-adic valued function* for $p = 2^n$:

$$f(\alpha) = \beta, \tag{2.2}$$

where $\alpha = \alpha_0\alpha_1\alpha_2\ldots$ and $\beta = \beta_0\beta_1\beta_2\ldots$ such that we have $f_{2^n}^{t=i}(\alpha_i) = \beta_i$ for each $i = 0, 1, 2, \ldots$ The numbers α and β are p-adic, because

$$\alpha = \sum_{i=0}^{\infty} \alpha_i \cdot p^i, \qquad \beta = \sum_{i=0}^{\infty} \beta_i \cdot p^i$$

and $\alpha_i, \beta_i \in \{0, \ldots, p-1\}$ for each $i = 0, 1, 2, \ldots$ Hence, the amoeboid long-time locomotion can be simulated by p-adic valued arithmetic functions of the form of (2.2).

In the paper [23], there was proposed the p-adic valued logic for simulating the locomotion of *Physarum polycephalum* plasmodia. The same logic can be used for simulating the *Amoeba proteus* locomotion as well as actin filament reactions of other cells. In this logic we can combine many trajectories of the form of (2.1) by which different amoebae have navigated.

Let us notice that arithmetic operations in p-adic valued logic can be defined *corecursively*. Assume that `[]` is an empty list and `a:s` is an infinite list of integers from $\{0, \ldots, p-1\}$ with a head `a` and a tail `s`. If the tail is a constant, it means that this constant repeats for ever. For example, `a:0` means that after `a` there is an infinite list of `0`. The list `a:s` can be defined as `a:b:s'`, the list `a:b:s'` as `a:b:c:s''`, etc. Meanwhile, `b` is the first element of the tail `s`, `c` is the first element of the tail `s'`, etc. So, each tail is an infinite list in turn: `s = b:s'`, `s' = c:s''`, etc. Let `next(a:b:s)= b:s`. Let us define the p-adic valued sum as follows:

```
sum (a,c):[] = sum (a:[], c:[])
next (sum (a:b:s, c:d:s')) = sum (b:s, d:s')
```

The p-adic valued product:

```
prod (a,c):[] = prod (a:[],  c:[])
next(prod (a:b:s, c:d:s')) = sum (prod (a:0, d:s'),
        prod(b:s, c:d:s'))
```

Now, let us assume that `0:0` is a minimal p-adic integer (it is an infinite list of 0) and `p-1:p-1` is a maximal p-adic integer (it is an infinite list of $p-1$). Then let us define the p-adic valued conjunction corecursively:

```
min (a,c):[] = min (a:[], c:[])
next (min (a:b:s, c:d:s')) = min (b:s, d:s')
```

as well as p-adic valued disjunction:

```
max (a,c):[] = max (a:[], c:[])
next (max (a:b:s, c:d:s')) = max (b:s, d:s')
```

The negation is defined as follows:

```
not(a):[] = p-1- (a:[])
next (not(a:b:s)) = p-1:p-1 - (b:s)
```

In this way we obtain the ring of p-adic integers, $\mathbf{Z}_p$, where $p = 2^n$, with the p-adic valued conjunction, disjunction, and negation. For $t \to \infty$ the amoeba or plasmodium motility implements arithmetic functions on $\mathbf{Z}_p$ by reversible logic gates as the FREDKIN gate. The field of p-adic numbers, $\mathbf{Q}_p$, exists just for the prime p, but our p is equal to an even number 2^n. So, we cannot obtain the field for $p = 2^n$, only the ring, if $n > 1$. Hence, for 2-adic numbers there is the field $\mathbf{Q}_2$. Another important feature is that $\mathbf{Z}_p$ contains infinite integers and the set $\mathbf{Z}_p$ is uncountable (see Table 2.3). Due to this fact $\mathbf{Z}_p$ differs from the ring of integers, $\mathbf{Z}$, a lot (the cardinal number of $\mathbf{Z}_p$ is larger, than the cardinal number of $\mathbf{Z}$, i.e. it is larger than $\aleph_0$). In particular, some arithmetic functions on $\mathbf{Z}_p$ are undecidable by definition. The matter is that $\mathbf{Z}_p$ is a codata set (non-inductive set, i.e. corecursive or coinductive data) with non-Archimedean properties. For $\mathbf{Z}_p$, there are no algorithms for calculating all arithmetic functions by definition, because their objects are defined coinductively, not inductively.

Table 2.3 The Cantor's diagonalization showing that the set of all outputs is uncountable in the p-adic universe. Let us take the diagonal $(p-1, p-1, p-3, \ldots)$ from the table and change it as follows: $(0, 0, 1, \ldots)$. This new p-adic integer does not occur in the enumeration

Input strings	The number of output string at t						
i_0	$\mathbf{p-1}$	$p-1$	$p-1$	$p-1$	$p-1$	$p-1$	...
i_1	$p-2$	$\mathbf{p-1}$	$p-2$	$p-1$	$p-2$	$p-1$	...
...	...	...	...	...	...	...	...
i_n	0	$p-1$	0	$p-1$	0	$p-1$	...
i_{n+1}	$p-1$	$p-2$	$p-1$	$p-2$	$p-1$	$p-2$	...
...	...	...	...	...	...	...	...

Table 2.4 The partition of the set of all p-adic valued arithmetic functions, F, into the subsets F_p^t, where $t = 0, 1, 2, \ldots, p = 2, 4, 8, 16, 32, \ldots$

	1-st n $(n = 1)$ 1-st p $(p = 2)$	2-nd n $(n = 2)$ 2-nd p $(p = 4)$	3-rd n $(n = 3)$ 3-rd p $(p = 8)$	...
$t = 0$	$f_{2^1}^{t=0} \in F_{p=2}^{t=0}$	$f_{2^2}^{t=0} \in F_{p=4}^{t=0}$	$f_{2^3}^{t=0} \in F_{p=8}^{t=0}$	...
$t = 1$	$f_{2^1}^{t=1} \in F_{p=2}^{t=1}$	$f_{2^2}^{t=1} \in F_{p=4}^{t=1}$	$f_{2^3}^{t=1} \in F_{p=8}^{t=1}$	...
$t = 2$	$f_{2^1}^{t=2} \in F_{p=2}^{t=2}$	$f_{2^2}^{t=2} \in F_{p=4}^{t=2}$	$f_{2^3}^{t=2} \in F_{p=8}^{t=2}$	...
⋮	⋮	⋮	⋮	⋮
$t \to \infty$	$f_{2^1}^{\infty} \in F_{p=2}^{\infty}$	$f_{2^2}^{\infty} \in F_{p=4}^{\infty}$	$f_{2^3}^{\infty} \in F_{p=8}^{\infty}$	...

Now, we can enumerate all the arithmetic functions which are implementable in actin filament networks as follows. Let us take a partition of all arithmetic functions on $\mathbf{Z}_p$ for different p in the way of Table 2.4. For each p there is the following enumeration of all arithmetic functions for the fixed t. Each function $f_{2^n}^t \in F_{p=2^n}^t$ is one of the possible t-th permutations of the numbers $0, 1, \ldots, 2^n - 1$ and can be distinguished by the following *code*:

$$\lceil f_{2^n}^t \rceil = \sum_{i=0}^{t} \left(\sum_{j=1}^{2^n-1} c_{ji} \cdot j! \right) \cdot ((2^n - 1)!)^i, \tag{2.3}$$

where c_{ji} counts the number of positions in the given i-th permutation that are to the right of value j and that contain a value less than j. For instance, for the FREDKIN gate we have the permutation (0, 1, 2, 3, 4, 6, 5, 7). This means that

$$\sum_{j=1}^{2^3-1} c_j \cdot j! = 0 \cdot 1! + 0 \cdot 2! + 0 \cdot 3! + 0 \cdot 4! + 0 \cdot 5! + 1 \cdot 6! + 0 \cdot 7! = 720.$$

Let us assume that the FREDKIN gate was applied two times and $t = 1$. As a consequence:

$$\sum_{i=0}^{1}\left(\sum_{j=1}^{2^3-1} c_{ji} \cdot j!\right) \cdot ((2^3 - 1)!)^i = 720 + 720 \cdot 7! = 3{,}629{,}520.$$

For the function $f_{2^n}^{\infty}$, its code is as follows:

$$\lceil f_{2^n}^{\infty} \rceil = \sum_{t=0}^{\infty}\left(\sum_{j=1}^{2^n-1} c_{jt} \cdot j!\right) \cdot ((2^n - 1)!)^t. \tag{2.4}$$

This code is $(p - 1)!$-adic.

Hence, each p-adic valued arithmetic function (2.2) coded by (2.4) denotes just an infinite trajectory of one amoeba or plasmodium of *Physarum polycephalum* under the conditions of not more than n inputs at each time step t. These trajectories can be combined by arithmetic and logical operations of the p-adic valued logic defined in [23]. In this way we can simulate a colony of amoebae, their common locomotion, their ongoing divisions and deaths, etc.

2.2.2 Undecidable Functions in Actin Filament Networks

Let us remind that a set A is called *computable* (*decidable* or *solvable*) if there exists a Turing machine M that behaves as follows:

$$M(x) = \begin{cases} 1, & \text{if } x \in A; \\ 0, & \text{otherwise.} \end{cases}$$

So, we assume that each F_p^t is decidable. Let $M_1, M_2, \ldots,$ be a standard list of Turing machines that includes all programs for F_p^t for a fixed p and from the index i it is possible to extract a code $i = \lceil f_{2^n}^t \rceil$ such that M_i decides that $f_{2^n}^t \in F_p^t$. It means, we suppose that there is a Turing machine M that takes an input $\langle i, x \rangle = \langle \lceil f_{2^n}^t \rceil, f_{2^n}^t \rangle$ and gives an output $M_i(x)$.

The set $K_0 = \{\langle i, x \rangle : M_i(x) \text{ halts}\}$ is called a *halting set*. If the computation halts, then we know that $\langle i, x \rangle \in K_0$.

Theorem 2.1 *The set K_0 is not decidable.*

Proof We can appeal to diagonalization to prove this statement. Let us assume that K_0 is decidable and let M_0 be a Turing machine that decides K_0. We can define M_0 as follows:

$$M_0(x) = \begin{cases} 1, & M_i \text{ halts}; \\ 0, & \text{otherwise.} \end{cases}$$

Since M_0 is a Turing machine, it has a code e, therefore $M_e = M_0$. Now, we can define $F_p''^t = \{i\colon M_i(i) \neq 1\}$. This set is undecidable for any i. Then the set $K_0 \supseteq \{\langle i, i\rangle : M_i(i) \neq 1\}$ is undecidable, too. □

Let us consider an example of $F_p''^t$ from Theorem 2.1. Take the following machine:

$$D(M) = \begin{cases} 1, & M(f_{2^n}^t) \neq M(\lceil f_{2^n}^t \rceil); \\ 0, & \text{otherwise.} \end{cases}$$

Then let us define the diagonalization:

$$D(D) = \begin{cases} 1, & D(M) \neq D(\lceil M \rceil); \\ 0, & \text{otherwise.} \end{cases}$$

The actin filament networks are too sensitive to the cellular surroundings. We have assumed that the amoeboid motility programmed by the actin filament networks is a kind of the reversible logic gates, i.e. for n inputs it gives just n outputs. However, the situation of the real amoeboid reactions is much more difficult. In reality, it looks like as follows. All the external signals have a scaling that is too different from the actin filament networks. The point is that the networks have a much better zooming than any outer stimulus. This fact allows them to react continuously to all possible signals at any point of the shape. The amoeba is an *analogue computer*.

The latter feature can be formulated as the situation when the number of outputs is larger, than the number of inputs. The amoeboid reactions with the n inputs and the $m > n$ outputs can be considered a *hybrid action* [116]. The set of hybrid actions can have an infinite set of labels [116]. So, it is unsolvable by definition. The hybrid actions are undecidable in the meaning of Theorem 2.1. Let us define them:

$$d(f_{2^n}^t) = \begin{cases} \textit{accept}, & \text{the amoeba accepts } g \text{ such that} \\ & g > f_{2^n}^t \text{ for any } f_{2^n}^t \\ & \text{at } t \text{ with the } n \text{ inputs;} \\ \textit{reject}, & \text{otherwise.} \end{cases}$$

Thus, we can implement undecidable arithmetic functions in the actin filament networks, too.

2.2.3 Formal Systems and Undecidability

Let us consider the following *formal system*: $\mathscr{F} = \langle K, V, P, I, L\rangle$, where $K = \{a, b, c, \dots\}$ is an alphabet consisting of signs, $V = \{x, y, z, \dots\}$ is a set of variables, $P = \{S_1, S_2, \dots\}$ is a set of predicates, I consists of comma (,) and implication ($\supset$), L is a set of axioms defining strings. The string is of the form $sA_1, \dots, A_n$, where

$A_1, \ldots, A_n \in K \cup V$. This string means that the sequence of symbols $A_1, \ldots, A_n$ belongs to S.

This system $\mathscr{F}$ generates some strings over signs, variables, and predicates by using axioms and the following two inference rules: (i) the *substitution rule*—we can replace the same variable by the same sequence of signs; (ii) *modus ponens*—if there are axioms A and $A \supset B$, then we obtain a string B, where A and B are some strings over K.

For instance, let $K = \{n, s, w, e\}$, where n denotes a north location, s is a south location, w is a west location, and e is an east location. Let V consist only of one variable x. Assume that the only predicate of the system, S, means an occupation by an organism (plasmodium or amoeba). So, an atomic string Sn means that the organism is located at the point denoted by n (north). Suppose that our axioms are as follows:

$$Sn; Ss; Sw; Se.$$

They mean that it is accessible that our organism is located at n, s, w, or e. The additional axioms:

$$Sns; Snw; Sne; Ssn; Ssw; Sse; Sen; Ses; Sew,$$

where each SAB means that the organism is located first at A, then at B.

Now the composite axioms:

$$Sxn \supset Sxne;$$

$$Sxs \supset Sxsw;$$

$$Sxe \supset Sxen;$$

$$Sxw \supset Sxws.$$

By using the substitution rule and modus ponens, we can define all the strings of that system simulating a propagation of organism in some chosen directions. If a string A is obtained by applying modus ponens t_m times and by applying the substitution rule t_s times, we say that this string A is obtained at the time step $t = t_m + t_s$. So, the strings Sn, Ss, Sw, Se, Sns, Snw, Sne, Ssn, Ssw, Sse, Sen, Ses, Sew are formed at the time $t = 0$. The string $Snsws$ is built at the time step $t = 4$, etc.

Another example is as follows. The FREDKIN gate of Figs. 2.16, 2.17, 2.18, 2.19, 2.20, 2.21, 2.22 and 2.23 has the following formal system: $\langle K, S, W, L \rangle$, where $K = \{u, r, d\}$ and u is a motion up right, r is a motion directly right, d is a motion down right. We have the following possible lines: (i) $A_1 := u, r, d$, Fig. 2.16; (ii) $A_2 := u, r$, Fig. 2.17; (iii) $A_3 := u, d$, Fig. 2.18; (iv) $A_4 := u$, Fig. 2.19; (v) $A_5 := r, d$, Fig. 2.20; (vi) $A_6 := d$, Fig. 2.21; (vii) $A_7 := r$, Fig. 2.22; (viii) $A_8 := \emptyset$, Fig. 2.23. So, let us consider not sequences of symbols, but lines of symbols. Let V consist of the two variables A_x and A_y for lines. The only predicate, S, means an occupation by plasmodium or amoeba. The axioms are as follows:

$$sA_1; sA_2; sA_3; sA_4; sA_5; sA_6; sA_7; sA_8. \tag{2.5}$$

For example, sA_3 means that we deal with a line Su, Sd. The next axiom

$$sA_x \supset (sA_y \supset sA_xA_y) \tag{2.6}$$

allows us to apply modus ponens for building new strings.

Let us show how we can construct new strings by applying the substitution rule and modus ponens. For instance, at formula (2.6), let sA_x be replaced by $sA_3(t = 1)$, then by modus ponens we have $(sA_y \supset sA_3A_y)$ $(t = 2)$. Then sA_y is replaced by $sA_5(t = 3)$. And by modus ponens we obtain sA_3A_5 $(t = 4)$. This sA_3A_5 means all the combinations of all signs from the line A_3 and A_5: Sur, Sdr, Sud, Sdd. In order to compare formal systems with 2^n-adic functions of (2.2), let us make the time step the same. For this purpose we should take just numbers $4t$ of time steps in formal systems and divide them by 4. The number $t < 4$ is taken as 0.

Generating an infinite string in a formal system can be considered as a 2^n-adic valued function (2.2), where n is a number of signs used at time t (e.g. $n = 3$ in the FREDKIN gate), with a $(2^n - 1)!$-adic valued code (2.4). At each time step t we can use different logic gates. So, axioms involved in generating strings can be different, too. For example, if a gate is FREDKIN, our axioms are (2.5) $\wedge$ (2.6). Assume that a string X is obtained by using the axioms $\text{Ax}_1, \ldots, \text{Ax}_k$. Then it is denoted by a string of the form $\text{Ax}_1 \star \cdots \star \text{Ax}_k \star X$. The formal system closed under such strings is denoted by $\mathscr{U}$.

We say that a predicate S represents a set of strings/lines, S', in the formal system $\mathscr{U}$ if for each sequence of signs/lines X we have:

$X \in S'$ if and only if $\text{Ax}_1 \star \cdots \star \text{Ax}_k \star SX$ and SX is inferable in $\mathscr{U}$ from axioms $\text{Ax}_1, \ldots, \text{Ax}_k$.

Axioms $\text{Ax}_1, \ldots, \text{Ax}_k$ used at time t are coded by $(2^n - 1)!$-adic valued number (2.3). Let us remember that each p-adic valued number $n = \sum_{t=0}^{\infty} n_t \cdot p^t$ has the notation $n = \ldots n_t \ldots n_1 n_0$. This notation is very useful for us. Let X and Y be two strings/lines in $\mathscr{U}$, and $\ulcorner X \urcorner$ and $\ulcorner Y \urcorner$ are their p-adic codes. Then a composite string/line XY has a p-adic code $\ulcorner XY \urcorner = \ulcorner Y \urcorner \ulcorner X \urcorner$. So, p-adic codes preserve concatenations of strings/lines.

We know that axioms $\text{Ax}_1, \ldots, \text{Ax}_k$ at t has a $(2^n - 1)!$-adic valued code $h = \ulcorner \text{Ax}_1, \ldots, \text{Ax}_k \urcorner$ calculated by formula (2.3). Let the code for SX be a $(2^n - 1)!$-adic valued number $h' = \ulcorner SX \urcorner$ that is given by an appropriate logic gate under conditions X. Then the $(2^n - 1)!$-adic valued code for $\text{Ax}_1 \star \cdots \star \text{Ax}_k \star SX$ is equal to $h'h$. Let us differ these codes for all t in the way: $h'_t h_t = h'_t h_t \ldots h'_1 h_1 h'_0 h_0$ and $h'_\infty h_\infty = \ldots h'_t h_t \ldots h'_1 h_1 h'_0 h_0$.

Let T be a set of all true strings/lines of $\mathscr{U}$ and T_0 be a set of $(2^n - 1)!$-adic valued codes for strings/lines from T.

Theorem 2.2 *The set T_0 is recursively enumerable and not decidable.*

Proof Let X be a string/line from $\mathcal{U}$ and A be a set of $(2^n - 1)!$-adic valued numbers. This X is called a *Gödel proposition* if:

$$X \in T \text{ if and only if } \ulcorner X \urcorner \in A.$$

Suppose that X is written in numbers of $(2^n - 1)!$-adic valued arithmetics and its $(2^n - 1)!$-adic code is X_0. The string X_0X is called a *diagonalization* of X. The axioms of our arithmetics:

$$Pa_1, \ulcorner a_1 \urcorner;$$

$$\ldots$$

$$Pa_k, \ulcorner a_k \urcorner;$$

$$\ldots,$$

where each a_k is $(2^n - 1)!$-adic valued number.

$$Px, \ulcorner x \urcorner \supset (Py, \ulcorner y \urcorner \supset Pyx, \ulcorner y \urcorner \ulcorner x \urcorner).$$

These axioms give for each X its diagonalization X_0X with a code $\ulcorner X_0 \urcorner \ulcorner X \urcorner$.

Assume that there is a set A' for each set of numbers A such that A' contains diagonalizations for all numbers from A. Now, let us show that if A is recursively enumerable, then A' is recursively enumerable. Let us introduce the new axiom:

$$Px, \ulcorner x \urcorner \supset (A \ulcorner x \urcorner x \supset A' \ulcorner x \urcorner).$$

Then if $A \ulcorner x \urcorner x$ is inferable, then $A' \ulcorner x \urcorner$ is inferable under conditions that A represents A and A' represents A'. From this it follows that if A is recursively enumerable, then A' is recursively enumerable, too.

Show that for each recursively enumerable set A there exists a Gödel proposition. Let A' be represented by a predicate H. Then for any number n

$$nH \text{ is true if and only if } n_0 \in A' \text{ if and only if } n_0 n \in A.$$

We know that n is a code $\ulcorner H \urcorner$, then:

$$\ulcorner H \urcorner H \text{ is true if and only if } \ulcorner \ulcorner H \urcorner \urcorner \ulcorner H \urcorner \in A.$$

Consequently, $\ulcorner H \urcorner H$ is a Gödel proposition for A.

Let $\neg T_0$ be a complement of T_0 in the set of all possible strings/lines in $\mathcal{U}$. Then there is no Gödel proposition for $\neg T_0$, because each Gödel proposition is defined for $A = T_0$. To be decidable, the set T_0 should be recursively enumerable and its complement, $\neg T_0$ should be recursively enumerable, too. However, we have shown that only T_0 is recursively enumerable. □

The main problem of designing the actin filament networks consists in controlling the signal transmission through the actin filaments [48, 59, 117, 118]. It is known that the actin filaments support propagation of voltage pulses (see [8] and compare to [119]) and, therefore, it is possible to explain the signal transmission through the actin filaments by an interaction between voltage pulses, where 1 ('true') is assigned to the presence of a voltage pulse in a given location of the actin filament, and 0 ('false') is assigned to the pulse's absence, so that Boolean logical gates and a one-bit half-adder with interacting voltage pulses can be constructed well [8]. In my research, I focus only on visible transformations of the cell shape to implement a formal arithmetic. Theorem 2.2 demonstrates that this arithmetic is implementable in the amoeboid motions with decidable and undecidable arithmetic functions. This statement supports the *hypothesis of orchestrated objective reduction* [37]. In order to prove this hypothesis, we should know how to control each signal transmission through the actin filaments, including voltage pulses.

I was grounded on the idea that each actin filament is a double chain of nodes, which take state 0 (resting) or 1 (excited), see Sect. 2.1.1. These states are updated in parallel in discrete time depending on states of two closest neighbours in the node chain and two closest neighbours in the complementary chain. In this way it is possible to represent the actin filaments as an automaton of finite states with transition rules that support traveling and mobile localizations [6, 120]. Also, we can assume that states of nodes depends not only on the current states of neighbouring node but also on their past states so that we assess the effect of memory of past states on the dynamics of acting automata [2].

The actin filament networks are responsible for cellular reactions to external stimuli. So, if it is possible to design a protein robot solving a complex of various tasks (learning, orientation in space, decision making about transitions etc.), then this robot will consist of actin filaments controlled by us. And taking into account the possibility to implement undecidable arithmetic functions in actin filament networks, this biological robot will present a kind of hypercomputation beyond any Turing machines.

The main result of the chapter is to show that reactions of *Amoeba proteus* to n external stimuli, expressed in transformations of the cell shape, can be represented as an arithmetic function $f_{2^n}(x) = y$, where $x, y \in \{0, 1, \ldots, 2^n - 1\}$. So, I have examined the FREDKIN gate (Figs. 2.16, 2.17, 2.18, 2.19, 2.20, 2.21, 2.22 and 2.23) dealing with the arithmetic function f_{2^3}, where the inputs and the outputs of Table 2.2 have been rewritten as natural numbers: $x_0x_1x_2 = \sum_{i=0}^{2} x_i \cdot 2^i$. Furthermore, we can distinguish time $t = 0, 1, 2, \ldots$ for amoeboid reactions, assuming that at each time step t, the *Amoeba proteus* faces different n external stimuli and reacts to them differently by implementing different arithmetic functions. It means that all the behaviour of *Amoeba proteus* is regarded as a realization of arithmetic functions $f_{2^n}^{t=0} f_{2^n}^{t=1} f_{2^n}^{t=2} \ldots$ at $t = 0, 1, 2, \ldots$ In this way we can show that a colony of amoebae implements a formal arithmetic with decidable and undecidable arithmetic functions (Theorem 2.2). This fact is a good argument supporting the *hypothesis of orchestrated objective reduction* [37], that is, we see that even at the level of unicellular organisms such as amoebae we observe a kind of intelligence. This intelligence can be explained by a feature of perceiving external signals by actin filament networks. The point is that the same signal can be perceived differently (see Fig. 2.14)—it depends on the cell shape and many other internal circumstances. As a consequence, we can have more outputs than inputs in possible reactions of actin filament networks. The latter case means that the organism can implement an undecidable arithmetic function (see the commentary after Theorem 2.1).

References

1. Adamatzky, A., Mayne, R.: Actin automata: Phenomenology and localizations. Int. J. Bifurc. Chaos **25**(2) (2015). https://doi.org/10.1142/S0218127415500303
2. Alonso-Sanz, R., Adamatzky, A.: Actin automata with memory. Int. J. Bifurc. Chaos **26**(1) (2016)
3. Schumann, A.: Toward a computational model of actin filament networks. In: Proceedings of the 9th International Joint Conference on Biomedical Engineering Systems and Technologies (BIOSTEC 2016)—Volume 4: BIOSIGNALS, Rome, Italy, 21–23 Feb 2016, pp. 290–297 (2016). https://doi.org/10.5220/0005828902900297
4. Schumann, A.: On arithmetic functions in actin filament networks. In: 10th EAI International Conference on Bio-inspired Information and Communications Technologies (formerly BIONETICS). ACM (2017). https://doi.org/10.4108/eai.22-3-2017.152402
5. Schumann, A.: Decidable and undecidable arithmetic functions in actin filament networks. J. Phys. D Appl. Phys. (2017). https://doi.org/10.1088/1361-6463/aa9d7b
6. Siccardi, S., Adamatzky, A.: Actin quantum automata: communication and computation in molecular networks. Nano Comm. Netw. **6**(1), 15–27 (2015). https://doi.org/10.1016/j.nancom.2015.01.002
7. Siccardi, S., Adamatzky, A.: Quantum actin automata and three-valued logics. IEEE J. Emerg. Sel. Topics Circuits Syst. **6**(1), 53–61 (2016)
8. Siccardi, S., Tuszynski, J.A., Adamatzky, A.: Boolean gates on actin filaments. Phys. Lett. A **380**(1), 88–97 (2016)
9. Adamatzky, A.: Physarum machine: implementation of a Kolmogorov-Uspensky machine on a biological substrate. Parallel Process. Lett. **17**(4), 455–467 (2007)
10. Adamatzky, A.: Physarum machines: encapsulating reaction-diffusion to compute spanning tree. Naturwisseschaften **94**, 975–980 (2007)
11. Adamatzky, A.: Physarum Machines: Computers from Slime Mould. Series on Nonlinear Science A, World Scientific (2010)
12. Adamatzky, A.: Slime mould logical gates: exploring ballistic approach. In: Applications, Tools and Techniques on the Road to Exascale Computing, vol. 1, pp. 41–56 (2010)
13. Adamatzky, A.: Slime mould computing. Int. J. Gen. Syst. **44**(3), 277–278 (2015)
14. Adamatzky, A.: A would-be nervous system made from a slime mold. Artif. Life **21**(1), 73–91 (2015)
15. Adamatzky, A., Schubert, T.: Slime mold microfluidic logical gates. Mater. Today **17**(2), 86–91 (2014)
16. Berzina, T., Dimonte, A., Cifarelli, A., Erokhin, V.: Hybrid slime mould-based system for unconventional computing. Int. J. Gen. Syst. **44**(3), 341–353 (2015). https://doi.org/10.1080/03081079.2014.997523
17. Jones, J., Mayne, R., Adamatzky, A.: Representation of shape mediated by environmental stimuli in Physarum polycephalum and a multi-agent model. JPEDS **32**(2), 166–184 (2017)

18. Jones, J.D., Adamatzky, A.: Towards Physarum binary adders. Biosystems **101**(1), 51–58 (2010)
19. Kalogeiton, V.S., Papadopoulos, D.P., Georgilas, I., Sirakoulis, G.C., Adamatzky, A.: Cellular automaton model of crowd evacuation inspired by slime mould. Int. J. Gen. Syst. **44**(3), 354–391 (2015). https://doi.org/10.1080/03081079.2014.997527
20. Mayne, R., Adamatzky, A.: Slime mould foraging behaviour as optically coupled logical operations. Int. J. Gen. Syst. **44**(3), 305–313 (2015). https://doi.org/10.1080/03081079.2014.997528
21. Ntinas, V.G., Vourkas, I., Sirakoulis, G.C., Adamatzky, A.: Oscillation-based slime mould electronic circuit model for maze-solving computations. IEEE Trans. Circuits Syst. **64-I**(6), 1552–1563 (2017)
22. Pancerz, K., Schumann, A.: Rough set models of Physarum machines. Int. J. Gen. Syst. **44**(3), 314–325 (2015)
23. Schumann, A.: *p*-adic valued logical calculi in simulations of the slime mould behaviour. J. Appl. Non-Class. Log. **25**(2), 125–139 (2015). https://doi.org/10.1080/11663081.2015.1049099
24. Schumann, A.: Towards slime mould based computer. New Math. Nat. Comput. **12**(2), 97–111 (2016). https://doi.org/10.1142/S1793005716500083
25. Schumann, A., Pancerz, K.: Logics for physarum chips. Studia Humana **5**(1), 16–30 (2016). https://doi.org/10.1515/sh-2016-0002
26. Shirakawa, T., Sato, H., Ishiguro, S.: Constrcution of living cellular automata using the Physarum polycephalum. Int. J. Gen. Syst. **44**, 292–304 (2015)
27. Shirakawa, T., Yokoyama, K., Yamachiyo, M., Gunji, Y.P., Miyake, Y.: Multi-scaled adaptability in motility and pattern formation of the Physarum plasmodium. Int. J. Bio-Inspir. Comput. **4**, 131–138 (2012)
28. Tsuda, S., Aono, M., Gunji, Y.P.: Robust and emergent Physarum-computing. BioSystems **73**, 45–55 (2004)
29. Whiting, J.G., de Lacy Costello, B.P., Adamatzky, A.: Slime mould logic gates based on frequency changes of electrical potential oscillation. Biosystems **124**, 21–25 (2014)
30. Adamatzky, A., Erokhin, V., Grube, M., Schubert, T., Schumann, A.: Physarum chip project: growing computers from slime mould. Int. J. Unconv. Comput. **8**(4), 319–323 (2012)
31. Pershin, Y.V., Di Ventra, M.: Memristive and memcapacitive models of Physarum learning. In: Advances in Physarum Machines, pp. 413–422. Springer (2016)
32. Traversa, F.L., Pershin, Y.V., Di Ventra, M.: Memory models of adaptive behavior. IEEE Trans. Neural Netw. Learn. Syst. **24**(9), 1437–1448 (2013)
33. Teplov, V.A., Romanovsky, Y.M., Latushkin, O.A.: A continuum model of contraction waves and protoplasm streaming in strands of Physarum polycephalum. Biosystems **24**, 269–289 (1991)
34. Teplov, V.A., Romanovsky, Y.M., Pavlov, D.A., Alt, W.: Auto-oscillatory processes and feedback mechanisms in Physarum plasmodium motility. In: Alt, W., Deutsh, A., Dunn, G. (eds.) Dynamics of Cell and Tissue Motion. Burkhauser, Basel, Switzerland (1997)
35. Forgacs, G.: On the possible role of cytoskeletal filamentous networks in intracellular signaling: an approach based on percolation. J. Cell Sci. **108**, 2131–2143 (1995)
36. Mayne, R., Adamatzky, A., Jones, J.: On the role of the plasmodial cytoskeleton in facilitating intelligent behavior in slime mold Physarum polycephalum. Commun. Integr. Biol. **8**(4), e1059007 (2015)
37. Hameroff, S., Penrose, R.: Consciousness in the universe: a review of the Orch OR theory. Phys. Life Rev. **11**, 39–78 (2014)
38. Maly, I.V., Borisy, G.G.: Self-organization of a propulsive actin network as an evolutionary process. Proc. Nat. Acad. Sci. U.S.A. **98**(20), 11324–11329 (2001)
39. Furukawa, R., Kundra, R., Fechheimer, M.: Formation of liquid crystals from actin filaments. Biochemistry **32**, 12346–12352 (1993)
40. Steinmetz, M., Goldie, K., Aebi, U.: A correlative analysis of actin filament assembly, structure, and dynamics. J. Cell Biol. **138**, 559–574 (1997)

41. Swezey, R., Somero, G.: Polymerization thermodynamics and structural stabilities of skeletal muscle actins from vertebrates adapted to different temperatures and hydrostatic pressures. Biochemistry **21**, 4496–4503 (1982)
42. Evans, E.: New physical concepts for cell amoeboid motion. Biophys. J. **64**, 1306–1322 (1993)
43. Fackler, O.T., Grosse, R.: Cell motility through plasma membrane blebbing. J. Cell Biol. **181**, 879–884 (2008)
44. Pollard, T.D., Borisy, G.G.: Cellular motility driven by assembly and disassembly of actin filaments. Cell **112**, 453–465 (2003)
45. Furuhashi, K., Ishigami, M., Suzuki, M., Titani, K.: Dry stress-induced phosphorylation of Physarum actin. Biochem. Biophys. Res. Commun. **242**, 653–658 (1998)
46. Nakagakia, T., Yamada, H., Ueda, T.: Interaction between cell shape and contraction pattern in the Physarum plasmodium. Biophys. Chem. **84**, 195–204 (2000)
47. Carlier, M.F.: Actin: protein structure and filament dynamics. J. Biol. Chem. **266**, 1–4 (1991)
48. Carlier, M.F., Pantaloni, D.: Control of actin dynamics in cell motility. J. Mol. Biol. **269**(4), 459–467 (1997)
49. Egelman, E.H.: The structure of f-actin. J. Muscle Res. Cell Motil. **6**(2), 129–151 (1985)
50. Etienne-Manneville, S.: Actin and microtubules in cell motility: which one is in control? Traffic **5**(7), 470–477 (2004)
51. Hill, T.L.: Microfilament or microtubule assembly or disassembly against a force. Proc. Natl. Acad. Sci. U.S.A. **78**(9), 5613–5617 (1981)
52. Le Clainche, C., Carlier, M.: Regulation of actin assembly associated with protrusion and adhesion in cell migration. Physiol. Rev. **88**(2), 489–513 (2008)
53. McGough, A.: F-actin-binding proteins. Curr. Opin. Struct. Biol. **8**(2), 166–176 (1998)
54. Mogilner, A., Oster, G.: Cell motility driven by actin polymerization. Biophys. J. **71**(6), 3030–3045 (1996)
55. Mooseker, M.S., Tilney, L.G.: Organization of an actin filament-membrane complex. Filament polarity and membrane attachment in the microvilli of intestinal epithelial cells. J. Cell Biol. **67**(3), 725–743 (1975)
56. Pollard, T.D., Cooper, J.A.: Actin, a central player in cell shape and movement. Science **326**(5957), 1208–1212 (2009)
57. Van Haastert, P.J., Devreotes, P.N.: Chemotaxis: signalling the way forward. Nat. Rev. Mol. Cell Biol. **5**, 626–634 (2004)
58. Holmes, K., Popp, D., Gebhard, W., Kabsch, W.: Atomic model of the actin filament. Nature **347**, 44–49 (1990)
59. Hu, J., Matzavinos, A., Othmer, H.G.: A theoretical approach to actin filament dynamics. J. Stat. Phys. **128**(1–2), 111–138 (2007)
60. Hu, K., Ji, L., Applegate, K.T., Danuser, G., Waterman-Storer, C.M.: Differential transmission of actin motion within focal adhesions. Science **315**, 111–115 (2007)
61. Tobacman, L.S., Korn, E.D.: The kinetics of actin nucleation and polymerization. J. Biol. Chem. **258**, 3207–3214 (1983)
62. Wegner, A.: Head to tail polymerization of actin. J. Mol. Biol. **108**, 139–150 (1976)
63. Wegner, A., Engel, J.: Kinetics of the cooperative association of actin to actin filaments. Biophys. Chem. **3**, 215–225 (1975)
64. Coppin, C., Leavis, P.: Quantitation of liquid-crystaline ordering in f-actin solutions. Biophysics **63**, 794–807 (1992)
65. Galbraith, C.G., Yamada, K.M., Galbraith, J.A.: Polymerizing actin fibers position integrins primed to probe for adhesion sites. Science **315**, 992–995 (2007)
66. Goldmann, W.H., Guttenberg, Z., Tang, J.X., Kroy, K., Isenberg, G., Ezzell, R.M.: Analysis of the f-actin binding fragments of vinculin using stopped-flow and dynamic lightscattering measurements. Eur. J. Biochem. **254**, 413–419 (1998)
67. Kabsch, W., Holmes, K.C.: The actin fold. FASEB J. **9**, 167–174 (1995)
68. Korn, E.D., Carlier, M., Pantaloni, D.: Actin polymerization and ATP hydrolysis. Science **238**, 638–644 (1987)

69. Coluccio, L.M., Tilney, L.G.: Under physiological conditions actin disassembles slowly from the nonpreferred end of an actin filament. J. Cell Biol. **97**, 1629–1634 (1983)
70. Svitkina, T.M., Borisy, G.G.: Arp2/3 complex and actin depolymerizing factor/cofilin in dendritic organization and treadmilling of actin filament array in lamellipodia. J. Cell Biol. **145**, 1009–1026 (1999)
71. Carlier, M.: Role of nucleotide hydrolysis in the dynamics of actin filaments and mictrotubules. Int. Rev. Cytol. **115**, 139–170 (1989)
72. Carlier, M.F., Valentin, R.C., Combeau, C., Fievez, S., Pantoloni, D.: Actin polymerization: regulation by divalent metal ion and nucleotide binding, ATP hydrolysis and binding of myosin. In: Advances in Experimental Medicine and Biology, vol. 358, pp. 71–81 (1994)
73. Chhabra, E.S., Higgs, H.N.: The many faces of actin: matching assembly factors with cellular structures. Nat. Cell Biol. **9**, 1110–1121 (2007)
74. Choi, C.K., Vicente-Manzanares, M., Zareno, J., Whitmore, L.A., Mogilner, A., Horwitz, A.R.: Actin and alpha-actinin orchestrate the assembly and maturation of nascent adhesions in a myosin II motor-independent manner. Nat. Cell Biol **10**, 1039–1050 (2008)
75. Meyer, R., Aebi, U.: Bundling of actin filaments by α-actinin depends on its molecular length. J. Cell Biol. **110**, 2013–2024 (1990)
76. Basaraba, R.J., Byerly, A.N., Stewart, G.C., Mosier, D.A., Fenwick, B.W., Chengappa, M.M., Laegreid, W.W.: Actin enhances the haemolytic activity of escherichia coli. Microbiology **144**(7), 1845–1852 (1998)
77. Iwasa, J.H., Mullins, R.D.: Spatial and temporal relationships between actin-filament nucleation, capping, and disassembly. Curr. Biol. **17**, 395–406 (2007)
78. Balaban, N.Q., Schwarz, U.S., Riveline, D., Goichberg, P., Tzur, G., Sabanay, I., Mahalu, D., Safran, S., Bershadsky, A., Addadi, L.: Force and focal adhesion assembly: a close relationship studied using elastic micropatterned substrates. Nat. Cell Biol. **3**, 466–472 (2001)
79. Brown, C.M., Hebert, B., Kolin, D.L., Zareno, J., Whitmore, L., Horwitz, A.R., Wiseman, P.W.: Probing the integrin-actin linkage using high-resolution protein velocity mapping. J. Cell Sci. **119**, 5204–5214 (2006)
80. Calderwood, D.A., Shattil, S.J., Ginsberg, M.H.: Integrins and actin filaments: reciprocal regulation of cell adhesion and signaling. J. Biol. Chem. **275**, 22607–22610 (2000)
81. Guo, W.H., Wang, Y.L.: Retrograde fluxes of focal adhesion proteins in response to cell migration and mechanical signals. Mol. Biol. Cell **18**, 4519–4527 (2007)
82. Condeelis, J.: Life at the leading edge: the formation of cell protrusions. Annu. Rev. Cell Biol. **9**, 411–444 (1993)
83. Mattila, P.K., Lappalainen, P.: Filopodia: molecular architecture and cellular functions. Nat. Rev. Mol. Cell Biol. **9**, 446–454 (2008)
84. Wachsstock, D.H., Schwarz, W.H., Pollard, T.D.: Cross-linker dynamics determine the mechanical properties of actin gels. Biophys. J. **66**, 801–809 (1994)
85. Elson, E.L.: Cellular mechanics as an indicator of cytoskeletal structure and function. Annu. Rev. Biophys. Chem. **17**, 397–430 (1988)
86. Isambert, H., Venier, P., Maggs, A.C., Fattoum, A., Kassab, R., Pantaloni, D., Carlier, M.F.: Flexibility of actin filaments derived from thermal fluctuations. Effect of bound nucleotide, phalloidin, and muscle regulatory proteins. J. Biol. Chem. **270**(11), 437–44 (1995)
87. Pollard, T.D.: Regulation of actin filament assembly by Arp2/3 complex and formins. Annu. Rev. Biophys. Biomol. Struct. **36**, 451–477 (2007)
88. Yoshigi, M., Hoffman, L.M., Jensen, C.C., Yost, H.J., Beckerle, M.C.: Mechanical force mobilizes zyxin from focal adhesions to actin filaments and regulates cytoskeletal reinforcement. J. Cell Biol. **171**, 209–215 (2005)
89. Vicente-Manzanares, M., Zareno, J., Whitmore, L., Choi, C.K., Horwitz, A.F.: Regulation of protrusion, adhesion dynamics, and polarity by myosins IIA and IIB in migrating cells. J. Cell Biol. **176**, 573–580 (2007)
90. Kas, J., Strey, H., Tang, J.X., Finger, D., Ezzell, R., Sackmann, E., Janmey, P.A.: F-actin, a model polymer for semiflexible chains in dilute, semidilute and liquid crystalline solutions. Biophys. J. **70**, 609–625 (1995)

91. Oosawa, F.: The flexibility of f-actin. Biophys. Chem. **11**, 443–446 (1980)
92. Asakura, S., Taniguchi, M., Oosawa, F.: Mechano-chemical behavior of f-actin. J. Mol. Biol. **7**, 55–69 (1963)
93. Hotulainen, P., Lappalainen, P.: Stress fibers are generated by two distinct actin assembly mechanisms in motile cells. J. Cell Biol. **173**, 383–394 (2006)
94. Janmey, P.A., Hvidt, S., Kas, J., Lerche, D., Maggs, A., Sackmann, E., Schliwa, M., Stossel, T.P.: The mechanical properties of actin gels. Elastic modulus and filament motions. J. Biol. Chem. **269**(32), 503–13 (1994)
95. Ridley, A.J., Schwartz, M.A., Burridge, K., Firtel, R.A., Ginsberg, M.H., Borisy, G., Parsons, J.T., Horwitz, A.R.: Cell migration: integrating signals from front to back. Science **302**, 1704–1709 (2003)
96. ben Avraham, D., Tirion, M.: Dynamic and elastic properties of f-actin: a normal-modes analysis. Biophys. J. **68**(4), 1231–1245 (1995)
97. Higuchi, H., Yanagida, T., Goldman, Y.E.: Compliance of thin filaments in skinned fibers of rabbit skeletal muscle. Biophys. J. **69**, 1000–1010 (1995)
98. Jen, C., McIntire, L., Bryan, J.: The viscoelastic properties of actin solutions. Arch. Biochem. Biophys. **216**, 126–132 (1982)
99. Xu, J.Y., Schwarz, W.H., Kas, J.A., Stossel, T.P., Janmey, P.A., Pollard, T.D.: Mechanical properties of actin filament networks depend on preparation, polymerization conditions, and storage of actin monomers. Biophys. J. **74**, 2731–2740 (1998)
100. Gimona, M., Mital, R.: The single CH domain of calponin is neither sufficient nor necessary for f-actin binding. J. Cell Sci. **111**(Pt 13), 1813–21 (1998)
101. Moore, P.B., Huxley, H.E., DeRosier, D.J.: Three-dimensional reconstruction of f-actin, thin filaments and decorated thin filaments. J. Mol. Biol. **50**(2), 279–295 (1970)
102. Schumann, A., Pancerz, K.: p-adic computation with Physarum. In: Adamatzky, A. (ed.) Advances in Physarum Machines: Sensing and Computing with Slime Mould. Emergence, Complexity and Computation, vol. 21, pp. 619–649. Springer International Publishing (2016)
103. Zadeh, L.: Fuzzy sets. Inform. Control **8**, 338–353 (1965)
104. Kornai, A.: Euclidean automata. In: Waser, M. (ed.) Proceedings of the AAAI Spring Symposium Implementing Selves with Safe Motivational Systems and Self-Improvement, pp. 25–30. AAAI Press (2014)
105. Ghosh, D., Chakraborty, D.: Analytical fuzzy plane geometry III. Fuzzy Sets Syst. **283**, 83–107 (2016)
106. Schumann, A.: Conventional and unconventional reversible logic gates on Physarum polycephalum. Int. J. Parallel, Emerg. Distrib. Syst. **32**(2), 218–231 (2017). https://doi.org/10.1080/17445760.2015.1068775
107. Toffoli, T.: Reversible computing. Tech. memo mit/lcs/tm-151. MIT (1980)
108. Jones, J.D.: Towards lateral inhibition and collective perception in unorganised non-neural systems. In: Pancerz, K., Zaitseva, E. (eds.) Computational Intelligence, Medicine and Biology: Selected Links. Springer (2015)
109. Schumann, A., Woleński, J.: Two squares of oppositions and their applications in pairwise comparisons analysis. Fundam. Inform. **144**(3–4), 241–254 (2016)
110. Dimonte, A., Berzina, T., Pavesi, M., Erokhin, V.: Hysteresis loop and cross-talk of organic memristive devices. Microelectron. J. **45**(11), 1396–1400 (2014). https://doi.org/10.1016/j.mejo.2014.09.009
111. Erokhin, V.: On the learning of stochastic networks of organic memristive devices. Int. J. Unconv. Comput. **9**(3–4), 303–310 (2013)
112. Erokhin, V., Howard, G.D., Adamatzky, A.: Organic memristor devices for logic elements with memory. Int. J. Bifurc. Chaos **22**(11) (2012). https://doi.org/10.1142/S0218127412502835
113. Pershin, Y.V., La Fontaine, S., Di Ventra, M.: Memristive model of amoeba learning. Phys. Rev. E **80**(2), 021926
114. Saigusa, T., Tero, A., Nakagak, T., Kuramoto, Y.: Amoebae anticipate periodic events. Phys. Rev. Lett. **100**(1), 018101 (2008)

115. Shirakawa, T., Gunji, Y.P., Miyake, Y.: An associative learning experiment using the plasmodium of Physarum polycephalum. Nano Commun. Netw. **2**, 99–105 (2011)
116. Schumann, A.: Towards context-based concurrent formal theories. Parallel Process. Lett. **25**, 1540008 (2015)
117. Taylor, B., Adamatzky, A., Greenman, J., Ieropoulos, I.: Physarum polycephalum: towards a biological controller. Biosystems **127**, 42–46 (2015)
118. Whiting, J.G.H., de Lacy Costello, B., Adamatzky, A.: Sensory fusion in physarum polycephalum and implementing multi-sensory functional computation. Biosystems **119**, 45–52 (2014)
119. Savtchenko, L.P., Poo, M.M., Rusakov, D.A.: Electrodiffusion phenomena in neuroscience: a neglected companion. Nat. Rev. Neurosci. **18**(10), 598–612 (2017)
120. Siccardi, S., Adamatzky, A.: Logical gates implemented by solitons at the junctions between one-dimensional lattices. Int. J. Bifurc. Chaos **26**(6), 1650,107 (2016)

Chapter 3
Unconventional Computers Designed on Swarm Behaviours

From the previous chapter we know that the opportunity of implementing decidable arithmetic functions on the amoeboid motions means that we can design an unconventional computer using amoeboid organisms. In this chapter, I am going to propose a kind of this computer designed on *Physarum polycephalum*. Designing computers on swarms (such as plasmodia or ant nests) would mean that *behaviourism* is valid indeed, that is, we can control the swarm behaviours by placing attractants and repellents. Nevertheless, this claim is only partly true, because we can implement undecidable arithmetic functions also. Hence, there are clear lines for behaviourism—we can control an animal behaviour only partly. In this chapter, I try to explain why. The matter is that chemical reactions responsible for sensing and motoring of swarm behaviours differ a lot from standard autocatalytic reactions, such as the Belousov-Zhabotinsky reaction. They assume a high variety of reactions to the same stimuli. See [1–11].

3.1 Reaction-Diffusion Computing

3.1.1 Prehistory of Unconventional Computing

The idea that each natural phenomenon is a computer came from Kabbalah, the esoteric teaching of Judaism.

In Kabbalah the illusiveness of the world is proved by that the world does not exist in itself. The world is a combination of the characters of sacred (Hebrew) language, in which the Holy Scripture is written. According to the *Sefer Yeẓirah* (*The Book of Creation*), the best known Kabbalistic source, the entire Universe was created from

A. Schumann, *Behaviourism in Studying Swarms: Logical Models of Sensing and Motoring*, Emergence, Complexity and Computation 33,
https://doi.org/10.1007/978-3-319-91542-5_3

the set of 22 Hebrew letters, each of which is also a number. God's holy words have created all:

בשלשים ושתים נתיבות פליאות חכמה חקק י״ה יהו״ה צבאות
את עולמו בשלשה ספרים בספר ספר וספור עשר ספירות
בלי מה במספר עשר אצבעות חמש כנגד חמש וברית יחוד
מכוונת באמצע במלת לשון ובמלת המעור

In two and thirty most occult and wonderful paths of wisdom did JAH the Lord of Hosts engrave his name: God of the armies of Israel, ever-living God, merciful and gracious, sublime, dwelling on high, who inhabiteth eternity. He created this universe by the three Sepharim, Number, Writing, and Speech. Ten are the numbers, as are the Sephiroth, and twenty-two the letters, these are the Foundation of all things. Of these letters, three are mothers, seven are double, and twelve are simple (*The Book of Creation*, translated by W. W. Wescott).

Many Kabbalists had attempts to recreate some samples of this Divine speech. They supposed that the Divine words may be generated by a purely mechanical process and have no meaning in human language. In fact, as we will see, the idea of *automaton*, a self-operating machine, flashed upon Kabbalists for the first time (for more details see [12, 13]).

As an example of Kabbalistic automata let us consider one of the automata simulating the Divine speech that has been proposed by Eleazar ben Judah ben Kalonymus of Worms (*Rokeaḥ*), the Talmudist and Kabbalist, born probably at Mayence about 1176 and died at Worms in 1238. He was a descendant of the great Kalonymus family of Mayence.

In the following example of Kabbalistic automaton, we can see, how the letters of word 'Talmud' emerge from *'alef* (the first letter of the Hebrew alphabet):

Example.
Halting pattern: {*t*, *l*, *m*, *w*, *d*} (Talmud)
Input string: *'a l f*
Generation1: *lf md 'a*
Generation2: *md 'a m lt lf*
Generation3: *m* ***lt lf m*** *md* ***w*** *m* ***d*** *'a*

All five letters of the word 'Talmud' appear after three generations. The letter *'alef* was first spelt out horizontally and then branches into a tree, see [1]:

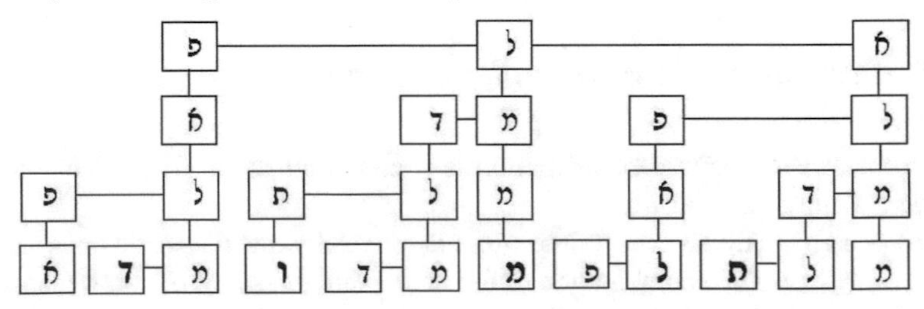

The idea of Kabbalistic automata in which operations over characters of the Hebrew alphabet should simulate physical and biological processes, suggested Leibniz an idea to work at a *characteristica universalis* or '*universal characteristic*', built on an alphabet of human thought in which each fundamental concept would be represented by a unique character:

> It is obvious that if we could find characters or signs suited for expressing all our thoughts as clearly and as exactly as arithmetic expresses numbers or geometry expresses lines, we could do in all matters insofar as they are subject to reasoning all that we can do in arithmetic and geometry. For all investigations which depend on reasoning would be carried out by transposing these characters and by a species of calculus (Leibniz, *Preface to the General Science*, 1677).

In universal characteristic, complex thoughts would be represented by combining characters for atomic thoughts. Characters of the Hebrew alphabet, which in Kabbalists' intention are atoms of any information, were prototypes of characters in Leibniz' meaning. For Leibniz such atoms should be presented by the system of coding of all elementary concepts by using prime numbers due to the uniqueness of prime factorization (also this idea was used in Gödel numbering).

Universal characteristic should have become an automaton simulating all physical processes, an automaton who knows the "Answer to the Ultimate Question of Life, the Universe, and Everything". This automaton should have become a universal checker which can check and verify any thought mechanically:

> The only way to rectify our reasonings is to make them as tangible as those of the Mathematicians, so that we can find our error at a glance, and when there are disputes among persons, we can simply say: Let us calculate [calculemus], without further ado, to see who is right (Leibniz, *The Art of Discovery*, 1685).

Due to such possibilities, universal characteristic may be regarded as sacral language, i.e. a unique language in which it is possible to think unmistakably, language that objectively reflects any physical and biological process, language which already coincides with reality. It is obvious that sacred language in the Kabbalistic meaning was a prototype of universal language in the Leibnizian meaning. The latter should have become a universal language for communications of all people:

> And although learnt men have long since thought of some kind of language or universal characteristic by which all concepts and things can be put into beautiful order, and with whose help different nations might communicate their thoughts and each read in his own language what another has written in his, yet no one has attempted a language or characteristic which includes at once both the arts of discovery and judgement, that is, one whose signs and characters serve the same purpose that arithmetical signs serve for numbers, and algebraic signs for quantities taken abstractly. Yet it does seem that since God has bestowed these two sciences on mankind, he has sought to notify us that a far greater secret lies hidden in our understanding, of which these are but the shadows [14].

Leibniz proposed some examples of how it is possible to design universal characteristic. In these examples Leibniz presented '*character*' as a certain number, assigned to a subject or a predicate as eigenvalue in the way that a derivated number corresponds to a combination (superposition) of numbers of initial terms. This combination can be presented, for example, by multiplication of numbers which are 'characters' of initial terms (atoms of universal characteristic). For instance, the expression 'rational animal', consisting of two terms: 'rational' with character a, and ‘animal’ with character r, obtains a new character ar, i.e. ‘man’ (‘man’ = ‘rational animal’). In particular, if $a = 3$, $r = 2$, then character of ‘man’ = 6.

In categorical propositions, relations between characters are defined by means of the following rules:

(1) The *universal affirmative proposition* is true if and only if the character of subject is divided by the character of predicate without remainder of division, e.g. in the proposition ‘man is a rational animal’ we have: ‘man’ (S) / ‘animal’ (P) = $ar/r = a$.
(2) The *particular affirmative proposition* is true if and only if the character of subject is divided by the character of predicate without remainder of division or the character of predicate is divided by the character of subject without remainder of division.
(3) The *universal negative proposition* is true if and only if the character of subject cannot be divided by the character of predicate without remainder of division and also the character of predicate cannot be divided by the character of subject without remainder of division.
(4) The *particular negative proposition* is true if and only if the character of subject cannot be divided by the character of predicate without remainder of division.

In case categorical propositions do not satisfy the above-mentioned rules, they are considered false. The favour of similar numerical interpretation of atomic propositions consists in that by means of simple operating over numbers it is possible to prove all syllogistic laws. At the same time, the evident assumption is as follows: characters $H = ABC$ and $J = BC$ for combinations of characters of simpler terms A, B, C are divided by the same number of term X without remainder of division if $X < H$ and $X < J$. For example, let the number ‘man’ be H and the number of ‘animal’ be A in the proposition ‘man is animal’. Then fraction H/A is reduced to prime numbers r/v such that r/v is an integer, i.e. $v = 1$. From this assumption it follows

1. ‘Every H is A’ iff $vH = rA$ and $v = 1$.
2. ‘Some A are H’ (or ‘Some H are A’) iff $rA = vH$ and ($r = 1$ or $v = 1$).
3. ‘No H is A’ (or ‘No A is H’) iff $vH = rA$ and ($r > 1$ and $v > 1$).
4. ‘Some A are not H’ iff $rA = vH$ and $r > 1$.

The given interpretation satisfies the square of opposition. Indeed, in (2) and (3) (resp. in (1) and (4)) the universal negative proposition and the particular affirmative proposition (resp. the universal affirmative proposition and the particular negative proposition) contradict each other (if one of them is true, another is false and vice versa). Further, from (1) we can infer (2) (resp. from (3) it follows that (4)). And so on.

Leibniz’ intention to construct universal characteristic was extremely grandiose. In his days people trusted in unlimited powers of reason. It seemed to them, a bit later and the whole world will be completely investigated.

An automaton (abstract machine) which would simulate any physical and biological process, can be said to be a *Kabbalistic-Leibnizian automaton*.

The main assumptions of Kabbalistic-Leibnizian automata are as follows:

- There is only information. Any chemical, physical or biological process is an unconventional computer. Thus, any chemical, physical or biological process can be simulated by an abstract machine.
- Any chemical, physical or biological process can be considered as transition in massive-parallel proofs. Therefore any chemical, physical or biological phenomenon may be completely reduced to logical relations, e.g. to logical inference rules.
- Any theoretical terms of chemistry, physics, biology can be eliminated if we replace them by logical inference rules.

3.1.2 *Unconventional Computing as Novel Paradigm in Natural Sciences*

In natural sciences there are many theoretical terms which as it seems to us allow to understand reality better, but actually they replace true reality by images made up. For example, in understanding the Belousov-Zhabotinsky reaction the following theoretical terms of chemistry are used: matter, chemical element, chemical reaction, autowave, valence etc. In understanding the dynamics of plasmodium of *Physarum polycephalum* the following theoretical terms of biology are assumed: contiguous living system, cell, protoplasm, physiology, morphology etc.

If we share positions of unconventional computing and agree that any natural process is a computation and nature as a whole is a computer, theoretical terms of physics, chemistry and biology will lose a sense. We can eliminate them from the language of scientific theories.

Theoretical terms cannot be verified, i.e. they are not reduced to observation terms. However, within the limits of unconventional computing we can substitute logical terms simulating the process of computation for theoretical terms. Logical terms do not require already verification. They form only the structure and algorithmization of observation terms.

Let us suppose that we created the universal language $\mathscr{L}_{\mathfrak{U}}$ of *Kabbalistic-Leibnizian automata*, i.e. the universal logical language of unconventional computing that allows us to consider any natural process as computation process. Then the language $\mathscr{L}_{\mathfrak{U}}$, first, eliminates all theoretical terms of natural sciences, replacing them by logical terms of massively parallel proof theory and, second, instead of huge number of scientific theories of physics, chemistry, biology we obtain an opportunity to have the universal logical theory that simulates the Universe.

Let us show, how we could eliminate theoretical terms in the language $\mathscr{L}_{\mathfrak{U}}$. Take n theoretical terms: $T_1, T_2, \ldots, T_n$, linked to observation terms by correspondence rules of an appropriate theory and m observation terms, i.e. their references are found out immediately: $O_1, O_2, \ldots, O_m$. The complete theoretical statement is a union consisting of T- and O-terms:

$$\langle T_1, T_2, \ldots, T_n; O_1, O_2, \ldots, O_m \rangle.$$

This formula can be replaced by another, identical to the first in the way that all theoretical terms are replaced by logical terms $U_1, U_2, \ldots, U_k$ which describe the process of computation of an appropriate automata with features $\langle O_1, O_2, \ldots, O_m \rangle$:

$$\langle U_1, U_2, \ldots, U_k; O_1, O_2, \ldots, O_m \rangle.$$

In such a statement, empirical values of theoretical terms are positioned, i.e. they obtain references. These references are expressed by observation terms. Such a representation of theoretical expressions has a number of advantages, e.g. sophisticated theoretical terms of natural sciences could find the classes of references. For example, it is obvious that the statements describing Earth's rotation have the set {the Earth} as a class of references. However, it is not clearly, which set of physical bodies realises the statements describing the concept 'electron'.

In a proposition of language $\mathcal{L}_{\mathfrak{U}}$ the sophisticated expression setting theoretical terms is transformed into an expression in which there are only observation terms as well as terms of logic and mathematics, therefore the concept 'electron' may be eliminated from theory expressions.

In the simulation of Belousov-Zhabotinsky reaction as observation terms the following states of an appropriate cellular automaton were used:

$$\{Ce^{3+}, HBrO_2, BrO_3^-, H^+, Ce^{4+}, H_2O, BrCH\ (COOH)_2, Br^-, HCOOH, CO_2, HOBr, Br_2, CH_2(COOH)_2\}.$$

It is enough for us to simulate this reaction. The theoretical terms needed for understanding the Belousov-Zhabotinsky reaction were replaced by inference rules of Figs. 3.4 and 3.5 or by inference rules (11.1)–(11.7).

This example shows that for designing the language $\mathcal{L}_{\mathfrak{U}}$ it is necessary to solve Leibniz' problem to find out characters (atoms of any information), i.e. to find out the minimum set of observation terms, sufficient for an elimination of any theoretical terms. This problem is similar to the problem of looking for Higgs boson. Its solution will allow to create the universal language of natural sciences, to construct a supercomputer "Earth" to come up with the Answer to The Ultimate Question of Life, the Universe, and Everything.

3.1.3 *Reaction-Diffusion Automata*

After an attempt to reconstruct actin filament networks theoretically in the previous chapter, we can assume that any animal behaviour (in any case the behaviour of unicellular organisms) can be explained by chemical reactions. There is an approach to design an unconventional computer on the basis of chemical reactions. This approach is called *reaction-diffusion computing*. Reaction-diffusion computers [7, 15–19] are spatially extended chemical systems, which process information using interacting growing patterns, of excitable and diffusive waves. In reaction-diffusion processors, both the data and the results of the computation are encoded as concentration profiles of the reagents. The computation is performed via the spreading and interaction of wave fronts. As specified in [17], an implementation of reaction-diffusion computers is based on three principles of the physics of computation [20]. The first principle states that the physical action measures an amount of information, i.e. a dynamics of the reaction-diffusion system is interpreted as computation. The second principle is that physical information travels only a finite distance, i.e. the natural computation is always local. The third principle says that the nature is governed by propagating patterns and traveling waves, i.e. an appropriate computation is spatial.

For the first time, Alan Turing provided mathematical models of pattern formation in reaction-diffusion media. In his famous paper [21] titled *The Chemical Basis of Morphogenesis* design patterns were considered as a conceptual framework for transferring knowledge from biology to computing. The importance of the idea of reaction-diffusion computing is that reaction-diffusion processes are observed throughout all the nature in all spatiotemporal pattern formation and self-organization far from equilibrium coupled with diffusion.

After Turing, the interest in controlling self-organized systems and in building up computational models based on them in order to achieve their specific desired behaviour has been increasing. The control of self-organization in reaction-diffusion frames is an ambitious goal allowing to simulate all massive-parallel natural processes: chemical reactions, morphogenesis, population dynamics, biological, economic and social evolution and other phenomena that produce a complex network of interactions among the component species.

In the initial meaning reaction-diffusion systems involve constituents locally transformed into each other by chemical reactions and transported in space by diffusion. The simplest kind of chemical wave emerges when reactants are converted into products as the front propagates through the reaction mixture. Usually, chemical reaction-diffusion systems eventually evolve to the state of chemical equilibrium, but there is a possibility of thermodynamic equilibrium like pattern formation in the *Belousov-Zhabotinsky reaction*. In the latter case, the equilibrium of self-organized patterns occurs due to an interaction of autocatalytic compensated by reaction step blocking the unbounded growth, see [22].

Wave is a key notion for analyzing reaction-diffusion processes. It is a form of propagating the spatial coupling of elements from the stable or the most stable state toward the unstable or the least stable one. Due to waves of reagent diffusions there is a chemical reaction. Furthermore, a wave can cause composite patterns in a regime of periodic or chaotic oscillations like that we observe in the Belousov-Zhabotinsky reaction. If a wave causes pattering and regulation of timing independently of input conditions and the letter are determined by the system itself, then this is called *auto-wave* [23, 24]. Its typical instance is described in the Belousov-Zhabotinsky reaction. Auto-waves may be detected in any self-organized systems, some examples of auto-waves are below. As a result, we can claim that the reaction-diffusion nature is observed not just within chemical reactions, but also in different pattering and interactions like as follows:

- *Auto-waves of morphogenesis* consisting in the processes that generate tissue organization and shape and are usually the downstream response to the timing and patterning.
- *Auto-waves in ecological systems* that change the network of interacting populations.
- *Auto-waves of epidemics*, they are observed in the rapid increase of the size of infected population due to the interaction between infected and susceptible individuals.
- *Auto-waves of opinion dynamics*, i.e. changes in opinions caused by interaction of people.
- *Auto-waves of aggregated individual plasmodium* of the cellular slime mold *Physarum polycephalum* to form a multicellular migrating slug, which moves toward a region suitable for culmination.

As we see, thanks to above-mentioned examples, the majority of physical, chemical, biological and sociological processes can be described in terms of auto-waves, i.e. propagating fronts of one nature or another.

The dynamics of complex reaction-diffusion frames can be described in terms of activator and inhibitor. The *activator*, A, is a short-range autocatalytic substance, and the *inhibitor*, I, is its long-range antagonist. A autocatalytically promotes its own synthesis. I inhibits synthesis of A, and I can diffuse faster than A. The pattern of peaks of A is dependent upon the reaction-diffusion frame. The dynamics of a self-organizing system is typically non-linear, because of circular or feedback relations between the components [25]. In other words, in complex reaction-diffusion systems there are circular cause-and-effect relations: each component affects the other components, but these components in turn affect the first component. This feedback can have two basic values: positive if the activator diffuses faster or negative if the inhibitor diffuses faster. In the positive feedback, the recurrent influence reinforces the initial change. In the negative feedback, the reaction is opposite to the initial action, so a change is counteracted. The aim of negative feedback is to stabilize the system, by returning to its original states. On the other hand, the positive feedback makes changes growing in an explosive manner. It leads to an accelerated development, which ends when all components have been absorbed into the new configuration, leaving the system in a stable state.

Diffusion waves of activator and inhibitor interact according to reaction rules. Solutions of those rules are represented by multisets of waves: this accounts for the associativity and commutativity of parallel composition, that is the implicit stirring mechanism. Setting diffusion waves by rules allows us to define reaction-diffusion processes as a kind of computing. This computing is fundamentally similar to parallel computing in that it takes advantage of the many different waves to try many different possibilities of diffusion at once.

Setting up reaction-diffusion computing is a way of simulating physics by a *universal computer* (Kabbalistic-Leibnizian automaton). This aim to build up a "life computer" is the most ambitious task attempted in unconventional computing, e.g. see for more details [26]. Its difficulties consist in that in conventional approach to computing a control is centralized. On the other hand, in self-organizing systems a control of organization is typically distributed over the whole system. Due to a centralized control, software is written for serial computation and a problem is broken into a discrete series of executed instructions so that only one instruction may execute at any moment in time. Due to a decentralized, distributed control in parallel computing, a computational problem is solved by the simultaneous use of multiple compute resources and a problem is broken into discrete parts that can be solved concurrently. We can add that reaction-diffusion computing is not just parallel, but massive-parallel. This means that parts a problem is broken into can be indiscrete and infinite many.

The *sequential model of computation* called von Neumann paradigm is unapplied, broken in the reaction-diffusion computing. The reason is that the latter computing may be presented as a massive-parallel locally-connected mathematical machine with circular and cyclic processes. These machines cannot have a single centralized source exercising a precise control over vast numbers of heterogeneous devices. The *interactive-computing paradigm* is able to describe concurrent (parallel) computations whose configuration may change during the computation and it is decentralized as well. Within the framework of this paradigm, one proposed a lot of so-called concurrency calculi also called *process algebras*. They are typically presented using systems of equations. These formalisms for concurrent systems are formal in the sense that they represent systems by expressions and then reason about systems by manipulating the corresponding expressions [27].

One of the unconventional, nature inspired models similar to reaction-diffusion computing is *chemical machine* in that molecules are viewed as computational processes supplemented with a minimal reaction kinetics. Berry and Boudol first built up a chemical abstract machine [28] as an example of how a chemical paradigm of the interactions between molecules can be utilized in concurrent computations (in algebraic process calculi).

There is a particular feature of reaction-diffusion chemical computers: the media are 'fully conductive' for chemical or excitation waves. Every point of a medium can be involved in the propagation of chemical waves and reactions between diffusing chemical species. Once a reaction is initiated in a point, it spreads all over the computing space by target and spiral waves. Such phenomena of wave-propagation, analogous to one-to-all broadcasting in massive-parallel systems, are employed to solve problems ranging from the Voronoi diagram construction to robot navigation [17, 18].

Every chemical non-stirred reaction-diffusion medium (micro-volume) can be represented as a finite state machine, or an automaton. Its states correspond to reagents which prevails in their concentration at any discrete time. For example, if there are two reagents, α and β, in a medium, then each micro-volume x can be represented by a finite automaton a_x such that if at the time step t the concentration of α in x exceeds the concentration of β in x, then a_x takes the state α, otherwise the stat β. Cellular automata [29–31] are best computational structures to represent space-time dynamics of spatially extended non-linear systems, including reaction-diffusion media. This is because a cellular automaton is a regular network, or a lattice, array, of locally connected finite automata updating their states in parallel [32]. Recall that a *cellular automaton* is a 4-tuple $\mathscr{A} = \langle \mathbf{Z}^d, S, u, f \rangle$, where (1) $d \in \mathbf{N}$ is a number of dimensions and the members of $\mathbf{Z}^d$ are cells, (2) S is a finite set of elements called the states of an automaton $\mathscr{A}$, the members of $\mathbf{Z}^d$ take their values in S, (3) $u \subset \mathbf{Z}^d \setminus \{0\}^d$ is a finite ordered set of n elements, $u(x)$ is said to be a neighbourhood for the cell x, (4) $f : S^{n+1} \rightarrow S$ that is f is the local transition function (or local rule). An automaton is regarded on the endless d-dimensional space of integers, i.e. on $\mathbf{Z}^d$. Discrete time is introduced for $t = 0, 1, 2, \ldots$ So, the sell x at time t is denoted by x^t. Each automaton calculates its next state depending on states of its closest neighbours.

To represent a chemical reaction-diffusion system in a cellular automaton [15] with local transition function f and cell-x neighbourhood $u(x) = \langle y_1, \ldots, y_n \rangle$, one needs to select a substrate state, let us call it s, such that

$$f(s, \ldots s) = s,$$

a set of reactants $\mathbf{Q} = \{r_1, \ldots, r_m\}$, and then to determine diffusion and reaction equations. In the most primitive form, the diffusion can be specified as follows:

$$x^{t+1} = \begin{cases} r_i \text{ if } x^t = s \text{ and } \mathbf{D_x^t} = \{s, r_i\} \\ s \text{ if } x^t \neq s \text{ and } |\mathbf{D_x^t}/\{s\}| > 1 \\ x^t \text{ othetwise,} \end{cases}$$

where $\mathbf{D_x^t} = \{y^t \in \mathbf{Q} : y^t \in u(x)\}$ is a set of states observed in a neighbourhood of x at time t.

Reactions between reactants of $\mathbf{Q}$ can be represented in cell-state transition rules by many different ways, the more generalized totalistic coding is suggested in [18, 19]: a cell's update depends on the number of different cell-states in its neighbourhood irrespective of the cell-states' positions. Thus, if there are $p = m + 1$ components in the chemical system then the update (transition) rule can be written as follows:

$$x^{t+1} = f(\delta_p(x)^t, \delta_{p-1}(x)^t, \ldots, \delta_0(x)^t),$$

where $\delta_a(x)^t$ is the number of cell x's neighbours with the cell-state $a \in \{s\} \cup \mathbf{Q}$ at time step t.

3.1.4 *Approximation of Voronoi Diagram and Calculations on Reaction-Diffusion Media*

Let us consider the most famous task—the approximation of Voronoi diagram in reaction-diffusion systems, see an overview in [17]. Let $\mathbf{P}$ be a nonempty finite set of planar points and $|\mathbf{P}| = n$. For points $p = (p_1, p_2)$ and $x = (x_1, x_2)$ let $d(p, x) = \sqrt{(p_1 - x_1)^2 + (p_2 - x_2)^2}$ denote their Euclidean distance. A planar *Voronoi diagram* of the set $\mathbf{P}$ is a partition of the plane into regions, such that for any element of $\mathbf{P}$, a region corresponding to a unique point p contains all those points of the plane which are closer to p in respect to the distance d than to any other node of $\mathbf{P}$. A unique region

$$vor(p) = \bigcap_{m \in \mathbf{P}, m \neq p} \{z \in \mathbf{R}^2 : d(p, z) < d(m, z)\}$$

assigned to the point p is called a *Voronoi cell* of the point p.

Voronoi cells of a planar set represent the natural or geographical neighbourhood of the set's elements. Therefore, the computation of a Voronoi diagram based on the spreading of some 'substance' from the data points is usually the first approach of those trying to design massively parallel algorithms, see an overview of experimental techniques in [33]. Let us consider a simplistic implementation of the Voronoi diagram construction in a reaction-diffusion medium with one diffusing reagent α and a substrate s.

The reagent α diffuses from sites introduced to correspond to the elements of a given planar set $\mathbf{P}$. When two diffusing wave fronts meet a super-threshold concentration of reagents prevents waves from spreading further. A cellular-automaton model represents this as follows.

Every cell has two possible states: s (resting state, or a substrate) and α (reagent). If the cell is in state α it remains in this state forever. If the cell is in state s and between one and three of its neighbours are in state α, then the cell takes the state α; otherwise, the cell remains in the state s (this reflects the 'super-threshold inhibition', or a 'self-inhibition' idea). A cell state transition rule is as follows:

$$x^{t+1} = \begin{cases} \alpha, & \text{if } x^t = s \text{ and } 1 \leq \delta(x)^t \leq a \\ x^t, & \text{otherwise} \end{cases} \tag{3.1}$$

where $\delta(x)^t = |\{y \in u(x) : y^t = \alpha\}|$, and $a = 3$ for a rectangular, 8-cell neighbourhood, and $a = 2$ for a hexagonal, 6-cell neighbourhood, cellular automaton.

Let X be the 2-dimensional Euclidean space $\mathbf{R}^2$. Let T denote the discrete time and Q be a finite set of states of the following three sorts: substrate s, activator a, and inhibitor i. At each point on the metric space X, we allocate an infinite sequence of state transitions. Let σ denote a function from X to $(Q \times T)^\omega$, i.e., for each point

$x \in X$, σ_x is a nonempty infinite sequence of pairs from $Q \times T$. Further, we will use some basic notions of stream calculus such as co-induction and bisimulation, for more details see [34–37].

The function σ_x is a kind of stream and will be said to be a *trajectory* of x. The set of all trajectories is denoted by $Tr(X, Q, T)$.

For a trajectory σ_x, we call $\sigma_x(0)$ the initial value of σ_x. We define the *derivative* of a trajectory σ_x, for all $n \geq 0$, by $\sigma_x'(n) = \sigma_x(n+1)$. For any $n \geq 0$, $\sigma_x(n)$ is called the n-th element of σ_x. It can also be expressed in terms of higher-order trajectory derivatives, defined, for all $k \geq 0$, by $\sigma_x^{(0)} = \sigma_x$; $\sigma_x^{(k+1)} = (\sigma_x^{(k)})'$. In this case the n-th element of a trajectory σ_x is given by $\sigma_x(n) = \sigma_x^{(n)}(0)$. Also, the trajectory is understood as an infinite sequence of derivatives. It will be denoted by an infinite sequence of values or by an infinite tuple: $\sigma_x = \sigma_x(0) :: \sigma_x(1) :: \sigma_x(2) :: \ldots :: \sigma_x(n-1) :: \sigma_x^{(n)}$, $\sigma_x = \langle \sigma_x(0), \sigma_x(1), \sigma_x(2), \ldots \rangle$.

A *constant trajectory* $\langle m, m, \ldots \rangle$ is denoted by $[m]$.

A *bisimulation* on $(Q \times T)^\omega$ is a relation $R \subseteq (Q \times T)^\omega \times (Q \times T)^\omega$ such that, for all σ_x and τ_y in $(Q \times T)^\omega$, if $\langle \sigma_x, \tau_y \rangle \in R$ then (i) $\sigma_x(0) = \tau_y(0)$ (this means that they have the same *initial value*) and (ii) $\langle \sigma_x', \tau_y' \rangle \in R$ (this means that they have the same *differential equation*).

If there *exists* a bisimulation relation R with $\langle \sigma_x, \tau_y \rangle \in R$ then we write $\sigma_x \sim \tau_y$ and say that σ_x and τ_y are *bisimilar*. In other words, the *bisimilarity* relation $\sim$ is the union of all bisimulations: $\sim := \bigcup \{R \subseteq (Q \times T)^\omega \times (Q \times T)^\omega : R$ is a bisimulation relation$\}$. Therewith, this relation $\sim$ is the greatest bisimulation. In addition, the bisimilarity relation is an equivalence relation.

Theorem 3.1 (Coinduction) *For all $\sigma_x, \tau_y \in (Q \times T)^\omega$, if there exists a bisimulation relation $R \subseteq (Q \times T)^\omega \times (Q \times T)^\omega$ with $\langle \sigma_x, \tau_y \rangle \in R$, then $\sigma_x = \tau_y$. In other words, $\sigma_x \sim \tau_y \Rightarrow \sigma_x = \tau_y$.* □

This proof principle is called *coinduction*, see [37–42]. It is a systematic way of proving the statement using bisimularity: instead of proving only the single identity $\sigma_x = \tau_y$, one computes the greatest bisimulation relation R that contains the pair $\langle \sigma_x, \tau_y \rangle$. By coinduction, it follows that $\sigma_x = \tau_y$ for all pairs $\langle \sigma_x, \tau_y \rangle \in R$.

Notice that $\sigma_x = \tau_y$ means that the points x and y of X have the same trajectory. Meanwhile, one point x can have different trajectories $\sigma_x \neq \tau_x$.

Let us assume that our metric space X has a partition on Voronoi cells of the form $vor(p)$ in accordance with planar points $p \in \mathbf{P}$. Each Voronoi cell $vor(p)$ has just one state. This means that it contains either a substrate (resting state), or only one reagent. Initial values of trajectories are substrate or reagents. In this case it is natural that if $x, y \in vor(p)$, then $\sigma_x(0) = \sigma_y(0)$ and if $x \in vor(q)$, $y \in vor(p)$ ($vor(p) \neq vor(q)$), then $\sigma_x(0) \neq \sigma_y(0)$. In other words, if points of X belong to the same Voronoi cell, then their trajectories have the same initial value and if points of X belong to different Voronoi cells, then their trajectories have different initial values. Trajectories depend on reactions among Voronoi cells (i.e. among substrate and reagents).

Let us distinguish two kinds of neighbourhood: for points of X (*p-neighbourhood*) and for cells of $\{vor(p) : p \in \mathbf{P}\}$ (*c-neighbourhood*). While the p-neighbourhood for *open* Voronoi cells consists of an infinite number of members that have the same initial state, the c-neighbourhood consists of a finite number of members (they are other Voronoi cells) which are necessarily in different initial states. Thus, the p-neighbourhood does not play a significant role in the transition of the whole system (it is important only for a transition within the framework of one Voronoi cell). Therefore we will consider c-neighbourhood more often.

A trajectory σ_x for every $x \in X$ depends on reactions among Voronoi cells, e.g. on a transition rule f characteristic for an appropriate cellular automaton of an appropriate Voronoi diagram.

Consider now a 3-adic state automaton, where every cell takes one of the cell-states of the following three sorts: the substrate $S = \{s_1, \dots, s_m\}$, the activator $A = \{a_1, \dots, a_l\}$, or the inhibitor $I = \{i_1, \dots, i_k\}$. The cardinality of the set of states $Q = S + A + I$ is not less than the number of members of the set $\mathbf{P}$: $|Q| = m + l + k \geq |\mathbf{P}|$ (we suppose that some states are superposition of basic states and the number of basic states are equal to $|\mathbf{P}|$, moreover $|Q| \geq |\mathbf{P}|$, because some reagents may be obtained as result of reactions of basic reagents of $\mathbf{P}$). The state of the first sort S is a dedicated substrate state: a point in state $s_j \in S$, whose c-neighbourhood is filled only with states $s_j \in S$, does not change its state (s_j are analogous to a quiescent state in cellular automaton models). The states of two other sorts, A and I, are assigned to be reactants. The cell-state transition rule f can be written as follows:

$$\sigma_x(n+1) = f(\Psi_{i_1}\sigma_x(n), \dots, \Psi_{i_k}\sigma_x(n), \Psi_{a_1}\sigma_x(n), \dots, \Psi_{a_l}\sigma_x(n), \Psi_{s_1}\sigma_x(n), \dots, \Psi_{s_m}\sigma_x(n)), \qquad (3.2)$$

where $k = |I|$, $l = |A|$, $m = |S|$, $|\Psi_p\sigma_x(n)|$ is the number of point x's c-neighbours with a state $p \in \{i_1, \dots, i_k, a_1, \dots, a_l, s_1, \dots, s_m\}$ at the time step t_n.

Let us consider a ternary state automaton based on a two-dimensional lattice with the hexagonal tiling. The automaton imitates the reaction-diffusion medium in a sub-excitable mode. In such a mode, a propagation of activator is limited and therefore spiral waves are formed by travelling a self-localizations emerge. See the detailed description and particular analysis in [18, 19].

The c-neighbourhood size is seven: the central cell and its six closest c-neighbours. To give a compact representation of the cell-state transition rule, we represent the cell-state transition rule as a matrix $M = (m_{ij})$, where $0 \leq i \leq j \leq 7$, $0 \leq i + j \leq 7$, and $m_{kl} \in \{s, a, i\}$. The output state of each c-neighbourhood is given by the row-index k (the number of c-neighbours in the cell-state i) and column-index l (the number of c-neighbours in the cell-state a). We do not have to count the number of c-neighbours in the cell-state s, because it is given by $7 - (k + l)$. A point with a c-neighbourhood represented by indexes k and l will update to cell-state m_{kl} which can be read off the matrix. In terms of the cell-state transition function this can be presented as follows: $\sigma_x(n+1) = M_{\Psi_a\sigma_x(n)\Psi_i\sigma_x(n)}$.

Here is the exact matrix structure, which corresponds to matrix M_3 (i.e. if $\Psi_a\sigma_x(n)\Psi_i\sigma_x(n) = 3$). This is a so-called 'spiral rule' [18]:

$$\left\| \begin{matrix} s\ a\ i\ a\ i\ i\ i\ i \\ s\ i\ i\ a\ i\ i\ i \\ s\ s\ i\ a\ i\ i \\ s\ i\ i\ a\ i \\ s\ s\ i\ a \\ s\ s\ i \\ s\ s \\ s \end{matrix} \right\| \tag{3.3}$$

This matrix represents an example of transition rule (3.2) for the ternary state automaton based on a two-dimensional lattice with the hexagonal tiling.

Now consider a propositional logic $\mathfrak{L}^\omega$ for the set of all trajectories $Tr(X, Q, T)$; its syntax and semantics are defined by coinduction and their objects are trajectories (i.e. streams of pairs of cell-states and time-steps). The syntax of $\mathfrak{L}^\omega$ is as follows:

Variables: $\mathbf{p} := p \mid q \mid r \ \ldots,$

where p, q, r are members of the product $Q \times T$ of the set of states Q for an appropriate reaction-diffusion cellular automaton and of the time line T.

Constants: $\mathbf{c} := \top \mid \bot$

where $\top$ means the truth and $\bot$ means the falsity.

Formulas: $\varphi, \psi := \mathbf{p} \mid \mathbf{c} \mid \neg\psi \mid \varphi \vee \psi \mid \varphi \wedge \psi \mid \varphi \supset \psi$

These definitions are coinductive. For instance,

- a variable $\mathbf{p}$ is of the form of a stream $\mathbf{p} = \mathbf{p}(0) :: \mathbf{p}(1) :: \mathbf{p}(2) :: \ldots :: \mathbf{p}(n-1) :: \mathbf{p}^{(n)}$, where $\mathbf{p}(i) \in \{p, q, r, \ldots\} = Q \times T$ for each $i \in \omega$;
- a constant $\mathbf{c}$ is of the form of a stream $\mathbf{c} = \mathbf{c}(0) :: \mathbf{c}(1) :: \mathbf{c}(2) :: \ldots :: \mathbf{c}(n-1) :: \mathbf{c}^{(n)}$, where $\mathbf{c}(i) \in \{\top, \bot\}$ for each $i \in \omega$, a particular case is $[\top] = [\top](0) :: [\top](1) :: [\top](2) :: \ldots :: [\top]^{(n)}$, where $[\top](i) = \top$ for each $i \in \omega$;
- a formula $\neg\varphi$ has the differential equation $(\neg\varphi)' = \neg(\varphi')$ and its initial value is $(\neg\varphi)(0) = \neg\varphi(0)$, this formula will be understood as $\varphi \supset [\bot]$;
- a formula $\varphi \vee \psi$ has the differential equation $(\varphi \vee \psi)' = \varphi' \vee \psi'$ and its initial value is $(\varphi \vee \psi)(0) = \varphi(0) \vee \psi(0)$;
- a formula $\varphi \wedge \psi$ has the differential equation $(\varphi \wedge \psi)' = \varphi' \wedge \psi'$ and its initial value is $(\varphi \wedge \psi)(0) = \varphi(0) \wedge \psi(0)$;
- a formula $\varphi \supset \psi$ has the differential equation $(\varphi \supset \psi)' = \varphi' \supset \psi'$ and its initial value is $(\varphi \supset \psi)(0) = \varphi(0) \supset \psi(0)$.

A variable $\mathbf{p} \in (Q \times T)^{\omega}$ defined as $\mathbf{p}(0) = \langle \mathbf{m}(0), \mathbf{t}(0) \rangle = \langle m, \mathbf{t}(0) \rangle \wedge \mathbf{p}' = \langle \mathbf{m}', \mathbf{t}' \rangle$ is denoted by $\langle [m], \mathbf{t} \rangle$ and is said to be a *semi-constant variable*. For instance, n-th element of semi-constant $\mathbf{p}$ is denoted by $\langle m, \mathbf{t}(n) \rangle$.

Further, let us consider semantics of $\mathfrak{L}^{\omega}$:

Truth-valuation of variables: $v(\mathbf{p}) := v(p) \mid v(q) \mid v(r) \ \ldots$

Truth-valuation of formulas: $v(\varphi), v(\psi) := v(\mathbf{p}) \mid v(\neg\psi) \mid v(\varphi \vee \psi) \mid v(\varphi \wedge \psi) \mid v(\varphi \supset \psi)$

These definitions are coinductive, too. For example, the truth-valuation of variables is a map from the set of variables onto the set of constants. The truth-valuation of formulas is defined as follows:

- the differential equation of $v(\neg\psi)$ is $(v(\neg\psi))' = [\top]' - v(\psi)'$ and its initial value is $(v(\neg\psi))(0) = [\top](0) - v(\psi)(0)$,
- the differential equation of $v(\varphi \vee \psi)$ is $(v(\varphi \vee \psi))' = \sup(v(\varphi)', v(\psi)')$ and its initial value is $(v(\varphi \vee \psi))(0) = \sup(v(\varphi)(0), v(\psi)(0))$,
- the differential equation of $v(\varphi \wedge \psi)$ is $(v(\varphi \wedge \psi))' = \inf(v(\varphi)', v(\psi)')$ and its initial value is $(v(\varphi \wedge \psi))(0) = \inf(v(\varphi)(0), v(\psi)(0))$,
- the differential equation of $v(\varphi \supset \psi)$ is $(v(\varphi \supset \psi))' = [\top]' - \sup(v(\varphi)', v(\psi)') + v(\psi)'$ and its initial value is $(v(\varphi \supset \psi))(0) = [\top](0) - \sup(v(\varphi)(0), v(\psi)(0)) + v(\psi)(0)$.

A formula φ is called logically *valid* (*satisfiable*) if for any (some) truth-valuation defined above, the equality $v(\varphi) = [\top]$ holds (i.e. $v(\varphi)$ and $[\top]$ are bisimilar).

By convention, we use Γ, Γ_1, Γ_2, etc., to range over sets of formulas, these may be empty. We write $\Gamma \vdash \varphi$ to indicate that φ is deducible (derivable) from Γ. Γ is called the set of premises and φ is called the conclusion.

The propositional logic $\mathfrak{L}^{\omega}$ has the standard inference rules for two-valued logic (see Fig. 3.1), but in the coinductive form; we can prove that it is a complete deductive system.

$$\frac{\Gamma(0) \vdash \varphi(0) \qquad \Gamma' \vdash \varphi'}{\Gamma \vdash \varphi}$$

$$\frac{\Gamma_1 \vdash \varphi \qquad \Gamma_2 \vdash \psi}{\Gamma_1, \Gamma_2 \vdash \varphi \wedge \psi} \qquad \frac{\Gamma \vdash \varphi}{\Gamma \vdash \varphi \vee \psi} \qquad \frac{\Gamma \vdash \psi}{\Gamma \vdash \varphi \vee \psi}$$

$$\frac{\Gamma \vdash \varphi \wedge \psi}{\Gamma \vdash \varphi} \qquad \frac{\Gamma \vdash \varphi \wedge \psi}{\Gamma \vdash \psi}$$

$$\frac{\varphi \in \Gamma}{\Gamma \vdash \varphi} \qquad \frac{\Gamma_1 \vdash \varphi \vee \psi \qquad \Gamma_2, \varphi \vdash \theta \qquad \Gamma_3, \psi \vdash \theta}{\Gamma_1, \Gamma_2, \Gamma_3 \vdash \theta}$$

$$\frac{\Gamma, \varphi \vdash \psi}{\Gamma \vdash \varphi \supset \psi} \qquad \frac{\Gamma_1 \vdash \varphi \supset \psi \qquad \Gamma_2 \vdash \varphi}{\Gamma_1, \Gamma_2 \vdash \psi}$$

$$\frac{\Gamma_1, \varphi \vdash \psi \qquad \Gamma_2, \varphi \vdash \neg\psi}{\Gamma_1, \Gamma_2 \vdash \neg\varphi} \qquad \frac{\Gamma \vdash \neg\neg\varphi}{\Gamma \vdash \varphi}$$

Fig. 3.1 Inference rules for the logic of trajectories. The entail relation $\vdash$ is the greatest fixed point of the rules above, or, equivalently, the conclusions of infinite derivation trees are built from these rules

Let us extend the language $\mathfrak{L}^{\omega}$ by adding a modal operator $\Diamond^k$ defined in the following way: a formula $\Diamond^k \varphi$ has the differential equation $(\Diamond^k \varphi)' = \Diamond^k(\varphi')$ and its initial value is $(\Diamond^k \varphi)(0) = \Diamond^k \varphi(0)$; the informal meaning of a formula $\Diamond^k \varphi$ is that in the c-neighbourhood, there exist k points belonging to different cells that satisfy φ. The new language will be denoted by $\mathfrak{L}^{\omega}_{\Diamond}$.

Now let us semantically define a transition rule for a cellular automaton of Voronoi diagram as follows:

$$\sigma_x(n) \hookrightarrow \sigma_x(n+1) \ \ \mathbf{if} \ \ \varphi, \tag{3.4}$$

where $\sigma_x(n), \sigma_x(n+1) \in Q \times T$ and φ is a well-formed formula of $\mathfrak{L}^{\omega}_{\Diamond}$. The cell-state transition rule (3.4) means that the state of point x at the time-step t_n is allowed to change from $\sigma_x(n)$ to $\sigma_x(n+1)$ if the condition expressed by formula $\varphi(n)$ is satisfied at the pair $\sigma_x(n+1) = \langle m_{n+1}, t_{n+1}\rangle_x \in Q \times T$.

The model, $\mathfrak{M}$, of logic $\mathfrak{L}^{\omega}_{\Diamond}$ is a cellular automaton of Voronoi diagram, σ_x is a trajectory of its point x. The notion of satisfaction for a formula φ at a pair $\sigma_x(n)$ of a model $\mathfrak{M}$ at time-step t_n is defined inductively thus:

$\mathfrak{M}, \sigma_x(n) \models \mathbf{p}$ iff for any i such that $0 \leq i \leq n-1$, $\sigma_x(i+1) = \mathbf{p}(i)$,
$\mathfrak{M}, \sigma_x(n) \models \neg\varphi$ if not $\mathfrak{M}, \sigma_x(i) \models \varphi$ for any i such that $0 \leq i \leq n$,
$\mathfrak{M}, \sigma_x(n) \models [\bot]$ never holds,
$\mathfrak{M}, \sigma_x(n) \models [\top]$ always holds,
$\mathfrak{M}, \sigma_x(n) \models \varphi \wedge \psi$ if both $\mathfrak{M}, \sigma_x(n) \models \varphi$ and $\mathfrak{M}, \sigma_x(n) \models \psi$,
$\mathfrak{M}, \sigma_x(n) \models \varphi \vee \psi$ if $\mathfrak{M}, \sigma_x(n) \models \varphi$ or $\mathfrak{M}, \sigma_x(n) \models \psi$,
$\mathfrak{M}, \sigma_x(n) \models \varphi \supset \psi$ if $\mathfrak{M}, \sigma_x(n) \models \varphi$ implies $\mathfrak{M}, \sigma_x(n) \models \psi$,
$\mathfrak{M}, \sigma_x(n) \models \Diamond^k \varphi$ if $\mathfrak{M}, \sigma_{x_1}(n) \models \varphi, \mathfrak{M}, \sigma_{x_2}(n) \models \varphi, \ldots, \mathfrak{M}, \sigma_{x_k}(n) \models \varphi$
and $vor(x_1), vor(x_2), \ldots, vor(x_k)$ are c-neighbours of $vor(x)$.

On the other hand, the notion of satisfaction for a formula φ at a trajectory σ_x of a model $\mathfrak{M}$ is defined coinductively:

$\mathfrak{M}, \sigma_x \models \mathbf{p}$ if and only if σ'_x and $\mathbf{p}$ are bisimilar,
$\mathfrak{M}, \sigma_x \models \neg\varphi$ if not $\mathfrak{M}, \sigma_x \models \varphi$,
$\mathfrak{M}, \sigma_x \models [\bot]$ never holds,
$\mathfrak{M}, \sigma_x \models [\top]$ always holds,
$\mathfrak{M}, \sigma_x \models \varphi \wedge \psi$ if both $\mathfrak{M}, \sigma_x \models \varphi$ and $\mathfrak{M}, \sigma_x \models \psi$,
$\mathfrak{M}, \sigma_x \models \varphi \vee \psi$ if $\mathfrak{M}, \sigma_x \models \varphi$ or $\mathfrak{M}, \sigma_x \models \psi$,
$\mathfrak{M}, \sigma_x \models \varphi \supset \psi$ if $\mathfrak{M}, \sigma_x \models \varphi$ implies $\mathfrak{M}, \sigma_x \models \psi$,
$\mathfrak{M}, \sigma_x \models \Diamond^k \varphi$ if $\mathfrak{M}, \sigma_{x_1} \models \varphi, \mathfrak{M}, \sigma_{x_2} \models \varphi, \ldots, \mathfrak{M}, \sigma_{x_k} \models \varphi$
and $vor(x_1), vor(x_2), \ldots, vor(x_k)$ are c-neighbours of $vor(x)$.

As we see, rule (3.4) means that φ and σ_x' are bisimilar, i.e. $\varphi = \sigma_x'$, if φ does not contain logical operations.

Proposition 3.1 *Let $x \in X$ and $\sigma_x \in (Q \times T)^{\omega}$. For each $n > 0$, there exists a transition rule of the form (3.4) such that $\mathfrak{M}, \sigma_x(n) \models \varphi$ holds.*

Proof By induction on the depth of formula. Let us suppose, φ is a variable $\mathbf{p}$, then there exists $\mathbf{p}$ such that $\mathbf{p}$ and σ_x' are bisimilar. If $\varphi = \neg\mathbf{p}$, there exists $\mathbf{p}$ such that $\mathbf{p}$ and σ_x' are not bisimilar. Suppose that $\varphi = \mathbf{p} \vee \mathbf{q}$ (resp. $\varphi = \mathbf{p} \wedge \mathbf{q}$ or $\varphi = \mathbf{p} \supset \mathbf{q}$), then there exist $\mathbf{p}, \mathbf{q}$ such that $\mathbf{p}$ or $\mathbf{q}$ are bisimilar with σ_x' (resp. $\mathbf{p}$ and $\mathbf{q}$ are bisimilar with σ_x' for $\varphi = \mathbf{p} \wedge \mathbf{q}$ and if $\mathbf{p}$ is bisimilar with σ_x', then $\mathbf{q}$ is bisimilar with σ_x' for $\varphi = \mathbf{p} \supset \mathbf{q}$), etc. □

Let us consider the case of an automaton with the cell-state transition rule (3.3). We have there only three states: s, a, i. Suppose m is a state-variable. Let us remember that we may obtain semi-constant variables of three sorts: $\langle [s], \mathbf{t}\rangle_x$, $\langle [a], \mathbf{t}\rangle_x$, $\langle [i], \mathbf{t}\rangle_x$ for any $x \in X$. The formula

$$\langle s, \mathbf{t}(n)\rangle_x \hookrightarrow \langle i, \mathbf{t}(n+1)\rangle_x \ \ \mathbf{if} \ \ \Diamond^2 \langle [a], \mathbf{t}\rangle \tag{3.5}$$

holds in the cellular automaton described by rule (3.3). This means a self-inhibition of activator if we observe this activator in two Voronoi cells of the c-neighbourhood. More precisely, according to formula $\Diamond^2\langle [a], \mathbf{t}\rangle$, we have a semi-constant trajectory $\sigma_x^{(n+1)} = \langle [i], \mathbf{t}\rangle_x$ after the time-step t_{n+1}. Also, in the cellular automaton described by rule (3.3) the following formula holds:

$$\langle s, \mathbf{t}(n)\rangle_x \hookrightarrow \langle i, \mathbf{t}(n+1)\rangle_x \ \ \mathbf{if} \ \ \Diamond^2 \langle a, \mathbf{t}(n)\rangle. \tag{3.6}$$

The difference between formulas (3.5) and (3.6) is that $\Diamond^2\langle [a], \mathbf{t}\rangle$ is a well-formed formula of $\mathfrak{L}^{\omega}_{\Diamond}$ and $\Diamond^2\langle a, \mathbf{t}(n)\rangle$ is an n-th element of a well-formed formula of $\mathfrak{L}^{\omega}_{\Diamond}$, therefore an uncountable infinite number of well-formed formulas of $\mathfrak{L}^{\omega}_{\Diamond}$ satisfies (3.6), namely there is an infinite number of formulas with n-th element $\Diamond^2\langle a, \mathbf{t}(n)\rangle$. This shows that there is no finite set of formulas describing the dynamics of cellular automaton (a finite set of transition rules). However, it is a normal and natural situation for coalgebraic methods. The basic notions are defined in coagebra as greatest fixed points, this means that they should contain an infinite number of objects.

The cell-state transition rule (3.3) we can define as the following set of logical formulas:

$$\frac{\sigma_x(0) \hookrightarrow \sigma_x(1) \text{ if } \varphi, \quad \ldots, \quad \sigma_x(i) \hookrightarrow \sigma_x(i+1) \text{ if } \varphi, \quad \ldots, \quad \vdash_{\sigma_x(0)} \varphi(0), \quad \vdash_{\sigma_x'} \varphi'}{\vdash_{\sigma_x} \varphi}$$

$$\frac{\Gamma_1 \vdash_{\sigma_x} \varphi \quad \Gamma_2 \vdash_{\sigma_x} \psi}{\Gamma_1, \Gamma_2 \vdash_{\sigma_x} \varphi \wedge \psi} \qquad \frac{\Gamma \vdash_{\sigma_x} \varphi}{\Gamma \vdash_{\sigma_x} \varphi \vee \psi} \qquad \frac{\Gamma \vdash_{\sigma_x} \psi}{\Gamma \vdash_{\sigma_x} \varphi \vee \psi}$$

$$\frac{\Gamma \vdash_{\sigma_x} \varphi \wedge \psi}{\Gamma \vdash_{\sigma_x} \varphi} \qquad \frac{\Gamma \vdash_{\sigma_x} \varphi \wedge \psi}{\Gamma \vdash_{\sigma_x} \psi}$$

$$\frac{\mathfrak{M}, \sigma_x \models \Gamma \quad \varphi \in \Gamma}{\Gamma \vdash_{\sigma_x} \varphi} \qquad \frac{\Gamma_1 \vdash_{\sigma_x} \varphi \vee \psi \quad \Gamma_2, \varphi \vdash_{\sigma_x} \theta \quad \Gamma_3, \psi \vdash_{\sigma_x} \theta}{\Gamma_1, \Gamma_2, \Gamma_3 \vdash_{\sigma_x} \theta}$$

$$\frac{\Gamma, \varphi \vdash_{\sigma_x} \psi}{\Gamma \vdash_{\sigma_x} \varphi \supset \psi} \qquad \frac{\Gamma_1 \vdash_{\sigma_x} \varphi \supset \psi \quad \Gamma_2 \vdash_{\sigma_x} \varphi}{\Gamma_1, \Gamma_2 \vdash_{\sigma_x} \psi}$$

$$\frac{\Gamma_1, \varphi \vdash_{\sigma_x} \psi \quad \Gamma_2, \varphi \vdash_{\sigma_x} \neg \psi}{\Gamma_1, \Gamma_2 \vdash_{\sigma_x} \neg \varphi} \qquad \frac{\Gamma \vdash_{\sigma_x} \neg\neg \varphi}{\Gamma \vdash_{\sigma_x} \varphi}$$

$$\frac{\Gamma_1 \vdash_{\sigma_{x_1}} \varphi \quad \Gamma_2 \vdash_{\sigma_{x_2}} \varphi \quad \ldots \quad \Gamma_k \vdash_{\sigma_{x_k}} \varphi \quad vor(x_1), vor(x_2), \ldots, vor(x_k) \in N(vor(x))}{\Gamma_1, \Gamma_2, \ldots, \Gamma_k \vdash_{\sigma_x} \Diamond^k \varphi}$$

$$\frac{\Gamma \vdash_{\sigma_x} \Diamond^k \varphi}{\Gamma \vdash_{\sigma_{x_i}} \varphi \quad vor(x_i) \in N(vor(x))} \; (1 \leq i \leq k)$$

Fig. 3.2 Inference rules for derivability $\vdash_{\sigma_x}$ of cell-state transitions rules on a trajectory σ_x. Γ is a set of well-formed formulas of $\mathfrak{L}^{\omega}_{\Diamond}$. If $\Gamma = \emptyset$, then $\Gamma \vdash_{\sigma_x} \varphi$ is denoted by $\vdash_{\sigma_x} \varphi$ and means that φ is an axiom for σ_x

$$\begin{array}{l}
\langle m, \mathbf{t}(n)\rangle_x \hookrightarrow \langle s, \mathbf{t}(n+1)\rangle_x \textbf{ if } \quad \Diamond^0\langle a, \mathbf{t}(n)\rangle \\
\langle m, \mathbf{t}(n)\rangle_x \hookrightarrow \langle i, \mathbf{t}(n+1)\rangle_x \textbf{ if } \quad \Diamond^2\langle a, \mathbf{t}(n)\rangle \\
\langle m, \mathbf{t}(n)\rangle_x \hookrightarrow \langle a, \mathbf{t}(n+1)\rangle_x \textbf{ if } \quad \Diamond^3\langle a, \mathbf{t}(n)\rangle \\
\langle m, \mathbf{t}(n)\rangle_x \hookrightarrow \langle i, \mathbf{t}(n+1)\rangle_x \textbf{ if } \quad \Diamond^4\langle a, \mathbf{t}(n)\rangle \\
\langle m, \mathbf{t}(n)\rangle_x \hookrightarrow \langle i, \mathbf{t}(n+1)\rangle_x \textbf{ if } \quad \Diamond^5\langle a, \mathbf{t}(n)\rangle \\
\langle m, \mathbf{t}(n)\rangle_x \hookrightarrow \langle i, \mathbf{t}(n+1)\rangle_x \textbf{ if } \quad \Diamond^6\langle a, \mathbf{t}(n)\rangle \\
\langle m, \mathbf{t}(n)\rangle_x \hookrightarrow \langle i, \mathbf{t}(n+1)\rangle_x \textbf{ if } \quad \Diamond^7\langle a, \mathbf{t}(n)\rangle \\
\langle m, \mathbf{t}(n)\rangle_x \hookrightarrow \langle a, \mathbf{t}(n+1)\rangle_x \textbf{ if } \Diamond^1\langle a, \mathbf{t}(n)\rangle \wedge \Diamond^0\langle i, \mathbf{t}(n)\rangle \\
\langle m, \mathbf{t}(n)\rangle_x \hookrightarrow \langle i, \mathbf{t}(n+1)\rangle_x \textbf{ if } \Diamond^1\langle a, \mathbf{t}(n)\rangle \wedge \Diamond^1\langle i, \mathbf{t}(n)\rangle \\
\langle m, \mathbf{t}(n)\rangle_x \hookrightarrow \langle s, \mathbf{t}(n+1)\rangle_x \textbf{ if } \Diamond^1\langle a, \mathbf{t}(n)\rangle \wedge \Diamond^2\langle i, \mathbf{t}(n)\rangle \\
\langle m, \mathbf{t}(n)\rangle_x \hookrightarrow \langle i, \mathbf{t}(n+1)\rangle_x \textbf{ if } \Diamond^1\langle a, \mathbf{t}(n)\rangle \wedge \Diamond^3\langle i, \mathbf{t}(n)\rangle \\
\langle m, \mathbf{t}(n)\rangle_x \hookrightarrow \langle s, \mathbf{t}(n+1)\rangle_x \textbf{ if } \Diamond^1\langle a, \mathbf{t}(n)\rangle \wedge \Diamond^4\langle i, \mathbf{t}(n)\rangle \\
\langle m, \mathbf{t}(n)\rangle_x \hookrightarrow \langle s, \mathbf{t}(n+1)\rangle_x \textbf{ if } \Diamond^1\langle a, \mathbf{t}(n)\rangle \wedge \Diamond^5\langle i, \mathbf{t}(n)\rangle \\
\langle m, \mathbf{t}(n)\rangle_x \hookrightarrow \langle s, \mathbf{t}(n+1)\rangle_x \textbf{ if } \Diamond^1\langle a, \mathbf{t}(n)\rangle \wedge \Diamond^6\langle i, \mathbf{t}(n)\rangle
\end{array} \tag{3.7}$$

Notice that rule (3.7) allows us to construct a semantical model of the cellular automaton described by rule (3.3).

For logic $\mathfrak{L}^{\omega}_{\Diamond}$, the derivability (deduction) relation is the greatest fixed point of the rules of Fig. 3.2.

Hence, we can program the medium of reaction-diffusion computing by the logic of trajectories. This fact takes place, because chemical reactions are quite deterministic. In the next section we examine a *process calculus* for reaction-diffusion computing. This calculus can be used for programming reaction-diffusion processes.

3.1.5 Process Calculus of Reaction-Diffusion Computing

A behaviour in reaction-diffusion systems is thought as the total of actions that a system can perform taking into account that we describe certain aspects of behaviour by introducing an abstraction or idealization of the 'real' behaviour. The simplest model of behaviour we will use is to consider behaviour as an input/output function. Evidently, it is a restriction, idealization we will deal with. Usually, such an input/output function approach is implemented in automata theory, where a process is modeled as an automaton. An automaton has a number of states and a number of transitions, going from one state to another state. A behaviour is a run, i.e. a path from the initial state to the final state. There are different ways of modeling the reaction-diffusion behaviour within this approach. The most popular is to simulate the behaviour by cellular automata.

In this Section, we will use a more general approach presented by *process algebra*—the study of the behaviour of parallel or distributed systems by algebraic means. It offers tools to describe or specify transition systems, and thus it has tools to talk about parallel composition in general. About transition systems, see [43–48].

Assume that the computational domain Ω is partitioned into computational cells $c_j = \overline{1, K}$ such that $c_i \cap c_j = \emptyset$, $i \neq j$ and $\bigcup_{j=1}^{K} c_j = \Omega$. Further, suppose that in the K cells, there are N chemically active species and the state of species i in cell j is denoted by p_{ij}, $i = \overline{1, N}$, $j = \overline{1, K}$. These states are time dependent and they are changed by reactions occurring between waves in the same cell and by diffusion where waves move to adjacent cells. In reaction-diffusion, the species interact concurrently and in a parallel manner.

Our process calculus contain the following basic operators: *Nil* (inaction), '•' (prefix), '|' (cooperation), '&' (reaction/fusion), '⊕' (choice). Let $N = \{a, b, \dots\}$ be a set of names and $L = \{a, \overline{a}\colon a \in N\}$, where a is considered as activator and $\overline{a}$ as inhibitor for a, be the set of labels built on N. Labels are said to be actions. We use the symbols α, β, etc., to range over labels (actions), with $\alpha = \overline{\overline{\alpha}}$, and the symbols P, Q, etc., to range over processes on states p_{ij}, $i = \overline{1, N}$, $j = \overline{1, K}$. The processes are given by the syntax:

$$P := Nil \mid \alpha \bullet P \mid (P|P) \mid P\&P \mid P \oplus P$$

Each label is a process, but not vice versa. An operational semantics for this syntax is defined in Fig. 3.3. The informal meanings of basic operations are as follows:

- *Nil* — This is the empty process.
- $\alpha \bullet P$ — A process P becomes active only after the action α has been performed.
- $P|Q$ — Actions P and Q are performed in parallel.
- $P\&Q$ — This is the fusion of P and Q; $P\&Q$ represents a system which may behave as both component P and Q. For instance, *Nil* behaves as $P\&\overline{P}$, where P is an activator and $\overline{P}$ an appropriate inhibitor respectively.

$$\textbf{Prefix}\ \frac{}{\alpha \bullet P \xrightarrow{\alpha} P}$$

$$\textbf{Choice}\ \frac{P \xrightarrow{\alpha} P'}{P \oplus Q \xrightarrow{\alpha} P'},\quad \frac{Q \xrightarrow{\alpha} Q'}{P \oplus Q \xrightarrow{\alpha} Q'}$$

$$\textbf{Cooperation}\ \frac{P \xrightarrow{\alpha} P'}{P|Q \xrightarrow{\alpha} P'|Q},\quad \frac{Q \xrightarrow{\alpha} Q'}{P|Q \xrightarrow{\alpha} P|Q'},\quad \frac{P \xrightarrow{\alpha} P' \quad Q \xrightarrow{\alpha} Q'}{P|Q \xrightarrow{\alpha} P'|Q'}$$

$$\textbf{Fusion}\ \frac{}{\alpha \bullet P \& \overline{P} \xrightarrow{\alpha} Nil},\quad \frac{P \xrightarrow{\alpha} P' \quad Q \xrightarrow{\alpha} P'}{P \& Q \xrightarrow{\alpha} P'},\quad \frac{P \xrightarrow{\alpha} P'}{P \& Q \xrightarrow{\alpha} P' \oplus Q'}$$

$$\frac{P \xrightarrow{\alpha} P' \quad Q \xrightarrow{\alpha} P'}{Q \& P \xrightarrow{\alpha} P'},\quad \frac{P \xrightarrow{\alpha} P'}{Q \& P \xrightarrow{\alpha} Q' \oplus P'}$$

Fig. 3.3 Operational semantics: inference rules for basic operations. The ternary relation $P \stackrel{\alpha}{\to} P'$ means that the initial action P is capable of engaging in action α and then behaving like P'

$P \oplus Q$ This is the choice between P and Q; $P \oplus Q$ represents a system which may behave either as component P or as Q.

Thus, in this *process calculus* we have two kinds of transitions between states: (1) the internal transitions $p_{mj} \stackrel{\alpha}{\to} p_{nj}$, i.e. a reaction α in a cell j is a transition from one state p_{mj} before the reaction to the state p_{nj} after the reaction, (2) the external transitions $p_{mk} \stackrel{\alpha}{\to} p_{nl}$ (diffusion), i.e. a reaction α in a cell l is a transition from one state p_{mk} in a cell k before the reaction to the state p_{nl} in a cell l after the reaction.

A behaviour in process calculus can be viewed as a *labelled transition system*, which consists of a collection of states $\mathscr{L} = \{p_{ij}\}_{i=\overline{1,N},j=\overline{1,K}}$ and a collection $\mathscr{T}$ of transitions (processes, actions) over them. Assume $\mathscr{T} : \mathscr{L} \mapsto \mathscr{P}(\mathscr{L})$, where $\mathscr{P}(\mathscr{L}) = \{T : T \subseteq \mathscr{L}\}$. This means that $\mathscr{T}(p)$ consists of all states that a reachable from p. The transition system is understood as a triple $\langle \mathscr{L}, \mathscr{T}, \longrightarrow \rangle$, where $\longrightarrow \subseteq \mathscr{L} \times \mathscr{T} \times \mathscr{L}$ is a transition relation that models how a state $p \in \mathscr{L}$ can evolve into another state $p' \in \mathscr{L}$ due to an interaction $\sigma \in \mathscr{T}$. Usually, $\langle p, \sigma, p' \rangle \in \longrightarrow$ is denoted by $p \stackrel{\sigma}{\to} p'$. So, a state p' is reachable from a state p if $p \stackrel{\sigma}{\to} p'$.

The finite word $\alpha_1\alpha_2 \ldots \alpha_n$ is a *finite trace of transition system* whenever there is a finite execution fragment of transition system

$$\rho = p_0\alpha_1 p_1\alpha_2 \ldots \alpha_n p_n \ \textit{such that}\ p_i \stackrel{\alpha_{i+1}}{\longrightarrow} p_{i+1}\ \textit{for all}\ 0 \le i < n.$$

The word $\alpha_1\alpha_2 \ldots \alpha_n$ is denoted by $trace(\rho)$. The infinite word $\alpha_1\alpha_2 \ldots$ is an *infinite trace* whenever there is an infinite execution fragment of transition system

$$\rho = p_0\alpha_1 p_1\alpha_2 p_2\alpha_3 p_3 \ldots \ \textit{such that}\ p_i \stackrel{\alpha_{i+1}}{\longrightarrow} p_{i+1}\ \textit{for all}\ 0 \le i.$$

The word $\alpha_1\alpha_2 \ldots$ is denoted by $trace(\rho)$ too. An infinite word $trace(\rho)$ is *circular* if

$$\rho = p_0\alpha_1 p_1\alpha_2 p_2 \ldots \alpha_n p_0\alpha_1 p_1\alpha_2 p_2 \ldots \alpha_n p_0\alpha_1 p_1\alpha_2 p_2 \ldots \alpha_n p_0 \ldots$$

such that $p_i \stackrel{\alpha_{i+1}}{\longrightarrow} p_{i+1}$ for all $0 \le i < n-1$ and $p_{n-1} \stackrel{\alpha_n}{\longrightarrow} p_0$.

Definition 3.1 An infinite (resp. finite) trace of state p denoted by $\rho(p)$ is the trace of an infinite (resp. finite) execution fragment starting in p.

Each trace can be regarded as a graph, where nodes represent states and edges transitions. In this way, the transition system is viewed as graph trees. On the other hand, in proof theory we can consider derivations as graph trees too. This will allow us to simulate transition system within the framework of proof theory, if we identify a process sequence $\alpha \bullet P$ with a derivation $\frac{\alpha}{P}$, where $\alpha, P \in L$, more precisely a transition $p \xrightarrow{\alpha} p'$ with a derivation $p \vdash^{\alpha} p'$, where $p, p' \in \mathscr{L}$.

By a *derivation tree*, we mean a possibly infinite tree of formulas in which each parent formula is obtained as the conclusion of an inference rule with its children as premises.

Definition 3.2 Let *For* denote the set of all well-formed formulas in the language $\langle L;$ $Nil, \&, \oplus\rangle$, where L is the set of propositional variables, *Nil* is an unary propositional connective, &, $\oplus$ are binary propositional connectives, and *Rul* denote some set of rules. Let $n \in \mathbf{N}$ be the maximum number of premises of any $R \in Rul$. Then a derivation graph is given by $\langle V, s, r, \tau\rangle$, where: V is a set of nodes, $s\colon V \mapsto For$, r is a partial mapping from V to *Rul*, and τ is a partial mapping from V to V^n (we write $\tau_j(v)$ for the j-th component of $\tau(v)$); for all $v \in V$, $\tau_j(v)$ is defined just in case $r(v)$ is a rule with m premises, $1 \leq j \leq m$ and:

$$\frac{s(\tau_1(v)) \ldots s(\tau_m(v))}{s(v)}$$

is an instance of rule $r(v)$.

A derivation graph $\mathscr{G}$ consists of the set V of nodes and the set of edges $E = \{\langle v, \tau_j, v'\rangle \colon v, v' \in V, \tau_j(v) = v'\}$. The finite word $\tau_1\tau_2 \ldots \tau_n$ is a *finite trace of derivation* whenever there is a finite execution fragment of transition system

$$\pi = v_0\tau_1v_1\tau_2 \ldots \tau_nv_n \text{ } \textit{such that } \tau_{i+1}(v_i) = v_{i+1} \textit{ for all } 0 \leq i < n.$$

The word $\tau_1\tau_2 \ldots \tau_n$ is denoted by $trace(\pi)$. The infinite word $\tau_1\tau_2 \ldots$ is an *infinite trace of derivation* whenever there is an infinite execution fragment of transition system

$$\pi = v_0\tau_1v_1\tau_2 \ldots \text{ } \textit{such that } \tau_{i+1}(v_i) = v_{i+1} \textit{ for all } 0 \leq i.$$

The word $\tau_1\tau_2 \ldots$ is denoted by $trace(\pi)$ too. An infinite word $trace(\pi)$ is *circular* if

$$\pi = v_0\tau_1v_1\tau_2 \ldots \tau_nv_0\tau_1v_1\tau_2 \ldots \tau_nv_0\tau_1v_1\tau_2 \ldots \tau_nv_0 \ldots$$

such that $\tau_{i+1}(v_i) = v_{i+1}$ for all $0 \leq i < n-1$ and $\tau_n(v_{n-1}) = v_0$. For more details about circular proofs see [49–51].

$$\frac{Ce^{3+}\&HBrO_2\&BrO_3^-\&H^+}{Ce^{4+}\oplus HBrO_2\oplus H_2O} \tag{3.8}$$

$$\frac{Ce^{4+}\&BrCH(COOH)_2\&H_2O}{Br^-\oplus Ce^{3+}\oplus HCOOH\oplus CO_2\oplus H^+} \tag{3.9}$$

$$\frac{HBrO_2}{HOBr\oplus BrO_3^-\oplus H^+} \tag{3.10}$$

$$\frac{BrO_3^-\&Br^-\&H^+}{HOBr\oplus HBrO_2} \tag{3.11}$$

$$\frac{HBrO_2\&Br^-\&H^+}{HOBr} \tag{3.12}$$

$$\frac{HOBr\&Br^-\&H^+}{Br_2\oplus H_2O} \tag{3.13}$$

$$\frac{Br_2\&CH_2(COOH)_2}{Br^-\oplus H^+\oplus BrCH(COOH)_2} \tag{3.14}$$

Fig. 3.4 Axioms of proof-theoretic simulation of Belousov-Zhabotinsky reaction

$$\textbf{Choice}: \frac{P\oplus Q}{P}\ (CL),\quad \frac{P\oplus Q}{Q}\ (CR),\quad \textbf{Empty process}: \frac{Nil}{}\ (Nil),$$

$$\textbf{Fusion}: \frac{P}{P\&Q}\ (FL),\quad \frac{Q}{P\&Q}\ (FR),\quad \frac{P\&\overline{P}}{}\ (\overline{F}),\quad \frac{P\&P}{P}\ (F).$$

Fig. 3.5 Inference rules of proof-theoretic simulation of Belousov-Zhabotinsky reaction. The variables P, Q run over the set $\{Ce^{3+}, HBrO_2, BrO_3^-, H^+, Ce^{4+}, H_2O, BrCH\ (COOH)_2, Br^-, HCOOH, CO_2, HOBr, Br_2, CH_2(COOH)_2\}$

Definition 3.3 A derivation graph $\mathscr{G}$ is a derivation tree if there is a distinguished node $v_0 \in V$ such that the trace of an infinite (resp. finite) execution fragment in $\mathscr{G}$ starts in v_0. This trace is denoted by $\pi(v_0)$.

Using the bisimulation principle [35, 36, 40], we can identify a transition system with a proof system, namely we can show that there exist a transition system and a proof system such that for any ρ and π (resp. p and v_0) $trace(\rho)$ and $trace(\pi)$ (resp. $\rho(p)$ and $\pi(v_0)$) are bisimilar.

Let us return to the process calculus of reaction-diffusion computing. Due to the bisimulation between a transition system and a proof system, we can set up proof-theoretic frameworks of this process calculus (Figs. 3.4 and 3.5). For example, we can define a proof-theoretic simulation of Belousov-Zhabotinsky reaction (Fig. 3.6).

Definition 3.4 Let $L = \{Ce^{3+}, HBrO_2, BrO_3^-, H^+, Ce^{4+}, H_2O, BrCH\ (COOH)_2, Br^-, HCOOH, CO_2, HOBr, Br_2, CH_2(COOH)_2\}$ be the set of value states. The set of initial transitions $\mathscr{T}_0$ is defined by axioms in Fig. 3.4. The two basic operations $\oplus$ (choice), & (fusion) are defined by inference rules in Fig. 3.5, which describe general properties of transitions. Then the proof-theoretic simulation of Belousov-Zhabotinsky reaction is a tuple $\langle L, \mathscr{T}_0, (CL), (CR), (FL), (FR), (\overline{F}), (F), (Nil)\rangle$.

$$\frac{Ce^{3+}\&HBrO_2\&BrO_3^-\&H^+\ \dagger}{Ce^{4+}\oplus HBrO_2\oplus H_2O}\ Axiom\ (3.8),$$

$$\frac{Ce^{4+}\oplus HBrO_2\oplus H_2O}{Ce^{4+}}\ Inf.rule\ 2\times(CL),$$

$$\frac{Ce^{4+}}{Ce^{4+}\&BrCH(COOH)_2\&H_2O}\ Inf.rule\ 2\times(FL),$$

$$\frac{Ce^{4+}\&BrCH(COOH)_2\&H_2O}{Br^-\oplus Ce^{3+}\oplus HCOOH\oplus CO_2\oplus H^+}\ Axiom\ (3.9),$$

$$\frac{Br^-\oplus Ce^{3+}\oplus HCOOH\oplus CO_2\oplus H^+}{Ce^{3+}}\ Inf.rules\ (CR), 3\times(LR),$$

$$\frac{Ce^{3+}}{Ce^{3+}\&HBrO_2\&BrO_3^-\&H^+\ \dagger}\ Inf.rule\ 3\times(FL),$$

Fig. 3.6 The derivation tree of proof-theoretic simulation of the Belousov-Zhabotinsky circular feedback $Ce^{3+}\longrightarrow Ce^{4+}\longrightarrow Ce^{3+}\longrightarrow\ldots$. We see that the pictured derivation tree has a *cycle*. This cycle presents the circular feedback in the Belousov-Zhabotinsky reaction (more precisely temporal oscillations in a well-stirred solution): cerium(III) is colourless and cerium(IV) is yellow. Under some conditions, this cycle will repeat several times: in the beginning the solution is colourless, later it is yellow, later it is colourless, etc.

In this system, transitions between states are identified with derivations of states. The example is given in Fig. 3.6. Each step of derivation means a transition. As a result, the circular trace of state Ce^{3+} (resp. Ce^{4+}) has a meaning of circular proof, where the state Ce^{3+} (resp. Ce^{4+}) is unfolded infinitely often among premisses and at the same time among derivable expressions. *This circular proof is an example of auto-wave.* Notice that the Belousov-Zhabotinsky reaction can be used in designing logic circuits, see [52].

Self-organization phenomena in nature assume circularity and cause-and-effect feedback relations: each component affects the other components, but these components in turn affect the first component. The most popular example of such self-organization is presented by Belousov-Zhabotinsky reaction. In this system we observe circularity in the interchange of solution colour: in the beginning the solution is colourless, then it becomes yellow, then it becomes colourless, etc. *In the logical simulation of Belousov-Zhabotinsky reaction we obtain circular proofs. This shows that the reaction-diffusion computing cannot dispense with logical circularity like cyclic proofs and feedback relations in state transitions.* We can suppose that logical circularity should be a key notion of "life computer", i.e. of each self-organized reaction-diffusion system.

3.2 Automata on Slime Mould

3.2.1 Slime Mould Cellular Automata

Let us show that the reaction-diffusion computing is realized on the media of swarms, too. And let us exemplify this fact by simulating the *slime mould* behaviour in the reaction-diffusion way. Plasmodium's active zones of growing pseudopodia are a key phenomenon of slime mould based computer. These zones interact concurrently. At these active zones, according to Adamatzky's experiments, three basic operations stimulated by nutrients (attractants) and some other conditions can be observed: fusion, multiplication, and direction operations [53]. The *fusion*, denoted *Fuse*, means that two active zones A_1 and A_2 either produce a new active zone A_3 (i.e. there is a collision of the active zones) or just a protoplasmic tube α: $Fuse(A_1, A_2) = A_3$ or $Fuse(A_1, A_2) = \alpha$. The *multiplication*, *Mult*, means that the active zone A_1 splits into two independent active zones A_2 and A_3 propagated according to their own trajectories: $Mult(A_1) = \{A_2, A_3\}$ or $Mult(\alpha) = \{A_2, A_3\}$. The *direction*, *Direct*, means that the active zone A is translated to a domain of an active space with certain initial velocity vector v: $Direct(A, v)$. These operations, *Fuse*, *Mult*, *Direct*, can be determined by the following stimuli:

- The set of attractants $\{N_1, N_2, \dots\}$—sources of nutrients, on which the plasmodium feeds. Each attractant N is characterized by its position and intensity. It is a function from one active zone to another.
- The set of repellents $\{R_1, R_2, \dots\}$. Plasmodium of *Physarum polycephalum* avoids light and some thermo- and salt-based conditions. Thus, domains of high illumination (or high grade of salt) are repellents such that each repellent R is characterized by its position and intensity (force of repelling). In other words, each repellent R is a function from one active zone to another.

The plasmodium behaviour is intelligent and can be represented as implementation of some abstract automata.

A cellular-automatic approach as the most natural way of presentation of plasmodium behaviours as computational media takes into account the fact that these behaviours are massively parallel. Recall again that a *cellular automaton* is a 4-tuple $\mathcal{A} = \langle \mathbf{Z}^d, S, u, f \rangle$, where (1) $d \in \mathbf{N}$ is a number of dimensions and the members of $\mathbf{Z}^d$ are referred as cells, (2) S is a finite set of elements called the states of an automaton $\mathcal{A}$, the members of $\mathbf{Z}^d$ take their values in S, (3) $u \subset \mathbf{Z}^d \setminus \{0\}^d$ is a finite ordered set of n elements, $u(x)$ is said to be a neighbourhood for the cell x, (4) $f : S^{n+1} \to S$, that is, f is the local transition function (or local rule). As we see, an automaton is considered in the endless d-dimensional space of integers, i.e. on $\mathbf{Z}^d$. Discrete time is introduced for $t = 0, 1, 2, \dots$ For instance, the cell x at time t is denoted by x^t. Each automaton calculates its next state depending on states of its closest neighbours. The cellular automata, thus, represent the locality of physics of information and massive parallelism in the space-time dynamics of natural systems [30, 31].

In abstract cellular automata, cells physically are identic. They can differ only by one of the possible states of S. In the case of *Physarum polycephalum*, cells can possess different topological properties. This depends on the intensity of chemo-attractants and chemo-repellents. The intensity entails the natural or geographical neighbourhood of the set's elements in accordance with the spreading of appropriate chemical signals (attractants or repellents). As a result, we obtain *Voronoi cells*. Let us remind what they are mathematically. Let $\mathbf{P}$ be a nonempty finite set of planar points and $|\mathbf{P}| = n$. For points $p = (p_1, p_2)$ and $x = (x_1, x_2)$ let $d(p, x) = \sqrt{(p_1 - x_1)^2 + (p_2 - x_2)^2}$ be their Euclidean distance. A *Voronoi diagram* of $\mathbf{P}$ is a partition of the plane into cells, such that for any element of $\mathbf{P}$, a cell corresponding to a unique point p contains all those points of the plane which are closer to p in respect to the distance d than to any other node of $\mathbf{P}$. As a consequence, a unique region

$$vor(p) = \bigcap_{m \in \mathbf{P}, m \neq p} \{z \in \mathbf{R}^2 : d(p, z) < d(m, z)\}$$

assigned to point p gives a *Voronoi cell* of p. Within one Voronoi cell, a reagent has a full power to attract or repel the plasmodium (or any other swarm). The distance d is defined by the intensity of reagent spreading like in other chemical reactions simulated by Voronoi diagrams [33]. The reagent attracts or repels the plasmodium and the distance is defined on the elements of a given planar set $\mathbf{P}$. When two spreading wave fronts of two reagents meet, this means that on the board of meeting the plasmodium cannot choose its further direction and it splits (see Fig. 3.7). Within one and the same Voronoi cell two active zones fuse.

The direction of protoplasmic tubes is defined by concentrations of chemo-attractants or chemo-repellents in the Voronoi neighbourhood. Let l, s, r be concentrations of reagents as measured by left, forward, and right *Physarum polycephalum* sensors, respectively [55]. Each step in the dynamics of protoplasmic tubes can be characterized by its current position x_t and the angle α_t at time step t. The angle is calculated as follows:

$$\alpha^{t+1} = \alpha^t + \begin{cases} 0, & \text{if } s^t > l^t \text{ and } s^t > r^t \\ -\psi, & \text{if } l^t > r^t \text{ and } l^t > s^t \\ \psi, & \text{if } r^t > l^t \text{ and } r^t > s^t \\ \text{random}(-\alpha, \alpha), & \text{otherwise} \end{cases}$$

Let $A_1, A_2, \ldots A_n$ denote growing pseudopodia. Then the Voronoi cell has the following state:

$$x^{t+1} = \begin{cases} \mathbf{D}(\alpha^{t+1}), & \text{if } A_i \in u(x^t) \\ x^t, & \text{otherwise} \end{cases}$$

where $\mathbf{D}(\alpha^{t+1})$ is a protoplasmic tube directed to the cell of x^{t+1} in accordance with the angle α^{t+1}.

Now we can define the *Physarum polycephalum* cellular automaton as a whole as follows:

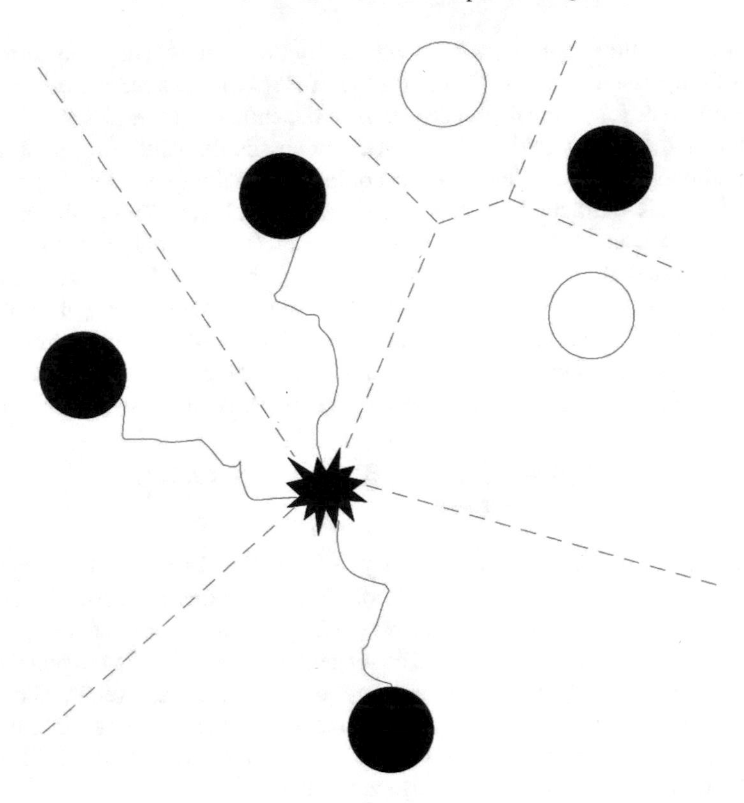

Fig. 3.7 The Voronoi diagram for *Physarum polycephalum*, where different attractants have different intensities and powers, see [54]

Definition 3.5 Consider a propositional language $\mathscr{L}$ with the only binary operation $\vee$, it is built in the standard way over the set of variables $S = \{0, A_1, A_2, \ldots, N_1, N_2, \ldots, R_1, R_2, \ldots, \mathbf{D}_1(\alpha_1), \mathbf{D}_2(\alpha_2), \ldots\}$, where 0 is inactive state, $A_1, A_2, \ldots$ are growing pseudopodia, $N_1, N_2, \ldots$ are attractants, $R_1, R_2, \ldots$ are repellents, $\mathbf{D}_1(\alpha_1)$, $\mathbf{D}_2(\alpha_2), \ldots$ are protoplasmic tubes. Let S be the set of states of proof-theoretic cellular automaton $\mathscr{A}$ on the *Physarum p.* Voronoi diagram described above. The inference rule of the automaton is as follows:

$$x^{t+1} = \begin{cases} X \vee A_j, & \text{if } x^t = X \vee \mathbf{D}_n(\alpha_n^t) \vee N_i \text{ and} \\ & \mathbf{D}_n(\alpha_n^t) \vee A_i \in u(x^t) \\ X \vee A_j, & \text{if } x^t = X \vee A_j \vee N_i \text{ and} \\ & \mathbf{D}_n(\alpha_n^t) \vee A_i \in u(x^t) \\ 0 \vee \mathbf{D}_n(\alpha_n^{t+1}), & \text{if } x^t = 0 \text{ and } R_i \notin u(x^t), \\ & \mathbf{D}_n(\alpha_n^t) \in u(x^t) \\ x^t, & \text{otherwise} \end{cases}$$

where X is a possibly empty metavariavle consisting of disjunctions. Then $\mathscr{A}$ can simulate the dynamics of plasmodium of *Physarum polycephalum*.

The automaton $\mathscr{A}$ is called *proof-theoretic* [2], because it is not a cellular automaton in the strict sense, it rather covers the class of automata by setting logical frameworks for their possible dynamics. This automaton defines basic features of *Physarum p.* computational behaviour which allow us to involve some abstract machines in analyzing *Physarum p.* computations. As a consequence, each *Physarum p.* proof-theoretic cellular automaton $\mathscr{A}$ serves as a logical metalanguage for an appropriate experiment with the *Physarum p.* plasmodium to fix experimental results. In other words, it is an experimental metalanguage containing observational terms and logical operations over them. Within this language, we can design abstract pointer machines and then implement them in real experiments with *Physarum polycephalum*.

3.2.2 Process Calculus of Slime Mould Computing

The *Physarum polycephalum* behaviour can be programmable within some abstract automata. It is easy to encode such programs into *process calculus* (π*-calculus*), developed by Milner [56] that describes concurrent computations whose network configuration may change during the computation. The *Physarum p.* version of π-calculus was proposed in [9]. This version is called *Physarum* spatial logic. Within π-calculus we can describe and simulate different intelligent processes such as business processes. In the meanwhile, *Physarum* spatial logic is a natural implementation of π-calculus. About logical aspects of concurrency and π-calculi, please see [43, 57, 58]. About spatial logic, please see [59–62].

This logic is built in the following way. The behaviour of *Physarum p.* plasmodium can be divided into the following elementary processes: inaction, fusion, cooperation, and choice, which could be interpreted as unconventional (spatial) falsity, conjunction, weak and strong disjunction respectively, denoted by *Nil*, &, |, and $+$. They may be described as special spatial transitions over states of *Physarum polycephalum* automata: inaction (*Nil*) means that pseudopodia has just stopped to behave; fusion (&) means that two pseudopodia come in contact one with another and then merge; cooperation (|) means that two pseudopodia behave concurrently (e.g. one tube splits); choice (+) means a competition between two pseudopodia in their behaviours.

A π-calculus for describing the dynamics of *Physarum p.* automata is presented as a labelled transition system with some logical relations.

Assume that there are n active species or growing pseudopodia within a *Physarum p.* Voronoi diagram and the state of species i is denoted by $p_i \in \Gamma$. These states are time dependent and they are changed by plasmodium's active zones interacting with each other and affected by attractants or repellents. Foraging plasmodium can be represented as a set of the following abstract entities (see also about possible transitions of slime mould in [63]):

- The set of elementary actions (growing pseudopodia), $T' = \{\alpha, \beta, \dots\}$, localized in *active zones*. The actions from T' are called the *simplest transitions*, the latter are defined as $\{p \overset{\alpha}{\rightarrow} q : p, q \in \Gamma, \alpha \in T'\}$. Notice that we also have transitions that do not belong to T'. Each action $\alpha \in T'$, starts on a state p_i, which is its current position, and says about a transition (propagation) of a state p_i to another state of the same or another Voronoi cell. Part of plasmodium feeding on a source of nutrients may not propagate, so its transition may be *Nil*, but this part can always start moving.
- The set of *attractants* $\{N_1, N_2, \dots\}$ are sources of nutrients, on which the plasmodium feeds. Each attractant N is a function from T' to T'.
- The set of *repellents* $\{R_1, R_2, \dots\}$. Each repellent R is a function from T' to T'.
- The set of *protoplasmic tubes* $\{C_1, C_2, \dots\}$. Typically plasmodium spans sources of nutrients with protoplasmic tubes/veins. $C(\alpha)$ means a diffusion of growing pseudopodia $\alpha \in T'$.

The set T of all transitions consists of members of T', $\{N_1, N_2, \dots\}$, $\{R_1, R_2, \dots\}$, $\{C_1, C_2, \dots\}$.

Our process calculus contains the following basic operators: *Nil* (inaction), . (prefix), | (cooperation), \ (hiding), & (reaction/fusion), + (choice), *a* (constant or restriction to a stable state), $A(\cdot)$ (attraction), $R(\cdot)$ (repelling), $C(\cdot)$ (spreading/diffusion). Let $T = \{a, b, \dots\}$, the set of all actions (evidently, this set is finite), be considered as a set of names. A name refers to a communication link or channel. With every $a \in T$ we associate a complementary action $\overline{a}$. Let us suppose that a designates an input port and $\overline{a}$ designates an output port. Any behaviour of *Physarum polycephalum* will be considered as outputs and any form of outside control and stimuli as appropriate inputs. For instance, T' is to be regarded just as the set of output ports and thereby $T \backslash T'$ contains ports that can be interpreted under different conditions as input ports or output ports. So, for each $X \in \{N_1, N_2, \dots\} \cup \{R_1, R_2, \dots\}$, we take that $X(\gamma) = \delta$ is a function from T' to T' such that X is a stimulus for γ and X makes an output $\delta \in T'$ along γ. Evidently that $X(Nil) = Nil$. Let $X(\gamma)$ denote an input and $\overline{X(\gamma)}$ an output.

Define $\mathscr{L}$ be the set of labels built on T (under this interpretation, $a = \overline{\overline{a}}$). Suppose that an action a communicates with its complement $\overline{a}$ to produce the internal action τ and τ belongs to $\mathscr{L}$, too.

We use now the symbols γ, δ, ..., etc., to range over labels (actions), with $a = \overline{\overline{a}}$, and the symbols P, Q, etc., to range over processes on states p_i. The processes are given by the syntax:

$$P, Q := Nil|\gamma \cdot P|N(\gamma) \cdot P|\overline{N(\gamma)} \cdot P|R(\gamma) \cdot P|\overline{R(\gamma)} \cdot P|C(\gamma) \cdot P|(P|Q)|P\backslash Q|P\&Q|P + Q|a$$

Each label is a process, but not vice versa, because a process may consists of many labels combined by the basic operators.

An operational semantics for this syntax is defined as follows:

$$\gamma := p_i, \text{ where } p_i \in \Gamma.$$

$$\textbf{Prefix}: \quad \frac{}{\gamma \cdot P \xrightarrow{\gamma} P},$$

$$\frac{}{N(\gamma) \cdot P \xrightarrow{\delta} P} \; (N(\gamma) = \delta), \quad \frac{}{\overline{N(\gamma)} \cdot P \xrightarrow{\delta} P} \; (\overline{N(\gamma)} = \delta),$$

$$\frac{}{R(\gamma) \cdot P \xrightarrow{\delta} P} \; (R(\gamma) = \delta), \quad \frac{}{\overline{R(\gamma)} \cdot P \xrightarrow{\delta} P} \; (\overline{R(\gamma)} = \delta),$$

(the conclusion states that the process of the form $\gamma \cdot P$ (resp. $N(\gamma) \cdot P$, $\overline{N(\gamma)} \cdot P$, $R(\gamma) \cdot P$, $\overline{R(\gamma)} \cdot P$) may engage in γ (resp. $N(\gamma)$, $\overline{N(\gamma)}$, $R(\gamma)$, $\overline{R(\gamma)}$) and thereafter they behave like P; in the presentations of behaviours as trees, $\gamma \cdot P$ (resp. $N(\gamma) \cdot P$, $\overline{N(\gamma)} \cdot P$, $R(\gamma) \cdot P$, $\overline{R(\gamma)} \cdot P$) is understood as an edge with two nodes: γ (resp. $N(\gamma)$, $\overline{N(\gamma)}$, $R(\gamma)$, $\overline{R(\gamma)}$) and the first action of P),

$$\textbf{Diffusion}: \quad \frac{P \xrightarrow{\gamma} P'}{P \xrightarrow{\gamma} C(\gamma)} \quad (C(\gamma) = P'),$$

$$\textbf{Constant}: \quad \frac{P \xrightarrow{\gamma} P'}{a \xrightarrow{\gamma} P'} \quad (a = P, a \in \Gamma),$$

$$\textbf{Choice}: \quad \frac{P \xrightarrow{\gamma} P'}{P + Q \xrightarrow{\gamma} P'}, \quad \frac{Q \xrightarrow{\gamma} Q'}{P + Q \xrightarrow{\gamma} Q'},$$

(these both rules state that a system of the form $P \mid Q$ saves the transitions of its subsystems P and Q),

$$\textbf{Cooperation}: \quad \frac{P \xrightarrow{\gamma} P'}{P|Q \xrightarrow{\gamma} P'|Q}, \quad \frac{Q \xrightarrow{\gamma} Q'}{P|Q \xrightarrow{\gamma} P|Q'},$$

(according to these rules, the cooperation | interleaves the transitions of its subsystems),

$$\frac{P \xrightarrow{\gamma} P' \quad Q \xrightarrow{\overline{\gamma}} Q'}{P|Q \xrightarrow{\tau} P'|Q'},$$

(i.e. subsystems may synchronize in the internal action τ on complementary actions γ and $\overline{\gamma}$),

$$\textbf{Hiding}: \quad \frac{P \xrightarrow{\gamma} P'}{P \backslash Q \xrightarrow{\gamma} P' \backslash Q} \quad (\gamma \notin Q, Q \subseteq \Gamma),$$

(this rule allows actions not mentioned in Q to be performed by $P\backslash Q$),

$$\textbf{Fusion}: \frac{}{\gamma \cdot P\&\breve{P} \xrightarrow{\gamma} Nil}$$

(the fusion of dual processes are to be performed into the inaction, e.g. a fusion of an attractant/repellent P and appropriate repellent/attractant $\breve{P}$),

$$\frac{P \xrightarrow{\gamma} P' \quad Q \xrightarrow{\gamma} P'}{P\&Q \xrightarrow{\gamma} P'}, \qquad \frac{P \xrightarrow{\gamma} P' \quad Q \xrightarrow{\gamma} P'}{Q\&P \xrightarrow{\gamma} P'}$$

(this means that if we obtain the same result P' that is produced by the same action γ and evaluates from two different processes P and Q, then P' may be obtained by that action γ started from the fusion $P\&Q$ or $Q\&P$),

$$\frac{P \xrightarrow{\gamma} P'}{P\&Q \xrightarrow{\gamma} Nil + C(\gamma) + P'}, \qquad \frac{P \xrightarrow{\gamma} P'}{Q\&P \xrightarrow{\gamma} Nil + C(\gamma) + P'}$$

(these rules state that if the result P' is produced by the action γ from the processes P, then a fusion $P\&Q$ (or $Q\&P$) is transformed by that same γ either into the inaction or diffusion or process P').

These are inference rules for basic operations. The ternary relation $P \xrightarrow{\gamma} P'$ means that the initial action P is capable of engaging in action γ and then behaving like P'.

Hence, the behaviour of plasmodium of *Physarum polycephalum* could be considered a kind of π-calculus and used as a programming language for *Physarum p.* automata or for simulating different intelligent processes. About our programming language, please see [64–70]. We have used the approach of object-programming [71].

3.2.3 Performative Propositions in Slime Mould π-Calculus

Involving *Physarum polycephalum* π-calculus as programming language into analyzing intelligent processes allows us to formalize many kinds of human interactions within *Physarum* automata. For instance, it is possible to consider simpler versions of *illocutionary logic* that was developed for explicating the logical nature of human speech acts [72–74]. This logic studies *performative propositions* which express our emotional and cognitive valuations to commit interactions. In *Physarum* π-calculus we can logically formulate some simple human performative propositions.

Let Ψ be any proposition that is built up by a superposition of standard propositional logical connectives ($\vee$, $\wedge$, $\neg$, $\Rightarrow$) in the conventional way. Let V is a valuation of each propositional variable p such that $V(p) \subseteq T$. We mean that $V(p)$ consists of possible worlds, where p is true. Now we can define whether a proposition Ψ is true in the event x of T. If it is true, it is denoted by $x \models \Psi$.

$$x \models p \text{ if } x \text{ belongs to } V(p);$$

$$x \not\models \textit{false}, \text{ where } \textit{false} \text{ is any contradiction;}$$

$$x \models \Phi \Rightarrow \Psi \text{ iff } x \not\models \Phi \text{ or } x \models \Psi;$$

$$x \models \neg\Phi \text{ iff } x \models (\Phi \Rightarrow \textit{false}).$$

Within *Physarum* spatial logic we can define the following three performative propositions:

- $\Box_{\text{eat}}\Psi$:= 'I would like to eat Ψ'
- $\Box_{\text{fear}}\Psi$:= 'I fear Ψ'
- $\Box_{\text{satisfy}}\Psi$:= 'I am almost satisfied by Ψ'

These propositions have the following semantics:

$x \models \Box_{\text{eat}}\Psi$ if for any process P containing $\overline{N(\gamma)}$, we have that if $x \models \Psi$, then P contains a transition $x \overset{\overline{N(\gamma)}}{\to} y$.

$x \models \Box_{\text{fear}}\Psi$ if for any process P containing $\overline{R(\gamma)}$, we have that if $x \models \Psi$, then P contains a transition $x \overset{\overline{R(\gamma)}}{\to} y$.

$x \models \Box_{\text{satisfy}}\Psi$ if for any process P containing $C(\gamma)$, we have that if P contains a transition $x \overset{C(\gamma)}{\to} y$, then $y \models \Psi$.

$\Diamond_{\text{eat}}\Psi = \neg\Box_{\text{eat}}(\neg\Psi)$.

$\Diamond_{\text{fear}}\Psi = \neg\Box_{\text{fear}}(\neg\Psi)$.

$\Diamond_{\text{satisfy}}\Psi = \neg\Box_{\text{satisfy}}(\neg\Psi)$.

As we see, we can consider the plasmodium activity as a verification of three basic performative propositions: $\Box_{\text{eat}}\Psi$, $\Box_{\text{fear}}\Psi$, $\Box_{\text{satisfy}}\Psi$.

3.3 Reversible Logic Gates on Swarm Behaviours

3.3.1 *Reaction-Diffusion Media of Topology of Attractants*

As we already know, the swarm behaviour is quite intelligent and with some restrictions it can be successfully managed by different outer stimuli: attractants and repellents [10, 53], and it may take place in accordance with *behaviourism*. This fact allows us to design some devices on the swarm motions assuming that logic variables $x_1, x_2, \ldots x_n$ are presented by attractants and their truth meanings either 0 or 1 is defined as follows: x_i has the value 1 (respectively, 0) if x_i is (respectively, is not) occupied by the swarm.

Each reaction to attractant is based on the reaction-diffusion computing defined in the previous sections. Hence, *the process calculus of reaction diffusion computing can be considered as a programming language for simulating chemo-biological reactions to attractants*. In this section, I show how we can design computers on topologies of attracting pieces. In order to design a biological computer on attracting the swarm behaviour we need to establish new ways of constructing logic circuits by involving the following simple actions of any swarm: fusion, multiplication, direction, and repelling. (i) The *fusion* means that two active zones A_1 and A_2 of a swarm produce new active zone A_3; (ii) the *multiplication* means that the active zone A_1 of the swarm splits into two independent active zones A_2 and A_3 propagating along their own trajectories for motions; (iii) the *direction* means that the active zone A is translated to a source of food with certain initial velocity vector $\overrightarrow{v}$; (iv) the *repelling* means that the swarm avoids some places under some conditions with certain initial opposite velocity vector $-\overrightarrow{v}$. The swarm behaviour can be considered a biological kind of labelled transition system [9]. It is the most fundamental type of behaviour managed by attractants and repellents.

In this section, let us exemplify the swarm behaviour by the plasmodium of *Physarum polycephalum* and let us show how we can design a biological computer on plasmodia. There are many significant results in designing logic circuits on plasmodia. For example, combinational logic circuits are proposed in [75] by using some external stimuli such as light and food to determine the outputs in the oscillation frequency. In that paper it is shown that basic gates OR, AND and NOT were correct 90%, 77.8% and 91.7% of the time respectively and derived logic circuits XOR, Half Adder and Full Adder were 70.8%, 65% and 58.8% accurate respectively. It is known [76] that the oscillations of electrical potential can also control the peristaltic activity of protoplasmic tubes, used for the distribution of nutrients in the *Physarum polycephalum* propagation. This feature can be engaged in designing logic gates $z = x \bigoplus y$ with two inputs x and y and one output z, where x, y, and z are tubes with diameter returns to a resting value 30 μm. For more details see [76].

In the object-oriented language for *Physarum polycephalum* computing we have constructed on the basis of the process calculus in [67, 68] to simulate the plasmodium propagation, we can design different logic gates on *Physarum polycephalum*—from classical to non-classical. In this section, I show, how we can define, first of all, reversible logic gates on *slime mould*. The point is that the plasmodium behaviour has some properties which are shared by quanta, e.g. we can perform the double-slit experiment [77] with *Physarum polycephalum* to prove that Kolmogorovian probabilities are unapplied in the case of slime mould as well as in the case of quanta [78]. Therefore we can assume that the evolution of plasmodium states can satisfy some properties of the evolution of quantum states, namely the evolution of *Physarum p.* states can be restricted by the unitarity property as well, i.e. every operation on a normalized plasmodium state must keep the sum of probabilities of all possible outcomes at exactly 1. Therefore any logic gate for *Physarum polycephalum* must be implemented as a unitary operator, too. So, it is reversible, also. Thus, it is better to appeal to reversible logic gates in the case of slime mould as well as in the case of quanta.

The main difference of states of plasmodium from the states of quanta is that the latter are described by the points of the *Bloch sphere*, while the first are described by the points of a disc around one active zone of plasmodium. For example, the quantum NOT gate has the effect of mapping a quantum state, $|\psi\rangle$, lying at any point on the surface of the Bloch sphere, into its antipodal state, $|\psi^{\perp}\rangle$, on the opposite side of the Bloch sphere. The *Physarum polycephalum* NOT gate maps a state of plasmodium on the disc around one active zone into its antipodal state on the opposite side of this disc. In other words, the fundamental meaning of *Physarum polycephalum* NOT is that if A denotes a direction of plasmodium with vector $\overrightarrow{v}$, then NOT A will denote a new direction of plasmodium with vector $-\overrightarrow{v}$ that is opposite to a given vector v with the same magnitude but the opposite direction.

The Bloch sphere is constructed as follows. Let us consider the spin-1/2 particle and the Stern-Gerlach apparatus with the upper and the lower detector plates for particles. While the upper detector plate is hit with probability P, the other detector plate is hit with probability $1 - P$. Then each proposition "The spin along the direction tilted at an angle θ ($0 \leq \theta \leq \pi$) from the z-axes is up" is probabilistic and belongs to a sphere.

However, the quantum reversible logic gates within the Bloch sphere are much easier to be designed than gates for *Physarum polycephalum* motions. The point is that the quantum behaviour is simpler than the behaviour of plasmodium. For quanta there are no lateral effects when quanta have a free will to choose a strategy of propagations. The plasmodium has such a free will. So, there are lateral inhibition and lateral excitation phenomena in the plasmodium networking. Let us recall that *lateral inhibition* (suppressing the activity of local neighbour neurons via inhibitory connections) and *lateral excitation* (*activation*) (activating local neighbour neurons via excitation mechanisms) are two basic phenomena of neural sensory modalities and it is a very substantial fact for designing logic circuits that both lateral inhibition and lateral excitation are also observed as two basic modalities in networking of *Physarum polycephalum* plasmodia [79]. The lateral inhibition means that the plasmodium prefers fusion of its tubes and the lateral excitation means that the plasmodium prefers multiplication of its tubes. Another problem is that the plasmodium can change its decisions and come back (compare to [80, 81]). As a result, we deal with non-well-founded sets [82].

In this section, first, we consider possibilities to implement asynchronous sequential logic gates for *Physarum polycephalum* motions and some examples of reversible logic gates within this approach, second, we show, how we can design some quantum-style conventional reversible logic gates on the medium of plasmodium behaviours within matrix multiplication groups, third, we propose an unconventional extension of matrix multiplication group theory to design reversible logic gates with an uncertain number of inputs and outputs. These results are a continuation of the work showing joint fundamentals of plasmodium behaviours and quantum behaviours [78].

Hence, we have used the following two approaches to designing reversible logic gates on plasmodia: (i) the approach by accepting the plasmodium propagation with controlling all directions and all ways of its motions; (ii) the approach by accepting the plasmodium propagation in all possible directions simultaneously without the controlling of each way of its behaviour. Both approaches have been applied by other scientists in designing logic gates on plasmodia, also. The first approach has been used in the design of logic circuits by appealing to the plasmodium motions in channels [83] or to the properties of plasmodium motions in the channels with direct and ballistic trajectories [84]. The same approach has been engaged in the optical coupling gates on plasmodia [85], in the plasmodium gates based on frequency changes of electrical potential oscillation [75], etc. The second approach has been used in the application of emergent computations on the medium of plasmodia [55].

For instance, in order to control all the ways of plasmodium propagation, we can design some channels plasmodia move through [83]. In this case the plasmodium motions can be interpreted as analogous to electric current in electric logic circuits. The presence of plasmodium in a given channel indicates ‘truth’ and absence ‘false’. Thus, in [83] there were proposed the two-input two-output gate $\langle x, y\rangle \rightarrow \langle xy, x + y\rangle$ and the three-input two-output gate $\langle x, y, z\rangle \rightarrow \langle \overline{x}yz, x + y + z\rangle$. Another mean is to use the difference of plasmodium behaviour in the channels with direct or ballistic trajectories [84]. In this way there were designed the following two-input two-output Boolean logic gates: $\langle x, y\rangle \rightarrow \langle xy, x + y\rangle$ and $\langle x, y\rangle \rightarrow \langle x, \overline{x}y\rangle$. Also, it is possible to oscillate the electrical activity of the protoplasmic tubes of plasmodia, creating a peristaltic like action within the tubes, to control the speed and direction of growth [75].

However, the most interesting way in designing logic circuits on plasmodia is presented by the second approach that applies emergent computations to the slime mould [55]. This way is more natural for a living system, because each living system possesses some emergent properties. Let us recall that emergence is a process whereby larger entities exhibit some new properties and patterns through interactions among smaller entities. In emergence there is a disequilibrium between local and global behaviours so that in emergent computations a whole system cannot be predicted locally. In [55] there is proposed a Boolean gate as an internally instable machine.

3.3.2 Sequential Logic Gates on Slime Mould and Some Examples of Reversible Logic Gates

In electronic circuits the two logical values 1 and 0 are represented by the high voltage at an input A (resp. at an output B) for $A = 1$ (resp. for $B = 1$) and low voltage at A (resp. at B) for $A = 0$ (resp. for $B = 0$). In the plasmodium circuits the value $A = 1$ (resp. $B = 1$) corresponds to the high concentration of plasmodium protoplasm at A (resp. at B) and the value $A = 0$ (resp. $B = 0$) corresponds to the low concentration of plasmodium protoplasm at A (resp. at B).

The plasmodium behaviour is managed by attractants and repellents. We can assume that logic variables $x_1, x_2, \ldots x_n$ are presented by attractants occupied or not by the plasmodium. Each variable x_i can be of a truth value either 0 or 1: it has the value 0 if x_i is not occupied by the plasmodium and it has the value 1 if x_i is occupied by the plasmodium. Let us denote an ordered collection consisting of all elements of a set $\{x_1, x_2, \ldots, x_n\}$ by $[x_1 x_2 \ldots x_n]$. It is a string. If we know a value of the i-sign x_i: say $x_i = 0$, we can write the string $[x_1 x_2 \ldots x_i \ldots x_n]$ as follows: $[x_1 x_2 \ldots \langle 0 \rangle \ldots x_n]$. We can replace all the variables with their values to obtain a binary combination. For example, variables $\{A, B, C, D\}$ of Fig. 3.8 have truth values $A = 0$, $B = 1$, $C = 1$, $D = 0$ at $t = 2$. In this case we receive the binary string and it is written as $[0110]$.

In Fig. 3.8, we see that one single variable can change its values. Moreover, the plasmodium can leave an occupied attractant B so that an appropriate value of B will be then 0. The sequence $\langle 011 \rangle$ means that the variable B is presented by three values fixed during time, since the step $t = 0$ and finishing at $t = 2$. The point at time $t = 0$ corresponds to the truth value $B(t_0) = 0$, the point at $t = 1$ to $B(t_1) = 1$, and the point at $t = 2$ to $B(t_2) = 1$. The time-depending variable $B(t)$ is defined by the sequence $\langle B(t_0)B(t_1)B(t_2) \rangle = \langle 011 \rangle$.

We assume that at each time step $t = 0, 1, 2, \ldots$ there is a change of meaning of at least one variable. This step is called the *clock edge*. The *set-up time* is to designate the time when the data inputs are valid before the clock transition (e.g. the 0 to 1 transition). The *hold time* is to designate the time when the data input remain valid after the clock edge.

Figure 3.8 can be regarded as a truth table structured as follows. There are horizontally located binary strings $X_0 = [0000]$, $X_1 = [0100]$, $X_2 = [0110]$ fixed at time t_0, t_1, t_2 respectively. All these strings are meanings of $[ABCD]$ at different time. Also, there are vertically located binary sequences $A(t) = \langle 000 \rangle$, $B(t) = \langle 011 \rangle$, $C(t) = \langle 001 \rangle$, $D(t) = \langle 000 \rangle$. These sequences can be presented as follows for short: $\langle X_0 X_1 X_2 \rangle$.

Now let us define a logical *switching* as an action to change the value from 0 to 1 or from 1 to 0 at time t. For instance, by passing from the binary string X_0 to X_1, variable B changes its value from logical zero to unity. At the same time, other variables do not change their truth values and, therefore, create a stable background for switching $B = 0/1$. In [86], the operation of *venjunction* is defined as the function $z = x \angle y$ which is read as follows: "switching x on the background y" and takes a unity value in the case of switching of variable x from zero to unity at the constant

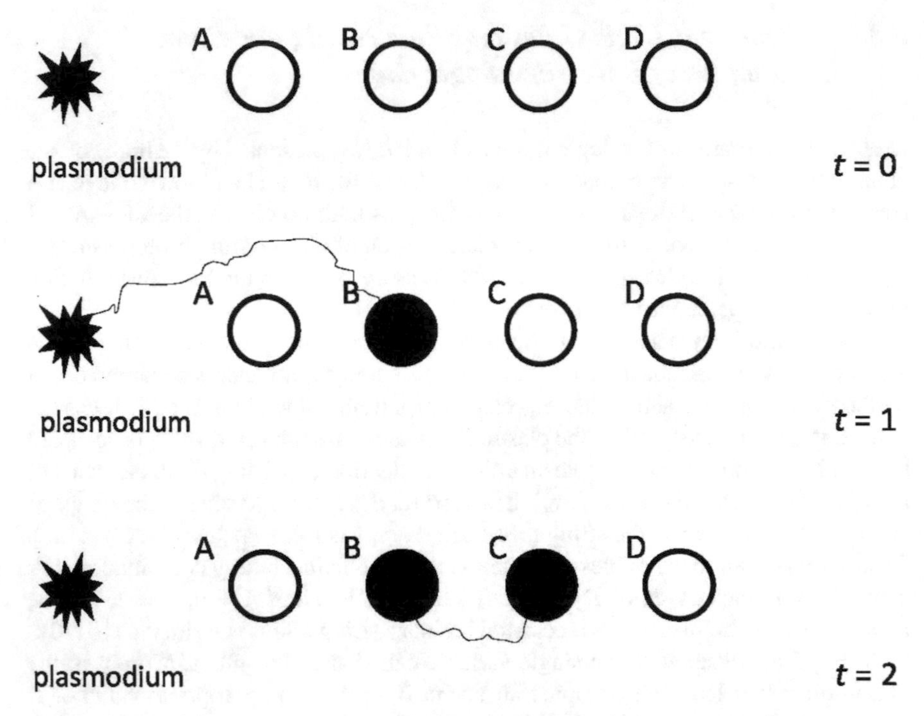

Fig. 3.8 One of the possible experiments when the plasmodium propagates a protoplasmic tube to attractants *A*, *B*, *C*, *D*. Let us pay attention that *A* is the nearest attractant and it is expected that it will be occupied first. However, the plasmodium occupies *B* at step $t = 1$ and then *C* at step $t = 2$ in this experiment, see [6]

unity value of variable *y*. It is supposed that the unity value of function remains until *x* or *y* reduces to zero. For example, $C \angle B$ takes the truth value 1 at $t = 2$.

On the basis of venjunction and some other Boolean functions, it is possible to implement *asynchronous sequential logic gates* [86, 87] on the medium of *Physarum polycephalum*. Notably that in asynchronous binary sequences there are no external control of time points $t_0, t_1, t_2, \ldots$, i.e. there is no external synchronizer which would set these points.

A *synchronous sequential circuit* is a logic circuit whose outputs are a function of their inputs, past inputs, and past outputs: *outputs* = *f*(*inputs, past inputs, past outputs*). In these sequential circuits we can deal with two outputs Q and $\neg Q$ one of which is the complement of the other. These circuits are called *bistable multivibrators* or *Flip Flops*. So, 1 as the first stable state at an output means rising on top ("flip") and 0 as the second stable state at an output means falling to ground ("flop"). The Flip Flop has one or two inputs and clock edge, as a result the current outputs are determined by the logic states at these inputs and the previous outputs. The commands formed by the inputs are as follows: (i) *set*: $Q = 1$; (ii) *reset*: $Q = 0$; (iii) *flip*: change

Table 3.1 The RS Flip Flop circuit with two inputs R ('reset') and S ('set') and two outputs $Q = \text{NAND}(S, \text{NOT}Q)$, $\neg Q = \text{NAND}(R, Q)$. Thus, when both R and S are 0, the Flip Flop is in a quiescent state and both outputs retain their value. If a positive pulse is applied to the S input, the Q output takes the 1 value (with $\neg Q = 0$) and if a positive pulse is applied to the R, he Q output takes the 0 value (with $\neg Q = 1$), see [6]

R	S	Q	$\neg Q$	Meaning
0	0	Q	$\neg Q$	No changes
0	1	1	0	Reset
1	0	0	1	Set
1	1	1/0	0/1	Undefined

the value from 0 to 1 or from 1 to 0; (iv) *hold value*: the value does not change (see Table 3.1).

The *Flip Flop circuit* can be implemented on *Physarum polycephalum* as it is shown in Fig. 3.9. This circuit is used for designing some arithmetic circuits within the sequential logic approach.

In sequential logic for formalizing the plasmodium propagation we can design some reversible gates: the *FEYNMAN gate* (see Fig. 3.10), the *TOFFOLI gate* (see Fig. 3.11), the *FREDKIN gate* (see Fig. 3.12).

The main disadvantage of sequential logic gates on slime mould is that there is an analogy between the uncertainty of plasmodium propagation and the uncertainty of position and momentum of quanta. So, in Fig. 3.8 we expect that the plasmodium occupies the nearest attractant first, but it can be different. Therefore we cannot predict, what will be occupied at once and what further. This means that it is difficult to manage switching of logic variables. Nevertheless, we can use a quantum-like reversible logic gate which does not erase any information when it acts, because we fix a transformation of the whole computation system. But we need too many repellents.

3.3.3 Conventional Reversible Logic Gates on Slime Mould

In *reversible logic gates* [88], there is always a unique input associated with a unique output and vice versa. So, any n-bit reversible gate specifies how to map each distinct bit string input of the length n into a distinct bit string output of the same length. For example, the *NOT gate* is reversible: it is a 1-input and 1-output gate that simply inverts the bit value 0 to 1 and 1 to 0 (see Table 3.2). In the case of the *CNOT gate* (the *2-bit controlled-NOT gate*), the four possible input bit strings are 00, 01, 10, 11 and these are mapped into 00, 01, 11, and 10 respectively (see Table 3.3). The CNOT gate can be generalized up to the *TOFFOLI gate* also called the *controlled-controlled-NOT gate* (see Table 3.4) or to the *FREDKIN gate* (see Table 3.5). It is a universal gate, where the values of the first two input bits control whether the third

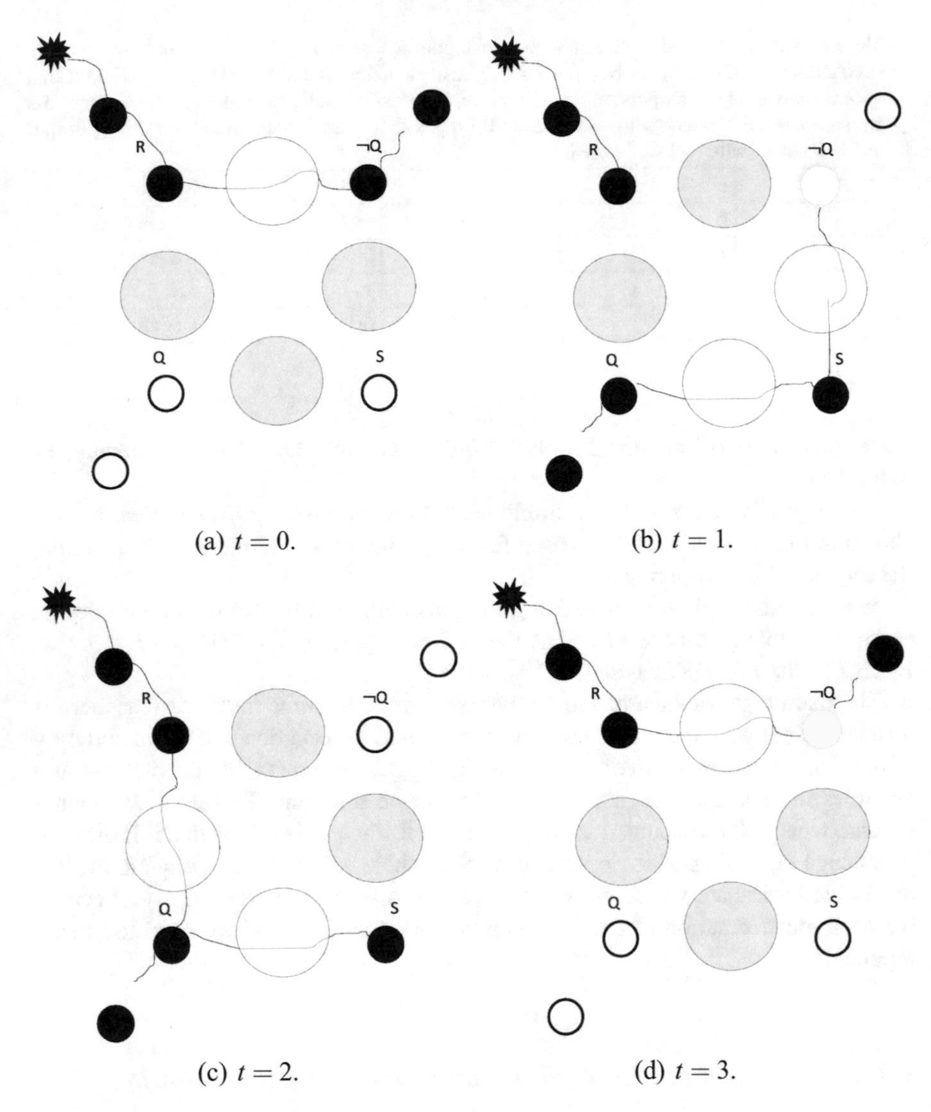

Fig. 3.9 The *Flip-Flop circuit*. We deal with two bit strings [RQ], [QS], [SQ], [QR]. The small black circles denote the attractants which are occupied. The small white circles denote the attractants which are not occupied. The big grey circles denote the zones with the activated light irradiation and the big white circles denote the zones with the deactivated light irradiation. The point is that the plasmodium avoids illumination with certain wavelengths of visible light and this feature can be used to repel the plasmodium from one place to another. The light irradiation is manipulated by the observer, see [6]

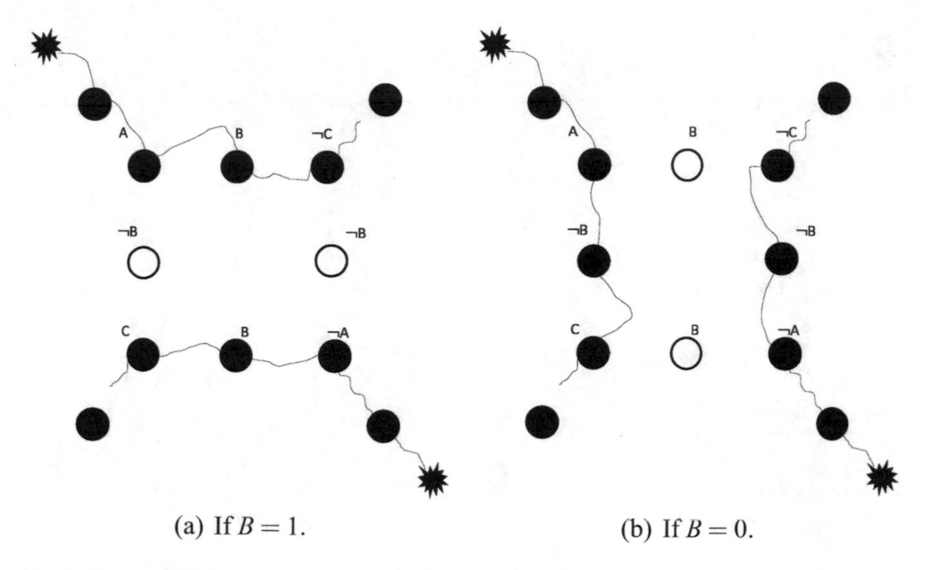

(a) If $B = 1$. (b) If $B = 0$.

Fig. 3.10 The *FEYNMAN gate* for *Physarum polycephalum*, i.e. the reversible logic gate $C = \text{XOR}(A, B)$ with two inputs A and $\neg A$ and two outputs C and $\neg C$. So, the plasmodium starts from both A and $\neg A$, see [6]

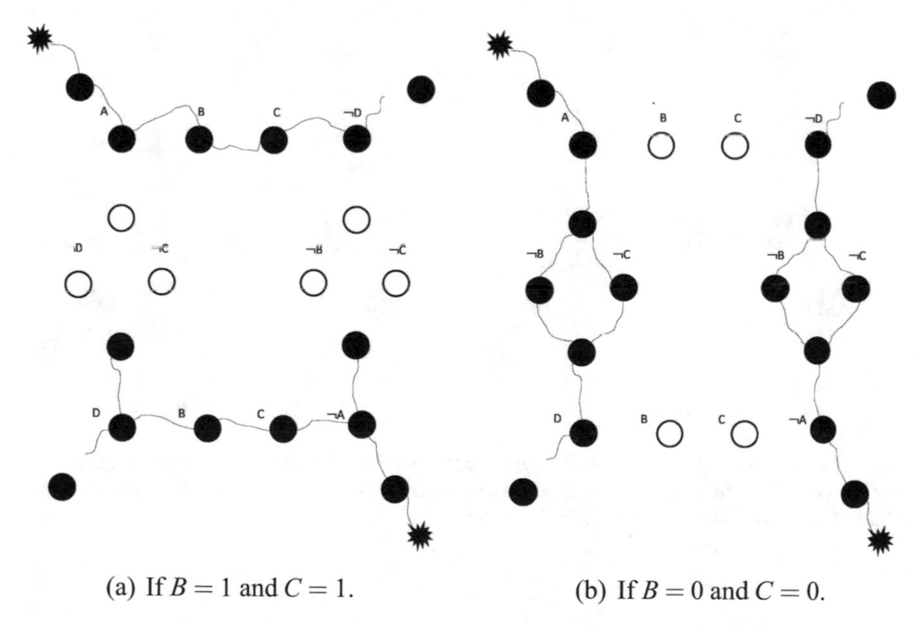

(a) If $B = 1$ and $C = 1$. (b) If $B = 0$ and $C = 0$.

Fig. 3.11 The *TOFFOLI gate* for *Physarum polycephalum*, i.e. the reversible logic gate $D = \text{XOR}(A, \text{AND}(B, C))$ with two inputs A and $\neg A$ and two outputs D and $\neg D$, see [6]

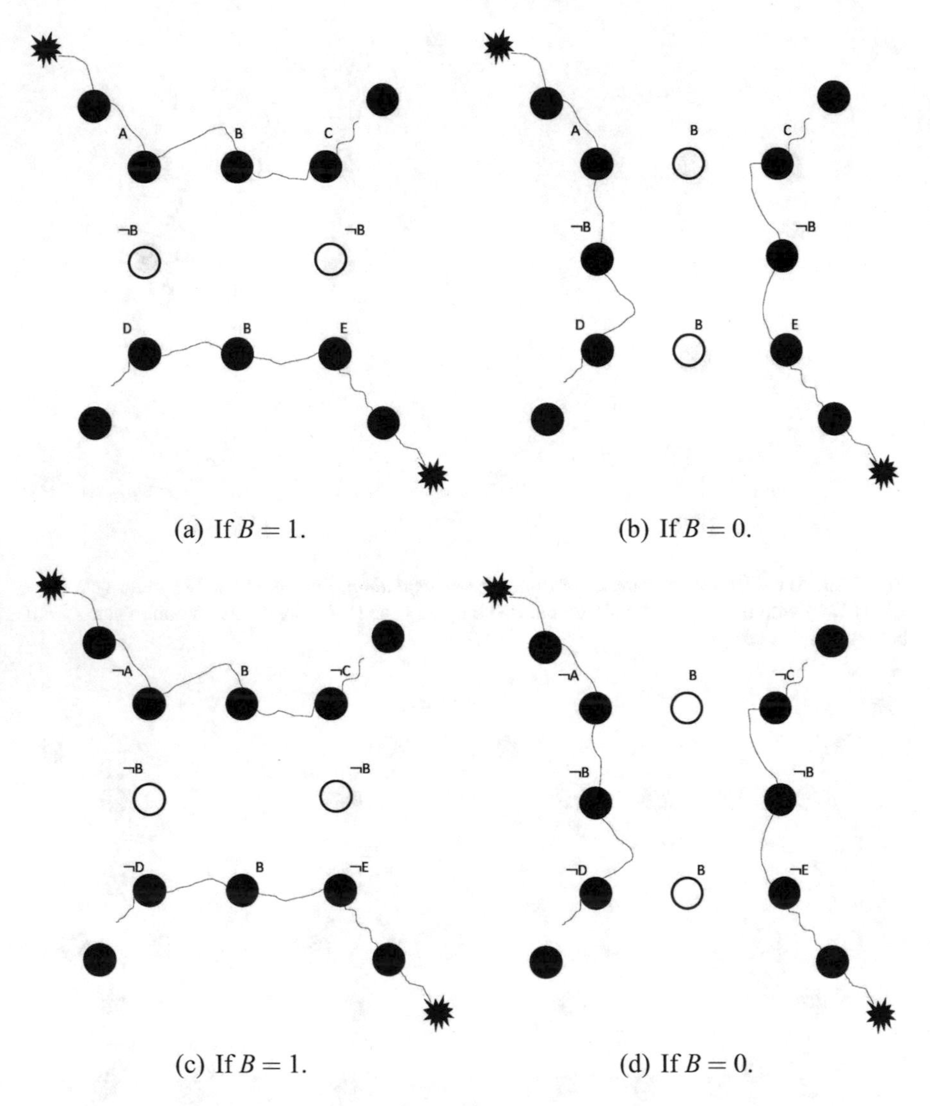

(a) If $B = 1$. (b) If $B = 0$.

(c) If $B = 1$. (d) If $B = 0$.

Fig. 3.12 The *FREDKIN gate* for *Physarum polycephalum*, i.e. the reversible logic gates $G = B, D = \text{OR}(\text{AND}(\text{NOT}B, A), \text{AND}(B, E))$ and $C = \text{OR}(\text{AND}(B, A), \text{AND}(\text{NOT}B, E))$ with three inputs A, B, E, and three outputs C, D, G, see [6]

Table 3.2 The NOT gate in the matrix form

	0	1
0	0	1
1	1	0

Table 3.3 The CNOT gate in the matrix form

	00	01	10	11
00	1	0	0	0
01	0	1	0	0
10	0	0	0	1
11	0	0	1	0

Table 3.4 The TOFFOLI gate in the matrix form

	000	001	010	011	100	101	110	111
000	1	0	0	0	0	0	0	0
001	0	1	0	0	0	0	0	0
010	0	0	1	0	0	0	0	0
011	0	0	0	1	0	0	0	0
100	0	0	0	0	1	0	0	0
101	0	0	0	0	0	1	0	0
110	0	0	0	0	0	0	0	1
111	0	0	0	0	0	0	1	0

input bit is switched, i.e. it is considered switching the third input bit if and only if the first two input bits are both 1. So, the TOFFOLI gate can be used as a logical gate by changing the control bits as inputs, e.g. for the input $[AB\ \langle 0\rangle]$ we have the output $[AB\ \langle \mathrm{AND}(A, B)\rangle]$, for the input $[AB\ \langle 1\rangle]$ we have the output $[AB\ \langle \mathrm{NAND}(A, B)\rangle]$, for the input $[\langle 1\rangle\ BC]$ we have the output $[\langle 1\rangle\ B\ \langle \mathrm{XOR}(B, C)\rangle]$, etc. The same situation holds for the FREDKIN gate, e.g. for the input $[A\ \langle 0\rangle\ C]$ we have the output $[A\ \langle \mathrm{AND}(A, C)\rangle\ \langle \mathrm{NOT}\,(\mathrm{OR}(\mathrm{NOT}\,C, A))\rangle]$, etc.

Thus, each reversible gate fixes a specification for how to permute the 2^n possible bit strings inputs expressible in n bits. In Tables 3.2, 3.3, 3.4 and 3.5, an n-bit reversible gate is considered an array whose rows and columns are indexed by the 2^n possible bit strings expressible in n bits. The (i, j)-th element of this array has the truth value 1 if and only if the input bit string corresponding to the i-th row is mapped to the output bit string corresponding to the j-th column. In this way we obtain a permutation matrix. So, logic gates combinations can be defined within a *permutation group* as multiplication of matrices.

The matrix of an n-bit circuit is called well-formed if it satisfies the following two conditions: (i) matrix elements can only be zeros or ones; (ii) each column or

Table 3.5 The FREDKIN gate in the matrix form

	000	001	010	011	100	101	110	111
000	1	0	0	0	0	0	0	0
001	0	1	0	0	0	0	0	0
010	0	0	1	0	0	0	0	0
011	0	0	0	1	0	0	0	0
100	0	0	0	0	1	0	0	0
101	0	0	0	0	0	0	1	0
110	0	0	0	0	0	1	0	0
111	0	0	0	0	0	0	0	1

Fig. 3.13 The three bit strings which are represented by column vectors

$$\text{(a) } [000] = \begin{pmatrix} 1 \\ 0 \\ 0 \\ 0 \\ 0 \\ 0 \\ 0 \\ 0 \end{pmatrix} \quad \text{(b) } [001] = \begin{pmatrix} 0 \\ 1 \\ 0 \\ 0 \\ 0 \\ 0 \\ 0 \\ 0 \end{pmatrix} \quad \text{(c) } [010] = \begin{pmatrix} 0 \\ 0 \\ 1 \\ 0 \\ 0 \\ 0 \\ 0 \\ 0 \end{pmatrix}$$

$$\text{(f) } [101] = \begin{pmatrix} 0 \\ 0 \\ 0 \\ 0 \\ 0 \\ 1 \\ 0 \\ 0 \end{pmatrix} \quad \text{(g) } [110] = \begin{pmatrix} 0 \\ 0 \\ 0 \\ 0 \\ 0 \\ 0 \\ 1 \\ 0 \end{pmatrix} \quad \text{(h) } [111] = \begin{pmatrix} 0 \\ 0 \\ 0 \\ 0 \\ 0 \\ 0 \\ 0 \\ 1 \end{pmatrix}$$

row has exactly one element with a value of 1, the other values are 0. This set of well-formed matrices is a group, i.e. it is closed under the matrix multiplication $\cdot$ and the operation $^{-1}$, and it contains the unit—an n-bit identity gate ($2^n \times 2^n$ identity matrix).

For example, the inputs of the TOFFOLI gate can be represented by a column vector consisting of $2^3 = 8$ slots, containing only once the number 1. So, each column vector corresponds to a three bit string, see Fig. 3.13.

Then the vector-matrix multiplication gives a transformation from an input to an output, for example:

Table 3.6 The Full Adder with three inputs C_i, A_i, and B_i and two outputs C_{i+1} and S_i. The input C_i is called carry-in and the output C_{i+1} is called carry-out. Sometimes C_i is denoted by C_{in} and C_{i+1} is denoted by C_{out}. C_{i+1} is needed, because by adding two binary numbers of the length i, we can obtain a binary number of the length $i+1$, e.g. by adding $1+1$ we obtain 10. According to this circuit, $S_i = \text{XOR}(C_i, A_i, B_i)$ and $C_{i+1} = \text{AND}(A_i, B_i) + \text{AND}(C_i, \text{XOR}(A_i, B_i))$, as a result $[C_{i+1}S_i] = [A_i] + [B_i]$.

C_i	A_i	B_i	C_{i+1}	S_i
0	0	0	0	0
0	0	1	0	1
0	1	0	0	1
0	1	1	1	0
1	0	0	0	1
1	0	1	1	0
1	1	0	1	0
1	1	1	1	1

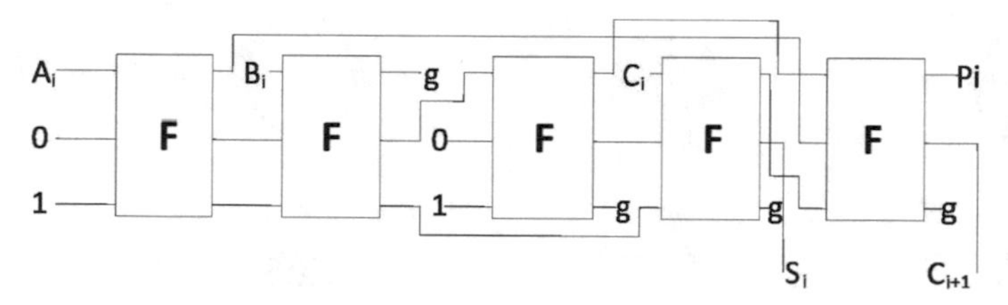

Fig. 3.14 The Full Adder circuit proposed in [89] and constructed by using five FREDKIN gates, where $P_i = \text{XOR}(A_i, B_i)$, see [6]

$$\text{TOFFOLI} \cdot [000] = \begin{pmatrix} 1&0&0&0&0&0&0&0 \\ 0&1&0&0&0&0&0&0 \\ 0&0&1&0&0&0&0&0 \\ 0&0&0&1&0&0&0&0 \\ 0&0&0&0&1&0&0&0 \\ 0&0&0&0&0&1&0&0 \\ 0&0&0&0&0&0&0&1 \\ 0&0&0&0&0&0&1&0 \end{pmatrix} \cdot \begin{pmatrix} 1\\0\\0\\0\\0\\0\\0\\0 \end{pmatrix} = \begin{pmatrix} 1\\0\\0\\0\\0\\0\\0\\0 \end{pmatrix} = [000].$$

By appealing to some reversible gates we can construct arithmetic circuits such as the *Full Adder* (see Table 3.6). The Full Adder computes the sum bit S_i and the carry output C_{i+1} based on addend inputs A_i and B_i and carry input C_i at time $t = i, i + 1, \ldots$. The output expressions for a ripple carry adder are (1) $S_i = \text{XOR}(A_i, B_i, C_i)$ (2) $C_{i+1} = A_iB_i + B_iC_i + C_iA_i$ $(i = 0, 1, 2, \ldots)$. The Full Adder can be designed by using FREDKIN gates (see Fig. 3.14).

On the *Physarum polycephalum* motions, logic variables of reversible gates are expressed by attractants, too. So, a string of the length n is a sequence of n attractants. The quantum-style NOT gate on slime mould is pictured in Fig. 3.15. In the

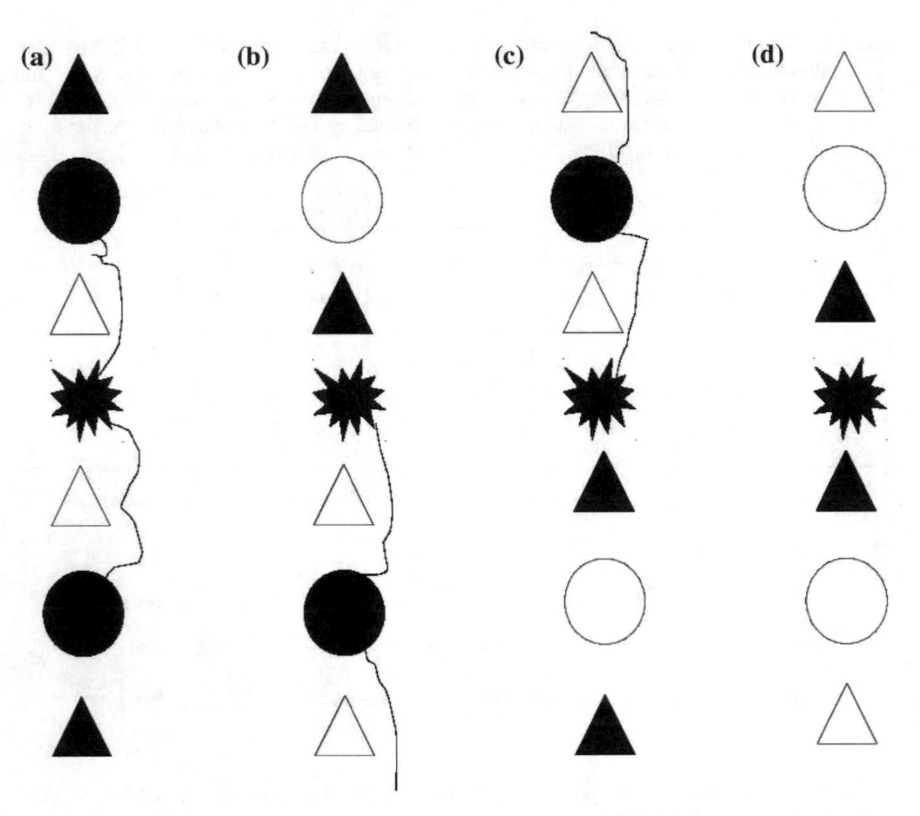

Fig. 3.15 The NOT gate on slime mould, where repellents are denoted by triangles, attractants by circles, thereby black circles are attractants occupied by the plasmodium, white circles are attractant which are not occupied by the plasmodium, black triangles are activated repellents, white triangles are deactivated repellents: **a** mapping 1 into 1, which is 0; **b** mapping 0 into 1, which is 1; **c** mapping 1 into 0, which is 1; **d** mapping 0 into 0, which is 0, see [6]

CNOT gate, we implement Table 3.3 and observe the four possible directions of plasmodium propagation (see Fig. 3.16), corresponding to the meaning 1 of Table 3.3. The transformation of 00 into 00, etc., which gives 1 (see Table 3.3), means that the plasmodium is propagated in both directions, where the two strings with the same meaning [00] are located (i.e. [*ab*] and [*cd*]). In the same way, the quantum-style TOFFOLI gate and the quantum-style FREDKIN gate on *Physarum* are designed.

Some other logic circuits for the slime mould propagation are proposed in [90].

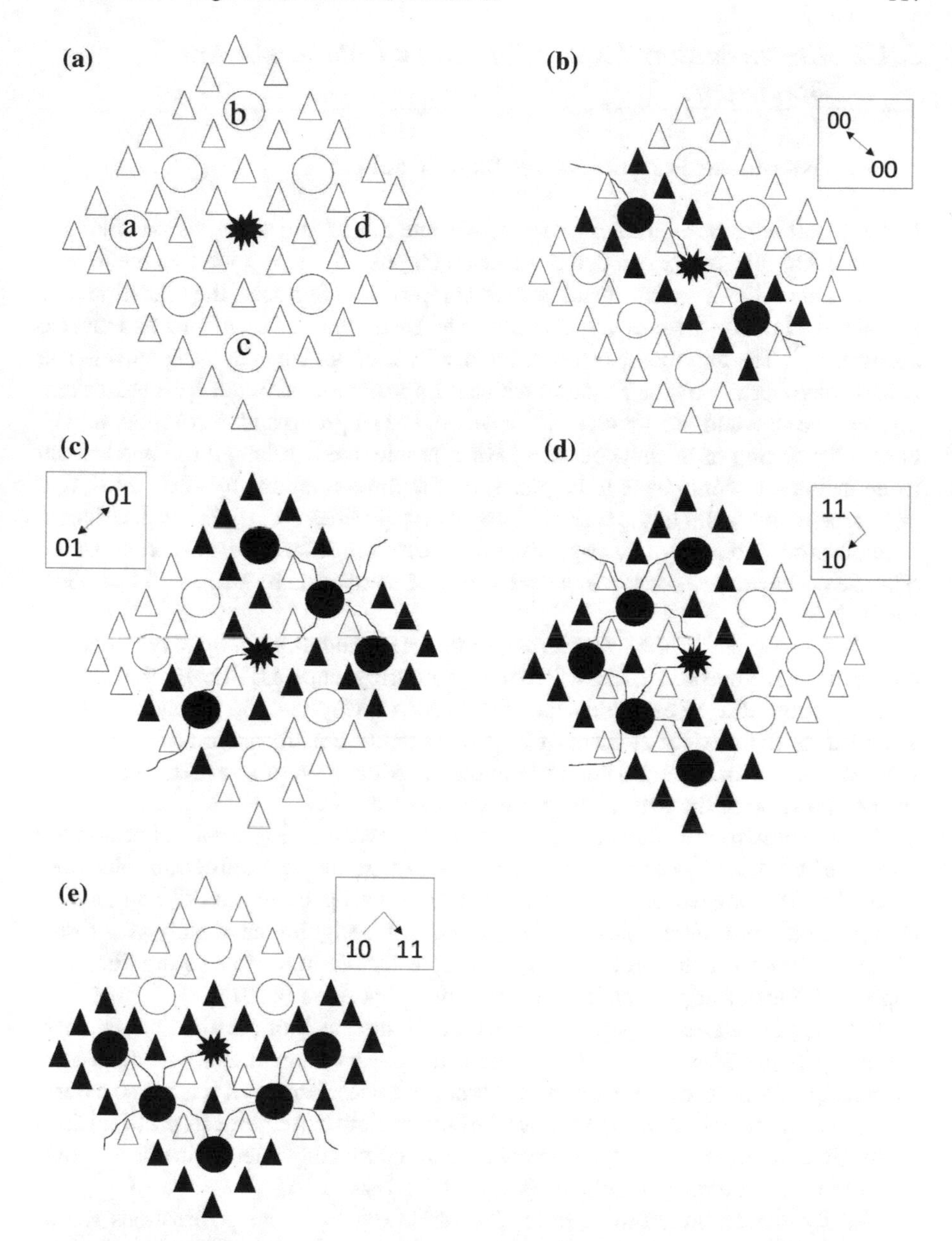

Fig. 3.16 The CNOT gate on slime mould, where black circles are attractants occupied by the plasmodium, white circles are attractant which are not occupied by the plasmodium, black triangles are activated repellents, white triangles are deactivated repellents: **a** the string [*ab*] is transformed into the string [*cd*] and vice versa; **b** mapping 00 into 00; **c** mapping 01 into 01; **d** mapping 11 into 10; **e** mapping 10 into 11. The line of repellents can be built up as a barrier, see [6]

3.3.4 *Unconventional Reversible Logic Gates on Swarm Behaviours*

3.3.4.1 Non-linear Permutations of Slime Mould

Let us recall what *permutation group* is. Assume that $\mathcal{G}$ is a set of various motions (actions). On this set we can define a linear composition: if x, y are two motions in $\mathcal{G}$, then the motion $x \cdot y$ is a result of their composition. Suppose, the operation $x \cdot y$ satisfies the law of associativity, $\mathcal{G}$ contains hereafter the unit member e and inverse member x^{-1}. The unit member means an identity motion, when nothing moves. The inverse member x^{-1} of x is a motion which is opposite to x. Obviously, it means that $e \cdot x = x \cdot e = x$ and $x^{-1} \cdot x = x \cdot x^{-1} = e$. So, $\mathcal{G}$ is a group and every motion in $\mathcal{G}$ causes the setting of linear sequences. For example, the equilateral triangle has the three motions to come back to its place, i.e. the three rotations by 120°, 240°, and 360° around the center of the triangle mapping every vertex (a_1, a_2, a_3) of the triangle to another one (to a_2, a_3, a_1, respectively). Indeed, for describing all motions of the triangle we have $x_1 \cdot x_2 \cdot x_3 = e$, where x_i is the i-th rotation by 120°, 240°, or 360°, $i = 1, 2, 3$.

If X is a set (e.g. it is the set of vertexes), then a *permutation group* may be defined as a group homomorphism h from $\mathcal{G}$ to the symmetric group on X. The motion assigns a permutation of X to each element of $\mathcal{G}$ in such a way that the permutation of X assigned to: the identity element of $\mathcal{G}$ is the identity transformation of X (e.g. the identity rotation of the equilateral triangle); a product $g \cdot h$ of two elements of $\mathcal{G}$ is the composition of the permutations assigned to g and h.

Permutation groups play an important role in reversible logic gates, because the latter can be defined by permutation matrices with a group operation of multiplication.

In this section, we are going to consider a strong extension of permutation groups, called *non-linear permutation groups*, for designing absolutely new forms of reversible logic gates. For more details about formal-logical and game-theoretic aspects of non-linear permutation group theory, please see [4, 91].

In the logic gates of Figs. 3.15 and 3.16, each next step of plasmodium is determined by its previous steps. In other words, the plasmodium chooses a direction of motions and its past choices determines its next choices. However, it is not so natural for the plasmodium behaviour, because the plasmodium can change its past decisions all the time. In order to avoid the plasmodium freedom and to determine its motions, we need to use too many repellents (see Figs. 3.15 and 3.16).

Usually, the plasmodium occupies the whole region in many directions simultaneously. Let us assume that all attractants stand in line (see Fig. 3.17). So, the plasmodium can move only to one of the two possible directions: either up or down. Suppose that we do not have repellents at all. Let us consider differences between the NOT gate of Fig. 3.15 and the logic device of Fig. 3.17. We can try to define the plasmodium motions as a group $\mathcal{G}$ of actions. Each attractant will be denoted by a member of $\mathcal{G}$ in the following way: (i) all attractants above are enumerated from a_0 to a_n and all attractants below are enumerated from b_0 to b_n; (ii) each b_i is equal a_i^{-1}

(this means that each a_i is equal b_i^{-1}), $i = 0, \ldots, n$; (iii) the member of $\mathcal{G}$ holds if it is occupied by the plasmodium. Hence, each motion a_i is opposite to the motion b_i. The linear composition $x \cdot y$ at time t is defined for all $x, y \in \mathcal{G}$ as follows: $x \cdot y$ occurs at time t if and only if both x and y are occupied by the plasmodium at this time.

The group $\mathcal{G}$ can be used for designing logic circuits on plasmodia. Let $A_i, A_j, \ldots$ contain not more than the same n members $x, y \in \mathcal{G}$ and each A_i represent a string of $\mathcal{G}$. Then the multiplication of $A_i, A_j, \ldots$ gives a logic circuit. We can understand strings of $\mathcal{G}$ as matrices such as ones given in Tables 3.2, 3.3, 3.4 and 3.5. In this case matrix multiplication gives a transformation from an input to an output.

Now, let us answer the question, whether $\mathcal{G}$ is a group indeed. To be a group, the following assumption has to hold on $\mathcal{G}$: the plasmodium chooses a direction of motions (up, down or up and down simultaneously) just at its start point. It cannot change its strategy at the points x, y, x^{-1}, y^{-1}. Nevertheless, this assumption is very strong and to realize it we must appeal to repellents, which block any come-back of plasmodium.

Thus, the plasmodium can permanently change its decisions at the points x, y, x^{-1}, y^{-1} of Fig. 3.17. In this case, x can become opposite to y^{-1} (i.e. y^{-1} to x as well) and y can become opposite to x^{-1} (x^{-1} to y), although first x was opposite to x^{-1} and y to y^{-1}. Let us assume that the plasmodium changes its choices all the time at all points. Then we obtain a strong extension of the group. Assume, $\mathcal{G}$ has $p - 1$ atoms a_0, a_1, ..., a_{p-2}. Then we can obtain all the p-adic streams over $\mathcal{G}$: $\mathcal{G}^\omega = \{\sigma : \mathbf{N} \rightarrow \mathcal{G}\}$. Each stream contains the history of decision changes. For instance, (i) let the stream $x = x \cdot y^{-1}$ mean that the plasmodium comes back from x to y^{-1}; (ii) let $y^{-1} = y^{-1} \cdot x$ mean that the plasmodium comes back from y^{-1} to x; (iii) $y = y \cdot x^{-1}$ that the plasmodium comes back from y to x^{-1}; (iv) $x^{-1} = x^{-1} \cdot y$ that the plasmodium comes back from x^{-1} to y, etc. In other words, we have several infinite streams as additional motions in $\mathcal{G} \cup \mathcal{G}^\omega$ for Fig. 3.17 like that:

$$x \cdot y^{-1} \cdot x \cdot y^{-1} \cdot x \cdot y^{-1} \cdot \ldots ;$$

$$y^{-1} \cdot x \cdot y^{-1} \cdot x \cdot y^{-1} \cdot x \cdot \ldots ;$$

$$y \cdot x^{-1} \cdot y \cdot x^{-1} \cdot y \cdot x^{-1} \cdot \ldots ;$$

$$x^{-1} \cdot y \cdot x^{-1} \cdot y \cdot x^{-1} \cdot y \cdot \ldots .$$

Furthermore, the plasmodium can move back at points x and y^{-1} (respectively, at points y and x^{-1}) simultaneously. This motions will be denoted by $()x, y^{-1}()$ (respectively, by $()y, x^{-1}()$), where $x = x \cdot y^{-1}$ and $y^{-1} = y^{-1} \cdot x$ (respectively, $y = y \cdot x^{-1}$ and $x^{-1} = x^{-1} \cdot y$). The objects $()x, y^{-1}()$, $()y, x^{-1}()$ will be called *wave sets*. They do not satisfy the set-theoretic foundation axiom [92]. These non-well-founded objects $()x, y^{-1}()$ and $()y, x^{-1}()$ are two additional members of $\mathcal{G} \cup \mathcal{G}^\omega$.

The informal meaning of wave sets is that any wave set contains just streams and we cannot say in accordance with which stream the wave set will behave. We face an

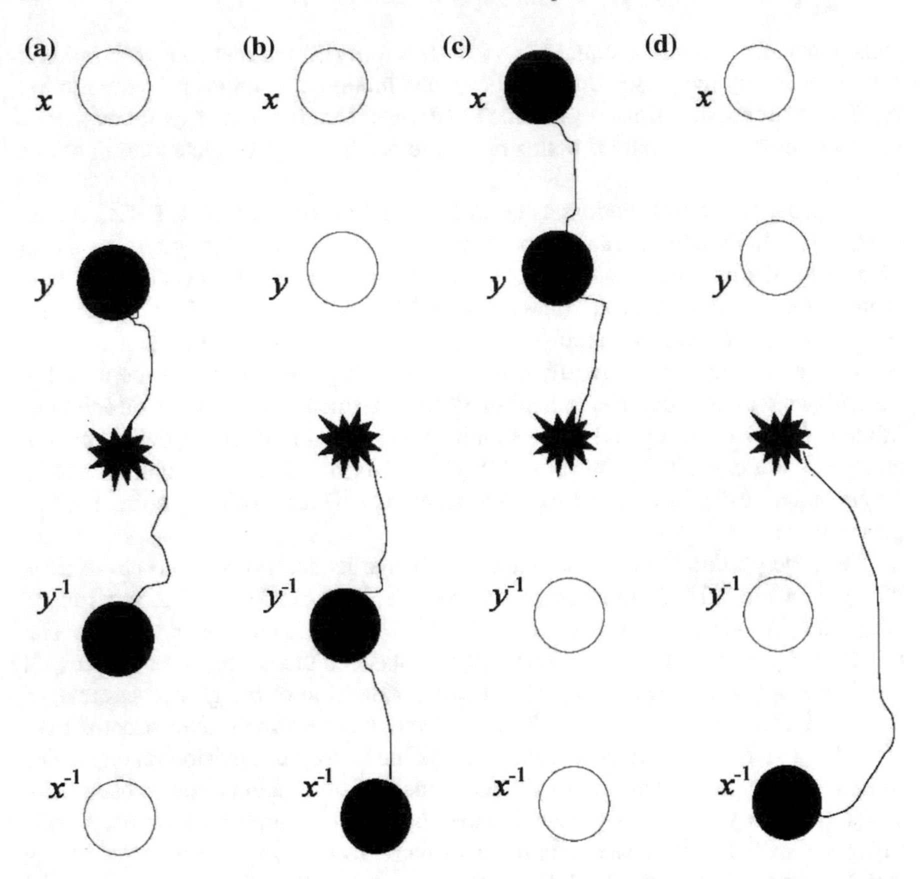

Fig. 3.17 The linear group $\{x, y, x^{-1}, y^{-1}\}$ on *Physarum polycephalum* at $t = 1$, black circles are attractants occupied by the plasmodium, white circles are attractant which are not occupied: **a** the plasmodium motion is expressed by the string $y \cdot y^{-1}$, which gives e; **b** the plasmodium motion is expressed by the string $x^{-1} \cdot y^{-1}$; **c** the plasmodium motion is expressed by the string $x \cdot y$; **d** the plasmodium motion is expressed by the string x^{-1}, see [6]

uncertainty of numbers of inputs and outputs. The gate can behave like any stream it contains but there is an uncertainty how exactly.

The gate as wave set includes as many versions of behaviour of the whole system as possible. In other words, the wave set is the greatest non-well-founded set containing alternative solutions. In different contexts, the wave set behaves differently, that is we apply a different solution contained in this wave set. This application is a clue for an appropriate context.

Thus, the composition of NOT gates pictured in Fig. 3.15 gives a permutation group $\mathscr{G}$. However, the composition of logic devices of Fig. 3.17 consisting just of attractants standing in line is not a conventional permutation group. This new group $\mathscr{G} \cup \mathscr{G}^{\omega}$ will be called *non-linear permutation group*. This group contains infinite streams and wave sets of radius 2, i.e. wave sets consisting only of two streams.

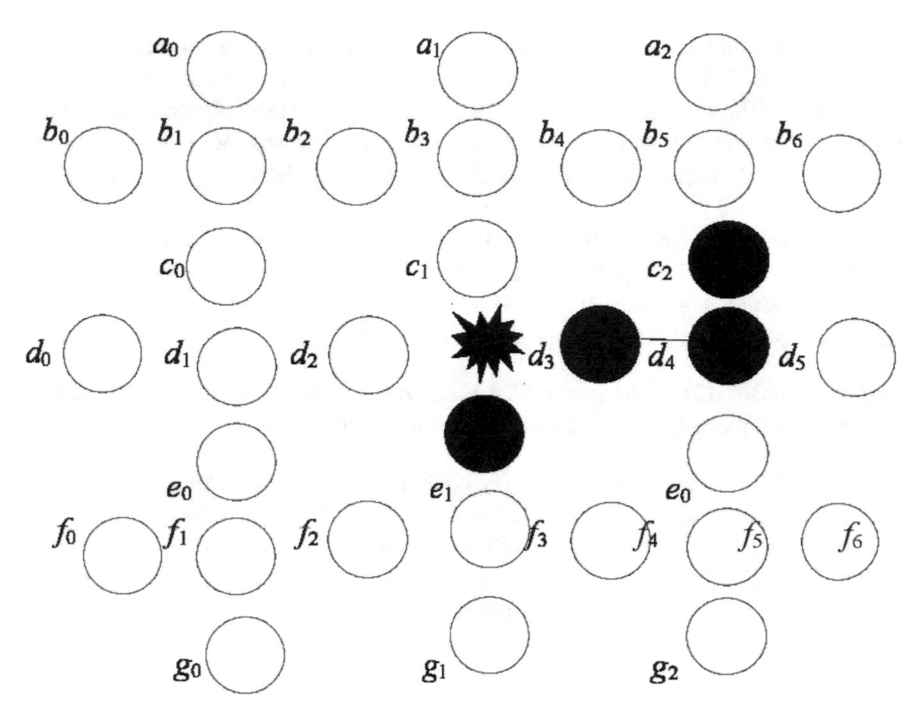

Fig. 3.18 The non-linear permutation group $\mathscr{G} \cup \mathscr{G}^{\omega}$ on *Physarum polycephalum* at $t = 1$, where black circles are attractants occupied by the plasmodium and white circles are attractant which are not occupied. At each point $b_1, b_3, b_5, d_1, d_4, f_1, f_3, f_5$ the plasmodium can make from one to four decisions, see [6]

In the same way, we can show that the composition of CNOT gates pictured in Fig. 3.16 gives a permutation group. Nevertheless, if we delete all repellents in the CNOT gate, we obtain a new logic device pictured in Fig. 3.18 that is not a conventional permutation group, too. This device means that the plasmodium can change its decisions at the points b_1, b_3, b_5, d_1, d_4, f_1, f_3, f_5. At each of this point, we can obtain an appropriate wave set of radius n, where $2 \leq n \leq 4$, i.e. this wave set can include from 2 to 4 infinite streams. For example, if the plasmodium changed its choices at b_1 and decided to come back in all four possible directions, then we have the wave set $()a_0, b_0, b_2, c_0()$. If the plasmodium changed its choices at d_1 and decided to come back only in the two possible directions towards d_0 and c_0, then we have the wave set $()d_0, c_0()$, etc.

The operations over $\mathscr{G} \cup \mathscr{G}^{\omega}$ are defined in Sect. 8.2.3. They can be used for designing logic circuits without repellents or barriers if the group multiplication is understood as a composition of circuits.

A non-linear (non-well-founded) composition in non-linear permutation group sets up not only sequences of symbols, defined by induction, but also their infinite streams and wave sets. We can state that for any simple motions x, $y \in \mathscr{G}$ there exists the absolute motion (*hyperunit*) $\overline{e}$ such that for any simple motions $x, y \in \mathcal{G}$, $x \cdot \overline{e} = y \cdot \overline{e} = \overline{e} \cdot x = \overline{e} \cdot y = \overline{e}$ and $e \cdot \overline{e} = \overline{e} \cdot e = \overline{e}$ (see Sect. 8.2.3).

As we see, the absolute motion gets all motions equal to itself. Hence, the absolute motion for any members of $\mathcal{G}$ determines the limits for any motions, i.e. it is the greatest possible motion in $\mathcal{G}$. For the absolute member there is no inverse motion, i.e. $\overline{e}^{-1} \cdot \overline{e} = \overline{e} \cdot \overline{e}^{-1} = \overline{e}$.

In multiplicative groups the absolute member plays the role of zero, but we can divide by zero as $\overline{e} \cdot \overline{e}^{-1} = \overline{e}$. In additive groups the absolute member plays the role of infinity, since $x + \overline{e} = y + \overline{e} = \overline{e} + x = \overline{e} + y = \overline{e}$. About formal properties of non-linear groups please see Proposition 8.1. One more property:

Proposition 3.2 *Each group $\mathcal{G}$ where basic operations ($\cdot$ and $^{-1}$) are extended to operations on wave sets is closed over the union of wave sets.*

Proof Let $\mathcal{G}$ contains $\{a, b, c, \dots\}$. By assumption, any non-empty subset $m \subseteq \mathcal{G}$ may be presented as wave set $f(m)$ such that $f(m)$ is the greatest set containing all members of m in the form of streams. Then the function f forms a natural order: $f(m_1) \subseteq f(m_2) \subseteq f(m_3) \subseteq \dots \subseteq f(\mathcal{G})$, where $f(\mathcal{G})$ is a maximal wave set that includes all members of $\mathcal{G}$. By properties of wave sets, we find $f(m_1), f(m_2), f(m_3), \dots, f(\mathcal{G}) \in \mathcal{G}$ and they are closed over union. □

Thus, in this chapter, I have shown (i) how to program the swarm behaviour on the example of *Physarum polycephalum* by using the process calculus (on the basis of this calculus we have designed an object-oriented language for simulating the slime mould) and (ii) how to implement sequential logic gates and reversible logic gates the swarm behaviour on the example of *Physarum polycephalum* behaviours. In the meanwhile, we can erase information in sequential logic because of uncertainty in the direction of plasmodium propagation. We can consider the standard motions of plasmodium in Figs. 3.17 and 3.18 as unconventional logic gates designed within an appropriate non-linear permutation group. In this case, we can avoid repellents as such. The logic gates of Figs. 3.9, 3.10, 3.11, 3.12, 3.13, 3.14, 3.15 and 3.16 need repellents to implement the standard permutation groups. Hence, the plasmodium without repellents demonstrates the more sophisticated behaviour, which can be described mathematically in strong extensions of standard groups of motions. It means that without repellents behaviourism as the idea of complete control of behaviours cannot be valid.

References

1. Schumann, A.: Kabbalistic-leibnizian automata for simulating the universe. In: Katz, E. (ed.) Molecular and Supramolecular Information Processing: From Molecular Switches to Logic Systems, pp. 259–280. Wiley-VCH, Weinheim (2012)
2. Schumann, A.: Proof-theoretic cellular automata as logic of unconventional computing. Int. J. Unconven. Comput. **8**(3), 263–280 (2012)

3. Schumann, A.: Unconventional logic for massively parallel reasoning. In: The 6th International Conference on Human System Interaction (HSI), June 2013, pp. 298–305. IEEE Xplore (2013)
4. Schumann, A.: Non-linear permutation groups on Physarum polycephalum. In: The 2014 2nd International Conference on Systems and Informatics (ICSAI 2014), pp. 246–251. Shanghai (2014)
5. Schumann, A.: Reversible logic gates on Physarum polycephalum. AIP Conf. Proc. **1648**(1), 580011 (2015). https://doi.org/10.1063/1.4912819
6. Schumann, A.: Conventional and unconventional reversible logic gates on Physarum polycephalum. Int. J. Parallel Emerg. Distrib. Syst. **32**(2), 218–231 (2017). https://doi.org/10.1080/17445760.2015.1068775
7. Schumann, A., Adamatzky, A.: Towards semantical model of reaction-diffusion computing. Kybernetes **38**(9), 1518–1531 (2009)
8. Schumann, A., Adamatzky, A.: Logical modelling of Physarum polycephalum. Analele Universitatii de Vest, Timisoara, Seria Matematica, Informatica XLVIII **3**, 175–190 (2010)
9. Schumann, A., Adamatzky, A.: Physarum spatial logic. New Math. Nat. Comput. **7**(3), 483–498 (2011)
10. Schumann, A., Akimova, L.: Simulating of schistosomatidae (trematoda: Digenea) behavior by Physarum spatial logic. In: Annals of Computer Science and Information Systems, Proceedings of the 2013 Federated Conference on Computer Science and Information Systems, vol 1. pp. 225–230. IEEE Xplore (2013)
11. Schumann, A., Akimova, L.: Process calculus and illocutionary logic for analyzing the behavior of schistosomatidae (trematoda: Digenea). In: Pancerz, K., Zaitseva, E. (eds.) Computational Intelligence, Medicine and Biology: Selected Links, pp. 81–101. Springer International Publishing, Cham (2015)
12. Schumann, A. (ed.): Judaic Logic. Gorgias Press (2010)
13. Schumann, A. (ed.): Modern Review of Judaic Logic. Special Issue of History and Philosophy of Logic (2011)
14. Leibniz, G.: Zur allgemeinen charakteristik. In: Hauptschriften zur Grundlegung der Philosophie. Philosophische Werke, vol. 1. Felix Meiner, Hamburg (1966)
15. Adamatzky, A.: Reaction-diffusion algorithm for constructing discrete generalized voronoi diagram. Neural Netw. World **6**, 635–643 (1994)
16. Adamatzky, A., De Lacy Costello, B.: Experimental logical gates in a reaction-diffusion medium: The xor gate and beyond. Phys. Rev. E **66**, 46–112 (2002)
17. Adamatzky, A., De Lacy Costello, B., Asai, T.: Reaction-Diffusion Computers. Elsevier (2005)
18. Adamatzky, A., Wuensche, A.: Computing in spiral rule reaction-diffusion hexagonal cellular automaton. Complex Syst. **16**, 4 (2007)
19. Adamatzky, A., Wuensche, A., Costello, B.D.L.: Glider-based computation in reaction-diffusion hexagonal cellular automata. Chaos, Solitons Fractals **27**, 287–295 (2006)
20. Margolus, N.: Physics-like models of computation. Physica D **10**, 81–95 (1984)
21. Turing, A.M.: The chemical basis of morphogenesis. Philos. Trans. Royal Soc. Lond. Series B Biol. Sci. **237**(641), 37–72 (1952)
22. Adamatzky, A., de Lacy Costello, B., Dittrich, P., Gorecki, J., Zauner, K.: On logical universality of Belousov-Zhabotinsky vesicles. Int. J. Gen. Syst. **43**(7), 757–769 (2014). https://doi.org/10.1080/03081079.2014.921000
23. Belousov, L.V.: Synergetics and Biological Morphogenesis, pp. 204–208. Springer, Berlin, Heidelberg (1984)
24. Boerlijst, M.C., Lamers, M.E., Hogeweg, P.: Evolutionary consequences of spiral waves in a host–parasitoid system. Proc. Royal Soc. Lond. B: Biol. Sci. **253**(1336), 15–18 (1993)
25. Adamatzky, A.: Computing in Nonlinear Media and Automata Collectives. Institute of Physics Publishing (2001)
26. Feynman, R.P.: Simulating physics with computers. Int. J. Theor. Phys. **21**(6–7), 467–488 (1982)
27. Hennessy, M., Milner, R.: Algebraic laws for nondeterminism and concurrency. JACM **32**(1), 137–161 (1985)

28. Berry, G., Boudol, G.: The chemical abstract machine. Theor. Comput. Sci. **96**, 217–248 (1992)
29. Boccara, N.: Modeling Complex Systems. Springer (2003)
30. Chopard, B., Droz, M.: Cellular Automata Modeling of Physical Systems. Cambridge University Press (2005)
31. Ilachinski, A.: Cellular Automata: A Discrete Universe. World Scientific (2001)
32. Wolfram, S.: Universality and complexity in cellular automata. Physica D **10**, 1–35 (1984)
33. de Lacy Costello., B.P.J., Adamatzky, A., Ratcliffe, N.M., Zanin, A., Purwins, H.G., Liehr, A.: The formation of voronoi diagrams in chemical and physical systems: experimental findings and theoretical models. Int. J. bifurc. chaos **14**, 2187–2210 (2004)
34. Gordon, A.D.: Bisimilarity as a theory of functional programming. Theor. Comput. Sci. **228**, 5–47 (1999)
35. Fiore, M.P.: A coinduction principle for recursive data types based on bisimulation. In: Proceedings 8th Conferences Logic in Computer Science (LICS'93), pp. 110–119 (1993)
36. Howe, D.J.: Proving congruence of bisimulation in functional programming languages. Inf. Comput. **124** (1996)
37. Pavlović, D., Escardó, M.H.: Calculus in coinductive form. In: Proceedings of the 13th Annual IEEE Symposium on Logic in Computer Science, pp. 408–417 (1998)
38. Jacobs, B., Rutten, J.: A tutorial on (co)algebras and (co)induction. EATCS Bull. **62**, 222–259 (1997)
39. Rutten, J.J.M.M.: Processes as terms: non-well-founded models for bisimulation. Math. Struct. Comput. Sci. **2**(3), 257–275 (1992)
40. Rutten, J.J.M.M.: A coinductive calculus of streams. Math. Struct. Comput. Sci. **15**(1), 93–147 (2005)
41. Rutten, J.J.M.M.: Behavioral differential equations: a coinductive calculus of streams, automata, and power series. Theor. Comput. Sci **308**, 1–53 (2003)
42. Rutten, J.J.M.M.: Universal coalgebra: a theory of systems. Theor. Comput. Sci. **249**(1), 3–80 (2000)
43. Keller, R.M.: Formal verification of parallel programs. Commun. ACM **19**(7), 371–384 (1976)
44. Nielsen, M., Rozenberg, G., Thiagarajan, P.S.: Elementary transition systems and refinement. Acta Inform. **29**(6), 555–578
45. Nielsen, M., Rozenberg, G., Thiagarajan, P.S.: Elementary transition systems. Theor. Comput. Sci. **96**(1), 3–33 (1992)
46. Henzinger, T.A., Manna, Z., Pnueli, A.: Timed transition systems. In: de Bakker, J., Huizing, C., de Roever, W., Rozenberg, G. (eds.) Real-Time: Theory in Practice. Lecture Notes in Computer Science, vol. 600, pp. 226–251. Springer, Berlin Heidelberg (1992)
47. Schumann, A., Pancerz, K.: Timed transition system models for programming Physarum machines: extended abstract. In: L. Popova-Zeugmann (ed.) Proceedings of the Workshop on Concurrency, Specification and Programming (CS&P'2014), pp. 180–183. Chemnitz, Germany (2014)
48. de Vink, E.P., Rutten, J.J.M.M.: Bisimulation for probabilistic transition systems: a coalgebraic approach. Theor. Comp. Sci. **221**, 271–293 (1999)
49. Brotherston, J.: Cyclic proofs for first-order logic with inductive definitions. In: 2005 Beckert, B. (ed.) TABLEAUX of LNAI, vol. 3702, pp. 78–92. Springer (2005)
50. Brotherston, J.: Sequent calculus proof systems for inductive definitions. Ph.D. thesis, University of Edinburgh (2006)
51. Santocanale, L.: A calculus of circular proofs and its categorical semantics. In: Proceedings of FoSSaCS 2002, LNCS 2303, pp. 357–371. Springer-Verlag, Grenoble (2002)
52. Dourvas, N.I., Sirakoulis, G.C., Adamatzky, A.: Cellular automaton Belousov-Zhabotinsky model for binary full adder. Int. J. Bifurc. Chaos **27**(6), 1–14 (2017)
53. Adamatzky, A.: Physarum Machines: computers from slime mould. World Scientific, Series on Nonlinear Science. Series A (2010)
54. Schumann, A.: Towards slime mould based computer. New Math. Nat. Comput. **12**(2), 97–111 (2016). https://doi.org/10.1142/S1793005716500083

55. Tsuda, S., Jones, J., Adamatzky, A.: Towards Physarum engines. Appl. Bionics Biomech. **9**, 221–240 (2012)
56. Milner, R.: Communicating and Mobile Systems: the π-calculus. Cambridge University Press, Cambridge (1999)
57. Dam, M.: Proof systems for pi-calculus logics. In: de Queiroz, R. (ed.) Logic for Concurrency and Synchronisation, pp. 145–212. Kluwer (2003)
58. Winskel, G., Nielsen, M.: Models for concurrency. In: Abramsky, S., Gabbay, D.M., Maibaum, T.S.E. (eds.) Handbook of Logic in Computer Science, vol. 4, pp. 1–148. Oxford University Press (1995)
59. Caires, L., Cardelli, L.: A spatial logic for concurrency: Part I. Inf. Comput. **186**(2), 194–235 (2003)
60. Caires, L., Cardelli, L.: A spatial logic for concurrency: Part II. Theor. Comput. Sci. **322**(3), 517–565 (2004)
61. Calcagno, C., Cardelli, L., Gordon, A.: Deciding validity in a spatial logic for trees. J. Funct. Progr. **15**, 543–572 (2005)
62. Cardelli, L., Gardner, P., Ghelli, G.: A spatial logic for querying graphs. In: P. Widmayer (ed.) Proc. ICALP'02, pp. 597–610. Springer (2002)
63. Dimonte, A., Berzina, T., Erokhin, V.: Basic transitions of Physarum polycephalum. In: 2015 Federated Conference on Computer Science and Information Systems, FedCSIS 2015, Lódz, Poland, Sept 13-16, pp. 599–606. https://doi.org/10.15439/2015F237
64. Pancerz, K., Schumann, A.: Principles of an object-oriented programming language for Physarum polycephalum computing. In: Proceedings of the 10th International Conference on Digital Technologies (DT'2014), pp. 273–280. Zilina, Slovak Republic (2014)
65. Pancerz, K., Schumann, A.: Some issues on an object-oriented programming language for Physarum machines. In: Bris, R., Majernik, J., Pancerz, K., Zaitseva, E. (eds.) Applications of Computational Intelligence in Biomedical Technology, Studies in Computational Intelligence, vol. 606, pp. 185–199. Springer International Publishing, Switzerland (2016)
66. Schumann, A., Pancerz: Physarumsoft—a software tool for programming Physarum machines and simulating Physarum games. In: Ganzha, M., Maciaszek, L., Paprzycki, M., (eds.) Proceedings of the 2015 Federated Conference on Computer Science and Information Systems (FedCSIS'2015), pp. 607–614. Lodz, Poland (2015)
67. Schumann, A., Pancerz, K.: Towards an object-oriented programming language for Physarum polycephalum computing. In: Szczuka, M., Czaja, L., Kacprzak, M. (eds.) Proceedings of the Workshop on Concurrency, Specification and Programming (CS&P'2013), pp. 389–397. Warsaw, Poland (2013)
68. Schumann, A., Pancerz, K.: Towards an object-oriented programming language for Physarum polycephalum computing: A Petri net model approach. Fundam. Inf. **133**(2–3), 271–285 (2014)
69. Schumann, A., Pancerz, K.: Petri net models of simple rule-based systems for programming Physarum machines: Extended abstract. In: Suraj, Z., Czaja, L. (eds.) Proceedings of the 24th International Workshop on Concurrency, Specification and Programming (CS&P'2015), vol. 2, pp. 155–160. Rzeszow, Poland (2015)
70. Schumann, A., Pancerz, K.: Physarumsoft: an update based on rough set theory. AIP Conf. Proc. **1863**(1), 360,005 (2017). https://doi.org/10.1063/1.4992534
71. Craig, I.: Object-Oriented Programming Languages: Interpretation. Springer, London (2007)
72. Searle, J.R.: Speech acts; an essay in the philosophy of language. Cambridge University Press, Cambridge (1969)
73. Searle, J.R.: Expression and meaning: studies in the theory of speech acts. Cambridge University Press, Cambridge (1979)
74. Searle, J.R., Vanderveken, D.: Foundations of Illocutionary Logic. Cambridge University Press, Cambridge (1984)
75. Whiting, J.G., de Lacy Costello, B.P., Adamatzky, A.: Slime mould logic gates based on frequency changes of electrical potential oscillation. Biosystems **124**, 21–25 (2014)
76. Adamatzky, A., Schubert, T.: Slime mold microfluidic logical gates. Mater. Today **17**(2), 86–91 (2014)

77. Möllenstedt, G., Jönsson, C.: Elektronen-mehrfachinterferenzen an regelmaessig hergestellten feinspalten. Zeitschrift fuer Physik A **155**(4), 472–474 (1959)
78. Khrennikov, A., Schumann, A.: Quantum non-objectivity from performativity of quantum phenomena. Phys. Scr. **T163** (2014)
79. Jones, J.D.: Towards lateral inhibition and collective perception in unorganised non-neural systems. In: Pancerz, K., Zaitseva, E. (eds.) Computational Intelligence. Selected Links. Springer, Medicine and Biology (2015)
80. Barwise, J., Etchemendy, J.: The Liar. Oxford University Press, New York (1987)
81. Barwise, J., Moss, L.: Vicious Circles. Stanford (1996)
82. Barwise, J., Moss, L.: Hypersets. Springer, New York (1992)
83. Jones, J.D., Adamatzky, A.: Towards Physarum binary adders. Biosystems **101**(1), 51–58 (2010)
84. Adamatzky, A.: Slime mould logical gates: exploring ballistic approach. Appl. Tools Tech. Road Exascale Comput. **1**, 41–56 (2010)
85. Mayne, R., Adamatzky, A.: Slime mould foraging behaviour as optically coupled logical operations. Int. J. Gen. Syst. **44**(3), 305–313 (2015). https://doi.org/10.1080/03081079.2014.997528
86. Vasyukevich, V.: Asynchronous Operators of Sequential Logic. Springer, Berlin, Heidelberg (2011)
87. Vasyukevich, V.: Asynchronous sequences decoding. ACCS J. **41**(2), 93–99 (2007)
88. Toffoli, T.: Reversible Computing. Tech. memo MIT/LCS/TM-151, MIT (1980)
89. Bruce, J.W., Thornton, M.A., Shivakumaraiah, L., Kokate, P.S., Li, X.: Efficient adder circuits based on a conservative reversible logic gate. In: Proceeding of the IEEE Computer Society Annual Symposium on VLSI, pp. 83–88. IEEE, Pittsburgh, Pennsylvania (2002)
90. Schumann, A., Pancerz, K., Jones, J.: Towards logic circuits based on Physarum polycephalum machines: The ladder diagram approach. In: Cliquet Jr A., Plantier, G., Schultz, T., Fred, A., Gamboa, H. (eds.) Proceedings of the International Conference on Biomedical Electronics and Devices (BIODEVICES'2014), pp. 165–170. Angers, France (2014)
91. Schumann, A.: Towards context-based concurrent formal theories. Parallel Process Lett **25**, 1540008 (2015)
92. Aczel, A.: Non-Well-Founded Sets. Stanford University Press (1988)

Chapter 4
Conventional and Unconventional Automata on Swarm Behaviours

In this chapter, I try to formulate swarms as labelled transition systems with the same set of labels (events or actions): direction, splitting, fusion, repelling. Then I show that these labelled transition systems can implement Kolmogorov-Uspensky machines, but with a low accuracy because of emergent patterns which occur if we have many states of appropriate transition system. Then I propose a p-adic arithmetic and logic to formalize emergent patterns of swarms. To sum up, I offer a general logical approach to swarm intelligence. See [1–6].

4.1 Formalizations of Swarm Transitions

4.1.1 From Emergent Computing to Swarm Computing

In bio-inspired computations there are many researches focused on swarm intelligence and different formalizations of swarm behaviours [7–9]. An intelligent swarm behaviour is demonstrated not only by animal groups, but also by groups of insects or even bacteria. For example, it was discovered experimentally that swarms of social insects [10] can solve complex computational problems in searching for food and in transporting sources and information due to massive-parallel behaviour with labour divisions.

The majority of researches in swarm intelligence are concentrated on algorithms in simulating and modelling swarms: the *Particle Swarm Optimization* (PSO) [11], the *Bacterial Foraging Optimization Algorithm* (BFOA) [12], the *Artificial Bee Colony* (ABC) [13], the *Cuckoo Optimization Algorithm* (COA) [14], the *Social Spider Optimization* (SSO) [15], the *Ant Colony Optimization* (ACO) [16], etc. These models combine different mathematical tools including probability theory to propose

A. Schumann, *Behaviourism in Studying Swarms: Logical Models of Sensing and Motoring*, Emergence, Complexity and Computation 33,
https://doi.org/10.1007/978-3-319-91542-5_4

artificial swarms. In this chapter, I try to consider swarm intelligence as a kind of computation.

In conventional logic circuits some electrical properties of transistors are used. In particular, the voltage is managed to be in only one of the two states: high (if the voltage runs the range from 2.8 to 5.0 V) or low (if the voltage is in the range from 0 to 0.8 V). The high state of voltage means '1' or logical true. The low state means '0' or logical false. Then the power of the circuit, P, is expressed as follows:

$$P = P_{static} \cdot P_{dynamic},$$

where $P_{static} = V_{CC} \cdot I_{CC}$ and $P_{dynamic} = [(C_{pd} + C_L) \cdot V_{CC}^2 \cdot f] \cdot N_{SW}$, V_{CC} is a supply voltage (V), I_{CC} is a power supply current (A), C_{pd} is a power dissipation capacitance (F), C_L is an external load capacitance (F), f is an operating frequency (Hz), N_{SW} is a total number of outputs switching.

The main idea of electric devices based on electrical properties of transistors is that the Boolean logic can be implemented with a very high accuracy. In this logic any complex logic expression is considered a composition of logical atoms (i.e. of simple logical propositions). In other words, there are no emergent phenomena. Let us recall that the emergence is detected, when new patterns of a highly structured collective behaviour appear and these patterns cannot be reduced to a linear composition of simple subsystems [17].

Notably, emergent phenomena are key phenomena in all self-organizing systems such as collective intelligent behaviours of animal groups: flocks of birds, colonies of ants, schools of fish [18], swarms of bees, etc. Emergence is observed in the economy as well: macroeconomic fluctuations, traffic jams, hierarchy of cities, motion picture industry and mass protest behaviour [17]. There are attempts to formalize the notion of emergence by algorithmic complexity theory. However, the Kolmogorov complexity function is not computable. There is no way to define the emergence by minimum linear compositions. Wolfram proposed a more useful approach in a mathematical definition of *emergency* [19]. He showed that the behaviour of one-dimensional cellular automata is divided into the following four cases:

- there are limit points of the system, i.e. we obtain a homogeneous state of the system;
- there are limit cycles of the system, i.e. we obtain separated periodic structures;
- there are chaotic attractors, i.e. we face chaotic patterns;
- there are complex localized structures.

In the last case we deal with an emergent phenomenon in the true sense. On the basis of the Wolfram's approach the so-called *emergent computing* has developed. In this kind of computing (i) the computation process is distributed over a set of parallel and autonomous processing units; (ii) each unit computes locally and can interact directly only with a small number of other units; (iii) there is a consistency among the processing units; (iv) there are more outputs, than inputs.

One of the most studied instances of emergent computing is represented by *swarm computing* [7, 11]. This computing is based on labor division in animal groups. Any swarm as a result of collective behaviour, such as birds flocking or fish schooling, is a self-organizing system, where, on the one hand, each unit responds to local stimuli individually and, on the other hand, all together they accomplish a global task (transporting, eating, self-protecting, etc.).

There were proposed many swarm algorithms to simulate the behaviour of insect or animal groups [11–16]. In the *Particle Swarm Optimization* (PSO) [11, 20] it is assumed that the particles (agents) know (i) their best position 'local best' (lb) and (ii) their neighbourhood's best position 'global best' (gb). The next position is determined by velocity. Let $x_i(t)$ denote the position of particle i in the search space at time step t, where t is discrete. Then the position x_i is changed by adding a velocity to the current position:

$$x_i(t+1) = x_i(t) + v_i(t+1),$$

where $v_i(t+1) = v_i(t) + c_1 r_1(lb(t) - x_i(t)) + c_2 r_2(gb(t) - x_i(t))$ and i is the particle index, c_1, c_2 are acceleration coefficients, such that $0 \leq c_1, c_2 \leq 2$, r_1, r_2 are random values (such that $0 \leq r_1, r_2 \leq 1$) regenerated every velocity update.

One of the possible PSO algorithms can be exemplified by the *bird flocking* [21, 22]. In flocks 'local best' and 'global best' of birds are defined by the following three rules: (i) collision avoidance (birds fly away before they crash into one another); (ii) velocity matching (birds fly about the same speed as their neighbours in the flock); and (iii) flock centering (birds fly toward the center of the flock as they perceive it). So, the position of a bird i at time t is given by its placement x_i at time $t-1$ shifted by its current velocity v_i. This v_i is determined by the rules (i)–(iii).

All the algorithms PSO, BFOA, ABC, COA, SSO, ACO are used to simulate swarms of different insects or animals. Let us try to answer the question, whether swarms can be considered an unconventional computer, i.e. whether swarms can calculate.

Each swarm is a natural transition system

$$TS = \langle S, E, T, I \rangle,$$

where:

- S is the non-empty set of states;
- E is the set of events;
- $T \subseteq S \times E \times S$ is the transition relation;
- $I \subseteq S$ is the set of initial states.

Any transition system is a labeled graph with nodes corresponding to states from S, edges representing the transition relation T, and labels of edges corresponding to events from E.

Let us consider some examples.

4.1.2 Ant Colony Transitions

Any *ant colony* is a swarm of ants localized first at the nest. This nest can be regarded as an initial state of ant colony transitions. Then ants use a special mechanism called *stigmergy* to build up all transitions. Stigmergy (stigma + ergon) means ‘stimulation by work’. This mechanism has the following steps [16]:

- At first ants are looking for food randomly, laying down pheromone trails.
- If ants find food, they return to the nest, leaving behind pheromone trails. So there is more pheromone on the shorter path than on the longer one.
- Ants prefer to go in the direction of the strongest pheromone smell. As a consequence, the concentration of pheromone is so strong on the shorter path, that all the ants prefer this path (it is experimentally proven in [16]).

Thus, stigmergy allows ants to transport food to their nest in a remarkably effective way. Food localizations are considered new states and ant roads to food places are regarded as transitions. Let $P_{ant} = \{n\}$ be an initial state of ant transitions (i.e. the nest), $A_{ant} = \{a_1, a_2, \ldots, a_j\}$ be a set of food pieces (attractants) localized at different places, $V_{ant} = \{r_1, r_2, \ldots, r_i\}$ be a set of ant roads. So the *ant colony transition system*, $TS_{ant} = \langle S_{ant}, E_{ant}, T_{ant}, I_{ant}\rangle$, can be defined as follows:

- σ: $P_{ant} \cup A_{ant} \rightarrow S_{ant}$ assigning a state to each original point of the ant colony as well as to each attractant;
- τ: $V_{ant} \rightarrow T_{ant}$ assigning a transition to each ant road;
- ι: $P_{ant} \rightarrow I_{ant}$ assigning an initial state to the nest.

Each event of the set of events E_{ant} is assigned to ant transitions in accordance with the following types of ant expansion:

- *direction* (the ants move from one state/attractant/initial point to another state/attractant),
- *fusion* (the ants move from different states/attractants to the same one state/attractant),
- *splitting* (the ants move from one state/attractant/initial point to different states/attractants),
- *repelling* (the ants stop to move in one direction).

The system TS_{ant} can be used to solve the *Travelling Salesman Problem* [16] formulated as follows: given a list of cities and the distances between each pair of cities, we must to define the shortest possible route that visits each city exactly once and returns to the origin city. This problem is NP-hard. Let P_{ant} be considered the origin city, A_{ant} be a set of all other cities, and V_{ant} be a set of all connections between cities. The Travelling Salesman Problem can be solved, because the ants lay down pheromone trails faster on the shortest path so that the shortest path gets reinforced with more pheromone to attract more future ants. As a result, pheromone trails on the edges between cities depend on the distance: more shorter, more attracting. This allows ants to find shorter tours of cities.

4.1.3 Bee Colony Transitions

A *bee colony* is another example of swarm intelligence [23]. The bee nest is an initial state of bee colony transitions. Any bee colony exploits a mechanism called *waggle dance* to optimize the food transporting to the nest. This mechanism is as follows. In the nest there is an area for communication among bees. At this area the bees knowing, where the food source is precisely, exchange the information about the direction, distance, and amount of nectar on the related food source by a waggle dance. The direction of waggle dancing bees shows the direction of the food source in relation to the Sun, the intensity of the waggles is associated to the distance, and the duration of the dance shows the amount of nectar. Due to this form of communication the bee colony transitions are built up by the following steps [13]:

- There are two kinds of bees: employed and unemployed. Employed bees know exactly, where a particular food source (nectar) is, and visit just this source. Unemployed bees do not know and seek a food source. The unemployed bees are divided into the following two groups: scouts and onlookers. A scout bee carries out search for new food sources without any guidance. An onlooker bee follows the instruction of a waggle dancing bee and visits the food source for the first time. An employed bee visits this source many times. So, the first step in constructing the bee transporting system is in sending scout bees.
- Then onlookers are sent.
- At the next step the food source is exploited by employed bees.
- An employed bee tests if the nectar amount of the new food source is higher than that of the previous one. If it is so, the bee memorizes the new place and forgets the old one. If the nectar amount decreased or exhausted and the employed bee does not know a new place, this bee become an unemployed bee. So, at this step the employed bees exchange the nectar information of the food sources to change their decision.

Assume that $P_{bee} = \{n\}$ is an initial state of bee transitions (i.e. the bee nest), $A_{bee} = \{a_1, a_2, \ldots, a_j\}$ is a set of food sources (attractants) localized at different places, $V_{onlooker} = \{r_1, r_2, \ldots, r_k\}$ is a set of onlooker bee roads, and $V_{employed} = \{r_1, r_2, \ldots, r_l\}$ is a set of employed bee roads. Then the *bee colony transition system*, $TS_{bee} = \langle S_{bee}, E_{bee}, T_{bee}, I_{bee}\rangle$, can be defined thus:

- σ: $P_{bee} \cup A_{bee} \rightarrow S_{bee}$ assigning a state to each original point of the bee colony as well as to each attractant;
- τ: $V_{onlooker} \cup V_{employed} \rightarrow T_{bee}$ assigning a transition to each bee road;
- ι: $P_{bee} \rightarrow I_{bee}$ assigning an initial state to the bee nest.

In the set of events E_{bee} there are the following types of labels for transitions:

- *direction* (the onlooker or employed bees move from one state/attractant/initial point to another state/attractant),
- *fusion* (the onlooker or employed bees move from different states/attractants to the same one state/attractant),

- *splitting* (the onlooker or employed bees move from one state/attractant/initial point to different states/attractants),
- *repelling* (the onlooker or employed bees stop to move in one direction).

The system TS_{bee}, if we use only $V_{employed}$ as a set of roads, can solve the *Travelling Salesman Problem*, also. Thereby, P_{bee} is examined as the origin city, A_{bee} is a set of all other cities, and $V_{employed}$ is a set of all connections between cities. The point is that the greater the number of iterations in sending onlooker or employed bees, the higher the influence of the distance in attracting the bees to appropriate food sources. As a result, the shorter distance seems to be more attracting for employed bees.

There is another NP-hard problem that can be solved by TS_{bee}, the so-called *Generalized Assignment Problem* formulated as follows: there are a number of agents and a number of tasks, each agent has a budget and each task assumes some cost and profit; we must find an assignment in which all agents do not exceed their budget and total profit of the assignment is maximized. In the case of the bee colony, the bees are regarded as agents, the nectar sources as tasks, the amount of nectar as profit, and the distance as cost. Hence, in this interpretation the bee colony can solve the Generalized Assignment Problem.

4.1.4 Escherichia Coli *Transitions*

The complex group behaviour can be observed in the bacterium life also. For instance, *Escherichia coli* bacteria form swarms in semisolid nutrient medium [24, 25]. In these swarms with high bacterial density, large-scale swirling and streaming motions of thousands or millions of cells are observed.

Let us suppose that $P_{one\ E.coli} = \{n_1, n_2, \ldots, n_m\}$ is a set of initial states of *Escherichia coli* transitions, $A_{one\ E.coli} = \{a_1, a_2, \ldots, a_j\}$ is a set of attractants (nutrient gradients), $V_{one\ E.coli} = \{r_1, r_2, \ldots, r_k\}$ is a set of paths for each *Escherichia coli* bacterium. Then the *Escherichia coli transition system*,

$$TS_{one\ E.coli} = \langle S_{one\ E.coli}, E_{one\ E.coli}, T_{one\ E.coli}, I_{one\ E.coli} \rangle,$$

can be defined in the following manner:

- σ: $P_{one\ E.coli} \cup A_{one\ E.coli} \rightarrow S_{one\ E.coli}$ assigning a state to each original point of the *Escherichia coli* population as well as to each attractant;
- τ: $V_{one\ E.coli} \rightarrow T_{one\ E.coli}$ assigning a transition to each path of each *Escherichia coli* bacterium;
- ι: $P_{one\ E.coli} \rightarrow I_{one\ E.coli}$ assigning initial states to initial positions of *Escherichia coli* bacteria.

There are the four types of labels for transitions in the set of events $E_{one\ E.coli}$:

- *Direction*: the *Escherichia coli* bacterium moves from one state/attractant/initial point to another state/attractant presented by a nutrient gradient. This locomotion is carried out by a set of tensile flagella which move in the counterclockwise direction helping the bacterium to swim very fast.

- *Tumbling*: the *Escherichia coli* bacterium can tumble to change its swim direction in the future. It is a Brownian motion of the bacterium. The tumbling is achieved by the flagella moving in the clockwise direction. In this direction each flagellum operates relatively independently of the others.
- *Repelling*: the *Escherichia coli* bacterium stops to move in one direction, because it avoids noxious environment.
- *Repropuction*: the health bacterium splits into two bacteria while the least healthy bacteria die.

In semisolid nutrient medium *Escherichia coli* bacteria make swarms: they become multinucleate and move across the surface in coordinated packs. However, swarms of *Escherichia coli* are not attracted by chemotaxis [24]. Their dynamics can be explained only mechanistically by collisions with neighbours. This means that *Escherichia coli* swarms can be simulated mathematically (by hydrodynamic interactions) on the basis of $TS_{one\ E.coli}$, but these swarms are not intelligent.

So we can add a set $E_{E.coli\ swarm}$ of swarming transitions defined mathematically to the transition system $TS_{one\ E.coli}$. There are the following four kinds of maneuvers during swarming [24, 25]:

- *Forward motion*: When the majority of *Escherichia coli* bacteria swim in the same way, their swarm moves forward.
- *Lateral motion*: Because of collisions with neighbouring cells or motor reversals there is a lateral motion in a swarm with propulsion angles of $>35°$.
- *Reversals*: Reversals occur every 1.5 s and require about 0.1 s for completion, but they do not have a large impact on the average cell behaviour.
- *Stalls*: When the bacteria pause, their flagella continue to spin and pump fluid over the agar in front of the swarm. As a result, stalls occur at the swarm edge.

Hence, *Escherichia coli* swarms do not calculate.

4.1.5 Paenibacillus Vortex *Transitions*

Notice that there are bacteria with intelligent and successful swarming strategies such as *Paenibacillus vortex* [26]. These strategies help *Paenibacillus vortex* bacteria to carry out a cooperative colonization of new territories. When *Paenibacillus vortex* is inoculated on hard agar surfaces with peptone, it develops complex colonies of vortices. When it is inoculated on soft agar surfaces, it organizes a special network of swarms with intricate internal traffic. In contrast to *Escherichia coli*, the *Paenibacillus vortex* swarms are sensitive to chemotaxis (attractants). As a consequence, the *Paenibacillus vortex* network is built up on the basis of interactions between swarms allowing them to transport nutrients, spores and other organisms [27, 28].

Let $P_{one\ P.vortex} = \{n_1, n_2, \ldots, n_m\}$ be a set of initial states of *Paenibacillus vortex* transitions, $A_{one\ P.vortex} = \{a_1, a_2, \ldots, a_j\}$ be a set of attractants (nutrient gradients), $V_{one\ P.vortex} = \{r_1, r_2, \ldots, r_k\}$ be a set of paths for each *Paenibacillus vortex* bacterium. Then the *Paenibacillus vortex transition system*, $TS_{one\ P.vortex} = \langle S_{one\ P.vortex}, E_{one\ P.vortex}, T_{one\ P.vortex}, I_{one\ P.vortex} \rangle$, is as follows:

- σ: $P_{one\ P.vortex} \cup A_{one\ P.vortex} \rightarrow S_{one\ P.vortex}$ assigning a state to each original point of the *Paenibacillus vortex* population as well as to each attractant;
- τ: $V_{one\ P.vortex} \rightarrow T_{one\ P.vortex}$ assigning a transition to each path of each *Paenibacillus vortex* bacterium;
- ι: $P_{one\ P.vortex} \rightarrow I_{one\ P.vortex}$ assigning initial states to initial positions of *Paenibacillus vortex* bacteria.

The four types of labels for transitions in the set of events $E_{one\ P.vortex}$:

- *direction*: the *Paenibacillus vortex* bacterium moves from one state/attractant/initial point to another state/attractant presented by a nutrient gradient;
- *tumbling*: the *Paenibacillus vortex* bacterium can tumble for a while;
- *repelling*: the *Paenibacillus vortex* bacterium stops to move in one direction, as it avoids some chemical concentrations;
- *repropuction*: the health bacterium splits into two bacteria.

Paenibacillus vortex bacteria can be organized in swarms. The locomotion in a swarm can be explained hydrodynamically by collisions among the bacteria and by the boundary of the layer of lubricant collectively generated by them. However, interactions among swarms can be considered intelligent. Each swarm has a snake-like formation. It looks for food and can cross each other's trail. When food is detected, swarms change their direction. The *Paenibacillus vortex* swarms can split and fuse in accordance with topology of nutrients.

So, for these swarms we can propose another transition system

$$TS_{P.vortex\ swarm} = \langle S_{P.vortex\ swarm}, E_{P.vortex\ swarm}, T_{P.vortex\ swarm}, I_{P.vortex\ swarm} \rangle,$$

where

- σ: $P_{P.vortex\ swarm} \cup A_{P.vortex\ swarm} \rightarrow S_{P.vortex\ swarm}$, where $P_{P.vortex\ swarm} = \{n_1, n_2, \ldots, n_l\}$ is a set of initial states of *Paenibacillus vortex* swarm transitions and $A_{P.vortex\ swarm} = A_{one\ P.vortex}$ is a set of attractants (nutrient gradients). The function σ assigns a state to each original point of the *Paenibacillus vortex* swarms as well as to each attractant;
- τ: $V_{P.vortex\ swarm} \rightarrow T_{P.vortex\ swarm}$, where $V_{P.vortex\ swarm} = \{r_1, r_2, \ldots, r_k\}$ is a set of paths for each *Paenibacillus vortex* swarm. The function τ is to assign a transition to each path of each *Paenibacillus vortex* swarm;
- ι: $P_{P.vortex\ swarm} \rightarrow I_{P.vortex\ swarm}$ is to assign initial states to initial positions of *Paenibacillus vortex* swarms.

Each event of the set of events $E_{P.vortex\ swarm}$ is assigned to swarm motions according to the following types of maneuvers:

- *direction*: the *Paenibacillus vortex* swarm moves from one state/attractant/initial point to another state/attractant,
- *fusion*: the *Paenibacillus vortex* swarms move from different states/attractants/ initial points to the same one state/attractant,
- *splitting*: the *Paenibacillus vortex* swarm moves from one state/attractant/initial point to different states/attractants (for the experimental details see [29]),
- *repelling*: the *Paenibacillus vortex* swarm stops to move in one direction if it faces a repellent.

As we see, the *Paenibacillus vortex* transition system $TS_{P.vortex\ swarm}$ can solve the Travelling Salesman Problem, too.

4.1.6 Towards Slime Mould Machines

Taking into account the fact that any plasmodium of *Physarum polycephalum* can be considered a typical intelligent swarm, we can use automata defined on plasmodia for demonstrating computational powers of different swarms: ant colonies, bee colonies, and *Paenibacillus vortex* swarms. Indeed, let

$$TS_{P.polycephalum} = \langle S_{P.polycephalum}, E_{P.polycephalum}, T_{P.polycephalum}, I_{P.polycephalum} \rangle$$

be a transition system for plasmodia and this system is defined standardly. Let f be a mapping from $S_{P.polycephalum}$ to $S_\star$ and from $I_{P.polycephalum}$ to $I_\star$, where $\star \in \{ant, bee, P.vortex\ swarm\}$. Assume that all transitions denoted by $\longrightarrow$ are the same for all $TS_{P.polycephalum}$, TS_{ant}, TS_{bee}, $TS_{P.vortex\ swarm}$. For example, we have the same direction, fusion, splitting, and repelling. The function f is a homomorphism if and only if

- for all $s \in (S_{P.polycephalum} \cup I_{P.polycephalum})$, if $s \longrightarrow s'$, for some $s' \in S_{P.polycephalum}$, then $f(s) \longrightarrow f(s')$;
- for all $s \in S_{P.polycephalum}$, if $f(s) \longrightarrow t$, for some $t \in S_\star$, then there exists $s' \in S_{P.polycephalum}$ with $s \longrightarrow s'$ and $f(s') = t$.

If f: $S_{P.polycephalum} \cup I_{P.polycephalum} \rightarrow S_\star \cup I_\star$ is a homomorphism as well as $f^{-1} : S_\star \cup I_\star \rightarrow S_{P.polycephalum} \cup I_{P.polycephalum}$ is a homomorphism, then f is an isomorphism. So, for any $TS_\star$, where $\star \in \{ant, bee, P.vortex\ swarm\}$, there is an isomorphism from $S_{P.polycephalum}$ to $S_\star$ and from $I_{P.polycephalum}$ to $I_\star$. This means that it is enough to study the *Physarum* machine $TS_{P.polycephalum}$ to know computational properties of different swarms: ant colonies, bee colonies, *Paenibacillus vortex* swarms, etc.

According to our previous study of $TS_{P.polycephalum}$ we know that it is impossible to define *Physarum* transitions as atomic acts [30]. For instance, under the same conditions, the plasmodium can follow splitting or direction, fusion or direction, etc. Nevertheless, with a low accuracy we can implement some conventional algorithms in $TS_{P.polycephalum}$.

Notice that $TS_{P.polycephalum}$ cannot be defined as an inductive set because of the absence of atomic acts. Let us define (i) the universe U of all transitions defined in $TS_{P.polycephalum}$ and (ii) functions, F, which take one or more elements from U as arguments and return an element of U. The universe is defined inductively (hence, it is an inductive set) if there is a base set $B \subseteq U$ such that unboundedly repeated compositions of F on B cover the whole U. So, we assume that B contains atomic acts (transitions). All functions $f \in F$ are defined *inductively* if (i) each f is defined on B; (ii) for each f we have $f(a_1, a_2, \dots, a_n) = f(b_1, b_2, \dots, b_n)$ for $a_1, a_2, \dots, a_n \in U$ and $b_1, b_2, \dots, b_n \in U$ iff $a_i = b_i$ for $i = 1, \dots, n$; (iii) each f has a range which is disjoint from the ranges of all other functions in F and from B. A function h is called *recursive* if (i) $h(a)$ is defined for all $a \in B$; (ii) for each $f \in F$ the value of $h(f(a_1, a_2, \dots, a_n))$ is defined in terms of $h(a_1)$, ..., $h(a_n)$.

In the very next section we will show how we can consider fragments of U such that there are some recursive functions in them.

4.2 Conventional Approach to Swarm Machines

It is known that, theoretically, Turing machines, Kolmogorov-Uspensky machines [31, 32], Schönhage's storage modification machines [33, 34], and random-access machines [35] have the same expressive power. In other words, the class of functions computable by these machines is the same. For the first time A. Adamatzky [36] experimentally showed that the *Physarum* machine $TS_{P.polycephalum}$ can be represented as a kind of Kolmogorov-Uspensky machines. Hence, we can implement conventional algorithms in $TS_{P.polycephalum}$.

4.2.1 Physarum *Kolmogorov-Uspensky Machines*

Let us show that $TS_{P.polycephalum}$ can be considered a *Kolmogorov-Uspensky machine*. Let $\Gamma = S_{P.polycephalum} \cup I_{P.polycephalum}$ be an alphabet, $k = |E_{P.polycephalum}|$ a natural number. We say that a tree is (Γ, k)-tree, if one of nodes is designated and it is called *root* and all edges are directed. Each node is labelled by one of signs of Γ and each edge from the same node is labelled by different numbers $\{1, \dots, k\}$ (so, each node has not more than k edges). We see that by this definition of (Γ, k)-tree, the plasmodium grows from the one active zone (so, we simulate the expansion from the one nest of ants or bees, or from the one inoculation of *Paenibacillus vortex* swarms), where all attractants are labelled by signs of Γ and protoplasmic tubes (roads of ants or roads of bees) are labelled by numbers of $\{1, \dots, k\}$. Thus, $TS_{P.polycephalum}$ (as well as TS_{ant}, TS_{bee}, or $TS_{P.vortex\ swarm}$) can be represented as a (Γ, k)-tree.

(Γ, k)-*Physarum complex* is any initial finite digraph which is connected (i.e. each vertex is accessible from the initial one by a directed path), each node is labelled by one of signs of Γ, and each edge from the same node is labelled by different numbers $\{1, \ldots, k\}$. The set of all vertices of (Γ, k)-*Physarum* complex U is denoted by $v(U)$.

The *r-neighbourhood* of (Γ, k)-complex is represented by a (Γ, k)-complex which consists of edges and vertices of initial complex that are accessible from initial vertex by a directed path that is not longer than r. Notice that r can be arbitrary. Any property of (Γ, k)-complex which is dependent just of r-neighbourhood is called *r-local property of* (Γ, k)*-complex*. Hence, we can ever project *Physarum* transitions (using attractants and repellents) for inducing different numbers r and appropriate local properties.

Definition 4.1 A program of *Physarum* Kolmogorov-Uspensky machine is any r-local action transforming some (Γ, k)-complexes of growing plasmodia into other (Γ, k)-complexes of growing plasmodia:

$$U \to \langle W, \gamma, \delta \rangle,$$

where U, W are (Γ, k)-*Physarum* complexes, γ is a mapping from $v(U)$ to $v(W)$, δ is an injection from $v(U)$ into $v(W)$. The algorithm of transformation complexes $S \to S^*$ is as follows [31, 32]:

- r-Neighbourhood of complex S is the same as of U.
- $v(S') = v(S \backslash U) \cup v(W)$.
- If $b \in U$, $a \in S \backslash U$, there is $\langle a, b \rangle$ in S and $\gamma(b)$ is defined, then $\langle a, \gamma(b) \rangle$ is an edge in S' with the same number as $\langle a, b \rangle$.
- If $a \in U$, $b \in S \backslash U$, there is $\langle a, b \rangle$ in S and $\delta(a)$ is defined, then $\langle \delta(a), b \rangle$ is an edge in S' with the same number as $\langle a, b \rangle$ (due to injectivity of δ we have different numbers for different edges from the same vertex).
- The initial vertex of W is an initial vertex of S' and we delete in S' all vertices (with appropriate edges) which are not accessible from the initial one. In this way we obtain S^*.

The simpler version of Kolmogorov-Uspensky machines is represented by Schönhage's storage modification machines [33, 34].

4.2.2 Physarum *Schönhage's Storage Modification Machines*

These machines consist of a fixed alphabet of input symbols, $\Gamma = S_{P.polycephalum} \cup I_{P.polycephalum}$, and a mutable directed graph with its arrows labelled by $E_{P.polycephalum}$ denoted by X for short. The set of nodes Γ, identifying with attractants, is finite. One fixed node $a \in I_{P.polycephalum}$ is identified as a distinguished center node of the graph. It is the first active zone of growing plasmodium (also associated with the one nest of ants or bees, just one inoculum of *Paenibacillus vortex* swarms). The distinguished node a has an edge x such that $x_\gamma(a) = a$ for all $\gamma \in X$. That is, all pointers from the distinguished center node point back to the center node. Each $\gamma \in X$ defines a mapping x_γ from Γ to Γ in accordance with directions of growing plasmodium; $x_\gamma(b)$ is the node found at the end of the edge starting at b labelled by γ. Each word of symbols in the alphabet Γ is a pathway through the machine from the distinguished center node. For example $ABBC$ would translate to taking path A from the start node, then path B from the resulting node, then path B, then path C. With respect to the word $ABBC$, the plasmodium moves.

Schönhage's machine modifies storage by adding new elements and redirecting edges. Its basic instructions are as follows:

- Creating a new node: **new** W. The machine reads the word W, following the path represented by the symbols of W until the machine comes to the last symbol in the word. It causes a new node y associated with the last symbol of W to be created and added to X; its location in relation to the other nodes and pointers is determined by W. If W is the empty string, this has the effect of creating a new center node a, linked to the old a. For example, **new** AB creates a new node that is reached by following the B pointer from the node designated by A. The growing plasmodium from active zone A to active zone B corresponds to this word AB. Adding a new node B means adding a new attractant denoted by B.
- A pointer redirection: **set** W **to** V. This instruction redirects an edge from the path represented by word W to a former node that represents word V. If W is the empty string, then this has the effect of renaming the center node a to be the node indicated by V. Notice that **set** W **to** V means removing nodes and the edges incident to $W\backslash V$. So, we can remove some attractants denoted by $W\backslash V$.
- A conditional instruction: **if** $V = W$ **then instruction** Z. It compares two paths represented by words W and V and if they end at the same node, then we jump to instruction Z else continue. This instruction serves to add edges between existing nodes. It corresponds to the splitting or fusion of *Physarum polycephalum*.

Definition 4.2 A program of *Physarum* Schönhage's storage modification machine is any action transforming sets X of nodes for growing plasmodia with the alphabet Γ into other sets X' of nodes for growing plasmodia with the same alphabet Γ which carries out by instructions **new** W; **set** W **to** V; **if** $V = W$ **then instruction** Z.

4.2.3 Physarum *Random-Access Machines*

In *random-access machines* there are registers (defined as simple locations with contents (single natural numbers) labelled by signs of $X = \{1, 2, \ldots, k\}$, where $k = |E_{P.polycephalum}|$. In case of *Physarum polycephalum*, the alphabet

$$\Gamma = S_{P.polycephalum} \cup I_{P.polycephalum}$$

consisting of of nodes (attractants) for growing plasmodia can be represented as set of registers. Their contents is defined by the number of protoplasmic tubes located at the place of appropriate register (i.e. by the number of protoplasmic tubes linked to this node). For example, $[\gamma]$ means 'the contents of register with location $\gamma \in \Gamma$'. So, $[\gamma]$ may be equal to 3. Then the node γ has three edges.

Instructions of *Physarum* random-access machines copy the contents of one registers and deposit them into other without destructing or changing registers. To do so we need repellents which stimulate plasmodium active zones to leave attractants.

Definition 4.3 A program of *Physarum* random-access machine is any action transforming contents of registers from X for the growing plasmodium with the alphabet Γ into other contents of the same registers in Γ for the growing plasmodium which carries out by the following instructions:

- Clear the content of register γ (set it to zero): $CLR([\gamma])$. All active zones are leaving γ due to repellents.
- Increment the contents of register γ: $INC([\gamma])$. The intensity of repellents at γ decreases.
- Decrement the contents of register γ: $DEC([\gamma])$. The intensity of repellents at γ increases.
- Copy the contents of register γ_j to register γ_k leaving the contents of γ_j intact: $CPY([\gamma_j], [\gamma_k])$. Using the *Physarum* transition, called direction, we can transmit active zones located at γ_j to add them to active zones located at γ_k.
- If register γ contains zero then jump to instruction Z else continue in sequence: $JZ([\gamma], Z)$.
- If the contents of register γ_j is equal to the contents of register γ_k then jump to instruction Z else continue in sequence: $JE([\gamma_j], [\gamma_k], Z)$.

Notice that in *Physarum* Kolmogorov-Uspensky machines and Schönhage's storage modification machines, the key role in computation models belongs to attractants, but in *Physarum* random-access machines, this role belongs to repellents.

4.2.4 Problems of Conventional Approach

Unfortunately, the computational complexity of implementations Kolmogorov-Uspensky machines, Schönhage's storage modification machines, and random-access machines on the *Physarum polycephalum* medium is so high. The point is that not every computable functions can be simulated by plasmodium behaviours properly (the more bit function, the higher complexity in its computation):

- first, the motion of plasmodia is too slow (several days are needed to compute simple functions such as 5-bit conjunction, 3-bit adder, etc., but the plasmodium stage of *Physarum polycephalum* is time-limited, therefore there is not enough time for realizations, e.g., of thousands-bit functions);
- second, the more attractants or repellents are involved in designing computable functions, the less accuracy of their implementation is, because the plasmodium

tries to be propagated in all possible directions and we will deal with indirected graphs, cycles, and other problems;

- third, the plasmodium is an adaptive organism that is very sensitive to environments, therefore it is very difficult to organize the same laboratory conditions for calculating the same k-bit functions, where k is large;
- fourth, the plasmodium has a free will and can make different decisions under the same conditions;
- fifth, the plasmodium follows emergent patterns which are fully eliminated in conventional automata such as Kolmogorov-Uspensky machines, although these patterns are natural for occupying many attractants.

Thus, swarm intelligence can be reduced to conventional automata, but with very low accuracy.

4.3 Unconventional Approach to Simulating the Swarm Behaviour

4.3.1 p*-Adic Integers*

As we already know, the intelligence of swarms is actively studied now, because swarms (ants, bees, some social bacteria, etc.) can solve logistic and transport problems very effectively [7–9, 11, 12, 21, 26, 28, 37–47]. For example, the slime mould can imitate route systems [48, 49]. So, as we know, the ant colony [16] can solve the Travelling Salesman Problem formulated in the following manner: given a list of "cities" (pieces of food scattered in the environment) and the distances between each pair of "cities", the ant nest can find the shortest possible route that visits each "city" exactly once and returns to the origin "city" (the nest). The bee colony [13] can solve the Generalized Assignment Problem formulated as follows: there are a number of "agents" (bees) and a number of "tasks" (nectar sources), each "agent" has a "budget" (possibilities in transporting nectars) and each "task" assumes some "cost" (distance in transporting nectars) and "profit" (delivered nectars); the bee colony can find an assignment in which all "agents" do not exceed their "budget" and total "profit" of the assignment is maximized. This problem is NP-hard, too.

Thus, the swarm behaviour is determined by attractants: any swarm looks for attractants (pieces of food), then the swarm builds a trasport system to logistically optimize the occupation of food and the absorption of all reachable pieces of food. Notice that the transport networks of swarms are not linear, but they hove many loops (circles) which change during the time.

The swarm behaviour can be modelled logically if we assume that at each time step of swarm propagation the swarm can find not more than $p-1$ neighbouring attractants which can be directly occupied. In this case we deal with the universe of p-adic integers for coding the swarm behaviour. Let us remember that the set of

p-adic integers is denoted by $\mathbf{Z}_p$ and each *p-adic integer* $n \in \mathbf{Z}_p$ has the following meaning [50]:

$$n = \sum_{i=0}^{\infty} a_i \cdot p^i,$$

where $a_i \in \{0, 1, \ldots, p-1\}$ and the following notation: $n = \ldots a_{i+1}a_i \ldots a_1 a_0$.

The set of p-adic integers includes all integers of $\mathbf{Z}$ as well as some rational numbers of $\mathbf{R}$. Let us see. Assume $p = 5$. Taking into account the fact that $|2/3|_5 = 1$, the rational number $2/3$ can be represented as a 5-adic integer. So, it has an expansion $\ldots a_{i+1}a_i \ldots a_1 a_0$ that starts with a_0. Note that $0 \leq a_0 < 5$ satisfies $a_0 \equiv 2/3 \mod 5$. Multiply by 3 gives $3a_0 \equiv 2 \mod 5$. The only residue of division by 5 that solves this is $a_0 = 4$. Then we have $(2/3 - 4) \equiv 5a_1 \mod 5^2$. Hence, $-10/3 \equiv 5a_1 \mod 5^2$. From this it follows that $-2/3 \equiv a_1 \mod 5$, i.e. $0 \equiv 3a_1 + 2 \mod 5$, and it is solved by $a_1 = 1$. For a_2, we have the equation $2/3 \equiv 4 + 1 \cdot 5 + 25 \cdot a_2 \mod 5^3$, i.e. $-1/3 \equiv a_2 \mod 5$, hence, $a_2 = 3$, etc. We see that $2/3 = 4 + 1 \cdot 5 + 3 \cdot 5^2 + 1 \cdot 5^3 + 3 \cdot 5^4 + 1 \cdot 5^5 + \cdots$ In the expansion form: $2/3 = \ldots 131314$.

In this chapter, I propose a way to code the swarm propagation by p-adic integers of $\mathbf{Z}_p$, see Sect. 4.3.2. All examples are examined for 5-adic integers and we concentrate just on motions of the plasmodium of *Physarum polycephalum* (slime mould) as an instance of swarm behaviour in looking for and occupying attractants with building an optimized transporting networks.

We know that each *swarm propagation* is divided into the following two stages: (i) the discover of localizations of neighbour attractants and repellents; (ii) the logistical optimization of the road system connecting all the reachable attractants and avoiding all the neighbour repellents. The first stage can be expressed by finite linear strings. However, the second stage can contain some loops. So, its appropriate strings can become infinite. This situation can be coded by infinite p-adic integers like $2/3 = \ldots 131314$ in $\mathbf{Z}_5$. But the strings of the second stage may be also non-linear, i.e. we can have several modifications of strings with an impossibility to differ them. Defining these non-linear strings is possible due to the notion of greatest fixed points giving rise to corecursion and codata. In this section, I show that there are ever some limits in any swarm propagation at the second/logistic stage. These limits can be taken as a hyperbolic space with patterns defined by corecursion or by an expanding of the classical group theory. I propose to follow the second way, i.e. to extend the group theory. I am going to introduce a p-adic valued logic for combining sets of linear and non-linear p-adic valued strings. Then I define a non-linear group theory that allows us to calculate processes over logistic stages in the swarm propagation.

4.3.2 *Synchronous Slime Mould Automata*

The plasmodium of *Physarum polycephalum*, as an agent of swarm behaviour, is sensitive to attractants and repellents and, therefore, its behaviour can be managed by distributing attractants and repellents [42, 51, 52]. So, we can illustrate p-adic valued models of swarm behaviour by the plasmodium. Motions of this organism are stimulated by a specific location of nutrients as attractants. For example, in our experiments we can use a Petri dish with dampened filter paper and scattered oat flakes attracting the plasmodium. As a result, the plasmodium migrates towards the oat flakes and builds a network of protoplasmic tubes (see Figs. 4.1, 4.2, 4.3 and 4.4). Under the conditions of *nutrient-poor concentration*, the plasmodium networks are well visible and can be represented as natural labelled transition systems, where states are attractants and transitions among states are protoplasmic tubes connecting attractants. In the meanwhile, under the conditions of *nutrient-rich concentration* of the plasmodium environment, e.g. under the conditions of an oatmeal rich agar spread on a Petri dish, the plasmodium propagates a wave front towards all directions (see Fig. 4.5). Then in order to adopt all found attractants the plasmodium builds a well visible network of protoplasmic tubes, too (see Fig. 4.6).

Without loss of generality, let us consider the universe of 5-adic integers $\mathbf{Z}_5$ for coding the swarm behaviour in occupying attractants. This means that at each step of propagation, the plasmodium faces not more than four neighbouring attractants (see Fig. 4.1). Let us construct a *synchronous Physarum automaton* for $t = 0, 1, 2, \ldots$ In these automata all plasmodium motions are synchronized and at each t the plasmodium can occupy only neighbouring attractants, i.e. this means that in Fig. 4.1 the plasmodium can occupy only attractants n_1, w_1, s_1, e_1 at $t = 0$ and if n_1 was occupied at $t = 0$, the plasmodium can occupy only attractants n_2, n_1w_1, n_1e_1 or go back to X at $t = 1$, etc. Hence, we assume that the plasmodium can go north (up), west (left), east (right), south (down) at each time step. The plasmodium can leave occupied attractants, e.g. it can leave its start point (inoculum). Therefore, if the plasmodium does not move forward after occupying some attractants and does not leave these attractants, this temporary stop is considered as a loop: the plasmodium does not stop, but go back and forward by circle along the existed road.

Now, let us code the swarm behaviour of the plasmodium by 5-adic integers. The states of synchronous *Physarum* automaton are presented by attractants n_1, w_1, s_1, e_1, ...(see Fig. 4.1). The transitions among states are presented by 5-adic integers obtained in the following manner: the 5-adic integer $\ldots a_n \ldots a_3a_2a_1a_0$ designates a transition started from the inoculum X and (i) if the plasmodium went north at $t = n$, then $a_n = 1$, (ii) if the plasmodium went west at $t = n$, then $a_n = 3$, (iii) if the plasmodium went east at $t = n$, then $a_n = 2$, (iv) if the plasmodium went south at $t = n$, then $a_n = 4$. Notice that in this way we can define the inverse motion as follows: let $a = a_n \ldots a_3a_2a_1a_0$ and $b = b_n \ldots b_3b_2b_1b_0$ be 5-adic integers, they are *inverse* if $a + b = \underbrace{111\ldots11}_{n+1}0$.

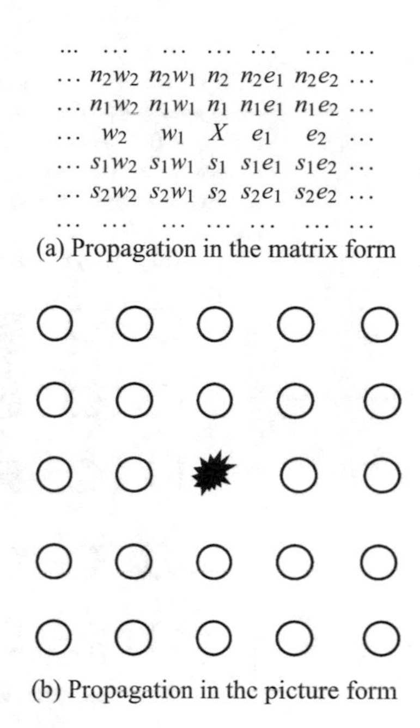

(a) Propagation in the matrix form

(b) Propagation in thc picture form

Fig. 4.1 The 5-adic valued universe of plasmodium propagation: **a** in the matrix form, where the inoculum of *Physarum polycephalum* is designated by X and at this point the plasmodium can grow north (n_1), west (w_1), south (s_1), east (e_1); at point n_1 the plasmodium can go up (n_2), left (n_1w_1), right (n_1e_1) or back (dawn); at point w_1 the plasmodium can go up (n_1w_1), left (w_2), down (s_1w_1) or back (right); at point s_1 the plasmodium can go down (s_2), left (s_1w_1), right (s_1e_1) or back (up); at point e_1 the plasmodium can go up (n_1e_1), down (s_1e_1), right (e_2) or back (left), etc.; **b** in the picture form, where the neighbouring attractants are designated by white circles, see [1]

The examples of 5-adic codes are given in Figs. 4.2, 4.3, 4.4 and 4.6. The p-adic integers are ascribed to all the non-empty strings $\gamma_0\gamma_1\ldots\gamma_k$ in the way: $[\gamma_0\gamma_1\ldots\gamma_k] \in \mathbf{Z}_p$. As we see in Figs. 4.2, 4.3, 4.4 and 4.6, the value $[\gamma_0\gamma_1\ldots\gamma_k]$ is different for the different $t = 0, 1, 2, \ldots$

We can define a p-adic distance $\rho_p(a, b)$ between different swarm paths $a = [\gamma_0\gamma_1\ldots\gamma_k]$ and $b = [\gamma'_0\gamma'_1\ldots\gamma'_m]$ as follows: $\rho_p(a, b) = |b - a|_p$. The norm $|\cdot|_p\colon \mathbf{Q}_p \to \mathbf{R}$ on $\mathbf{Q}_p$, where $\mathbf{Q}_p$ is the feld of all the p-adic numbers, is defined thus:

$$|n = \sum_{k=N}^{\infty} \alpha_k \cdot p^k|_p := p^{-N},$$

where N is an index of the first number distinct from zero in p-adic expansion of n.

Note that $|0|_p := 0$. The function $|\cdot|_p$ has values 0 and $\{p^l\}_{l\in\mathbf{Z}}$ on $\mathbf{Q}_p$. Finally, $|x|_p \geq 0$ and $|x|_p = 0 \equiv x = 0$.

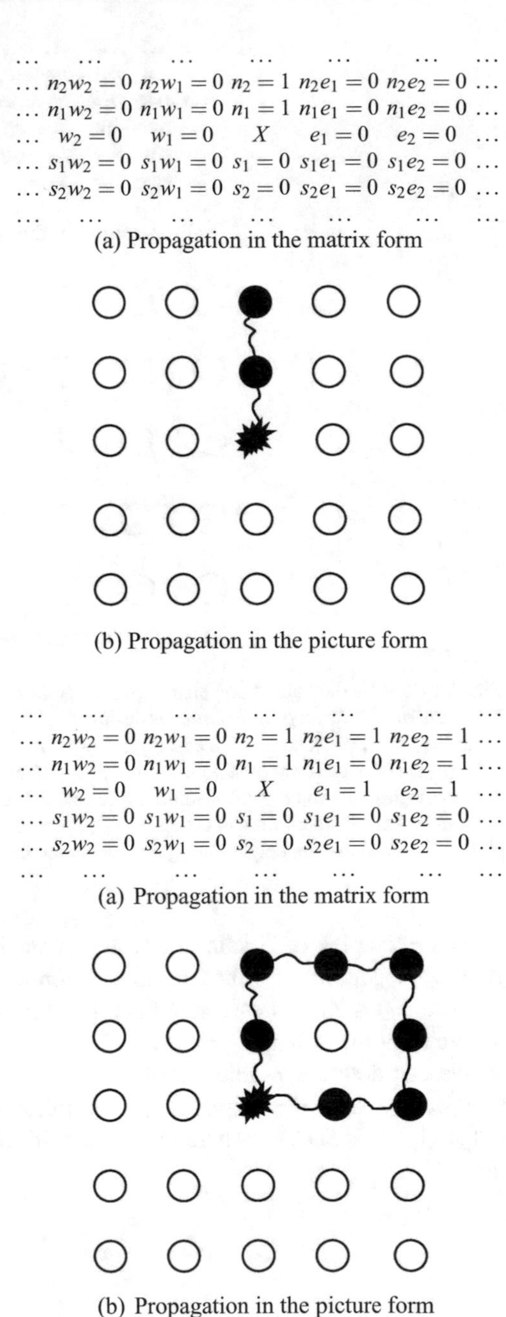

Fig. 4.2 The 5-adic valued string $[Xn_1n_2] = 11$ at time $t = 1$, the 5-adic valued string $[Xn_1n_2] = 411$ at time $t = 2$, the 5-adic valued string $[Xn_1n_2] = 4411$ at time $t = 3$, the 5-adic valued string $[Xn_1n_2] = 14411$ at time $t = 4$, the 5-adic valued string $[Xn_1n_2] = 114411$ at time $t = 5$, the 5-adic valued string $[Xn_1n_2] = 11441144114411441 1$ at time $t = 17$, etc. In **a** the occupied (activated) attractants are equal 1 (all others are equal 0). In **b** the occupied attractants are designated by black circles connected by protoplasmic tubes, see [1]

Fig. 4.3 The 5-adic valued string $[Xn_1n_2(n_2e_1)(n_2e_2)(n_1e_2)e_2e_1] = 33442211$ at time $t = 7$, the 5-adic valued string $[Xn_1n_2(n_2e_1)(n_2e_2)(n_1e_2)e_2e_1] = 1133442211$ at time $t = 9$, the 5-adic valued string $[Xn_1n_2(n_2e_1)(n_2e_2)(n_1e_2)e_2e_1] = 3344221133442211$ at time $t = 15$, etc. In **a** the occupied (activated) attractants are equal 1 (all others are equal 0). In **b** the occupied attractants are designated by black circles connected by protoplasmic tubes, see [1]

$$\begin{array}{ccccccc}
\dots & \dots & \dots & \dots & \dots & \dots & \dots \\
\dots & n_2w_2=0 & n_2w_1=0 & n_2=0 & n_2e_1=0 & n_2e_2=0 & \dots \\
\dots & n_1w_2=0 & n_1w_1=0 & n_1=1 & n_1e_1=1 & n_1e_2=0 & \dots \\
\dots & w_2=0 & w_1=1 & X & e_1=0 & e_2=0 & \dots \\
\dots & s_1w_2=0 & s_1w_1=0 & s_1=0 & s_1e_1=0 & s_1e_2=0 & \dots \\
\dots & s_2w_2=0 & s_2w_1=0 & s_2=0 & s_2e_1=0 & s_2e_2=0 & \dots \\
\dots & \dots & \dots & \dots & \dots & \dots & \dots
\end{array}$$

(a) Propagation in the matrix form

(b) Propagation in the picture form

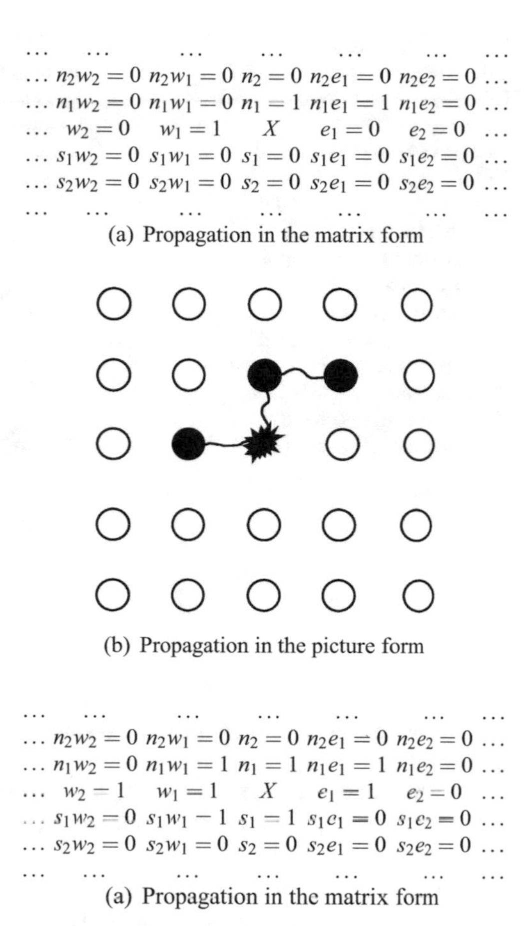

Fig. 4.4 The 5-adic valued strings $[Xn_1(n_1e_1)] = 21$ and $[Xw_1] = 23$ at time $t = 1$, the 5-adic valued strings $[Xn_1(n_1e_1)] = 212121212121$ and $[Xw_1] = 232323232323$ at time $t = 11$, etc. In **a** the occupied (activated) attractants are equal 1 (all others are equal 0). In **b** the occupied attractants are designated by black circles connected by protoplasmic tubes, see [1]

$$\begin{array}{ccccccc}
\dots & \dots & \dots & \dots & \dots & \dots & \dots \\
\dots & n_2w_2=0 & n_2w_1=0 & n_2=0 & n_2e_1=0 & n_2e_2=0 & \dots \\
\dots & n_1w_2=0 & n_1w_1=1 & n_1=1 & n_1e_1=1 & n_1e_2=0 & \dots \\
\dots & w_2=1 & w_1=1 & X & e_1=1 & e_2=0 & \dots \\
\dots & s_1w_2=0 & s_1w_1=1 & s_1=1 & s_1e_1=0 & s_1e_2=0 & \dots \\
\dots & s_2w_2=0 & s_2w_1=0 & s_2=0 & s_2e_1=0 & s_2e_2=0 & \dots \\
\dots & \dots & \dots & \dots & \dots & \dots & \dots
\end{array}$$

(a) Propagation in the matrix form

(b) Propagation in the picture form

Fig. 4.5 The wave front on the nutrient-rich substrate with the occupation of 8 attractants, see [1]

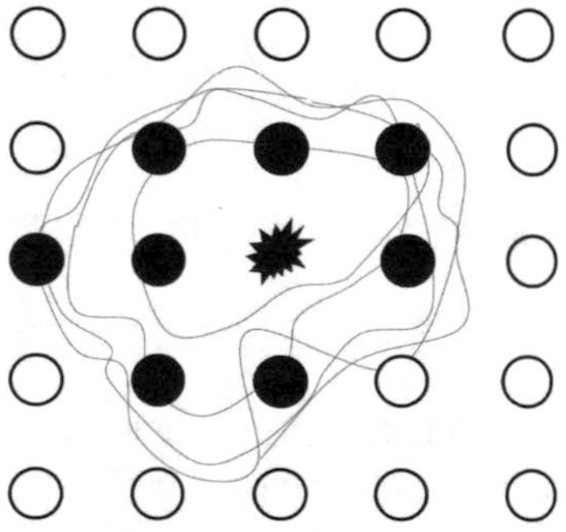

$$\begin{array}{cccccc}
\ldots & \ldots & \ldots & \ldots & \ldots & \ldots \quad \ldots \\
\ldots\ n_2w_2=0 & n_2w_1=0 & n_2=0 & n_2e_1=0 & n_2e_2=0 & \ldots \\
\ldots\ n_1w_2=0 & n_1w_1=1 & n_1=1 & n_1e_1=1 & n_1e_2=0 & \ldots \\
\ldots\ \ w_2=1 & w_1=1 & X & e_1=1 & e_2=0 & \ldots \\
\ldots\ s_1w_2=0 & s_1w_1=1 & s_1=1 & s_1e_1=0 & s_1e_2=0 & \ldots \\
\ldots\ s_2w_2=0 & s_2w_1=0 & s_2=0 & s_2e_1=0 & s_2e_2=0 & \ldots \\
\ldots & \ldots & \ldots & \ldots & \ldots & \ldots \quad \ldots
\end{array}$$

(a) Propagation in the matrix form

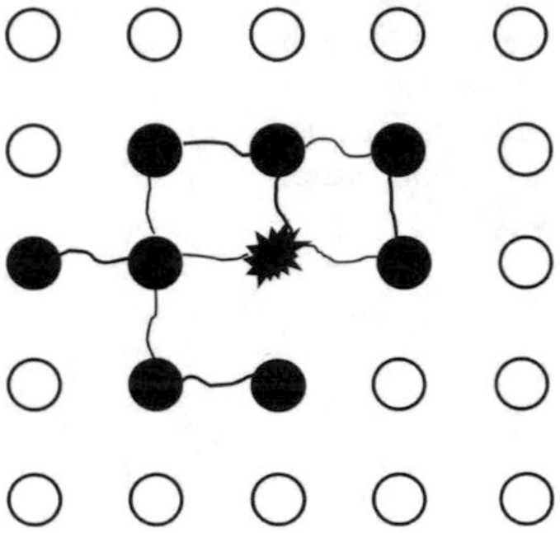

(b) Propagation in the picture form

Fig. 4.6 The 5-adic valued strings $[Xw_1(s_1w_1)s_1] = 243$, $[Xw_1w_2] = 233$, $[Xn_1(n_1w_1)w_1] = 431$, and $[Xe_1(n_1e_1)n_1] = 312$ at time $t = 2$. So, we have the three branches at point X and the two branches at point w_1. In **a** the occupied (activated) attractants are equal 1 (all others are equal 0). In **b** the occupied attractants are designated by black circles connected by protoplasmic tubes, see [1]

The value $|b - a|_p = p^l$ means that since $t = l$ the swarm paths $a = [\gamma_0\gamma_1 \ldots \gamma_k]$ and $b = [\gamma'_0\gamma'_1 \ldots \gamma'_m]$ are different.

The function $\rho_p(x, y) = |x - y|_p$ is a metric on $\mathbf{Q}_p$. It is a translation invariant metric, i.e. $\rho_p(x + h, y + h) = \rho_p(x, y)$. As usual in metric spaces we define closed and open balls in $\mathbf{Q}_p$:

$$B[a, p^l] := B_l[a] = \{x \in \mathbf{Q}_p \colon \rho_p(x, a) \leq p^l\},$$

$$B(a, p^l) := B_l(a) = \{x \in \mathbf{Q}_p \colon \rho_p(x, a) < p^l\},$$

where $p^l \in \mathbf{R}_+$.

The metric ρ_p satisfies the strong triangle inequality:

$$\rho_p(x, y) \leq \max(\rho_p(x, z), \rho_p(z, y)).$$

Such a kind of metrics is called an *ultrametrics* or *non-Archimedean metrics*. We note that any open or closed ball in an ultrametric space is a simultaneously closed and open subset, because $B_{l-1}[a] = B_l(a)$.

The balls $B_l(0)$ are additive subgroups of $\mathbf{Q}_p$: if $|x|_p, |y|_p \leq r$, then $|x + y|_p \leq \max(|x|_p, |y|_p) \leq r$. Furthermore, the ball $B[0, 1]$ is a ring: if $|x|_p, |y|_p \leq 1$, then $|x \cdot y|_p \leq |x|_p \cdot |y|_p \leq 1$. Moreover, the ball $B[0, 1]$ is the ring of p-adic integers $\mathbf{Z}_p$.

The ball $B_l(a)$ with the center $a = [\gamma_0\gamma_1 \dots \gamma_k]$ contains all the values for all the strings which differ from a after the time step $t = l$.

Hence, p-adic metrics gives us a partition of varieties of string values into the balls.

4.3.3 Non-linear Strings and the Limits in the Swarm Propagation

In Fig. 4.6 we see the situation when we have at once at the same zone several 5-adic valued strings $Xw_1(s_1w_1)s_1$, Xw_1w_2, $Xn_1(n_1w_1)w_1$, and $Xe_1(n_1e_1)n_1$ since $t = 2$. This situation is called the logistic stage in the propagation. At this stage it is difficult to distinguish one strings from other strings, because they are overlapped and interconnected among themselves.

Let us consider an example of two stages in the mould propagation, see Fig. 4.7. At both pictures we see two species of mould (*fungi* that grow in the form of multicellular filaments called hyphae), conditionally called by us here the 'green' and 'white' mould. They are in a competition for the food and at the discovering stage they are trying to sketch main directions of their propagations. At the logistic stage they are constructing networks, called a mycelium, to product spores (conidia) formed by differentiation at the ends of hyphae.

The same pattern with distinguishing the two stages in the propagation can be seen everywhere in the swarm behaviour, including a phenomenon of diffusion given by the growth of colonies of bacteria, see [53–55]. In [55, 56] there was proposed a simulation of the logistic stage by celular automata in the hyperbolic plane, where there are innitely many different tilings. One of the examples of the space for these cellular automata is given by the tiling when three copies of polygon can be put around a vertex in order to cover a neighbourhood of the vertex with no overlapping. This tiling is called the ternary heptagrid or, simply, the heptagrid, see Fig. 4.8.

On the one hand, in Fig. 4.8 we see infinitely many cells of automata. On the other hand, they are limited in the space by a circle and they are contained in an open interval.

Margenstern defines a cellular automaton on the heptagrid by a local transition function in form of a table, where each row defines a rule and the table has nine columns numbered from 0 to 8 so that each entry of the table containing a state of the automaton. On each row, column 0 contains the state of the cell to which the rule applies. Columns from 1 to 8 contain the states of neighbours. For the central cell, its neighbour 1 is fixed once and for all. For another cell, its neighbour 1 is its father. In all cases, the other neighbours are increasingly numbered from 2 to 7 while counter-clockwise turning around the cell starting from side 1. For more details, see [56]. In this way, the logistic stage of the swarm propagation is simulated by a possibly infinite behaviour within a rigid limits. These limits are given as a greatest fixed point—the circle in the case of heptagrid.

Fig. 4.7 The competition between the 'green' and 'white' mould in the same unwashed cup of coffee. In **a** we see a discovering stage of both species of mould to occupy the food space as prompt as possible. In **b** we see a logistic stage to effectively connect the own occupied attractants, see [1]

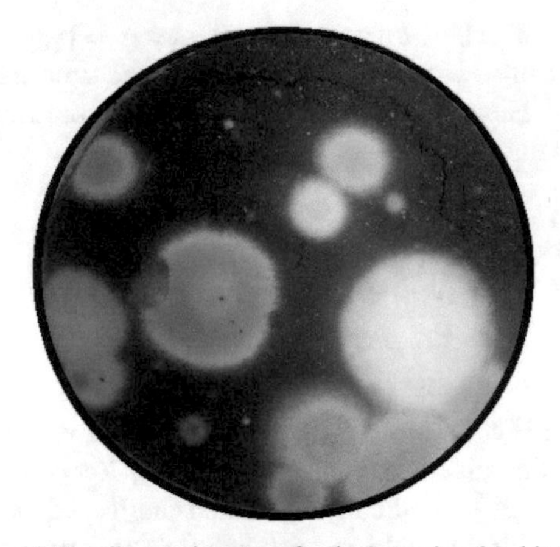

(a) The discovering stage for the 'green' and 'white' mould, ca.4 weeks of the unwashed cup of coffee

(b) The logistical stage for the 'green' and 'white' mould, ca.7 weeks of the unwashed cup of coffee

I support this Margenstern's idea to regard the logistic stage in the hyperbolic plane, but his approach has some disadvantages: (i) the cells of the heptagrid are taken before any calculations and cannot change in the calculating process, but the swarm behaviour is more context-based and assumes different own modifications in course of time; (ii) we cannot even imagine a transmission from one hyperbolic plane (one logistic stage) to another, but such things take place in the reality of swarms, because swarms can migrate.

It is better to consider the limits in the swarm propagation as limits in possible permutations, see [30]. Let $\mathcal{G}$ be a set of various motions with a linear composition: if x, y are two motions in $\mathcal{G}$, then the motion $x \cdot y$ is a result of their composition. Suppose that (i) the operation $x \cdot y$ satisfies the law of associativity, (ii) $\mathcal{G}$ contains the

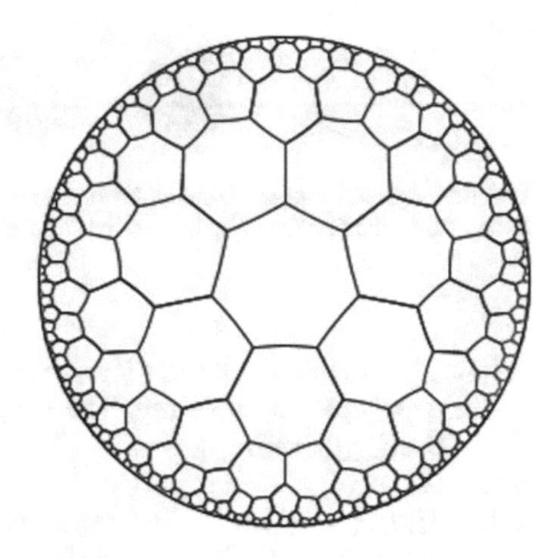

Fig. 4.8 The heptagrid as a space for cellular automata simulating the growth of bacterial colonies, see [55]

unit member e, and (iii) $\mathcal{G}$ contains the inverse member x^{-1}. The unit member means an identity motion, when nothing moves. The inverse member x^{-1} of x is a motion which is opposite to x. Consequently, $e \cdot x = x \cdot e = x$ and $x^{-1} \cdot x = x \cdot x^{-1} = e$. Thus, $\mathcal{G}$ is a group and every motion in $\mathcal{G}$ causes the setting of linear sequences. This group is called a permutation group.

A non-linear composition in non-linear permutation group assumes the existence of their limit represented as the absolute motion (*hyperunit*) $\overline{e}$ such that for any simple motions $x, y \in \mathcal{G}, x \cdot \overline{e} = y \cdot \overline{e} = \overline{e} \cdot x = \overline{e} \cdot y = \overline{e}$ and $e \cdot \overline{e} = \overline{e} \cdot e = \overline{e}$. So, the absolute motion gets all motions equal to itself. Hence, the absolute motion for any members of $\mathcal{G}$ determines the limits for any motions, i.e. it is the greatest possible motion in $\mathcal{G}$. For the absolute member there is no inverse motion, i.e. $\overline{e}^{-1} \cdot \overline{e} = \overline{e} \cdot \overline{e}^{-1} = \overline{e}$.

In multiplicative groups the absolute member plays the role of zero, but we can divide by zero as $\overline{e} \cdot \overline{e}^{-1} = \overline{e}$. In additive groups the absolute member plays the role of infinity, since $x + \overline{e} = y + \overline{e} = \overline{e} + x = \overline{e} + y = \overline{e}$.

The non-linear groups are formally defined at the end of Sect. 4.3.5.

4.3.4 p-*Adic Valued Logic for Non-linear Strings*

Let us consider non-empty strings $\gamma_0\gamma_1 \ldots \gamma_k$ of Sect. 4.3.2 and let each string $\gamma_0\gamma_1 \ldots \gamma_k$ have a numeric value $[\gamma_0\gamma_1 \ldots \gamma_k] \in \mathbf{Z}_p$. Let us define a weight $[[\gamma_i]]$ for each point/node γ_i occupied by a swarm.

- Let $[[\gamma_0]]$ denote an integer $\leq p - 1$ for the point $\gamma_0 \in \gamma_0\gamma_1 \ldots \gamma_k$. This integer is equal to the number of closest neighbours for γ_0 occupied by a swarm. The weight of γ_0 is coded by a p-adic integer $\ldots 0000[[\gamma_0]]$.
- Let $[[\gamma_0]]$ and $[[\gamma_1]]$ denote some integers $\leq p - 1$ for the points $\gamma_0, \gamma_1 \in \gamma_0\gamma_1 \ldots \gamma_k$. The integer $[[\gamma_0]]$ is equal to the number of occupied closest neigh-

Fig. 4.9 At each step of plasmodium propagation, there are only two attractants which can be occupied. At the first time step the plasmodium expansion is coded by the 3-adic integer ...000001, see [1]

bours for γ_0 and the integer $[[\gamma_1]]$ is equal to the number of closest neighbours for γ_1 occupied by a swarm. The weight of $\gamma_0\gamma_1$ is coded by a p-adic integer $\ldots 0000[[\gamma_1]][[\gamma_0]]$.

- ...
- Let $[[\gamma_0]]$, $[[\gamma_1]]$, ..., $[[\gamma_k]]$ denote some integers $\leq p-1$ for the points γ_0, γ_1, ..., $\gamma_k \in \gamma_0\gamma_1 \ldots \gamma_k$. The integer $[[\gamma_0]]$ is equal to the number of occupied neighbours for γ_0, the integer $[[\gamma_1]]$ is equal to the number of neighbours for γ_1 occupied by a swarm, ..., the integer $[[\gamma_k]]$ is equal to the number of neighbours for γ_k occupied by a swarm. The weight of $\gamma_0\gamma_1 \ldots \gamma_k$ is coded by a p-adic integer $\ldots 0000[[\gamma_k]] \ldots [[\gamma_1]][[\gamma_0]]$. We see that $0000[[\gamma_k]] \ldots [[\gamma_1]][[\gamma_0]] = 0000[[\gamma_k \ldots \gamma_1\gamma_0]] = [[\gamma_k \ldots \gamma_1\gamma_0]] = [[\gamma_0\gamma_1 \ldots \gamma_k]]$.

Evidently that according to this definition if in the string $\gamma_0\gamma_1 \ldots \gamma_k$ each γ_i $(0 \leq i \leq k)$ has no neighbouring attractants occupied by a swarm, then $[[\gamma_0\gamma_1 \ldots \gamma_k]] = 0 \in \mathbf{Z}_p$ and if in the string $\gamma_0\gamma_1 \ldots \gamma_k$ each γ_i $(0 \leq i \leq k)$ has all neighbouring attractants occupied by a swarm, then $[[\gamma_0\gamma_1 \ldots \gamma_k]] = \sum_{i=0}^{k}(p-1) \cdot p^i \in \mathbf{Z}_p$. The strings $[[\gamma_0\gamma_1 \ldots \gamma_k]] \neq 0$ are called non-empty.

Some examples of weights are given in Figs. 4.9, 4.10, 4.11 and 4.12.

Let us analyze the case when we have two strings $\gamma_0\gamma_1 \ldots \gamma_k$ and $\gamma_0\gamma_1' \ldots \gamma_m'$ started from the same state γ_0. Suppose that γ_{i_k} $(0 \leq i_k \leq k)$ and γ_{i_m}' $(0 \leq i_m \leq m)$ have some neighbouring attractants occupied by the plasmodium. This means that we face a splitting of the plasmodium at the node γ_0. Assume that there is not more splitting for nodes from $\gamma_0\gamma_1 \ldots \gamma_k$ and $\gamma_0\gamma_1' \ldots \gamma_m'$ and only $\gamma_0\gamma_1 \ldots \gamma_k$ and $\gamma_0\gamma_1' \ldots \gamma_m'$ are non-empty. Then the transition system at the logistic stage is coded by the set

$$\{[[\gamma_0\gamma_1 \ldots \gamma_k]], [[\gamma_0\gamma_1' \ldots \gamma_m']]\}.$$

Another similar situation is observed when we have two strings $\gamma_0\gamma_1\gamma_2 \ldots \gamma_k$ and $\gamma_0\gamma_1\gamma_2' \ldots \gamma_m'$ started from the same state γ_0 and with the splitting at the node γ_1. If only $\gamma_0\gamma_1\gamma_2 \ldots \gamma_k$ and $\gamma_0\gamma_1\gamma_2' \ldots \gamma_m'$ are non-empty, then the transition system is coded by the set

$$\{[[\gamma_0\gamma_1\gamma_2 \ldots \gamma_k]], [[\gamma_0\gamma_1\gamma_2' \ldots \gamma_m']]\}.$$

Let A_{γ_0} be a non-empty set of all non-empty strings started from γ_0 and this set be coded by $[A_{\gamma_0}]$, i.e. by a set of p-adic integers obtained as weights for all strings from A_{γ_0}. If we are going to distinguish a set of *all* possible strings started from γ_0 in

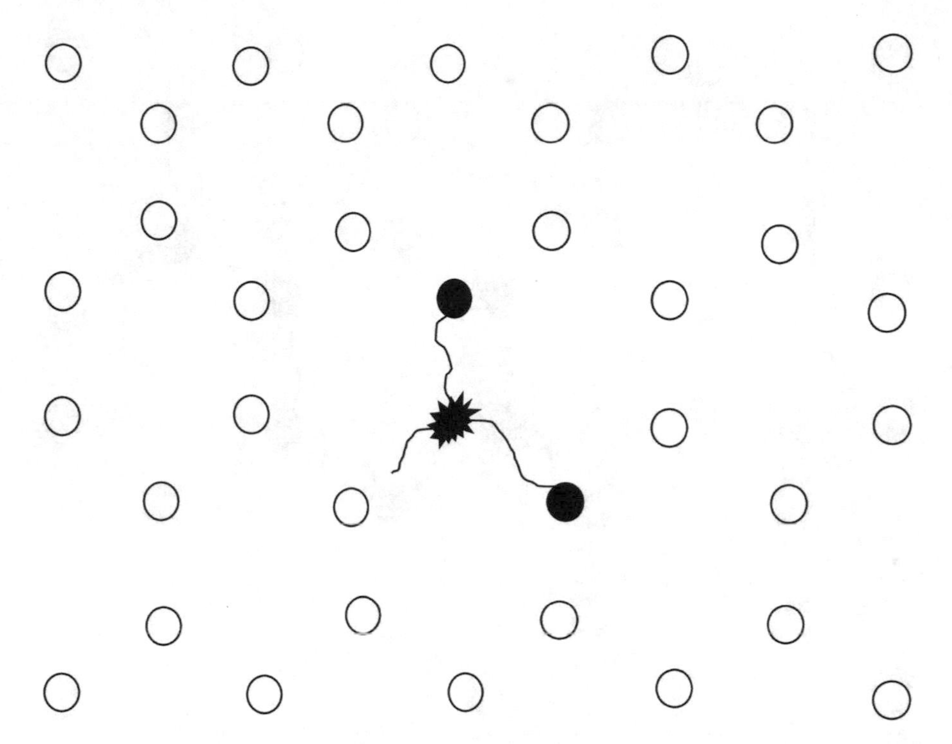

Fig. 4.10 At each step of plasmodium propagation, there are only three attractants which can be occupied. At the first time step the plasmodium expansion is coded by the 4-adic integer ...000002, see [1]

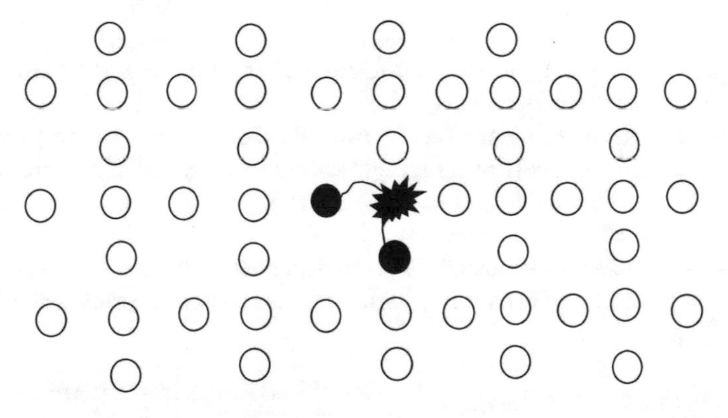

Fig. 4.11 At each step of plasmodium propagation, there are not more than four attractants which can be occupied. At the first time step the plasmodium expansion is coded by the 5-adic integer ...000002, see [1]

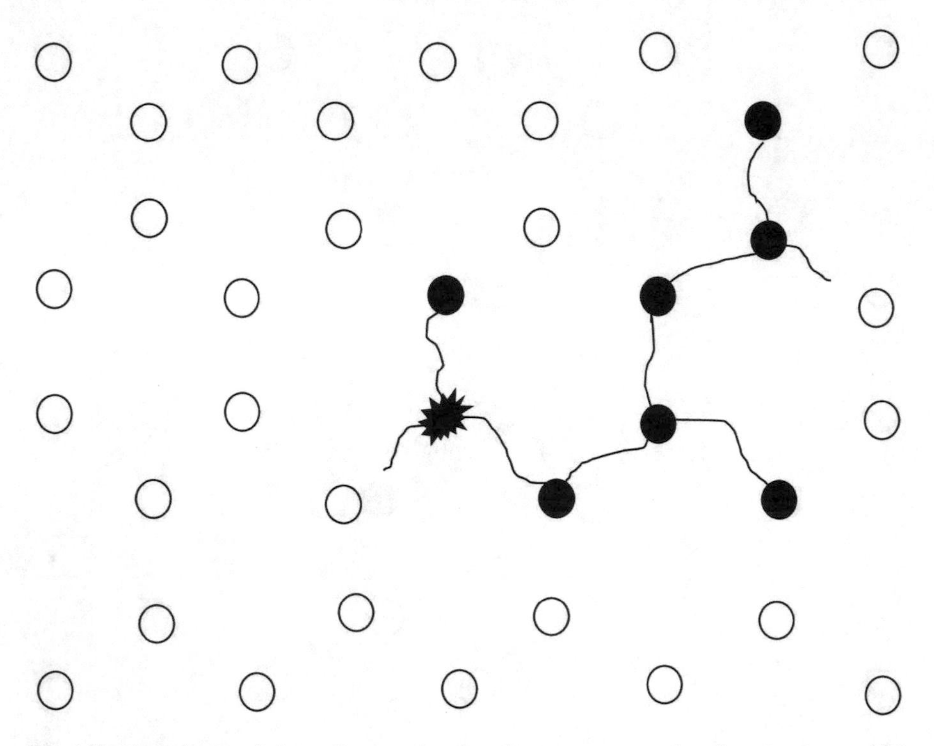

Fig. 4.12 At each step of plasmodium propagation, there are not more than four attractants which can be occupied. At the time step $t > 0$ the plasmodium expansion is coded by the 4-adic integer ...00000232, see [1]

a local space from a set A_{γ_0} of *some* strings started from γ_0, we will use the notation Ω_{γ_0}. Obviously, for each A_{γ_0}, $A_{\gamma_0} \subseteq \Omega_{\gamma_0}$.

Now we can define compositions of two sets A_{γ_0} and $A_{\gamma'_0}$, where $\gamma_0 \neq \gamma'_0$: we have A_{γ_0,γ'_0} and this set contains all strings started from γ_0 and started from γ'_0. The system A_{γ_0,γ'_0} is coded by $[A_{\gamma_0,\gamma'_0}] = [A_{\gamma_0}] \cup [A_{\gamma'_0}]$.

By induction, we can define sets $A_{\gamma_0,\gamma'_0,\ldots,\gamma''_0}$.

Notably, the swarm expansion is time dependent. So, we can consider sets $A^t_{\gamma_0,\gamma'_0,\ldots,\gamma''_0}$ at $t = 0, 1, \ldots$ Let us define logical operations over the same sets $A^t_{\gamma_0,\gamma'_0,\ldots,\gamma''_0}$ with different t:

conjunction $A^{t=k}_{\gamma_0,\gamma'_0,\ldots,\gamma''_0} \wedge A^{t=l}_{\gamma_0,\gamma'_0,\ldots,\gamma''_0}$: Notice that strings from different $A^{t=k}_{\gamma_0,\gamma'_0,\ldots,\gamma''_0}$ and $A^{t=l}_{\gamma_0,\gamma'_0,\ldots,\gamma''_0}$, where $k \neq l$, are the same, but they can be coded by different p-adic integers at $t = k$ and $t = l$. Let us consider each string $\gamma_0\gamma_1\gamma_2\ldots\gamma_m$. Let $[[\gamma_0\gamma_1\gamma_2\ldots\gamma_m]]_k$ be a p-adic numerical weight of $\gamma_0\gamma_1\gamma_2\ldots\gamma_m$ at $t = k$ and $[[\gamma_0\gamma_1\gamma_2\ldots\gamma_m]]_l$ be a p-adic numerical weight of $\gamma_0\gamma_1\gamma_2\ldots\gamma_m$ at $t = l$. Then we define $\min([[\gamma_0\gamma_1\gamma_2\ldots\gamma_m]]_k, [[\gamma_0\gamma_1\gamma_2\ldots\gamma_m]]_l)$ digit by digit:

$$\min(\ldots 000\gamma_{m,k}\ldots\gamma_{2,k}\gamma_{1,k}\gamma_{0,k};\ \ldots 000\gamma_{m,l}\ldots\gamma_{2,l}\gamma_{1,l}\gamma_{0,l}) =$$

$$\ldots 000 \min(\gamma_{m,k}, \gamma_{m,l}) \ldots \min(\gamma_{2,k}, \gamma_{2,l}) \min(\gamma_{1,k}, \gamma_{1,l}) \min(\gamma_{0,k}, \gamma_{0,l}).$$

The set $A^{t=k}_{\gamma_0,\gamma'_0,\ldots,\gamma''_0} \wedge A^{t=l}_{\gamma_0,\gamma'_0,\ldots,\gamma''_0}$ contains such minimum for each string.

disjunction $A^{t=k}_{\gamma_0,\gamma'_0,\ldots,\gamma''_0} \vee A^{t=l}_{\gamma_0,\gamma'_0,\ldots,\gamma''_0}$: Let us consider each string $\gamma_0\gamma_1\gamma_2\ldots\gamma_m$. Let $[[\gamma_0\gamma_1\gamma_2\ldots\gamma_m]]_k$ be a p-adic numerical value of $\gamma_0\gamma_1\gamma_2\ldots\gamma_m$ at $t = k$ and $[[\gamma_0\gamma_1\gamma_2\ldots\gamma_m]]_l$ be a p-adic numerical value of $\gamma_0\gamma_1\gamma_2\ldots\gamma_m$ at $t = l$. Then we define $\max([[\gamma_0\gamma_1\gamma_2\ldots\gamma_m]]_k, [[\gamma_0\gamma_1\gamma_2\ldots\gamma_m]]_l)$ digit by digit:

$$\max(\ldots 000\gamma_{m,k}\ldots\gamma_{2,k}\gamma_{1,k}\gamma_{0,k}; \ldots 000\gamma_{m,l}\ldots\gamma_{2,l}\gamma_{1,l}\gamma_{0,l}) =$$

$$\ldots 000 \max(\gamma_{m,k}, \gamma_{m,l}) \ldots \max(\gamma_{2,k}, \gamma_{2,l}) \max(\gamma_{1,k}, \gamma_{1,l}) \max(\gamma_{0,k}, \gamma_{0,l}).$$

The set $A^{t=k}_{\gamma_0,\gamma'_0,\ldots,\gamma''_0} \vee A^{t=l}_{\gamma_0,\gamma'_0,\ldots,\gamma''_0}$ contains such maximum for each string.

negation $\neg A^{t=k}_{\gamma_0,\gamma'_0,\ldots,\gamma''_0}$: Let us define the universe $\Omega_{\gamma_0,\gamma'_0,\ldots,\gamma''_0}$ as a set of all possible strings started from the nodes γ_0, γ'_0, ..., γ''_0. A numerical value of each string $\gamma_0\gamma_1\ldots\gamma_m$ from $\Omega_{\gamma_0,\gamma'_0,\ldots,\gamma''_0}$ is maximal: $[[\gamma_0\gamma_1\ldots\gamma_m]] = \sum_{i=0}^{m}(p-1)\cdot p^i$. Then $\neg A^{t=k}_{\gamma_0,\gamma'_0,\ldots,\gamma''_0} = \Omega_{\gamma_0,\gamma'_0,\ldots,\gamma''_0} \backslash A^{t=k}_{\gamma_0,\gamma'_0,\ldots,\gamma''_0}$ and it is coded by a set of p-adic integers $\sum_{i=0}^{m}(p-1)\cdot p^i - [[\gamma_0\gamma_1\gamma_2\ldots\gamma_m]]_k$ for each string $\gamma_0\gamma_1\gamma_2\ldots\gamma_m \in A^{t=k}_{\gamma_0,\gamma'_0,\ldots,\gamma''_0}$.

We can define logical connectives connecting different sets $A_{\gamma_0,\ldots,\gamma'_0}$ and $B_{\gamma''_0,\ldots,\gamma'''_0}$:

conjunction $A_{\gamma_0,\ldots,\gamma'_0} \wedge B_{\gamma''_0,\ldots,\gamma'''_0}$: The sets $A_{\gamma_0,\ldots,\gamma'_0}$ and $B_{\gamma''_0,\ldots,\gamma'''_0}$ contain different strings. Let $\gamma_0\gamma_1\gamma_2\ldots\gamma_m$ be a string from $A_{\gamma_0,\ldots,\gamma'_0}$ and $\gamma''_0\gamma''_1\gamma''_2\ldots\gamma''_k$ be a string from $B_{\gamma''_0,\ldots,\gamma'''_0}$. Let $[[\gamma_0\gamma_1\gamma_2\ldots\gamma_m]]$ be a p-adic numerical weight of $\gamma_0\gamma_1\gamma_2\ldots\gamma_m$ and $[[\gamma''_0\gamma''_1\gamma''_2\ldots\gamma''_k]]$ be a p-adic numerical weight of $\gamma''_0\gamma''_1\gamma''_2\ldots\gamma''_k$. Then we define $\min([[\gamma_0\gamma_1\gamma_2\ldots\gamma_m]], [[\gamma''_0\gamma''_1\gamma''_2\ldots\gamma''_k]])$ digit by digit:

$$\min(\ldots 000[[\gamma_0\gamma_1\gamma_2\ldots\gamma_m]]; \ldots 000[[\gamma''_0\gamma''_1\gamma''_2\ldots\gamma''_k]]) =$$

$$\ldots 000 \ldots \min([[\gamma_1]], [[\gamma''_1]]) \min([[\gamma_0]], [[\gamma''_0]]).$$

The set $A_{\gamma_0,\ldots,\gamma'_0} \wedge B_{\gamma''_0,\ldots,\gamma'''_0}$ contains such minimum for each string.

disjunction $A_{\gamma_0,\ldots,\gamma'_0} \vee B_{\gamma''_0,\ldots,\gamma'''_0}$: The sets $A_{\gamma_0,\ldots,\gamma'_0}$ and $B_{\gamma''_0,\ldots,\gamma'''_0}$ contain different strings. Let $\gamma_0\gamma_1\gamma_2\ldots\gamma_m$ be a string from $A_{\gamma_0,\ldots,\gamma'_0}$ and $\gamma''_0\gamma''_1\gamma''_2\ldots\gamma''_k$ be a string from $B_{\gamma''_0,\ldots,\gamma'''_0}$. Let $[[\gamma_0\gamma_1\gamma_2\ldots\gamma_m]]$ be a p-adic numerical weight of $\gamma_0\gamma_1\gamma_2\ldots\gamma_m$ and $[[\gamma''_0\gamma''_1\gamma''_2\ldots\gamma''_k]]$ be a p-adic numerical weight of $\gamma''_0\gamma''_1\gamma''_2\ldots\gamma''_k$. Then we define $\max([[\gamma_0\gamma_1\gamma_2\ldots\gamma_m]], [[\gamma''_0\gamma''_1\gamma''_2\ldots\gamma''_k]])$ digit by digit:

$$\max(\ldots 000[[\gamma_0\gamma_1\gamma_2\ldots\gamma_m]]; \ldots 000[[\gamma''_0\gamma''_1\gamma''_2\ldots\gamma''_k]]) =$$

$$\ldots 000 \ldots \max([[\gamma_1]], [[\gamma''_1]]) \max([[\gamma_0]], [[\gamma''_0]]).$$

The set $A_{\gamma_0,\ldots,\gamma'_0} \wedge B_{\gamma''_0,\ldots,\gamma'''_0}$ contains such maximum for each string.

negation $\neg A_{\gamma_0,\gamma'_0,\ldots,\gamma''_0}$: Let us define the universe $\Omega_{\gamma_0,\gamma'_0,\ldots,\gamma''_0}$ as a set of all possible strings started from the nodes γ_0, γ'_0, ..., γ''_0. A numerical value of each string

$\gamma_0\gamma_1 \ldots \gamma_m$ from $\Omega_{\gamma_0,\gamma_0',\ldots,\gamma_0''}$ is maximal: $[[\gamma_0\gamma_1 \ldots \gamma_m]] = \sum_{i=0}^{m}(p-1)\cdot p^i$. Then $\neg A_{\gamma_0,\gamma_0',\ldots,\gamma_0''} = \Omega_{\gamma_0,\gamma_0',\ldots,\gamma_0''} \backslash A_{\gamma_0,\gamma_0',\ldots,\gamma_0''}$ and it is coded by a set of p-adic integers $\sum_{i=0}^{m}(p-1)\cdot p^i - [[\gamma_0\gamma_1\gamma_2 \ldots \gamma_m]]$ for each string $\gamma_0\gamma_1\gamma_2 \ldots \gamma_m \in A_{\gamma_0,\gamma_0',\ldots,\gamma_0''}$.

By using this logic we can define an expansion strategy of a swarm on $\gamma_0\gamma_1\gamma_2 \ldots \gamma_m$ in the universe $A_{\gamma_0,\gamma_0',\ldots,\gamma_0''}$:

$$\mathscr{S}(\gamma_0\gamma_1\gamma_2 \ldots \gamma_m \in A_{\gamma_0,\gamma_0',\ldots,\gamma_0''}) = \bigwedge_{k=0}^{k=m} \sum_{t=0}^{\infty} (max([[\gamma_k]]_t, [[\gamma_k]]_{t+1})) \cdot p^t.$$

$$\mathscr{S}(A_{\gamma_0,\gamma_0',\ldots,\gamma_0''}) = \{\mathscr{S}(\gamma_0\gamma_1\gamma_2 \ldots \gamma_m) \colon \gamma_0\gamma_1\gamma_2 \ldots \gamma_m \in A_{\gamma_0,\gamma_0',\ldots,\gamma_0''}\}.$$

We know that $A_{\gamma_0,\gamma_0',\ldots,\gamma_0''} \subseteq \Omega_{\gamma_0,\gamma_0',\ldots,\gamma_0''}$. We can define the *p-adic cardinality*, $\lceil A_{\gamma_0,\gamma_0',\ldots,\gamma_0'',\ldots} \rceil$, in the following manner:

$$\lceil A_{\gamma_0,\gamma_0',\ldots,\gamma_0'',\ldots} \rceil = \min_{\gamma_0\gamma_1\gamma_2 \ldots \gamma_m \in A_{\gamma_0,\gamma_0',\ldots,\gamma_0'',\ldots}} \mathscr{S}(\gamma_0\gamma_1\gamma_2 \ldots \gamma_m).$$

Evidently, $\lceil A_{\gamma_0,\gamma_0',\ldots,\gamma_0''} \rceil \leq \lceil \Omega_{\gamma_0,\gamma_0',\ldots,\gamma_0''} \rceil$ even if $A_{\gamma_0,\gamma_0',\ldots,\gamma_0''} = \Omega_{\gamma_0,\gamma_0',\ldots,\gamma_0''}$, as for each $\gamma_0\gamma_1\gamma_2 \ldots \gamma_m \in \Omega_{\gamma_0,\gamma_0',\ldots,\gamma_0''}$, $\mathscr{S}(\gamma_0\gamma_1\gamma_2 \ldots \gamma_m) = \bigwedge_{k=0}^{k=m} \sum_{t=0}^{\infty} (max([[\gamma_k]]_t, [[\gamma_k]]_{t+1})) \cdot p^t = \sum_{t=0}^{\infty}(p-1)\cdot p^t = -1$. Thus,

$$\lceil \Omega_{\gamma_0,\gamma_0',\ldots,\gamma_0''} \rceil = -1$$

Let Ω be a set of all strings including infinite ones. Then

$$\mathscr{S}(\Omega) = \{\mathscr{S}(\gamma_0\gamma_1\gamma_2 \ldots \gamma_m \ldots) \in \mathbf{Z}_p \colon \gamma_0\gamma_1\gamma_2 \ldots \gamma_m \ldots \in \Omega\}$$

and

$$\lceil \Omega \rceil = \min(\mathscr{S}(\Omega)) = -1.$$

For any $A_{\gamma_0,\gamma_0',\ldots,\gamma_0''} \subseteq \Omega_{\gamma_0,\gamma_0',\ldots,\gamma_0''}$ and $A_{\gamma_0,\gamma_0',\ldots,\gamma_0''\ldots} \subseteq \Omega$, a *p-adic valued probability measure*, $P(A_{\gamma_0,\gamma_0',\ldots,\gamma_0''})$ and $P(A_{\gamma_0,\gamma_0',\ldots,\gamma_0''\ldots})$, is defined as follows:

- $P(A_{\gamma_0,\gamma_0',\ldots,\gamma_0''}) = \frac{\lceil A_{\gamma_0,\gamma_0',\ldots,\gamma_0''} \wedge \Omega_{\gamma_0,\gamma_0',\ldots,\gamma_0''} \rceil}{\lceil \Omega_{\gamma_0,\gamma_0',\ldots,\gamma_0''} \rceil}$ and $P(A_{\gamma_0,\gamma_0',\ldots,\gamma_0'',\ldots}) = \frac{\lceil A_{\gamma_0,\gamma_0',\ldots,\gamma_0'',\ldots} \wedge \Omega \rceil}{\lceil \Omega \rceil}$.
- $P(\Omega_{\gamma_0,\gamma_0',\ldots,\gamma_0''}) = 1$, $P(\Omega) = 1$, and $P(\emptyset) = 0$.
- if $A_{\gamma_0,\ldots,\gamma_0'} \subseteq \Omega_{\gamma_0,\ldots,\gamma_0'}$ and $B_{\gamma_0'',\ldots,\gamma_0'''} \subseteq \Omega_{\gamma_0'',\ldots,\gamma_0'''}$ are disjoint, i.e.

$$\inf(P(A_{\gamma_0,\ldots,\gamma_0'}), P(B_{\gamma_0'',\ldots,\gamma_0'''})) = 0,$$

then $P(A_{\gamma_0,\ldots,\gamma_0'} \vee B_{\gamma_0'',\ldots,\gamma_0'''}) = P(A_{\gamma_0,\ldots,\gamma_0'}) + P(B_{\gamma_0'',\ldots,\gamma_0'''})$. Otherwise,

$$P(A_{\gamma_0,\ldots,\gamma_0'} \vee B_{\gamma_0'',\ldots,\gamma_0'''}) = P(A_{\gamma_0,\ldots,\gamma_0'}) + P(B_{\gamma_0'',\ldots,\gamma_0'''}) -$$

$$\inf(P(A_{\gamma_0,\dots,\gamma_0'}), P(B_{\gamma_0'',\dots,\gamma_0'''})) = \sup(P(A_{\gamma_0,\dots,\gamma_0'}), P(B_{\gamma_0'',\dots,\gamma_0'''})).$$

All these equalities hold for infinite sets $A_{\gamma_0,\dots,\gamma_0',\dots}$ and $B_{\gamma_0'',\dots,\gamma_0''',\dots}$, also.

- $P(\neg A_{\gamma_0,\gamma_0',\dots,\gamma_0''}) = 1 - P(A_{\gamma_0,\gamma_0',\dots,\gamma_0''})$ for all finite $A_{\gamma_0,\gamma_0',\dots,\gamma_0''} \subseteq \Omega_{\gamma_0,\gamma_0',\dots,\gamma_0''}$, where $\neg A_{\gamma_0,\gamma_0',\dots,\gamma_0''} = \Omega_{\gamma_0,\gamma_0',\dots,\gamma_0''} \backslash A_{\gamma_0,\gamma_0',\dots,\gamma_0''}$. $P(\neg A_{\gamma_0,\gamma_0',\dots,\gamma_0'',\dots}) = 1 - P(A_{\gamma_0,\gamma_0',\dots,\gamma_0'',\dots})$ for all infinite $A_{\gamma_0,\gamma_0',\dots,\gamma_0'',\dots} \subseteq \Omega$, where $\neg A_{\gamma_0,\gamma_0',\dots,\gamma_0'',\dots} = \Omega \backslash A_{\gamma_0,\gamma_0',\dots,\gamma_0'',\dots}$.
- Relative probability functions

$$P(A_{\gamma_0,\dots,\gamma_0'} | B_{\gamma_0'',\dots,\gamma_0'''}) \in \mathbf{Q}_p \text{ and } P(A_{\gamma_0,\dots,\gamma_0',\dots} \mid B_{\gamma_0'',\dots,\gamma_0''',\dots}) \in \mathbf{Q}_p$$

are defined as follows:

$$P(A_{\gamma_0,\dots,\gamma_0'} | B_{\gamma_0'',\dots,\gamma_0'''}) = \frac{P(A_{\gamma_0,\dots,\gamma_0'} \wedge B_{\gamma_0'',\dots,\gamma_0'''})}{P(B_{\gamma_0'',\dots,\gamma_0'''})},$$

where $P(B_{\gamma_0'',\dots,\gamma_0'''}) \neq 0$ and

$$P(A_{\gamma_0,\dots,\gamma_0'} \wedge B_{\gamma_0'',\dots,\gamma_0'''}) = \inf(P(A_{\gamma_0,\dots,\gamma_0'}), P(B_{\gamma_0'',\dots,\gamma_0'''})).$$

All these equalities hold for infinite sets $A_{\gamma_0,\dots,\gamma_0',\dots}$ and $B_{\gamma_0'',\dots,\gamma_0''',\dots}$, too.

For deductive properties of p-adic valued logic for simulating the slime mould behaviour and the behaviour of other swarms, please see [57].

4.3.5 *Relations and Functions on Non-linear Strings and the Non-linear Group Theory*

Let us continue to consider only sets with non-empty strings:

$$A_{\gamma_0,\dots,\gamma_0'} = \{\gamma_0\gamma_1\gamma_2\cdots\gamma_m : [[\gamma_0\gamma_1\gamma_2\cdots\gamma_m]] = \sum_{i=0}^{m}[[\gamma_i]]\cdot p^i \neq 0 \wedge [[\gamma_j]] = 0 \text{ for all } j > m\}.$$

Then the set $R_n(\underbrace{A_{\gamma_0,\dots,\gamma_0'}, \dots, B_{\gamma_0'',\dots,\gamma_0'''}}_{n}) \in \mathscr{P}(\Omega)^n$ is called an n-ary relation among the sets of p-adic valued strings.

Let Φ and Ψ be two binary relations on sets of p-adic valued strings. Then their composition $\Phi \circ \Psi$ is defined as follows:
$\Phi \circ \Psi = \{\langle A_{\gamma_0,\dots,\gamma_0'}, B_{\gamma_0'',\dots,\gamma_0'''}\rangle : \langle A_{\gamma_0,\dots,\gamma_0'}, C_{\gamma_0'''',\dots,\gamma_0'''''}\rangle \in \Phi$ and $\langle C_{\gamma_0'''',\dots,\gamma_0'''''}, B_{\gamma_0'',\dots,\gamma_0'''}\rangle \in \Psi$ for some $C_{\gamma_0'''',\dots,\gamma_0'''''} \subseteq \Omega\}$.

This product has the property of associativity:

$$\Phi \circ (\Psi \circ \Theta) = (\Phi \circ \Psi) \circ \Theta.$$

Also, we can define a diagonal relation in $\mathscr{A} \subset \mathscr{P}(\Omega)$:

$$1_{\mathscr{A}} = \{\langle A_{\gamma_0,\ldots,\gamma_0'}, A_{\gamma_0,\ldots,\gamma_0'}\rangle : A_{\gamma_0,\ldots,\gamma_0'} \in \mathscr{A}\},$$

and an inverse relation:

$$\Phi^{-1} = \{\langle A_{\gamma_0,\ldots,\gamma_0'}, B_{\gamma_0'',\ldots,\gamma_0'''}\rangle : \langle B_{\gamma_0'',\ldots,\gamma_0'''}, A_{\gamma_0,\ldots,\gamma_0'}\rangle \in \Phi\}.$$

In other words, while $\Phi \subseteq \mathscr{A} \times \mathscr{B}$, $\Phi^{-1} \subseteq \mathscr{B} \times \mathscr{A}$, where $\mathscr{A}, \mathscr{B} \subseteq \mathscr{P}(\Omega)$
If Φ is a relation from $\mathscr{A}$ to $\mathscr{B}$, then

$$\Phi \circ 1_{\mathscr{B}} = 1_{\mathscr{A}} \circ \Phi = \Phi;$$

$$\Phi \circ \Phi^{-1} = 1_{\mathscr{A}};$$

$$\Phi^{-1} \circ \Phi = 1_{\mathscr{B}}.$$

Let $\Phi_1, \ldots, \Phi_n$ be relations, $\circ$ their product, $^{-1}$ their inverse relation, 1 their diagonal. Then we have a group.

The *horizontal equivalence* on the set $\mathscr{A}$ is said to be any reflexive, symmetric, and transitive relation Φ in $\mathscr{A}$. This means that we have respectively:

1. $\Phi \supseteq 1_{\mathscr{A}}$,
2. $\Phi^{-1} = \Phi$,
3. $\Phi \circ \Phi \subseteq \Phi$.

Let us define an absolute member $0_{\mathscr{A}}$ for each set $\mathscr{A} \subseteq \mathscr{P}(\Omega)$ as follows:

$$0_{\mathscr{A}} = \{\langle \Omega_{\gamma_0,\ldots,\gamma_0'}, \Omega_{\gamma_0,\ldots,\gamma_0'}\rangle : A_{\gamma_0,\ldots,\gamma_0'} \in \mathscr{A}\}.$$

It means that for any $\mathscr{A} \subseteq \mathscr{P}(\Omega)$, $1_{\mathscr{A}} \subseteq 0_{\mathscr{A}}$.

A *diagonalization* of a relation Φ has the following meaning:

$$\overline{\Phi^{-1}} = \{\langle \Omega_{\gamma_0'',\ldots,\gamma_0'''}, \Omega_{\gamma_0,\ldots,\gamma_0'}\rangle : \langle A_{\gamma_0,\ldots,\gamma_0'}, B_{\gamma_0'',\ldots,\gamma_0'''}\rangle \in \Phi\}$$

Continuing in the same way, we say that a *diagonalization* of a relation Φ^{-1} has the following meaning:

$$\overline{\Phi} = \{\langle \Omega_{\gamma_0,\ldots,\gamma_0'}, \Omega_{\gamma_0'',\ldots,\gamma_0'''}\rangle : \langle B_{\gamma_0'',\ldots,\gamma_0'''}, A_{\gamma_0,\ldots,\gamma_0'}\rangle \in \Phi^{-1}\}$$

The features of diagonalizations are as follows:

1. $\overline{\Phi} \circ \overline{\Phi^{-1}} = 0_{\mathscr{A}}$;
2. $\overline{\Phi^{-1}} \circ \overline{\Phi} = \underline{0_{\mathscr{B}}}$;
3. $\Phi \circ \overline{\Phi^{-1}} = \overline{\Phi^{-1}}$;
4. $\Phi^{-1} \circ \overline{\Phi} = \overline{\Phi}$;

5. $\overline{\Phi} \circ \Phi^{-1} = \Phi^{-1}$;
6. $\overline{\Phi^{-1}} \circ \Phi = \Phi$;
7. if $\overline{\Phi^{-1}}$ is a diagonalization from $\mathscr{B}$ to $\mathscr{A}$, then $\overline{\Phi^{-1}} \circ 0_{\mathscr{A}} = 0_{\mathscr{B}} \circ \overline{\Phi^{-1}} = \overline{\Phi^{-1}}$;
8. if $\overline{\Phi}$ is a diagonalization from $\mathscr{A}$ to $\mathscr{B}$, then $\overline{\Phi} \circ 0_{\mathscr{B}} = 0_{\mathscr{A}} \circ \overline{\Phi} = \overline{\Phi}$.

From this it follows that

1. $1_{\mathscr{A}} \circ 0_{\mathscr{A}} = 0_{\mathscr{A}}$,
2. $0_{\mathscr{A}} \circ 1_{\mathscr{A}} = 1_{\mathscr{A}}$.

Let Φ_1, ..., Φ_n be relations, $\overline{\Phi_1^{-1}}$, ..., $\overline{\Phi_n^{-1}}$ their diagonalizations, $\circ$ their product extended to diagonalizations, $^{-1}$ their inverse relation extended to diagonalizations, 1 their diagonal, 0 their absolute member. Then we have a non-linear group.

In non-linear groups we have absolute members $\overline{\Phi_1^{-1}}$, ..., $\overline{\Phi_n^{-1}}$ which give the rigid limits, respectively, for the behaviours Φ_1, ..., Φ_n, and the greatest absolute member 0 that gives the rigid limits for all possible behaviours.

A relation Φ is a function from $\mathscr{A}$ to $\mathscr{B}$ if and only if $\Phi \circ \Phi^{-1} \supseteq 1_{\mathscr{A}}$ and $\Phi^{-1} \circ \Phi \subseteq 1_{\mathscr{B}}$. A diagonalization $\overline{\Phi^{-1}}$ is a function if and only if $\overline{\Phi} \circ \overline{\Phi^{-1}} \subseteq 0_{\mathscr{A}}$ and $\overline{\Phi^{-1}} \circ \overline{\Phi} \supseteq 0_{\mathscr{B}}$.

A *mapping* is said to be the triple $\langle \mathscr{A}, \mathscr{B}, f \rangle$, where the set $\mathscr{A}$ is called origin of f, $\mathscr{B}$ is called its goal, f is a function from $\mathscr{A}$ to $\mathscr{B}$. If for mappings $f\colon \mathscr{A} \to \mathscr{B}$, $g\colon \mathscr{B} \to \mathscr{C}$, and $h\colon \mathscr{A} \to \mathscr{C}$ the equality $f \circ g = h$ takes place, then we say that an appropriate diagram is commutative.

The triple $\langle \mathscr{A}, \mathscr{B}, \overline{f^{-1}} \rangle$ is called a *mapping diagonaization*.

A mapping f is *surjective* if $f^{-1} \circ f = 1_{\mathscr{B}}$. In this case, the range of values of f is equal to $\mathscr{B}$. On the other hand, a mapping f is *injective* if $f \circ f^{-1} = 1_{\mathscr{A}}$. In this case the identity of images tells us about the identity of inverse images. If a mapping is both surjective and injective, then it is called a *bijection* or one-to-one correspondence.

A mapping diagonalization $\overline{f^{-1}}$ is *surjective* if $\overline{f} \circ \overline{f^{-1}} = 0_{\mathscr{A}}$. In this case, the range of values of $\overline{f^{-1}}$ is equal to the diagonalization of $\mathscr{A}$. On the other hand, a mapping $\overline{f^{-1}}$ is *injective* if $\overline{f^{-1}} \circ \overline{f} = 0_{\mathscr{B}}$. In this case the identity of images tells us about the identity of inverse images. If a mapping diagonalization is both surjective and injective, then it is *bijective*.

Let $q = f \circ f^{-1}$. Then q ia a horizontal equivalence. Indeed, first, $f \circ f^{-1} \supseteq 1_{\mathscr{A}}$, second, $(f \circ f^{-1})^{-1} = (f^{-1})^{-1} \circ f^{-1} = f \circ f^{-1}$, third, $f \circ f^{-1} \circ f \circ f^{-1}$ belongs to $f \circ 1_{\mathscr{B}} \circ f^{-1}$, because $f^{-1} \circ f \subseteq 1_{\mathscr{B}}$, thus, from $f \circ f^{-1} \circ f \circ f^{-1}$ it follows that $f \circ f^{-1}$. The equivalence $q = f \circ f^{-1}$ is called *kernel equivalence* or *core of mapping* f and it is denoted by $\mathbf{Ker}\, f$.

If q is an equivalence on $\mathscr{A}$, then for any $X \in \mathscr{A}$ there exists a subset $|X|_q$ of set $\mathscr{A}$ defined by equality:

$$|X|_q = \{Y \in \mathscr{A}\colon \langle X, Y \rangle \in q\}.$$

Instead of $\langle X, Y \rangle \in q$ very often the notation $X \equiv Y (\text{mod } q)$ or only $X \equiv Y$ is used. The set $|X|_q$ is called *q-class* of the member X or *coset* of the member X if fixing $q_1, q_2, \ldots q_n$ does not play any role. From properties of horizontal equivalence it follows that if q-classes are intersected (have joint members), then these classes form the one joint q-class.

The intersection of any system of equivalence relations on the set $\mathscr{A}$ is an equivalence relation on $\mathscr{A}$. The residue classes (classes of equivalences) either are mutually disjoint or coincide.

Therefore q-classes make partition of set $\mathscr{A}$, partitioning of $\mathscr{A}$ onto non-empty disjoint subsets. A subset of power set $\mathscr{P}(\mathscr{A})$ consisting of all q-classes is called *horizontal quotient set* of set $\mathscr{A}$ by the equivalence relation q and it is denoted by $\mathscr{A}/q$. If q is kernel equivalence, then an appropriate quotient set is denoted by $\mathscr{A}/\mathbf{Ker}\, f$. If each $X \in \mathscr{A}$ is associated with $|X|_q$, then we have a mapping from $\mathscr{A}$ to $\mathscr{A}/q$ which is called *natural mapping* or *canonical mapping* and is denoted by **nat** q. Obviously, this mapping is surjective.

The *vertical equivalence* on the set $\mathscr{A}$ is said to be any reflexive, symmetric, and supertransitive relation $\overline{\Phi^{-1}}$ in $\mathscr{A}$, i.e. such that:

1. $\overline{\Phi^{-1}} \supseteq 1_{\mathscr{A}}$,
2. $(\overline{\Phi^{-1}})^{-1} = \overline{\Phi^{-1}}$,
3. $\overline{\Phi^{-1}} \circ \overline{\Phi^{-1}} \supseteq \overline{\Phi^{-1}}$.

Let $\overline{q} = \overline{f} \circ \overline{f^{-1}}$. Then $\overline{q}$ is a vertical equivalence. First, $\overline{f} \circ \overline{f^{-1}} \subseteq 0_{\mathscr{A}}$. Second, $(\overline{f} \circ \overline{f^{-1}})^{-1} = \overline{f} \circ \overline{f^{-1}}$. Third, $\overline{f} \circ \overline{f^{-1}} \circ \overline{f} \circ \overline{f^{-1}} \supseteq \overline{f} \circ \overline{f^{-1}}$. This equivalence is called *atom equivalence* and it is denoted by **Atm** $\overline{f^{-1}}$.

Let $\overline{q}$ be a vertical equivalence on $\mathscr{A}$. Then for any X such that $X \in \mathscr{A}$ there exists the class $\|X\|_{\overline{q}}$ defined by equality:

$$\|X\|_{\overline{q}} = \{Y : \langle X, Y\rangle \in \overline{q} \text{ and } Y \in \mathscr{A}\}.$$

The union of any system of vertical equivalences on the set $\mathscr{A}$ is a vertical equivalence on $\mathscr{A}$. The residue classes of vertical equivalences can be always united. From this proposition it follows that any two classes of vertical equivalence on the same set $\mathscr{A}$ have always a non-empty intersection that may appear not a class of vertically equivalent members.

In the class of p-adic valued strings we can define the vertical equivalence $\overline{q}$ as follows:

$$\overline{q}(A_{\gamma_0,\ldots,\gamma_0'}, B_{\gamma_0'',\ldots,\gamma_0'''}) \; if \; and \; only \; if \; P(A_{\gamma_0,\ldots,\gamma_0'}|B_{\gamma_0'',\ldots,\gamma_0'''}) > 0.$$

We can check that it has the following properties:

- *identity*: $\overline{q}(A_{\gamma_0,\ldots,\gamma_0'}, A_{\gamma_0'',\ldots,\gamma_0'''})$ for any $A_{\gamma_0,\ldots,\gamma_0'} \subseteq \Omega$, as

$$P(A_{\gamma_0,\ldots,\gamma_0'}|A_{\gamma_0,\ldots,\gamma_0'}) > 0;$$

- *symmetry*: if $\overline{q}(A_{\gamma_0,\ldots,\gamma_0'}, B_{\gamma_0'',\ldots,\gamma_0'''})$, then $\overline{q}(B_{\gamma_0'',\ldots,\gamma_0'''}, A_{\gamma_0,\ldots,\gamma_0'})$ for any $A_{\gamma_0,\ldots,\gamma_0'}$,

$B_{\gamma_0'',\ldots,\gamma_0'''} \subseteq \Omega$, since from $P(A_{\gamma_0,\ldots,\gamma_0'}|B_{\gamma_0'',\ldots,\gamma_0'''}) > 0$ it follows that

$$P(B_{\gamma_0'',\ldots,\gamma_0'''}|A_{\gamma_0,\ldots,\gamma_0'}) > 0.$$

- *supertransitivity*: if $\overline{q}(A_{\gamma_0,\ldots,\gamma_0'}, C_{\gamma_0'''',\ldots,\gamma_0'''''})$, then there exists $B_{\gamma_0'',\ldots,\gamma_0'''} \subseteq \Omega$ such that $\overline{q}(B_{\gamma_0'',\ldots,\gamma_0'''}, C_{\gamma_0'''',\ldots,\gamma_0'''''})$ and $\overline{q}(A_{\gamma_0,\ldots,\gamma_0'}, B_{\gamma_0'',\ldots,\gamma_0'''})$, because from

$P(A_{\gamma_0,\ldots,\gamma_0'}|C_{\gamma_0'''',\ldots,\gamma_0'''''}) > 0$ it follows that there exists $B_{\gamma_0'',\ldots,\gamma_0'''}$ such that

$P(B_{\gamma_0'',\ldots,\gamma_0'''}|C_{\gamma_0'''',\ldots,\gamma_0'''''}) > 0$ and $P(A_{\gamma_0,\ldots,\gamma_0'}|B_{\gamma_0'',\ldots,\gamma_0'''}) > 0$. It gives a vertical partition. Let us show that we have no transitivity. Assume that $P(A_{\gamma_0,\ldots,\gamma_0'}) = \ldots 01011$, $P(B_{\gamma_0'',\ldots,\gamma_0'''}) = \ldots 10010$, $P(C_{\gamma_0'''',\ldots,\gamma_0'''''}) = \ldots 01000$. Then

$P(A_{\gamma_0,\ldots,\gamma_0'}|B_{\gamma_0'',\ldots,\gamma_0'''}) > 0$ and $P(A_{\gamma_0,\ldots,\gamma_0'}|C_{\gamma_0'''',\ldots,\gamma_0'''''}) > 0$, but

$P(B_{\gamma_0'',\ldots,\gamma_0'''})|C_{\gamma_0'''',\ldots,\gamma_0'''''}) = 0$.

Notably that the supertransitivity is a feature of p-adic valued probabilities. In real probabilities with values in the interval $[0, 1]$ of real numbers the conditional probability $P(A|B) > 0$ defines a standard equivalence relation (called by us horizontal equivalence) that gives a partition of Ω into disjoint subsets. In particular, $P(A|B) > 0$ satisfies the transitivity: there exists B such that if $P(B|C) > 0$ and $P(A|B) > 0$, then $P(A|C) > 0$. It gives a (horizontal) partition.

In the class of p-adic valued strings we can define also the horizontal equivalence q as follows:

$$q(A_{\gamma_0,\ldots,\gamma_0'}, B_{\gamma_0'',\ldots,\gamma_0'''}) \; if \; and \; only \; if \; \frac{\sup(P(A_{\gamma_0,\ldots,\gamma_0'}), P(B_{\gamma_0'',\ldots,\gamma_0'''}))}{P(B_{\gamma_0'',\ldots,\gamma_0'''})} > 0.$$

Let us check that it has the next properties:

- *identity*: $q(A_{\gamma_0,\ldots,\gamma_0'}, A_{\gamma_0'',\ldots,\gamma_0'''})$ for any $A_{\gamma_0,\ldots,\gamma_0'} \subseteq \Omega$, as

$$\frac{\sup(P(A_{\gamma_0,\ldots,\gamma_0'}), P(A_{\gamma_0,\ldots,\gamma_0'}))}{P(A_{\gamma_0,\ldots,\gamma_0'})} > 0;$$

- *symmetry*: if $q(A_{\gamma_0,\ldots,\gamma_0'}, B_{\gamma_0'',\ldots,\gamma_0'''})$, then $q(B_{\gamma_0'',\ldots,\gamma_0'''}, A_{\gamma_0,\ldots,\gamma_0'})$ for any $A_{\gamma_0,\ldots,\gamma_0'}$,

$B_{\gamma_0'',\ldots,\gamma_0'''} \subseteq \Omega$, since from $\frac{\sup(P(A_{\gamma_0,\ldots,\gamma_0'}), P(B_{\gamma_0'',\ldots,\gamma_0'''}))}{P(B_{\gamma_0'',\ldots,\gamma_0'''})} > 0$ it follows that

$$\frac{\sup(P(B_{\gamma_0'',\ldots,\gamma_0'''}), P(A_{\gamma_0,\ldots,\gamma_0'}))}{P(A_{\gamma_0,\ldots,\gamma_0'})} > 0.$$

- *transitivity*: if $q(B_{\gamma_0'',\ldots,\gamma_0'''}, C_{\gamma_0'''',\ldots,\gamma_0'''''})$ and $q(A_{\gamma_0,\ldots,\gamma_0'}, B_{\gamma_0'',\ldots,\gamma_0'''})$, then

$q(A_{\gamma_0,\dots,\gamma_0'}, C_{\gamma_0'''',\dots,\gamma_0'''''})$, indeed, from $\frac{\sup(P(B_{\gamma_0'',\dots,\gamma_0'''}), P(C_{\gamma_0'''',\dots,\gamma_0'''''}))}{P(C_{\gamma_0'''',\dots,\gamma_0'''''})} > 0$ and $\frac{\sup(P(A_{\gamma_0,\dots,\gamma_0'}), P(B_{\gamma_0'',\dots,\gamma_0'''}))}{P(B_{\gamma_0'',\dots,\gamma_0'''})} > 0$ it follows that $\frac{\sup(P(A_{\gamma_0,\dots,\gamma_0'}), P(C_{\gamma_0'''',\dots,\gamma_0'''''}))}{P(C_{\gamma_0'''',\dots,\gamma_0'''''})} > 0$.

Now, we can define *non-linear permutation groups* by appealing to the p-adic probabilities, see Sect. 4.3.4.

The standard permutation group, $\mathcal{G}$, on sets of p-adic valued strings is defined as follows. Let us consider some zones $A, B, C, \dots$ with scattered attractants. Each zone can be active if it is occupied by a swarm and logistically optimized for some needs of this swarm. If a zone X became active, it is denoted by γ_X. If γ_A, γ_B are two active zones which contain joint occupied nodes, then $\gamma_A \cdot \gamma_B$ is a result of composition. We can assume that the operation $\gamma_A \cdot \gamma_B$ satisfies the law of associativity. Let us define the unit member e as stopping in the swarm propagation, when nothing moves. And let us define the inverse member γ_X^{-1} as deactivating the zone of X. Hence,

$$e \cdot \gamma_X = \gamma_X \cdot e = \gamma_X \tag{4.1}$$

$$\gamma_X^{-1} \cdot \gamma_X = \gamma_X \cdot \gamma_X^{-1} = e \tag{4.2}$$

A non-linear extension of $\mathcal{G}$ is being obtained in the following manner. Let $\overline{\gamma_X^{-1}}$ be an activization at the zone of X such that it is a class of members which are vertical equivalent to γ_X, i.e. $\overline{\gamma_X^{-1}} = \{\gamma_Y \colon P(\gamma_X|\gamma_Y) > 0 \text{ and } P(\gamma_X|\gamma_Y) \in \mathbf{Q}_p\}$. Let $\overline{\gamma_X}$ be a deactivization at the zone of X such that it is a class of members which are vertical equivalent to γ_X^{-1}, i.e. $\overline{\gamma_X} = \{\gamma_Y^{-1} \colon P(\gamma_X^{-1}|\gamma_Y^{-1}) > 0 \text{ and } P(\gamma_X^{-1}|\gamma_Y^{-1}) \in \mathbf{Q}_p\}$. Finally, let $\overline{e}$ be an absolute member or *hyperunit*.

We have the following equalities:

$$\overline{\gamma_X} \cdot \overline{\gamma_X^{-1}} = \overline{e} \tag{4.3}$$

$$\overline{\gamma_X^{-1}} \cdot \overline{\gamma_X} = \overline{e} \tag{4.4}$$

$$\gamma_X \cdot \overline{\gamma_X^{-1}} = \overline{\gamma_X^{-1}} \tag{4.5}$$

$$\overline{\gamma_X^{-1}} \cdot \gamma_X = \overline{\gamma_X^{-1}} \tag{4.6}$$

$$\gamma_X^{-1} \cdot \overline{\gamma_X} = \overline{\gamma_X} \tag{4.7}$$

$$\overline{\gamma_X} \cdot \gamma_X^{-1} = \overline{\gamma_X} \tag{4.8}$$

$$e \cdot \overline{e} = \overline{e} \tag{4.9}$$

$$\overline{e} \cdot e = \overline{e} \tag{4.10}$$

which give us a *non-linear permutation group*. See Proposition 8.1.

I have offered to distinguish the following two stages in the swarm propagation: the discovering stage and the logistic stage. At the discovering stage we can ascribe a p-adic valued code to each string in the way of Sect. 4.3.2. At the logistic stage we can ascribe a p-adic valued weight to each string in the way of Sect. 4.3.4. On sets of p-adic valued strings we can define relations and functions (Sect. 4.3.5). And by using p-adic valued probabilities we can introduce a non-linear group by adding axioms (4.3)–(4.10) to standard group-theoretic axioms (4.1), (4.2), see Sect. 4.3.5. The new axioms allow us to introduce the absolute member $\overline{e}$ that holds the limits for the swarm propagation (Sect. 4.3.3).

Hence, in any intelligent swarm behaviour there are emergent patterns that cannot be reduced to linear combinations of subsystems. Therefore conventional algorithms such as Kolmogorov-Uspensky machines have very low accuracy of their implementations on swarm systems. Nevertheless, we can define a p-adic valued logic that can describe a massive-parallel behaviour of swarms.

References

1. Schumann, A.: Group theory and P-adic valued models of swarm behaviour. Math. Methods Appl. Sci. https://doi.org/10.1002/mma.4540
2. Schumann, A.: From swarm simulations to swarm intelligence. In: BICT 2015, Proceedings of the 9th EAI International Conference on Bio-inspired Information and Communications Technologies (formerly BIONETICS), December 3–5, pp. 461–468. New York City, United States (2015)
3. Schumann, A.: Towards slime mould based computer. New Math. Nat. Comput. **12**(2), 97–111 (2016). https://doi.org/10.1142/S1793005716500083
4. Schumann, A.: Conventional and unconventional approaches to swarm logic. In: Adamatzky, A. (ed.) Advances in Unconventional Computing: Volume 1: Theory, pp. 711–734. Springer International Publishing, Cham (2017)
5. Schumann, A.: P-adic valued models of swarm behaviour. AIP Conf. Proc. **1863**(1), 360009 (2017). https://doi.org/10.1063/1.4992538
6. Schumann, A., Pancerz, K., Szelc, A.: The swarm computing approach to business intelligence. Stud. Hum. **4**(3), 41–50 (2015). https://doi.org/10.1515/sh-2015-0019
7. Bonabeau, E., Dorigo, M., Theraulaz, G.: Swarm Intelligence: from natural to artificial systems. Oxford University Press (1999)
8. Kassabalidis, I., El-Sharkawi, M.A., Marks, R.J., Arabshahi, P., Gray, A.A.: Swarm intelligence for routing in communication networks. In: 2001 IEEE Global Telecommunications Conference, GLOBECOM '01, vol. 6, pp. 3613–3617. IEEE (2001)
9. Wang, Y., Li, B., Weise, T., Wang, J., Yuan, B., Tian, Q.: Self-adaptive learning based particle swarm optimization. Inf. Sci. **181**(20), 4515–4538 (2011)
10. Gordon, D.: The organization of work in social insect colonies. Complexity **8**(1), 43–46 (2003)
11. Kennedy, J., Eberhart, R.: Swarm Intelligence. Morgan Kaufmann Publishers, Inc. (2001)
12. Passino, K.M.: Biomimicry of bacterial foraging for distributed optimization and control. Control Syst. **22**(3), 52–67 (2002)
13. Karaboga, D.: An idea based on honey bee swarm for numerical optimization. Technical Report-tr06, Engineering Faculty, Computer Engineering Department, Erciyes University (2005)
14. Rajabioun, R.: Cuckoo optimization algorithm. Appl. Soft Comput. **11**, 5508–5518 (1987)
15. Cuevas, E., Cienfuegos, M., Zaldivar, D., Perez-Cisneros, M.: A swarm optimization algorithm inspired in the behavior of the social-spider. Expert Syst. Appl. **40**(16), 6374–6384 (2013)

16. Dorigo, M., Stutzle, T.: Ant Colony Optimization. MIT Press (2004)
17. Lee, C.: Emergence and universal computation. Metroeconomica **55**(2–3), 219–238 (2004)
18. Hafner, G.S., Tokarski, T.R.: Morphogenesis and pattern formation in the retina of the crayfish Procambarus clarkii. Cell Tissue Res. **293**, 535–550 (1998)
19. Wolfram, S.: Universality and complexity in cellular automata. Physica D **10**, 1–35 (1984)
20. Kennedy, J., Eberhart, R.C.: Particle swarm optimization. In: Proceedings of the 1995 IEEE International Conference on Neural Networks, vol. 4, pp. 1942–1948. Perth, Australia, IEEE Service Center, Piscataway, NJ (1995)
21. Reynolds, C.W.: Flocks, herds, and schools: a distributed behavioral model. Comput. Gr. **21**, 25–34 (1987)
22. Reynolds, R.G.: An introduction to cultural algorithms. In: Proceedings of the 3rd Annual Conference on Evolutionary Programming, pp. 131–139 (1994)
23. Karaboga, D., Akay, B.: A comparative study of artificial bee colony algorithm. Appl. Math. Comput. **214**(1), 108–132 (2009)
24. Darnton, N.C., Turner, L., Rojevsky, S., Berg, H.C.: Dynamics of bacterial swarming. Biophys. J. **98**, 2082–2090 (2010)
25. Turner, L., Zhang, R., Darnton, N.C., Berg, H.C.: Visualization of flagella during bacterial swarming. J. Bacteriol. **192**, 3259–3267 (2010)
26. Ariel, G., Shklarsh, A., Kalisman, O., Ingham, C., Ben-Jacob, E.: From organized internal traffic to collective navigation of bacterial swarms. New J. Phys. **15**, 12,501 (2013)
27. Ingham, C.J., Kalisman, O., Finkelshtein, A., Ben-Jacob, E.: Mutually facilitated dispersal between the nonmotile fungus aspergillus fumigatus and the swarming bacterium paeni bacillus vortex. Proc. Nat. Acad. Sci. U.S.A. **108**(49), 19731–19736 (2011)
28. Shklarsh, A., Finkelshtein, A., Ariel, G., Kalisman, O., Ingham, C., Ben-Jacob, E.: Collective navigation of cargo-carrying swarms. Interface Focus **2**, 689–692 (2012)
29. Ingham, C.J., Ben-Jacob, E.: Swarming and complex pattern formation in paenibacillus vortex studied by imaging and tracking cells. BMC Microbiology **36**, 8 (2008)
30. Schumann, A.: Towards context-based concurrent formal theories. Parallel Process. Lett. **25**, 1540,008 (2015)
31. Kolmogorov, A.N.: On the concept of algorithm. Uspekhi Mateaticheskich Nauk **8**(4), 175–176 (1953)
32. Uspensky, V.U.: Kolmogorov and mathematical logic. J. Symb. Logic **57**, 385–412 (1992)
33. Schoenhage, A.: Real-time simulation of multi-dimensional turing machines by storage modification machines. Project MAC technical memorandum 37, MIT (1973)
34. Schoenhage, A.: Storage modification machines. SIAM J. Comput. **9**, 490–508 (1980)
35. Tarjan, R.E.: Reference machines require non-linear time to maintain disjoint sets. Stan-cs-77–603 (1977)
36. Adamatzky, A.: Physarum machine: implementation of a Kolmogorov-Uspensky machine on a biological substrate. Parallel Process. Lett. **17**(4), 455–467 (2007)
37. Kalogeiton, V.S., Papadopoulos, D.P., Georgilas, I., Sirakoulis, G.C., Adamatzky, A.: Cellular automaton model of crowd evacuation inspired by slime mould. Int. J. Gen. Syst. **44**(3), 354–391 (2015). https://doi.org/10.1080/03081079.2014.997527
38. Nakagaki, T., Yamada, H., Toth, A.: Maze-solving by an amoeboid organism. Nature **407**, 470–470 (2000)
39. Nakagaki, T., Yamada, H., Tothm, A.: Path finding by tube morphogenesis in an amoeboid organism. Biophys. Chem. **92**, 47–52 (2001)
40. Nakagaki, T., Iima, M., Ueda, T., Nishiura, Y., Saigusa, T., Tero, A., Kobayashi, R., Showalter, K.: Minimum-risk path finding by an adaptive amoeba network. Phys. Rev. Lett. **99**, 68–104 (2007)
41. Ntinas, V.G., Vourkas, I., Sirakoulis, G.C., Adamatzky, A.: Oscillation-based slime mould electronic circuit model for maze-solving computations. IEEE Trans. Circuits Syst. **64-I**(6), 1552–1563 (2017)
42. Shirakawa, T., Yokoyama, K., Yamachiyo, M., Y-p, G., Miyake, Y.: Multi-scaled adaptability in motility and pattern formation of the Physarum plasmodium. Int. J. Bio-Inspir. Comput. **4**, 131–138 (2012)

43. Tero, A., Nakagaki, T., Toyabe, K., Yumiki, K., Kobayashi, R.: A method inspired by Physarum for solving the Steiner problem. Int. J. Unconv. Comput. **6**(2), 109–123 (2010)
44. Tsuda, S., Aono, M., Gunji, Y.P.: Robust and emergent Physarum-computing. BioSystems **73**, 45–55 (2004)
45. Watanabe, S., Tero, A., Takamatsu, A., Nakagaki, T.: Traffic optimization in railroad networks using an algorithm mimicking an amoeba-like organism. Physarum plasmodium. Biosystems **105**(3), 225–232 (2011)
46. Westendorf, C., Gruber, C., Grube, M.: Quantitative comparison of plasmodial networks of different slime molds. In: BICT 2015, Proceedings of the 9th EAI International Conference on Bio-inspired Information and Communications Technologies (formerly BIONETICS), December 3-5, pp. 611–612. New York City, United States (2015). http://dl.acm.org/citation.cfm?id=2954754
47. Whiting, J.G.H., de Lacy Costello, B., Adamatzky, A.: Transfer function of protoplasmic tubes of Physarum polycephalum. Biosystems **128**, 48–51 (2015)
48. Adamatzky, A., Yang, X., Zhao, Y.: Slime mould imitates transport networks in china. Int. J. Intell. Comput. Cybern. **6**(3), 232–251 (2013). https://doi.org/10.1108/IJICC-02-2013-0005
49. Adamatzky, A., Ilachinski, A.: Slime mold imitates the united states interstate system. Complex Syst. **21**(1) (2012)
50. Koblitz, N.: P-adic numbers, P-adic analysis and zeta functions, 2nd edn. Springer (1984)
51. Adamatzky, A.: Physarum Machines: computers from slime mould. World Scientific, Series on Nonlinear Science Series A (2010)
52. Shirakawa, T., Sato, H., Ishiguro, S.: Constrcution of living cellular automata using the Physarum polycephalum. Int. J. Gen. Syst. **44**, 292–304 (2015)
53. Ben-Jacob, E.: Social behavior of bacteria: from physics to complex organization. Eur. Phys. J. B **65**(3), 315–322 (2008)
54. Ivanitsky, G.R., Kunisky, A.S., Tzyganov, M.A.: Study of 'target patterns' in a phage-bacterium system. In: Krinsky, V. (ed.) Self-organization: Autowaves and Structures Far From Equilibrium, pp. 214–217. Springer, Heidelberg (1984)
55. Margenstern, M.: Bacteria inspired patterns grown with hyperbolic cellular automata. In: HPCS, pp. 757–763 (2011)
56. Margenstern, M.: An algorithmic approach to tilings of hyperbolic spaces: universality results. Fundam. Inform. **138**(1–2), 113–125 (2015)
57. Schumann, A.: p-adic valued logical calculi in simulations of the slime mould behaviour. J. Appl. Non-Class. Log. **25**(2), 125–139 (2015). https://doi.org/10.1080/11663081.2015.1049099

Chapter 5
Non-Archimedean Valued Fuzzy and Probability Logics

In this chapter, I concern some basic logical aspects of non-Archimedean valued probabilities and non-Archimedean valued fuzziness involved in simulating swarm behaviours. So, this chapter contains general results concerning non-Archimedean probabilities and fuzziness, i.e. logical values which run over an uncountable infinite, non-well-ordered and non-well-founded set. This kind of probabilities and fuzziness generalizes the notion of p-adic valued probabilities defined in [1–7]. This chapter contains results published in [8–22]. About p-adic numbers used in modeling cognitions, see also [23].

5.1 Non-Archimedean Numbers as Logical Values

Let us remember that the *Archimedean axiom* affirms that for any positive real or rational number ε, there exists a positive integer n such that $\varepsilon \geq \frac{1}{n}$ or $n \cdot \varepsilon \geq 1$. The field that satisfies all properties of $\mathbf{R}$ of real numbers without Archimedes' axiom is called the field of *hyperreal numbers* and it is denoted by $^*\mathbf{R}$. The field that satisfies all properties of $\mathbf{Q}$ of rational numbers without Archimedes' axiom is called the field of *hyperrational numbers* and it is denoted by $^*\mathbf{Q}$. By definition of field, if $\varepsilon \in \mathbf{R}$ (resp. $\varepsilon \in \mathbf{Q}$), then $1/\varepsilon \in \mathbf{R}$ (resp. $1/\varepsilon \in \mathbf{Q}$). Therefore $^*\mathbf{R}$ and $^*\mathbf{Q}$ contain simultaneously *infinitesimals* and *infinitely large integers*: for an infinitesimal ε, we have $N = \frac{1}{\varepsilon}$, where N is an infinitely large integer.

In this chapter, I will consider logical values defined on the following three sets: (i) the set $^*\mathbf{R}$ of hyperreal numbers, (ii) the set $^*\mathbf{Q}$ of hyperrational numbers, (iii) the set $\mathbf{Z}_p$ of p-adic integers. It is well known that sets $^*\mathbf{R}$, $^*\mathbf{Q}$ satisfy properties of field and set $\mathbf{Z}_p$ properties of ring. All those sets contain infinitely large numbers and in addition sets $^*\mathbf{R}$, $^*\mathbf{Q}$ contain infinitesimals (infinitely small numbers). For more details on infinitesimal and p-adic analysis see: [24–29].

A. Schumann, *Behaviourism in Studying Swarms: Logical Models of Sensing and Motoring*, Emergence, Complexity and Computation 33,
https://doi.org/10.1007/978-3-319-91542-5_5

Probability theory first axiomatized by Andrey Kolmogorov is closed over properties of a probability space $\langle \Omega, \mathscr{F}, \mathbf{P} \rangle$ consisting of: (i) an arbitrary non-empty set Ω called the *sample space*; (ii) the *σ-field* $\mathscr{F}$, i.e. a set of subsets of Ω containing the sample space (that is, $\Omega \in \mathscr{F}$) such that $\mathscr{F}$ is closed under complements (that is, if $A \in \mathscr{F}$, then also $(\neg A) \in \mathscr{F}$) and it is closed under countable unions (that is, if $A_i \in \mathscr{F}$ for $i = 1, 2, \ldots$, then also $(\bigcup_i A_i) \in \mathscr{F}$); (iii) the *probability measure* $\mathbf{P}$: $\mathscr{F} \to [0, 1]$, i.e. a function on $\mathscr{F}$ such that $\mathbf{P}$ on Ω is equal to one (that is, $\mathbf{P}(\Omega) = 1$) and $\mathbf{P}$ is countably additive (that is, if $A_i \in \mathscr{F}$ is a countable collection of pairwise disjoint sets, then $\mathbf{P}(\bigcup_i A_i) = \sum_i \mathbf{P}(A_i)$).

As we see, the probability measure $\mathbf{P}$ takes real numbers of $[0, 1]$ as its values. Nevertheless, according to *Alexander Ostrowski's theorem* proved in 1916, any non-trivial norm on the field $\mathbf{Q}$ of rational numbers is equivalent to either the usual real absolute value $|\cdot|_\infty$ giving the field $\mathbf{R}$ of real numbers as completion of $\mathbf{Q}$ or a p-adic norm $|\cdot|_p$ for some prime number p giving the field $\mathbf{Q}_p$ of p-adic numbers as completion of $\mathbf{Q}$, see [27]. In other words, we can construct a completion of the field $\mathbf{Q}$ with respect to a norm just in two ways: either as $\mathbf{R}$ or $\mathbf{Q}_p$. This means that we can try to define a probability measure taking values on $\mathbf{Q}_p$ as well.

For the first time, Andrei Khrennikov proposed an approach to *p-adic probability theory*, i.e. to probability theory with values not in $[0, 1]$, but in $\mathbf{Q}_p$. This theory was involved by him in solving some quantum paradoxes [30–33] and in modelling cognitive abilities [34, 35]. Several significant generalizations of his approach are contained in [3, 4, 36].

By his works Khrennikov suggested two ways of studies of p-adic probability logics: first, conventional logics closed over propositions such as "the probability of α belongs to the p-adic ball with the center r and the radius ρ", "the p-adic distance between the probabilities of α and β is less than or equal to ρ", etc. (see [1, 2, 7, 37]), second, unconventional logics closed over propositions whose truth values run over all p-adic numbers or hypernumbers (see [11–15, 17]).

In this chapter, I am constructing non-Archimedean valued fuzzy logics $Ł\Pi\forall_\infty$ and $Ł\Pi\frac{1}{2}\forall_\infty$ and a p-adic valued fuzzy logic $BL\forall_\infty$ that are built as ω-order extensions of the logics $Ł\Pi\forall$, $Ł\Pi\frac{1}{2}\forall$, and $BL\forall$ respectively. Recall that the logics $Ł\Pi\forall$, $Ł\Pi\frac{1}{2}\forall$, $BL\forall$ are considered in [38–41]. About fuzzy logic see [42]. The non-Archimedean fuzziness is considered in the framework of the t-norm based approach. The idea of non-Archimedean multiple-validity and non-Archimedean fuzziness is that (1) the set of values for the fuzziness is uncountable infinite and (2) this set isn't well-ordered and well-founded. This idea can have a lot of applications in probabilistic reasoning (see [5, 6, 15]).

Further, I get non-Archimedean probabilities on a class of non-Archimedean valued fuzzy subsets and build a non-Archimedean probability logic on the basis of $BL\forall_\infty$.

In this chapter, I cover the main essentials of the t-norm based approach, defining fuzzy logics as logics based on t-norms and their residua. Recall that t-*norm* is an operation $*\colon [0, 1]^2 \to [0, 1]$ which is commutative and associative, non-decreasing in both arguments and have 1 as unit element and 0 as zero element, i.e.

$$x * y = y * x,$$

$$(x * y) * z = x * (y * z),$$

$$x \leq x' \text{ and } y \leq y' \text{ implies } x * y \leq x' * y',$$

$$1 * x = x, \; 0 * x = 0.$$

We shall use only continuous t-norms as truth-functions of a conjunction. Each t-norm determines uniquely its corresponding implication $\Rightarrow$ (*residuum*) satisfying for all $x, y, z \in [0, 1]$,

$$z \leq x \Rightarrow y \text{ iff } x * z \leq y$$

The following are important examples of continuous t-norms and their residua:

1. *Łukasiewicz's logic*:
   - $x * y = \max(x + y - 1, 0)$,
   - $x \Rightarrow y = 1$ for $x \leq y$ and $x \Rightarrow y = 1 - x + y$ otherwise.

   In this logic $*$ and $\Rightarrow$ are denoted by $\&_L$ and $\rightarrow_L$ respectively.
2. *Gödel's logic*:
   - $x * y = \min(x, y)$,
   - $x \Rightarrow y = 1$ for $x \leq y$ and $x \Rightarrow y = y$ otherwise.

   In this logic $*$ and $\Rightarrow$ are denoted by $\&_G$ and $\rightarrow_G$ respectively.
3. *Product logic*:
   - $x * y = x \cdot y$
   - $x \Rightarrow y = 1$ for $x \leq y$ and $x \Rightarrow y = y/x$ otherwise.

   In this logic $*$ and $\Rightarrow$ are denoted by $\&_\Pi$ and $\rightarrow_\Pi$ respectively.

A *regular residuated lattice* (or a *BL-algebra*) is an algebra $\mathfrak{L}_L = \langle L, \wedge, \vee, *, \Rightarrow, 0, 1\rangle$ such that (1) $\langle L, \wedge, \vee, 0, 1\rangle$ is a lattice with the largest element 1 and the least element 0, (2) $\langle L, *, 1\rangle$ is a commutative semigroup with the unit element 1, i.e. $*$ is commutative, associative, and $1 * x = x$ for all x, (3) the following conditions hold

$$z \leq (x \Rightarrow y) \text{ iff } x * z \leq y \text{ for all } x, y, z;$$

$$x \wedge y = x * (x \Rightarrow y);$$

$$x \vee y = ((x \Rightarrow y) \Rightarrow y) \wedge ((y \Rightarrow x) \Rightarrow x),$$

$$(x \Rightarrow y) \vee (y \Rightarrow x) = 1.$$

It is known the following result. Let $*$ be a continuous t-norm with residuum $\Rightarrow$, then $\mathscr{A} = \langle [0, 1], \min, \max, *, \Rightarrow, 0, 1\rangle$ is a BL-algebra, called a standard BL-algebra;

if $*$ is the Łukasiewicz, Gödel or Product t-norm then $\mathscr{A}$ is an **L**-algebra, **G**-algebra or Π-algebra respectively, called the standard **L**-algebra, **G**-algebra or Π-algebra.

5.1.1 Hypernumbers

Let $\boldsymbol{\Theta}$ be a set and I an infinite set of indices. We consider, in a standard way, the family $\boldsymbol{\Theta}^I$, i.e. the set of all functions: $f: I \mapsto \boldsymbol{\Theta}$. Now we define a *filter* $\mathscr{F}$ on I as a family of sets $\mathscr{F} \subset 2^I$ for which: (1) $A \in \mathscr{F}, A \subset B \to B \in \mathscr{F}$; (2) $A_1, \ldots, A_n \in \mathscr{F} \to \bigcap_{k=1}^{n} A_k \in \mathscr{F}$ for any $n \geq 1$; (3) $\emptyset \notin \mathscr{F}$. The set of all complements for finite subsets of I is a filter and it is called a *Fréchet filter on I* and it is denoted by $\mathscr{U}_I$.

Further, define a relation $\backsim$ on the set $\boldsymbol{\Theta}^I$ by $f \backsim g \equiv \{\alpha \in I: f(\alpha) = g(\alpha)\} \in \mathscr{U}_I$. It is easily be proved that the relation $\backsim$ is an equivalence. For each $f \in \boldsymbol{\Theta}^I$ let $[f]$ denote the equivalence class of f under $\backsim$. The *ultrapower* $\boldsymbol{\Theta}^I/\mathscr{U}_I$ is then defined to be the set of all equivalence classes $[f]$ as f ranges over $\boldsymbol{\Theta}^I$: $\boldsymbol{\Theta}^I/\mathscr{U}_I := \{[f]: f \in \boldsymbol{\Theta}^I\}$.

The ultrapower $\boldsymbol{\Theta}^I/\mathscr{U}_I$ is said to be a *proper nonstandard extension* of $\boldsymbol{\Theta}$ and it is denoted by ${}^*\boldsymbol{\Theta}$. Recall that each element of ${}^*\boldsymbol{\Theta}$ is an equivalence class $[f]$ where $f: I \to \boldsymbol{\Theta}$. There exist two groups of members of ${}^*\boldsymbol{\Theta}$: (1) equivalence classes of constant functions, e.g. $f(\alpha) = m \in \boldsymbol{\Theta}$ for all $\alpha \in I$. Such equivalence class is denoted by ${}^*m$ or $[f = m]$, (2) equivalence classes of functions that aren't constant.

The set ${}^\sigma\boldsymbol{\Theta} = \{{}^*m: m \in \boldsymbol{\Theta}\}$ is called *standard set*. The members of ${}^\sigma\boldsymbol{\Theta}$ are called *standard*. It is readily seen that ${}^\sigma\boldsymbol{\Theta}$ and $\boldsymbol{\Theta}$ are isomorphic: ${}^\sigma\boldsymbol{\Theta} \simeq \boldsymbol{\Theta}$.

If $\boldsymbol{\Theta}$ is a number system, then members of ${}^*\boldsymbol{\Theta}$ will be called *hypernumbers*. We can define operations on them:

$$[f] \odot [g] = [h] \quad \equiv \quad \{\alpha \in I: f(\alpha) \odot g(\alpha) = h(\alpha)\} \in \mathscr{U}_I,$$

where $\odot \in \{+, -, \cdot, /\}$

In this way we can obtain hyperreal numbers of ${}^*\mathbf{R}$ and hyperrational numbers of ${}^*\mathbf{Q}$ satisfying properties of field.

5.1.2 p-Adic Numbers

Now we are trying to compare p-adic numbers with hypernumbers. For this let us consider a particular case of nonstandard extension when $\boldsymbol{\Theta}$ is a finite set such that $|\boldsymbol{\Theta}| = p$, where p is a prime number. Since we stopped to consider the general case there is no need to excuse that we lose generality. Let the set **N** of natural numbers be the index set and let $\mathscr{U}_N$ be a Fréchet filter on **N**. Then there exists a nonstandard extension ${}^*\boldsymbol{\Theta} := \boldsymbol{\Theta}^{\mathbf{N}}/\mathscr{U}_N$.

As usual, every considered function $f : \mathbf{N} \to \boldsymbol{\Theta}$ can be meant as an infinite-tuple $\langle f(0), f(1), f(2), \ldots \rangle$, where $f(j) \in \boldsymbol{\Theta}$ for any $j = 0, 1, \ldots$

It is obvious that if $n \leq m$ for $n, m \in \boldsymbol{\Theta}$, then we can set $^*n \leq {}^*m$ for $^*n, {}^*m \in {}^*\boldsymbol{\Theta}$. In other words, the order relation on the members of $\boldsymbol{\Theta}$ can be extended to the order relation on the constant functions of $^*\boldsymbol{\Theta}$.

Now consider f, such that there is no $k \in \boldsymbol{\Theta}$ for which $[f] = {}^*k$.

Let $f_0, \ldots, f_{p-1}, f_p, \ldots, f_{2p-1}, \ldots, f_{p^2-p}, \ldots, f_{p^2-1}, \ldots$ be functions that satisfy the following condition:

$[f_m] = [\langle n_0, n_1, n_2, \ldots, n_k, 0, 0, \ldots \rangle]$, where $n_0, n_1, n_2, \ldots, n_k$ are such that $m = n_0 + n_1 p + n_2 p^2 + \cdots + n_k p^k$.

For instance,

1. $[f_k] = [\langle k, 0, 0, \ldots \rangle]$, for any $k = 0, 1, \ldots, p - 1$,
2. $[f_k] = [\langle k \mod p, 1, 0, \ldots \rangle]$, for any $k = p, \ldots, 2p - 1$,
3. $[f_k] = [\langle k \mod p, p - 1, 0, \ldots \rangle]$, for any $k = p^2 - p, \ldots, p^2 - 1$,
4. $\ldots$

We can extend the ordering relation on the members of $\boldsymbol{\Theta}$ to the ordering relation $\leq_*$ on the members $[f_0], \ldots, [f_{p-1}], [f_p], \ldots, [f_{2 \cdot p-1}], \ldots, [f_{p^2-p}], \ldots, [f_{p^2-1}], \ldots$ of $^*\boldsymbol{\Theta}$. Define this order structure as follows:

$$\begin{cases} [f_k] \leq_* [f_l] \text{ iff } k \leq l, & \text{if } k, l \in \mathbf{N}; \\ [f], [f'] \text{ are incompatible under } \leq_*, & \text{if there is no } k \in \mathbf{N} \text{ such that} \\ & [f] = [f_k] \text{ or } [f'] = [f_k]. \end{cases}$$

Notice that $\leq_*$ is partial, because $^*\boldsymbol{\Theta}$ contains an uncountable number of its members, therefore there exists $[f]$ such that there is no $k \in \mathbf{N}$ for which $[f] = [f_k]$, this $[f]$ is said to be infinitely large integer. We can assign the natural number to each member $[f_k] \in {}^*\boldsymbol{\Theta}$, were k is finite; namely to each $[f_k]$ as $k \to \infty$, we can assign the following expansion

$$f_k(0) + f_k(1) \cdot p + \ldots + f_k(n) \cdot p^n + \ldots = \sum_{n=0}^{\infty} f_k(n) \cdot p^n,$$

where $f_k(n) \in \{0, 1, \ldots, p - 1\}, \forall n \in \mathbf{N}$. This expansion is called the p-adic integer. This number sometimes has the following notation:

$$\ldots \beta_n \ldots \beta_3 \beta_2 \beta_1 \beta_0,$$

where $\beta_0 = f_k(0), \beta_1 = f_k(1), \beta_2 = f_k(2), \ldots$.

Thus, we have shown that the nonstandard extension $\boldsymbol{\Theta}^{\mathbf{N}}/\mathscr{U}_N$, where $\boldsymbol{\Theta} = \{0, \ldots, p - 1\}$, is isomorphic with the set $\mathbf{Z}_p$ of p-adic integers. Usual denary operations (/, +, −, ·) can be extrapolated to the case of them, [27, 28]. For example, for 5-adic integers $\ldots 02324$ and $\ldots 003$ we obtain: $\frac{\ldots 02324}{\ldots 003} = \ldots 0423$ (the operation of division is not defined for all p-adic integers), $\ldots 02324 + \ldots 003 = \ldots 02332, \ldots 02324 - \ldots 003 = \ldots 02321, \ldots 02324 \cdot \ldots 003 = \ldots 013032$. Finite numbers of $\mathbf{Z}_p$ can be regarded as positive integers, namely each $[f_k]$ considered above such that $k \in \mathbf{N}$ can be identified with k. In this way we can identify $\ldots 02324$ with 339 and $\ldots 003$ with 3.

5.2 Hyper-valued and p-Adic Valued Logical Matrices

5.2.1 Hyper-valued ŁΠ-Matrix

Let $\mathbf{Q}_{[0,1]} = \mathbf{Q} \cap [0, 1]$. We can extend the usual order structure on $\mathbf{Q}_{[0,1]}$ to a partial order structure on ${}^*\mathbf{Q}_{[0,1]} := \mathbf{Q}^{\mathbf{N}}/\mathscr{U}_N$:

1. for any ${}^*x, {}^*y \in {}^\sigma\mathbf{Q}_{[0,1]}$ we have ${}^*x \preceq_* {}^*y$ iff $x \leq y$ in $\mathbf{Q}_{[0,1]}$,
2. if ${}^*x \neq {}^*0$, then $[f] \preceq_* {}^*x$, i.e. each positive rational number ${}^*x \in {}^\sigma\mathbf{Q}_{[0,1]}$ is greater than any number $[f] \in {}^*\mathbf{Q}_{[0,1]} \backslash {}^\sigma\mathbf{Q}_{[0,1]}$.

These conditions have the following informal sense: (1) the sets ${}^\sigma\mathbf{Q}_{[0,1]}$ and $\mathbf{Q}_{[0,1]}$ have an isomorphic order structure; (2) the set ${}^*\mathbf{Q}_{[0,1]}$ contains actual infinities that are less than any positive rational number of ${}^\sigma\mathbf{Q}_{[0,1]}$. Define this partial order structure on ${}^*\mathbf{Q}_{[0,1]}$ as follows:

$\mathscr{O}_{*\mathbf{Q}}$ (i) For any $[x], [y] \in {}^*\mathbf{Q}_{[0,1]}$ we have $[x] \preceq_* [y]$ iff $\{\alpha \in \mathbf{N} \colon x(\alpha) \leq y(\alpha)\} \in \mathscr{U}_N$. (ii) For any ${}^*z \in {}^\sigma\mathbf{Q}_{[0,1]}$ and $[y] \in {}^*\mathbf{Q}_{[0,1]} \backslash {}^\sigma\mathbf{Q}_{[0,1]}$ if ${}^*z \neq {}^*0$, then $[y] \prec_* {}^*z$. Notice that we have $[x] \prec_* [y]$ iff $[x] \neq [y]$ and $[x] \preceq_* [y]$.

Introduce two operations sup, inf in the partial order structure $\mathscr{O}_{*\mathbf{Q}}$:

$$\inf([x], [y]) := [\min(x, y)];$$

$$\sup([x], [y]) := [\max(x, y)].$$

Note there exist the maximal number ${}^*1 \in {}^*\mathbf{Q}_{[0,1]}$ and the minimal number ${}^*0 \in {}^*\mathbf{Q}_{[0,1]}$ under the meaning expressed in the condition $\mathscr{O}_{*\mathbf{Q}}$. Indeed, $\{\alpha \in \mathbf{N} \colon x(\alpha) \leq 1\} \in \mathscr{U}_N$ for any $x \in \mathbf{Q}^{\mathbf{N}}$ and $\{\alpha \in \mathbf{N} \colon 0 \leq x(\alpha)\} \in \mathscr{U}_N$ for any $x \in \mathbf{Q}^{\mathbf{N}}$.

Let $+, -, \cdot, /$ be the addition, subtraction, multiplication and division defined on $\mathbf{Q}$; we can extend them on ${}^*\mathbf{Q}$ as follows: $[x] \odot [y] = [z]$ iff $\{\alpha \in \mathbf{N} \colon x(\alpha) \odot y(\alpha)) = z(\alpha)\} \in \mathscr{U}_N$, where $\odot \in \{+, -, \cdot, /\}$.

Now introduce the following new operations defined for all $[x], [y] \in {}^*\mathbf{Q}$ in the partial order structure $\mathscr{O}_{*\mathbf{Q}}$:

- $[x] \rightarrow_L [y] = {}^*1 - \sup([x], [y]) + [y]$,

- $[x] \rightarrow_{\Pi} [y] = \begin{cases} {}^*1 & \text{if } [x] \preceq_* [y], \\ \inf\left({}^*1, \frac{[y]}{[x]}\right) & \text{otherwise,} \end{cases}$

 notice that we have $\inf\left({}^*1, \frac{[y]}{[x]}\right) = [h]$ iff $\{\alpha \in \mathbf{N}\colon \min\left(1, \frac{y(\alpha)}{x(\alpha)}\right) = h(\alpha)\} \in \mathscr{U}_N$, let us also remember that the members $[x]$, $[y]$ can be incompatible under $\mathscr{O}_{{}^*\mathbf{Q}}$,
- $\neg_L[x] = {}^*1 - [x]$, i.e. $[x] \rightarrow_L {}^*0$,
- $\neg_{\Pi}[x] = \begin{cases} {}^*1 & \text{if } [x] = {}^*0, \\ {}^*0 & \text{otherwise,} \end{cases}$ i.e. $\neg_{\Pi}[x] = [x] \rightarrow_{\Pi} {}^*0$,
- $\Delta[x] = \begin{cases} {}^*1 & \text{if } [x] = {}^*1, \\ {}^*0 & \text{otherwise,} \end{cases}$ i.e. $\Delta[x] = \neg_{\Pi}\neg_L[x]$,
- $[x]\&_L[y] = \sup([x], {}^*1 - [y]) + [y] - {}^*1$, i.e. $[x]\&_L[y] = \neg_L([x] \rightarrow_L \neg_L[y])$,
- $[x]\&_{\Pi}[y] = [x] \cdot [y]$,
- $[x] \oplus [y] := \neg_L[x] \rightarrow_L [y]$,
- $[x] \ominus [y] := [x]\&_L\neg_L[y]$,
- $[x] \wedge [y] = \inf([x], [y])$, i.e. $[x] \wedge [y] = [x]\&_L([x] \rightarrow_L [y])$,
 Let us show that $\wedge$ is really derived from $\&_L$ and $\rightarrow_L$. Recall that inf and sup are defined digit by digit. This means that if $[x] = [\langle x_0, x_1, x_2, \ldots\rangle]$ and $[y] = [\langle y_0, y_1, y_2, \ldots\rangle]$ then

$$\inf([x], [y]) = [\langle \min(x_0, y_0), \min(x_1, y_1), \min(x_2, y_2), \ldots\rangle].$$

 For each $i = 0, 1, 2, \ldots$ we have $x_i \leq y_i$ or $x_i > y_i$. At the same time, if $x_i \leq y_i$, then $x_i\&_L(x_i \rightarrow_L y_i) = \max(x_i, \max(x_i, y_i) - y_i) + y_i - \max(x_i, y_i) = x_i$ and if $x_i > y_i$, then $x_i\&_L(x_i \rightarrow_L y_i) = \max(x_i, \max(x_i, y_i) - y_i) + y_i - \max(x_i, y_i) = y_i$.
- $[x] \vee [y] = \sup([x], [y])$, i.e. $[x] \vee [y] = ([x] \rightarrow_L [y]) \rightarrow_L [y]$,
 Indeed, $([x] \rightarrow_L [y]) \rightarrow_L [y] = {}^*1 - \sup({}^*1 - \sup([x], [y]) + [y], [y]) + [y] = \sup([x], [y])$.
- $[x] \leftrightarrow [y] := ([x] \rightarrow_L [y]) \wedge ([y] \rightarrow_L [x])$,
- $[x]\&_G[y] := \inf([x], [y])$,
- $[x] \rightarrow_G [y] = \begin{cases} {}^*1 & \text{if } [x] \preceq_* [y], \\ [y] & \text{otherwise,} \end{cases}$ i.e. $[x] \rightarrow_G [y] = \Delta([x] \rightarrow_L [y]) \vee [y]$.

Proposition 5.1 *A structure* $\mathfrak{L}_{{}^*\mathbf{Q}} = \langle {}^*\mathbf{Q}_{[0,1]}, \oplus, \neg_L, \rightarrow_{\Pi}, \&_{\Pi}, {}^*0, {}^*1\rangle$ *is a hyperrational valued* ŁΠ*-matrix.*

Proof We should show that

1. $\langle {}^*\mathbf{Q}_{[0,1]}, \oplus, \neg_L, {}^*0\rangle$ is an MV-algebra, i.e. (1) $\langle {}^*\mathbf{Q}_{[0,1]}, \oplus, {}^*0\rangle$ is a commutative monoid, (2) $[x] \oplus {}^*1 = {}^*1$, (3) $\neg_L\neg_L[x] = [x]$, (4) $([x] \ominus [y]) \oplus [y] = ([y] \ominus [x]) \oplus [x]$;
2. $\langle {}^*\mathbf{Q}_{[0,1]}, \vee, \wedge, \rightarrow_{\Pi}, \&_{\Pi}, {}^*0, {}^*1\rangle$ is a Π-algebra, i.e. (1) $\langle {}^*\mathbf{Q}_{[0,1]}, \vee, \wedge\rangle$ is a bounded lattice with the order $\preceq_*$, with the top element ^{*1} and the bottom element

^{*0}, (2) $\langle {}^*\mathbf{Q}_{[0,1]}, \&_{\Pi}, {}^*1\rangle$ is a commutative semigroup with the unit element ^{*1}, (3) $\rightarrow_{\Pi}$ and $\&_{\Pi}$ form an adjoint pair, i.e. $[z] \preceq_* [x] \rightarrow_{\Pi} [y]$ iff $[x]\&_{\Pi}[z] \preceq_* [y]$ for all $[x], [y], [z] \in {}^*\mathbf{Q}_{[0,1]}$;
3. $[x]\&_{\Pi}([y] \ominus [z]) = ([x]\&_{\Pi}[y]) \ominus ([x]\&_{\Pi}[z])$.

However, all three items are readily checked. For instance,

1. $([x] \ominus [y]) \oplus [y] = {}^*1 - \sup[{}^*1 - (\sup([x], [y]) - [y]), [y]] + [y] = \sup([x], [y]) = ([y] \ominus [x]) \oplus [x]$.
2. Show that $[z] \preceq_* [x] \rightarrow_{\Pi} [y]$ iff $[x]\&_{\Pi}[z] \preceq_* [y]$ for all $[x], [y], [z] \in {}^*\mathbf{Q}_{[0,1]}$.
(i) Suppose that $[x] \preceq_* [y]$. Then $[z] \preceq_* [x] \rightarrow_{\Pi} [y] = {}^*1$ and $[x] \cdot [z] \preceq_* [y]$.
(ii) Otherwise $[x] \rightarrow_{\Pi} [y] = \inf\left({}^*1, \frac{[y]}{[x]}\right) \preceq_* \frac{[y]}{[x]}$. In this case $[z] \preceq_* \frac{[y]}{[x]}$ iff $[x] \cdot [z] \preceq_* [y]$.
3. $[x]\&_{\Pi}([y] \ominus [z]) = [x] \cdot (\sup([y], [z]) - [z]) = (\sup([x] \cdot [y], [x] \cdot [z]) - [x] \cdot [z]) = ([x]\&_{\Pi}[y]) \ominus ([x]\&_{\Pi}[z])$. □

A *hyperrational valued* Ł$\Pi\frac{1}{2}$*-matrix* is a structure $\mathfrak{L}_{{}^*\mathbf{Q}} = \langle {}^*\mathbf{Q}_{[0,1]}, \oplus, \neg_L, \rightarrow_{\Pi}, \&_{\Pi}, {}^*0, {}^*1, {}^*\frac{1}{2}\rangle$, where the reduct $\langle {}^*\mathbf{Q}_{[0,1]}, \oplus, \neg_L, \rightarrow_{\Pi}, \&_{\Pi}, {}^*0, {}^*1\rangle$ is a hyperrational valued ŁΠ-matrix and the identity ${}^*\frac{1}{2} = \neg_L {}^*\frac{1}{2}$ holds.

The truth value $^*0 \in {}^*\mathbf{Q}_{[0,1]}$ of a hyperrational valued ŁΠ- (resp. Ł$\Pi\frac{1}{2}$)-matrix is falsity, the truth value $^*1 \in {}^*\mathbf{Q}_{[0,1]}$ is truth, and other truth values $x \in {}^*\mathbf{Q}_{[0,1]}\backslash\{{}^*0, {}^*1\}$ are called neutral.

If we replace the set $\mathbf{Q}_{[0,1]}$ by $\mathbf{R}_{[0,1]}$ and the set ${}^*\mathbf{Q}_{[0,1]}$ by ${}^*\mathbf{R}_{[0,1]}$ in all above definitions, then we obtain *hyperreal valued* ŁΠ*-matrix* (resp. *hyperreal valued* Ł$\Pi\frac{1}{2}$*-matrix*) $\mathfrak{L}_{{}^*\mathbf{R}}$.

5.2.2 p-*Adic Valued* BL-*Matrix*

Extend the standard order structure on $\mathbf{N}$ to a partial order structure on $\mathbf{Z}_p$. We know that each finite number of $\mathbf{Z}_p$ can be identified with a positive integer.

1. for any finite numbers $x, y \in \mathbf{Z}_p$ we have $x \preceq_p y$ iff $x \leq y$ in $\mathbf{N}$,
2. each finite natural number x is less than any infinite number y, i.e. $x \prec_p y$ for any $x \in \mathbf{N}$ and $y \in \mathbf{Z}_p\backslash\mathbf{N}$, $y \neq 0$. Notice that we have $x \prec_p y$ iff $x \neq y$ and $x \preceq_p y$.

Define this partial order structure on $\mathbf{Z}_p$ as follows:

$\mathcal{O}_{\mathbf{Z}_p}$ Let $x = \ldots x_n \ldots x_1 x_0$ and $y = \ldots y_n \ldots y_1 y_0$ be the canonical expansions of two p-adic integers $x, y \in \mathbf{Z}_p$. (1) We set $x \prec_p y$ if the following three conditions hold: (i) there exists n such that $x_n < y_n$; (ii) $x_k \leq y_k$ for all $k > n$; (iii) x is a finite integer, i.e. there exists l such that $x_m = 0$ for all $m \geq l$. (2) We set $x = y$ if $x_n = y_n$ for each $n = 0, 1, \ldots$ (3) Suppose that both x and y are infinite integers. We set $x \preceq_p y$ if we have $x_n \leq y_n$ for each $n = 0, 1, \ldots$

and we set $x \prec_p y$ if we have $x_n \leq y_n$ for each $n = 0, 1, \dots$ and there exists n_0 such that $x_{n_0} < y_{n_0}$.

Now introduce two operations sup, inf in the partial order structure on $\mathbf{Z}_p$: $\sup(x, y) = y$ and $\inf(x, y) = x$ iff $x \preceq_p y$. Let $x = \dots x_n \dots x_1 x_0$ and $y = \dots y_n \dots y_1 y_0$ be the canonical expansions of two p-adic integers $x, y \in \mathbf{Z}_p$ and x, y are incompatible in $\mathcal{O}_{\mathbf{Z}_p}$. We get $\inf(x, y) = z = \dots z_n \dots z_1 z_0$, where, for each $n = 0, 1, \dots$, we set (1) $z_n = y_n$ if $x_n \geq y_n$, (2) $z_n = x_n$ if $x_n \leq y_n$, (3) $z_n = x_n = y_n$ if $x_n = y_n$. We get $\sup(x, y) = z = \dots z_n \dots z_1 z_0$, where, for each $n = 0, 1, \dots$, we set (1) $z_n = y_n$ if $x_n \leq y_n$, (2) $z_n = x_n$ if $x_n \geq y_n$, (3) $z_n = x_n = y_n$ if $x_n = y_n$.

It is important to remark that there exists the maximal number $N_{max} \in \mathbf{Z}_p$ in $\mathcal{O}_{\mathbf{Z}_p}$. It is easy to see: $N_{max} = -1 = (p-1) + (p-1) \cdot p + \dots + (p-1) \cdot p^k + \dots$[1]

Further, consider the following new operations defined for all $x, y \in \mathbf{Z}_p$ in the partial order structure $\mathcal{O}_{\mathbf{Z}_p}$:

- $x \rightarrow_L y = N_{max} - \sup(x, y) + y$,
- $x \rightarrow_\Pi y = \begin{cases} N_{max} & \text{if } x \preceq_p y, \\ \text{integral part of } \dfrac{y}{x} & \text{otherwise,} \end{cases}$
- $\neg_L x = N_{max} - x$, i.e. $x \rightarrow_L 0$,
- $\neg_\Pi x = \begin{cases} N_{max} & \text{if } x = 0, \\ 0 & \text{otherwise,} \end{cases}$ i.e. $\neg_\Pi x = x \rightarrow_\Pi 0$,
- $\Delta x = \begin{cases} N_{max} & \text{if } x = N_{max}, \\ 0 & \text{otherwise,} \end{cases}$ i.e. $\Delta x = \neg_\Pi \neg_L x$,
- $x \&_L y = \sup(x, N_{max} - y) + y - N_{max}$, i.e. $x \&_L y = \neg_L (x \rightarrow_L \neg_L y)$,
- $x \&_\Pi y = x \cdot y$,
- $x \oplus y := \neg_L x \rightarrow_L y$,
- $x \ominus y := x \&_L \neg_L y$,
- $x \wedge y = \inf(x, y)$, i.e. $x \wedge y = x \&_L (x \rightarrow_L y)$,
- $x \&_G y := \inf(x, y)$,
- $x \vee y = \sup(x, y)$, i.e. $x \vee y = (x \rightarrow_L y) \rightarrow_L y$,
- $x \leftrightarrow y := (x \rightarrow_L y) \wedge (y \rightarrow_L x)$,
- $x \rightarrow_G y = \begin{cases} N_{max} & \text{if } x \preceq_p y, \\ y & \text{otherwise,} \end{cases}$ i.e. $x \rightarrow_G y = \Delta(x \rightarrow_L y) \vee y$.

Proposition 5.2 *A structure $\langle \mathbf{Z}_p, \oplus, \neg_L, 0 \rangle$ is a p-adic valued MV-algebra.* □

Proposition 5.3 *A structure $\langle \mathbf{Z}_p, \oplus, \neg_L, \rightarrow_\Pi, \&_\Pi, 0, N_{max} \rangle$ is not a p-adic valued ŁΠ-matrix.*

Proof Indeed, it can be easily shown that $\langle \mathbf{Z}_p, \vee, \wedge, \rightarrow_\Pi, \&_\Pi, 0, N_{max} \rangle$ is not a Π-algebra. □

---

[1] If $x = \sum_{i=n}^{\infty} x_i \cdot p^i$ then $-x = \sum_{i=n}^{\infty} y_i \cdot p^i$, where $y_n = p - x_n$ and $y_i = (p-1) - x_i$ for $i > n$. For example, $-0 = 0$ and if $p = 5$ we have $\frac{1}{3} = \dots 1313132$ and $-\frac{1}{3} = \dots 131313$.

Proposition 5.4 *A structure $\mathfrak{L}_{\mathbf{Z}_p} = \langle \mathbf{Z}_p, \wedge, \vee, *, \Rightarrow, 0, N_{max} \rangle$, where $* \in \{\&_L, \&_G\}$ and $\Rightarrow \in \{\rightarrow_L, \rightarrow_G\}$ is a p-adic valued BL-matrix. If $* = \&_L$ and $\Rightarrow = \rightarrow_L$, it is called a p-adic valued* **L**-*algebra. If $* = \&_G$ and $\Rightarrow = \rightarrow_G$, it is called a p-adic valued* **G**-*algebra.*

Proof We can show that (1) $\langle \mathbf{Z}_p, \wedge, \vee, 0, N_{max} \rangle$ is a lattice with the largest element N_{max} and the least element 0, (2) $\langle \mathbf{Z}_p, *, N_{max} \rangle$ is a commutative semigroup with the unit element N_{max}, i.e. $*$ is commutative, associative, and $N_{max} * x = x$ for all $x \in \mathbf{Z}_p$, (3) the following conditions hold

$$z \preceq_p (x \Rightarrow y) \text{ iff } x * z \preceq_p y \text{ for all } x, y, z \in \mathbf{Z}_p;$$

$$x \wedge y = x * (x \Rightarrow y);$$

$$x \vee y = ((x \Rightarrow y) \Rightarrow y) \wedge ((y \Rightarrow x) \Rightarrow x);$$

$$(x \Rightarrow y) \vee (y \Rightarrow x) = N_{max}.$$

□

The truth value $0 \in \mathbf{Z}_p$ of a p-adic valued BL-matrix is called falsity, the truth value N_{max} is called truth, and other truth values $x \in \mathbf{Z}_p \backslash \{0, N_{max}\}$ are called neutral.

We can dualize the order $\preceq_p$ in the following natural way: $x \succeq_p^N y$ iff $x \preceq_p y$ and $x \neq 0$. As we see, 1 was the least positive p-adic integer due to $\preceq_p$ and became the maximal number due to $\preceq_p^N$ (respectively, -1 was the largest p-adic integer and became the least positive integer). Let us set 0 as the minimal.

Reintroduce two operations sup, inf in the new partial order structure on $\mathbf{Z}_p$: $\sup(x, y) = y$ and $\inf(x, y) = x$ iff $x \preceq_p^N y$. If two p-adic integers x, y are incompatible, their maximum and minimum are defined digit by digit too. Consider the following new operations defined for all $x, y \in \mathbf{Z}_p$ in the partial order structure ordered by $\preceq_p^N$:

- $x \rightarrow_L y = 1 - \sup(x, y) + y$,
- $x \rightarrow_\Pi y = \begin{cases} 1 & \text{if } x \preceq_p^N y, \\ \text{integral part of } \frac{y}{x} & \text{otherwise,} \end{cases}$
- $\neg_L x = 1 - x$, i.e. $x \rightarrow_L 0$,
- $\neg_\Pi x = \begin{cases} 1 & \text{if } x = 0, \\ 0 & \text{otherwise,} \end{cases}$ i.e. $\neg_\Pi x = x \rightarrow_\Pi 0$,
- $\Delta x = \begin{cases} 1 & \text{if } x = 1, \\ 0 & \text{otherwise,} \end{cases}$ i.e. $\Delta x = \neg_\Pi \neg_L x$,
- $x \&_L y = \sup(x, 1 - y) + y - 1$, i.e. $x \&_L y = \neg_L (x \rightarrow_L \neg_L y)$,
- $x \&_\Pi y = x \cdot y$,
- $x \oplus y := \neg_L x \rightarrow_L y$,
- $x \ominus y := x \&_L \neg_L y$,

- $x \wedge y = \inf(x, y)$, i.e. $x \wedge y = x \&_L (x \to_L y)$,
- $x \vee y = \sup(x, y)$, i.e. $x \vee y = (x \to_L y) \to_L y$,
- $x \leftrightarrow y := (x \to_L y) \wedge (y \to_L x)$,
- $x \to_G y = \begin{cases} 1 & \text{if } x \preceq_p^N y, \\ y & \text{otherwise,} \end{cases}$ i.e. $x \to_G y = \Delta(x \to_L y) \vee y$.

Proposition 5.5 *A structure* $\mathfrak{L}'_{\mathbf{Z}_p} = \langle \mathbf{Z}_p, \wedge, \vee, *, \Rightarrow, 0, 1 \rangle$, *where* $* \in \{\&_L, \&_G\}$ *and* $\Rightarrow \in \{\to_L, \to_G\}$ *is a p-adic valued* BL*-matrix.* □

The truth value $0 \in \mathbf{Z}_p$ of a p-adic valued BL-matrix is called falsity, the truth value $1 \in \mathbf{Z}_p$ is called truth, and other truth values $x \in \mathbf{Z}_p \backslash \{0, 1\}$ are called neutral.

5.2.3 Non-Archimedean Approach to Higher-Order Fuzzy Classes

Now consider fuzzy predicates with infinite-order (ω-order) fuzziness and show that they can have a non-Archimedean valued interpretation. For the first time the higher-order fuzziness was modeled degree-theoretically in the framework of higher-order many-valued logic in the paper [43]. Notice that fuzzy sets of higher level are defined in fuzzy class theory considered in [38, 44, 45]. Fuzzy relations of higher level are introduced in Henkin-style fuzzy type theory and higher-order fuzzy logic proposed in [46].

Consider fuzzy relations of higher level in order to get later quantified fuzzy relations of infinite level.

Begin with fuzzy relations of first level. Let us remember that every relation R of D^n can be represented by its characteristic function $c_R : D^n \to \{0, 1\}$ defined by setting $c_R(x_1, \ldots, x_n) = 1$ if $\langle x_1, \ldots, x_n \rangle \in R$ and $c_R(x_1, \ldots, x_n) = 0$ if $\langle x_1, \ldots, x_n \rangle \notin R$. We can also represent the extension of a vague predicate by a generalized characteristic function assuming values in a set $[0, 1]$, e.g. either in the set $\mathbf{Q}_{[0,1]}$ or in $\mathbf{R}_{[0,1]}$. A *fuzzy relation* or *fuzzy subset* P of D^n is any map $P : D^n \to [0, 1]$. For instance, the subset $\emptyset \subset D^n$ (resp. $D^n \subseteq D^n$) can be represented by a fuzzy relation P that is considered as the constant map $P : D^n \ni x \mapsto 0$ (resp. $P : D^n \ni x \mapsto 1$), where 0 is the least element of $[0, 1]$ and 1 is the largest element of $[0, 1]$. We denote by $\mathscr{F}(D^n)$ the class of fuzzy subsets of D^n. It is obvious that if we identify sets with their characteristic functions, then $\mathscr{P}(D^n) \subset \mathscr{F}(D^n)$, where $\mathscr{P}(D^n)$ is the powerset.

Let 1 be the top element of $[0, 1]$. Observe that, if P and P' are fuzzy subsets of D^n, then we can define,

- the *inclusion* by setting $P \subseteq P'$ iff $P(x_1, \ldots, x_n) \leq P'(x_1, \ldots, x_n)$ for every $\langle x_1, \ldots, x_n \rangle \in D^n$,
- *Łukasiewicz's relative complement* $P \to_L P'$ by setting $(P \to_L P')(x_1, \ldots, x_n) = 1 - \max(P(x_1, \ldots, x_n), P'(x_1, \ldots, x_n)) + P'(x_1, \ldots, x_n)$ for every $\langle x_1, \ldots, x_n \rangle \in D^n$,

- *Łukasiewicz's complement* $\neg_L P$ of P by setting $(\neg_L P)(x_1, \ldots, x_n) = 1 - P(x_1, \ldots, x_n)$ for every $\langle x_1, \ldots, x_n \rangle \in D^n$,
- *Łukasiewicz's intersection* $P \&_L P'$ as $(P \&_L P')(x_1, \ldots, x_n) = \max(P(x_1, \ldots, x_n), 1 - P'(x_1, \ldots, x_n)) + P'(x_1, \ldots, x_n) - 1$ for every $\langle x_1, \ldots, x_n \rangle \in D^n$,
- the *Product relative complement* $P \rightarrow_\Pi P'$ by setting $(P \rightarrow_\Pi P')(x_1, \ldots, x_n) = 1$ if $P(x_1, \ldots, x_n) \leq P'(x_1, \ldots, x_n)$ for every $\langle x_1, \ldots, x_n \rangle \in D^n$ and $(P \rightarrow_\Pi P')(x_1, \ldots, x_n) = \frac{P(x_1, \ldots, x_n)}{P'(x_1, \ldots, x_n)}$ otherwise,
- the *Product complement* $\neg_\Pi P$ as $\neg_\Pi P(x_1, \ldots, x_n) := P(x_1, \ldots, x_n) \rightarrow_\Pi 0$ for every $\langle x_1, \ldots, x_n \rangle \in D^n$,
- the *Product intersection* $P \&_\Pi P'$ as $(P \&_\Pi P')(x_1, \ldots, x_n) = P(x_1, \ldots, x_n) \cdot P'(x_1, \ldots, x_n)$ for every $\langle x_1, \ldots, x_n \rangle \in D^n$,
- the *union* $P \vee P'$ as $(P \vee P')(x_1, \ldots, x_n) = \max(P(x_1, \ldots, x_n), P'(x_1, \ldots, x_n))$ for every $\langle x_1, \ldots, x_n \rangle \in D^n$,
- the *intersection* $P \wedge P'$ by setting $(P \wedge P')(x_1, \ldots, x_n) = \min(P(x_1, \ldots, x_n), P'(x_1, \ldots, x_n))$ for every $\langle x_1, \ldots, x_n \rangle \in D^n$,
- the *union* $\bigvee\limits_{i \in I} P_i$, given a family $(P_i)_{i \in I}$ of fuzzy subsets of D^n, as

$$\bigvee_{i \in I} P_i(x_1, \ldots, x_n) = \max\{P_i(x_1, \ldots, x_n) \colon i \in I\}$$

  for every $\langle x_1, \ldots, x_n \rangle \in D^n$,
- the *intersection* $\bigwedge\limits_{i \in I} P_i$, given a family $(P_i)_{i \in I}$ of fuzzy subsets of D^n, as

$$\bigwedge_{i \in I} P_i(x_1, \ldots, x_n) = \min\{P_i(x_1, \ldots, x_n) \colon i \in I\}$$

  for every $\langle x_1, \ldots, x_n \rangle \in D^n$.

The structure $\langle \mathscr{F}(D^n), \wedge, \vee, \neg_L, \bot(D^n), \top(D^n) \rangle$ is a complete lattice with an involution (see the proof in [47]), where $\bot(D^n)$ and $\top(D^n)$ are the constant maps $\bot(D^n) \colon D^n \ni x \mapsto 0 \in V$ and $\top(D^n) \colon D^n \ni x \mapsto 1 \in V$ respectively. The structure $\mathfrak{L}_V = \langle \mathscr{F}(D^n), \wedge, \vee, *, \Rightarrow, \bot(D^n), \top(D^n) \rangle$ is a BL-algebra. It is the direct power, with index set D^n, of the structure $\langle V, \wedge, \vee, *, \Rightarrow, 0, 1 \rangle$.

We define an $i+1$-*order class* $\mathscr{F}_{i+1}(D^n)$ *of fuzzy subsets* of D^n by recursion: $\mathscr{F}_1(D^n) = \mathscr{F}(D^n)$ and $\mathscr{F}_{i+1}(D^n) = \mathscr{F}(\mathscr{F}_i(D^n))$, where $i \geq 1$. The members $P_{i+1} \in \mathscr{F}_{i+1}(D^n)$ are called *fuzzy relations/subsets of* $i+1$-th *level* (or $i+1$-*order fuzzy relations/subsets*). It is possible to define an *infinite-order class* $\mathscr{F}_\infty(D^n)$ *of fuzzy subsets* by setting $\mathscr{F}_\infty(D^n) := \bigcup_{i \in \mathbf{N}} \mathscr{F}_i(D^n)$. The members $P_\infty \in \mathscr{F}_\infty(D^n)$ are called *fuzzy relations/subsets of infinite level* (or *infinite-order fuzzy relations/subsets*).

Let V be any nonempty set such that $|V| > 2$ and $\mathscr{F}_i(D)$ is the class of all maps $\underbrace{V^{V^{\cdots V^D}}}_{i+1} \rightarrow V$. The following sets are not isomorphic:

$$\underbrace{V^{V^{\cdots V^{D^n}}}}_{i+2} \not\cong (\underbrace{V \times \ldots \times V}_{i})^{D^n}.$$

We can build the following map $\boldsymbol{\rho}$ of the class of all *constant mappings* of $(V^i)^{D^n}$ to the set $\mathscr{F}_i(D^n)$:

- if a fuzzy set $P_i \in \mathscr{F}_i(D^n)$ has the range $\mathbf{rg}(P_i) = \{k\}$ (i.e. P_i is the constant mapping such that it takes each member $P_{i-1} \in \mathscr{F}_{i-1}(D^n)$ to the value k), so the map
$$P_i^{\mathrm{Q}} \colon D^n \to \{\langle \underbrace{k, \ldots, k}_{i} \rangle\}$$
belongs to the set $\boldsymbol{\rho}^{-1}(P_i)$ (it is evident that in this case P_i^{Q} is the only member of $\boldsymbol{\rho}^{-1}(P_i)$),
- suppose (1) a fuzzy set $P_i \in \mathscr{F}_i(D^n)$ has the value $k_i \in \mathbf{rg}(P_i)$ for some valuations (i.e. P_i is the mapping such that it takes *some* members $P_{i-1} \in \mathscr{F}_{i-1}(D^n)$ to the value k_i, these members will be denoted by $P_{i-1}^{k_i}$), (2) a fuzzy set $P_{i-1}^{k_i} \in \mathscr{F}_{i-1}(D^n)$ has the value $k_{i-1} \in \mathbf{rg}(P_{i-1}^{k_i})$ for some valuations, ..., (i) a fuzzy set $P_1^{k_2} \in \mathscr{F}_1(D^n)$ has the value $k_1 \in \mathbf{rg}(P_1^{k_2})$ for some valuations (i.e. $P_1^{k_2}$ is the mapping such that it takes *some* members of D^n to the value k_1), then the map
$$P_i^{\mathrm{Q}} \colon D^n \ni x \mapsto \langle k_1, k_2, \ldots, k_i \rangle$$
belongs to the set $\boldsymbol{\rho}^{-1}(P_i)$ (it is obvious that in this case $\boldsymbol{\rho}^{-1}(P_i)$ contains more than one member).

(1) In the first case (if a fuzzy set $P_i \in \mathscr{F}_i(D^n)$ has the range $\mathbf{rg}(P_i) = \{k\}$), we will denote P_i^{Q} by $\forall^{\langle k,\ldots,k\rangle} P_{i-1}$. This means that at the level i the membership degree of a relation P_{i-1} in P_i is equal to k for any valuations. (2) In the second case (if a fuzzy set $P_i \in \mathscr{F}_i(D^n)$ is not a constant function), we will denote P_i^{Q} by $\exists^{\langle k_1,\ldots,k_i\rangle} P_{i-1}$. This means that at the level $r \in \{2, \ldots, i\}$ the membership degree of a relation P_{r-1} in P_r is equal to k_r for some valuations.

Predicates of the form $\forall^{\langle k,\ldots,k\rangle} P_{i-1}$ and $\exists^{\langle k_1,\ldots,k_i\rangle} P_{i-1}$ are called *i-order quantified fuzzy relations/predicates* (or quantified fuzzy relations/predicates of i-th level). For instance, we can represent a quantified fuzzy relation P_{i+1}^{Q} of $i+1$-th level as any constant map $P_{i+1}^{\mathrm{Q}} : D^n \to V^{i+1}$. The *class of all quantified fuzzy relations of i- th* level we will denote by $\mathscr{F}_i^{\mathrm{Q}}(D^n)$. Define the *class of all quantified fuzzy relations of infinite level* by setting $\mathscr{F}_\infty^{\mathrm{Q}}(D^n) := \bigcup_{i \in \mathbf{N}} \mathscr{F}_i^{\mathrm{Q}}(D^n)$. The members $P_\infty \in \mathscr{F}_\infty^{\mathrm{Q}}(D^n)$ we can consider as the following constant maps $P_\infty : D^n \to {}^*V$, i.e. we set functions of the form $P_\infty(x_1, \ldots, x_n) = \langle y_1, \ldots, y_{i+1}, \ldots \rangle = [f]$, where $\langle x_1, \ldots, x_n \rangle \in D^n$ and $[f] \in {}^*V$.

Let us construct an extension $\mathbf{F}_V = \langle \mathscr{F}_\infty^{\mathrm{Q}}(D^n), \wedge, \vee, *, \Rightarrow, {}^*\bot(D^n), {}^*\top(D^n) \rangle$ of the BL-algebra on $\mathscr{F}(D^n)$ as the direct power, with index set D^n (and restricted to the set of all constant functions), of non-Archimedean valued BL-matrix $\langle {}^*V,$

$\wedge, \vee, *, \Rightarrow, {}^*0, {}^*1\rangle$. Note that in the case $V = \mathbf{Q}_{[0,1]}$, we have the hyperrational multiple-validity and in the case $V = \mathbf{R}_{[0,1]}$, the hyperreal multiple-validity.

Thanks to $\mathbf{F}_V$, we can define i-order (ω-order) quantified predicates of a higher-order extension of the logic $BL\forall$ by some i-order (ω-order) quantified fuzzy relations. (1) The $i+1$-order predicate $\forall^{\langle k,\ldots,k\rangle} P_i \in \rho^{-1}(P'_{i+1})$ regarded as membership function is the constant map

$$\forall^{\langle k,\ldots,k\rangle} P_i \colon D^n \ni x \mapsto \underbrace{\langle k, \ldots, k\rangle}_{i+1},$$

where $k \in V$, this means that for any $P_i \in \mathscr{F}_i(D^n)$ we have $\langle P_i, k\rangle \in P'_{i+1}$. (2) The ω-order predicate $\forall^{*k} P_\infty$ regarded as membership function is the constant map $\forall^{*k} P_\infty : D^n \ni x \mapsto {}^*k$, where ${}^*k \in {}^*V$. (3) The $i+1$-order predicate $\exists^{\langle k_1,\ldots,k_{i+1}\rangle} P_i \in \rho^{-1}(P'_{i+1})$ regarded as membership function is the constant map

$$\exists^{\langle k_1,\ldots,k_{i+1}\rangle} P_i \colon D^n \ni x \mapsto \langle k_1, k_2, \ldots, k_{i+1}\rangle \in V^{i+1},$$

this means that for some $P_i \in \mathscr{F}_i(D^n)$ we have $\langle P_i, k_{i+1}\rangle \in P'_{i+1}$, for some $P_{i-1} \in \mathscr{F}_{i-1}(D^n)$ we have $\langle P_{i-1}, k_i\rangle \in P_i$, etc. (4) The ω-order predicate $\exists^{[f]} P_\infty$ regarded as membership function is the constant map

$$\exists^{[f]} P_\infty \colon D^n \ni x \to [f] \in {}^*V.$$

Also, the structure $\mathbf{F}_V$ allows to set an infinite-order predicate language and its semantics.

5.2.4 Non-Archimedean Valued Predicate Logic $\mathbf{BL}\forall_\infty$

An *infinite-order predicate logical language* $\mathscr{L}_V^\infty$ (with the set of truth values V) consists of the following symbols: (1) Variables: (i) denumerable set of free variables $a_0, a_1, a_2, \ldots$; (ii) denumerable set of bound variables $x_0, x_1, x_2, \ldots$; (iii) infinite set of i-order predicate variables of arity n of the form P_i^n ($i \in \omega$); (iv) infinite set of infinite-order (ω-order) predicate variables of arity n of the form P_∞^n. (2) Constants: (i) denumerable set of constant symbols $c_0, c_1, c_2, \ldots$; (ii) denumerable set of function symbols of arity n: $F_0^n, F_1^n, F_2^n, \ldots$; (iii) first-order predicate signs of arity n: $N_0^n, N_1^n, N_2^n, \ldots$. (3) Logical symbols: (i) the truth constant ^{*0}, (ii) various order propositional connectives of arity n_j: $\square_0^{n_0}, \square_1^{n_1}, \ldots, \square_r^{n_r}$, which are built by superposition of $\to_\star$, $\&_\star$, where $\star \in \{L, G, \Pi\}$; (iii) first-order quantifiers $\forall, \exists$; (iv) infinite set of i-order vertical quantifiers (or predicate quantifiers) of the form $\forall^{\langle k,\ldots,k\rangle}$ and $\exists^{\langle k_1,\ldots,k_i\rangle}$ ($i \in \omega$); (v) infinite set of infinite-order (ω-order) vertical quantifiers of the form $\forall^{*k}$ and $\exists^{[f]}$. (4) Auxiliary symbols (,).

If $V = \{0, 1, \ldots, p - 1\}$, where p is a prime number, then we have p universal vertical quantifiers at the level i and p existential vertical quantifiers at the level i. If $V = \mathbf{Q}_{[0,1]}$, then we have $\aleph_0$ universal vertical quantifiers at the level i and $\aleph_0$ existential vertical quantifiers at the level i. If $V = \mathbf{R}_{[0,1]}$, then we have $2^{\aleph_0}$ universal vertical quantifiers at the level i and $2^{\aleph_0}$ existential vertical quantifiers at the level i.

Terms, well-formed atomic formulas, and well-formed first-order formulas of $\mathscr{L}_V^\infty$ are defined by standard way. *Well-formed higher-order formulas* of $\mathscr{L}_V^\infty$ are inductively defined as follows:

1. If $\boldsymbol{\Psi}$ is a first-order or $i + 1$-order formula not containing the i-order predicate variable P_i, N is a first-order predicate, $\mathrm{Q} \in \{\forall^{\langle k,\ldots,k\rangle}, \exists^{\langle k_1,\ldots,k_{i+1}\rangle}\}$ is an $i + 1$-order vertical quantifier, then $\mathrm{Q}P_i\ \boldsymbol{\Psi}(P_i)$, where $\boldsymbol{\Psi}(P_i)$ is obtained from $\boldsymbol{\Psi}$ by replacing N by P_i at every occurrence of N in $\boldsymbol{\Psi}$, is an $i + 1$-order formula. Its outermost logical symbol is Q.
2. If $\boldsymbol{\Psi}$ is a first-order or ω-order formula not containing the ω-order predicate variable P_∞, N is a first-order predicate, $\mathrm{Q} \in \{\forall^{*k}, \exists^{[f]}\}$ is an infinite-order vertical quantifier, then $\mathrm{Q}P_\infty\ \boldsymbol{\Psi}(P_\infty)$, where $\boldsymbol{\Psi}(P_\infty)$ is obtained from $\boldsymbol{\Psi}$ by replacing N by P_∞ at every occurrence of N in $\boldsymbol{\Psi}$, is an infinite-order formula. Its outermost logical symbol is Q.

A higher-order formula is called *vertical bounded* if it does not contain first-order predicates.

Let $\mathfrak{L}_{*V}$ be a non-Archimedean valued BL-matrix. Notice that ${}^*V$ is identified with $\mathbf{Z}_p$ for $\{0, 1, \ldots, p - 1\}$. An $\mathfrak{L}_{*V}$-structure $\mathbf{M} = \langle D, I\rangle$ for first-order formulas of $\mathscr{L}_V^\infty$ consists of the following: (1) a non-empty set D; (2) a mapping I such that each constant symbol c is mapped to an element c_I of D, each n-place function symbol F^n to a function $F_I^n : D^n \to D$, each n-place predicate symbol N^n to a *fuzzy* subset N_I^n of D^n. Let $A = \{a_0, a_1, a_2, \ldots\}$ be the set of all free variable symbols. A function $s : A \to D$ is said to be a variable assignment for $\mathbf{M} = \langle D, I\rangle$. Let T be the set of all terms. Then we can extend a variable assignment s to a denotation assignment $\mathbf{s} : T \to D$ as follows: (1) for each free variable a, $\mathbf{s}(a) = s(a)$; (2) for each constant symbol c, $\mathbf{s}(c) = c_I$; (3) for every n-place function symbol F^n and terms t_1, ..., t_n, $\mathbf{s}(F(t_1, \ldots, t_n)) = F_I(\mathbf{s}(t_1), \ldots, \mathbf{s}(t_n))$, (4) for every n-place predicate symbol N^n and terms t_1, ..., t_n, $\mathbf{s}(N(t_1, \ldots, t_n)) = N_I(\mathbf{s}(t_1), \ldots, \mathbf{s}(t_n))$.

We define a *predicate assignment* S as follows: (1) for each n-place first-order predicate variable P_1, ${}^S P_1$ is a fuzzy subset of D^n; (2) for each n-place i-order predicate variable P_i, ${}^S P_i$ is a member of $\mathscr{F}_i(D^n)$; (3) for each n-place ω-order predicate variable P_∞, ${}^S P_\infty$ is a member of $\mathscr{F}_\infty(D^n)$. Now extend an $\mathfrak{L}_{*V}$-structure $\langle\mathbf{M}, \mathbf{s}\rangle$ to a *higher-order $\mathfrak{L}_{*V}$-structure* $\mathbf{M}_\infty = \langle(\mathscr{F}_\infty(D^n))_{n\in\omega}, \mathbf{s}_\infty\rangle$ as follows: (1) for each n-place i-order predicate variable P_i and terms t_1, ..., t_n, $\mathbf{s}_\infty(P_i(t_1, \ldots, t_n)) = {}^S P_i(\mathbf{s}(t_1), \ldots, \mathbf{s}(t_n))$; (2) for each n-place ω-order predicate variable P_∞ and terms t_1, ..., t_n, $\mathbf{s}_\infty(P_\infty(t_1, \ldots, t_n)) = {}^S P_\infty(\mathbf{s}(t_1), \ldots, \mathbf{s}(t_n))$.

Let $\mathfrak{L}_{*V}$ be a BL-matrix and $\mathbf{M}_\infty$ be an $\mathfrak{L}_{*V}$-structure for $\mathscr{L}_V^\infty$.

An *i-order truth assignment* on a higher-order $\mathfrak{L}_{*V}$-structure $\mathbf{M}_\infty$ is a function $||\cdot||^i$ whose domain is the set of all i-order formulas of $\mathscr{L}_V^\infty$ and whose range is the set V^i of truth values such that:

1. For any first-order formula Φ, $||\Phi||^1$ is a truth assignment of the basic fuzzy logic $BL\forall$ (e.g. of Łukasiewicz's predicate logic **L**, Gödel's predicate logic **G** or Product predicate logic Π). For instance,

$$||N(t_1,\ldots,t_n)||^1 = \mathbf{s}(N(t_1,\ldots,t_n)) = N_I(\mathbf{s}(t_1),\ldots,\mathbf{s}(t_n)),$$

$$||P_1(t_1,\ldots,t_n)||^1 = \mathbf{s}_\infty(P_1(t_1,\ldots,t_n)) = {}^S P_1(\mathbf{s}(t_1),\ldots,\mathbf{s}(t_n)).$$

   Extend $||\cdot||^1$ as follows:

$$||P_i(t_1,\ldots,t_n)||^1 = \mathbf{s}_\infty(P_i(t_1,\ldots,t_n)) = {}^S P_i(\mathbf{s}(t_1),\ldots,\mathbf{s}(t_n)).$$

2. For any formula Φ without vertical quantifiers,

$$||\Phi||^i = \langle\underbrace{y_1,\ldots y_1}_{i}\rangle \text{ iff } ||\Phi||^1 = y_1.$$

3. For any first- or i-order formula Φ, $||\star\Phi||^i = \star||\Phi||^i$, where $\star \in \{\neg_L, \neg_G, \neg_\Pi\}$, the symbols $\neg_L, \neg_G, \neg_\Pi$ on the left-hand side are operations in logic **L**, **G**, and Π respectively, and on the right-hand side they are appropriate logical connectives in algebra **L**, **G**, and Π respectively.
4. For any first- or i-order formulas Φ and $\boldsymbol{\Psi}$,

$$||\Phi\star\boldsymbol{\Psi}||^i = ||\Phi||^i\star||\boldsymbol{\Psi}||^i = \langle\star(x_1,y_1),\star(x_2,y_2),\ldots,\star(x_i,y_i)\rangle,$$

   where $\star \in \{\rightarrow_L, \rightarrow_\Pi, \rightarrow_G, \&_L, \&_\Pi, \&_G\}$, $||\Phi||^i = \langle x_1, x_2,\ldots,x_i\rangle$, and $||\boldsymbol{\Psi}||^i = \langle y_1, y_2,\ldots,y_i\rangle$.
5. For any i-order vertical bounded formula $\forall^{\langle k,\ldots,k\rangle}P_{i-1}\boldsymbol{\Psi}(P_{i-1})$,

$$||\forall^{\langle k,\ldots,k\rangle}P_{i-1}\boldsymbol{\Psi}(P_{i-1})||^i = \langle\underbrace{k,\ldots,k}_{i}\rangle$$

   if for any member ${}^S P_{i-1}$ of $\mathscr{F}_{i-1}(D^n)$, we have $||\boldsymbol{\Psi}(P_{i-1})||^1 = \boldsymbol{\Psi}({}^S P_{i-1}) = k$, i.e. if there exists the member ${}^S\forall^{\langle k,\ldots,k\rangle}P_{i-1}$ of $\mathscr{F}^{\mathrm{Q}}_{i-1}(D^n)$ such that

$$||\forall^{\langle k,\ldots,k\rangle}P_{i-1}\boldsymbol{\Psi}(P_{i-1})||^i = \boldsymbol{\Psi}({}^S\forall^{\langle k,\ldots,k\rangle}P_{i-1}) = \langle\underbrace{k,\ldots,k}_{i}\rangle.$$

   Otherwise $||\forall^{\langle k,\ldots,k\rangle}P_{i-1}\boldsymbol{\Psi}(P_{i-1})||^i = \langle\underbrace{0,\ldots,0}_{i}\rangle$.
6. For any i-order vertical bounded formula $\exists^{\langle k_1,\ldots,k_{i-1},k_i\rangle}P_{i-1}\boldsymbol{\Psi}(P_{i-1})$,

$$||\exists^{\langle k_1,\ldots,k_{i-1},k_i\rangle}P_{i-1}\boldsymbol{\Psi}(P_{i-1})||^i = \langle k_1,\ldots,k_{i-1},k_i\rangle$$

if (i) for some members ${}^S P_{i-1}$ of $\mathscr{F}_{i-1}(D^n)$, we have

$$||\exists^{\langle k_1,\dots,k_{i-1},k_i\rangle} P_{i-1}\boldsymbol{\Psi}(P_{i-1})||_1 = \boldsymbol{\Psi}({}^S P_{i-1}) = k_i,$$

(ii) there exists ${}^S\exists^{\langle k_1,\dots,k_{i-1}\rangle} P_{i-2} \in \boldsymbol{\rho}^{-1}({}^S P_{i-1})$ such that

$$||\exists^{\langle k_1,\dots,k_{i-1}\rangle} P_{i-2}\boldsymbol{\Psi}(P_{i-2})||^{i-1} = \langle k_1, \dots, k_{i-1}\rangle.$$

In other words, if there exists the member ${}^S\exists^{\langle k_1,\dots,k_i\rangle} P_{i-1}$ of $\mathscr{F}^{\mathrm{Q}}_{i-1}(D^n)$ such that

$$||\exists^{\langle k_1,\dots,k_i\rangle} P_{i-1}\, \boldsymbol{\Psi}(P_{i-1})||^i = \boldsymbol{\Psi}({}^S\exists^{\langle k_1,\dots,k_i\rangle} P_{i-1}) = \langle k_1, \dots, k_i\rangle.$$

Otherwise $||\exists^{\langle k_1,\dots,k_i\rangle} P_{i-1}\boldsymbol{\Psi}(P_{i-1})||^i = \langle\underbrace{0, \dots, 0}_{i}\rangle$.

7. For any i-order vertical bounded formula $\forall x\, \mathrm{Q}P_{i-1}\, \Phi(P_{i-1}(x))$,

$$||\forall x\, \mathrm{Q}P_{i-1}\, \Phi(P_{i-1}(x))||^i = ||\mathrm{Q}P_{i-1}\, \forall x\, \Phi(P_{i-1}(x))||^i = \\ = \inf_{a\in D} ||\Phi(P_{i-1}(x))||^1_{s[x\to a]} \wedge \Phi({}^S\mathrm{Q}P_{i-1}),$$

where $\mathrm{Q} \in \{\forall^{\langle k,\dots,k\rangle}, \exists^{\langle k_1,\dots,k_i\rangle}\}$ and $s[x \to a]$ is a variable assignment such that $s[x \to a](x) = a$ and $s[x \to a](y) = s(y)$ for each individual variable y different from x.

An *infinite-order* (ω-*order*) *truth assignment* on a higher-order $\mathfrak{L}_{{}^*V}$-structure $\mathbf{M}_\infty$ is a function $||\cdot||^\infty$ whose domain is the set of all infinite-order formulas of $\mathscr{L}_V^\infty$ and whose range is the set ${}^*V$ of truth values such that:

1. Extend $||\cdot||^1$ as follows:

$$||P_\infty(t_1, \dots, t_n)||^1 = \mathbf{s}_\infty(P_\infty(t_1, \dots, t_n)) = {}^S P_\infty(\mathbf{s}(t_1), \dots, \mathbf{s}(t_n)).$$

2. For any formula Φ without vertical quantifiers,

$$||\Phi||^\infty = {}^*y_1 = \langle y_1, y_1, \dots\rangle \text{ iff } ||\Phi||^1 = y_1.$$

3. For any first- or ω-order formula Φ, $||\star\,\Phi||^\infty = \star||\Phi||^\infty$, where $\star \in \{\neg_L, \neg_G, \neg_\Pi\}$, the symbols $\neg_L, \neg_G, \neg_\Pi$ on the left-hand side are operations in ω-order logic, and on the right-hand side they are appropriate logical connectives in BL-algebra $\mathfrak{L}_{{}^*V}$.
4. For any first- or ω-order formulas Φ and $\boldsymbol{\Psi}$,

$$||\Phi \star \boldsymbol{\Psi}||^\infty = ||\Phi||^\infty \star ||\boldsymbol{\Psi}||^\infty,$$

where $\star \in \{\to_L, \to_\Pi, \to_G, \&_L, \&_\Pi, \&_G\}$, the symbol $\star$ on the right-hand side is an appropriate operation in BL-algebra $\mathfrak{L}_{{}^*V}$.

5. For any ω-order vertical bounded formula $\forall^{*k} P_\infty \boldsymbol{\Psi}(P_\infty)$,

$$||\forall^{*k} P_\infty \boldsymbol{\Psi}(P_\infty)||^\infty = {}^*k \in {}^*V$$

if there exists ${}^S\forall^{*k} P_\infty \in \boldsymbol{\rho}^{-1}({}^S P_\infty)$ such that

$$||\forall^{*k} P_\infty \boldsymbol{\Psi}(P_\infty)||^\infty = \boldsymbol{\Psi}({}^S\forall^{*k} P_\infty) = {}^*k.$$

Otherwise $||\forall^{*k} P_\infty \boldsymbol{\Psi}(P_\infty)||^\infty = {}^*0$.

6. For any ω-order vertical bounded formula $\exists^{[f]} P_\infty \boldsymbol{\Psi}(P_\infty)$,

$$||\exists^{[f]} P_\infty \boldsymbol{\Psi}(P_\infty)||^\infty = [f] \in {}^*V$$

if there exists ${}^S\exists^{[f]} P_\infty \in \boldsymbol{\rho}^{-1}({}^S P_\infty)$ such that

$$||\exists^{[f]} P_\infty \boldsymbol{\Psi}(P_\infty)||^\infty = \boldsymbol{\Psi}({}^S\exists^{[f]} P_\infty) = [f].$$

Otherwise $||\exists^{[f]} P_\infty \boldsymbol{\Psi}(P_\infty)||^\infty = {}^*0$.

7. For any ω-order vertical bounded formula $\forall x\, \mathrm{Q} P_\infty\, \Phi(P_\infty(x))$,

$$||\forall x\, \mathrm{Q} P_\infty\, \Phi(P_\infty(x))||^\infty = ||\mathrm{Q} P_\infty\, \forall x\, \Phi(P_\infty(x))||^\infty = \\ = \inf_{a \in D} ||\Phi(P_\infty(x))||^1_{s[x \to a]} \wedge \Phi({}^S\mathrm{Q} P_\infty),$$

where $\mathrm{Q} \in \{\forall^{*k}, \exists^{[f]}\}$ and $s[x \to a]$ is a variable assignment such that $s[x \to a](x) = a$ and $s[x \to a](y) = s(y)$ for each individual variable y different from x.

5.2.4.1 Hyper-valued Predicate Logic $BL\forall_\infty$

Let us concentrate on the case if ${}^*V = {}^*[0, 1]$.

We say that an i-order formula Φ_i (resp. ω-order formula Φ_∞) is *logically valid*/is an $\mathfrak{L}_{^*V}$*-tautology* if $||\Phi_i||^i = \langle 1, \dots, 1\rangle$ (resp. $||\Phi_\infty||^\infty = {}^*1$) for each higher-order $\mathfrak{L}_{^*V}$-structure $\mathbf{M}_\infty$. An i-order formula Φ_i (resp. ω-order formula Φ_∞) is *logically satisfiable* if $||\Phi_i||^i \neq \langle 0, \dots, 0\rangle$ (resp. $||\Phi_\infty||^\infty \neq {}^*0$) for some $\mathfrak{L}_{^*V}$-structures $\mathbf{M}_\infty$.

We say that an $\mathfrak{L}_{^*V}$-structure $\mathbf{M}_\infty$ is an i-order $\mathfrak{L}_{^*V}$*-model* (resp. an ω-order $\mathfrak{L}_{^*V}$*-model*) of an $\mathscr{L}_V^\infty$-theory T iff $||\Phi||^i = \langle 1, \dots, 1\rangle$ (resp. $||\Phi||^\infty = {}^*1$) on $\mathbf{M}_\infty$ for each $\Phi \in T$.

Denote by Π_1 (resp. by Σ_1) the class of vertical bounded higher-order formulas $\boldsymbol{\Psi}$ such that all vertical quantifiers of $\boldsymbol{\Psi}$ are universal (existential) with the same upper indices.

Theorem 5.1 *1. Suppose that $\boldsymbol{\Psi}_i$ (resp. $\boldsymbol{\Psi}_\infty$) is an i-order (resp. ω-order) formula that belongs to Π_1 and $\boldsymbol{\Psi}$ is a first-order formula in that every i-order (resp.*

ω-order) predicate variable of $\boldsymbol{\Psi}_i$ (resp. of $\boldsymbol{\Psi}_\infty$) is replaced by a first-order predicate constant, then $\boldsymbol{\Psi}_i$ (resp. $\boldsymbol{\Psi}_\infty$) is logically valid iff $\boldsymbol{\Psi}$ is logically valid.

2. *Suppose that $\boldsymbol{\Psi}_i$ (resp. $\boldsymbol{\Psi}_\infty$) is an i-order (resp. infinite-order) formula that belongs to Σ_1 and $\boldsymbol{\Psi}$ is a first-order formula in that every i-order (resp. infinite-order) predicate variable of $\boldsymbol{\Psi}_i$ (resp. of $\boldsymbol{\Psi}_\infty$) is replaced by a first-order predicate constant, then $\boldsymbol{\Psi}_i$ (resp. $\boldsymbol{\Psi}_\infty$) is logically satisfiable iff $\boldsymbol{\Psi}$ is logically satisfiable.*

Proof It follows from the semantic rules of $\mathscr{L}_V^\infty$. □

As a result, the class of higher-order formulas belonging to Π_1 and Σ_1 is effectively axiomatizable. In this section we will consider just formulas of these both classes.

Let us construct an infinite-order extension of $BL\forall$ denoted by $BL\forall_\infty$. Remember that the logic $BL\forall$ has just two propositional operations: &, $\rightarrow$, which are understood as t-norm and its residuum respectively.

In $BL\forall$ we can define the following new operations:

- $\Phi \wedge \boldsymbol{\Psi} := \Phi \& (\Phi \rightarrow \boldsymbol{\Psi})$,
- $\Phi \vee \boldsymbol{\Psi} := ((\Phi \rightarrow \boldsymbol{\Psi}) \rightarrow \boldsymbol{\Psi}) \wedge ((\boldsymbol{\Psi} \rightarrow \Phi) \rightarrow \Phi)$,
- $\neg\Phi := \Phi \rightarrow 0$,
- $\Phi \leftrightarrow \boldsymbol{\Psi} := (\Phi \rightarrow \boldsymbol{\Psi}) \& (\boldsymbol{\Psi} \rightarrow \Phi)$,
- $\Phi \oplus \boldsymbol{\Psi} := \neg\Phi \rightarrow \boldsymbol{\Psi}$,
- $\Phi \ominus \boldsymbol{\Psi} := \Phi \& \neg\boldsymbol{\Psi}$,

The logic $BL\forall_\infty$ is given by the following axioms, where all subformulas are either first-order or i-order (ω-order) belonging to Π_1 and Σ_1:

$$(\Phi \rightarrow \boldsymbol{\Psi}) \rightarrow ((\boldsymbol{\Psi} \rightarrow \Gamma) \rightarrow (\Phi \rightarrow \Gamma)) \tag{5.1}$$

$$(\Phi \& \boldsymbol{\Psi}) \rightarrow \Phi \tag{5.2}$$

$$(\Phi \& \boldsymbol{\Psi}) \rightarrow (\boldsymbol{\Psi} \& \Phi) \tag{5.3}$$

$$(\Phi \& (\Phi \rightarrow \boldsymbol{\Psi})) \rightarrow (\boldsymbol{\Psi} \& (\boldsymbol{\Psi} \rightarrow \Phi)) \tag{5.4}$$

$$(\Phi \rightarrow (\boldsymbol{\Psi} \rightarrow \Gamma)) \rightarrow ((\Phi \& \boldsymbol{\Psi}) \rightarrow \Gamma) \tag{5.5}$$

$$((\Phi \& \boldsymbol{\Psi}) \rightarrow \Gamma) \rightarrow (\Phi \rightarrow (\boldsymbol{\Psi} \rightarrow \Gamma)) \tag{5.6}$$

$$((\Phi \to \Psi) \to \Gamma) \to (((\Psi \to \Phi) \to \Gamma) \to \Gamma) \tag{5.7}$$

$$^*0 \to \Psi \tag{5.8}$$

$$\forall x\, \Psi(x) \to \Psi(t), \tag{5.9}$$

where t is substitutable for x in $\Psi(x)$.

$$\Psi(t) \to \exists x\, \Psi(x), \tag{5.10}$$

where t is substitutable for x in $\Psi(x)$.

$$\forall x(\Gamma \to \Psi) \to (\Gamma \to \forall x\, \Psi), \tag{5.11}$$

where x is not free in Γ.

$$\forall x(\Psi \to \Gamma) \to (\exists x\, \Psi \to \Gamma), \tag{5.12}$$

where x is not free in Γ.

$$\forall x(\Gamma \vee \Psi) \to (\Gamma \vee \forall x\, \Psi), \tag{5.13}$$

where x is not free in Γ.

These axioms are said to be *horizontal*. Introduce also some new axioms that show basic properties of non-Archimedean ordered structures. These express a connection between formulas of various level. It is well known that there exist infinitesimals that are less than any positive number of [0, 1]. This property can be expressed by means of the following logical axiom:

$$(\neg(\Psi_1 \leftrightarrow \Psi_\infty)\&\neg(\Phi_1 \leftrightarrow {}^*0)) \to (\Psi_\infty \to \Phi_1), \tag{5.14}$$

where Ψ_1, Φ_1 are first-order formulas and Ψ_∞ is an ω-order formula of classes Π_1 or Σ_1 that has such a logical matrix as Ψ_1. Axiom (5.14) is said to be *vertical*.

The deduction rules of $BL\forall_\infty$ are *modus ponens* (from Φ, $\Phi \to \Psi$ infer Ψ) and *generalization* (from Φ infer $\forall x\, \Phi$).

The notions of proof, derivability $\vdash$, theorem, and theory over $BL\forall_\infty$ are defined as usual.

Theorem 5.2 *(Completeness) Let Φ be an i-order (resp. ω-order) formula of $\mathscr{L}_V^\infty$, T an i-order (resp. ω-order) $\mathscr{L}_V^\infty$-theory. Then the following conditions are equivalent:*

- $T \vdash \Phi$;
- $||\Phi||^i = \langle \underbrace{1, \ldots, 1}_{i} \rangle$ *(resp. $||\Phi||^\infty = {}^*1$) for each $\mathfrak{L}_{*V}$-model $\mathbf{M}_\infty$ of T;*

Proof The first-order case is proved in [41] and the higher-order case follows from Theorem 5.1. □

5.2.4.2 p-Adic Valued Logic $BL'\forall_\infty$

Let us consider the case $^*V = \{0, 1, \dots, p-1\}$.

The p-adic valued logic $BL'\forall_\infty$ is given by the following axioms, where all subformulas are either first-order or i-order (ω-order):

$$(\Phi \to \Psi) \to ((\Psi \to \Gamma) \to (\Phi \to \Gamma)) \tag{5.15}$$

$$(\Phi \& \Psi) \to \Phi \tag{5.16}$$

$$(\Phi \& \Psi) \to (\Psi \& \Phi) \tag{5.17}$$

$$(\Phi \& (\Phi \to \Psi)) \to (\Psi \& (\Psi \to \Phi)) \tag{5.18}$$

$$(\Phi \to (\Psi \to \Gamma)) \to ((\Phi \& \Psi) \to \Gamma) \tag{5.19}$$

$$((\Phi \& \Psi) \to \Gamma) \to (\Phi \to (\Psi \to \Gamma)) \tag{5.20}$$

$$((\Phi \to \Psi) \to \Gamma) \to (((\Psi \to \Phi) \to \Gamma) \to \Gamma) \tag{5.21}$$

$$0 \to \Psi \tag{5.22}$$

$$\forall x \, \Psi(x) \to \Psi(t), \tag{5.23}$$

where t is substitutable for x in $\Psi(x)$.

$$\Psi(t) \to \exists x \, \Psi(x), \tag{5.24}$$

where t is substitutable for x in $\Psi(x)$.

$$\forall x (\Gamma \to \Psi) \to (\Gamma \to \forall x \, \Psi), \tag{5.25}$$

where x is not free Γ.

$$\forall x(\boldsymbol{\Psi} \to \Gamma) \to (\exists x\, \boldsymbol{\Psi} \to \Gamma), \tag{5.26}$$

where x is not free Γ.

$$\forall x(\Gamma \vee \boldsymbol{\Psi}) \to (\Gamma \vee \forall x\, \boldsymbol{\Psi}), \tag{5.27}$$

where x is not free Γ.

These axioms are said to be *horizontal*.

There is a well known theorem according to that every equivalence class a for which $|a|_p \leq 1$ (this means that a is a p-adic integer) has exactly one representative Cauchy sequence for which:

1. $0 \leq a_i < p^i$ for $i = 1, 2, 3, \ldots$;
2. $a_i \equiv a_{i+1} \mod p^i$ for $i = 1, 2, 3, \ldots$

This property can be expressed by means of the following logical axioms:

$$\begin{aligned}&((\overline{p^{i+1}-1} \ominus \overline{p^i - 1}) \to_L \boldsymbol{\Psi}_{i+1}) \to_L \\ &(\boldsymbol{\Psi}_{i+1} \leftrightarrow (\underbrace{\overline{p-1} \oplus \ldots \oplus \overline{p-1}}_{p^i} \oplus \boldsymbol{\Psi}_i),\end{aligned} \tag{5.28}$$

$$\begin{aligned}&(\boldsymbol{\Psi}_{i+1} \leftrightarrow (\overbrace{\overline{k} \oplus \ldots \oplus \overline{k}}^{p^i}) \oplus \boldsymbol{\Psi}_i) \to_L \\ &((((\ldots(\underbrace{\overline{p^{i+1}-1} \ominus \overline{p^i - 1}) \ominus \ldots) \ominus \overline{p^i - 1}}_{0<p-k\leq p}) \ominus \neg_L \overline{k}) \to_L \boldsymbol{\Psi}_{i+1}),\end{aligned} \tag{5.29}$$

$$\begin{aligned}&(\boldsymbol{\Psi}_{i+1} \leftrightarrow (\overbrace{\overline{k} \oplus \ldots \oplus \overline{k}}^{p^i}) \oplus \boldsymbol{\Psi}_i) \to_L \\ &(\boldsymbol{\Psi}_{i+1} \to_L (((\ldots(\overline{p^{i+1}-1} \ominus \underbrace{\overline{p^i - 1}) \ominus \ldots) \ominus \overline{p^i - 1}}_{0\leq(p-1)-k\leq p-1}) \ominus \neg_L \overline{k})),\end{aligned} \tag{5.30}$$

$$(\boldsymbol{\Psi}_{i+1} \to_L \overline{p^i - 1}) \to_L (\boldsymbol{\Psi}_{i+1} \leftrightarrow \boldsymbol{\Psi}_i), \tag{5.31}$$

$$\begin{aligned}&(\boldsymbol{\Psi}_{i+1} \leftrightarrow \boldsymbol{\Psi}_i) \vee (\boldsymbol{\Psi}_{i+1} \leftrightarrow (\boldsymbol{\Psi}_i \oplus p^i \cdot \overline{1})) \vee \ldots \\ &\vee(\boldsymbol{\Psi}_{i+1} \leftrightarrow (\boldsymbol{\Psi}_i \oplus p^i \cdot \overline{p-1})),\end{aligned} \tag{5.32}$$

where $\overline{p-1}$ is a tautology at the first-order level and $\overline{p^i - 1}$ (respectively $\overline{p^{i+1} - 1}$) a tautology at i-th order level (respectively at $(i+1)$-th order level); $i+1$-order formula Ψ_{i+1} and i-order formula Ψ_i have the same logical matrix; $\neg_L \overline{k}$ is a first-order formula that has the truth value $((p-1)-k) \in \{0, \ldots, p-1\}$ for any its interpretations and $\overline{k}$ is a first-order formula that has the truth value $k \in \{0, \ldots, p-1\}$ for any its interpretations; $\overline{1}$ is a first-order formula that has the truth value 1 for any its interpretations, etc. The denoting $p^i \cdot \overline{k}$ means $\underbrace{\overline{k} \oplus \ldots \oplus \overline{k}}_{p^i}$.

Axioms (5.28)–(5.32) are said to be *vertical*.

The inference rules of $BL'\forall_\infty$ are *modus ponens* (from Φ, $\Phi \to \Psi$ infer Ψ) and *generalization* (from Φ infer $\forall x\, \Phi$).

Theorem 5.3 *(Completeness) Let Φ be an i-order (resp. ω-order) formula of $\mathcal{L}^{\infty}_{*\forall}$, T an i-order (resp. ω-order) $\mathcal{L}^{\infty}_{*\forall}$-theory. Then the following conditions are equivalent:*

- $T \vdash \Phi$;
- $\left[\Phi\right]^i = \langle \underbrace{\top, \ldots, \top}_{i} \rangle$ *(resp.* $\left[\Phi\right]^\infty = {}^*\top$*) for each* $\mathfrak{L}_{\mathbf{Z}_p}$*-model* $\mathbf{M}_\infty$ *of* T;

Proof The first-order case is proved in [41] and the higher-order case follows from Theorem 5.1. □

5.3 Non-Archimedean Valued Probability Logics

5.3.1 Hyper-valued Probabilities

Now set up the problem to get non-Archimedean probability logic on the basis of $BL\forall_\infty$.

In the standard way, probabilities are defined on an algebra of subsets. Recall that an *algebra* $\mathscr{A}$ *of subsets* $A \subset X$ consists of the following: (1) union, intersection, and difference of two subsets of X; (2) $\emptyset$ and X. Then a *finitely additive probability measure* is a nonnegative set function $\mathbf{P}(\cdot)$ defined for sets $A \in \mathscr{A}$ that satisfies the following properties:

1. $\mathbf{P}(A) \geq 0$ for all $A \in \mathscr{A}$,
2. $\mathbf{P}(X) = 1$ and $\mathbf{P}(\emptyset) = 0$,
3. if $A \in \mathscr{A}$ and $B \in \mathscr{A}$ are disjoint, then $\mathbf{P}(A \cup B) = \mathbf{P}(A) + \mathbf{P}(B)$. In particular $\mathbf{P}(\neg A) = 1 - \mathbf{P}(A)$ for all $A \in \mathscr{A}$.

The algebra $\mathscr{A}$ is called a σ-*algebra* if it is assumed to be closed under countable union (or equivalently, countable intersection), i.e. if for every n, $A_n \in \mathscr{A}$ causes $A = \bigcup_n A_n \in \mathscr{A}$.

A set function $\mathbf{P}(\cdot)$ defined on a σ-algebra is called a *countable additive probability measure* (or a σ-*additive probability measure*) if in addition to satisfying equations of the definition of finitely additive probability measure, it satisfies the following countable additivity property: for any sequence of pairwise disjoint sets A_n, $\mathbf{P}(A) = \sum_n \mathbf{P}(A_n)$. The ordered system $\langle X, \mathscr{A}, \mathbf{P}\rangle$ is called a *probability space*.

Now consider non-Archimedean probabilities.

Define a structure $\langle \mathscr{P}(^*\boldsymbol{\Theta}), \wedge, \vee, \neg, {}^*\boldsymbol{\Theta}\rangle$ as follows (1) each $A \in \mathscr{P}(^*\boldsymbol{\Theta})$ is identified with the set of points of the form $f(\alpha)$, (2) for any $A, B \in \mathscr{P}(^*\boldsymbol{\Theta})$, $A \wedge B = \{f(\alpha)\colon f(\alpha) \in A \wedge f(\alpha) \in B\}$, (3) for any $A, B \in \mathscr{P}(^*\boldsymbol{\Theta})$, $A \vee B = \{f(\alpha)\colon f(\alpha) \in A \vee f(\alpha) \in B\}$, (4) for any $A \in \mathscr{P}(^*\boldsymbol{\Theta})$, $\neg A = \{f(\alpha)\colon f(\alpha) \in {}^*\boldsymbol{\Theta}\backslash A\}$. Then a structure $\langle \mathscr{P}(^*\boldsymbol{\Theta}), \wedge, \vee, \neg, {}^*\boldsymbol{\Theta}\rangle$ is not a Boolean algebra if $|\boldsymbol{\Theta}| \geq 2$, because the set $\mathscr{P}(^*\boldsymbol{\Theta})$ is not closed under intersection. Indeed, by definition of non-standard extension, some elements of ${}^*\boldsymbol{\Theta}$ have a non-empty intersection. Therefore some subsets of ${}^*\boldsymbol{\Theta}$ have a non-empty intersection that does not belong to $\mathscr{P}(^*\boldsymbol{\Theta})$, i.e. they aren't atoms of $\mathscr{P}(^*\boldsymbol{\Theta})$ of the form $[f]$.

Consequently, there is a problem how it is possible to define probabilities on non-Archimedean structures if we have not an opportunity to put them on an algebra of subsets. A. Khrennikov who considered non-Archimedean probability theory on the set of p-adic numbers [6] proposed a semi-algebra that is closed only with respect to a finite unions of sets, which have empty intersections. However there is a better way if we get non-Archimedean probabilities on a class $\mathscr{F}^{\mathrm{Q}}_{\infty}(X)$ of fuzzy subsets $A \subset X$.

In the next section we shall survey non-Archimedean probability logic in that probabilities are defined on a class $\mathscr{F}^{\mathrm{Q}}_{\infty}(X)$ of fuzzy subsets $A \subset X$. In this case a *finitely additive probability measure* is a nonnegative set function $\mathbf{P}(\cdot)$ defined for sets $A \in \mathscr{F}^{\mathrm{Q}}_{\infty}(X)$ that runs the set Q and satisfies the following properties:

1. $\mathbf{P}(A) \geq {}^*0$ for all $A \in \mathscr{F}^{\mathrm{Q}}_{\infty}(X)$,
2. $\mathbf{P}(X) = {}^*1$ and $\mathbf{P}(\emptyset) = {}^*0$,
3. if $A \in \mathscr{F}^{\mathrm{Q}}_{\infty}(X)$ and $B \in \mathscr{F}^{\mathrm{Q}}_{\infty}(X)$ are disjoint, then $\mathbf{P}(A \cup B) = \mathbf{P}(A) + \mathbf{P}(B)$.
4. $\mathbf{P}(\neg A) = {}^*1 - \mathbf{P}(A)$ for all $A \in \mathscr{F}^{\mathrm{Q}}_{\infty}(X)$,

where ^{*1} is the largest member of $Q = {}^*[0, 1]$ and ^{*0} is the least member of $Q = {}^*[0, 1]$.

This probability measure is called *non-Archimedean fuzzy probability*. The main originality of fuzzy probabilities is that conditions 3, 4 are independent. As a result, in a probability space $\langle X, \mathscr{F}^{\mathrm{Q}}_{\infty}(X), \mathbf{P}\rangle$ some Bayes' formulas do not hold in the general case (see Theorem 5.4).

5.3.2 *Hyper-valued Probability Logic*

A function $\mathbf{P}(\Phi)$ is said to be a *non-Archimedean absolute fuzzy probability function* for a statement Φ in $\mathscr{L}^{\infty}_{V}$ if $\mathbf{P}(\Phi)$ ranges over the number set ${}^*V$ (${}^*V = {}^*\mathbf{Q}_{[0,1]}$ or ${}^*V = {}^*\mathbf{R}_{[0,1]}$) and satisfies the following constraints:

$$\mathbf{P}(\Phi) = ||\Phi||^{\infty}, \tag{5.33}$$

where $||\Phi||^{\infty}$ is a truth valuation of Φ;

$$\mathbf{P}(\Phi) + \mathbf{P}(\neg\Phi) = {}^{*}1; \tag{5.34}$$

if $\Phi \wedge \Psi$ is logically false, then

$$\mathbf{P}(\Phi \vee \Psi) = \mathbf{P}(\Phi) + \mathbf{P}(\Psi); \tag{5.35}$$

if Φ and Ψ are logically equivalent, then

$$\mathbf{P}(\Phi) = \mathbf{P}(\Psi); \tag{5.36}$$

$$\mathbf{P}(\forall x\Phi \wedge \Psi) = \lim_{n\to\infty} \mathbf{P}((\Phi(t_1/x) \wedge \ldots \wedge \Phi(t_n/x)) \wedge \Psi), \tag{5.37}$$

where $\Phi(t_1/x)$ is the result of replacing the variable x everywhere in Φ by a term t_1, $\Phi(t_n/x)$ is the result of replacing the variable x everywhere in Φ by a term t_n, etc;

$$\mathbf{P}(t_i = t_i) = {}^{*}1; \tag{5.38}$$

$$\mathbf{P}(t_i = t_j \supset (\Phi \equiv \Phi(t_j//t_i))) = {}^{*}1, \tag{5.39}$$

where $\Phi(t_j//t_i)$ is the result of replacing t_i by t_j at zero or more places in Φ;

if t_i and t_j are terms of $\mathscr{L}_V^{\infty}$, then

$$\mathbf{P}(t_i = t_j) = ||t_i = t_j||^{\infty}, \tag{5.40}$$

where $||t_i = t_j||^{\infty}$ is a truth valuation of $t_i = t_j$.

From (5.33) it follows that

$$\mathbf{P}(\bot) = {}^{*}0,$$

where $\bot$ is a contradiction of $BL\forall_{\infty}$ (i.e. it has the truth value *0 for all valuations);

$$\mathbf{P}(\top) = {}^{*}1,$$

where $\top$ is a tautology of $BL\forall_{\infty}$ (i.e. it has the truth value *1 for all valuations).

Under the condition of (5.34), we obtain $\mathbf{P}(\neg\Phi) = {}^{*}1 - \mathbf{P}(\Phi)$.

Non-Archimedean relative fuzzy probability functions $\mathbf{P}(\Phi/\Psi) \in {}^{*}V$ for the language $\mathscr{L}_V^{\infty}$ are characterized by the following constraint:

$$\mathbf{P}(\Phi/\Psi) = \frac{\mathbf{P}(\Phi \wedge \Psi)}{\mathbf{P}(\Psi)} = \frac{||\Phi \wedge \Psi||^{\infty}}{||\Psi||^{\infty}} = \frac{\inf(||\Phi||^{\infty}, ||\Psi||^{\infty})}{||\Psi||^{\infty}}, \tag{5.41}$$

when $\boldsymbol{\Psi}$ is not logically false, where $||\Phi||^{\infty}$ is a truth valuation of Φ and $||\boldsymbol{\Psi}||^{\infty}$ is a truth valuation of $\boldsymbol{\Psi}$.

From this constraint we deduce the following properties of non-Archimedean relative fuzzy probability functions:

$$\mathbf{P}(\Phi/\Phi) = {}^*1, \text{ when } \Phi \text{ is not logically false;} \tag{5.42}$$

if $\Phi \wedge \boldsymbol{\Psi}$ is logically false, then

$$\mathbf{P}(\Phi \vee \boldsymbol{\Psi}/\Gamma) = \mathbf{P}(\Phi/\Gamma) + \mathbf{P}(\boldsymbol{\Psi}/\Gamma), \tag{5.43}$$

when Γ is not logically false;

$$\mathbf{P}(\Phi \wedge \boldsymbol{\Psi}/\Gamma) = \mathbf{P}(\Phi/\boldsymbol{\Psi} \wedge \Gamma) \cdot \mathbf{P}(\boldsymbol{\Psi}/\Gamma), \tag{5.44}$$

when Γ and $\boldsymbol{\Psi} \wedge \Gamma$ are not logically false;

if Φ and $\boldsymbol{\Psi}$ are logically equivalent, then

$$\mathbf{P}(\Phi/\Gamma) = \mathbf{P}(\boldsymbol{\Psi}/\Gamma), \tag{5.45}$$

when Γ is not logically false;

if Φ and $\boldsymbol{\Psi}$ are logically equivalent, then

$$\mathbf{P}(\Gamma/\Phi) = \mathbf{P}(\Gamma/\boldsymbol{\Psi}), \tag{5.46}$$

when Φ and $\boldsymbol{\Psi}$ are not logically false;

$$\mathbf{P}(\Phi/\boldsymbol{\Psi} \wedge \Phi) = {}^*1, \tag{5.47}$$

when $\boldsymbol{\Psi} \wedge \Phi$ is not logically false;

if $\Phi_1, \Phi_2, \ldots, \Phi_N$ are mutually exclusive, then

$$\mathbf{P}(\Phi_1 \vee \Phi_2 \vee \ldots \vee \Phi_N/\Gamma) = \sum_{n=1}^{N} \mathbf{P}(\Phi_n/\Gamma), \tag{5.48}$$

when Γ is not logically false;

$$\mathbf{P}(\forall x \Phi/\boldsymbol{\Psi}) = \lim_{n \to \infty} \mathbf{P}(\Phi(t_1/x) \wedge \ldots \wedge \Phi(t_n/x)/\boldsymbol{\Psi}), \tag{5.49}$$

where $\boldsymbol{\Psi}$ is not logically false, $\Phi(t_1/x)$ is the result of replacing the variable x everywhere in Φ by a term t_1, $\Phi(t_n/x)$ is the result of replacing the variable x everywhere in Φ by a term t_n, etc;

$$\mathbf{P}(t_i = t_i/\boldsymbol{\Psi}) = {}^*1, \tag{5.50}$$

when $\boldsymbol{\Psi}$ is not logically false;

$$\mathbf{P}(t_i = t_j \supset (\Phi \equiv \Phi(t_j//t_i))/\boldsymbol{\Psi}) = {}^*1, \tag{5.51}$$

when $\boldsymbol{\Psi}$ is not logically false, where $\Phi(t_j//t_i)$ is the result of replacing t_i by t_j at zero or more places in Φ;

if t_i and t_j are terms of $\mathscr{L}_V^\infty$, then

$$\mathbf{P}(t_i = t_j/\boldsymbol{\Psi}) = \frac{||(t_i = t_j) \wedge \boldsymbol{\Psi}||^\infty}{||\boldsymbol{\Psi}||^\infty}, \tag{5.52}$$

when $\boldsymbol{\Psi}$ is not logically false.

We postulate that

$$\text{if } \boldsymbol{\Psi} \text{ is logically false, then } \mathbf{P}(\Phi/\boldsymbol{\Psi}) = {}^*1, \text{ for every statement } \Phi \text{ of } \mathscr{L}_V^\infty. \tag{5.53}$$

From (5.33) and (5.53) it follows that

$$\text{if } \mathbf{P}(\Phi/\boldsymbol{\Psi}) = {}^*1 \text{ for all sentences } \boldsymbol{\Psi}, \text{ then } \Phi \text{ is a tautology;} \tag{5.54}$$

$$\text{if } ||\boldsymbol{\Psi} \supset \Phi||^\infty = {}^*1 \text{ for any truth valuations, then } \mathbf{P}(\Phi/\boldsymbol{\Psi}) = {}^*1. \tag{5.55}$$

Notice that all formulas of real inductive logic are also valid in non-Archimedean inductive logic if we restrict the set ${}^*V$ of hyperrational (hyperreal) numbers to the set ${}^*\{0, 1\} = \{0, 1\}^{\mathbf{N}}/\mathscr{U}$, i.e. to the set that contains only hypernumbers satisfying the following property: if $[f] = \langle \alpha_1, \alpha_2, \alpha_3, \ldots \rangle$, then for each $j = 1, 2, \ldots$ we have $\alpha_j \in \{0, 1\}$.

Under this condition, we have the Boolean complement. For example, the formula $\mathbf{P}(\Phi/\boldsymbol{\Psi}) + \mathbf{P}(\neg\Phi/\boldsymbol{\Psi}) = 1$ is valid for hypernumbers of ${}^*\{0, 1\}$, but in the general case, it isn't valid in non-Archimedean probability logic. Therefore non-Archimedean inductive logic is more general probability logic than real one. Show that Bayes' formulas are also valid for hypernumbers of the set ${}^*\{0, 1\}$.

Theorem 5.4 *(Bayes' theorem) Let the propositions $\Gamma_1, \ldots, \Gamma_N$ have truth valuations such that all truth values have the form of infinite tuple $\langle \alpha_1, \alpha_2, \alpha_3, \ldots \rangle$, where for any $j = 1, 2, \ldots$ we have $\alpha_j \in \{0, 1\}$. If proposition $\boldsymbol{\Psi}$ implies that the propositions $\Gamma_1, \ldots, \Gamma_N$ are mutually disjoint (i.e. $\Gamma_1 \vee \ldots \vee \Gamma_N$ has the truth value ^{*1} for all truth valuations), then*

$$\mathbf{P}(\Phi/\Psi) = \sum_{n=1}^{N} \mathbf{P}(\Phi \wedge \Gamma_n/\Psi), \tag{5.56}$$

$$\mathbf{P}(\Phi/\Psi) = \sum_{n=1}^{N} \mathbf{P}(\Phi/\Gamma_n \wedge \Psi) \cdot \mathbf{P}(\Gamma_n \wedge \Psi), \tag{5.57}$$

$$\mathbf{P}(\Gamma_n/\Phi \wedge \Psi) = \frac{\mathbf{P}(\Phi/\Gamma_k \wedge \Psi) \cdot \mathbf{P}(\Gamma_k \wedge \Psi)}{\sum_{n=1}^{N} \mathbf{P}(\Phi/\Gamma_n \wedge \Psi) \cdot \mathbf{P}(\Gamma_n \wedge \Psi)}, \tag{5.58}$$

where $k \in \{1, \ldots, N\}$.

Proof Since Ψ implies that the proposition $\Gamma_1 \vee \ldots \vee \Gamma_N$ is true, we see that $||\Psi||^{\infty} = ||\Psi||^{\infty} \wedge ||\Gamma_1 \vee \ldots \vee \Gamma_N||^{\infty}$. Thus, from (5.47) we obtain $\mathbf{P}(\Gamma_1 \vee \ldots \vee \Gamma_N/\Psi) = {}^*1$ and $\mathbf{P}(\Phi \wedge (\Gamma_1 \vee \ldots \vee \Gamma_N)/\Psi) = \mathbf{P}(\Phi/(\Gamma_1 \vee \ldots \vee \Gamma_N) \wedge \Psi) \cdot \mathbf{P}(\Gamma_1 \vee \ldots \vee \Gamma_N/\Psi)$.

Since Ψ implies that $\Gamma_1, \ldots, \Gamma_N$ are mutually exclusive, we have that $\Phi \wedge \Gamma_n$ and $\Phi \wedge \Gamma_m$ are also mutually exclusive ($n \neq m$ and $n, m \in \{1, \ldots, N\}$). Finally, from (5.34) we get (5.56).

Formula (5.57) follows from (5.56). Statement (5.58) follows from (5.56) and from rule (5.43). □

5.3.3 p-*Adic Valued Probabilities*

Let us remember that for each fixed prime number p, we can define a *p-adic norm* $|\cdot|_p$ on $\mathbf{Q}$ as follows. Let $a \in \mathbf{Q}$, $a \neq 0$, and $a = p^n q/r$, where q, r, and p are all co-prime to each other. Then $|a|_p = |p^n q/r|_p = p^{-n}$ and $|0|_p = 0$. By this definition, we understand the distance between two points $a, b \in \mathbf{Q}$ as $d(a, b) = |a - b|_p$. So, the larger power of a prime p, the closer distance. For example, in 7-adic metric, $d(2, 51) < d(1, 2)$, because $d(2, 51) = |51 - 2|^7 = |49|^7 = |7^2|^7 = 7^{-2} = 1/49$ and $d(1, 2) = |2 - 1|^7 = |1|^7 = |7^0|^7 = 7^0 = 1$.

The basic properties of p-adic norm/distance are as follows: (i) $|x|_p = 0$ iff $x = 0$; (ii) $|x \cdot y|_p = |x|_p \cdot |y|_p$; (iii) $|x + y|_p \leq \max\{|x|_p, |y|_p\}$. Property (iii) says that we deal with the strong triangle inequality that gives a *non-Archimedean metrics*.

Assume that $a \in \mathbf{Q}_p$, $r \in \{p^n : n \in \mathbf{Z}\} \cup \{0\}$. Then we can define the closed *p-adic ball* with center a and radius r: $B[a, r] := \{x \in \mathbf{Q}_p : |x - a|_p \leq r\}$, the open p-adic ball with center a and radius r: $B(a, r) := \{x \in \mathbf{Q}_p : |x - a|_p < r\}$, and the p-adic sphere with center a and radius r: $S(a, r) := \{x \in \mathbf{Q}_p : |x - a|_p = r\}$. Notice that the closed ball $B[0, 1]$ is the ring of p-adic integers denoted by $\mathbf{Z}_p$.

There are many interesting topological properties of p-adic metrics like that: (1) two p-adic balls are disjoint or one is contained in another; (2) every point of p-adic ball can be its center (if $b \in B[a, r]$, then $B[a, r] = B[b, r]$), (3) $B[a, r_1] \cap$

$B[b, r_2] = \emptyset$ iff $|a - b|_p > \max\{r_1, r_2\}$; (4) the field of p-adic numbers is not ordered, etc.

Khrennikov defines the *p-adic probability measure* as follows. Let us fix $r = p^n$ for some $n \in \mathbf{Z}$. Suppose, Ω is an arbitrary set of events and $\mathscr{F}$ is a field of subset of Ω. Suppose that there is an additive measure $\mathbf{P} : \mathscr{F} \to B[0, r]$ such that $\mathbf{P}(\Omega) = 1$. Then $\langle \Omega, \mathscr{F}, \mathbf{P} \rangle$ is called *p-adic r-probability space* and $\mathbf{P}$ is called *p-adic r-probability*. The main problem of this definition is that we cannot define the standard σ-additivity in general case, because all σ-additive p-adic valued measures defined on σ-fields are discrete, although there are many other measures. Therefore Khrennikov proposes to consider just p-adic valued measures which are bounded, continuous, and normalized (see [3]). This idea is his substantial restriction in defining p-adic valued measures.

In papers of [1, 2, 7, 37], there are constructed p-adic valued probability logics, where propositions take values in p-adic balls in accordance with Khrennikov's bounded, continuous, and normalized p-adic valued measures.

Another approach based on Khrennikov's attempts to define p-adic valued (or more generally, non-Archimedean valued or stream-valued) probabilities is offered in the papers: [11–15, 17]. Finitely additive p-adic measures presented in those papers are essential for our programming of the behaviour of a slime mould, so we shall state some basic facts about them.

The ring $\mathbf{Z}_p$ cannot be linearly ordered, but there are many possibilities to define a *partial ordering relation* on the set $\mathbf{Z}_p$. For example, we can assume that (i) for any finite p-adic integers $\sigma, \tau \in \mathbf{N}$, we have $\sigma \leq \tau$ in $\mathbf{Z}_p$ iff $\sigma \leq \tau$ in $\mathbf{N}$; (ii) $\sigma < \tau$ for any $\sigma \in \mathbf{N}$ and $\tau \in \mathbf{Z}_p \setminus \mathbf{N}$; (iii) for any pair of infinite $\sigma, \tau \in \mathbf{Z}_p$ let $\sigma \leq \tau$ iff $\sigma_t \leq \tau_t$ for all $t \in \mathbf{N}$. Let us denote this ordering relation by $\mathscr{O}_{\mathbf{Z}_p}$. It is easy to see that $\mathscr{O}_{\mathbf{Z}_p}$ is not a linear ordering. For example, let $p = 2$ and let σ represents the p-adic integer $-1/3 = \ldots 10101 \ldots 101$ and τ the p-adic integer $-2/3 = \ldots 01010 \ldots 010$. Then the p-adic numbers σ and τ are incomparable. Now we can define sup and inf digit by digit. It is easy to see that for any pair of infinite p-adic integers σ and τ, the corresponding infimum and supremum are determined by $\inf(\sigma, \tau) = \ldots \alpha_2 \alpha_1 \alpha_0$, where $\alpha_i = \min(\sigma_i, \tau_i)$, and $\sup(\sigma, \tau) = \ldots \beta_2 \beta_1 \beta_0$, where $\beta_i = \max(\sigma_i, \tau_i)$. The greatest p-adic integer according to our definition is $-1 = \ldots xxxxxx$, where $x = p - 1$, and the smallest is $0 = \ldots 00000$.

Let Ω^* denote all infinite streams built from members of Ω at each $t = 0, 1, 2, \ldots, \infty$. Its subsets will be denoted by $A^*, B^* \subseteq \Omega^*$. Assume that the cardinality of Ω is equal to $p - 1$. Now, we can define p-adic fuzziness as follows: a *p-adic fuzzy measure* is a set function $F_{Z_p}(\cdot)$ defined for sets $A^*, B^* \subseteq \Omega^*$, it runs over the set $\mathbf{Z}_p$ and satisfies the following properties:

- $F_{Z_p}(\Omega^*) = -1$ and $F_{Z_p}(\emptyset^*) = 0$.
- If $A^* \subseteq \Omega^*$ and $B^* \subseteq \Omega^*$ are disjoint, i.e. $\inf(F_{Z_p}(A^*), F_{Z_p}(B^*)) = 0$, then $F_{Z_p}(A^* \cup B^*) = F_{Z_p}(A^*) + F_{Z_p}(B^*)$. Otherwise, $F_{Z_p}(A^* \cup B^*) = F_{Z_p}(A^*) + F_{Z_p}(B^*) - \inf(F_{Z_p}(A^*), F_{Z_p}(B^*)) = \sup(F_{Z_p}(A^*), F_{Z_p}(B^*))$.
- If $A^*, B^* \subseteq \Omega^*$, then $F_{Z_p}(A^* \cap B^*) = \inf(F_{Z_p}(A^*), F_{Z_p}(B^*))$.
- $F_{Z_p}(\neg A^*) = -1 - F_{Z_p}(A^*)$ for all $A^* \subseteq \Omega^*$, where $\neg A^* = \Omega^* \setminus A^*$.

Hence, $F_{Z_p}(\cdot)$ can be interpreted as a p-adic valued cardinal number of members of $\mathscr{P}(\Omega^*)$.

A *p-adic probability measure* is a set function $\mathbf{P}_{Z_p}(\cdot)$ defined for sets A^*, $B^* \subseteq \Omega^*$ thus:

- $\mathbf{P}_{Z_p}(A^*) = -F_{Z_p}(A^*) \in \mathbf{Z}_p$
- $\mathbf{P}_{Z_p}(A^* \,|B^*) \in \mathbf{Q}_p$ is characterized by the following constraint:

$$\mathbf{P}_{Z_p}(A^* \,|B^*) = \frac{\mathbf{P}_{Z_p}(A^* \cap B^*)}{\mathbf{P}_{Z_p}(B^*)} = \frac{F_{Z_p}(A^* \cap B^*)}{F_{Z_p}(B^*)},$$

where $\mathbf{P}_{Z_p}(B^*) \neq 0$, $\mathbf{P}_{Z_p}(A^* \cap B^*) = \inf(\mathbf{P}_{Z_p}(A^*), \mathbf{P}_{Z_p}(B^*))$.

The measure $\mathbf{P}_{Z_p}(\cdot)$ runs over the set $\mathbf{Q}_p$ of all p-adic numbers (not only integers). Notice that while $\mathbf{Z}_p$ is the ring of p-adic integers, $\mathbf{Q}_p$ is the field of p-adic numbers.

For non-Archimedean valued probabilities, Bayes' theorem does not hold in general case (Theorem 5.4). This theorem holds only on 2-adic numbers. There are the following two principal versions of non-Archimedean valued predicate logical calculi proposed in [17]: p-adic valued extension of BL and hyper-valued extension of ŁΠ. So, a structure $\langle \mathbf{Z}_p, \oplus, \neg_L, 0 \rangle$ is a p-adic valued MV-algebra. Furthermore, a structure $\langle \mathbf{Z}_p, \wedge, \vee, *, \Rightarrow, 0, -1 \rangle$, where $* \in \{\&_L, \&_G\}$ and $\Rightarrow \in \{\rightarrow_L, \rightarrow_G\}$ is a p-adic valued BL-matrix. And if $* = \&_L$ and $\Rightarrow = \rightarrow_L$, it is called a p-adic valued **L**-algebra. If $* = \&_G$ and $\Rightarrow = \rightarrow_G$, it is called a p-adic valued **G**-algebra. However, a structure $\langle \mathbf{Z}_p, \oplus, \neg_L, \rightarrow_\Pi, \&_\Pi, 0, -1 \rangle$ is not a p-adic valued ŁΠ-matrix. Only a structure $\langle {}^*[0, 1], \oplus, \neg_L, \rightarrow_\Pi, \&_\Pi, {}^*0, {}^*1 \rangle$ is a hyperrational valued ŁΠ-matrix, where ${}^*[0, 1]$ is a non-standard (non-Archimedean) extension of $[0, 1]$.

Using non-Archimedean (stream-valued) probabilities we can define non-Archimedean evolution of logical cellular automata [48], disprove the Aumann's agreement theorem [49], and simulate the dynamics of slime mould [19].

Let us show how we can simulate a swarm behaviour in p-aic valued logics.

5.4 *p*-Adic Valued Logic for Simulating the Swarm Behaviours

5.4.1 p-*Adic Logical Values in Experiments on Swarms*

Assume, the cardinal number of Ω is equal to $p - 1$, i.e. $\lceil \Omega \rceil = p - 1$. Let us assume that all attractants used in an experiment with a swarm are located in such a way that at each time step $t = 0, 1, 2, \ldots, \infty$ the swarm cannot achieve more than $p - 1$ attractants and for some $t = i$ the swarm can occupy exactly $p - 1$ attractants. Then the family Ω^* of all infinite streams which can be constructed, i.e. $\Omega^* = \prod_{t=0}^{\infty} \Omega_t$, has the cardinal number $-1 = F_{Z_p}(\Omega^*) = \lceil \Omega^* \rceil = \lceil \mathbf{Z}_p \rceil$, the largest integer of $\mathbf{Z}_p$ (indeed, if $p = 3$, then $\lceil \Omega^* \rceil = \ldots 22222 = 0 - 1 = -1$ and if $p = 5$, then $\lceil \Omega^* \rceil = \ldots 44444 = 0 - 1 = -1$).

Hence, Ω^* is understood as the universe for the experiment with the swarm. Suppose that A_t := 'Attractants, which can be occupied by the swarm i at time t'; B_t := 'Attractants accessible for the attractant N_i by a logistic system of the swarm at time t'; C_t := 'If there are attractants accessible for the attractant N_i by a logistic system of the swarm i at time t, then there are attractants accessible for the attractant N_j by a logistic system of the swarm j at time t'. Then we obtain $A^* = \prod_{t=0}^{\infty} A_t$; $B^* = \prod_{t=0}^{\infty} B_t$; $C^* = \prod_{t=0}^{\infty} C_t$ such that $A^*, B^*, C^* \subseteq \Omega^*$. In other words, all statements about the experiment can be considered subsets of Ω^*.

The meanings of A^*, B^*, C^* are presented by cardinal numbers $F_{Z_p}(A^*) = \lceil A^* \rceil$, $F_{Z_p}(B^*) = \lceil B^* \rceil$, $F_{Z_p}(C^*) = \lceil C^* \rceil$, respectively. These numbers run over the set $\mathbf{Z}_p$.

Thus, by performing the experiment we can study some ensembles $\Omega^* = \prod_{t=0}^{\infty} \Omega_t$, such that $\lceil \Omega^* \rceil = -1$. Let us consider a sequence of ensembles Ω_t having volumes $\lceil \Omega_t \rceil$, $t = 0, 1, \ldots$ We may imagine an ensemble Ω^* as being the population of a tower $T = T_\Omega$, which has an infinite number of floors with the following distribution of population through floors: population of t-th floor is Ω_t. Set $T_k = \prod_{t=0}^{k} \Omega_t \times \prod_{m=k+1}^{\infty} \emptyset_m$. This is a population of the first $k+1$ floors. Let $A^* \subset \Omega^*$ and let us assume that $n(A^*) = \lim_{t \to \infty} n_t(A^*)$, where $n_t(A^*) = \lceil A^* \cap T_t \rceil$. The quantity $n(A^*)$ is said to be a *non-well-founded volume of the set* A^*.

Now we can define the probability of A^* by the standard proportional relation:

$$\mathbf{P}(A^*) := \mathbf{P}_{\Omega^*}(A^*) = \frac{n(A^*)}{-1},$$

where $\lceil \Omega^* \rceil = -1$, $n(A^*) = \lceil A^* \cap \Omega^* \rceil = \lceil A^* \rceil$. In this case $\mathbf{P}(\emptyset^*) = 0$ and $\mathbf{P}(\Omega^*) = \frac{-1}{-1} = 1$.

We denote the family of all $A^* \subseteq \Omega^*$, for which $\mathbf{P}(A^*)$ exists, by $\mathcal{G}_\Omega$. The sets $A^* \in \mathcal{G}_\Omega$ are said to be events. The ordered system $\langle \Omega^*, \mathcal{G}_\Omega, \mathbf{P}_{\Omega^*} \rangle$ is called a *non-well-founded ensemble probability space for the ensemble* Ω. These probabilities are p-adic valued.

The swarm behaviour is intelligent. This means that the swarm makes decisions at each $t = 0, 1, 2, \ldots$ how to occupy as many attractants as possible. For making decisions the swarm i appeals to a *knowledge structure*, a function $\mathbf{P}_i$ which assigns to each attractant $\omega_t \in \prod_t \Omega_t$ at t a non-empty subset of Ω^*, so that each thing ω_t belongs to one or more elements of each $\mathbf{P}_i$, i.e. Ω^* is contained in a union of $\mathbf{P}_i$, but $\mathbf{P}_i$ are not mutually disjoint (see the previous subsections).

Suppose, we have a set of N swarms starting to move, call them i = 1, ..., N. Then $\mathbf{P}_i(\omega_t)$ is called i's *knowledge state* at the attractant ω_t. This means that if the actual state is ω_t, the swarm only knows that the actual state is in $\mathbf{P}_i(\omega_t)$.

We can interpret $\mathbf{P}_i(\omega_t)$ probabilistically as follows:

$$\mathbf{P}_i(\omega_t) = \{\omega' : P^i_{Z_p}(\omega'|\omega_t) = \frac{P_{Z_p}(\omega' \wedge \omega_t)}{P_{Z_p}(\omega_t)} > 0\},$$

where (i) $P^i_{Z_p}(\omega_t) = 1$ iff ω_t is occupied by the swarm for all t; (ii) $P^i_{Z_p}(\omega_t) = 0$ iff ω_t is not occupied by the swarm for all t; (iii) $P^i_{Z_p}(\omega_t) = -\ldots\alpha_t\ldots\alpha_1\alpha_0$, where $\alpha_t = p-1$ if ω_t is occupied at t, $\alpha_t = 0$ if ω_t is not occupied at t, $\alpha_t = p-2 > 0$ if $p-2$ neighbour attractants of ω_t are occupied at t, etc.; (iv) $P_{Z_p}(\omega' \wedge \omega_t) = \inf(P_{Z_p}(\omega'), P_{Z_p}(\omega_t))$. Evidently that $P^i_{Z_p}(\omega_t|\omega_t) > 0$ for all $\omega_t \in \prod_t \Omega_t$, therefore for all $\omega_t \in \prod_t \Omega_t$, $\omega_t \in \mathbf{P}_i(\omega_t)$.

Now we consider the relation $A^* \subseteq \mathbf{P}_i(\omega_t)$, where $A^* \subseteq \Omega^*$, as the statement that at ω_t agent i *accepts the performance* A^*:

$$K_i A^* = \{\omega_t : A^* \subseteq \mathbf{P}_i(\omega_t)\}.$$

Some basic properties of *knowledge operators* K_i:

$$K_i \Omega \subseteq \Omega^*; \tag{5.59}$$

$$(K_i A^* \cap K_i B^*) \Rightarrow K_i(A^* \cap B^*); \tag{5.60}$$

$$K_i(A^* \cup B^*) \Rightarrow (K_i A^* \cup K_i B^*); \tag{5.61}$$

$$K_i(A^* \cup B^*) = (K_i A^* \cap K_i B^*); \tag{5.62}$$

$$A^* \subseteq B^* \Rightarrow K_i A^* \subseteq K_i B^*; \tag{5.63}$$

$$A \subseteq K_i A; \tag{5.64}$$

$$K_i K_i A^* = K_i A^*. \tag{5.65}$$

Using relations (5.59)–(5.65), we can disprove the Aumann's agreement theorem for swarms (see Theorems 10.2 and 10.3).

Let B_i^* mean 'Attractants, which can be occupied by agent i'. After several steps, we expect fusions of all roads built by the swarm so that all accessible attractants are being occupied by all agents. Does it mean that we observe a union of B_i^*? No, it does not. We face just the situation that since a time step $t = k$ the sets B_i^* are intersected. Let C_i^* mean 'Attractants accessible for the attractant N_i by a road system'. Assume, $\omega \in B_i^*$ and $\omega' \in C_i^*$. Evidently, $P^i_{Z_p}(\omega'|\omega) > 0$. As a consequence, we assume according to our definitions that each agent i knows ω at ω' and knows ω' at ω, i.e. agent i accepts the performance B_i^* at ω' and i accepts the performance C_i^* at ω, i.e. $K_i B_i^* = K_i C_i^*$.

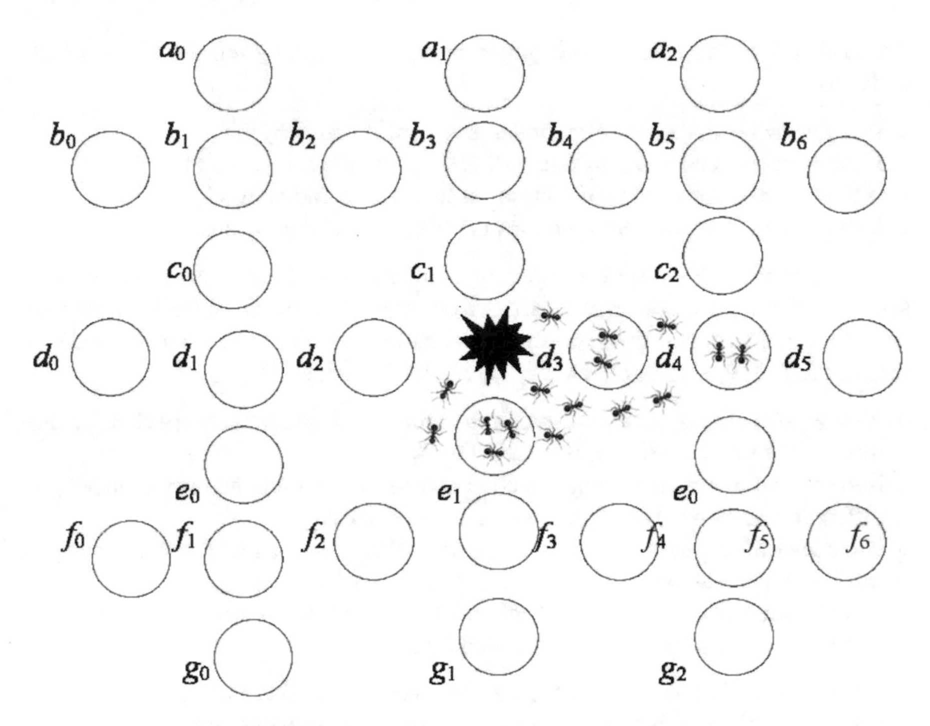

Fig. 5.1 The propagation of ant nest at $t = 1$, where circles are attractants and attractants e_1, d_3, and d_4 are occupied by ants. We see that at each step the ant nest can detect not more than 4 attractants. Furthermore, at points $b_1, b_3, b_5, d_1, d_4, f_1, f_3, f_5$ the ant nest can make from one to four decisions

5.4.2 p-*Adic Valued Adder and* p-*Adic Valued Subtracter*

If the environment of a swarm contains many attractants, the swarm can simultaneously propagate its networks in many different directions. Let us assume that the swarm cannot see more than $p - 1$ attractants at time $t = 0, 1, 2, \ldots$ and for some $t = i$ the swarm can occupy exactly $p - 1$ attractants (see Fig. 5.1). Without loss of generality, let us consider the case $p = 5$. This means that the swarm has four possible directions in its motions. These directions can be coded by integers from 1 to 4 as follows:

- if the swarm moves just in one direction at time t, it is coded by 1_t;
- if the swarm moves in two different directions at time t, it is coded by 2_t;
- if the swarm moves in three directions at time t, it is coded by 3_t;
- if the swarm moves in four directions at time t, it is coded by 4_t.

For example, the swarm can move to the north, to the south, to the east, or to the west. Nevertheless, in these motions the swarm can change the direction at the same t and start to move to the northwest, to the northeast, to the southwest, or to the southeast. In this case the new directions can be coded by integers 5, 10, 15, 20 as follows:

- if the swarm changes one direction at time t, it is coded by 5_t;
- if the swarm changes two different directions at time t, it is coded by 10_t;
- if the swarm changes three directions at time t, it is coded by 15_t;
- if the swarm changes four directions at time t, it is coded by 20_t.

Furthermore, the swarm can move in standards directions simultaneously with switching to non-standard directions, e.g. it can move to the north and to the northwest or to the southwest and to the north, etc. Then the new directions can be coded by integers 6, 7, 8, 9, 11, 12, 13, 14, 16, 17, 18, 19, 21, 22, 23, 24, e.g.:

- if the swarm moves just in one direction at time t and simultaneously changes one direction at the same time t, it is coded by 6_t;
- if the swarm moves just in one direction at time t and simultaneously changes two different directions at the same time t, it is coded by 11_t;
- if the swarm moves just in one direction at time t and simultaneously changes three directions at the same time t, it is coded by 16_t;
- if the swarm moves just in one direction at time t and simultaneously changes four directions at the same time t, it is coded by 21_t, etc.

Thus, at one time step t we consider 24 codes of possible motions. This means that we are limited by 5-adic two bit strings $[\alpha_0\alpha_1]$ with values $\alpha_0 + 5 \cdot \alpha_1$. These two bit strings can be represented by the following 2-adic matrices:

$$\begin{vmatrix} 1 & 1 & 1 & 1 & b_1 = 1 \\ a_{21} & a_{22} & a_{23} & a_{24} & b_2 = 0 \\ a_{31} & a_{32} & a_{33} & a_{34} & b_3 = 0 \\ a_{41} & a_{42} & a_{43} & a_{44} & b_4 = 0 \\ c_1 = 0 & c_2 = 0 & c_3 = 0 & c_4 = 0 & \end{vmatrix} = 1_t;$$

$$\begin{vmatrix} 1 & 1 & 1 & 1 & b_1 = 1 \\ 1 & 1 & 1 & 1 & b_2 = 1 \\ a_{31} & a_{32} & a_{33} & a_{34} & b_3 = 0 \\ a_{41} & a_{42} & a_{43} & a_{44} & b_4 = 0 \\ c_1 = 0 & c_2 = 0 & c_3 = 0 & c_4 = 0 & \end{vmatrix} = 2_t;$$

$$\begin{vmatrix} 1 & 1 & 1 & 1 & b_1 = 1 \\ 1 & 1 & 1 & 1 & b_2 = 1 \\ 1 & 1 & 1 & 1 & b_3 = 1 \\ a_{41} & a_{42} & a_{43} & a_{44} & b_4 = 0 \\ c_1 = 0 & c_2 = 0 & c_3 = 0 & c_4 = 0 & \end{vmatrix} = 3_t;$$

$$\left| \begin{array}{ccccc} 1 & 1 & 1 & 1 & b_1 = 1 \\ 1 & 1 & 1 & 1 & b_2 = 1 \\ 1 & 1 & 1 & 1 & b_3 = 1 \\ 1 & 1 & 1 & 1 & b_4 = 1 \\ c_1 = 0 & c_2 = 0 & c_3 = 0 & c_4 = 0 & \end{array} \right| = 4_t;$$

$$\left| \begin{array}{ccccc} 1 & a_{12} & a_{13} & a_{14} & b_1 = 1 \\ 1 & a_{22} & a_{23} & a_{24} & b_2 = 0 \\ 1 & a_{32} & a_{33} & a_{34} & b_3 = 0 \\ 1 & a_{42} & a_{43} & a_{44} & b_4 = 0 \\ c_1 = 1 & c_2 = 0 & c_3 = 0 & c_4 = 0 & \end{array} \right| = 5_t;$$

$$\left| \begin{array}{ccccc} 1 & 1 & 1 & 1 & b_1 = 1 \\ 1 & a_{22} & a_{23} & a_{24} & b_2 = 0 \\ 1 & a_{32} & a_{33} & a_{34} & b_3 = 0 \\ 1 & a_{42} & a_{43} & a_{44} & b_4 = 0 \\ c_1 = 1 & c_2 = 0 & c_3 = 0 & c_4 = 0 & \end{array} \right| = 6_t;$$

$$\left| \begin{array}{ccccc} 1 & 1 & 1 & 1 & b_1 - 1 \\ 1 & 1 & 1 & 1 & b_2 = 1 \\ 1 & a_{32} & a_{33} & a_{34} & b_3 = 0 \\ 1 & a_{42} & a_{43} & a_{44} & b_4 = 0 \\ c_1 = 1 & c_2 = 0 & c_3 = 0 & c_4 = 0 & \end{array} \right| = 7_t, etc.,$$

$$\left| \begin{array}{ccccc} 1 & 1 & 1 & 1 & b_1 = 1 \\ 1 & 1 & 1 & 1 & b_2 = 1 \\ 1 & 1 & 1 & 1 & b_3 = 1 \\ 1 & 1 & 1 & 1 & b_4 = 1 \\ c_1 = 1 & c_2 = 1 & c_3 = 1 & c_4 = 1 & \end{array} \right| = 24_t,$$

where $a_{ij} = 1$ (respectively, $b_i = 1$ or $c_j = 1$) if the attractant denoted by a_{ij} (respectively, by b_i or c_j) is occupied by the swarm. Otherwise $a_{ij} = 0$ (respectively, $b_i = 0$ or $c_j = 0$). In the matrix coded by 1_t, there exists i such that $a_{i1} = 0, a_{i2} = 0, a_{i3} = 0$, $a_{i4} = 0$ and there exists j such that $a_{2j} = 0$, $a_{3j} = 0$, $a_{4j} = 0$. In the matrix coded by 2_t, there exists i such that $a_{i1} = 0$, $a_{i2} = 0$, $a_{i3} = 0$, $a_{i4} = 0$ and there exists j such that $a_{3j} = 0$, $a_{4j} = 0$, etc. Notice that b_i are outputs for rows and c_j are outputs for columns.

The *p-adic valued adder* for calculating $x + y$, where $x, y \in \{1_t, 2_t, \dots, 24_t\}$, is represented by the fusion of two plasmodia coded by x and y at t.

The swarm can leave some attractants at time step t under specific conditions such as there are neighbour repellents or in case an appropriate attractant was not sufficient for feeding. Then the *p-adic valued subtracter* for calculating $x - y$, where $x, y \in \{1_t, 2_t, \dots, 24_t\}$ and $x > y$, is represented by leaving attractants coded by y at t.

5.4.3 p-Adic Valued Towers of Higher-Order Sequents

Thus, different locations of $p - 1 = \lceil \Omega \rceil$ attractants cause different outcomes of experiments with swarms. The wave sets A^*, $B^* \subseteq \Omega^*$, such that $0 \leq \lceil A^* \rceil, \lceil B^* \rceil \leq -1$, are properties of the swarm expansions: how quickly the swarm moves, in what direction, how many attractants it occupies and how many attractants it leaves. The point is that by different locations of attractants, we can affect on swarm motions differently. These locations combined with different intensity of attractants are outer stimuli for swarm motions. Hence, appropriate descriptions $A^*, B^*, \ldots$ of the experiment can be understood as propositions, such that $F_{Z_p}(A^*), F_{Z_p}(B^*), \ldots$ are p-adic valued truth values of $A^*, B^*, \ldots$, respectively.

Assume that we have a standard logical language $Ł_p$ with standard definitions of well-formed propositions (formulas) ${}^1\Phi$, ${}^1\Psi, \ldots$ Let these formulas be p-valued and denote the propositions describing experiments with swarms at time step $t = 0$. Examples of ${}^1\Phi$, ${}^1\Psi, \ldots$ at $t = 0$ are as follows: 'Attractants, which can be occupied by the swarm i', 'Attractants accessible for the attractant N_i by swarm networks', 'If there are attractants accessible for the attractant N_i by networks of the swarm i, then there are attractants accessible for the attractant N_j by networks of the swarm j' etc.

Let us build up an extension of $Ł_p$, where we can use the formulas ${}^i\Phi$, ${}^i\Psi$, ..., $i > 0$, i.e. the propositions describing experiments with swarms at time step $t = i - 1$, as well as we can use the formulas ${}^\infty\Phi$, ${}^\infty\Psi$, ..., i.e. the propositions describing experiments with swarms at an infinite time.

The natural deduction system for p-valued logic $Ł_p$, where p is a finite natural number, is considered in [50]. A *p-valued sequent* is regarded as a p-tuple of finite sets Γ_i $(1 \leq i \leq p)$ of formulas, denoted by $\Gamma_1 \mid \Gamma_2 \mid \ldots \mid \Gamma_p$. It is defined to be *satisfied by an interpretation* iff for some $i \in \{1, \ldots, p\}$ at least one formula in Γ_i takes the truth value $i - 1 \in \{0, \ldots, p - 1\}$, where $\{0, \ldots, p - 1\}$ is the set of truth values for $Ł_p$.

By this approach, a two-valued sequent with Gentzen's standard notation $\Gamma_1 \vdash \Gamma_2$, where Γ_1 and Γ_2 are finite sequences of formulas, is interpreted truth-functionally: either one of the formulas in Γ_1 is false or one of the formulas in Γ_2 is true. In other words, we can denote a sequent $\Gamma_1 \vdash \Gamma_2$ by $\Gamma_1 \mid \Gamma_2$ and we can define it to be satisfied by an interpretation iff for some $i \in \{1, 2\}$ at least one formula in Γ_i takes the truth value $i - 1 \in \{0, 1\}$.

We can extend this approach of [50] to the infinite-order case. In other words, we can obtain a p-adic valued extension of the natural deduction system for $Ł_p$. Notably, the deductive systems for non-Archimedean many-valued logics (including p-adic many-valued logics) are proposed in [17, 19]. In this subsection, we consider another system, which is built as an extension of p-valued sequent logic constructed in [50]. So, we will appeal to

- formulas $\langle {}^1\Phi, {}^2\Phi, \ldots, {}^i\Phi \rangle$, $\langle {}^1\Psi, {}^2\Psi, \ldots, {}^i\Psi \rangle$, ..., as towers with i floors, these formulas are interpreted as i-order formulas belonging to Π_1 or Σ_1 (see the next subsection),

- formulas $\langle {}^1\Phi, {}^2\Phi, \ldots, {}^i\Phi, \ldots, {}^\infty\Phi\rangle$, $\langle {}^1\Psi, {}^2\Psi, \ldots, {}^i\Psi, \ldots, {}^\infty\Psi\rangle$, ..., as towers with infinite floors, these formulas are interpreted as infinite-order formulas belonging to Π_1 or Σ_1 (see the next subsection).

A higher-order formula is called vertical bounded if it does not contain predicate constants and all predicate variables are quantified. Denote by Π_1 (resp. by Σ_1) the class of vertical bounded higher-order formulas of the form Ψ such that all predicate quantifiers of Ψ are universal (existential).

By ${}^j\Gamma_i$ denote the set of j-order formulas belonging to Π_1 or Σ_1. We obtain these formulas by j-order quantification of all formulas of Γ_i. We set ${}^1\Gamma_i := \Gamma_i$. Now we can get a j-order sequent ${}^j\Gamma_1 \mid {}^j\Gamma_2 \mid \ldots \mid {}^j\Gamma_p$ regarded as a *tower of higher-order sequents with the height equal to j*

$$\begin{array}{c} {}^j\Gamma_1 \mid {}^j\Gamma_2 \mid \ldots \mid {}^j\Gamma_p \\ \ldots \\ {}^3\Gamma_1 \mid {}^3\Gamma_2 \mid \ldots \mid {}^3\Gamma_p \\ {}^2\Gamma_1 \mid {}^2\Gamma_2 \mid \ldots \mid {}^2\Gamma_p \\ {}^1\Gamma_1 \mid {}^1\Gamma_2 \mid \ldots \mid {}^1\Gamma_p \end{array}$$

and we can get an infinite-order sequent ${}^\infty\Gamma_1 \mid {}^\infty\Gamma_2 \mid \ldots \mid {}^\infty\Gamma_p$ regarded as an infinite tower of higher-order sequents

$$\begin{array}{c} \ldots \\ {}^3\Gamma_1 \mid {}^3\Gamma_2 \mid \ldots \mid {}^3\Gamma_p \\ {}^2\Gamma_1 \mid {}^2\Gamma_2 \mid \ldots \mid {}^2\Gamma_p \\ {}^1\Gamma_1 \mid {}^1\Gamma_2 \mid \ldots \mid {}^1\Gamma_p \end{array}$$

Higher-order and infinite-order sequents have interpretations in the set $\mathbf{Z}_p$ of p-adic integers. Thus, we obtain a p-adic multiple validity for an infinite-order extension of first-order p-valued sequent logic of [50].

For instance, let us consider a second-order extension of a two-valued sequent $\Gamma_1 \mid \Gamma_2$. Suppose that ${}^2\Gamma_i$ $(i = 1, 2)$ denote the set of second-order formulas belonging to Π_1 or Σ_1 such that they were obtained by second-order quantification of all formulas of Γ_i. Then each formula in ${}^2\Gamma_1, {}^2\Gamma_2$ has one of the forms $\forall P\ \Psi(P)$, $\forall P\ \neg\Psi(P), \exists P\ \Psi(P), \exists P\ \neg\Psi(P)$. Assume that these formulas are vertical bounded. Consequently, the formula $\forall P\ \Psi(P)$ can be considered as a statement that the formula $\Psi(P)$ is general valid, the formula $\forall P\ \neg\Psi(P)$ as a statement that $\Psi(P)$ is non-satisfiable, the formula $\exists P\ \Psi(P)$ as a statement that $\Psi(P)$ is satisfiable, and the formula $\exists P\ \neg\Psi(P)$ as a statement that $\Psi(P)$ is non-general valid. Also, we can say that in the second-order case, we have a four-valued logic with the following truth values: $\langle 1, 1\rangle$ ('general validity'), $\langle x, 1\rangle$ for $x \in \{0, 1\}$ ('satisfiability'), $\langle 0, 0\rangle$ ('non-satisfiability'), $\langle x, 0\rangle$ for $x \in \{0, 1\}$ ('non-general validity'). This implies that if we build a matrix logic for infinite-order extensions of formulas $\Psi(P)$, then the infinite-order extension of binary logic has $D = \{(2^i - 1)_{i=1}^{\infty}\} = \{\langle 1, 1, 1, \ldots\rangle\}$ as the set of designated truth values and it has $V = \{0, 1, 2, \ldots, (2^i - 1)_{i=1}^{\infty}\}$ as the set of all truth values. Thus, we obtain a 2-adic valued logic.

A second-order sequent ${}^2\Gamma_1 \mid {}^2\Gamma_2$ is defined to be *satisfied by an interpretation* iff for some $i \in \{1, 2\}$ at least one formula in ${}^2\Gamma_i$ takes the truth value $\langle x, i - 1\rangle$ (where $x \in \{0, 1\}$). A second-order sequent

$$\frac{{}^2\Gamma_1 \mid {}^2\Gamma_2}{{}^1\Gamma_1 \mid {}^1\Gamma_2}$$

is defined to be satisfied by an interpretation iff for some

$$\langle i, j\rangle \in \{\langle 1, 1\rangle, \langle 1, 2\rangle, \langle 2, 1\rangle, \langle 2, 2\rangle\}$$

at least one formula in ${}^1\Gamma_i, {}^2\Gamma_j$ takes the truth value $\langle i-1, j-1\rangle$.

5.4.4 p-*Adic Valued Natural Deductions*

Let us assume that Ω^* is a family of p-adic streams describing an experiment with the swarm. Let us build a logical language for propositions such as 'Attractants, which can be occupied by the swarm i at time t', 'Attractants accessible for the attractant N_i by networks at time t', 'If there are attractants accessible for the attractant N_i by networks of the swarm i at time t, then there are attractants accessible for the attractant N_j by networks of the swarm j at time t', etc. These propositions do not have quantifiers or terms and say about the experiment at time $t = 0, 1, 2, \ldots$ The truth valuation of these propositions runs over the set $\{0, 1, \ldots, p-1\}$ at each $t = 0, 1, 2, \ldots$ Let us denote the propositions at $t = 0$ by ${}^1\Phi, {}^1\Psi, \ldots$and the propositions at time $t = i > 0$ by ${}^{i+1}\Phi, {}^{i+1}\Psi, \ldots$Also, we can assume the existence of propositions at $t = \infty$ and denote them by ${}^{\infty}\Phi, {}^{\infty}\Psi, \ldots$

Now, let us identify each formula ${}^k\Phi$ with a finite stream $\langle {}^1\Phi, \ldots, {}^k\Phi\rangle$ with the same propositional content Φ and each formula ${}^{\infty}\Phi$ with an infinite stream $\langle {}^1\Phi, \ldots, {}^k\Phi, \ldots\rangle$ with the same propositional content Φ. In this case ${}^k\Phi$ is $(p^i - 1)_{i=1}^k$-valued and ${}^{\infty}\Phi$ is p-adic valued. These propositions describe the whole experiment up to the time $t = 1, 2, \ldots, \infty$.

We associate each formula ${}^k\Phi$ or ${}^{\infty}\Phi$ with a statement $||{}^k\Phi||$ or $||{}^{\infty}\Phi||$ describing the experiment with the swarm in the universe $T_{j-1} = \prod_{t=0}^{j-1} \Omega_t \times \prod_{m=j}^{\infty} \emptyset_m$ or ${}^*\Omega$, respectively. The truth value of $||{}^k\Phi||$ or $||{}^{\infty}\Phi||$ is considered a cardinal number $\lceil ||{}^k\Phi|| \rceil$ or $\lceil ||{}^{\infty}\Phi|| \rceil$.

Suppose that each attractant involved into the experiment Ω^* is denoted by one of the symbols $d_1, \ldots, d_m$. Also, let us add some constants $a_1, \ldots, a_m$ to our logical language. By $\Omega^*(d_1/a_1, \ldots, d_m/a_m)$ let us denote valuations of constants in Ω^*. Further, we can introduce individual variables $x, y, \ldots$, running over all the attractants of Ω, and the following four terms: $+$ (addition of p-adic integers), $\cdot$ (multiplication of p-adic integers), $-$ (subtraction of p-adic integers), $=$. Then, we can introduce predicate symbols ${}^{t+1}A_1, \ldots, {}^{t+1}A_n$ with meanings describing some properties of the experiment Ω^* with the swarm at time t. Examples of these properties are as follows: 'Occupation by the swarm i at time t', 'Accessibility for the attractant N_i by networks at time t', etc. Hence, ${}^{t+1}A_l(a)$ can mean that 'Occupation by the swarm i at time t' holds on an attractant a. The formula $\exists x^{t+1}A_l(x)$ can mean that there exist attractant x such that 'Accessibility for the attractant N_i by networks at time t' holds on x. The formula $\forall x^{t+1} A_l(x)$ can mean that for all attractants x 'Neighbourhood for the attractant N_i at time t' holds on x. In the same way as for the propositional logic we can identify atomic propositions ${}^{t+1}A_l(a)$, $\exists x^{t+1} A_l(x)$, $\forall x^{t+1} A_l(x)$ with finite streams:

$$\langle {}^1A_l(a), \ldots, {}^{t+1}A_l(a) \rangle,$$

$$\langle \exists x^1 A_l(x), \ldots, \exists x^{t+1} A_l(x) \rangle,$$

$$\langle \forall x^1 A_l(x), \ldots, \forall x^{t+1} A_l(x) \rangle,$$

respectively, and generalize them up to infinite streams ${}^{\infty}A_l(a)$, $\exists x^{\infty} A_l(x)$, $\forall x^{\infty} A_l(x)$ as follows:

$${}^{\infty}A_l(a) := \langle {}^1A_l(a), \ldots, {}^{t+1}A_l(a), \ldots \rangle,$$

$$\exists x^{\infty} A_l(x) := \langle \exists x^1 A_l(x), \ldots, \exists x^{t+1} A_l(x), \ldots \rangle,$$

$$\forall x^{\infty} A_l(x) := \langle \forall x^1 A_l(x), \ldots, \forall x^{t+1} A_l(x), \ldots \rangle.$$

Thus, at each $t = 0, 1, 2, \ldots$ we deal with $(p^i - 1)_{i=1}^{t+1}$-valued logic. The natural deductive system for p-valued logic $Ł_p$ is considered in [50]. A *p-valued sequent* is regarded as a p-tuple of finite sets Γ_i $(1 \leq i \leq p)$ of formulas, denoted by $\Gamma_1 \mid \Gamma_2 \mid \ldots \mid \Gamma_p$. It is defined to be *satisfied by an interpretation* iff for some $i \in \{1, \ldots, p\}$ at least one formula in Γ_i takes the truth value $i - 1 \in \{0, \ldots, p - 1\}$, where $\{0, \ldots, p - 1\}$ is the set of truth values for $Ł_p$.

We can extend this approach of [50] to $(p^i - 1)_{i=1}^{t+1}$-valued or p-adic valued case. By ${}^j\Gamma_i$ denote the set of ${}^j\Psi$ formulas understood as streams. We set ${}^1\Gamma_i := \Gamma_i$. Now we can get a j-order sequent ${}^j\Gamma_1 \mid {}^j\Gamma_2 \mid \ldots \mid {}^j\Gamma_p$ regarded as a *tower of higher-order sequents with the height equal to j*

$$\begin{array}{c} {}^j\Gamma_1 \mid {}^j\Gamma_2 \mid \ldots \mid {}^j\Gamma_p \\ \ldots \\ {}^3\Gamma_1 \mid {}^3\Gamma_2 \mid \ldots \mid {}^3\Gamma_p \\ {}^2\Gamma_1 \mid {}^2\Gamma_2 \mid \ldots \mid {}^2\Gamma_p \\ {}^1\Gamma_1 \mid {}^1\Gamma_2 \mid \ldots \mid {}^1\Gamma_p \end{array}$$

and we can get an infinite-order sequent ${}^{\infty}\Gamma_1 \mid {}^{\infty}\Gamma_2 \mid \ldots \mid {}^{\infty}\Gamma_p$ regarded as an infinite tower of higher-order sequents

$$\begin{array}{c} \ldots \\ {}^3\Gamma_1 \mid {}^3\Gamma_2 \mid \ldots \mid {}^3\Gamma_p \\ {}^2\Gamma_1 \mid {}^2\Gamma_2 \mid \ldots \mid {}^2\Gamma_p \\ {}^1\Gamma_1 \mid {}^1\Gamma_2 \mid \ldots \mid {}^1\Gamma_p \end{array}$$

Higher-order and infinite-order sequents have interpretations in the set $\mathbf{Z}_p$ of p-adic integers. Thus, we obtain a p-adic multiple validity for an infinite-order extension of first-order p-valued sequent logic of [50].

For instance, a second-order sequent ${}^2\Gamma_1 \mid {}^2\Gamma_2$ is defined to be *satisfied by an interpretation* iff for some $i \in \{1, 2\}$ at least one formula in ${}^2\Gamma_i$ takes the truth value $\langle x, i-1\rangle$ (where $x \in \{0, 1\}$). A second-order sequent

$$\begin{array}{c} {}^2\Gamma_1 \mid {}^2\Gamma_2 \\ {}^1\Gamma_1 \mid {}^1\Gamma_2 \end{array}$$

is defined to be satisfied by an interpretation iff for some

$$\langle i, j\rangle \in \{\langle 1, 1\rangle, \langle 1, 2\rangle, \langle 2, 1\rangle, \langle 2, 2\rangle\}$$

at least one formula in ${}^1\Gamma_i, {}^2\Gamma_j$ takes the truth value $\langle i-1, j-1\rangle$.

So, each formula $\boldsymbol{\Psi} \in {}^j\Gamma_i$ is interpreted as a statement $\|\boldsymbol{\Psi}\|$ describing the experiment with the swarm in the universe $T_{j-1} = \prod_{t=0}^{j-1} \Omega_t \times \prod_{m=j}^{\infty} \emptyset_m$. A model Ω^*, describing a real experiment with the swarm, is said to *satisfy* a sequent ${}^j\Gamma_1 \mid {}^j\Gamma_2 \mid \ldots \mid {}^j\Gamma_p$ (resp. ${}^\infty\Gamma_1 \mid {}^\infty\Gamma_2 \mid \ldots \mid {}^\infty\Gamma_p$) if there is a p-adic integer $\ldots 0 \ldots 0 i \alpha_{j-2} \ldots \alpha_1\alpha_0$ (resp. a p-adic integer $\ldots \alpha_{j-1}\alpha_{j-2} \ldots \alpha_1\alpha_0$) and a formula ${}^j\boldsymbol{\Psi} \in {}^j\Gamma_i$ (resp. ${}^\infty\boldsymbol{\Psi} \in {}^\infty\Gamma_i$) such that $\ulcorner \|{}^j\boldsymbol{\Psi}\| \urcorner = \ldots 00 \ldots 00 i \alpha_{j-2} \ldots \alpha_1\alpha_0$ (resp. $\ulcorner \|{}^\infty\boldsymbol{\Psi}\| \urcorner = \ldots \alpha_{j-1}\alpha_{j-2} \ldots \alpha_1\alpha_0$) on Ω^*. A sequent is called *logically valid* if it is satisfied on each model Ω^*, i.e. on each real experiment with the slime mould.

Each formula of the set ${}^j\Gamma_i$ $(j > 1)$ contains an information about formulas of the order smaller than j in the tower:

$$\begin{array}{c} {}^j\Gamma_1 \mid {}^j\Gamma_2 \mid \ldots \mid {}^j\Gamma_p \\ \ldots \\ {}^3\Gamma_1 \mid {}^3\Gamma_2 \mid \ldots \mid {}^3\Gamma_p \\ {}^2\Gamma_1 \mid {}^2\Gamma_2 \mid \ldots \mid {}^2\Gamma_p \\ {}^1\Gamma_1 \mid {}^1\Gamma_2 \mid \ldots \mid {}^1\Gamma_p \end{array}$$

A model Ω^* satisfies this sequent if there is a p-adic integer

$$\ldots 00 \ldots 00 \alpha_{j-1}\alpha_{j-2} \ldots \alpha_1\alpha_0$$

and formulas ${}^j\boldsymbol{\Psi} \in {}^j\Gamma_{\alpha_{j-1}}, {}^{j-1}\boldsymbol{\Psi} \in {}^{j-1}\Gamma_{\alpha_{j-2}}, \ldots, {}^2\boldsymbol{\Psi} \in {}^2\Gamma_{\alpha_1}, {}^1\boldsymbol{\Psi} \in {}^1\Gamma_{\alpha_0}$ such that

$$\ulcorner \|{}^j\boldsymbol{\Psi}\| \urcorner = \ldots 00 \ldots 00 \alpha_{j-1}\alpha_{j-2} \ldots \alpha_1\alpha_0.$$

A *j-order introduction rule for a connective R at place i* is a schema of the form:

$$\frac{\langle {}^j\Gamma_1^k, {}^j\Delta_1^k \mid \ldots \mid {}^j\Gamma_p^k, {}^j\Delta_p^k\rangle_{k \in I}}{{}^j\Gamma_1 \mid \ldots \mid {}^j\Gamma_i, R(A_1, \ldots, A_n) \mid \ldots \mid {}^j\Gamma_p}$$

where the arity of R is n, I is a finite set, ${}^j\Gamma_l = \bigcup_{k \in I} {}^j\Gamma_l^k$, ${}^j\Delta_l^k \subseteq \{A_1, \ldots, A_n\}$ and the following condition holds: Let Ω^* be a model. Then the following are equivalent:

(1) $\ulcorner \| R(A_1, \ldots, A_n) \| \urcorner = \ldots 00 \ldots 00 i \alpha_{j-2} \ldots \alpha_1 \alpha_0$ on Ω^*; (2) Ω^* satisfies the sequents ${}^j\Delta_1^k \mid \ldots \mid {}^j\Delta_p^k$ for all $k \in I$.

An infinite-order introduction rule for a connective R at place i is defined for infinite-order sequents in the same way.

If a p-adic sequent is regarded as a tower, then a j-order introduction rule for a connective R is considered not as one at place i, but as one at place

$$\ldots 00 \ldots 00 \alpha_{j-1} \alpha_{j-2} \ldots \alpha_1 \alpha_0.$$

Thus, the following condition holds: Let Ω^* be a model. Then the following are equivalent:

(1) $\ulcorner \| R(A_1, \ldots, A_n) \| \urcorner = \ldots 00 \ldots 00 \alpha_{j-1} \alpha_{j-2} \ldots \alpha_1 \alpha_0$ on Ω^*; (2) for all $k \in I$, Ω^* satisfies the sequents ${}^j\Delta_1^k \mid \ldots \mid {}^j\Delta_p^k$ regarded as towers.

A *j-order introduction rule for a first-order quantifier* Q *at place i* is a schema of the form:

$$\frac{\langle {}^j\Gamma_1^k, {}^j\Delta_1^k \mid \ldots \mid {}^j\Gamma_p^k, {}^j\Delta_p^k \rangle_{k \in I}}{{}^j\Gamma_1 \mid \ldots \mid {}^j\Gamma_i, \mathrm{Q}x A(x) \mid \ldots \mid {}^j\Gamma_p}$$

where I is a finite set, ${}^j\Gamma_l = \bigcup_{k \in I} {}^j\Gamma_l^k$, ${}^j\Delta_l^k \subseteq \{A(a_1), \ldots, A(a_m)\} \cup \{A(t_1), \ldots, A(t_q)\}$. The a_l are metavariables for free variables satisfying the condition that they do not occur in the lower sequent, the t_n are metavariables for arbitrary terms, and the following condition holds: Let Ω^* be a model. Then the following are equivalent:

- $\ulcorner \| \mathrm{Q}x A(x) \| \urcorner = \ldots 00 \ldots 00 i \alpha_{j-2} \ldots \alpha_1 \alpha_0$ in Ω^*;
- for all $d_1, \ldots, d_m \in \Omega$, there are terms $t_1', \ldots, t_q'$ such that for all $k \in I$, $\Omega^*(d_1/a_1, \ldots, d_m/a_m)$ satisfies ${}^j\Delta_1^{\prime k} \mid \ldots \mid {}^j\Delta_p^{\prime k}$ where ${}^j\Delta_l^{\prime k}$ is obtained from ${}^j\Delta_l^k$ by instantiating the eigenvariable t_m with t_m' $(1 \leq m \leq q)$.

An infinite-order introduction rule for a first-order quantifier Q at place i is defined for infinite-order sequents in the same way.

If a p-adic sequent is regarded as a tower, then a j-order introduction rule for a first-order quantifier Q is considered not as one at place i, but as one at place $\ldots 00 \ldots 00 \alpha_{j-1} \alpha_{j-2} \ldots \alpha_1 \alpha_0$. Then the following condition holds: Let Ω^* be a model. Then the following are equivalent:

- $\ulcorner \| \mathrm{Q}x A(x) \| \urcorner = \ldots 00 \ldots 00 \alpha_{j-1} \alpha_{j-2} \ldots \alpha_1 \alpha_0$ in Ω^*;
- for all $d_1, \ldots, d_m \in \Omega$, there are terms $t_1', \ldots, t_q'$ such that for all $k \in I$, $\Omega^*(d_1/a_1, \ldots, d_m/a_m)$ satisfies ${}^j\Delta_1^{\prime k} \mid \ldots \mid {}^j\Delta_p^{\prime k}$ regarded as towers, where ${}^j\Delta_l^{\prime k}$ is obtained from ${}^j\Delta_l^k$ by instantiating the eigenvariable t_m with t_m' $(1 \leq m \leq q)$.

So, a (p-adic) *infinite-order sequent calculus* is given by:

1. axioms of the form: $A \mid A \mid \ldots \mid A$, where A is any j-order (infinite-order) formula;

2. for every connective R and every truth value $\alpha \in \mathbf{Z}_p$, there is a j-order (infinite-order) introduction rule at place α;
3. for every quantifier Q and every truth value $\alpha \in \mathbf{Z}_p$ a j-order (infinite-order) introduction rule at place α;
4. j-order and infinite-order weakening rules for every place i:

$$\frac{{}^j\Gamma_1|...|^j\Gamma_i|...|^j\Gamma_p}{{}^j\Gamma_1|...|^j\Gamma_i, A|...|^j\Gamma_p}, \qquad \frac{{}^\infty\Gamma_1|...|^\infty\Gamma_i|...|^\infty\Gamma_p}{{}^\infty\Gamma_1|...|^\infty\Gamma_i, A|...|^\infty\Gamma_p};$$

5. cut rules for every pair of truth values $\alpha, \beta \in \mathbf{Z}_p$ such that $n \neq m$ in the expansions of p-adic integers $\alpha = \ldots \alpha_j n \alpha_{j-2} \ldots \alpha_0$, $\beta = \ldots \beta_j m \beta_{j-2} \ldots \beta_0$:

$$\frac{{}^j\Gamma_1 \mid \ldots \mid {}^j\Gamma_n, A \mid \ldots \mid {}^j\Gamma_p \quad {}^j\Delta_1 \mid \ldots \mid {}^j\Delta_m, A \mid \ldots \mid {}^j\Delta_p}{{}^j\Gamma_1, {}^j\Delta_1 \mid \ldots \mid {}^j\Gamma_p, {}^j\Delta_p}.$$

A j-order (infinite-order) sequent is *provable* if there is an upward tree of sequents such that every topmost j-order (infinite-order) sequent is an axiom and every other j-order (infinite-order) sequent is obtained from the ones standing immediately above it by an application of one of the rules.

Theorem 5.5 *(Soundness and Completeness) For every (p-adic) infinite-order sequent calculus the following holds: a j-order (infinite-order) sequent is logically valid iff it is provable without cuts from atomic axioms.*

Proof It is an infinite generalization of all discrete cases of [50]. □

Any intelligent behaviour consists in achieving a maximum efficiency and successful control in looking for attractants and in reaching them. For different animals including human beings these attractants are different, too. If we assume that at any time step t, I can find out or reach not more than $p - 1$ my attractants, then my behaviour can be coded by p-adic numbers. As a result, my behaviour implements some p-adic valued arithmetic circuits and verifies some p-adic valued logical propositions. In this chapter, I have proposed an appropriate logical language for describing the behaviour for searching for and reaching not more than $p - 1$ attractants at any time step.

References

1. Ilić-Stepić, A., Ognjanović, Z.: Logics for reasoning about processes of thinking with information coded by p-adic numbers. Studia Log. (2014)
2. Ilić-Stepić, A., Ognjanović, Z., Ikodinović, N., Perović, A.: A p-adic probability logic. Math. Log. Quarter. **58**(4–5), 263–280 (2012)
3. Khrennikov, A.: Toward theory of p-adic valued probabilities. Stud. Log. Gramm. Rhetor. **14**(27), 137–154 (2008)
4. Khrennikov, A., Yamada, S., van Rooij, A.: The measure-theoretical approach to p-adic probability theory. Ann. Math. Blaise Pascal **6**(1), 21–32 (1999)

5. Khrennikov, A.Y.: p-adic quantum mechanics with p-adic valued functions. J. Math. Phys. **32**(4), 932–937 (1991)
6. Khrennikov, A.Y.: Interpretations of Probability. VSP International Scientific Publishers, Utrecht/Tokyo (1999)
7. Milošević, M.: A propositional p-adic probability logic. Publications de l'Institut Mathématique, Nouvelle série **87**(101), 75–83 (2010)
8. Khrennikov, A., Schumann, A.: p-adic physics, non-well-founded reality and unconventional computing. p-Adic Num. Ultrametr. Anal. Appl. **1**(4), 297–306 (2009)
9. Khrennikov, A.Y., Schumann, A.: Logical approach to p-adic probabilities. Bullet. Sect. Log. **35**(1), 49–57 (2006)
10. Schumann, A.: Group theory and p-adic valued models of swarm behaviour. Math. Methods Appl. Sci. https://doi.org/10.1002/mma.4540
11. Schumann, A.: Dsm models and non-archimedean reasoning. In: Smarandache, F., Dezert, J. (eds.) Advances and Applications of DSmT (Collected works), vol. 2, pp. 183–204. American Research Press, Rehoboth (2006)
12. Schumann, A.: Non-Archimedean valued sequent logic. In: Eighth International Symposium on Symbolic and Numeric Algorithms for Scientific Computing (SYNASC'06), pp. 89–92. IEEE Press (2006)
13. Schumann, A.: Non-Archimedean valued predicate logic. Bullet. Sect. Log. **36**(1–2), 67–78 (2007)
14. Schumann, A.: p-adic multiple-validity and p-adic valued logical calculi. J. Multi. Val. Log. Soft Comput. **13**(1–2), 29–60 (2007)
15. Schumann, A.: Non-Archimedean fuzzy and probability logic. J. Appl. Non-Class. Log. **18**(1), 29–48 (2008)
16. Schumann, A.: Non-Archimedean valued and p-adic valued fuzzy cellular automata. J. Cell. Automata **3**(4), 337–354 (2008)
17. Schumann, A.: Non-archimedean valued extension of logic $l\pi$ and p-adic valued extension of logic bl. J. Uncertain Syst. **4**(2), 99–115 (2010)
18. Schumann, A.: p-adic valued fuzzyness and experiments with Physarum polycephalum. In: 11th International Conference on Fuzzy Systems and Knowledge Discovery (FSKD), pp. 466–472. IEEE Xplore (2014)
19. Schumann, A.: p-adic valued logical calculi in simulations of the slime mould behaviour. J. Appl. Non-Classic. Log. **25**(2), 125–139 (2015). https://doi.org/10.1080/11663081.2015.1049099
20. Schumann, A.: p-adic valued models of swarm behaviour. AIP Conference Proceedings **1863**(1), 360–369 (2017). https://doi.org/10.1063/1.4992538
21. Schumann, A., Adamatzky, A.: The double-slit experiment with Physarum polycephalum and p-adic valued probabilities and fuzziness. Int. J. Gen. Syst. **44**(3), 392–408 (2015)
22. Schumann, A., Pancerz, K.: p-adic computation with Physarum. In: A. Adamatzky (ed.) Advances in Physarum Machines: Sensing and Computing with Slime Mould, Emergence, Complexity and Computation, vol. 21, pp. 619–649. Springer International Publishing (2016)
23. Khrennikov, A.Y.: Modeling of Processes of Thinking in p-Adic Coordinates. Nauka, Fizmatlit, Moscow (2004). (in Russian)
24. Bachman, G.: Introduction to p-adic Numbers and Valuation Theory. Academic Press (1964)
25. Hasse, H.: Über-adische schiefkörper und ihre bedeutung für die arithmetik hyper-komplexen zahlsysteme. Math. Ann. **104**, 495–534 (1931)
26. Hasse, H.: Zahlentheorie. Akademie-Verlag, Berlin (1949)
27. Koblitz, N.: p-Adic Numbers, p-Adic Analysis and Zeta Functions, 2 edn. Springer (1984)
28. Mahler, K.: Introduction to p-Adic Numbers and Their Functions, 2 edn. Cambridge University Press (1981)
29. Robinson, A.: Non-Standard Analysis. North-Holland, Studies in Logic and the Foundations of Mathematics (1966)
30. Khrennikov, A.: Mathematical methods of the non-archimedean physics. Uspekhi Matematicheskikh Nauk **45**(4), 79–110 (1990)

31. Khrennikov, A.: Non-Archimedean Analysis: Quantum Paradoxes. Dynamical Systems and Biological Models, Springer (1997)
32. Khrennikov, A.: p-adic probability interpretation of Bell's inequality. Phys. Lett. A **200**(3–4), 219–223 (1995)
33. Khrennikov, A.Y.: p-Adic Valued Distributions in Mathematical Physics. Kluwer Academic Publishers, Dordrecht (1994)
34. Khrennikov, A.: Human subconscious as a p-adic dynamical system. J. Theor. Biol. **193**, 179–196 (1998)
35. Khrennikov, A.: p-Adic discrete dynamical systems and collective behaviour of information states in cognitive models. Discr. Dynam. Nature Soc. **5**, 59–69 (2000)
36. Khrennikov, A.: Interpretations of probability and their p-adic extensions. Theor. Probab. Appl. **46**(2), 256–273 (2001)
37. Ilić-Stepić, A., Ognjanović, Z., Ikodinović, N.: Conditional p-adic probability logic. Int. J. Approx. Reason. **55**, 9
38. Běhounek, L., Cintula, P.: Fuzzy class theory. Fuzzy Sets Syst. **154**(1), 34–55 (2005)
39. Cintula, P.: The $l\pi$ and $l\pi\frac{1}{2}$ propositional and predicate logics. Fuzzy Sets Syst. **124**(3), 21–34 (2001)
40. Cintula, P.: Advances in the $l\pi$ and $l\pi\frac{1}{2}$ logics. Arch. Math. Log. **42**(5), 449–468 (2003)
41. Hájek, P.: Metamathematics of Fuzzy Logic. Springer, Netherlands, Dordrecht (1998)
42. Novák, V., Perfilieva, I., Močkoř, J.: Mathematical Principles of Fuzzy Logic. Kluwer, Boston/Dordrecht (1999)
43. Gottwald, S.: Set theory for fuzzy sets of higher level. Fuzzy Sets Syst. **2**, 25–51 (1979)
44. Gottwald, S.: Fuzzy Sets and Fuzzy Logic: Foundations of Application–from A Mathematical Point of View. Vieweg, Wiesbaden (1993)
45. Gottwald, S.: Universes of fuzzy sets and axiomatizations of fuzzy set theory. part i: Model-based and axiomatic approaches. Studia Logica **82**, 211–244 (2006)
46. Novak, V.: On fuzzy type theory. Fuzzy Sets Syst. **149**, 235–273 (2004)
47. Gerla, G.: Fuzzy logic. In: Mathematical Tools for Approximate Reasoning, Kluwer Academic Publishers (2001)
48. Schumann, A.: Proof-theoretic cellular automata as logic of unconventional computing. Int. J. Unconvention. Comput. **8**(3), 263–280 (2012)
49. Schumann, A.: Reflexive games and non-Archimedean probabilities. P-Adic Num. Ultrametr. Anal. Appl. **6**(1), 66–79 (2014)
50. Baaz, M., Fermüller, C., Zach, R.: Systematic construction of natural deduction systems for many-valued logics. In: Proceedings of the 23rd International Symposium on Multiple Valued Logic, pp. 208–213. Sacramento, USA (1993)

Chapter 6
Individual-Collective Duality in Swarm Behaviours

The main assumption of behaviourism is that any animal behaviour can be controlled by attractants and repellents. It is quite true just for bacteria, but evidently it is untrue for swarms. The matter is that any swarm with a coordination of its members can behave according to an *individual-collective duality*—it means, it can perform only one action or many concurrent actions simultaneously and there is a fundamental uncertainty which mode will by chosen by the swarm right now (to behave as a big individual or a collective of distributed members). In this chapter, I am going to explicate this problem logically. See [1–4].

6.1 Quantum Double-Slit Experiment

6.1.1 Non-additive Measures

The fuzzy measures allow us to consider the so-called *spatial uncertainties*, when a phenomenon exists here and now, but lexicographically it has cloud boundaries, i.e. there is no possibility to define additivity for measuring this phenomenon. In this chapter, we introduce the so-called *temporal uncertainties*, when a phenomenon continuously changes, but in these changes there cannot be additivity, too, i.e. there are always uncertainties in its dynamics. Non-additivity of some dynamical phenomena does not mean that they cannot be studied mathematically. There are some rigorous approaches such as p-adic probability theory [5] which allow us to do it. There has been developed even p-adic quantum mechanics started from the publication of Igor Volovich in [6], about quantum mechanics see [7, 8]. In p-adic quantum mechanics [9, 10], p-adic probabilities are applied instead of real ones (about some logical aspects of p-adic physics see [11]).

A. Schumann, *Behaviourism in Studying Swarms: Logical Models of Sensing and Motoring*, Emergence, Complexity and Computation 33,
https://doi.org/10.1007/978-3-319-91542-5_6

The most significant feature of p-adic probabilities (or more generally, non-Archimedean probabilities or probabilities on infinite streams) is that they do not satisfy additivity. On the one hand, the p-adic analogies of the central limit theorem in real numbers face the problem that the normalized sums of independent and i.d. random variables do not converge to a unique distribution, there are many limit points [12], therefore in p-adic distributions we cannot build up the Gauss curve. On the other hand, the powerset over infinite streams like p-adic numbers is not a Boolean algebra in general case. In particular, there is no additivity—we cannot obtain a partition for any set into disjoint subsets whose sum gives the whole set (see Proposition 10.1).

The main feature of any swarm behaviour is that, on the one hand, the swarm motions are intelligent, but, on the other hand, they do not verify the *induction principle* (when the minimal set satisfying appropriate properties is given). This means that they can implement Kolmogorov-Uspensky machines only in a form of approximation, because the swarm performs much more, than just conventional calculations (the set realized is not minimal). So, if we perform the *double-slit experiment* for a swarm, we detect self-inconsistencies showing that we cannot approximate atomic individual acts of swarms as well as it is impossible to approximate single photons. From the standpoint of measure theory, it means that we cannot define additive measures for swarm actions.

In my opinion, the claim that there are no additive measures on swarm actions is a fundamental result for many behavioural sciences. Non-additivity of actions can be expressed in different ways: (i) *natural transition systems, such as swarm behaviour, cannot be reduced to Kolmogorov-Uspensky machines*, although their actions are intelligent, (ii) *there is an individual-collective duality, when we cannot approximate atomic individual acts* (an individual, such as plasmodium of *Physarum polycephalum*, can behave like a collective and a collective, such as collective of plasmodia, can behave like an individual).

In this chapter, we propose to use p-adic valued fuzziness and probabilities for measuring behaviours which cannot be measured additively.

6.1.2 Performativity in Quantum Experiments

In order to check whether there is an additivity for quantum probabilities, i.e. whether quanta are logical atoms indeed, the double-slit experiments have been performed with different particles: photons (this experiment was earliest and done by Thomas Young between 1801 and 1805), electrons [13], neutrons [14–16], etc. See also [17–20]. In these experiments a fluorescent screen has been used, so that the arrival of each electron is registered as a flash of light. On the one hand, the flash of a single electron on the screen occurs totally at random. On the other hand, if both slits are open and the experiment is continued for a sufficiently long period of time for thousands single electrons, there is an accumulation of flashes in some regions according to an interference pattern. As a result, the sum of the patterns with one slit covered at a time is not equal to the interference pattern with both slits opened: $P_{12}(x) = P_1(x) + P_2(x) + 2\sqrt{P_1(x)P_2(x)}\cos\delta$, where $\delta = 2\pi d \sin\theta/\lambda$ and $\lambda = h/p$, where p is the momentum of the passing electrons and λ is de Broglie wavelength. This means that the electrons cannot be considered the logical atoms, as they can behave like waves (collectives, i.e. composite sets), not particles in the conventional meaning.

The fact of that we cannot obtain a partition of sets for quantum probabilities just means that quanta cannot be regarded as logical atoms. Therefore Kolmogorovian probabilities are unapplied for quantum phenomena. The set-theoretic universes, where there are no atoms are called *non-well-founded*. They are studied in non-well-founded set theory [21]. Appropriate number systems, where numbers are defined as infinite streams, have a non-Archimedean structure. As a consequence, probabilities for non-well-founded sets may be defined on non-Archimedean number systems such as p-adic numbers [22–24] or infinitesimals [25].

In this section, I argue that the matter of quantum non-objectivity is that, on the one hand, the formalism of quantum mechanics constructed as a mathematical theory is self-consistent, but, on the other hand, quantum phenomena as results of experimenter's performances are not self-consistent. This self-inconsistency is an effect of that the language of quantum mechanics differs much from the language of human performances. The first is the language of a mathematical theory which uses some Aristotelian and Russellian assumptions (e.g. the assumption that there are logical atoms). The second language consists of performative propositions which are self-inconsistent only from the viewpoint of conventional mathematical theory, but they satisfy another logic which is non-Aristotelian. Hence, the representation of quantum reality in linguistic terms may be different: from a mathematical theory to a logic of performative propositions. At the level of mathematical theory, we deal with linguistic terms, satisfying the Aristotelian assumptions. At the level of logic of experimenter's performances, we deal with linguistic terms, not satisfying the Aristotelian assumptions.

Thus, it is possible to avoid the "quantum inconsistency" by applying modern tools of symbolic logic for studying intelligent behaviour (performances) and we will show that the quantum behaviour satisfies all the basic properties of performances. Logical tools for studying human behaviour were first proposed in the 20th-century language philosophy. Notice that in philosophy of language since Ludwig Wittgenstein [26], John Searle [27–29], and John Langshaw Austin [30] the ideas of non-objectivity of our everyday reality have actively developed within the so-called paradigm of *linguistic solipsism*. According to this paradigm, we deal just with linguistic reality if we think or act and cannot go out of language and return to things themselves. Any fact is seeable and understandable if and only if this is speakable and the fact can be described in a language [31]. So, in any thinking we are limited by our possible speech acts and in any activity by speech interactions. Language is a part of our behaviour and the way we are interacting with others, for example by commanding, requesting, pleading, joking, debating, etc. Philosophers of language distinguish performative propositions designating and expressing our behaviour from informative propositions denoting facts. While the *informative propositions* are truth-functions of the elementary propositions whose meanings are presented by facts, therefore they always have references in the real world, the *performative propositions* are non-objective in principle, we cannot find out any real references for them. They are self-referent and their meanings are just their utterances [26, 30]. It is possible to claim that *some quantum statements should be considered as performative propositions, as well.*

Notably, the informative propositions are ever *referential*, i.e. their meanings do not depend upon contexts and directly refer to facts. For example, it rains or it does not. Therefore ‘it rains’ is a true or false proposition. Simple proposition refer to simple facts (‘it rains’, ‘the wind blows’, etc.). Composite propositions refer to composite facts (‘it is raining now and the wind blows hard’). So, we can ever suppose the existence of logical atoms: first, logical atoms as simple propositions to build theories by composition rules and, second, logical atoms as simple facts to build models by composition rules. And in mathematical logic, we study relations between theories and models.

Let us consider now the proposition ‘I sell asset X’. Its meaning varies in the contexts of different situations and, therefore, it is a performative proposition. For example, at a loss relative to the price at which assets were purchased, the rational investors are likely to sell assets. On the other hand, the individual investors (they are usually irrational) prefer to sell assets which have gone up in value relative to their purchase price. Meanings of performative propositions depend on contexts and can change in time. For example, investors in their performative propositions (e.g. forecasts) might be optimistic, while at other times, they might be pessimistic. So, we cannot find out facts as logical atoms for the proposition ‘I sell asset X’. The point is that with the lapse of time investors can build different families of intensions as the meaning of phenomenon ‘selling asset X’ which can always be self-inconsistent, as a result. The *self-reference* means here that the proposition ‘I sell asset X’ refers not to facts, but to different intensions which ever change and can satisfy self-inconsistency. Hence, the self-referentiality just means that there are no logical atoms for both performative propositions and their meanings. In this chapter, I will show how the self-referentiality could be understood for quantum phenomena and propositions about them.

By emphasizing the role of performative propositions in quantum mechanics, we cannot avoid a discussion on the role of *free will.* I will show that in quantum mechanics the problem of free will is involved into considerations—how quantum performances can be thought and treated as appropriate performative propositions for which there are no real references, because they (as well as performative propositions about human interactions) have a non-objective status, see [32, 33].

For example, let us define truth-valuations of quantum mechanics conventionally, in the way of Aristotle and Russell:

- (i) the property E is actual (true) in a given state S, whenever a test of E on any physical object x in S would show that $E(x)$ is true for every x in the state S;
- (ii) the property E is nonactual (false) in a given state S whenever a complementary property $\neg E$ on any physical object x in S would show that $\neg E(x)$ is true for every x in the state S.

Objects x are interpreted as individuals (logical atoms). Assume that x mean quanta, $E, \neg E$ properties, discovered in the double-slit experiment, with the following meanings:

$E :=$ "*the non-detection of the position of x on the first screen in one of the two slits and the detection of the position of x on the registration screen corresponding to the momentum representation with the interference picture*";

$\neg E :=$ "*the detection of the position of x on the first screen in one of the two slits or the non-detection of the position of x on the registration screen corresponding to the momentum representation with the interference picture*".

According to the double-slit tests, we face that $E(x)$ and $\neg E(x)$ are true for the same state S. Obviously, that is self-inconsistent.[1]

But we *can deny our assumptions that x are logical atoms*, i.e. we can assume that x are not exclusive individuals. For instance, we can put forward the following self-referent definition of x: $\{x\} = \{a, b\}$, i.e., x is both a and b, where $a = (b, (a))$ and $b = (a, (b))$, i.e., $a = (b(a(b(a(b\ldots)))))$ and $b = (a(b(a(b(a\ldots)))))$ are two mutually depended infinite streams. Let $E(a)$ be true, $\neg E(b)$ be true, $E(b)$ be false, and $\neg E(a)$ be false. In this case, $E(x)$ and $\neg E(x)$ are true for the same state S and we cannot logically divide x into a and b, because x is a simple object, although it is not an individual. In this chapter, I will show, how we can deal with these strange non-Aristotelian objects logically.

Notice that the language of any physical theory consists of two different languages: a *theoretical language* (formal theory with axioms and inference rules) and an *observational language* (semantics for the theoretical language). The first language contains theoretical terms, which are understood as expressions that refer to non-observable entities or properties. The second language contains observational terms (observables).

---

[1]Notice that in this chapter, we deal just with the first-order logic to show that its non-classical version proposed by us, where there are no logical atoms, is a more applicable basic logical theory for quantum mechanics. Evidently, some disadvantages of first-order logic and its Aristotelian-Russellian semantics for quantum mechanics can be avoided in logics of higher orders. For example, the statement that the property E is actual (true) in a state S may assume quantifying on S in an appropriate second-order logic. But it is better to avoid the disadvantages of Aristotelian-Russellian semantics right at the level of first-order logic what we are doing actually. So, in classical logic the properties $\neg E(x)$ and $E(x)$ defined above cannot be true for the same S. Obviously, there are properties describing the double-slit experiment which are not self-inconsistent such as $E' :=$ "*if there is a non-detection of the position of x on the first screen in one of the two slits, then there is a detection of the position of x on the registration screen corresponding to the momentum representation with the interference picture*"; $E'' :=$ "*there is a detection of the position of x on the first screen in one of the two slits, then there is a non-detection of the position of x on the registration screen corresponding to the momentum representation with the interference picture*". Thus, we face the theoretical impossibility of assigning simultaneously truth values to all statements of the form $E(x)$ describing the double-slit experiment, because some assignments would be self-inconsistent formally, such as $\neg E(x)$ and $E(x)$, although experimentally we can suppose a verification of both $\neg E(x)$ and $E(x)$ at the same S. Our logic proposed in the chapter extends these limits to describe more properties for the same state, than the classical logic can allow.

A logic for observables is constructed in the observational language and a logic for theoretical entities in the theoretical language. In the early 20th century, there was a philosophical movement of logicism and its followers claimed that it is possible to construct a general logic for both observables and theoretical terms. This general logic could be called "*logical physics*". It is a part of logic, where logical properties of terms and propositions in relation to space, time, motion, causality, etc. are studied [34]. Usually, logical physics has been presented by logical tools for reducing propositions with theoretical entities to propositions with observables. In the accordance with this task, the theoretical terms are understood as follows: a term t is theoretical if and only if it holds, for all methods m of determining its extension, that m rests upon some axioms of some theory T and otherwise it is observational [35]. For example, in classical mechanics all methods of determining the force acting upon a particle appeal to some axioms of classical mechanics (CM), therefore force is a theoretical term for CM. The entity of spatial distance does not depend upon the Newtonian axioms. Hence, it is an observable of CM. Thus, the reduction procedure, which eliminates theoretical terms of an axiomatic theory by means of observables, is considered a set of semantic rules for interpreting propositions with theoretical terms on propositions with observables.

In the way proposed by R. Carnap and C.-O. Hempel, we can reduce theoretical entities by the following schemata: $c \Rightarrow (h \Rightarrow e)$, where h is a proposition in theoretical terms (hypothesis), c and e are propositions in terms of observables such that c expresses certain observational conditions, which are satisfied, e presents suitable detecting devices, which then have to show observable responses.

It was proven that there are ever theoretical terms which cannot be reduced to observational terms by any logical schemata. This circumstance of the existence of irreducible theoretical terms shows the rigorous limits of logical physics and logicism in physical sciences at all. For example, in *quantum mechanics* (QM) there are, first, a formal physical theory formulated in a theoretical language and, second, its semantics formulated in an observational language describing quantum experiments. There is also a logical way to reduce theoretical entities to observables. This way is presented by *quantum logic* (QL). Its logical schemas of reduction are as follows:

$$p, q := \text{'observable } o \text{ has a value in a Borel set } \Delta\text{'}$$

and such propositions are represented by closed subspaces of a Hilbert space, $\mathscr{H}$. The set of all such subspaces forms an ortholattice, $\mathscr{L}(\mathscr{H})$, with $p \leq q$ defined by 'p is a subspace of q'. The logical operations of 'and', 'or', and 'not' are modelled respectively by the operations of meet (infimum), join (supremum) and orthocomplement on $\mathscr{L}(\mathscr{H})$. The lattice $\mathscr{L}(\mathscr{H})$ is atomistic, complete, and orthomodular (non-distributive).

So, as we see, another concept of truth is defined in QL and this concept is radically different from the classical (Aristotelian-Russelian) concept of truth [36], because the QL schemas of reducing theoretical terms are different a lot from the Carnap and Hempel's classical manner. The main problem of QL is that even in non-classical means of interpreting QM there are irreducible theoretical statements. For example, the QM explanations of the double-slit experiment cannot be directly interpreted in QL. Nevertheless,

> quantum logics can be interpreted as a pragmatic language of pragmatically decidable assertive formulas, which formalize statements about physical systems that are empirically justified or unjustified in the framework of quantum mechanics. According to this interpretation, quantum logic formalizes properties of the metalinguistic concept of empirical justification within quantum mechanics rather than properties of a quantum concept of truth [36].

The Garola's pragmatic extension of QL allows him to define justifications of theoretical statements which cannot be reduced directly, e.g. within this extension it is possible to justify the QM explanations of the double-slit experiment. We can remember that the irreducibility of theoretic entities can imply even scientific anarchism:

> Science is an essentially anarchic enterprise: theoretical anarchism is more humanitarian and more likely to encourage progress than its law-and-order alternatives [37].

Therefore, the pragmatic approach can explicate many presuppositions of quantum physicists and their way of reasoning as one of the possible ways.

My approach to QL is different from the conventional QL with propositions defined on members of $\mathcal{L}(\mathcal{H})$ and the Garola's pragmatic extension of this QL. First of all, we would like to follow the pure logicism that has been reanimated by unconventional computing recently. In unconventional computing, we appeal to the following schemata of logical reductions: $I \Rightarrow (h \Rightarrow O)$, where h is a theoretical proposition, I are inputs of an unconventional computer (quantum computer, DNA-computer, *Physarum polycephalum* computer, etc.) and O are outputs of this computer. In these schemata, h is interpreted as a processor of suitable unconventional computer.

Unconventional computing is not so ambitious as physical theories such as QM. This new approach to computations completely ignores theoretical entities if they cannot be applied in designing an appropriate unconventional (abstract or real) processor. Hence, it deals just with reducible theoretical terms. In my research, I found out that the behavioural logic constructed on the observables of *Physarum polycephalum* and parasites of *Schistosomatidae* (Trematoda: Digenea) can be directly applied in the double-slit experiment with quanta. The basic idea of this behavioural logic is in the individual-collective dualism that there are no logical atoms in behaviours. Notice that logical theories for unconventional computing are always constructed in an observational language. In my opinion, the propagation of photons has some similarities with an intelligent propagation of *Physarum polycephalum*, parasites of *Schistosomatidae* (Trematoda: Digenea), and may other living organisms. Perhaps, we can claim about a new version of pantheism and idealism that the same patterns of intelligent behaviours are observed everywhere—from quanta to one-cell organisms and human beings.

6.1.3 Different Syllogistics for Classical Mechanics and Quantum Mechanics

Let us recall that the Kolmogorov's main assumption in probability theory is that there exists a set partition into disjoint subsets and, respectively, the probability measure defined on the given set is calculated as the addition of appropriate probabilities defined on subsets. However, we have just exemplified in the previous subsection (by the quantum double-slit experiment) that there are observables, where the additivity for probabilities is falsified if we deal with behaviours of quanta.

The *intuition of objectivity* that has been felt by the majority of physicists since the Ancient times till now was first formulated by Aristotle. According to him, there is '*hypokeimenon*' as substratum of any predicates. *Hypokeimenon* is a family of singular events or singular facts ('atoms' in the first meaning proposed by Democritus). For quantum physicists, *hypokeimenon* is given by smallest particles and all the world is described by predicates in relation to these particles, like that: 'the quantum has the property A', 'the complex B of quanta has properties $A_x, A_y, A_z, \ldots$, which explore physical phenomena $x, y, z, \ldots$, respectively', ..., etc. By Kolmogorov, probabilities should be involved in our reasoning just on particles (singular events) or their Boolean compositions. However, the double-slit experiment means that photons cannot be considered the Aristotelian *hypokeimenon* and there are no singular events at all.

According to Aristotle, *hypokeimenon*, the 'first subject', underlying things a, b, $c, \ldots$, present an objective reality. Every underlying thing possesses unique properties. It means that $a, b, c, \ldots$ are atoms of our database. There is nothing less than them. Due to properties, we can group atoms within different classes $P, Q, R, \ldots$ The more general property of thing, the more extensive class to which it belongs by this property.

Hence, the idea of *hypokeimenon*, the underlying things, allowed Aristotle to build up formal databases as well-founded trees of data (i.e. these trees are finite and without cycles or loops). He started with underlying things as primary descendants of trees in constructing ontological (syllogistic) databases. Let us notice that, by Aristotle, different sciences have different syllogistic databases, because they use different means for obtaining predicates for *hypokeimenon*. The quantum physics follows this Aristotelian understanding of objectivity and differs from the Ancient physics only by different ways of creating predicates for underlying things which are understood now as smallest particles.

Syllogistic trees contain genus-species relations among items. We know that in genus-species relations we can consider a branch (a relation between a genus and species) as implication, where the top of branch (genus) is regarded as consequent of implication and the bottom of branch (species) as antecedent of implication. Then for each node of the genera-species tree, we may define an intension as all reachable genera (all higher nodes) and an extent as all reachable species (all lower nodes). It is known that the greater extent, the smaller intension and the greater intension, the smaller extent.

Thus, the first logical database was invented by Aristotle. It is designed in his syllogistics. He suggested using this database as a logical frame for different sciences. Therefore if we claim that a science is a database constructed on the basis of empirical observations by applying logical inference rules, then we can claim that the history of exact science has started since Aristotle.

Let us assume that such a database is closed under all logical operations. Then in this database the following relations take place. Let P be a property. Then there is also a property non-P. Further let P be more extensive than Q (i.e. an appropriate class P is more extensive, than Q). Then we obtain the following relations:

- Q and non-P are properties which can be together false, but they cannot be together true in relation to any atom of our database;
- P and non-Q are properties which can be together true, but cannot be together false in relation to any atom of our database;
- P and non-P are properties which cannot be together true and cannot be together false in relation to any atom of our database;
- Q and non-Q are properties which cannot be together true and cannot be together false in relation to any atom of our database;
- if Q is true in relation to some atoms of our database, then also P is true in relation to the same atoms of our database;
- if P is not true in relation to some atoms of our database, then also Q is not true in relation to the same atoms of our database.

Thus, in the Aristotelian database we deal with the Boolean algebra due to the assumption of existence of *logical atoms* (singular events). Since Aristotle the objectivity has been understood as a possibility to construct databases when there is 'subject', the underlying things, the family of atoms grouped in classes, so that these classes are closed under all logical operations. In different sciences we choose different properties of atoms and as a consequence we group atoms differently. Such an intuition of objectivity holds in quantum physics till now.

Notably, *classical mechanics* (CM) can be readily presented as a semantics for the Aristotelian logic closed over logical superpositions of syllogistic propositions of the following kind: "All S are P", "Some S are P", "No S are P", "Some S are not P". The point is that in CM, first, we have Aristotelian atoms or individuals defined as particles, second, in CM the state of a system $\mathcal{S}$ consisting of N particles is defined by giving the $3N$ position coordinates and the $3N$ momentum coordinates. Hence, according to CM, any state of $\mathcal{S}$ is fully determined by three values for position and three values for momentum of all particles of $\mathcal{S}$. These two circumstances allow us to define a verification of syllogistic propositions as follows:

$S\mathbf{a}_{CM}P$: "All particles of $\mathcal{S}$ with positions S have momentums P" means that the system $\mathcal{S}$ is not empty (i.e. it contains some particles) and for all particles of $\mathcal{S}$ if we know their position S, then we know their momentum P;

$S\mathbf{i}_{CM}P$: "Some particles of $\mathcal{S}$ with positions S have momentums P" means that for some particles of the system $\mathcal{S}$ we know their position S and we know their momentum P;

$S\mathbf{e}_{CM}P$: "No particles of $\mathscr{S}$ with positions S have momentums P" means that for all particles of the system $\mathscr{S}$ we do not know their position S or we do not know their momentum P;

$S\mathbf{o}_{CM}P$: "Some particles of $\mathscr{S}$ with positions S do not have momentums P" means that all particles do not belong to the system $\mathscr{S}$ or there are particles of $\mathscr{S}$ such that we know their position S and we do not know their momentum P.

Formally:

$$S\mathbf{a}_{CM}P := (\exists A(A\varepsilon S) \wedge \forall A(A\varepsilon S \Rightarrow A\varepsilon P)); \tag{6.1}$$

$$S\mathbf{i}_{CM}P := \exists A(A\varepsilon S \wedge A\varepsilon P); \tag{6.2}$$

$$S\mathbf{e}_{CM}P := \neg(S\mathbf{i}_{CM}P); \tag{6.3}$$

$$S\mathbf{o}_{CM}P := \neg(S\mathbf{a}_{CM}P). \tag{6.4}$$

All other propositions of Aristotelian logic are defined thus: (i) each syllogistic proposition defined in (6.1)–(6.4) is a proposition, (ii) if X, Y are propositions, then $\neg X$, $\neg Y$, $X \star Y$, where $\star \in \{\vee, \wedge, \Rightarrow\}$, are propositions, too. Now, the Aristotelian logic can describe properties of our knowledge on systems $\mathscr{S}$ of CM.

Nevertheless, we can assume reality without objectivity, i.e. without logical atoms of databases. In modern logic universes in which there are no atoms are studied as well. However, in modern sciences the intuition that logical atoms exist has been used till now, and Aristotle's reasoning has been intuitively applied.

Notice that any context-based reasoning can be realized only in a universe without atoms. For instance, let us consider the following two propositions from the Bible: 'bestow that money for sheep' and 'bestow for whatsoever thy soul desireth' (*Deut.* 14:26). Syntactically, if we assume the existence of logical atoms, 'bestow for whatsoever thy soul desireth' is a universal affirmative proposition ($S\mathbf{a}P$) and 'bestow that money for sheep' is a particular affirmative proposition ($S\mathbf{i}P$), i.e. the first is more general, than the second. However, for example, I do not desire sheep and I do not know people who desire it. Perhaps, such people exist, but I do not know. Then we cannot plot the classical square of opposition, because 'bestow that money for sheep' is not included into 'bestow for whatsoever thy soul desireth', e.g. maybe my soul does not desire sheep, but desire many other things. The matter is that 'thy soul desireth' is a performative proposition and it has different meanings at different situations (there are no atoms for that proposition). This means that in this Biblical example the implication $S\mathbf{a}P \Rightarrow S\mathbf{i}P$ is false in general case. Thence, we could assume another semantics, where $S\mathbf{a}P$ and $S\mathbf{i}P$ are different viewpoints of the same level. Therefore, at one and the same situation of utterance both statements ('bestow for whatsoever thy soul desireth', $S\mathbf{a}P$, and 'bestow that money for sheep', $S\mathbf{i}P$) may be simultaneously false, but cannot be simultaneously true. In this way we obtain the unconventional square of opposition.

The same situation takes place for photons and other quanta. The conventional square of opposition does not hold for them, because logical atoms do not exist for all situations (for the double-slit experiment when both slits are open). Let us consider the propositions 'the photon can be detected passing through all slits' ($S\mathbf{a}_{photon}P$), 'the photon cannot be detected passing through the slits' ($S\mathbf{e}_{photon}P$), 'the photon can be detected passing through some slits (probably through the one slit)' ($S\mathbf{i}_{photon}P$), 'the photon cannot be detected passing through some slits (probably through the one slit)' ($S\mathbf{o}_{photon}P$). If it is possible to say that the following implication is valid: if 'the photon can be detected passing through all slits', then 'the photon can be detected passing through some slits (probably through the one slit)'? In other words, if the photon is a wave and described in a proposition in theoretical terms of 'wave', then the same photon is a particle and described in another proposition in theoretical terms of 'particle'? Rather there should be the disjunction: $S\mathbf{a}_{photon}P$ or $S\mathbf{i}_{photon}P$. But not implication. Indeed, if $S\mathbf{a}_{photon}P$ is true, then the photon is not a particle in common meaning, and if $S\mathbf{i}_{photon}P$ is true, then maybe the photon is a particle in common meaning.

In QM, it is impossible to define logical atoms. Indeed, if there were logical atoms in QM, our propositions would not be performative and would not depend upon the context of quantum experiments. It means that the reality exists, but it does not correspond to the Aristotelian intuition of objectivity when the underlying things exist, i.e. any behaviour is reduced to an individual behaviour. Instead of this intuition, we can propose another intuition, the *non-well-funded objectivity* [11], when there are no logical atoms, i.e. there is no 'first subject' in the Aristotelian meaning. I can add that in this picture of the world there are no things themselves ('*Dinge an sich*' in the Kantian meaning), nothing, only behavioural complexes described by self-referent performative propositions. It is a kind of idealism proposed in the *linguistic solipsism* of Wittgenstein and accepted by me. A similar idealistic approach was proposed by E. Husserl. He states that phenomena are nothing more than our consciousness, and pure phenomenology is the science of pure consciousness:

> Natural objects, for example, must be experienced before any theorizing about them can occur. Experiencing is consciousness that intuits something and values it to be actual; experiencing is intrinsically characterized as consciousness of the natural object in question and of it as the original: there is consciousness of the original as being there 'in person' [38].

Thus,

> the concept 'phenomenon' carries over, furthermore, to the changing modes of being conscious of something—or example, the clear and the obscure, evident and blind modes in which one and the same relation or connection, one and the same state of affairs, one and the same logical coherency, etc., can be given to consciousness [38].

Notice that Gestalt psychology is based on these ideas of Husserl.

Thus, in non-well-founded objectivity there are no logical atoms. What does it mean? We have properties (classes) P, Q, R, ... Some of these classes have non-empty intersections, and some others do not. Logical atoms are classes which cannot be intersected at all, they are singletons. Their intersection is always empty. This fact can be considered the definition of logical atom. Their logical combination (disjunction, conjunction, complement) gives any class $P, Q, R, \ldots$ Accordingly, the universe in which there are no logical atoms is universe in which the intersection of classes is not empty. Different combinations of these intersections give different contexts of performative propositions. For such a universe instead of the Aristotelian square of opposition, another square takes place, e.g.:

- *In the double-slit experiment with photons*: 'The photon can be detected in all slits' ($S\mathbf{a}_{photon}P$), 'the photon can be detected in no slits' ($S\mathbf{e}_{photon}P$), 'the photon can be detected in some slits (in QM, it means, just in one)' ($S\mathbf{i}_{photon}P$), 'the photon cannot be detected in some slits (just in one)' ($S\mathbf{o}_{photon}P$).
- *Formally*:

1. $S\mathbf{a}_{photon}P$ and $S\mathbf{i}_{photon}P$ are properties which can be together unjustified, but they cannot be together justified in relation to any performative situation of our database;
2. $S\mathbf{e}_{photon}P$ and $S\mathbf{o}_{photon}P$ are properties which can be together justified, but cannot be together unjustified in relation to any performative situation of our database;
3. $S\mathbf{a}_{photon}P$ and $S\mathbf{o}_{photon}P$ are properties which cannot be together justified and cannot be together unjustified in relation to any performative situation of our database;
4. $S\mathbf{e}_{photon}P$ and $S\mathbf{i}_{photon}P$ are properties which cannot be together justified and cannot be together unjustified in relation to any performative situation of our database;
5. if $S\mathbf{a}_{photon}P$ (respectively, $S\mathbf{i}_{photon}P$) is justified in relation to some performative situations of our database, then also $S\mathbf{e}_{photon}P$ (respectively, $S\mathbf{o}_{photon}P$) is justified in relation to the same situations of our database[2];
6. if $S\mathbf{e}_{photon}P$ (respectively, $S\mathbf{o}_{photon}P$) is not justified in relation to some performative situations of our database, then also $S\mathbf{a}_{photon}P$ (respectively, $S\mathbf{i}_{photon}P$) is not justified in relation to the same situations of our database.

Notice that in this version of QL, performative propositions expressing observables are not true or false in a conventional meaning of Russellian-Tarskian semantics, but they are justified or unjustified. Really, they have a pragmatic rather than a semantic interpretation. In the Austian semantics (also called the situation semantics), they are evaluated as successful or unsuccessful in the given situation of utterances.

So, instead of atoms in the quantum universe we deal with performative situations, i.e. with different intersections of classes (properties) in a collective behaviour. The logical theory of performative propositions was proposed in [39]. In this theory we

[2]I should remark that in logic, the implication does not mean a causal relation or deep semantic relationship between antecedent and consequent. For example, the sentence "2 + 2 = 4" implies that "We are born in the USSR", because both sentences are true. In case of syllogistic, we group quantified propositions into some classes according to their truth-conditions. So, this relationship between antecedent and consequent just formally follows from our formal definitions of $S\mathbf{a}_{photon}P$, $S\mathbf{e}_{photon}P$, $S\mathbf{i}_{photon}P$, $S\mathbf{o}_{photon}P$, see (6.5)–(6.8) and their interpretations on quantum observables below. In an informal interpretation, this relationship between antecedent and consequent means that the class of events of non-detecting in both slits is largest. The class of detecting in one slit or in both slits is smallest. It directly follows from our formal definitions below.

obtain non-well-founded syllogistic trees for which there cannot be underlying things (*hypokeimenon*). Thus, there is no objectivity in classical meaning. Indeed, we can always define intersections $A\&B$ for some situations A and B such that $A\&B$ is an infimum of A and B. Therefore there are no atoms which can be used for building trees-molecules as their superpositions. Instead of underlying things, we suppose situations that can always be intersected.

The Aristotelian logic with syllogistic propositions defined in (6.1)–(6.4) is self-inconsistent on quantum observables, although CM plays the role of semantics for this logic, as we said. Nevertheless, we can offer a non-Aristotelian system without logical atoms, where syllogistic propositions have the following meanings:

$S\mathbf{a}_{QM}P$: "All quanta of $\mathcal{S}$ with positions S have momentums P" means that the system $\mathcal{S}$ is not empty (i.e., it contains some quanta) and for all experiments with quanta of $\mathcal{S}$ their position S is absolutely uncertain for all possibilities and we know their momentum P; *in the double-slit experiment*: the system $\mathcal{S}$ is not empty and for all experiments with quanta of $\mathcal{S}$ these quanta pass through both slits (S) and their momentum has an interference picture (P);

$S\mathbf{i}_{QM}P$: "Some quanta of $\mathcal{S}$ with positions S have momentums P" means that for all quanta of the system $\mathcal{S}$ we know their position S and we do not know their momentum P; *in the double-slit experiment*: for all experiments with quanta of $\mathcal{S}$ these quanta pass through one slit (S) and their momentum does not have an interference picture (P);

$S\mathbf{e}_{QM}P$: "No quanta of $\mathcal{S}$ with positions S have momentums P" is justified iff "Some quanta of $\mathcal{S}$ with positions S have momentums P" is not justified;

$S\mathbf{o}_{QM}P$: "Some quanta of $\mathcal{S}$ with positions S do not have momentums P" is justified iff "All quanta of $\mathcal{S}$ with positions S have momentums P" is not justified.

Formally:

$$S\mathbf{a}_{QM}P := (\exists A(A\varepsilon S) \wedge \forall A(A\varepsilon S \wedge A\varepsilon P)); \tag{6.5}$$

$$S\mathbf{i}_{QM}P := \forall A(\neg(A\varepsilon S) \wedge \neg(A\varepsilon P)); \tag{6.6}$$

$$S\mathbf{e}_{QM}P := \neg(S\mathbf{i}P); \tag{6.7}$$

$$S\mathbf{o}_{QM}P := \neg(S\mathbf{a}P). \tag{6.8}$$

All other propositions of non-Aristotelian quantum logic are defined as follows: (i) each syllogistic proposition defined in (6.5)–(6.8) is a proposition, (ii) if X, Y are propositions, then $\neg X, \neg Y, X \star Y$, where $\star \in \{\vee, \wedge, \Rightarrow\}$, are propositions, also. This non-Aristotelian logic is a very simple version of QL without logical atoms. All its propositions are performative and depend upon contexts.

Hence, we could claim that from the viewpoint of performativity and logical theories studying performative propositions, there is no objective reality in the classical (Aristotelian) meaning. In QM, scientists try to appeal to the objective reality with logical atoms of quantum systems, which causes self-inconsistencies. Therefore, the only outcome is in appealing to non-well-founded reality [39] and performative propositions in QM. Self-inconsistency occurs only in cases of applying classical logic and classical semantics. In our logic, there are no contradictions. The same situation is in the so-called paraconsistent logics, where there are contradictions as new truth-values. Self-inconsistency is just in that we avoid logical atoms and even contrary statements in the classical logic may have non-empty intersections in our new semantics. Thus, the term self-inconsistency only concerns logical properties of our version of QL and not things themselves. According to Kant, we know nothing about things themselves.

6.2 Non-additivity of Swarm Behaviour

6.2.1 *Double-Slit Experiment with* Physarum Polycephalum

In the Kolmogorov's definition of *finitely additive probability measure*, it is assumed that for any set A there is always its partition into disjoint subsets so that their union gives A. We can illustrate this property by a thought *double-slit experiment with particles*. Let us imagine a machine gun spraying small particles at a screen in which there are two slits which may or may not be covered. Behind the first screen there is another one called observation screen where we detect particles that pass through the slits. Suppose that the following three experiments are carried out: (i) slit 1 is opened, slit 2 is covered; (ii) slit 1 is covered, slit 2 is opened; (iii) both slit 1 and 2 are opened. In the first (second) experiment particles strike the screen at random in a region somewhere opposite the position of slit 1 (slit 2). This outcome can be presented as a curve $P_1(x)$ (respectively, $P_2(x)$) which is understood in the following manner: $P_i(x)\delta x$ is equal the probability of a particle striking the screen in some region $(x, x + \delta x)$, where $i = 1, 2$.

Let A mean a set of all particles landing at the second screen. Its partition is set up as follows: A_1 means particles that pass through slit 1 and A_2 means particles that pass through slit 2. In the third experiment both slits are opened. Then we observe the particles varying between the two possibilities in a random way, if they would pass either through slit 1 or through slit 2, i.e. we obtain a new curve $P_{12}(x) = P_1(x) + P_2(x)$. In other words, we have $P(A) = P(A_1) + P(A_2)$. Thus, for the Kolmogorov's classical approach, the probability of set A is calculated as the sum of its two disjoint subsets A_1 and A_2 which are a partition of the given set A.

In the definition of probability measures, Kolmogorov appeals to a Boolean algebra of subsets and assumes that any set is a union of *logical atoms*—singletons. Thus, additivity means the existence of logical atoms (*elementary events* in probability theory) from the logical point of view.

Let us consider now the plasmodium behaviour of *Physarum polycephalum* as an example of swarming. This behaviour, on the one hand, is intelligent, but, on the other hand, the variety of plasmodium responses to stimuli is large enough. After performing many and many experiments we can start iterating the experiment we expect, but it is unreal to obtain an iteration of the same complex experiment forever. The plasmodium verifies the following three basic operations stimulated by attractants: (i) The *fusion*, denoted *Fuse*, means that two active zones A_1 and A_2 produce a new active zone A_3. (ii) The *multiplication*, *Mult*, means that the active zone A_1 splits into two independent active zones A_2 and A_3 propagating along their own trajectories. (iii) The *direction*, *Direct*, means that the active zone A is translated to a source of nutrients with certain initial velocity vector v. The basic operation stimulated by repellents: (iv) The *repelling*, *Repel*, means that the active zone A avoids a place of repellent with certain initial velocity vector v.

The operations *Fuse*, *Mult*, *Direct*, *Repel* can be examined as the most basic forms of intelligent behaviour of living organisms. For example, in [40] and [41], we showed that the behaviour of collectives of the genus *Trichobilharzia* Skrjabin & Zakharov, 1920 (*Schistosomatidae* Stiles & Hassall, 1898) can be simulated by the *Physarum* operations *Fuse*, *Mult*, *Direct* (without *Repel*). This means that, first, a local group of *Schistosomatidae* can behave as a programmable biological computer, second, a biologized kind of process calculus, such as *Physarum* transition system [42], can describe concurrent biological processes at all.

Thus, we showed [41] that the behaviour of one-cell organism of *Physarum* and the behaviour of local group of cercariae have identical patterns in the same transition systems. Other studies of simple population collectives behaving as a single entity as equivalent to the behaviour of the *Physarum* plasmodium is carried out in [43]. These joint patterns may be considered an effect of the *individual-collective duality*. In some situations, living agents may behave individually, but in the same situations at another time, they may behave collectively.

To approximate atomic acts of *Physarum* we can carry out the *double-slit experiment with plasmodia* to show that nevertheless the propagation of protoplasmic tubes can be considered a collective behaviour at a time, too. As a consequence, *Fuse*, *Mult*, *Direct*, *Repel* are simple actions indeed, but they are not atomic. Therefore, we cannot define additive measures on frames composite by *Fuse*, *Mult*, *Direct*, and *Repel*.

Let us take the first screen with two slits which are covered or opened and the second screen behind the first at which attractants are distributed evenly. Before the first screen there is an active zone of plasmodium. Then let us perform the following three experiments: (i) slit 1 is opened, slit 2 is covered; (ii) slit 1 is covered, slit 2 is opened; (iii) both slit 1 and 2 are opened (see Fig. 6.1). In the first (second) experiment, protoplasmic tubes arrive at the screen at random in a region somewhere opposite the position of slit 1 (slit 2). We have a curve $P_1(x)$ (respectively, $P_2(x)$) which is interpreted probabilistically: $P_i(x)\delta x$ is equal the probability of tubes arriving at the screen in some region $(x, x + \delta x)$, where $i = 1, 2$. The difference from the experiments with particles consists in that tubes split before the second screen and

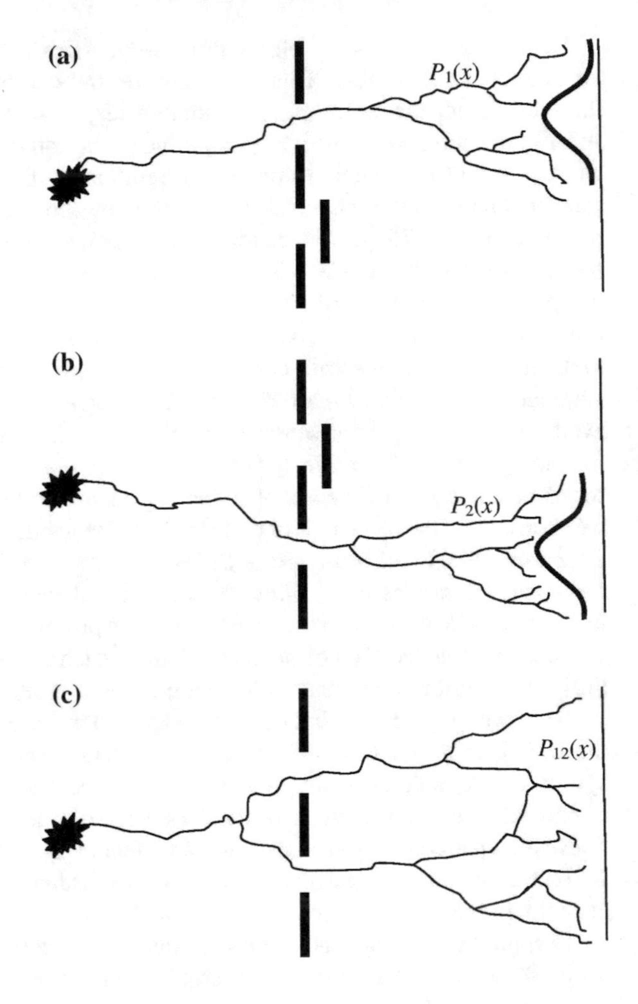

Fig. 6.1 The result of reaching plasmodium protoplasmic tubes at a screen when **a** only slit 1 is open; **b** only slit 2 is open; **c** both slits are open. The curves $P_1(x)$, $P_2(x)$, $P_{12}(x)$ represent the intensity of the tubes passing through the slits, see [4]

we always have several tubes split from the one and reaching the screen in some region simultaneously.

Let us denote all tubes landing at the second screen by A, thereby all tubes that pass through slit 1 by A_1 and all tubes that pass through slit 2 by A_2. Now we can check if either there is a partition of set A in case of *Physarum polycephalum* into sets A_1 and A_2 or there is a self-inconsistency. We open both slits. Then we see that the plasmodium behaves like photons, namely it can propagate just one tube passing through either slit 1 or slit 2 or it can propagate two tubes passing through both slits simultaneously. In the second case, these tubes split before the second screen and do not always occur at the same place, i.e. they appear to occur randomly across the whole screen (Fig. 6.1c). Thus, the total probability $P(A)$ corresponding to the intensity of plasmodium reaching the screen is not just the sum of the probabilities $P(A_1)$ and $P(A_2)$. This means that the plasmodium has the fundamental property of photons discovered in the double-slit experiment, i.e. we face a self-inconsistency, too, according to that it is impossible to approximate not only single photons, but also single actions of *Physarum polycephalum*. Indeed, in Fig. 6.1c we observe the one act of *Physarum* that is partly *Direct*, partly *Fuse*, and partly *Mult*. So, this one act of plasmodium is not individual, but collective. This allows us to state that we face the *individual-collective duality* of *Physarum polycephalum* plasmodium. In other words, there are no atomic acts for *Physarum polycephalum*. Its acts are parallel: in the one act, the plasmodium can do many things at once (as in Fig. 6.1c) and it can always perform many and many acts concurrently.

In Fig. 6.1, the ideal experiment is described. In reality, we can face more varieties of plasmodium activity, than it can be expected.

To show what we mean, let us consider the results of the three double-slit laboratory experiments with *Physarum* performed for me by Andrew Adamatzky, when only slit 2 is open, Figs. 6.2, 6.3, 6.4. In each experiment *Physarum* was inoculated near the center of one side of the Petri dish and the oat flakes, acting as attractants of *Physarum*, were placed on the opposite side of the dish. In the experiment shown in Fig. 6.2, *Physarum* propagates from the site of initial inoculation, explores its environment and determines gradients of chemo-attractant diffusion (Fig. 6.2a). It then locates widest opening between obstacles (Fig. 6.2b) and propagates towards the oat flakes and occupies them (Fig. 6.2c). In the experiments shown in Figs. 6.3 and 6.4, *Physarum* explores the space and even reaches the widest opening between the obstacles yet does not go through the opening but instead propagates around the chain of obstacles and then towards the source of nutrients.

We see the evidences that the plasmodium can achieve attractants by the shortest and longest paths. This property of plasmodium to choose different distances and not to follow only straight paths (as it would be expected in Kolmogorov-Uspensky machines) may be metaphorically called that *the plasmodium prefers Chinese strategies instead of Caesarian strategies*. The name 'Caesarian' is to refer to his famous motto 'veni, vidi, vici', which shows that the best way in decision-making is to follow algorithms, i.e. the shortest solutions.

While the foundation axiom of set theory [21] states that we can always build up *inductive sets* (i.e. straight paths in the shortest distance), its analogue in physics is a statement that any physical phenomenon may be measured by inductive sets. This statement is called the Archimedean axiom. The simplest case of the latter statement is that any physical phenomenon may be measured by a ruler (e.g. by rational numbers). The Chinese strategic thinking means that in paths of decision making, the Archimedean axiom can be invalid.

The plasmodium behaviour of Figs. 6.2, 6.3, and 6.4 shows that we cannot approximate single actions even in the situation when only one slit is open. It is an evidence for *non-additivity of measurement of Physarum activity*. Indeed, in Figs. 6.3, 6.4, the plasmodium follows partly *Fuse*, partly *Direct*, partly *Mult*, and partly *Repel*. So, according to this experiment, *Fuse*, *Direct*, *Mult*, and *Repel* are not atomic actions. Therefore we cannot obtain additivity on the plasmodium behaviours.

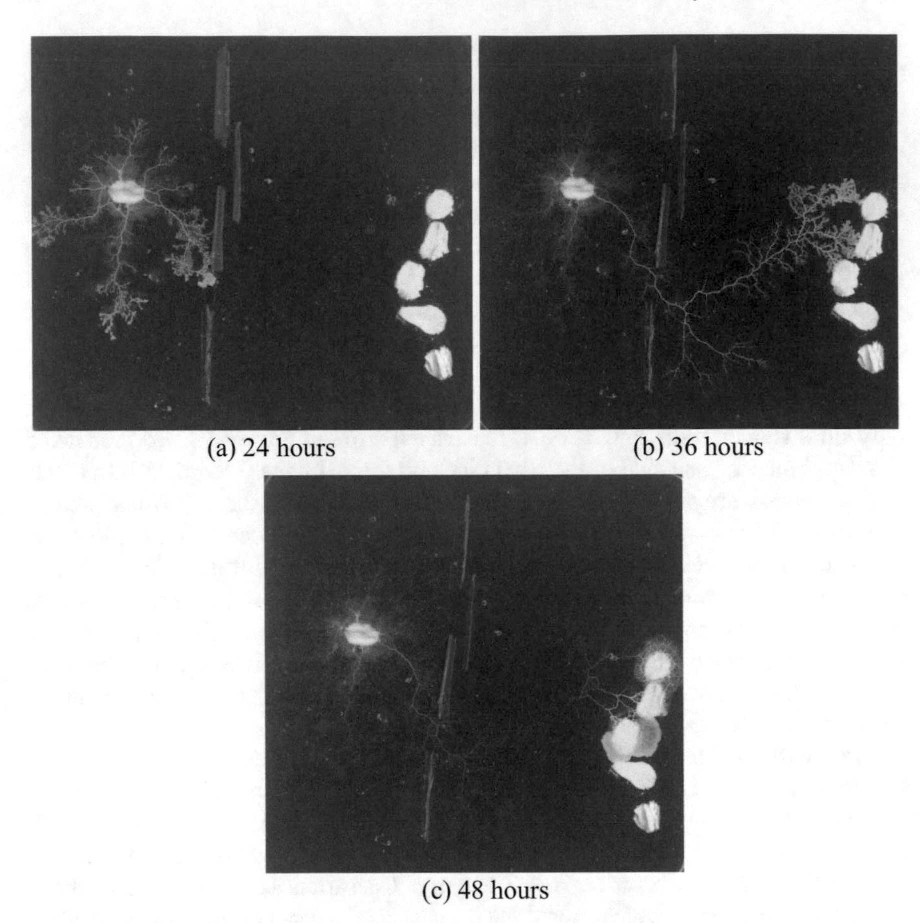

Fig. 6.2 The first experiment of reaching plasmodium protoplasmic tubes at a screen when only slit 2 is open. It has the form predicted in the thought experiment. The experiment was performed by Andrew Adamatzky, the picture was taken from [4] by his courtesy

6.2.2 *Growing Sample Space*

In probability theory, the universe is understood as a sample space Ω such that every member of $\mathscr{P}(\Omega)$ is called *event*. Conventionally, probability measures run over real numbers of the unit [0, 1] and their domain is a Boolean algebra of $\mathscr{P}(\Omega)$ with atoms (singletons) denoting simple (atomic) events. In the previous subsection, we have just shown that in the *Physarum* behaviour we cannot find out atomic events/actions, therefore we cannot define additive measures in the universe Ω. This means that if we wanted to define a domain of probability measures to calculate probabilities on the plasmodium, we could not appeal to the Boolean algebra of $\mathscr{P}(\Omega)$. Otherwise, our measures would be additive.

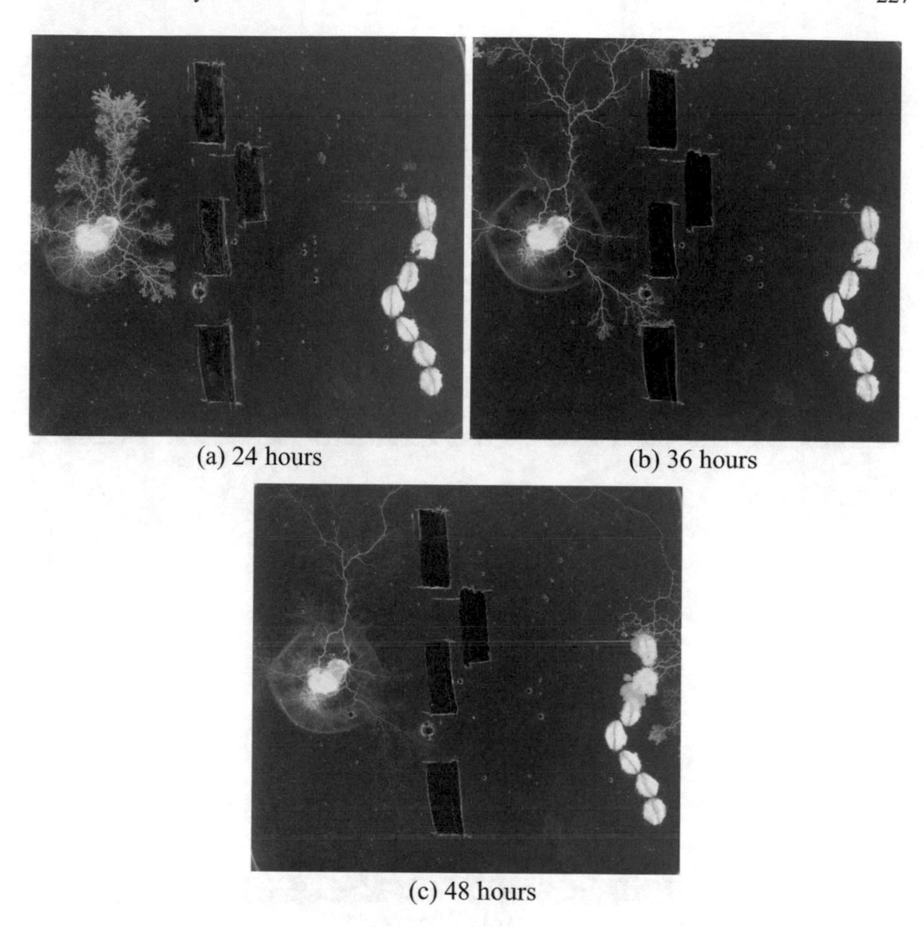

(a) 24 hours (b) 36 hours

(c) 48 hours

Fig. 6.3 The second experiment of reaching plasmodium protoplasmic tubes at a screen when only slit 2 is open. It has the form that was not predicted in the thought experiment. The plasmodium avoids the shortest distance to attractants. The experiment was performed by Andrew Adamatzky, the picture was taken from [4] by his courtesy

So, in our case, the powerset of $\mathcal{P}(\Omega)$ for the universe Ω must be non-atomistic. It would be possible if the sample space Ω was not fixed, but it changed, continuously at $t = 0, 1, 2 \ldots$ In other words, let us assume that Ω is a set at $t = 0$, which can grow, be expanded, decrease or just change in itself at next steps. As a consequence, we will deal not with atoms as members of Ω, but with streams of the form $\langle a_0, a_1, a_2, \ldots \rangle$, where $a_0 \in \Omega$ at $t = 0$, $a_1 \in \Omega$ at $t = 1$, $a_2 \in \Omega$ at $t = 2$, etc. Let us denote this instable set, i.e. the set of all these infinite streams, by Ω^* and call it a *wave set*. The powerset $\mathcal{P}(\Omega^*)$ is not a Boolean algebra, e.g. it is not atomistic (see Proposition 10.1).

Any swarm starts to move within a certain topology of attractants. Let this start point of motions be a sample set Ω at $t = 0$. This set contains information about

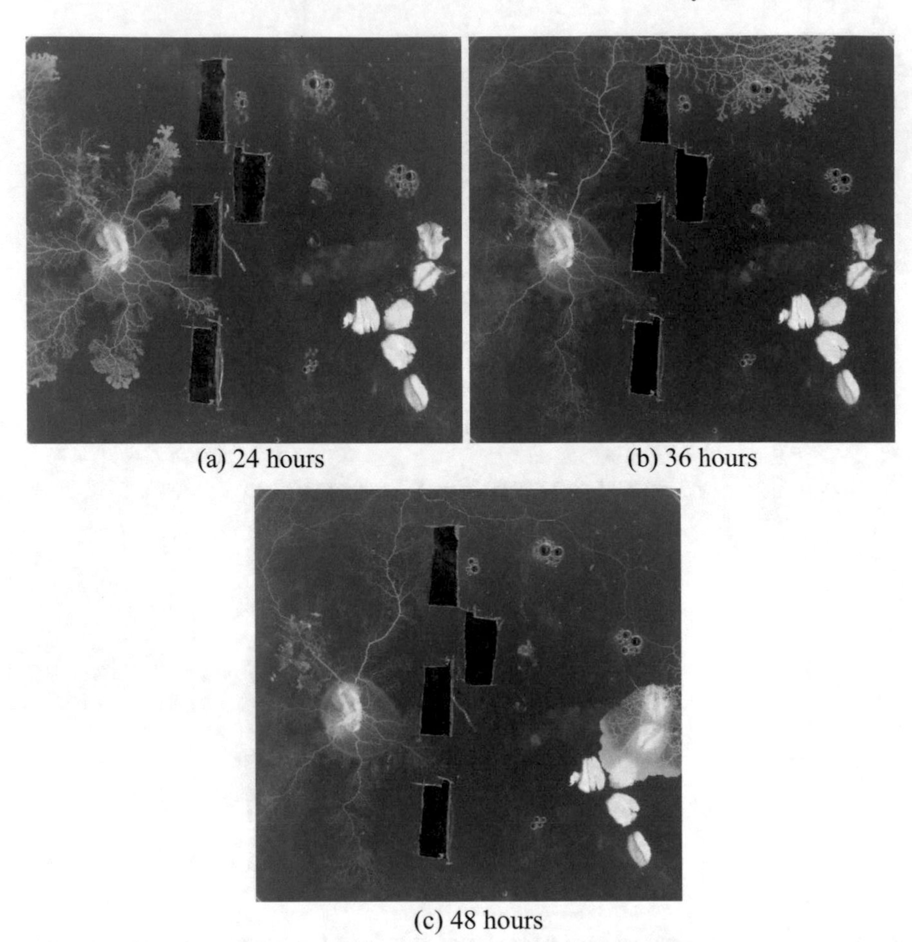

Fig. 6.4 The third experiment of reaching plasmodium protoplasmic tubes at a screen when only slit 2 is open. It has the form that was not predicted in the thought experiment, too. The plasmodium again avoids the shortest distance to attractants. The experiment was performed by Andrew Adamatzky, the picture was taken from [4] by his courtesy

positions of attractants. At next steps $t = 1, 2, 3, \ldots$, Ω contains information about the ways of occupying attractants by the swarm. Namely, let us assume that there are two neighbour attractants a and b. We say that there is a string ab or ba if both attractants a and b are occupied by the swarm. As a result, the dynamics of the swarm is observed as a continuous expansion of the set of strings at $t = 1, 2, 3, \ldots$

Suppose that Ω consists of $p - 1$ attractants and $A, B, \ldots \subseteq \Omega$. Let $A :=$ 'Attractants accessible for the attractant N_1 by swarm networks' and $B :=$ 'Neighbours for the attractant N_1', etc. (see Fig. 6.5). In other words, let $\mathscr{P}(\Omega)$ consist of sets which are defined by propositions describing our experiment with the swarm at $t = 0$. Some conditions of the experiment, fixed by subsets of Ω, do not change for different time

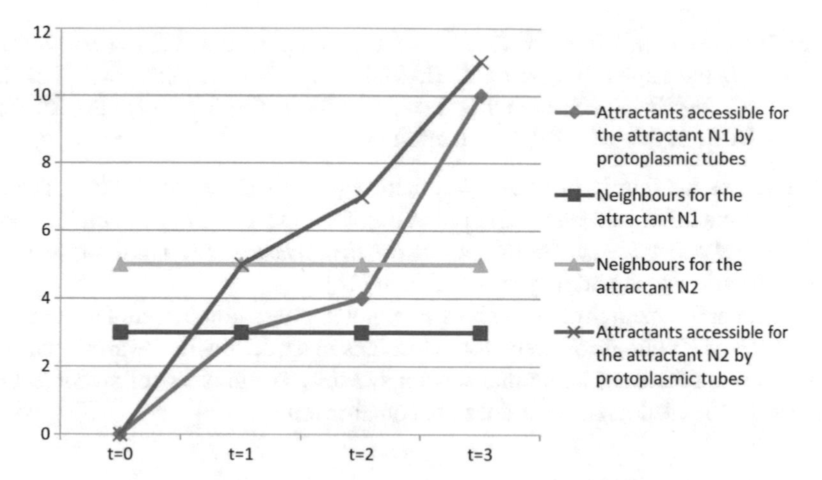

Fig. 6.5 The number of members of Ω which satisfy properties 'Attractants accessible for the attractant N_1 by swarm networks', 'Attractants accessible for the attractant N_2 by swarm networks', 'Neighbours for the attractant N_1' and 'Neighbours for the attractant N_2' for time $t = 0, 1, 2, 3$, see [4]

$t = 0, 1, 2, \ldots$ Some other conditions change for different time $t = 0, 1, 2, \ldots$ So, we can see that the property B is verified on the same number of members of Ω for any time $t = 0, 1, 2, \ldots$ Nevertheless, the property A is verified on a different number of members of Ω for different time $t = 0, 1, 2, \ldots$ (see Fig. 6.5).

Let $\lceil X \rceil$ mean a cardinal number of $X \subseteq \Omega$. So, $0 \leq \lceil X \rceil \leq p - 1$. Now we define cardinalities of wave sets as follows:

$$\lceil X^* \rceil := \langle \lceil X \rceil \text{ for } t = 0; \lceil X \rceil \text{ for } t = 1; \lceil X \rceil \text{ for } t = 2, \ldots \rangle$$

In particular, $\lceil \emptyset^* \rceil = \langle 0, 0, 0, \ldots \rangle$ and $\lceil \Omega^* \rceil = \langle p-1, p-1, p-1, \ldots \rangle$. Then $\langle 0, 0, 0, \ldots \rangle \leq \lceil X^* \rceil \leq \langle p-1, p-1, p-1, \ldots \rangle$ covers all p-adic integers, i.e. it belongs to the set $\mathbf{Z}_p$. In the example of Fig. 6.5, the cardinal numbers of all sets are infinite integers (i.e. they are larger than any natural number).

Hence, we can study some ensembles $\Omega^* = \prod_{t=0}^{\infty} \Omega_t$, such that $\lceil \Omega^* \rceil = \sigma$, where σ is an infinite stream. Let us consider a sequence of ensembles Ω_t having volumes $\lceil \Omega_t \rceil$, $t = 0, 1, \ldots$ We may imagine an ensemble Ω^* as being the population of a tower $T = T_\Omega$, which has an infinite number of floors with the following distribution of population through floors: population of t-th floor is Ω_t. Set $T_k = \prod_{t=0}^{k} \Omega_t \times \prod_{m=k+1}^{\infty} \emptyset_m$. This is a population of the first $k+1$ floors. Let $A^* \subset \Omega^*$ and let us assume that $n(A^*) = \lim_{t \to \infty} n_t(A^*)$, where $n_t(A^*) = \lceil A^* \cap T_t \rceil$. The quantity $n(A^*)$ is said to be a *non-well-founded volume of the set* A^*.

Now we can define the probability of A^* by the standard proportional relation:

$$\mathbf{P}(A^*) := \mathbf{P}_{\Omega^*}(A^*) = \frac{n(A^*)}{\sigma},$$

where $\lceil \Omega^* \rceil = \sigma$, $n(A^*) = \lceil A^* \cap \Omega^* \rceil = \lceil A^* \rceil$. Suppose that $\lceil \Omega \rceil = p - 1$. Then $\lceil \Omega^* \rceil = -1$, the largest integer of $\mathbf{Z}_p$ (indeed, if $p = 3$, then $\lceil \Omega^* \rceil = \ldots 22222 = 0 - 1 = -1$ and if $p = 5$, then $\lceil \Omega^* \rceil = \ldots 44444 = 0 - 1 = -1$). So, $\mathbf{P}(A^*) = \frac{\lceil A^* \rceil}{-1} = -\lceil A^* \rceil$. In this case $\mathbf{P}(\emptyset^*) = 0$ and $\mathbf{P}(\Omega^*) = \frac{-1}{-1} = 1$.

We denote the family of all $A^* \subseteq \Omega^*$, for which $\mathbf{P}(A^*)$ exists, by $\mathscr{G}_\Omega$. The sets $A^* \in \mathscr{G}_\Omega$ are said to be events. The ordered system $\langle \Omega^*, \mathscr{G}_\Omega, \mathbf{P}_{\Omega^*} \rangle$ is called a *non-well-founded ensemble probability space for the ensemble* Ω. These probabilities are non-Archimedean and they are studied in [44].

As we see, the main problem, why we cannot apply spatial presentations of algorithms such as Kolmogorov-Uspensky machines in modelling the swarm behaviour, consists in that there are no atomic acts for swarms. The universe of swarm actions is non-well-founded [21], i.e. it does not contain atoms.

6.2.3 p-*Adic Valued Probability Measure*

From the point of view of measure theory, the double-slit experiment with *Physarum polycephalum* means that the probability measure of Fig. 6.1 is not additive, although any probability measure must be additive according to definition. Usually, the additive property of probability measures is replaced by the weaker property of monotonicity in fuzzy measures. Thus, the fuzzy measure is considered an extension of probability measure, where additivity is not valid [45]. Let us recall that a *fuzzy measure* in space Ω is defined as follows: a function $\mu\colon \mathscr{P}(\Omega) \to [0, \infty)$ is a fuzzy measure if it satisfies the following properties: (i) $\mu(\emptyset) = 0$; (ii) if $A, B \in \mathscr{P}(\Omega)$ and $A \subseteq B$, then $\mu(A) \leq \mu(B)$. A fuzzy measure is called normalized if $\mu(\Omega) = 1$.

However, the fuzzy measure is too weak for our task to model any swarm behaviour. Therefore we define a stronger non-additive measure—*p-adic valued measure*. It is weaker, than the conventional probability measure, because it is not additive, but stronger than the conventional normalized fuzzy measure, because the maxitivity and minitivity axioms taken together have non-trivial solutions. From the standpoint of logic, it is called *p-adic valued fuzzy measure*. From the standpoint of measure theory, it is called *p-adic valued probability measure*. But it is the same.

So, in order to define probability measures with a domain on actions, we should appeal to the so-called non-well founded sets which do not have atoms at all (instead of atoms, they have streams). The main problem of these sets is that we cannot obtain a partition of sets in general case. Therefore we can preserve measurability without additivity. These new fuzzy and probability measures may be defined on non-Archimedean numbers, in particular on p-adic integers. In particular, if $\lceil \Omega^* \rceil = \lceil \mathbf{Z}_p \rceil$, where $\lceil \Omega \rceil = p - 1$, then for any $A^* \subseteq \Omega^*$, $\mathbf{P}(A^*) = \mathbf{P}_{\Omega^*}(A^*) = -\lceil A^* \rceil$.

Let us extend the standard order structure on $\mathbf{N}$ to a partial order structure on $\mathbf{Z}_p$: (i) for any p-adic integers $\sigma, \tau \in \mathbf{N}$, we have $\sigma \leq \tau$ in $\mathbf{N}$ iff $\sigma \leq \tau$ in $\mathbf{Z}_p$; (ii) each finite p-adic integer $n = \ldots \alpha_3\alpha_2\alpha_1\alpha_0$ (i.e. such that $\alpha_i = 0$ for any $i > j$) is less than any infinite number τ, i.e. $\sigma < \tau$ for any $\sigma \in \mathbf{N}$ and $\tau \in \mathbf{Z}_p \backslash \mathbf{N}$; (iii) if $\sigma = \ldots \sigma_3\sigma_2\sigma_1\sigma_0$

and $\tau = \ldots \tau_3\tau_2\tau_1\tau_0$ are infinite p-adic integers, then $\sigma \leq \tau$ iff $\sigma_n \leq \tau_n$ for all $n = 0$, 1, 2, ...

Now we can define sup and inf digit by digit. Then if $\sigma \leq \tau$, so $\inf(\sigma, \tau) = \sigma$ and $\sup(\sigma, \tau) = \tau$. The greatest p-adic integer according to our definition is $-1 = \ldots xxxxxx$, where $x = p - 1$, and the smallest is $0 = \ldots 00000$, i.e. $\lceil \mathbf{Z}_p \rceil = -1$.

Let us define p-adic fuzziness as follows: a *p-adic fuzzy measure* is a set function $F_{Z_p}(\cdot)$ defined for sets $A^*, B^* \subseteq \Omega^*$, it runs over the set $\mathbf{Z}_p$ and satisfies the following properties:

- $F_{Z_p}(\Omega^*) = -1$ and $F_{Z_p}(\emptyset^*) = 0$.
- If $A^* \subseteq \Omega^*$ and $B^* \subseteq \Omega^*$ are disjoint, i.e. $\inf(F_{Z_p}(A^*), F_{Z_p}(B^*)) = 0$, then $F_{Z_p}(A^* \cup B^*) = F_{Z_p}(A^*) + F_{Z_p}(B^*)$. Otherwise, $F_{Z_p}(A^* \cup B^*) = F_{Z_p}(A^*) + F_{Z_p}(B^*) - \inf(F_{Z_p}(A^*), F_{Z_p}(B^*)) = \sup(F_{Z_p}(A^*), F_{Z_p}(B^*))$.
- If $A^*, B^* \subseteq \Omega^*$, then $F_{Z_p}(A^* \cap B^*) = \inf(F_{Z_p}(A^*), F_{Z_p}(B^*))$.
- $F_{Z_p}(\neg A^*) = -1 - F_{Z_p}(A^*)$ for all $A^* \subseteq \Omega^*$, where $\neg A^* = \Omega^* \backslash A^*$.

A *p-adic probability measure* is a set function $\mathbf{P}_{Z_p}(\cdot)$ defined for sets $A^*, B^* \subseteq \Omega^*$ thus:

- $\mathbf{P}_{Z_p}(A^*) = -F_{Z_p}(A^*) \in \mathbf{Z}_p$
- $\mathbf{P}_{Z_p}(A^* | B^*) \in \mathbf{Q}_p$ is characterized by the following constraint:

$$\mathbf{P}_{Z_p}(A^* | B^*) = \frac{\mathbf{P}_{Z_p}(A^* \cap B^*)}{\mathbf{P}_{Z_p}(B^*)} = \frac{F_{Z_p}(A^* \cap B^*)}{F_{Z_p}(B^*)},$$

where $\mathbf{P}_{Z_p}(B^*) \neq 0$, $\mathbf{P}_{Z_p}(A^* \cap B^*) = \inf(\mathbf{P}_{Z_p}(A^*), \mathbf{P}_{Z_p}(B^*))$. The measure $\mathbf{P}_{Z_p}(\cdot)$ runs over the set $\mathbf{Q}_p$ of all p-adic numbers (not only integers).

6.2.4 p-*Adic Valued Fuzzy Syllogistic*

Let us recall that *classical syllogism* is a deductive inference schema based on the quantity relations $\mathbf{Q}$ among sets S and P and these relations are presented by the propositions of the following form: '$\mathbf{Q}S$ are P', where $\mathbf{Q}$ stands for one of the four classical crisp quantifiers: 'All', 'No', 'Some', 'Some... not', S is a subject-term denoting an appropriate crisp set, P is a predicate-term denoting an appropriate crisp set, too. Nevertheless, we can extend the syllogistic reasoning process to an optimization problem where the quantifiers can be: (1) proportional quantifiers like 'most', 'few', 'almost all', etc. with definitions on fuzzy sets; (2) interval imprecise quantifiers with well-defined bounds like 'between 50 and 70%', etc., (3) fuzzy quantifiers with imprecise-defined bounds: 'something more than 50%', etc.

In this subsection, we define quantifiers 'All', 'No', 'Some', 'Some... not' unconventionally as p-adic valued fuzzy quantifiers expressing the massive-parallel behaviour of swarm. The matter is that classically, these four quantifiers are defined on well-founded sets, which are constructed by means of atoms (individuals, things).

However, as we have demonstrated in the previous subsections, the universe of swarm actions (in particular, the universe of plasmodia of *Physarum polycephalum*) have no atoms (atomic acts) and, as a result, it is non-well-founded in the meaning of set theory without foundation axiom [21].

While Aristotelian syllogisms assume well-founded sets, i.e. finite crisp sets built up from atoms constructively, in the most cases of swarm behaviour we observe a spatial expansion of the swarm in all directions with many cycles without possibilities to define atoms (atomic actions). Under these circumstances it is more natural to define all the basic syllogistic propositions 'All S are P', 'No S are P', 'Some S are P', 'Some S are not P' in the way they would satisfy the inverse relationship, when all converses are valid: (1) 'All S are P' $\Rightarrow$ 'All P are S', 'No S are P' $\Rightarrow$ 'No P are S', 'Some S are P' $\Rightarrow$ 'Some P are S', 'Some S are not P' $\Rightarrow$ 'Some P are S'. In other words, then we can draw more natural conclusions for swarm networks which are decentralized and have some cycles. The formal syllogistic system over propositions with such properties is constructed in [39]. This system is called the *performative syllogistic*. In this section, we propose an extension of that system, where syllogistic propositions are temporally bounded. This system has semantics on p-adic integers.

The alphabet of our system for simulation of swarm behaviour contains as descriptive signs the syllogistic letters $S, P, M, \ldots$, as logical-semantic signs the syllogistic connectives $\mathbf{a}_t$, $\mathbf{e}_t$, $\mathbf{i}_t$, $\mathbf{o}_t$ at time t, $\mathbf{a}_\infty$, $\mathbf{e}_\infty$, $\mathbf{i}_\infty$, $\mathbf{o}_\infty$ for infinite time, and the propositional connectives: $\neg$, $\vee$, $\wedge$, $\Rightarrow$. Simple propositions are defined as follows: SxP, where $x \in \{\mathbf{a}_t, \mathbf{e}_t, \mathbf{i}_t, \mathbf{o}_t, \mathbf{a}_\infty, \mathbf{e}_\infty, \mathbf{i}_\infty, \mathbf{o}_\infty\}$. All other propositions are defined thus: (i) each simple proposition is a proposition, (ii) if X, Y are propositions, then $\neg X$, $\neg Y$, $X \star Y$, where $\star \in \{\vee, \wedge, \Rightarrow\}$, are propositions, too.

Syllogistic letters $S, P, M, \ldots$ are interpreted as attractants as follows: a data point S is considered empty if and only if an appropriate attractant denoted by S is not occupied by the swarm. Let us define syllogistic strings of the form SP at time t with the following notation: 'S is$_t$ P,' and with the following meaning: SP at t is true if and only if S and P are neighbour cells and both S and P are not empty at t, otherwise SP is false. Thus, this definition represents syllogistic reasoning as swarm labelled transition system $\langle$States, Edg$\rangle$, where $S, P, M, \ldots \in$ States and true propositions S is$_t$ $P, \ldots \in$ Edg. Using the definition of syllogistic strings, we can define simple syllogistic propositions as follows:

1. 'All S are P at time t' ($S\mathbf{a}_tP$): there is a string AS at time t and for any A which is a neighbour for S and P, there are strings AS and AP. This means that we have a massive-parallel occupation of region at t, where the cells S and P are located.
2. 'Some S are P at time t' ($S\mathbf{i}_tP$): for any A which is a neighbour for S and P at t, there are no strings AS and AP. This means that the swarm cannot reach S from P or P from S immediately at t.
3. 'No S are P at time t' ($S\mathbf{e}_tP$): there exists A at time t which is a neighbour for S and P such that there is a string AS or there is a string AP. This means that the swarm occupies S or P, but surely not the whole region at time t, where the cells S and P are located.

4. 'Some S are not P at time t' ($S\mathbf{o}_tP$): for any A which is a neighbour for S and P at time t there is no string AS or there exists A which is a neighbour for S and P such that there is no string AS or there is no string AP. This means that at time t the swarm does not occupy S or there is a neighbour cell which is not connected with S or P by a network.

Formally:

$$S\mathbf{a}_tP := (\exists A(A \text{ is}_t S) \wedge (\forall A(A \text{ is}_t S \wedge A \text{ is}_t P))); \tag{6.9}$$

$$S\mathbf{i}_tP := \forall A(\neg(A \text{ is}_t S) \wedge \neg(A \text{ is}_t P)); \tag{6.10}$$

$$\begin{aligned} &S\mathbf{e}_tP := \neg\forall A(\neg(A \text{ is}_t S) \wedge \neg(A \text{ is}_t P)), \textit{i.e.} \\ &\exists A(A \text{ is}_t S \vee A \text{ is}_t P). \end{aligned} \tag{6.11}$$

$$\begin{aligned} &S\mathbf{o}_tP := \neg(\exists A(A \text{ is}_t S) \vee (\forall A(A \text{ is}_t P \wedge A \text{ is}_t S))), \\ &\textit{i.e.}(\forall A\neg(A \text{ is}_t S) \wedge \exists A(\neg(A \text{ is}_t P) \vee \neg(A \text{ is}_t S))); \end{aligned} \tag{6.12}$$

Among simple syllogistic propositions of different time there are the following relations:

$$S\mathbf{a}_tP \Rightarrow S\mathbf{e}_{t+1}P. \tag{6.13}$$

From this it follows that

$$S\mathbf{i}_{t+1}P \Rightarrow S\mathbf{o}_tP. \tag{6.14}$$

The formulas $S\mathbf{a}_tP \Rightarrow S\mathbf{a}_{t+1}P$ and $S\mathbf{i}_{t+1}P \Rightarrow S\mathbf{i}_tP$ are not valid in general case, because the swarm can leave attractants which were occupied earlier. Now we can define simple syllogistic propositions for infinite time: $S\mathbf{a}_\infty P := \lim_{t\to\infty} S\mathbf{a}_tP$. Consequently, we may generalize Eqs. (6.9) and (6.10) in the following manner:

$$S\mathbf{a}_\infty P \Rightarrow \exists t.S\mathbf{a}_tP; \tag{6.15}$$

$$S\mathbf{i}_\infty P \Rightarrow \forall t.S\mathbf{i}_tP. \tag{6.16}$$

Other axioms are as follows:

$$S\mathbf{a}_tP \Rightarrow S\mathbf{e}_tP; \quad S\mathbf{a}_\infty P \Rightarrow S\mathbf{e}_\infty P; \tag{6.17}$$

$$S\mathbf{a}_tP \Rightarrow P\mathbf{a}_tS; \quad S\mathbf{a}_\infty P \Rightarrow P\mathbf{a}_\infty S; \tag{6.18}$$

$$S\mathbf{i}_tP \Rightarrow P\mathbf{i}_tS; \quad S\mathbf{i}_\infty P \Rightarrow P\mathbf{i}_\infty S; \tag{6.19}$$

$$S\mathbf{a}_tM \Rightarrow S\mathbf{e}_tP; \quad S\mathbf{a}_\infty M \Rightarrow S\mathbf{e}_\infty P; \tag{6.20}$$

$$M\mathbf{a}_tP \Rightarrow S\mathbf{e}_tP; \quad M\mathbf{a}_\infty P \Rightarrow S\mathbf{e}_\infty P; \tag{6.21}$$

$$(M\mathbf{a}_t P \wedge S\mathbf{a}_t M) \Rightarrow S\mathbf{a}_t P; \tag{6.22}$$

$$(M\mathbf{i}_t P \wedge S\mathbf{i}_t M) \Rightarrow S\mathbf{i}_t P; \tag{6.23}$$

$$(M\mathbf{a}_\infty P \wedge S\mathbf{a}_\infty M) \Rightarrow S\mathbf{a}_\infty P; \tag{6.24}$$

$$(M\mathbf{i}_\infty P \wedge S\mathbf{i}_\infty M) \Rightarrow S\mathbf{i}_\infty P. \tag{6.25}$$

Some basic formal properties of that axiomatic system are considered in the next Chapter (see also [39]). In the p-adic valued fuzzy syllogistic we can analyze the collective dimension of behaviour for different time. Within this system we can study, how the swarm occupies all possible attractants in any direction at time t, if it can only see them. So, this system shows logical properties of a massive-parallel behaviour.

Now let us show that, on the one hand, the Aristotelian syllogistic is unapplied for describing the double-slit experiment with photons or *Physarum polycephalum* (as well as with any other swarm) and, on the other hand, the new syllogistic closed over Eqs. (6.9)–(6.25) can do it. The *conventional logical square of opposition* is not valid within the double-slit experiment with photons or *Physarum polycephalum*. The problem is that the photon behaves both individually (as a particle) and collectively (as a wave). The same problem is with swarms: they behave both individually (according to one of the basic actions: *Direct*, *Fuse* or *Mult*) and collectively (as a mixing of *Direct*, *Fuse*, and *Mult* in the one simple act). In other words, we cannot find out logical atoms in the logical reconstruction of behaviours of photons or swarms.

For the universe of swarms as well as for the universe of quanta, instead of the Aristotelian square of opposition, another square of opposition takes place:

- *In the double-slit experiment*:
  'The coincidence clicks will always occur at t' ($S\mathbf{a}_t P$), 'the coincidence clicks will never occur at t' ($S\mathbf{e}_t P$), 'the coincidence clicks will somewhen occur at t' ($S\mathbf{i}_t P$), 'the coincidence clicks will not somewhen occur at t' ($S\mathbf{o}_t P$). Among these simple propositions, the following relations hold true in the p-adic valued fuzzy syllogistic:

  1. $S\mathbf{a}_t P$ and $S\mathbf{i}_t P$ are properties which can be together false, but they cannot be together true;
  2. $S\mathbf{e}_t P$ and $S\mathbf{o}_t P$ are properties which can be together true, but cannot be together false;
  3. $S\mathbf{a}_t P$ and $S\mathbf{o}_t P$ are properties which cannot be together true and cannot be together false;
  4. $S\mathbf{e}_t P$ and $S\mathbf{i}_t P$ are properties which cannot be together true and cannot be together false;
  5. if $S\mathbf{a}_t P$ (respectively, $S\mathbf{i}_t P$) is true, then also $S\mathbf{e}_t P$ (respectively, $S\mathbf{o}_t P$) is true,
  6. if $S\mathbf{e}_t P$ (respectively, $S\mathbf{o}_t P$) is not true, then also $S\mathbf{a}_t P$ (respectively, $S\mathbf{i}_t P$) is not true.

In any experiment with a swarm we deal with attractants which can be placed differently to obtain different topologies and to induce different transitions of the swarm. Let Ω, consisting of attractants, be the universe of our experiments such that every member of $\mathscr{P}(\Omega)$ is called *property of an appropriate experiment*. Let us exemplify one of the possible properties by the following proposition: "How many attractants are occupied by the swarm". This property is not constant, but it changes during the time, $t = 0, 1, 2, \ldots$ Therefore the universe of our experiments must depend on time, too. Let us assume that our time of experiments is infinite. So, the universe of our experiments should be regarded as a set of infinite streams, Ω^*, and then properties are members of $\mathscr{P}(\Omega^*)$. Notice that the powerset $\mathscr{P}(\Omega^*)$ cannot be a Boolean algebra (see Proposition 10.1).

The probability space $\langle \Omega^*, \mathscr{G}_\Omega, \mathbf{P}_{Z_p} \rangle$ may be considered semantics for the p-adic valued fuzzy syllogistic system. Let $A_{t,\{S,P\}}$ mean "attractants that are neighbours of attractants S and P at time t" and $O_{t,\{S,P\}}$ mean "attractants that are neighbours of attractants S and P and are occupied by the swarm at time t". Suppose, $A_{\{S,P\}} := \lim_{t\to\infty} A_{t,\{S,P\}}$ and $O_{\{S,P\}} := \lim_{t\to\infty} O_{t,\{S,P\}}$. We know that $\lceil A_{\{S,P\}} \rceil = \sigma$ and $\lceil O_{\{S,P\}} \rceil = \tau$ are infinite p-adic integers and we can define their p-adic valued probabilities as non-well-founded volumes: $\mathbf{P}_{Z_p}(A_{\{S,P\}}) = \lceil A_{\{S,P\}} \rceil$ and $\mathbf{P}_{Z_p}(O_{\{S,P\}}) = \lceil O_{\{S,P\}} \rceil$. Consequently, $\lceil A_{t,\{S,P\}} \rceil = \sigma$ and $\lceil O_{t,\{S,P\}} \rceil = \tau$ are finite p-adic integers (natural numbers) which are defined by induction as follows:

- $\lceil A_{t=0,\{S,P\}} \rceil = i_0 \in \{0, 1, \ldots, p-1\}$,
  $\lceil A_{t=1,\{S,P\}} \rceil = \ldots 00000 i_1 i_0$,
  where $i_0, i_1 \in \{0, 1, \ldots, p-1\}$,
  $\lceil A_{t=2,\{S,P\}} \rceil = \ldots 00000 i_2 i_1 i_0$,
  where $i_0, i_1, i_2 \in \{0, 1, \ldots, p-1\}$,
  $\ldots$,
  $\lceil A_{t=j,\{S,P\}} \rceil = \ldots 00000 i_j \ldots i_2 i_1 i_0$,
  where $i_0, i_1, i_2, \ldots, i_j \in \{0, 1, \ldots, p-1\}$.
- $\lceil O_{t=0,\{S,P\}} \rceil = i_0 \in \{0, 1, \ldots, p-1\}$,
  $\lceil O_{t=1,\{S,P\}} \rceil = \ldots 00000 i_1 i_0$,
  where $i_0, i_1 \in \{0, 1, \ldots, p-1\}$,
  $\lceil O_{t=2,\{S,P\}} \rceil = \ldots 00000 i_2 i_1 i_0$,
  where $i_0, i_1, i_2 \in \{0, 1, \ldots, p-1\}$,
  $\ldots$,
  $\lceil O_{t=j,\{S,P\}} \rceil = \ldots 00000 i_j \ldots i_2 i_1 i_0$,
  where $i_0, i_1, i_2, \ldots, i_j \in \{0, 1, \ldots, p-1\}$.

We define the p-adic valued probabilities of $A_{t=j,\{S,P\}}$ and $O_{t=j,\{S,P\}}$ as their non-well-founded volumes: $\mathbf{P}_{Z_p}(A_{t=j,\{S,P\}}) = \lceil A_{t=j,\{S,P\}} \rceil$ and $\mathbf{P}_{Z_p}(O_{t=j,\{S,P\}}) = \lceil O_{t=j,\{S,P\}} \rceil$. They are natural numbers.

The *semantics for p-adic valued fuzzy reasoning* is as follows:

- $S\mathbf{a}_t P$ is true if and only if $\mathbf{P}_{Z_p}(A_{t=j,\{S,P\}}) = \mathbf{P}_{Z_p}(O_{t=j,\{S,P\}})$;
- $S\mathbf{i}_t P$ is true if and only if $\mathbf{P}_{Z_p}(O_{t=j,\{S,P\}}) = 0$;
- $S\mathbf{e}_t P$ is true if and only if $\mathbf{P}_{Z_p}(O_{t=j,\{S,P\}}) > 0$;

- $S\mathbf{o}_t P$ is true if and only if $\mathbf{P}_{Z_p}(A_{t=j,\{S,P\}}) > \mathbf{P}_{Z_p}(O_{t=j,\{S,P\}})$;
- $S\mathbf{a}_\infty P$ is true if and only if $\mathbf{P}_{Z_p}(A_{\{S,P\}}) = \mathbf{P}_{Z_p}(O_{\{S,P\}})$;
- $S\mathbf{i}_\infty P$ is true if and only if $\mathbf{P}_{Z_p}(O_{\{S,P\}}) = 0$;
- $S\mathbf{e}_\infty P$ is true if and only if $\mathbf{P}_{Z_p}(O_{\{S,P\}}) > 0$;
- $S\mathbf{o}_\infty P$ is true if and only if $\mathbf{P}_{Z_p}(A_{\{S,P\}}) > \mathbf{P}_{Z_p}(O_{\{S,P\}})$.

In the space $\langle \Omega^*, \mathcal{G}_\Omega, \mathbf{P}_{Z_p} \rangle$, it is possible to define many other fuzzy quantifiers such as conventional quantifiers 'All', 'No', 'Some', 'Some... not' in the Aristotelian meaning, but defined on fuzzy subsets of Ω^ω. Notice that these subsets may be called fuzzy, because the powerset $\mathcal{P}(\Omega^*)$ is not a Boolean algebra (see Proposition 10.1). The semantics for the Aristotelian quantifiers on members S, P of $\mathcal{P}(\Omega^*)$ are as follows:

- 'All S are P' is true if and only if $\mathbf{P}_{Z_p}(P\,|S) = -1$;
- 'No S are P' is true if and only if $\mathbf{P}_{Z_p}(P\,|S) = 0$;
- 'Some S are P' is true if and only if $\mathbf{P}_{Z_p}(P\,|S) > 0$;
- 'Some S are not P' is true if and only if $\mathbf{P}_{Z_p}(P\,|S) < -1$.

Some other p-adic valued fuzzy quantifiers on members S, P of $\mathcal{P}(\Omega^*)$:

- 'Most S are P' is true if and only if $-1 - \Delta \leq \mathbf{P}_{Z_p}(P\,|S) < -1$, where Δ is small;
- 'Few S are P' is true if and only if $0 < \mathbf{P}_{Z_p}(P\,|S) \leq \Delta$.

The double-slit experiment with *Physarum polycephalum* shows that, first, we cannot extract atomic actions of a swarm from all the kinds of the swarm behaviour, second, probability measures used in describing this experiment are not additive. In order to simulate the swarm behaviour, we construct the p-adic valued fuzzy syllogistic, where quantifiers 'All', 'No', 'Some', 'Some... not' are interpreted in the massive-parallel way to describe the swarm propagation in different directions. Then we define p-adic probability (fuzzyness) to fix all properties (conditions) for any experiment with the swarm.

6.2.5 p-Adic Valued Fuzzy Logic Controllers

In p-adic many-valued logic, we can study the wave sets $A^*, B^*, \ldots \subseteq \Omega^*$ and their logical combinations as propositions, where truth values are p-adic integers $F_{Z_p}(A^*)$, $F_{Z_p}(B^*), \ldots$ and their appropriate logical compositions. This logic can be engaged in designing p-adic valued fuzzy controllers on the medium of a swarm behaviour.

Let us assume that all attractants used for p-adic valued fuzzy controllers are located in the way that at each time step t the swarm cannot achieve more than $p - 1$ attractants and for some t_i the swarm can occupy exactly $p - 1$ attractants. This means that, first, at any step t there are $p - 1$ or less neighbour attractants, second, our sample space is, on the one hand, expended spatially by adding new attractants, but, on the other hand, can be interpreted as p-adic valued still. If the swarm moves in n directions simultaneously, then it is understood as a union of n wave sets. So, one wave set is a tunnel with the diameter consisting of $p - 1$ attractants.

Notice that p-adic many-valued logic can study the sample space of swarm motions for an infinite time $t \to \infty$. However, in controllers it is enough to deal with a finite $t = k$. For this purpose, we will appeal to the following p-adic towers with k floors: $T_k = \prod_{t=0}^{k} \Omega_t \times \prod_{m=k+1}^{\infty} \emptyset_m$; $T_k^A = \prod_{t=0}^{k} A_t \times \prod_{m=k+1}^{\infty} \emptyset_m$; $T_k^B = \prod_{t=0}^{k} B_t \times \prod_{m=k+1}^{\infty} \emptyset_m$, such that $T_k^A, T_k^B \subseteq T_k$. In the fragment of p-adic valued logic closed over simple propositions $T_k^A, T_k^B \subseteq T_k$ and their logical combinations, the set of designated truth values is presented by $D = \{(p^i - 1)_{i=1}^{k}\} = \{\langle \underbrace{p-1, p-1, p-1, \ldots, p-1}_{k} \rangle\}$ and the set of all truth values by $V = \{0, 1, 2, \ldots, (p^i - 1)_{i=1}^{k}\}$.

Thus, each proposition of the $(p^i - 1)_{i=1}^{k}$-valued fragment of p-adic valued logic describes one of the forms of swarm propagations at time step $t = k$. This propagation can be treated as a *p-adic valued fuzzy logic controller* (pFLC). It is a device computing an output vector from an input vector by some *linguistic rules*. In these rules, first, we use some linguistic labels, such as 'high', 'low', 'medium', to designate input ports of pFLC, second, we use some linguistic labels to designate output ports of pFLC. Input ports are called *premisses of linguistic rules*. Output ports are called *conclusions of linguistic rules*. The pFLC is a labyrinth of $(p-1)$-valued tunnels with n input ports and m output ports. This labyrinth is designed as a model for appropriate propositions of $(p^i - 1)_{i=1}^{k}$-valued logic.

The algorithm of constructing pFLC at $t = k$ is as follows (Fig. 6.6):

- *Linguistic variables*: definition of input and output linguistic variables; selection of meaningful linguistic states ('high', 'low', 'medium', etc.) for each variable.
- *Fuzzification*: identification of input ports of pFLC with input linguistic variables; construction of fuzzy membership function for input variables (how strongly they belong to linguistic states); identification of fuzzy membership function for input variables with concentration of swarm at an appropriate input port (0% of membership means that there is no swarm at the input port, 100% means that all attractants at $t = 0$ are occupied, $0\% < x\% < 100\%$ means that $x\%$ of attractants at $t = 0$ are occupied).
- *Rule base*: construction of rule base for transforming input ports of pFLC into its output ports. This rule base is a family of $(p^i - 1)_{i=1}^{k}$-valued propositions. Appropriate experiments with the swarm are models for these propositions. We can simulate the swarm motions on the basis of $(p^i - 1)_{i=1}^{k}$-valued propositions.
- *Inference engine*: conversion of states for input variables into states of output variables in accordance with the p-adic valued rule base; simulation of the swarm motions in an object-oriented language (e.g. in the object-oriented language for *Physarum*); performing of experiment with the swarm in accordance with the given simulation.
- *Defuzzification*: combination of results of output ports of pFLC; identification of output ports with output linguistic variables; detection of swarm concentration at each tunnel of appropriate output port at $t = 0, 1, \ldots, k$; conversion of these concentrations into p-adic valued wave sets (see Fig. 6.5) defined by the motions through the tunnels of output variables; conversion of p-adic fuzzy values into a single real number (as center of maxima, mean of maxima, etc).

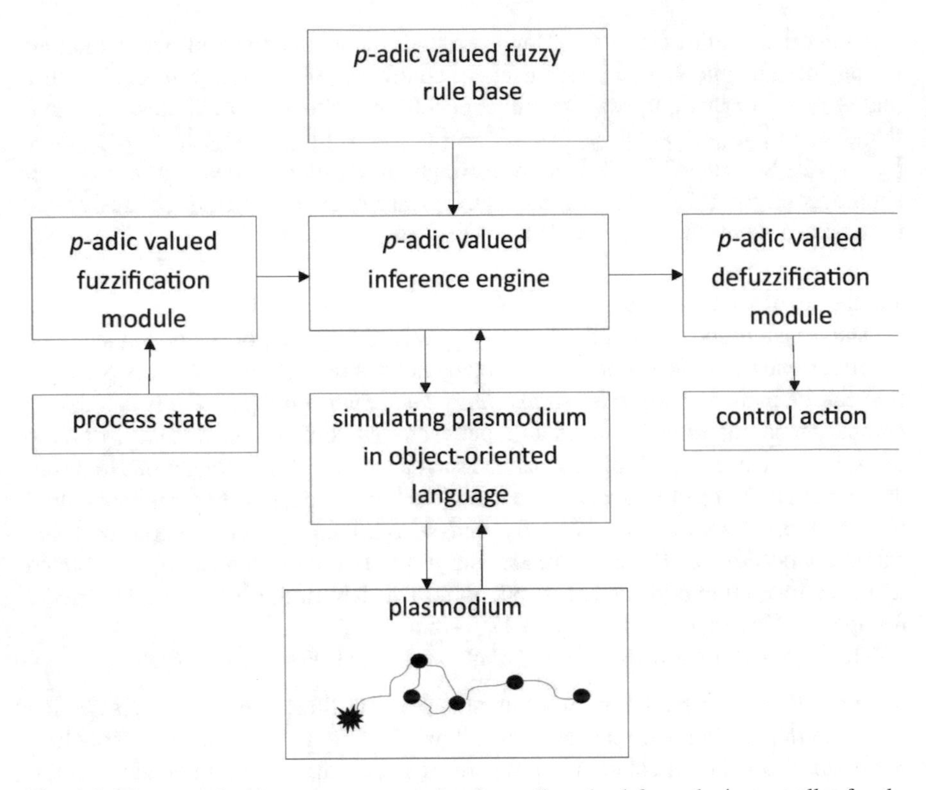

Fig. 6.6 The steps for the output computation in p-adic valued fuzzy logic controller for the plasmodium of *Physarum polycephalum*, see [4]

Let $0 < m < k$. Then pFLC can be used for prognoses what action will be chosen at $t = m$. In other words, pFLC is understood as the set of the following rules: *if agent i performs an action A , then at t = m (s)he will perform an action B with an uncertainty (meaning)* $F_{Z_p}(B)$, where A is an input variable, B is an output variable at time $t = m$. Meanings of output variables run over finite p-adic integers $F_{Z_p}(T^A_m), F_{Z_p}(T^B_m) \leq F_{Z_p}(T_m)$. This understanding of pFLC allows us to define the following repeating operative cycle of pFLC (see Fig. 6.7):

- *Inputs at time t*: measurements of all input linguistic variables at time t.
- *Fuzzification*: conversion of the measurements at time t into fuzzy sets to express their uncertainties; these fuzzy sets are used to identify the swarm concentration at input ports.
- *Inference engine*: Data of input ports at time t are transformed by the inference engine into output ports of time $t + 1$ in accordance with the control rules formulated in $(p^i - 1)^k_{i=1}$-valued logic with a simulation of a swarm.
- *Defuzzification*: outputs at time $t + 1$ are converted into single crisp values (the best representatives of the p-adic valued wave sets).

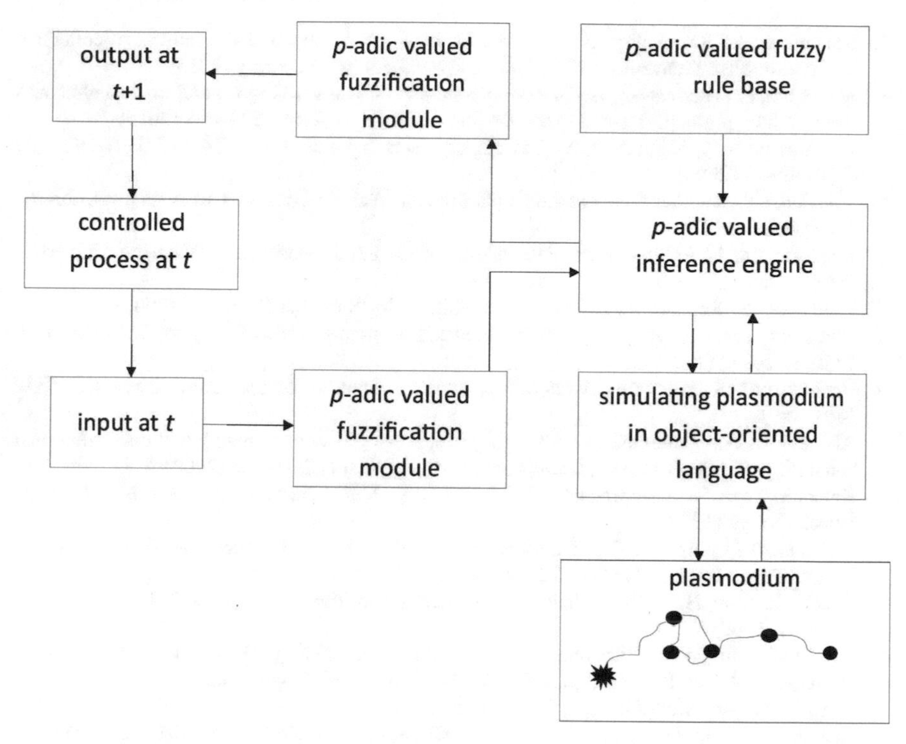

Fig. 6.7 The operative cycle of p-adic valued fuzzy logic controller for the plasmodium of *Physarum polycephalum*, see [4]

- *Inputs at time* $t + 1$: measurements of all input linguistic variables at time $t + 1$, etc.
- …

The double-slit experiment with *Physarum* as well as with any other swarm shows that, first, we cannot approximate its single actions and, second, we can appeal to p-adic fuzzy and probability measures which are not additive. For describing experiments with the swarm, we have constructed p-adic many-valued logic. On the basis of this logic it is possible to design p-adic valued fuzzy logic controllers.

References

1. Khrennikov, A., Schumann, A.: Quantum non-objectivity from performativity of quantum phenomena. Phys. Scr. **T163** (2014)
2. Schumann, A.: p-adic valued fuzzyness and experiments with Physarum polycephalum. In: 11th International Conference on Fuzzy Systems and Knowledge Discovery (FSKD), pp. 466–472. IEEE Xplore (2014)

3. Schumann, A.: Rationality in the behaviour of slime moulds and the individual-collective duality. In: AISB Convention 2015. University of Kent in Canterbury (2015)
4. Schumann, A., Adamatzky, A.: The double-slit experiment with Physarum polycephalum and p-adic valued probabilities and fuzziness. Int. J. Gen. Syst. **44**(3), 392–408 (2015)
5. Khrennikov, A.Y., Schumann, A.: Logical approach to p-adic probabilities. Bull. Sect. Logic **35**(1), 49–57 (2006)
6. Volovich, I.V.: Number theory as the ultimate theory. Technical Report, CERN preprint, CERN-TH.4791/87 (1987)
7. von Neuman, J.: Mathematical Foundations of Quantum Mechanics. Princeton University Press, Princeton (1955)
8. Plotnitsky, A.: Reading Bohr: Physics and Philosophy. Springer, Dordrecht (2006)
9. Khrennikov, A.Y.: p-adic quantum mechanics with p-adic valued functions. J. Math. Phys. **32**(4), 932–937 (1991)
10. Vladimirov, V.S., Volovich, I.V.: p-adic quantum mechanics. Commun. Math. Phys. **123**, 659–676 (1989)
11. Khrennikov, A., Schumann, A.: p-adic physics, non-well-founded reality and unconventional computing. P-Adic Numbers, Ultrametric. Anal. Appl. **1**(4), 297–306 (2009)
12. Khrennikov, A.Y.: Interpretations of Probability. VSP International Science Publishers, Utrecht/Tokyo (1999)
13. Möllenstedt, G., Jönsson, C.: Elektronen-mehrfachinterferenzen an regelmaessig hergestellten feinspalten. Z. Phys. A **155**(4), 472–474 (1959)
14. Rauch, H., Kurz, H.: Beugung thermischer neutronen an einem strichgitter. Z. Phys. A **220**(5), 419–426 (1969)
15. Shull, C.G.: Single-slit difraction of neutrons. Phys. Rev. **179**(3), 752–754 (1969)
16. Zeilinger, A., Ghler, R., Shull, C.G., Treimer, W., Mampe, W.: Single- and double-slit diffraction of neutrons. Rev. Mod. Phys. **60**, 1067–1073 (1988)
17. Bartell, L.S.: Complementarity in the double-slit experiment: On simple realizable systems for observing intermediate particle-wave behaviour. Phys. Rev. D **21** (1980)
18. Bell, J.S.: Speakable and Unspeakable in Quantum Mechanics. Cambridge University Press, Cambridge (1987)
19. Grangier, P.: Etude expérimentale de propriétés non-classiques de la lumière: interférence à un seul photon. Université de Paris-Sud, Centre D'Orsay (1986)
20. Greenberger, D., Yasin, A.: Simultaneous wave and particle knowledge in a neutron interferometer. Phys. Lett. A **128** (1988)
21. Aczel, A.: Non-Well-Founded Sets. Stanford University Press (1988)
22. Bachman, G.: Introduction to p-adic Numbers and Valuation Theory. Academic Press (1964)
23. Koblitz, N.: p-adic Numbers, p-adic Analysis and Zeta Functions, 2nd edn. Springer (1984)
24. Mahler, K.: Introduction to p-Adic Numbers and their Functions, 2nd edn. Cambridge University Press (1981)
25. Robinson, A.: Non-Standard Analysis. North-Holland, Studies in Logic and the Foundations of Mathematics (1966)
26. Wittgenstein, L.: Philosophische untersuchungen (1953)
27. Searle, J.R.: Speech Acts; an Essay in the Philosophy of Language. Cambridge University Press, Cambridge (1969)
28. Searle, J.R.: Expression and Meaning: Studies in the Theory of Speech Acts. Cambridge University Press, Cambridge (1979)
29. Searle, J.R., Vanderveken, D.: Foundations of Illocutionary Logic. Cambridge University Press, Cambridge (1984)
30. Austin, J.L.: How to do things with words: the William James lectures delivered at Harvard University in 1955. In: Urmson, J.O., Sbisa, M. (eds.). Clarendon Press, Oxford (1962)
31. Weinberg, J.R.: An Examination of Logical Positivism. Mason Press (2007)
32. Khrennikov, A.: Contextual Approach to Quantum Formalism. Springer, Berlin, Heidelberg, New York (2009)

33. Khrennikov, A.Y.: Information Dynamics in Cognitive. Psychological and Anomalous Phenomena. Kluwer Academic Publishers, Dordrecht (2004)
34. Zinov'ev, A.: Logical physics. In: Boston Studies in the Philosophy and History of Science. Springer (1983)
35. Sneed, J.: The Logical Structure of Mathematical Physics. D. Reidel Publishing Company, Dordrecht (1971)
36. Garola, C.: Pragmatic interpretation of quantum logic. CoRR (2014). arXiv:abs/1409.0194
37. Feyerabend, P.: Against Method, 4th edn. Verso (2010)
38. Husserl, E.: Pure Phenomenology (1917)
39. Schumann, A.: On two squares of opposition: the Lesniewski's style formalization of synthetic propositions. Acta Anal. **28**, 71–93 (2013)
40. Schumann, A., Akimova, L.: Simulating of Schistosomatidae (Trematoda: Digenea) behavior by Physarum spatial logic. In: Annals of Computer Science and Information Systems, vol. 1. Proceedings of the 2013 Federated Conference on Computer Science and Information Systems, pp. 225–230. IEEE Xplore (2013)
41. Schumann, A., Akimova, L.: Syllogistic system for the propagation of parasites. The case of Schistosomatidae (Trematoda: Digenea). Stud. Logic Grammar Rhetor. **40**(53), 303–319 (2015)
42. Schumann, A., Adamatzky, A.: Physarum spatial logic. New Math. Nat. Comput. **7**(3), 483–498 (2011)
43. Ozasa, K., Aono, M., Maeda, M., Hara, M.: Simulation of neurocomputing based on the photophobic reactions of euglena with optical feedback stimulation. BioSyst. **100**(2), 101–107 (2010)
44. Schumann, A.: Non-archimedean fuzzy and probability logic. J. Appl. Non-Class. Logics **18**(1), 29–48 (2008)
45. Zadeh, L.: Fuzzy sets. Inf. Control **8**, 338–353 (1965)

Chapter 7
Syllogistic Systems of Swarm Propagation

In the p-adic valued universe of stimuli for controlling the swarm behaviour (including swarm sensing and motoring), we can define syllogistic propositions with two quantifiers: 'all neighbours of an attractant/repellent' and 'some neighbours of an attractant/repellent'. In this chapter, I examine Aristotelian and non-Aristotelian syllogistics for simulating the swarms. In the Aristotelian syllogistic the models verifying swarming are well-founded and in the non-Aristotelian syllogistic the models verifying swarming are non-well-founded (i.e. they have no logical atoms). See [1–8].

7.1 Aristotelian and Non-Aristotelian Syllogistics for Modelling the Swarm Behaviour

7.1.1 Leśniewski's Ontology Without Logical Atoms

Leśniewski's ontology (see [9–11]) is based on propositional logic that is built in the standard way. We shall use axioms of *Łukasiewicz's propositional calculus* $\mathscr{S}_{PL}$ as the input set of provable propositions [12]:

$$(p \Rightarrow q) \Rightarrow ((q \Rightarrow r) \Rightarrow (p \Rightarrow r)), \tag{7.1}$$

$$(\neg p \Rightarrow p) \Rightarrow p, \tag{7.2}$$

$$p \Rightarrow (\neg p \Rightarrow q). \tag{7.3}$$

The implication and complement are given there as basic operations. Other operations are derivable, e.g. the conjunction and disjunction are defined as follows:

A. Schumann, *Behaviourism in Studying Swarms: Logical Models of Sensing and Motoring*, Emergence, Complexity and Computation 33,
https://doi.org/10.1007/978-3-319-91542-5_7

$$p \wedge q := \neg(p \Rightarrow \neg q), \tag{7.4}$$

$$p \vee q := \neg p \Rightarrow q, \tag{7.5}$$

$$p \equiv q := (p \Rightarrow q) \wedge (q \Rightarrow p). \tag{7.6}$$

By combining axioms (7.1)–(7.3) and using inference rules, we could obtain all other provable propositions for the system $\mathscr{S}_{PL}$.

Leśniewski's ontology is an extension of propositional logic.

Definition 7.1 The *alphabet of Leśniewski's ontology* is the ordered system $\mathscr{A}_{LO} = \langle V, Q, L_1, L_2, L_3, L_4, K \rangle$, where

1. V is the set of propositional variables $p, q, r, \ldots$;
2. Q is the set of ontological variables $A, B, C, \ldots$;
3. L_1 is the set of unary propositional connectives consisting of one element $\neg$ called the symbol of negation;
4. L_2 is the set of binary propositional connectives containing three elements: $\wedge$, $\vee$, $\Rightarrow$ called the symbols of conjunction, disjunction, and implication respectively;
5. L_3 is the set of binary ontological connectives containing the only element ε called the functor "…is…";
6. L_4 is the set consisting of two quantifiers: existential ($\exists$) and universal ($\forall$);
7. K is the set of auxiliary symbols containing two brackets: (,).

The sets V and Q are denumerable.

Definition 7.2 The *language of Leśniewski's ontology* is the ordered system $\mathscr{L}_{LO} = \langle \mathscr{A}_{LO}, \mathscr{F}_{LO} \rangle$, where

1. $\mathscr{A}_{LO}$ is the alphabet of Leśniewski's ontology;
2. $\mathscr{F}_{LO}$ is the set of all formulas formed by means of symbols in $\mathscr{A}_{LO}$; this set $\mathscr{F}_{LO}$ contains all formulas defined by the following rules:
   a. every propositional variable p, q, r, … or their Boolean combination is a formula of Leśniewski's ontology;
   b. if A and B are ontological variables, then an expression $A \varepsilon B$ is a formula of Leśniewski's ontology;
   c. if α is a formula of Leśniewski's ontology, where there is a free ontological variable A, then $\mathrm{Q}A\alpha$, where $\mathrm{Q} \in L_4$, is a formula of Leśniewski's ontology too;
   d. if α, β are formulas of Leśniewski's ontology, then expressions $\neg\alpha$, $\alpha \wedge \beta$, $\alpha \vee \beta$, $\alpha \Rightarrow \beta$ are formulas of Leśniewski's ontology too;
   e. a finite sequence of symbols of $\mathscr{A}_{LO}$ is called a formula of propositional logic if that sequence satisfies above mentioned conditions.

Thus, an expression that is derivable by rules of this definition is called a formula of Leśniewski's ontology. Formulas that do not contain propositional variables are called *formulas of Leśniewski's ontology in the restricted sense*.

Definition 7.3 *Leśniewski's ontology* is the ordered system $\mathscr{S}_{LO} = \langle \mathscr{A}_{LO}, \mathscr{F}_{LO}, \mathscr{C} \rangle$, where

1. $\mathscr{A}_{LO}$ is the alphabet of Leśniewski's ontology;
2. $\mathscr{F}_{LO}$ is the set of all formulas formed by means of symbols in $\mathscr{A}_{LO}$;
3. $\mathscr{C}$ is the inference operation in $\mathscr{F}_{LO}$.

The inference rules of Leśniewski's ontology are as follows:

1. *the substitution rule*: we replace a propositional variable p_j of formula $\alpha(p_1, \ldots, p_n)$, containing propositional variables $p_1, \ldots, p_n$, by a formula $\beta(q_1, \ldots, q_k)$, containing propositional variables $q_1, \ldots, q_k$ (resp. by a formula $\beta(A_l, B_m)$, containing ontological variables A_l, B_m), and we obtain a new propositional formula $\alpha'(p_1, \ldots, p_{j-1}, \beta(q_1, \ldots, q_k), p_{j+1}, \ldots, p_n)$ (resp. a new ontological formula $\alpha'(p_1, \ldots, p_{j-1}, \beta(A_l, B_m), p_{j+1}, \ldots, p_n)$):

$$\frac{\alpha(p_1, \ldots, p_j, \ldots, p_n)}{\alpha'(p_1, \ldots, p_{j-1}, \beta(q_1, \ldots, q_k), p_{j+1}, \ldots, p_n)}$$

   or

$$\frac{\alpha(p_1, \ldots, p_j, \ldots, p_n)}{\alpha'(p_1, \ldots, p_{j-1}, \beta(A_l, B_m), p_{j+1}, \ldots, p_n)},$$

   In the same way, from an ontological formula $\alpha(A_j, B_i)$ we can infer a new formula $\alpha'(A_k, B_i)$ or $\alpha'(A_j, B_l)$ if we replace an ontological variable A_j by an ontological variable A_k or B_i by B_l:

$$\frac{\alpha(A_j, B_i)}{\alpha'(A_k, B_i)}$$

   or

$$\frac{\alpha(A_j, B_i)}{\alpha'(A_j, B_l)};$$

2. *modus ponens*: according to that if two formulas of Leśniewski's ontology α and $\alpha \Rightarrow \beta$ hold, then we deduce a formula β:

$$\frac{\alpha, \alpha \Rightarrow \beta}{\beta}.$$

3. *the universal generalization*: if an ontological formula α, where there is no free variable A, implies an ontological formula β, where there is a free variable A, then from formula β we infer a formula $\forall A \beta$:

$$\frac{\alpha}{\frac{\beta}{\forall A\beta}};$$

4. *the universal restriction*: an ontological formula $\forall A \alpha$ entails a formula α:

$$\frac{\forall A\alpha}{\alpha};$$

5. *the existential generalization*: an ontological formula α is followed by $\exists A\alpha$:

$$\frac{\alpha}{\exists A\alpha};$$

6. *the existential restriction*: from an ontological formula $\exists A\alpha$ we deduce α, where the variable A is replaced:

$$\frac{\exists A\alpha}{\alpha}.$$

The axioms of Leśniewski's ontology include axioms of propositional logic (e.g. axioms (7.1) – (7.3) of the propositional system $\mathscr{S}_{PL}$), and the following expression:

$$A\varepsilon B \equiv (\exists C(C\varepsilon A) \wedge \forall C\forall D((C\varepsilon A \wedge D\varepsilon A) \Rightarrow C\varepsilon D) \wedge \forall C(C\varepsilon A \Rightarrow C\varepsilon B)), \tag{7.7}$$

where the expression $A\varepsilon B$ is read "A is B" and we are defining three properties of the connective "…is…": (i) subject (A) is not empty, i.e. $\exists C(C\varepsilon A)$, (ii) subject ($A$) is a singleton (consists of the only member), i.e. $\forall C\forall D((C\varepsilon A \wedge D\varepsilon A) \Rightarrow C\varepsilon D)$, (iii) any member of subject (of A) belongs to predicate (B) as well, i.e. $\forall C(C\varepsilon A \Rightarrow C\varepsilon B)$. The third property (transitivity) means that if we have a proposition $A\varepsilon B$ and a predicate B has another predicate X, then this X is a predicate of a subject A, too. There is ever a predicate of predicate. In fact, the third property says that *the predicate is a genus for the subject*. This is the main property of analytic propositions [2].

Let us extend Leśniewski's ontology by adding the new axiom:

$$\exists C(C\varepsilon A). \tag{7.8}$$

The deductive system (7.1)–(7.8) is called *non-empty Leśniewski's ontology*.

Aristotle offered the first formal theory called syllogistic [13–15], where there are four logical connectives: **a** ("every + noun 1 + is + noun 2"), **i** ("some + noun 1 + is + noun 2"), **e** ("no + noun 1 + is + noun 2") and **o** ("some + noun 1 + is not + noun 2").

The axiomatic system of *Aristotelian syllogistic* was created first by Łukasiewicz [12]. His axioms are as follows [(they are added to axioms (7.1)–(7.3)]:

$$S\mathbf{a}S, \tag{7.9}$$

$$S\mathbf{i}S, \tag{7.10}$$

$$(M\mathbf{a}P \wedge S\mathbf{a}M) \Rightarrow S\mathbf{a}P, \textit{ i.e., Barbara}, \tag{7.11}$$

$$(M\mathbf{a}P \wedge M\mathbf{i}S) \Rightarrow S\mathbf{i}P, \textit{ i.e., Datisi}. \tag{7.12}$$

The functors **a** and **i** are basic and two others are defined as follows:

$$S\mathbf{e}P := \neg(S\mathbf{i}P), \tag{7.13}$$

$$S\mathbf{o}P := \neg(S\mathbf{a}P). \tag{7.14}$$

By using axioms (7.1)–(7.3), (7.9)–(7.12), and definitions (7.4)–(7.6), (7.13), (7.14), we may obtain all tautologies of Aristotle's syllogistic.

Proposition 7.1 *Aristotelian syllogistic is contained in non-empty Leśniewski's ontology.*

Proof See [11]. □

The Aristotelian syllogistic describes *analytic propositions* in the Kantian terminology or *informative propositions* in the Austian terminology. For them there ever exist denotations (things in reality) verifying syllogistic propositions. In other words, there are Venn diagrams to demonstrate the relationships among concepts A and B: intersection ($A \cap B \neq \emptyset$), inclusion ($A \subseteq B$)), and independence ($A \cap B = \emptyset$). Nevertheless, there are *synthetic propositions* in the Kantian terminology or *performative propositions* in the Austian terminology, also. For them it is impossible to build Venn diagrams, because appropriate concepts have no denotations (logical atoms).

Let us consider the following two examples

(1) Each love is a gift of the Lord; Some homosexuals have respect to love; Then, Some homosexuals have respect to the gift of the Lord.

(2) All whites are being covered by flecks; Socrates became white; Then, Socrates became covering by flecks.

Both statements have no intuitive meanings in the Aristotelian syllogistic. Therefore they are considered sophisms. The point is that in (1) there is no way to define the extensions of the concepts 'love' and 'gift of the Lord'. What are denotations for them? And the conclusion that 'Some homosexuals have respect to the gift of the Lord' is wrong. According to this conclusion, it is possible even to read that 'Homosexuals who have respect to love are religious'. But they can be religious and have no respect to love or they can have respect to love and not be religious. In (2) 'covering by flecks' is not a genus for 'white'. Here we cannot use Venn diagrams also. In the same way, we cannot say that 'becoming white' is a genus for 'Socrates', although the form of both premises says that they are general. Thus, there are concepts for defining relations among which we cannot use Venn diagrams. In the first premise of (1) we cannot use Venn diagrams, because extensions (logical scopes) of 'love' and 'gift of the Lord' cannot be defined at all. In both premises of (2) we cannot use Venn diagrams, because predicates cannot be considered genus for subjects. The fact that there are propositions like (1) and (2) for which the formulas of Aristotelian syllogistic have no intuitive meanings have been known since Aristotle (please see his own counterexamples in *Topics* and *Sophistical Refutations*). Kant was one of the first philosophers who tried to explain why there are such propositions. Propositions for which predicates are regarded genus for subjects are called by him analytic. General propositions (like the first premise of (1) and (2)) for which predicates cannot be regarded genus for subjects are called by him synthetic.

Performative or synthetic propositions have non-well-founded models. These propositions can be defined by replacing implication by conjunction.

The novel ontology is said to be *synthetic* or *performative*. It is built up by adding the following new axiom to axioms (7.1)–(7.3):

$$A\mathfrak{ist}B \equiv (\exists C(C\mathfrak{ist}A) \wedge \forall C \forall D((C\mathfrak{ist}A \wedge D\mathfrak{ist}A) \Rightarrow C\mathfrak{ist}D) \wedge \\ \forall C(C\mathfrak{ist}A \wedge C\mathfrak{ist}B)), \quad (7.15)$$

The formula $A\mathfrak{ist}B$ designates any perforative/synthetic propositions. Let us compare formula (7.15) with (7.7). We see that instead of the property that the predicate is a genus of the subject, we have the property that the subject has ever a non-empty intersection with the predicate.

Inference rules in synthetic ontology are the same as in Leśniewski's ontology.

Non-empty performative/synthetic ontology is obtained by adding the new axiom

$$\exists C(C\mathfrak{ist}A). \quad (7.16)$$

Proposition 7.2 *All theorems of (non-empty) Leśniewski's ontology are theorems of (non-empty) performative/synthetic ontology as well.* □

Let us sketch now the syllogistic system formalizing performative/synthetic propositions. This system is said to be *synthetic syllogistic* or *performative syllogistic*, while we are assuming that Aristotelian syllogistic is analytic. The basic logical connectives of performative/synthetic syllogistic are as follows: $\mathfrak{a}$ ("every + noun + is + adjective"), $\mathfrak{i}$ ("some + noun + is + adjective"), $\mathfrak{e}$ ("no + noun + is + adjective") and $\mathfrak{o}$ ("some + noun + is not + adjective") that are defined in performative/synthetic ontology in the following way:

$$S\mathfrak{a}P := (\exists A(A\mathfrak{ist}S) \wedge (\forall A(A\mathfrak{ist}S \wedge A\mathfrak{ist}P))); \quad (7.17)$$

$$S\mathfrak{i}P := \forall A(\neg(A\mathfrak{ist}S) \wedge \neg(A\mathfrak{ist}P)); \quad (7.18)$$

$$S\mathfrak{o}P := \neg(\exists A(A\mathfrak{ist}S) \wedge (\forall A(A\mathfrak{ist}S \wedge A\mathfrak{ist}P))), \ i.e. \\ (\forall A\neg(A\mathfrak{ist}S) \vee \exists A(\neg(A\mathfrak{ist}S) \vee \neg(A\mathfrak{ist}P))); \quad (7.19)$$

$$S\mathfrak{e}P := \neg\forall A(\neg(A\mathfrak{ist}S) \wedge \neg(A\mathfrak{ist}P)), \ i.e. \ \exists A(A\mathfrak{ist}S \vee A\mathfrak{ist}P). \quad (7.20)$$

Now let us formulate axioms of performative/synthetic syllogistic:

$$S\mathfrak{a}P \Rightarrow S\mathfrak{e}P; \tag{7.21}$$

$$S\mathfrak{a}P \Rightarrow P\mathfrak{a}S; \tag{7.22}$$

$$S\mathfrak{i}P \Rightarrow P\mathfrak{i}S; \tag{7.23}$$

$$S\mathfrak{a}M \Rightarrow S\mathfrak{e}P; \tag{7.24}$$

$$M\mathfrak{a}P \Rightarrow S\mathfrak{e}P; \tag{7.25}$$

$$(M\mathfrak{a}P \wedge S\mathfrak{a}M) \Rightarrow S\mathfrak{a}P; \tag{7.26}$$

$$(M\mathfrak{i}P \wedge S\mathfrak{i}M) \Rightarrow S\mathfrak{i}P. \tag{7.27}$$

Proposition 7.3 *Performative/synthetic syllogistic is a deductive part of non-empty performative/synthetic ontology.* □

In this syllogistic we have no logical atoms. Hence, we can use non-well-founded models.

7.1.2 *Lateral Inhibition and Lateral Activation*

As we know, any biological activity can be controlled by placing attractants and repellents—some items which are programmed to attract and repel the behaviour. So, any swarm can be represented as a computation medium, because the behaviour of any swarm can be programmed by localization of attractants presented as food pieces and repellents presented as dangerous places. Nevertheless, we can program the behaviour of many unicellular organisms in the same way, such as behaviours of *Amoeba proteus* or plasmodia of *Physarum plycephalum*. At the level of one cell, this controlling is explained by the appearance and disappearance of *actin filaments* or *F-actin*, i.e. of the protein which is organized into higher-order structures, forming linear bundles, two-dimensional networks, and three-dimensional semisolid gels to react to all the external stimuli. Actin filaments are connected to the plasma membrane to provide a mechanical support by an actin cortex. If there is an attractant before the cell, actin filaments form a wave to change the cell shape to allow the movement of the cell surface to build a pseudopodium by cross-linked filaments to catch the attractant. If there is a repellent before the cell, actin filaments form a wave to change the cell shape to avoid the repellent.

In swarm computing we use real organisms like ants or slime mould with completely controlling their behaviour. But we may expect that in the future we can control all the chemical reactions responsible for assembling and disassembling the actin filament networks. It means that we would have an “artificial protein monster” whose reactions are programmed by us even at the molecular level.

Conventionally, any device for computations has been regarded as a “mechanical” calculator—a machine designed from inanimate-nature objects and used to perform automatically all the computations in the way of mechanical (later electronic) simulations of calculating processes. The first calculating machine was designed by Blaise Pascal (1623–1662) to mechanize calculations. This attempt gave many inspirations for some logicians at that time. So, Gottfried Wilhelm von Leibniz (1646–1716) introduced the idea of *characteristica universalis*—the universal computer to mechanize all thinking processes, not only calculations:

> ...a kind of general algebra in which all truths of reason would be reduced to a kind of calculus. At the same time, this would be a kind of universal language or writing, though infinitely different from all such languages which have thus far been proposed; for the characters and the words themselves would direct the mind, and the errors – excepting those of fact – would only be calculation mistakes. It would be very difficult to form or invent this language or characteristic, but very easy to learn it without any dictionaries (Leibniz, *Letter to Nicolas Remond, 10 January 1714*, Tr. by Leroy Loemker).

Nobody has thought of building up computers from the animate nature. Later, the Leibniz's idea of *characteristica universalis* or universal “mechanical” calculator to calculate everything was continued and developed within *logical empiricism* (or *logical positivism*)—a branch of analytic philosophy, in which all the true pictures of reality were considered as logical compositions of atomic sentences which in turn consist only of observational terms—terms for which there are correspondence rules, showing appropriate facts as the empiric values of terms, see [16]. Hence, a true picture of reality is a composition constructed by logical connectives to combine facts expressed in observational terms. Otto Neurath (1882–1945) called this Leibniz-inspirited program ‘*physicalism*’—a universal system which would comprehend each positive verifiable knowledge furnished by all the various empirical sciences, see [17, 18]. It is a recent analogous for the older Leibnizian *characteristica universalis.* In 1928 Rudolf Carnap (1891–1970) published the book *The Logical Structure of the World* (*Der logische Aufbau der Welt*), in which he developed a logical formalism to define all scientific terms by means only of observational terms and symbolic-logical rules. This system is a “mechanical” calculator to draw all the true conclusions from the empirical knowledge expressed in observational terms.

Ludwig Josef Johann Wittgenstein (1889–1951) introduced this physicalism as follows:

- The world, i.e. the picture of reality, is a combination of facts.
- A fact is expressed by an atomic proposition and a combination of facts is expressed by a logical composition of appropriate atomic proposition verified on these facts (Wittgenstein, *The Tractatus Logico-Philosophicus*).

Hence, knowledge was understood by the logical empiricists as a process that can be simulated by logical machines drawing conclusions automatically from observations. These logical machines were theoretically explicated concurrently in the three ways: (i) mathematically by Kurt Friedrich Gödel (1906–1978); (ii) from the point of view of programming by Alonzo Church (1903–1995); and (iii) from the point of view of engineering by Alan Mathison Turing (1912–1954).

(i) So, Gödel proposed the idea of μ-recursive functions to explicate a class of partial functions from natural numbers to natural numbers that are 'computable' in an intuitive sense. These functions are defined by inductive sets. Now, there is a notion of the so-called *corecursive functions* defined by coinductive sets. This notion allows us to formally describe any behaviour, even not-algorithmic.

(ii) Church introduced λ-calculus—a formal system for expressing computation based on function abstraction, variable binding, and substitution. This calculus is an abstract programming language expressing computation as such. Now, there are *process calculi* like π*-calculus* applying corecursuive functions for programming instead of recursive functions. These new calculi are used for simulating different behavioural systems including not-algorithmic and concurrent. Notice, λ-calculus cannot do this.

(iii) Finally, Turing defined a Turing machine—an abstract machine that manipulates symbols on a strip of tape according to a table of rules to calculate all functions, 'computable' in an intuitive sense. This machine is inanimate in principle. In unconventional computing, designing computers from swarms or designing an "*artificial protein monster*" (in the future) is an attempt to explain the animal behaviour as such and this attempt is parallel to mathematical theories on corecursive functions and programming languages involving process calculi. The latter try to explain the behaviour, as well.

Thus, conventionally there was a philosophical presupposition that the human being is unique who possesses intelligence and all computers can be made just as mechanical (electric) devices simulating the human algorithmic thinking.

However, there is an old tradition of *panpsychism*—a view that all animate things bear a mind or a mind-like quality, too. So, Johann Wolfgang von Goethe (1749–1832) stated that nothing exists without an internal intelligence called by him *Seele* (spirit):

> Since, however, matter can never exist and act without spirit [Seele], nor spirit without matter, matter is also capable of undergoing intensification, and spirit cannot be denied its attraction and repulsion (Goeth, *Commentary on the Aphoristic Essay 'Nature' (1828)*", in: D. Miller (ed.), *Goethe: Scientific Studies*. New York: Suhrkamp, 1988).

The panpsychistic idea of internal intelligence of all things is well expressed in Qaballah, the Judaic mysticism. The Bible verse 'And the spirit of God moved upon the face of the waters' (*Genesis* 1:2) was interpreted as affirming that there exists a spirit (*ruaḥ*, רוח) of the Messiah or a pure man before the world creation (*Genesis Rabbah* 8:1). This spirit is named *'Adam Qadmon* (אדם קדמון). He is the cosmic man or Self and represents 'crown' (*keter*, כתר), the divine will to create everything. From *'Adam Qadmon* emerge the following four worlds: (i) the divine light or pure emanation (*'aẓilut*, אצילות); (ii) the creation or divine waters (*briy'ah*, בריאה); (iii) the formation or internal essence of all things (*yeẓirah*, יצירה); and (iv) the action and all the forms of behaviour (*'aśiyah*, עשיה).

We find out almost the same description of internal intelligence of all things in the Hindu tradition, as well. The cosmic man or Self is named *Puruṣa* and from him emerge also the same four worlds: (i) the divine light or pure emanation ('the *Agni* [A.Sch.: divine fire] whose fuel is the sun'); (ii) the creation or divine waters ('*parjánya*' or 'clouds whose fuel is the moon'); (iii) the formation or internal essence of all things (contained in 'medicinal plant'); and (iv) the action and all the forms of behaviour (actions started from 'the male which sheds the semen on woman'):

> From him the *Agni* (*dyú loka*) whose fuel is the sun; from the moon in the *dyú loka*, *parjánya* (clouds); from the clouds, the medicinal plant that grows on earth; from these, the male (fire) which sheds the semen on woman, thus gradually many living beings such as Brahmins, etc., are born of the *Puruṣa* (*Muṇḍaka Upaniṣad* 2, 1:5; Tr. by S. Sitarama Sastri).

It is quite mysterious why in Judaism and Hinduism (the religions, not connected at all between themselves) there are the similar notions of cosmic men *'Adam Qadmon* and *Puruṣa* with the same four emanations from them.

Hence, according to some religious traditions, such as Judaism and Hinduism, panpsychism holds indeed—it is assumed that intelligence is everywhere. Therefore, their believers suppose that there are many non-human (or even over-human) forms of intelligence in natural processes.

In accordance with panpsychism, each animate thing is a kind of computer. So, an "artificial protein monster" (*Golem* in Qabbalah) is possible, too. The idea of this monster was used in a science fiction television series, *Lexx*, created by Paul Donovan, Lex Gigeroff, and Jeffrey Hirschfield, originally released from 18 April 1997 to 26 April 2002. *Lexx* is a narration about the adventures of a group of mismatched individuals aboard the organic space craft called Lexx. All the devices and technics in this series are organic and look like protein monsters called 'insects.'

The panpsychist idea cannot be scientific because of its religious roots, but it can be inspiriting for us. It is so surprising that in swarm computing there are some evidences supporting panpsychism. We know that in the neural networks there are the following two mechanisms responsible for perceiving signals: (i) increasing the intensity of the signal by *lateral inhibition*, when inhibitory interneurons inhibit neighbouring cells in the neural network to make the contrast of the signal more visible; (ii) decreasing the intensity of the signal by *lateral activation*, when activation interneurons activate neighbouring cells to make the contrast of the signal less visible. Due to both mechanisms, we deal with some illusions such as the Müller-Lyer one—a geometric illusion in which the perceived length of a line depends on whether the line terminates in an arrow tail (when we face the lateral inhibition effect) or arrowhead (the lateral activation effect).

Hence, the lateral inhibition and lateral activation are two mechanisms of our mind in perceiving signals (in the case of the Müller-Lyer illusion the signals are visual). Nevertheless, the same mechanisms of transmitting signals are discovered (i) on the level of Amoeboid organisms and (ii) on the level of swarms optimizing their transport networks. For the first time, Sakiyama and Gunji have shown in [19] that both neuronal effects are observed in the collective patterns of garden ants (*Lasius niger*) which optimize the own logistics by local interactions (the analogue of the neuronal lateral inhibition) and global explorations (the analogue of the neuronal lateral activation). Later Jones has paid attention in [20] that plasmodia of *Physarum polycephalum* verify both effects also in their adaptive behaviours. It is quite strange taking into account the fact that plasmodia are unicellular and, therefore, they have no nervous systems at all.

The matter is that both effects are basic for the actin filament networks: (i) among actin filaments, neighbouring bundles can be inhibited to increase the intensity of the signal to make just one zone of actin filament polymerization active; (ii) neighbouring bundles can be activated to decrease the intensity of the signal to make several zones of actin filament polymerization active.

Thus, the lateral inhibition and lateral activation can be detected in any forms of swarm networking including social bacteria and plasmodia of *Physarum polycephalum*. The same effects are observed even in the swarm behaviour of alcohol-dependent people [21], i.e. on the level of collective patterns of the human beings. It means that on the level of actin filament networks we have a kind of intelligence that is enough for the adaptation and optimization of logistics. So, the "artificial protein monster" (*Golem*) consisting of actin filament networks and solving many computational tasks connected to orientation and locomotion is absolutely real. The basic logic for this monster is proposed in [22].

To sum up, panpsychism in computer science means that we design bioinspired robots by assuming scale-invariant mechanisms that have been conserved across species. In particular, lateral inhibition and lateral activation are ubiquitous events that occur over many scales including within the cell during cell polarization, between groups of neuron within the visual cortex to process visual cues, and between active zones of swarms to react to their environments. Conventional logics, such as Aristotelian syllogistic, μ-recursive functions, λ-calculus, Turing machines, can describe the behaviour just under the conditions of lateral inhibition. But the behaviour under the conditions of lateral activation is a substantial part of the swarming, too. Let us consider then, how we can define the Aristotelian syllogistic on laterally inhibited actions and later how we can define the performative/synthetic syllogistic on laterally activated actions.

7.1.3 Lateral Inhibition and Aristotelian Syllogistic

In the swarm implementation of Aristotelian syllogistic, all data points are denoted by appropriate syllogistic letters as attractants. A data point S is considered empty if and only if an appropriate attractant denoted by S is not occupied by the swarm. We have syllogistic strings of the form SP with the following interpretation: 'S is P,' and with the following meaning: SP is true if and only if S and P are reachable by the swarm and both S and P are not empty, otherwise SP is false. By this definition of syllogistic strings, we can define atomic syllogistic propositions as follows:

'All S are P' ($S\mathbf{a}P$): *In the formal syllogistic*: there exists A such that A is S and for any A, if A is S, then A is P. *In the swarm model*: there is a swarm A and for any A, if A is located at S, then A is located at P.

'Some S are P' ($S\mathbf{i}P$): *In the formal syllogistic*: there exists A such that both 'A is S' is true and 'A is P' is true. *In the swarm model*: there exists a swarm A such that A is located at S and A is located at P.

'No S are P' ($S\mathbf{e}P$): *In the formal syllogistic*: for all A, 'A is S' is false or 'A is P' is false. *In the swarm model*: for all swarms A, A is not located at S or A is not located at P.

'Some S are not P' ($S\mathbf{o}P$): *In the formal syllogistic*: for any A, 'A is S' is false or there exists A such that 'A is S' is true and 'A is P' is false. *In the swarm model*: for any swarms A, A is not located at S or there exists A such that A is located at S and A is not located at P.

Formally, this semantics is defined as follows. Let M be a set of attractants. Take a subset $\|X\| \subseteq M$ of attractants occupied by the swarm as a meaning for each syllogistic variable X. Next, define an ordering relation $\subseteq$ on subsets $\|S\|, \|P\| \subseteq M$ as: $\|S\| \subseteq \|P\|$ iff attractants from $\|P\|$ are reachable for the swarm located at the attractants from $\|S\|$. Hence, $\|S\| \cap \|P\| \neq \emptyset$ means that some attractants from $\|P\|$ are reachable for the swarm located at the attractants from $\|S\|$ and $\|S\| \cap \|P\| = \emptyset$ means that no attractants from $\|P\|$ are reachable for the swarm located at the attractants from $\|S\|$. This gives rise to models $\mathcal{M} = \langle M, \| \cdot \| \rangle$ such that

- $\mathcal{M} \models S\mathbf{a}P$ iff $\|S\| \subseteq \|P\|$;
- $\mathcal{M} \models S\mathbf{i}P$ iff $\|S\| \cap \|P\| \neq \emptyset$;
- $\mathcal{M} \models S\mathbf{e}P$ iff $\|S\| \cap \|P\| = \emptyset$;
- $\mathcal{M} \models p \wedge q$ iff $\mathcal{M} \models p$ and $\mathcal{M} \models q$;
- $\mathcal{M} \models p \vee q$ iff $\mathcal{M} \models p$ or $\mathcal{M} \models q$;
- $\mathcal{M} \models \neg p$ iff it is false that $\mathcal{M} \models p$.

Proposition 7.4 *The Aristotelian syllogistic is sound and complete relatively to $\mathcal{M}$ if we understand $\subseteq$ as an inclusion relation.*

Proof It is a well-known result of [23]. □

However, relatively to all possible swarm behaviours the Aristotelian syllogistic is not complete. *This syllogistic simulates the lateral inhibition effect, when just one possible direction of propagation is chosen at each move.*

7.1.4 Lateral Activation and Non-Aristotelian Syllogistic

While in Aristotelian syllogisms we are concentrating on one direction of many swarm motions, and dealing with acyclic directed graphs with fusions of many swarm networks toward one data point, in most cases of swarm behaviour, not limited by repellents, we observe a spatial expansion of swarm in all directions with many cycles. Under these circumstances it is more natural to define all the basic syllogistic propositions $S\mathfrak{a}P$, $S\mathfrak{i}P$, $S\mathfrak{e}P$, $S\mathfrak{o}P$ in a way they satisfies the inverse relationship when all converses are valid: $S\mathfrak{a}P \Rightarrow P\mathfrak{a}S$, $S\mathfrak{i}P \Rightarrow P\mathfrak{i}S$, $S\mathfrak{e}P \Rightarrow P\mathfrak{e}S$, $S\mathfrak{o}P \Rightarrow P\mathfrak{o}S$. In other words, we can draw more natural conclusions for swarm networks which are decentralized and have some cycles. This system is called the *performative/synthetic syllogistic* defined above. The alphabet of this system contains as descriptive signs the syllogistic letters S, P, M, ..., as logical-semantic signs the syllogistic connectives $\mathfrak{a}$, $\mathfrak{e}$, $\mathfrak{i}$, $\mathfrak{o}$ and the propositional connectives $\neg$, $\vee$, $\wedge$, $\Rightarrow$. Atomic propositions are defined as follows: $S\mathfrak{x}P$, where $\mathfrak{x} \in \{\mathfrak{a}, \mathfrak{e}, \mathfrak{i}, \mathfrak{o}\}$. All other propositions are defined thus: (i) each atomic proposition is a proposition, (ii) if X, Y are propositions, then $\neg X$, $\neg Y$, $X \star Y$, where $\star \in \{\vee, \wedge, \Rightarrow\}$, are propositions, too.

In order to implement the performative/synthetic syllogistic in the behaviour of swarms, we will interpret all data points denoted by appropriate syllogistic letters as attractants. A data point S is considered empty if and only if an appropriate attractant denoted by S is not occupied by the swarm. Let us define syllogistic strings of the form SP with the following interpretation: 'S is P,' and with the following meaning: SP is true if and only if S and P are reachable for each other by the swarm and both S and P are not empty, otherwise SP is false. Using this definition of syllogistic strings, we can define atomic syllogistic propositions as follows:

'All S are P' ($S\mathfrak{a}P$): *In the formal performative syllogistic*: there exists A such that A is S and for any A, A is S and A is P. *In the swarm model*: there is a string AS and for any A which is a neighbour for S and P, there are strings AS and AP. This means that we have a massive-parallel occupation of the region where the cells S and P are located.

'Some S are P' ($S\mathfrak{i}P$): *In the formal performative syllogistic*: for any A, both 'A is S' is false and 'A is P' is false. *In the swarm model*: for any A which is a neighbour for S and P, there are no strings AS and AP. This means that the swarm cannot reach S from P or P from S immediately.

'No S are P' ($S\mathfrak{e}P$): *In the formal performative syllogistic*: there exists A such that if 'A is S' is false, then 'A is P' is true. *In the swarm model*: there exists A which is a neighbour for S and P such that there is a string AS or there is a string AP. This means that the swarm occupies S or P, but not the whole region where the cells S and P are located.

'Some S are not P' ($S\mathfrak{o}P$): *In the formal performative syllogistic*: for any A, 'A is S' is false or there exists A such that 'A is S' is false or 'A is P' is false. *In the swarm model*: for any

A which is a neighbour for S and P there is no string AS or there exists A which is a neighbour for S and P such that there is no string AS or there is no string AP. This means that the swarm does not occupy S or there is a neighbouring cell which is not connected to S or P by a swarm network.

A model $\mathscr{M}' = \langle M', \| \cdot \|_x \rangle$ for the performative syllogistic, where M' is the set of attractants and $\|X\|_x \subseteq M'$ is a meaning of syllogistic letter X which is understood as all attractants reachable for the swarm from the point x, is defined as follows:

- $\mathscr{M}' \models$ *All x are y* iff $\|X\|_x \neq \emptyset$, $\|X\|_y \neq \emptyset$, and $\|X\|_x \cap \|X\|_y \neq \emptyset$, i.e. the swarm can move from neighbours of y to x and it can move from neighbours of x to y;
- $\mathscr{M}' \models$ *Some x are y* iff $y \notin \|X\|_x$ and $x \notin \|X\|_y$, i.e. the swarm cannot move from neighbours of y to x and it cannot move from neighbours of x to y;
- $\mathscr{M}' \models$ *No x are y* iff $y \in \|X\|_x$ or $x \in \|X\|_y$, i.e. the swarm can move from neighbours of y to x or it can move from neighbours of x to y;
- $\mathscr{M}' \models$ *Some x are not y* iff $y \notin \|X\|_x$ or $x \notin \|X\|_y$, i.e. the swarm cannot move from neighbours of y to x or it cannot move from neighbours of x to y;
- $\mathscr{M}' \models p \wedge q$ iff $\mathscr{M}' \models p$ and $\mathscr{M}' \models q$;
- $\mathscr{M}' \models p \vee q$ iff $\mathscr{M}' \models p$ or $\mathscr{M}' \models q$;
- $\mathscr{M}' \models \neg p$ iff it is false that $\mathscr{M}' \models p$.

Proposition 7.5 *The performative syllogistic is sound and complete in* $\mathscr{M}'$. □

In the performative syllogistic we simulate the lateral activation effect. In this system we have no logical atoms.

7.2 Spatial Diagrams for Syllogistic Propositions

One of the first logicians who proposed a spatial implementation of Aristotelian syllogistic reasoning was Lewis Carroll [24, 25]. He used three kinds of syllogistic propositions: (i) the universal affirmative ('all members of its subject are members of its predicate'), (ii) the universal negative ('no members of its subject are members of its predicate'), (iii) and the particular affirmative ('some members of its subject are members of its predicate'). His examples are as follows: 'All red apples are ripe', 'No red apples are ripe', 'Some red apples are ripe'. For verifying syllogistic propositions he proposed the following biliteral diagram:

xy	xy'
$x'y$	$x'y'$

that plays the role of 'universe of discourse' for all syllogistic propositions over adjuncts x, y, non-x (which is denoted by x'), non-y (which is denoted by y'). For example, let x mean 'old,' so that x' will mean 'new'. Let y mean 'English,' so that y' will mean 'foreign'. Assume that 'books' are an appropriate universe of discourse. Then we can divide this universe into the following four classes: xy ('old English books'), xy' ('old foreign books'), $x'y$ ('new English books'), $x'y'$ ('new foreign books').

Now let us take two kinds of counters: grey and black. If a black counter is placed within a cell, this means that "this cell is occupied" (i.e. "there is at least one thing in it"). If a grey counter is placed within a cell, this means that "this cell is empty" (i.e. "there is nothing in it"). Thus, using grey and black counters we can verify all the basic syllogistic propositions.

In this section, I develop Carroll's ideas, but our diagrams will be represented by living organisms, e.g. by the slime mould. One of the most interesting outcomes of this research is that, first, the universe of discourse is considered a continuously growing living organism (*Physarum polycephalum* plasmodium), second, any motions of this organism are considered as inferring syllogistic conclusions. Carroll's diagrams allow us to define all possible directions in any motion of the slime mould.

7.2.1 Strings in the Slime Mould Growing Universe

The universe, where *Physarum polycephalum* lives, consists of cells possessing different topological properties according to the intensity of chemo-attractants and chemo-repellents. The intensity entails the natural or geographical neighbourhood of the set's elements in accordance with the spreading of attractants or repellents. As a result, we obtain Voronoi cells. A planar Voronoi diagram of the set P is a partition of the plane into cells, such that for any element of P, a cell corresponding to a unique point p contains all those points of the plane which are closer to p in respect to the distance d than to any other node of P. A unique region

$$vor(p) = \bigcap_{m \in P, m \neq p} \{z \in \mathbf{R}^2 : d(p, z) < d(m, z)\}$$

assigned to the point p is called the *Voronoi cell* of the point p. Within one Voronoi cell a reagent has the full power to attract or repel the plasmodium. The distance d is defined by intensity of reagent spreading like in other chemical reactions simulated by *Voronoi diagrams*. When two spreading wave fronts of two reagents meet, this means that the plasmodium cannot choose any further direction, and splits. Within the same Voronoi cell two active zones will fuse.

If a Voronoi center is presented by an attractant a that is activated and occupied by the plasmodium, this means that there exists a string a. This string has the meaning "a exists". If a Voronoi center is presented by a repellent $[a]$ that is activated and avoided by the plasmodium, this means that there exists a string $[a]$. This string has the meaning "a does not exist". If two neighbour Voronoi cells contain activated attractants a and b, which are occupied by the plasmodium, and between both centers there are protoplasmic tubes, then we say that there exists a string ab and a string ba. The meaning of those strings is equal and it is as follows: "ab exist", "ba exist", "some a is b", "some b is a".

If one neighbour Voronoi cell contains an activated attractant a which is occupied by the plasmodium and another neighbour Voronoi cell contains an activated repellent

$[b]$ which is avoided by the plasmodium, then we say that there exists a string $a[b]$ and a string $[b]a$. The meaning of those strings is equal and it is as follows: "*ab* do not exist, but *a* exists without *b*", "there exists *a* and no *a* is *b*", "no *b* is *a* and there exists *a*", "*a* exists and *b* does not exist".

If two neighbour Voronoi cells contain activated repellents $[a]$ and $[b]$ which are avoided by the plasmodium, then there exists a string $[ab]$ and a string $[ba]$. The meaning of those strings is equal and it is as follows: "*ab* do not exist together", "there are no *a* and there are no *b*", "no *b* is *a*", "no *a* is *b*".

Thus, the family of strings between nearest attractants presents a proximity graph which continuously grows from one attractant to another. This expansion could be demonstrated as a *Toussaint hierarchy* [26], where the family of strings starts from a nearest-neighbourhood graph and any next graph in the hierarchy is produced from previous graph by adding some edges between non-adjacent nodes.

Each string in the Toussaint hierarchy could be interpreted as a syllogistic proposition. At the beginning, we have just data points without strings. Then some first strings have grown up and in some cases we see the first syllogistic conclusions, when three or more points are connected by protoplasmic tubes. At the end, we observe all possible syllogistic conclusions in the topology of attractants and repellents we have set up for *Physarum polycephalum*. As a result, the growth of plasmodia is considered syllogistic conclusions. Due to different stimuli, we can manage this growth in directions we want, therefore we can foresee all possible syllogistic conclusions, which can be implemented within the certain topology of attractants and repellents. Moreover, we can deal with different syllogistic systems in managing the slime mould behaviour. In particular, we can implement the Aristotelian syllogistic and performative syllogistic. For denoting all possible logic circuits of syllogistic systems implemented in the *Physarum* behaviour, we will use the so-called *Physarum diagrams* which are a modification of the well-known Lewis Carroll's diagrams [24, 25].

7.2.2 *Aristotelian Syllogistic for* Physarum *Plasmodia with Repellents*

In the *Physarum* diagrams for verifying all the basic syllogistic propositions, we will use the following four cells: x, y, x', y', where x' means all cells which differ from x, but they are neighbours for y, and y' means all cells which differ from y and are neighbours for x. These cells express appropriate meanings of syllogistic letters. The corresponding universe of discourse will be denoted by means of the following diagram:

x	y'
y	x'

.

Assume that a black counter denotes an attractant and if it is placed within a cell x, this means that "this Voronoi cell contains an attractant N_x activated and occupied by the plasmodium." It is a verification of the syllogistic letter S_x at cell x. A grey counter denotes a repellent and if it is placed within a cell x, this means that "this Voronoi cell contains a repellent R_x activated and there is no plasmodium in it."

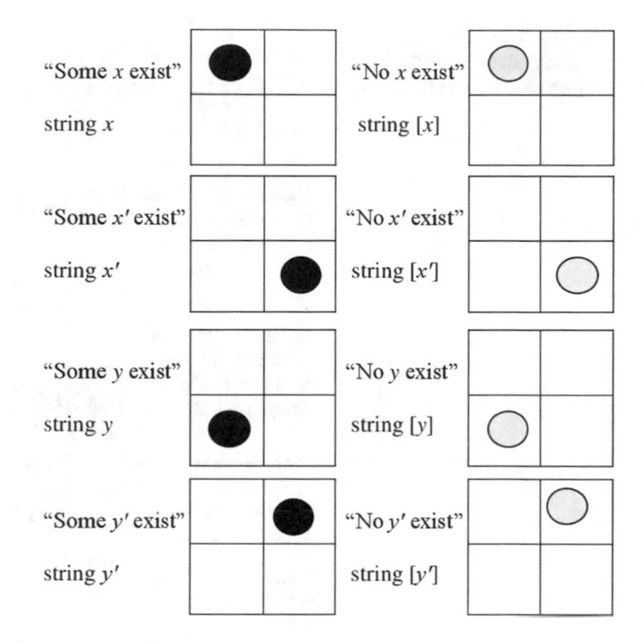

Fig. 7.1 The *Physarum* diagrams for the basic existence strings, see [5]

It is a verification of a new syllogistic letter $[S_x]$. For the sake of convenience, we will denote S_x by x and $[S_x]$ by $[x]$. Using these counters, we can verify all the basic existence syllogistic propositions in a way analogous, though different to Carroll's diagrams (see Fig. 7.1).

Physarum strings of the form xy, yx are interpreted as particular affirmative propositions "Some x are y" and "Some y are x" respectively, strings of the form $[xy]$, $[yx]$, $x[y]$, $y[x]$ are interpreted as universal negative propositions "No x are y" and "No y are x." A universal affirmative proposition "All x are y" are presented by a complex string $xy\&x[y']$. The sign & means that we have strings xy and $x[y']$ simultaneously and they are considered the one complex string. All these strings are verified on the basis of the diagrams of Fig. 7.2.

For verifying syllogisms we will use the following diagrams symbolizing some neighbour cells:

	m	m'	
m'	x	y'	m
m	y	x'	m'
	m'	m	

The motion of plasmodium starts from one of the central cells (x, y, x', y') and goes towards one of the four directions (northwest, southwest, northeast, southeast). The syllogism shows a connection between two not-neighbour cells on the basis of its joint neighbour and says if there was either multiplication or fusion. As a syllogistic conclusion, we obtain another diagram: $\begin{array}{|c|c|}\hline x & m' \\ \hline m & x' \\ \hline\end{array}$. Different syllogistic conclusions

Fig. 7.2 The *Physarum* diagrams for syllogistic propositions, see [5]

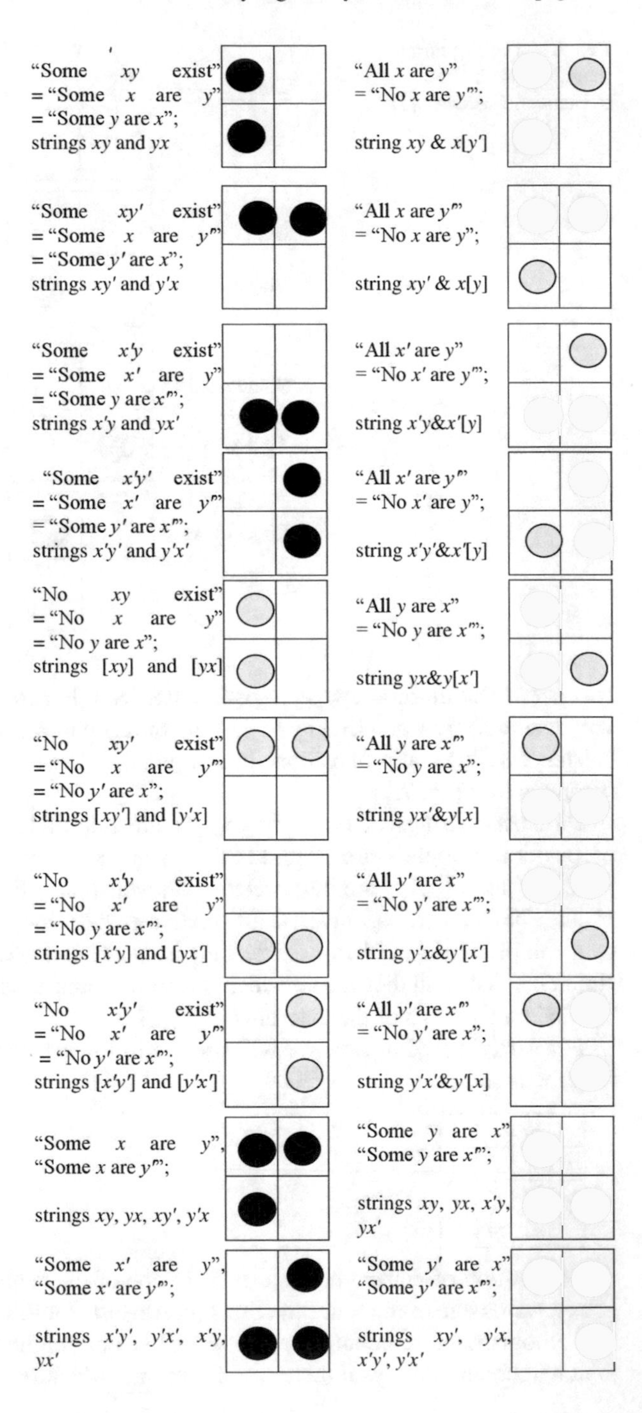

derived show directions of plasmodium's propagation. Some examples are provided in Figs. 7.3, 7.4 and 7.5:

Continuing in the same way, we can construct a syllogistic system, where conclusions are derived from three premises. The motion of plasmodium starts from one of the central cells (x, y, x', y') and goes towards one of the four directions (northwest, southwest, northeast, southeast), then towards one of the eight directions (north-northwest, west-northwest, south-southwest, west-southwest, north-northeast, east-northeast, south-southeast, east-southeast), etc.

Hence, a spatial expansion of plasmodium is interpreted as a set of syllogistic propositions. The universal affirmative proposition $xy\&x[y']$ means that the plasmodium at the place x goes only to y and all other directions are excluded. The universal negative proposition $x[y]$ or $[xy]$ means that the plasmodium at the place x cannot go to y and we know nothing about other directions. The particular affirmative proposition xy means that the plasmodium at the place x goes to y and we know nothing about other directions. Syllogistic conclusions allow us to mentally reduce the number of syllogistic propositions showing plasmodium's propagation.

For the implementation of Aristotelian syllogistic we appeal to repellents to delete some possibilities in the plasmodium propagation. So, model $\mathcal{M}$ defined above should be understood as follows:

- $\mathcal{M} \models$ *All x are y* iff $xy\&x[y']$, i.e. the plasmodium is located at x and can move only to y and cannot move towards all other directions;
- $\mathcal{M} \models$ *Some x are y* iff xy, i.e. the plasmodium is located at x and can move to y;
- $\mathcal{M} \models$ *No x are y* iff $x[y]$ or $[xy]$, i.e. the plasmodium cannot move to y in any case.

It is evident in this formulation that the Aristotelian syllogistic is so unnatural for plasmodia. Without repellents, this syllogistic system cannot be verified in the medium of plasmodium propagations. In other words, we can prove the next proposition:

Proposition 7.6 *The Aristotelian syllogistic is not sound and complete on the plasmodium without repellents.* □

7.2.3 *Non-Aristotelian Syllogistic for* Physarum *Plasmodia Without Repellents*

In the *Physarum* diagrams for the performative syllogistic, the 'universe of discourse' covers the cells x, y, non-x (which be denoted by x'), non-y (which be denoted by y'): $\begin{array}{|c|c|}\hline x & y' \\ \hline y & x' \\ \hline\end{array}$, where x, y, x', y' are neighbour cells containing attractants for *Physarum*, x' are all neighbours for y which differ from x, and y' are all neighbours for x which differ from y. Suppose that we have black, white, and grey counters and (i) if a black counter is placed within a cell, this means that "this cell is occupied" (i.e. "there is

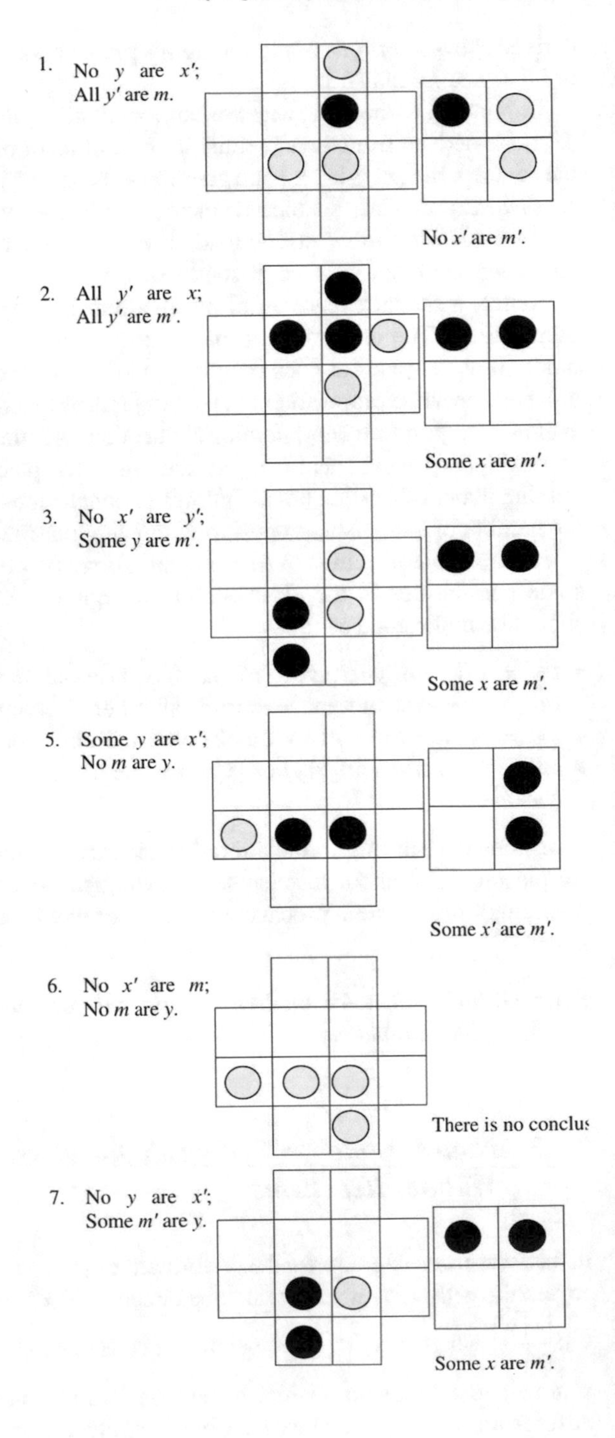

Fig. 7.3 *Physarum* diagrams for syllogisms (part 1), see [5]

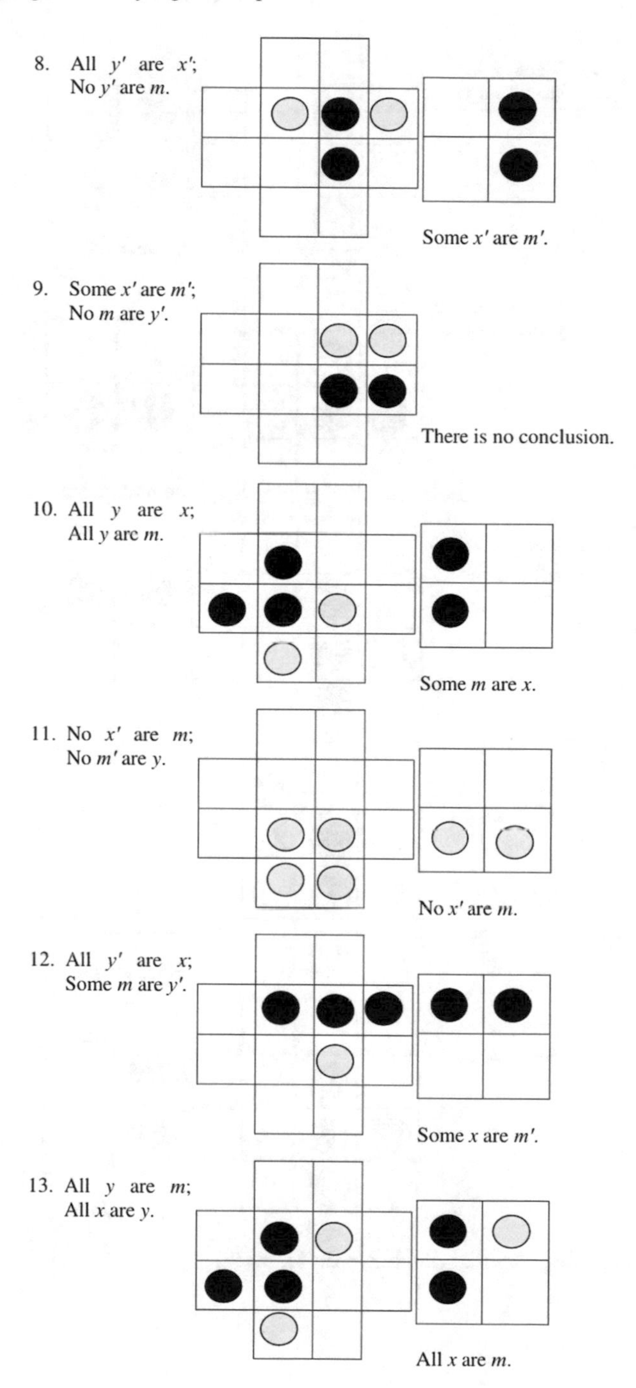

Fig. 7.4 *Physarum* diagrams for syllogisms (part 2), see [5]

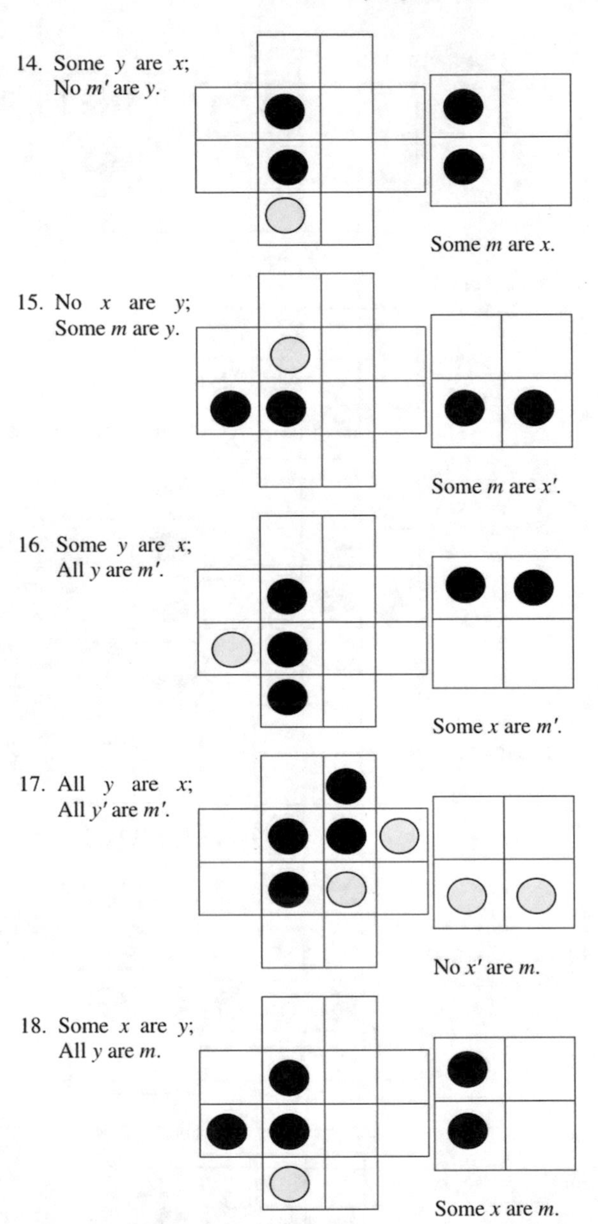

Fig. 7.5 *Physarum* diagrams for syllogisms (part 3), see [5]

at least one thing in it"), (ii) if a white counter is placed within a cell, this means that "this cell is not occupied" (i.e. "there is not thing in it"), (ii) if a grey counter is placed within a cell, this means that "it is not known if this cell is occupied". All possible combinations of *Physarum* diagrams for atomic propositions within our universe of discourse are pictured in Fig. 7.6.

The universe of discourse for simulating performative syllogisms by means of *Physarum* behaviours covers cells x, y, m, x', y', m' in the following manner:

y'	m	m'	x'
m'	x	y'	m
m	y	x'	m'
x	m'	m	y

The motion of plasmodium starts from one of the central cells (x, y, x', y') and goes towards one of the four directions (northwest, southwest, northeast, southeast). The *Physarum* diagram for syllogistic conclusions is as follows:

x	m'
m	x'

Some examples of performative syllogistic conclusions are regarded in Fig. 7.7.

Thus, the performative syllogistic allows us to study different zones containing attractants for *Physarum* if they are connected by protoplasmic tubes homogenously.

A model $\mathcal{M}' = \langle M', \|\cdot\|_x\rangle$ for the performative syllogistic, where M' is the set of attractants and $\|X\|_x \subseteq M'$ is a meaning of syllogistic letter X which is understood as all attractants reachable for the plasmodium from the point x, is defined as follows:

- $\mathcal{M}' \models$ *All x are y* iff $\|X\|_x \neq \emptyset$, $\|X\|_y \neq \emptyset$, and $\|X\|_x \cap \|X\|_y \neq \emptyset$, more precisely both $(x'\&y')x$ and $(x'\&y')y$ hold in $\mathcal{M}'$, i.e. the plasmodium can move from neighbours of y to x and it can move from neighbours of x to y;
- $\mathcal{M}' \models$ *Some x are y* iff $y \notin \|X\|_x$ and $x \notin \|X\|_y$, more precisely neither $(x'\&y')x$ nor $(x'\&y')y$ hold in $\mathcal{M}'$, i.e. the plasmodium cannot move from neighbours of y to x and it cannot move from neighbours of x to y;
- $\mathcal{M}' \models$ *No x are y* iff $y \in \|X\|_x$ or $x \in \|X\|_y$, more precisely $(x'\&y')x$ or $(x'\&y')y$ hold in $\mathcal{M}'$, i.e. the plasmodium can move from neighbours of y to x or it can move from neighbours of x to y;
- $\mathcal{M}' \models$ *Some x are not y* iff $y \notin \|X\|_x$ or $x \notin \|X\|_y$, more precisely $(x'\&y')x$ or $(x'\&y')y$ do not hold in $\mathcal{M}'$, i.e. the plasmodium cannot move from neighbours of y to x or it cannot move from neighbours of x to y;
- $\mathcal{M}' \models p \wedge q$ iff $\mathcal{M}' \models p$ and $\mathcal{M}' \models q$;
- $\mathcal{M}' \models p \vee q$ iff $\mathcal{M}' \models p$ or $\mathcal{M}' \models q$;
- $\mathcal{M}' \models \neg p$ iff it is false that $\mathcal{M}' \models p$.

Proposition 7.7 *The performative syllogistic is sound and complete in* $\mathcal{M}'$. □

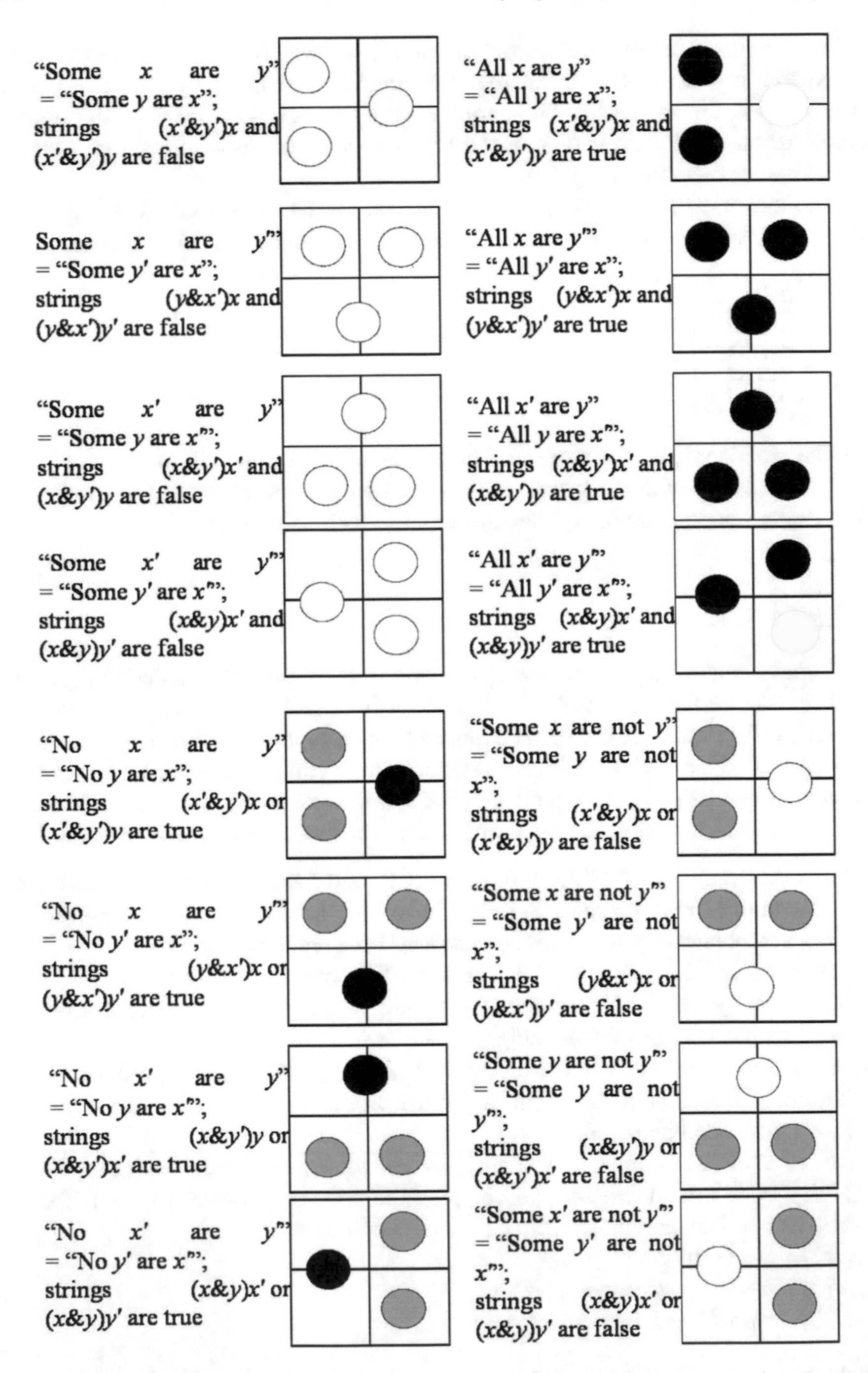

Fig. 7.6 The *Physarum* diagrams for premises of performative syllogisms. Strings of the form $(x'\&y')x$ mean that in cells x' and y' there are neighbours A for x such that Ax, i.e. $(x'\&y')$ is a metavariable in $(x'\&y')x$ that is used to denote all attractants of x' and y' which are neighbours for the attractant of x, see [5]

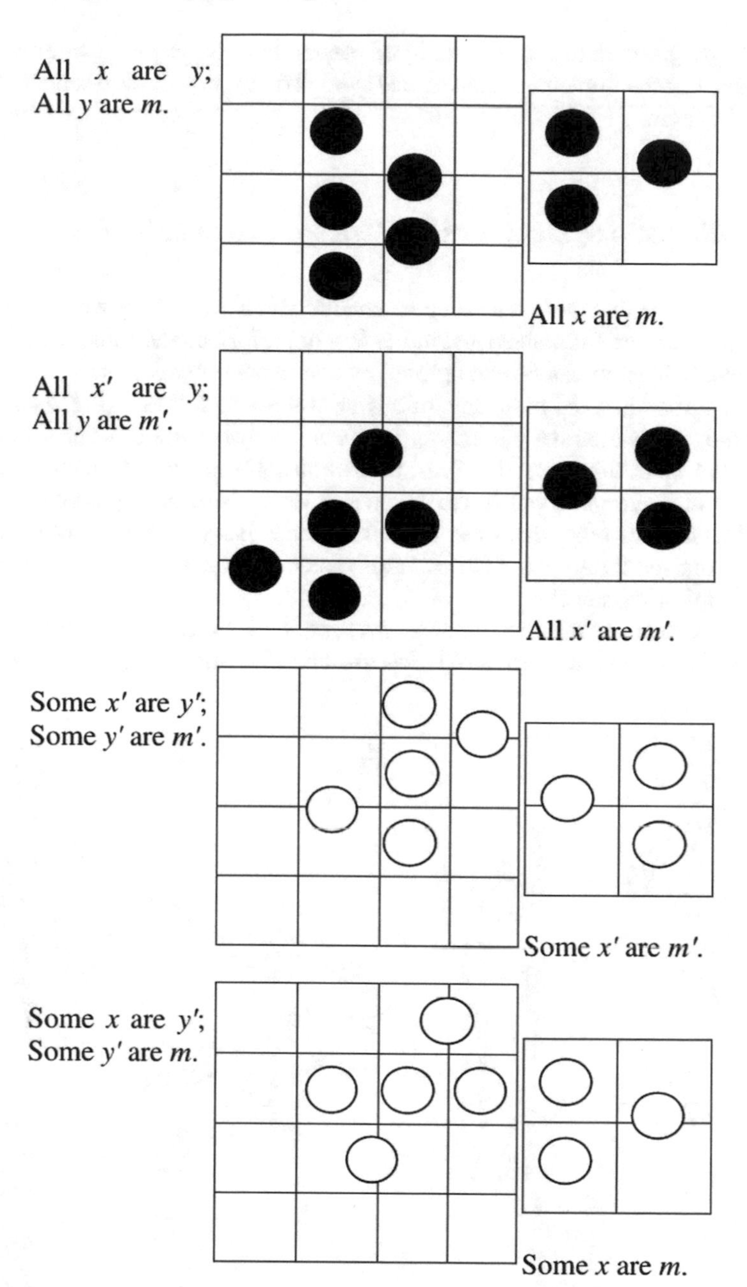

Fig. 7.7 The *Physarum* diagrams for performative syllogisms with true conclusions, see [5]

This syllogistic describes the logic of plasmodium propagation in all possible directions. For the implementation of this syllogistic we do not need repellents. It is a natural system.

7.3 Syllogistic System for the Parasite Propagation

Now, let us show that there is a group behaviour which satisfies only the lateral activation. So, it can be formalized just in the way of performative/synthetic syllogistic. This group behaviour can be exemplified by a *parasite propagation*.

As we remember, the basic acts of any swarm are as follows: (i) *Direction*: the swarm moves towards the attractant; (ii) *Fuse*: the two parts of swarm fuse after meeting the same attractant; (iii) *Split*: the swarm splits in front of many attractants. The same acts are observed in the behaviour of *Schistosomatidae* cercariae (see Fig. 7.8) and many other parasites. The difference is just in different attractants: for example, plasmodia are attracted by nutrients and cercariae are attracted by fatty acids of bird or human skin.

Syllogistically, we can model the behaviour of collectives of the genus *Trichobilharzia* Skrjabin and Zakharov, 1920 (*Schistosomatidae* Stiles and Hassall, 1898) and

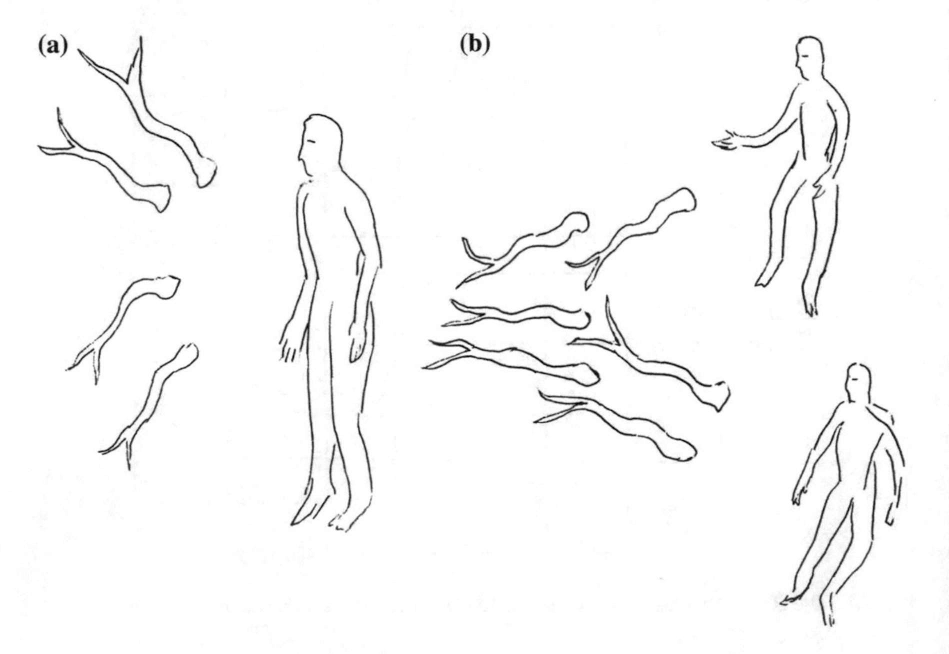

Fig. 7.8 The stimulation of the following operations in cercariae motions: **a** the fusion of cercariae collectives, **b** the multiplication of cercariae collective, where the human beings are attractants, see [7]

the behaviour of many other parasites, as well. This research was made together with Ludmila Akimova. About chemotaxis of parasites, please see [27–30].

Schistosomatidae have been studied recently because of cercarial dermatitis which they cause in humans by the secretion of penetration glands. Notice that the cercaria begins the incorporation into the human skin approximately in 8 seconds (the range from 0 to 80 s) just after first contacts [31]. The process of full penetration into the skin takes about 4 min (the range from 83 s till 13 min 37 s). In human kids, the cercariae can be brought by the venous blood into the lungs, causing hemorrhages and inflammations [32]. Many other complications can come because of repeated infestations [32].

The meaning of the behaviour of parasites such as cercariae consists in the propagation of their collectives in all possible directions in looking for hosts and then their further propagation according to their life cycle. That behaviour is concurrent and not linear (it continuously splits and fuses). In this system, attractants for parasites are considered syllogistic letters. Any two neighbour attractants which can be immediately reached by the one collective of parasites are considered a syllogistic string that is true if both attractants are occupied by parasites and false otherwise. The good science-fiction example of the propagation of parasites is presented by the movie *World War Z* (2013) directed by Marc Forster. Zombies in this movie are attracted by the chemotaxis from able-bodied humans, sounds and motions, which causes the propagation of zombies in all possible directions. Cercariae are attracted by the chemotaxis from a skin of potential hosts, turbulence of water and lights. Notice that the syllogistic system for modelling the propagation of parasites is proposed by me for the first time.

7.3.1 Trichobilharzia szidati *(Diginea: Schistosomatidae) and Their Life Cycle*

All representatives of subclass *Digenea* Carus, 1863 (Platyhelminthes: Trematoda) including *Trichobilharzia szidati* are endoparasites of animals. Their life cycle has the form of heterogony: there are amphimictic and parthenogenetic stages. At these stages, digeneans have different outward, different ways of reproduction and different adaptation to different specific and incompetent hosts. The majority of them have the life cycle with participation of three hosts: intermediate, additional (metacercarial) and final. Molluscs are always the first intermediate hosts, while different classes of vertebrate animals are final hosts.

Among digeneans, there are parasites belonging to the family of *Schistosomatidae*, which have adapted to parasitizing in the circulatory system of vertebrate animals. This family includes the following three subfamilies: *Schistosomatinae* Stiles and Hassall, 1898, they parasitize a variety of birds and mammals, including human beings; *Bilharziellinae* Price, 1929 and *Gigantobilharziinae* Mehra, 1940, they par-

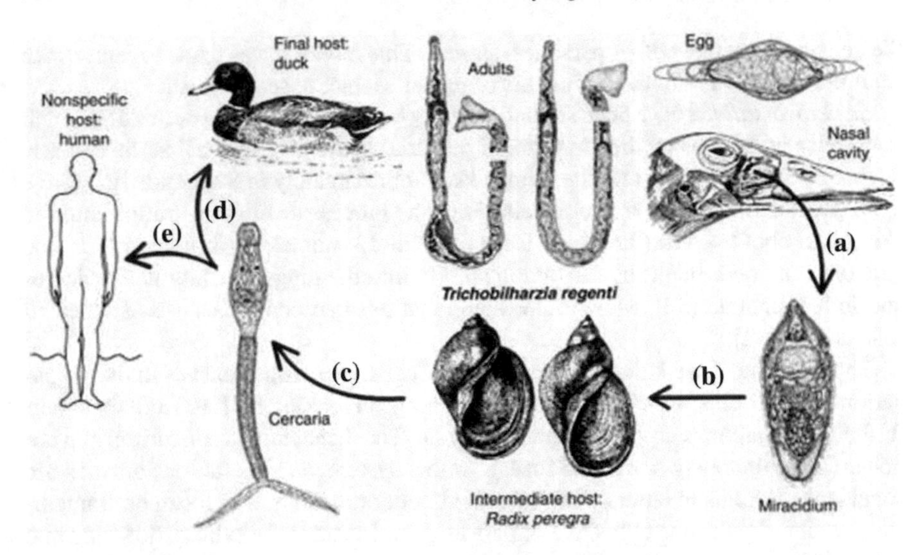

Fig. 7.9 The life cycle of *Trichobilharzia regenti*: **a** mature worms lay the eggs in the nasal mucosa of ducks, then the eggs get to water and from these eggs miracidia hatch; **b** miracidia infect the intermediate hosts; **c** from these hosts cercariae are released; **d** they penetrate the skin of a specific host or **e** the skin of an incompetent mammalian host such as human being (from [33]), see [7]

asitize birds. *Schistosomatidae* can penetrate the human skin, where they perish and, therefore, invoke allergic dermatitis.

The life cycle of all representatives of the family *Schistosomatidae* is identical (see Fig. 7.9). Its members have the following two free-swimming stages: miracidia and cercariae, which actively search for their hosts (intermediate and final, respectively). Miracidial and cercarial host-finding is initiated mainly by response to some gravitational, light and chemical attractants.

7.3.2 *Miracidia (Genus* Trichobilharzia*): Morphology and Behaviour*

Miracidia are free-swimming larval stage (Fig. 7.10). Their body surface is covered with four rows of epithelial plates that carry a multitude of cilia involved in active motions of miracidia (Fig. 7.10b). The anterior part of the body contains terminal openings of the following penetration unicellular glands: two cells of apical glands and two cells of lateral glands. The posterior part of the body contains clusters of germ cells from which daughter sporosysts are formed subsequently. Miracidia lack any digestive system and they cannot feed.

Miracidia of *Trichobilharzia szidati* [34] hatch from eggs within a short time 5–10 min. In order to survive, miracidia must infect a snail host within 20 h at 24 °C before they die [34]. The positive photo- and negative geotactic orientation is

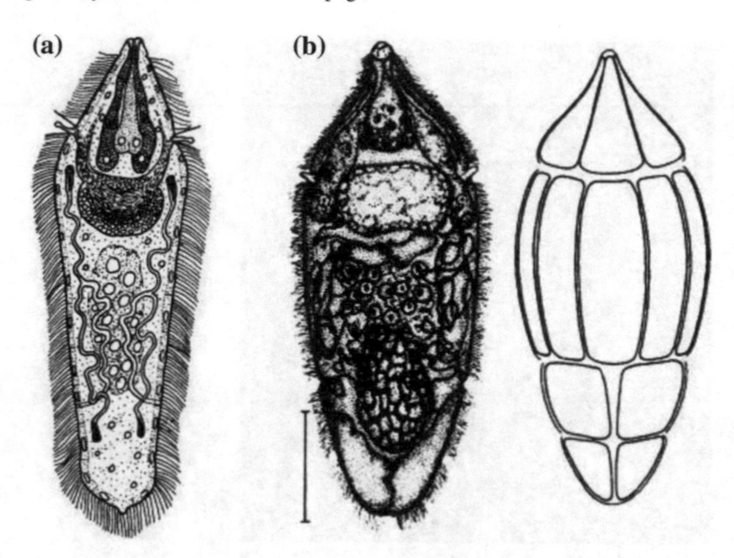

Fig. 7.10 The morphology of miracidia. **a** The general morphology of *Trichobilharzia szidati* (from [34]); **b** the general morphology of *Trichobilharzia regenti* (left) and the arrangement of ciliated plates (right), scale bar = 25 mm (from [33]), see [7]

an adaptation to reach the preferred habitats of their host-snail species [35–39]. Also, it is now confirmed that miracidia are attracted by some chemical host signals [34, 40–42]. Snail-hosts release various secreted/excreted products into water. Miracidia are activated by the macromolecular components of these products, which consist of glycoproteins larger than 30 kDa, termed "miracidia attracting glycoproteins" [43]. These glycoproteins produced by different species of snails have similar peptide-based, but differ by saccharide component. Miracidia can differentiate between these glycosylation patterns in order to find and infect mainly specific snail species [43, 44]. Thus, the chemical cues secreted by the freshwater gastropod *Lymnaea stagnalis* stimulate behavioural responses of *Trichobilharzia szidati* miracidia. In other words, a miracidium moves towards an appropriate chemical signal. Other kinds of attractants for miracidia are presented by light (there is a positive phototaxis) and gravitation (negative geotaxis). We will designate all miracidian attractants by syllogistic letters $S_{m_1}, S_{m_2}, \ldots, S_{m_n}, P_{m_1}, P_{m_2}, \ldots, P_{m_n}, M_{m_1}, M_{m_2}, \ldots, M_{m_n}$. They can differ by their power and intensity.

7.3.3 *Cercariae (Genus* Trichobilharzia*): Morphology and Behaviour*

Cercariae are free-swimming larvae of pubertal generation parasitizing vertebrate animals. Their length is about 1.0 mm. They are capable to insinuate into the skin of

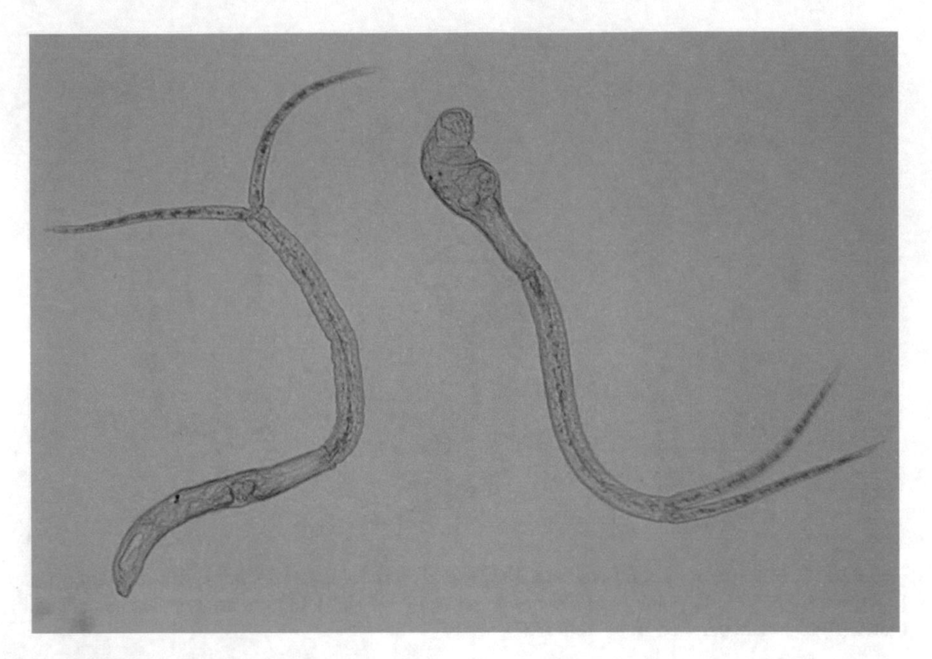

Fig. 7.11 The appearance of *Trichobilharzia szidati*. A detection place: the lake of Naroch (the Minsk region, the Mjadelsky destrict, Belarus). *Lymnaea stagnalis* is an intermediate host. By courtesy of Ludmila Akimova, see [7]

human being who is for them an incompetent host. As a result, they cause an allergic reaction, the so-called *cercarial dermatitis*. This term was proposed by Cort [45], who for the first time correlated this disease with molluscs of certain kinds, and then with cercariae.

Cercariae of the genus *Trichobilharzia* belong to the bunch of furcocercariae, their posterior tail part consists of two branches (furcae), and the length of furca is approximately a half of length of tail. Even in a small zoom its pigmented eye-spots are well visible. The cercarial body is translucent. During motion it is strongly reduced, receiving various forms (Fig. 7.11).

All European species of *Trichobilharzia* possess the five pairs of penetration glands (Fig. 7.12). The two pairs are presented by circumacetabular glands located round the ventral sucker, and the three other by postacetabular glands located sequentially one after another below. The secreta of penetration glands helps larvae to break a dermal barrier of vertebrate hosts.

Cercariae of the genus *Trichobilharzia* after leaving a mollusc actively swimming in the water for an hour. Such an active behaviour of larvae after leaving a mollusc provides a cercarial distribution in the water space. Then cercariae pass to a passive behaviour. Free-swimming cercariae need to insinuate into a final host during the limited time interval (1–1.5 days at temperature 24 °C) [34]. In a resting state, cercariae are attached to a vascular wall or on a water film by means of acetabulum.

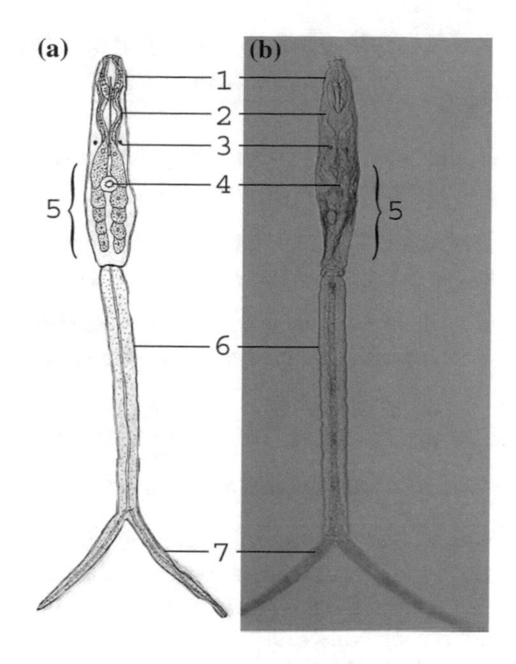

Fig. 7.12 The constitution of *Trichobilharzia szidati*. **a** The schematic structure of cercaria (from [46]), **b** The photo of cercaria. 1. Penetration organ, 2. Penetration gland ducts, 3 Pigmented eye-spots, 4. Ventral sucker, 5. Penetration glands (5 pairs), 6. Tail stem, 7. Furcae. By courtesy of Ludmila Akimova, see [7]

Active motions are characteristic only by the strong shaking of pot or by the water interfusion.

The cercarial behaviour of bird schistosomes (family *Schistosomatidae*) is well studied due to representatives of the genus *Trichobilharzia* [33]. Their behaviour is characterized by the specific taxis implying their looking for specific hosts, their affixion to a surface of host body as well as their incorporation into a host cutaneous covering and their penetration into a circulatory system, where a parasite reaches sexual maturity. Thus, taxis is presented by an enough large family of attractants: larvae of digeneans of *Trichobilharzia* possess a positive phototaxis, negative geotaxis, chemotaxis, and also actively react to turbulence of water [47].

Larvae have chemoreceptors which receive appropriate chemical signals proceeding from a skin of potential host. The similarity of compound of fatty acids of bird and human skin leads to that cercariae equally react to the bird and human appearance in water: they move in their direction, and then they are attached to the skin and begin penetration into it [48]. In experimental researches, it has been shown that any attachment of cercariae of *Trichobilharzia* to the skin is stimulated by cholesterol and ceramides, and incorporation into the skin by linoleic and linolenic acids, all these materials are present on the skin of both the bird and the human being [31, 49]. Thereby surface lipids of human skin invoke higher frequency of cercarial incorporations into the skin, than surface lipids of birds [48].

Hence, the chemotaxis from a skin of potential hosts, the positive phototaxis, the negative geotaxis and the water turbulence should be considered cercarial attractants, which will be designated by $S_{c_1}, S_{c_2}, \ldots, S_{c_n}, P_{c_1}, P_{c_2}, \ldots, P_{c_n}, M_{c_1}, M_{c_2}, \ldots, M_{c_n}$.

7.3.4 *Syllogistic System for the Propagation of* Schistosomatidae

So, in order to formalize the propagation of *Schistosomatidae*, we will use the *performative* or *synthetic syllogistic*.

Definition 7.4 The alphabet of performative-syllogistic system for modelling the propagation of *Schistosomatidae* contains

- as descriptive signs the syllogistic letters S_x, P_x, M_x, ..., where $x \in \{m, c\}$;
- as logical-semantic signs the syllogistic connectives
  - $\mathbf{a}_t$, $\mathbf{e}_t$, $\mathbf{i}_t$, $\mathbf{o}_t$ at time $t = 0, 1, \ldots, n, \ldots$,
  - $\mathbf{a}_\infty$, $\mathbf{e}_\infty$, $\mathbf{i}_\infty$, $\mathbf{o}_\infty$ for infinite time;
- and the propositional connectives $\neg$, $\vee$, $\wedge$, $\Rightarrow$.

Simple propositions are defined as follows: $S_x \star P_x$, where $\star \in \{\mathbf{a}_t, \mathbf{e}_t, \mathbf{i}_t, \mathbf{o}_t, \mathbf{a}_\infty, \mathbf{e}_\infty, \mathbf{i}_\infty, \mathbf{o}_\infty\}$. All other propositions are defined thus: (i) each simple proposition is a proposition, (ii) if X, Y are propositions, then $\neg X$, $\neg Y$, $X \star Y$, where $\star \in \{\vee, \wedge, \Rightarrow\}$, are propositions, too.

Syllogistic letters S_m, P_m, M_m, ... are interpreted as attractants for miracidia as follows: a data point S_m is evaluated as empty if and only if an appropriate attractant for miracidia denoted by S_m is not occupied by any miracidium. Syllogistic letters S_c, P_c, M_c, ...are interpreted as attractants for cercariae in the following manner: an item S_c is evaluated as empty if and only if an appropriate attractant for cercariae denoted by S_c is not occupied by any cercaria.

Let us define syllogistic strings of the form $S_m P_m$ and $S_c P_c$ at time t with the following notation: 'S_m is$_t$ P_m' and 'S_c is$_t$ P_c,' and with the following meaning:

- $S_m P_m$ at time $t = 0, 1, 2, \ldots$ is true if and only if S_m and P_m are cells occupied by miracidia at t and there is a path between S_m and P_m and this path consists of cells also occupied by miracidia at t (i.e. both S_m and P_m are not empty at t and between them there is a path of non-empty cells), otherwise $S_m P_m$ is false (e.g. in Fig. 7.13 there are paths $S_4 S_5$, $S_4 S_{10}$ at $t = 0$);
- $S_c P_c$ at t is true if and only if both S_c and P_c are not empty cells at t and between them there is a path of non-empty cells at t, otherwise $S_c P_c$ is false.

Now, we define syllogistic propositions 'All ... are ...', 'No ... are ...', 'Some ... are ...', 'Some ... are not ...' in the non-Aristotelian way to construct a syllogistic of propagation. The meaning of the proposition 'All S are P' is that between points S and P we observe a propagation in all possible directions. The meaning of the proposition 'Some S are P' is that between points S and P we cannot observe a propagation in all possible directions and the propagation is just contingent and casual. So, while in the Aristotelian syllogistic the propositions 'All ... are ...' and 'No ... are ...' are contrary, in the syllogistic of propagation the propositions 'All ... are ...' and 'Some ... are ...' are contrary. Other propositions are understood conventionally:

'No ... are ...' := It is false that 'Some ... are ...'; 'Some ... are not ...' := It is false that 'All ... are ...'. The point of this non-Aristotelian interpretation is that we cannot exclude propagation at all. There are always some forms of propagation from massive-parallel (when 'All ... are ...' is true) to casual (when 'Some ... are ...' is true).

Thus, using the definition of syllogistic strings, we can define simple performative-syllogistic propositions as follows:

1. 'All S_m are P_m at time t' ($S_m \mathbf{a}_t P_m$): there is a string $A_m S_m$ at time t and for any A_m which is a neighbour for S_m or P_m, there are strings $A_m S_m$ and $A_m P_m$ at t. This means that we have a massive-parallel occupation of the region at t, where the cells S_m and P_m are located, i.e. the propagation holds in all possible directions.
2. 'All S_c are P_c at time t' ($S_c \mathbf{a}_t P_c$): there is a string $A_c S_c$ at time t and for any A_c which is a neighbour for S_c or P_c, there are strings $A_c S_c$ and $A_c P_c$ at t.
3. 'Some S_m are P_m at time t' ($S_m \mathbf{i}_t P_m$): for any A_m which is a neighbour for S_m or P_m at t, there are no strings $A_m S_m$ and $A_m P_m$. This means that the collective of miracidia cannot reach S_m from P_m or P_m from S_m immediately at t, but it does not mean that there are no propagating miracidia. Some forms of their propagation ever exist.
4. 'Some S_c are P_c at time t' ($S_c \mathbf{i}_t P_c$): for any A_c which is a neighbour for S_c or P_c at t, there are no strings $A_c S_c$ and $A_c P_c$.
5. 'No S_m are P_m at time t' ($S_m \mathbf{e}_t P_m$): there exists A_m at time t which is a neighbour for S_m or P_m such that there is a string $A_m S_m$ or there is a string $A_m P_m$. This means that the collective of miracidia occupies S_m or P_m, but surely not the whole region at time t, where the cells S_m and P_m are located.
6. 'No S_c are P_c at time t' ($S_c \mathbf{e}_t P_c$): there exists A_c at time t which is a neighbour for S_c or P_c such that there is a string $A_c S_c$ or there is a string $A_c P_c$.
7. 'Some S_m are not P_m at time t' ($S_m \mathbf{o}_t P_m$): for any A_m which is a neighbour for S_m or P_m at time t there is no string $A_m S_m$ or there exists A_m which is a neighbour for S_m and P_m such that there is no string $A_m S_m$ or there is no string $A_m P_m$. This means that at time t the collective of miracidia does not occupy S_m or there is a neighbour cell which is not connected with S_m or P_m by the same propagated collective of miracidia.
8. 'Some S_c are not P_c at time t' ($S_c \mathbf{o}_t P_c$): for any A_c which is a neighbour for S_c or P_c at time t there is no string $A_c S_c$ or there exists A_c which is a neighbour for S_c or P_c such that there is no string $A_c S_c$ or there is no string $A_c P_c$.

Formally:

$$S_m \mathbf{a}_t P_m := (\exists A_m (A_m \text{ is}_t\ S_m) \wedge (\forall A_m (A_m \text{ is}_t\ S_m \wedge A_m \text{ is}_t\ P_m))); \tag{7.28}$$

$$S_c \mathbf{a}_t P_c := (\exists A_c (A_c \text{ is}_t\ S_c) \wedge (\forall A_c (A_c \text{ is}_t\ S_c \wedge A_c \text{ is}_t\ P_c))); \tag{7.29}$$

$$S_m \mathbf{i}_t P_m := \forall A_m (\neg (A_m \text{ is}_t\ S_m) \wedge \neg (A_m \text{ is}_t\ P_m)); \tag{7.30}$$

$$S_c \mathbf{i}_t P_c := \forall A_c(\neg(A_c \text{ is}_t S_c) \wedge \neg(A_c \text{ is}_t P_c)); \tag{7.31}$$

$$S_m \mathbf{e}_t P_m := \neg\forall A_m(\neg(A_m \text{ is}_t S_m) \wedge \neg(A_m \text{ is}_t P_m)), i.e. \\ \exists A_m(A_m \text{ is}_t S_m \vee A_m \text{ is}_t P_m). \tag{7.32}$$

$$S_c \mathbf{e}_t P_c := \neg\forall A_c(\neg(A_c \text{ is}_t S_c) \wedge \neg(A_c \text{ is}_t P_c)), i.e. \\ \exists A_c(A_c \text{ is}_t S_c \vee A_c \text{ is}_t P_c). \tag{7.33}$$

$$S_m \mathbf{o}_t P_m := \neg(\exists A_m(A_m \text{ is}_t S_m) \vee (\forall A_m(A_m \text{ is}_t P_m \wedge A_m \text{ is}_t S_m))), \\ i.e.(\forall A_m \neg(A_m \text{ is}_t S_m) \wedge \exists A_m(\neg(A_m \text{ is}_t P_m) \vee \neg(A_m \text{ is}_t S_m))); \tag{7.34}$$

$$S_c \mathbf{o}_t P_c := \neg(\exists A_c(A_c \text{ is}_t S_c) \vee (\forall A_c(A_c \text{ is}_t P_c \wedge A_c \text{ is}_t S_c))), \\ i.e.(\forall A_c \neg(A_c \text{ is}_t S_c) \wedge \exists A_c(\neg(A_c \text{ is}_t P_c) \vee \neg(A_c \text{ is}_t S_c))); \tag{7.35}$$

Notably, this system is essentially non-Aristotelian (i.e. performative or synthetic), in particular $S_m \mathbf{a}_t P_m \not\Rightarrow S_m \mathbf{e}_t P_m$; $S_m \mathbf{i}_t P_m \not\Rightarrow S_m \mathbf{o}_t P_m$; $S_m \mathbf{a}_t P_m \not\Rightarrow P_m \mathbf{a}_t S_m$; $S_m \mathbf{e}_t P_m \not\Rightarrow P_m \mathbf{e}_t S_m$; $S_m \mathbf{i}_t P_m \not\Rightarrow P_m \mathbf{i}_t S_m$; $S_m \mathbf{o}_t P_m \not\Rightarrow P_m \mathbf{o}_t S_m$.

The topology of attractants changes permanently (see Fig. 7.13), because different attractants become occupied in due course. Therefore for different time $t = 0, 1, 2, \ldots$, we observe different true syllogistic propositions (see Fig. 7.13).

Also, we can assume that the attractants can move at $t = 0, 1, 2\ldots$ Let us postulate that their motions at different time are limited by the following rules, where $t + 1$ means the next step of life cycle:

$$S_m \mathbf{a}_t P_m \Rightarrow S_m \mathbf{e}_{t+1} P_m. \tag{7.36}$$

$$S_c \mathbf{a}_t P_c \Rightarrow S_c \mathbf{e}_{t+1} P_c. \tag{7.37}$$

From this we have

$$S_m \mathbf{i}_{t+1} P_m \Rightarrow S_m \mathbf{o}_t P_m. \tag{7.38}$$

$$S_c \mathbf{i}_{t+1} P_c \Rightarrow S_c \mathbf{o}_t P_c. \tag{7.39}$$

Attractants for miracidia and cercariae are different, but they live in the same lake. We can postulate that if the whole region is occupied by miracidia (cercariae) at time t, then surely the part of this region is occupied by cercariae (miracidia) at time $t + 1$. So, the propagation of collectives of miracidia at t causes the propagation of collectives of cercariae at $t + 1$:

$$S_m \mathbf{a}_t P_m \Rightarrow S_c \mathbf{e}_{t+1} P_c. \tag{7.40}$$

$$S_m \mathbf{i}_{t+1} P_m \Rightarrow S_c \mathbf{o}_t P_c. \tag{7.41}$$

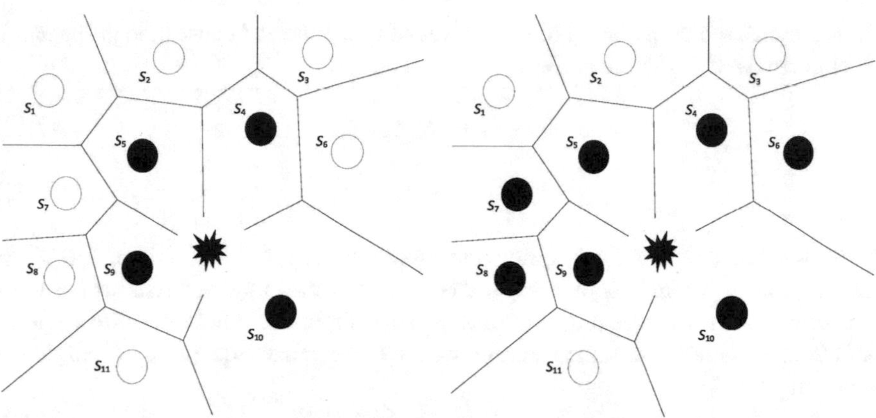

(a) $t = 0$; the four attractants denoted by S_4, S_5, S_9, S_{10} are occupied by the parasites

(b) $t = 1$; the seven attractants denoted by $S_4, S_5, S_6, S_7, S_8, S_9, S_{10}$ are occupied by the parasites

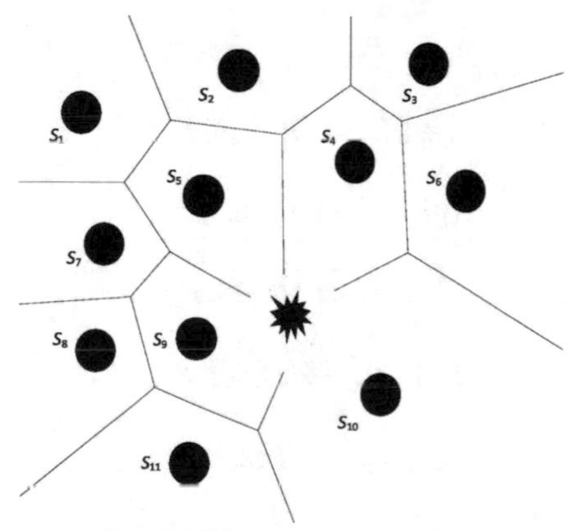

(c) $t = 2$; the eleven attractants denoted by $S_1, S_2, S_3, S_4, S_5, S_6, S_7, S_8, S_9, S_{10}, S_{11}$ are occupied by the parasites

Fig. 7.13 Example of propagation of parasites. First of all, parasites are located in the center of pictures (**a**), (**b**), (**c**). Attractants which are not occupied by parasites are denoted by white circles. Attractants which are occupied by parasites are denoted by black circles. Each attractant is a center of suitable Voronoi cell. The radius of Voronoi cell is a maximal distance of power to attract parasites in one direction. At time $t = 0, 1, 2$, different syllogistic propositions are true: **a** $S_7\mathbf{i}_0S_{11}$, $S_{11}\mathbf{i}_0S_7$, $S_1\mathbf{i}_0S_3$, $S_3\mathbf{i}_0S_1$, $S_1\mathbf{e}_0S_{10}$, $S_{10}\mathbf{e}_0S_1$, $S_1\mathbf{e}_0S_9$, $S_9\mathbf{e}_0S_1$, $S_1\mathbf{e}_0S_4$, $S_4\mathbf{e}_0S_1$, etc.; **b** $S_1\mathbf{i}_1S_{11}$, $S_{11}\mathbf{i}_1S_1$, $S_2\mathbf{o}_1S_6$, $S_6\mathbf{o}_1S_2$, etc.; **c** $S_7\mathbf{a}_2S_{11}$, $S_{11}\mathbf{a}_2S_7$, $S_1\mathbf{a}_2S_3$, $S_3\mathbf{a}_2S_1$, $S_1\mathbf{a}_2S_{10}$, $S_{10}\mathbf{a}_2S_1$, $S_1\mathbf{a}_2S_9$, $S_9\mathbf{a}_2S_1$, $S_1\mathbf{a}_2S_4$, $S_4\mathbf{a}_2S_1$, etc., see [7]

Furthermore, the propagation of collectives of cercariae at t causes the propagation of collectives of miracidia at $t + 1$:

$$S_c \mathbf{a}_t P_c \Rightarrow S_m \mathbf{e}_{t+1} P_m. \quad (7.42)$$

$$S_c \mathbf{i}_{t+1} P_c \Rightarrow S_m \mathbf{o}_t P_m. \quad (7.43)$$

Now, we can define simple syllogistic propositions for infinite time: $S\mathbf{x}_\infty P := \lim_{t \to \infty} S\mathbf{x}_t P$, where $\mathbf{x} \in \{\mathbf{a}, \mathbf{e}, \mathbf{i}, \mathbf{o}\}$. The informal meaning of these propositions is that we put forward syllogistic propositions about the whole one life cycle of *Schistosomatidae* in the same lake. So, we may generalize Eqs. (7.36)–(7.43) in the following way:

$$S_m \mathbf{a}_\infty P_m \Rightarrow \exists t. S_m \mathbf{a}_t P_m; \quad (7.44)$$

$$S_c \mathbf{a}_\infty P_c \Rightarrow \exists t. S_c \mathbf{a}_t P_c; \quad (7.45)$$

$$S_m \mathbf{a}_\infty P_m \Rightarrow \exists t. S_c \mathbf{a}_t P_c; \quad (7.46)$$

$$S_c \mathbf{a}_\infty P_c \Rightarrow \exists t. S_m \mathbf{a}_t P_m; \quad (7.47)$$

$$S_m \mathbf{i}_\infty P_m \Rightarrow \forall t. S_m \mathbf{i}_t P_m; \quad (7.48)$$

$$S_c \mathbf{i}_\infty P_c \Rightarrow \forall t. S_c \mathbf{i}_t P_c; \quad (7.49)$$

$$S_m \mathbf{i}_\infty P_m \Rightarrow \forall t. S_c \mathbf{i}_t P_c; \quad (7.50)$$

$$S_c \mathbf{i}_\infty P_c \Rightarrow \forall t. S_m \mathbf{i}_t P_m. \quad (7.51)$$

Formula (7.44) means that if the whole region is occupied by miracidia for one life cycle of *Schistosomatidae*, then there exists time t such that the whole region is occupied by miracidia. Formula (7.45) means that if the whole region is occupied by cercariae for one life cycle of *Schistosomatidae*, then there exists time t such that the whole region is occupied by cercariae. Formula (7.46) means that if the whole region is occupied by miracidia for one life cycle of *Schistosomatidae*, then there exists time t such that the whole region is occupied by cercariae. Formula (7.47) means that if the whole region is occupied by cercariae for one life cycle of *Schistosomatidae*, then there exists time t such that the whole region is occupied by miracidia.

Formula (7.48) means that if there is just a casual propagation of miracidia for one life cycle of *Schistosomatidae*, then for all time t of this life cycle there is just a casual propagation of miracidia. Formula (7.49) means that if there is just a casual propagation of cercariae for one life cycle of *Schistosomatidae*, then for all time t of this life cycle there is just a casual propagation of cercariae. Formula (7.50) means that if there is just a casual propagation of miracidia for one life cycle of *Schistosomatidae*, then for all time t of this life cycle there is just a casual propagation of cercariae. Formula (7.51) means that if there is just a casual propagation of cercariae for one life cycle of *Schistosomatidae*, then for all time t of this life cycle there is just a casual propagation of miracidia.

Other axioms are as follows (for $x \in \{m, c\}$):

$$S_x \mathbf{a}_t P_x \Rightarrow S_x \mathbf{e}_t P_x; \quad S_x \mathbf{a}_\infty P_x \Rightarrow S_x \mathbf{e}_\infty P_x; \tag{7.52}$$

$$S_x \mathbf{a}_t P_x \Rightarrow P_x \mathbf{a}_t S_x; \quad S_x \mathbf{a}_\infty P_x \Rightarrow P_x \mathbf{a}_\infty S_x; \tag{7.53}$$

$$S_x \mathbf{i}_t P_x \Rightarrow P_x \mathbf{i}_t S_x; \quad S_x \mathbf{i}_\infty P_x \Rightarrow P_x \mathbf{i}_\infty S_x; \tag{7.54}$$

$$S_x \mathbf{a}_t M_x \Rightarrow S_x \mathbf{e}_t P_x; \quad S_x \mathbf{a}_\infty M_x \Rightarrow S_x \mathbf{e}_\infty P_x; \tag{7.55}$$

$$M_x \mathbf{a}_t P_x \Rightarrow S_x \mathbf{e}_t P_x; \quad M_x \mathbf{a}_\infty P_x \Rightarrow S_x \mathbf{e}_\infty P_x; \tag{7.56}$$

$$(M_x \mathbf{a}_t P_x \wedge S_x \mathbf{a}_t M_x) \Rightarrow S_x \mathbf{a}_t P_x; \tag{7.57}$$

$$(M_x \mathbf{i}_t P_x \wedge S_x \mathbf{i}_t M_x) \Rightarrow S_x \mathbf{i}_t P_x; \tag{7.58}$$

$$(M_x \mathbf{a}_\infty P_x \wedge S_x \mathbf{a}_\infty M_x) \Rightarrow S_x \mathbf{a}_\infty P_x; \tag{7.59}$$

$$(M_x \mathbf{i}_\infty P_x \wedge S_x \mathbf{i}_\infty M_x) \Rightarrow S_x \mathbf{i}_\infty P_x. \tag{7.60}$$

Some basic formal properties of that axiomatic system closed over axioms (7.28)–(7.35) and (7.52)–(7.60), where **a**, **e**, **i**, **o** do not depend on time, are considered in Sects. 7.1.1 and 7.2.3. Probabilistic semantics for the time-depended version of that system for the slime mould behaviour is proposed in Sect. 6.2.4.

Notice that formulas (7.58) and (7.60) are axioms of the syllogistic system of propagation, while they are not valid in the Aristotelian system. Indeed, $(M_x \mathbf{i}_t P_x \wedge S_x \mathbf{i}_t M_x) \Rightarrow S_x \mathbf{i}_t P_x$ means that

$$[\forall A_x(\neg(A_x \text{ is}_t\ M_x) \wedge \neg(A_x \text{ is}_t\ P_x)) \wedge \forall A_x(\neg(A_x \text{ is}_t\ S_x) \wedge \neg(A_x \text{ is}_t\ M_x))]$$

$$\Rightarrow \forall A_x(\neg(A_x \text{ is}_t\ S_x) \wedge \neg(A_x \text{ is}_t\ P_x)),$$

the latter formula is valid.

In this chapter, I have constructed the syllogistic system for the swarm propagation on the example of *Schistosomatidae*. In syllogistics for the explication of propagations, the inverse relations hold true for all syllogistic connectives:

$$S\mathbf{a}_t P \Rightarrow P\mathbf{a}_t S;$$

$$S\mathbf{e}_t P \Rightarrow P\mathbf{e}_t S;$$

$$S\mathbf{i}_t P \Rightarrow P\mathbf{i}_t S;$$

$$S\mathbf{o}_t P \Rightarrow P\mathbf{o}_t S.$$

It is inferred from definitions (7.28)–(7.35). These properties fix propagations in all possible directions. *Such propagations correspond to lateral activation and can be described only by the performative/synthetic syllogistic*. It is a good example that there is possible a group behaviour that cannot be laterally inhibited, such as the behaviour of *Schistosomatidae*.

References

1. Schumann, A.: Two squares of opposition: for analytic and synthetic propositions. Bull. Sect. Log. **40**(3–4), 165–178 (2011)
2. Schumann, A.: On two squares of opposition: the Lesniewski's style formalization of synthetic propositions. Acta Anal. **28**, 71–93 (2013)
3. Schumann, A.: Physarum syllogistic L-systems. In: Future Computing 2014, The Sixth International Conference on Future Computational Technologies and Applications. ThinkMind (2014)
4. Schumann, A.: Physarum polycephalum syllogistic L-systems and Judaic roots of unconventional computing. Stud. Log. Gramm. Rhetor. **44**(1), 181–201 (2016). https://doi.org/10.1515/slgr-2016-0011
5. Schumann, A.: Syllogistic versions of go games on Physarum. In: Adamatzky, A. (ed.) Advances in Physarum Machines: Sensing and Computing with Slime Mould, pp. 651–685. Springer International Publishing, Cham (2016)
6. Schumann, A., Adamatzky, A.: Physarum polycephalum diagrams for syllogistic systems. IfCoLog J. Log. Appl. **2**(1), 35–68 (2015)
7. Schumann, A., Akimova, L.: Syllogistic system for the propagation of parasites. The case of Schistosomatidae (Trematoda: Digenea). Stud. Log. Gramm. Rhetor. **40**(53), 303–319 (2015)
8. Schumann, A., Woleński, J.: Two squares of oppositions and their applications in pairwise comparisons analysis. Fundam. Inf. **144**(3–4), 241–254 (2016)
9. Leśniewski, S.: O podstawach matematyki. Przeglạnd Filozoficzny **30, 31, 32, 33, 34**, 164–206, 261–291, 60–101, 77–105, 142–170 (1927–1931)
10. Leśniewski, S.: Über die Grundlagen der Ontologie. Comptes Rendus des Séances de la Société des Sciences et des Lettres de Varsovie, Classe **III**(23), 111–132 (1930)
11. Słupecki, J.: St. Leśniewski's calculus of classes. Stud. Log. **3**, 7–71 (1953)
12. Łukasiewicz, J.: Aristotle's Syllogistic from the Standpoint of Modern Formal Logic. Garland (1951)
13. Bocheński, I.M.: Formale Logik. Karl Alber, Freiburg-München (1956)
14. Ross, W.D. (ed.): The Works of Aristotle, Volume 1: Logic. Oxford University Press (1928)
15. Rose, L.E.: Aristotle's Syllogistic. Charles C. Thomas Publisher (1968)
16. Weinberg, J.R.: An Examination of Logical Positivism. Mason Press (2007)
17. Sneed, J.: The Logical Structure of Mathematical Physics. D. Reidel Publishing Company, Dordrecht (1971)
18. Zinov'ev, A.: Logical Physics. Boston Studies in the Philosophy and History of Science, Springer (1983)
19. Sakiyama, T., Gunji, Y.P.: The Müller-Lyer illusion in ant foraging. PLOS ONE **12**(8), 1–12 (2013)
20. Jones, J.D.: Towards lateral inhibition and collective perception in unorganised non-neural systems. In: Pancerz, K., Zaitseva, E. (eds.) Computational Intelligence, Medicine and Biology. Selected Links. Springer, (2015)

21. Schumann, A., Fris, V.: Swarm intelligence among humans—the case of alcoholics. In: Proceedings of the 10th International Joint Conference on Biomedical Engineering Systems and Technologies—Volume 4: BIOSIGNALS (BIOSTEC 2017), pp. 17–25. ScitePress (2017)
22. Schumann, A.: Towards context-based concurrent formal theories. Parallel Process. Lett. **25**, 1540,008 (2015)
23. Smith, R.: Completeness of an ecthetic syllogistic. Notre Dame J. Form. Log. **24**(2), 224–232 (1983)
24. Carroll, L.: The Game of Logic. Macmillan and Co., London (1886)
25. Carroll, L.: Symbolic Logic. Part I. Elementary. Macmillan and Co., London (1897)
26. Adamatzky, A.: Developing proximity graphs by Physarum polycephalum: does the plasmodium follow the Toussaint hierarcy? Parallel Process. Lett. **19**, 105 (2009)
27. Feiler, W., Haas, W.: *Trichobilharzia ocellata*: chemical stimuli of duck skin for cercarial attachment. Parasitology **96**, 507–517 (1988)
28. Haas, W.: Host finding mechanisms–a physiological effect. In: Mehlhorn, Y. (ed.) Biology, pp. 382–383. Encyclopedic Reference of Parasitology, Structure, Function (1988)
29. Smyth, J.D., Halton, D.W.: The Physiology of Trematodes, 2nd edn. Cambridge University Press, Cambridge (1983)
30. Haas, W., Haberl, B.: Host Recognition by Trematode Miracidia and Cercariae, pp. 197–227. CRC Press, Boca Raton, Florida (1997)
31. Haas, W., Haeberlein, S.: Penetration of cercariae into the living human skin: *Schistosoma mansoni* vs. *Trichobilharzia szidati*. Parasitol. Res. **105**(4), 1061–1066 (2009)
32. Bear, S.A., Voronin, M.V.: Cercariae in Urbanized Ecosystems. Nauka, Moskow (2007). (in Russian)
33. Horák, P., Kolárová, L., Adema, C.M.: Biology of the schistosome genus Trichobilharzia. Adv. Parasitol. **52**, 155–233 (2002)
34. Saladin, K.S.: Behavioral parasitology and perspectives on miracidial host-finding. Z. Parasitenkd. **60**(3), 197–210 (1979)
35. Haas, W.: Parasitic worms: strategies of host finding, recognition and invasion. Zoology **106**, 349–364 (2003)
36. Haas, W., Haberl, B., Kalbe, M., Körner, M.: Snail-host-finding by miracidia and cercariae: chemical host cues. Parasitol. Today **11**, 468–472 (1995)
37. Hertel, J., Holweg, A., Haberl, B., Kalbe, M., Haas, W.: Snail odour-clouds: spreading and contribution to the transmission success of *Trichobilharzia ocellata* (trematoda, digenea) miracidia. Oecologia **147**, 173–180 (2006)
38. Takahashi, T., Mori, K., Shigeta, Y.: Phototactic, thermotactic and geotactic responses of miracidia of *Schistosoma japonicum*. Jpn. J. Parasitol. **110**, 686–691 (1961)
39. Podhorsky, M., Huzova, Z., Mikes, L., Horak, P.: Cercarial dimensions and surface structures as a tool for species determination of *Trichobilharzia spp*. Acta Parasitol. **50**, 343–365 (2009)
40. MacInnis, A.J.: How parasites find their hosts: some thoughts on the inception of host-parasite integration. In: Cennedy, C. (ed.) Ecological Aspects of Parasitology, pp. 3–20. North-Holand, Amsterdam (1976)
41. Sukhdeo, M.V.K., Mettrick, D.F.: Parasite behaviour: understanding platyhelminth responses. Adv. Parasitol. **26**, 73–144 (1987)
42. Sukhdeo, M.V.K., Sukhdeo, S.C.: Trematode behaviours and the perceptual worlds of parasites. Can. J. Zool. **82**, 292–315 (2004)
43. Kalbe, M., Haberl, B., Haas, W.: Finding of the snail host by *Fasciola hepatica* and *Trichobilharzia ocellata*: compound analysis of 'miracidia attracting glycoprotein'. Exp. Parasitol. **96**, 231–242 (2000)
44. Kock, S.: Investigations on intermediate host specificity help to elucidate the taxonomic status of *Trichobilharzia ocellata* (digenea: Schistosomatidae). Parasitology **123**, 67–70 (2001)
45. Cort, W.W.: Schistosome dermatitis in the United States (Michigan). J. Am. Med. Assoc. **90**, 1027–1029 (1928)
46. Ginetsinskaya, T.A.: Trematodes, Their Life Cycles, Biology and Evolution (in Russian). Nauka, Leningrad (1968)

47. Haas, W.: Physiological analysis of cercarial behavior. J. Parasitol. **78**, 243–255 (1992)
48. Haas, W., Roemer, A.: Invasion of the vertebrate skin by cercariae of *Trichobilharzia ocellata*: penetration processes and stimulating host signals. Parasitol. Res. **84**(10), 787–795 (1998)
49. Mikes, L., Zìdková, L., Kasný, M., Dvorák, J., Horák, P.: In vitro stimulation of penetration gland emptying by *Trichobilharzia szidati* and *T. regenti* (schistosomatidae) cercariae. Quantitative collection and partial characterization of the products. Parasitol. Res. **96**(4), 230–241 (2005)

Chapter 8
Context-Based Games of Swarms

The behaviour of *Schistosomatidae* cannot realize the lateral inhibition. As a consequence, it cannot realize different pictures of reality depending on a context. The point is that the parasites try to be propagated in all possible directions, no matter what conditions are at the moment. Nevertheless, true swarms, such as the ant nests or plasmodia of *Physarum polycephalum*, react differently under different conditions. It means that they can produce different pictures of reality or even compete with their own kind. The latter situation of their competitions for the same food allows us to define their behaviour as a game. In this chapter, I will define a *bio-inspired game theory*. See [1–10].

8.1 Bio-inspired Game Theory

8.1.1 Symbolic Values and Non-additive Measures

Let us start with regarding presuppositions of classical game theory. The majority of expert systems used to predict behaviours appeal to appropriate statistical and econometric tools [11–14]. In their possible applications they are extremely limited by some fundamental assumptions about the characters of material laws. First of all, it is assumed that the system of the material universe consists of primary bodies (atoms) and their combinations and relationships described by mathematical equalities, in particular it is supposed that each atom bears its own separate and independent effect so that the total state is being compounded of a number of separate effects detected in the proceeding state. In other words, in order to explore the total state we should present an appropriate proceeding state as a machine:

> And in this matter the example of several bodies made by art was of great service to me: for I recognize no difference between these and natural bodies beyond this, that the effects of machines depend for the most part on the agency of certain instruments, which, as they

A. Schumann, *Behaviourism in Studying Swarms: Logical Models of Sensing and Motoring*, Emergence, Complexity and Computation 33,
https://doi.org/10.1007/978-3-319-91542-5_8

> must bear some proportion to the hands of those who make them, are always so large that their figures and motions can be seen; in place of which, the effects of natural bodies almost always depend upon certain organs so minute as to escape our senses. And it is certain that all the rules of mechanics belong also to physics, of which it is a part or species, [so that all that is artificial is withal natural]: for it is not less natural for a clock, made of the requisite number of wheels, to mark the hours, than for a tree, which has sprung from this or that seed, to produce the fruit peculiar to it. Accordingly, just as those who are familiar with automata, when they are informed of the use of a machine, and see some of its parts, easily infer from these the way in which the others, that are not seen by them, are made; so from considering the sensible effects and parts of natural bodies, I have essayed to determine the character of their causes and insensible parts (René Descartes, *Principles of Philosophy*, 1644; translated by John Veith).

René Descartes was one of the first thinkers who have put forward the assumption that wholes can be studied due to laws of connection between their individual parts described my maths, i.e. wholes are subject to different laws in proportion to the differences of their parts and these proportions can be analyzed mathematically. This one of the main presuppositions of mathematical tools in science is called *measurability* and *additivity* of reality. Due to this assumption modern physics can have obtained all its results. For discovering the material universe it has appealed to *additive measures* such as mass, force, energy, temperature, etc. Economics and conventional game theory try to continue this empiricist tradition and in statistical and econometric tools they deal only with the measurable aspects of reality. They try to obtain additive measures in economics and in studies of real intelligent behaviour, also.

Nevertheless, there is always the possibility that there are important variables of economic systems which are unobservable and non-additive in principle. We should understand that statistical and econometric methods can be rigorously applied in economics just after the presupposition that the phenomena of our social world are ruled by stable causal relations between variables. However, let us assume that we have obtained a fixed parameter model with values estimated in specific spatio-temporal contexts. Can it be exportable to totally different contexts? Are real social systems governed by stable causal mechanisms with atomistic and additive features?

In the 19th century there was a causal relation between power demand and good and service consumption: the increase of good and service consumption has implied the increase of power demand. But now this relation is untrue, because power demand does not increase and good consumption does. Hence, the same causal relation was true in the industrial society and false in the post-industrial society. In other words, that fact shows that in real social systems there is *no ergodicity*. Recall that in case of ergodicity we can describe a dynamical system which has the same behaviour averaged over time as averaged over the space for all states. Therefore it is sophisticated to find out additive measures in economics at all.

One of the additive measures that have been widely applied in economies is *money*. Due to money we can compare goods and services as well as capitals. *Economic capital* is the term to describe already-produced goods or any asset that is used in production of goods or services. There is also its part, *financial capital*, to denote money used to buy what is needed to provide services to the sector of the economy upon which an appropriate operation is based. Money allows us to evaluate material welfare, goods, and services. Nevertheless, we can face non-additivity there too. The matter is that some welfare is not additive. For example, two oil-paintings with the same parameters can have so different surplus exchange values: they can become cheap, expansive or precious.

Goods and services have a high dollar surplus exchange value if they are produced as a part of *symbolic capital* [15, 16] which denotes a non-economic capital (such as education, networking, power, publicity, image) allowing us to aid social exchange. Economic capital consists of any resources which can be used for producing goods or services to obtain profit. Symbolic capital consists of cultural values of goods and services which increase their surplus exchange values extremely.

The Karl Marx's economic theory [17, 18] tried to describe causal connections of industrial society that was concentrated on producing goods. But the modern society is post-industrial and it is concentrated on producing services, where symbolic capital plays more significant role than it took place in the industrial society. According to Marx, any society has the following two levels: (i) the base (relations of production, relations of production forces) and (ii) superstructure (cultural, symbolic relations). The superstructure is derivable from the base. In the industrial society there were not enough places for symbolic capital. The transition from the feudal formation to the capitalist one is, first of all, a reduction of symbolic capital, its depreciation. Public statuses, titles of noble families were not as important as the economic capitals.

The role of symbolic capital has mainly increased in the post-industrial society. It is caused by a priority which services have over goods now in earning money and obtaining profits. In services there is always an appreciable share of symbolic capital and symbolic values. In sausages or tooth-brushes there is no *symbolic value* (as well as in other consumer goods), but if we take fashion shows or cinema there is already nothing more than symbolic values. Accordingly, surplus values can be so different. In the modern society the Marx's scheme about the base domination over the superstructure is not true. Nowadays the superstructure already determines the base. Symbolic capital dominates over economic capital. Any development of information technologies only strengthens this domination. Money and goods are connected now with social exchanges mediated by information technologies. Such a revaluation began to transform promptly all societies towards increasing the importance of publicity and openness. Any society with the higher role of symbolic capital becomes transparent.

Symbolic values which are involved now in producing goods and services cannot be additive measures. However, they can be studied within *symbolic interactionism*, the theory developed since George Herbert Mead [19–21] and Herbert Blümer [22, 23]. They have stated that people act toward things based on *symbolic meanings* they ascribe to those things. In turn, these meanings are derived from social interactions and transformed through their interpretations. Symbolic meanings are defined and studied by qualitative research methods. Symbolic meanings have no equilibria in the classical meaning, see examples in [24, 25]. About symbolic meanings, see [15, 16, 26].

Thus, in mathematical tools of economics and conventional game theory we accept only phenomena with causal connections measured by additive measures. Nevertheless, in the social world we deal with symbolic interactions studied by *non-additive labels* (symbolic meanings or symbolic values). For accepting the variety of such phenomena we should avoid additivity of basic labels.

In Chap. 6, I have shown that the swarm behaviour cannot be described by additive measures. It means, on the one hand, *we cannot use classical game theory built on additive measures for simulating the swarm behaviour*, but, on the other hand, *the bio-inspired game theory we are going to obtain can be used for describing symbolic interactionism in modern societies in turn*. A primitive kind of symbolic interaction (a *proto-symbolic interaction*) takes place even among swarms. The point is that swarms can be laterally inhibited or laterally activated and these conditions change a proto-symbolic value of the item perceived by the swarm: (i) its proto-symbolic value can be increased (the lateral inhibition effect); (ii) its proto-symbolic value can be decreased (the lateral activation effect).

8.1.2 Basic Assumptions of Game Theory and Non-additivity in Symbolic Interactions

Presuppositions of conventional probability theory, including the idea of additivity (partition into mutually disjoint subsets), are continued in game theory, where the human behaviour is understood as a step-by-step interaction of decision-makers for own payoffs. It is supposed that each player has a certain objective called *payoff* and takes actions deliberately in an attempt to achieve that objective. For this purpose, each player takes into account knowledge or expectations of other decision-makers, because payoffs of different players can stay in conflicts. So, the basic entity of game theory is a *player* who may be interpreted as an individual or as a group of individuals. A player is everyone who has an effect on others' payoffs. However, we can ever assume that a player participates in a symbolic interaction with others, when a payoff is laterally inhibited or laterally activated (i.e. its proto-symbolic or symbolic value is increased or decreased respectively). Later I will show that this proto-symbolic or symbolic interaction can be evaluated by non-Archimedean probability measures (see Chap. 6).

In game theory, possible actions of individual players are considered primitives (logical atoms). Therefore we deal with a set $\Omega_{\mathcal{G}}$ of all *strategies* (actions available to each player) in a game $\mathcal{G}$ such that $\Omega_{\mathcal{G}}$ has a partition into mutually disjoint subsets E_1, E_2, ..., where each E_i contains actions of player i, $i = 1, 2, \ldots$ Each member $\langle e_1, e_2, \ldots \rangle$ of a set $E_1 \times E_2 \times \ldots$ is an outcome of the game and it is associated with *payoffs* $\langle a_1, a_2, \ldots \rangle$, where a_i is a payoff of player i after using a strategy e_i, $i = 1, 2, \ldots$ So, each player has own strategies and combinations with strategies of others give payoffs. Thus, the number of players is fixed and known to all parties. It is the first assumption of game theory, corresponding to the Descartes' hypothesis of additivity of labels in scientific investigations. We have mutually disjoint subsets E_1, $E_2 \ldots$, of $\Omega_{\mathcal{G}}$ and each player knows such a partition. In other words, each player chooses among two or more possible strategies and knows how each strategy chosen by him/her or by another player determines the whole play.

Nevertheless, in symbolic interactions of human beings very often we face the situations when we do not know all players (e.g. lobbyists in politic games can be hidden), so we do not know an appropriate partition of $\Omega_{\mathcal{G}}$ and it is possible that we do not know all strategies of $\Omega_{\mathcal{G}}$ as such. In this case we can appeal to sets with a non-Archimedean ordering structure.

The second assumption of game theory, corresponding to the Descartes' hypothesis of additivity, is that all players are considered *fully rational*. It is understood as follows:

1. Players know *all the rules of the game*, but it can appear that some rules change during the game. In proto-symbolic interactions of swarms and in symbolic interactions of humans all depends on players and (proto-)symbolic meanings they produce within concrete interactions. For example, a swarm can be laterally inhibited and laterally activated through the same game at different times and it is caused just by a topology of payoffs.
2. Players assume other parties to be fully rational. This means that all players have a *zero reflexion*: they know each other and know everything about each other including all strategies. Evidently, this assumption does not hold for (proto-)symbolic interactions. I can cheat to hide my true motives and utter false announcements to lie. Therefore I cannot trust others. In the case of swarms, there is the same uncertainty: we cannot predict fully the swarm behaviour.
3. All players attempt to *maximize their utility*. The latter means some ranking of the subjective welfare, when (s)he is ready to change something. As a result, all players *accept the highest payoffs*. Nevertheless, in (proto-)symbolic interactions I can avoid the highest payoffs for the sake of some symbolic values, e.g. I can be altruistic or even sacrifice my life for somebody. In the case of swarms, it means that the same payoffs can be different: they can be laterally inhibited or laterally activated.
4. All players have *resistance points*, i.e. they can accept only solution's that are at or greater than their security levels. In symbolic interactions I can avoid this item, too, for the same reasons as in the previous item.
5. All players *know utilities and preferences* (preference relations) *of other players* and develop tangible preferences among those options. *Preferences remain constant* throughout the game. But in (proto-)symbolic interactions including swarm behaviours, players can change their preferences through the interaction (for instance, the human beings can lie and hide their true preferences).
6. For any game there is *Pareto efficiency*. All players can take maximally efficient decisions which maximize each player's own interests. Let us recall that a distribution of utility A is called *Pareto superior* over another distribution B if from state B there is a possible redistribution of utility to A such that at least one player has the better payoff in A than in B and no player has the worse payoffs. In the situations of (proto-)symbolic interactions when preferences may change through the game there is no Pareto efficiency in general case.

Due to all the assumptions of game theory mentioned above there are always common game solutions giving an endogenously stable or equilibrated state. These solutions are called *equilibria*. This term is extrapolated from physics, where it means a stable state in which all the causal forces internal to the system balance each other out unless it is perturbed by the intervention of some external force. So, game-theorists consider economic systems as mutually constraining causal relations, just like physical systems. These equilibria can be found out just by using the math tools of computations over payoffs. For (proto-)symbolic interactions including swarm behaviours there are no equilibria in that meaning, but it is ever possible to reach a consensus that will be called a *performative equilibrium*.

If we use p-adic probability measures, we can appeal to other game-theoretic assumptions (for more details see [7]):

1. Each *game* can be assumed *infinite*, because its rules can change.
2. Players can have *different levels of reflexion*: one player can know everything about another, but the second can know just false announcements from the first.
3. Some utilities can have *proto-symbolic meanings* or *symbolic meanings*. These meanings are results of accepting (proto-)symbolic values by some players, e.g. swarms can consider the same item being laterally inhibited or laterally activated. The higher symbolism of payoffs, the higher level of reflexion of appropriate players. On the zero level of reflexion, the payoffs do not have symbolic meanings at all. For consensus the players are looking for joint symbolic meanings.
4. *Resistance points* for players are reduced to the payoffs of the *zero level of reflexion*.
5. The *joint* (proto-)*symbolic meanings can change* through the game if a player increases his/her level of reflexion.
6. For any game there is *performative efficiency*, when all (proto-)symbolic meanings of one player are shared by other players.

In the case of these new game-theoretic assumptions we can calculate some aspects of (proto-)symbolic interactions by probabilistic tools in non-Archimedean numbers [27], see Chap. 10. These new assumptions correspond to bio-inspired game theory as well as to theory of reflexive games. So, on the one hand, in bio-inspired game theory all the game moves are performed under the lateral inhibition or lateral activation conditions, therefore the swarm behaviour is not forecasted by additive measures. On the other hand, in reflexive games players can lie to each other, therefore their behaviours are not predictable by additive measures, too.

8.1.3 Concurrent Go Games on the Slime Mould

The behaviour of swarms can be represented as a bio-inspired game, i.e. within the experimental game theory, where, on the one hand, all basic definitions are verified in the experiments with swarms and, on the other hand, all basic algorithms are implemented in an object-oriented language for simulations of swarms (e.g. the object-oriented language for plasmodia is introduced in [28–31]).

Let us consider an example of bio-inspired game. There are two species of the slime mould: *Physarum polycephalum* and *Badhamia utricularis*. The first can move faster than the second, but the second has bigger strands, see [7]. The main feature of gamic competition between *Physarum polycephalum* and *Badhamia utricularis* is that they move concurrently. Hence, we deal with a concurrent game, not sequential. So, we can show that the slime mould can be a model for concurrent games.

In context-based games, we try to define in this chapter, players can move concurrently as well as in concurrent games, but the set of actions is ever infinite. In our experiments, we adopt the following interpretations of basic entities: (i) attractants as payoffs; (ii) attractants occupied by the swarm as states of the game; (iii) active zones of swarm as players; (iv) logic gates for behaviours as moves (available actions) for the players; (v) propagation of the swarm as the transition table which associates, with a given set of states and a given move of the players, the set of states resulting from that move.

In this *bio-inspired game theory* we can demonstrate creativity of swarms. The point is that swarms do not strictly follow spatial algorithms like Kolmogorov-Uspensky machines, but perform many additional actions (see Chap. 4). So, the swarm behaviour can be formalized within strong extensions of spatial algorithms, e.g. within concurrent games or context-based games [3].

Now, let us show how we can represent the plasmodium behaviour as a *concurrent Go game*. Remember that the Go game is a board game with two players (called Black and White) who alternately place black and white stones, accordingly, on the vacant intersections (called points) of a board with a 19×19 grid of lines [32]. Black moves first. Stones are placed until they reach a point where stones of another color are located. There are the following two basic rules of the game: (i) each stone must have at least one open point (called liberty) directly next to it (up, down, left, or right), or must be part of a connected group that has at least one such open point; stones which lose their last liberty are removed from the board; (ii) the stones must never repeat a previous position of stones. The aim of the game is in surrounding more empty points by player's stones. At the end of game, the number of empty points player's stones surround are counted, together with the number of stones the player captured. This number determines who the winner is.

In our version of Go game for the slime mould, both players move concurrently, not sequentially. In this view the black stones are considered attractants occupied by the plasmodium and the white stones are regarded as repellents. By this interpretation, we have two players, also: Black (this player places attractants) and White (this player places repellents). The winner is determined by the number of empty points player's stones surround.

Notice that the number of possible Go games is too large, 10,761. Therefore it is better to focus just on games, where locations of black and white stones simulate spatial reasoning.

The plasmodium of *Physarum polycephalum* moves to attractants to connect them and in the meanwhile it avoids places, where repellents are located. The radius, where

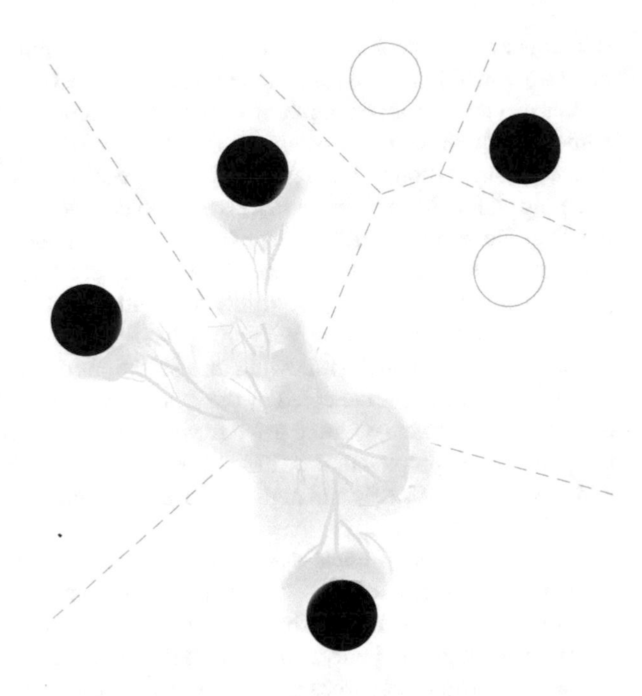

Fig. 8.1 The six Voronoi cells in accordance with the four attractants denoted by the black stones and with the two repellents denoted by the white stones. The plasmodium located in the center of the picture connects the three attractants by the three protoplasmic tubes. It cannot see the fourth attractant because of the two repellents

chemical signals from attractants (repellents) can be detected by the plasmodium to attract (repel) the latter, determines the structure of natural *Voronoi cells*, where each Voronoi cell is a place, where a chemical signal holds (see Fig. 8.1).

For the plasmodium of Fig. 8.1 we have just the four neighbour cells. Notably that in Go games at each point we have only four neighbours everywhere. In other words, we deal with calculations in the 5-adic universe. So, we can design the space for plasmodia in the way to have just four neighbours at each point of moves. We associate black stones with attractants occupied by plasmodia and white stones with repellents. *For the sake of convenience and more analogy with Voronoi cells, let us consider cells (not intersections of lines) as points for stone locations*. Then we can use all rules of Go games to simulate plasmodium motions (see Fig. 8.2).

8.1.4 Aristotelian Go Game on the Slime Mould

In the concurrent Go implementation of *Aristotelian syllogistic*, the syllogistic letters S, P, M, ... are interpreted as cells of the board with the 19×19 grid of lines. The letter S is understood as empty if and only if the white stone is located on an appropriate cell denoted by S. This letter is treated as non-empty if and only if the black stone is located on an appropriate cell denoted by S. If a cell does not contain any stone, this means that this cell is out of the game.

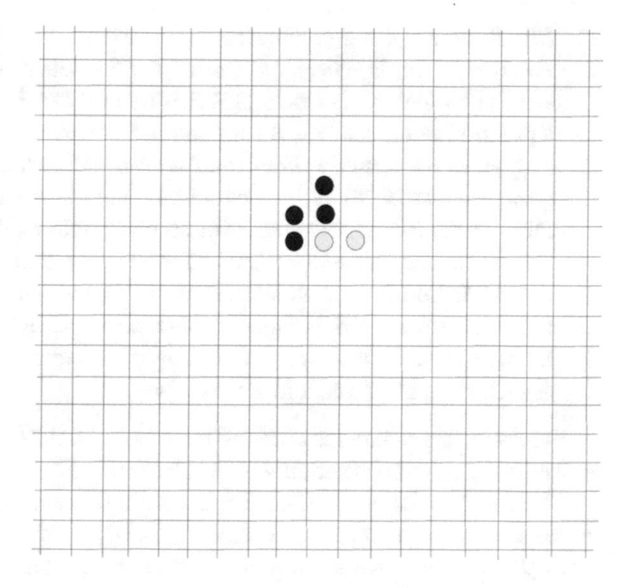

Fig. 8.2 The Go game board with two white stones designating repellents and four black stones designating attractants, see [4]

The Go implementation of Aristotelian syllogistic can be embodied in the behaviour of *Physarum* plasmodium. Let us design cells of *Physarum* syllogistic which will designate classes of terms. We can suppose that cells can possess different topological properties. This depends on intensity of chemo-attractants and chemo-repellents. The intensity entails the natural or geographical neighbourhood of the set's elements in accordance with the spreading of attractants or repellents. As a result, we obtain Voronoi cells.

Thus, in the *Physarum* interpretation of the Go game, the non-empty syllogistic letters S, P, M, ..., i.e. the cells denoted by S, P, M, ... containing black stones are considered attractants and the empty syllogistic letters S, P, M, ..., i.e. the cells denoted by S, P, M, ... containing white stones are considered repellents. So, a data point S is regarded as non-empty if and only if an appropriate attractant located in S is occupied by a plasmodium. This data point S is regarded as empty if and only if an appropriate repellent located in S repels the plasmodium.

Thus, in the Aristotelian version of the Go game we have syllogistic strings of the form SP with the following interpretation: 'S is P', and with the following meaning: SP is true if and only if S and P are neighbours and both S and P are not empty, otherwise SP is false. We can extend this meaning as follows: SP is true if and only if S and P are not empty and there is a line of non-empty cells between points S and P, otherwise SP is false. By the definition of true syllogistic strings, we can define atomic syllogistic propositions as follows:

1. $Sa^{A}_{Go}P$. *In the formal syllogistic*: there exists A such that A is S and for any A, if A is S, then A is P. *In the Go game model*: there is a cell A containing the black stone and for any A, if AS is true, then AP is true. *In the Physarum model*: there is a plasmodium in the cell A and for any A, if AS is true, then AP is true.

2. $Si^{A}_{Go}P$. *In the formal syllogistic*: there exists A such that both AS is true and AP is true. *In the Go game model*: there exists a cell A containing the black stone such that AS is true and AP is true. *In the Physarum model*: there exists a plasmodium in the cell A such that AS is true and AP is true.
3. $Se^{A}_{Go}P$. *In the formal syllogistic*: for all A, AS is false or AP is false. *In the Go game model*: for all cells A containing the black stones, AS is false or AP is false. *In the Physarum model*: for all plasmodia A, AS is false or AP is false.
4. $So^{A}_{Go}P$. *In the formal syllogistic*: for any A, AS is false or there exists A such that AS is true and AP is false. *In the Go game model*: for all cells A containing the black stones, AS is false or there exists A such that AS is true and AP is false. *In the Physarum model*: for any plasmodia A, AS is false or there exists A such that AS is true and AP is false.

Formally, this semantics is defined as follows. Let M be a set of attractants. Take a subset $|X| \subseteq M$ of cells containing the black stones (i.e. of cells containing attractants and occupied by the plasmodium) as a meaning for each syllogistic variable X. Next, define an ordering relation $\subseteq$ on subsets $|S|, |P| \subseteq M$ as: $|S| \subseteq |P|$ iff all attractants from $|P|$ are reachable for the plasmodium located at the attractants from $|S|$, i.e. iff for all cells of $|S|$ with black stones there are lines of black stones connecting them to cells of $|P|$ also containing black stones. Hence, $|S| \cap |P| \neq \emptyset$ means that some attractants from $|P|$ are reachable for the plasmodium located at the attractants from $|S|$ and $|S| \cap |P| = \emptyset$ means that no attractants from $|P|$ are reachable for the plasmodium located at the attractants from $|S|$. In the Go game model $|S| \cap |P| \neq \emptyset$ means that some cells from $|P|$ occupied by the black stones are connected by the lines of black stones with the cells from $|S|$ occupied by the black stones and $|S| \cap |P| = \emptyset$ means that there are no lines of black stones from the cells of $|P|$ to the cells of $|S|$.

This allows us to define models $\mathscr{M} = \langle M, |\cdot|\rangle$ such that

- $\mathscr{M} \models Sa^{A}_{Go}P$ iff $|S| \subseteq |P|$;
- $\mathscr{M} \models Si^{A}_{Go}P$ iff $|S| \cap |P| \neq \emptyset$;
- $\mathscr{M} \models Se^{A}_{Go}P$ iff $|S| \cap |P| = \emptyset$;
- $\mathscr{M} \models p \wedge q$ iff $\mathscr{M} \models p$ and $\mathscr{M} \models q$;
- $\mathscr{M} \models p \vee q$ iff $\mathscr{M} \models p$ or $\mathscr{M} \models q$;
- $\mathscr{M} \models \neg p$ iff it is false that $\mathscr{M} \models p$.

Let us consider a concurrent game of Go at time step 10, i.e. when White and Black players have placed the 10 white stones and the 10 black stones respectively. Let this game be pictured in Fig. 8.3. Each Voronoi cell is denoted from $S_{1,1}$ to $S_{18,18}$. So, in Fig. 8.3 syllogistic letters $S_{6,4}$, $S_{7,5}$, $S_{7,6}$, $S_{8,7}$, $S_{8,8}$, $S_{7,9}$, $S_{8,10}$, $S_{6,11}$, $S_{4,9}$, $S_{4,10}$ are understood as non-empty and syllogistic letters $S_{4,6}$, $S_{5,6}$, $S_{6,8}$, $S_{6,9}$, $S_{6,10}$, $S_{4,11}$, $S_{7,10}$, $S_{7,12}$, $S_{11,8}$, $S_{12,8}$ as empty. As a result, we can build some true syllogistic propositions in this universe like that: 'Some $S_{7,5}$ are $S_{7,6}$', 'Some $S_{8,7}$ are $S_{8,8}$', 'Some $S_{4,9}$ are $S_{4,10}$', 'No $S_{4,6}$ are $S_{5,6}$', 'No $S_{6,8}$ are $S_{6,9}$', 'No $S_{6,9}$ are $S_{6,10}$', 'No $S_{6,10}$ are $S_{7,10}$', 'No $S_{11,8}$ are $S_{12,8}$', etc. Let us notice that in the universe of Fig. 8.3 we do not have universal affirmative propositions. But we can draw some syllogistic

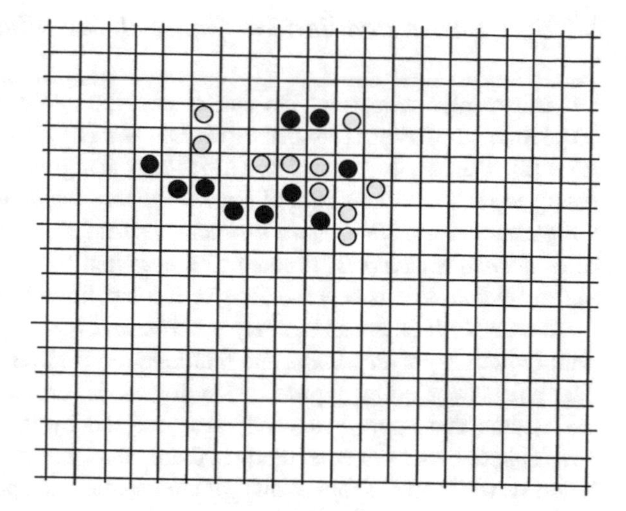

Fig. 8.3 The Aristotelian concurrent Go game 1 at time step 10. The black stones are attractants occupied by the plasmodium, the white stones are repellents, see [4]

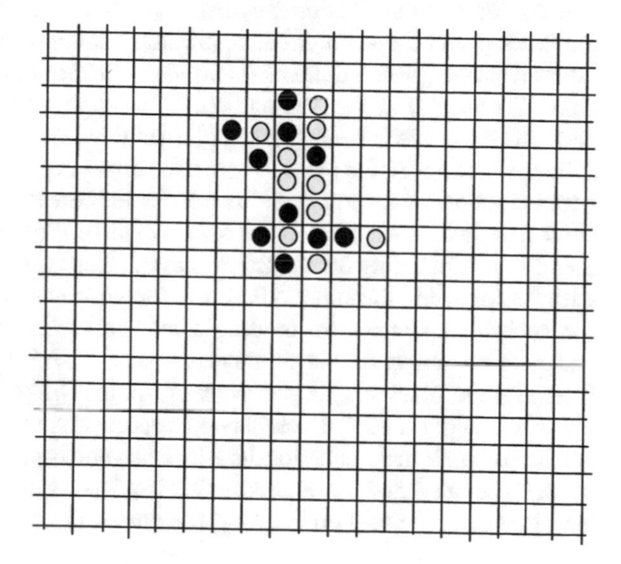

Fig. 8.4 The Aristotelian concurrent Go game 2 at time step 10. The black stones are attractants occupied by the plasmodium, the white stones are repellents, see [4]

conclusions such as 'If $S_{4,10}[S_{4,11}]$ and $S_{4,9}S_{4,10}$, then $S_{4,9}[S_{4,11}]$' (i.e. 'If no $S_{4,10}$ are $S_{4,11}$ and some $S_{4,9}$ are $S_{4,10}$, then no $S_{4,9}$ are $S_{4,11}$').

On the contrary, in the universe pictured in the concurrent Go game of Fig. 8.4 we have universal affirmative propositions such as 'All $S_{8,10}$ are $S_{8,11}$' and 'All $S_{4,9}$ are $S_{3,9}$'. Some possible conclusions: 'If no $S_{8,11}$ are $S_{8,12}$ and all $S_{8,10}$ are $S_{8,11}$, then no $S_{8,10}$ are $S_{8,12}$' and 'If no $S_{3,9}$ are $S_{3,10}$ and all $S_{4,9}$ are $S_{3,9}$, then no $S_{4,9}$ are $S_{3,10}$'.

8.1.5 Non-Aristotelian Go Game on the Slime Mould

Let us consider concurrent Go games with the two different kinds of plasmodia: (i) plasmodia of *Physarum polycephalum* and (ii) plasmodia of *Badhamia utricularis* [7]. They try to occupy free attractants antagonistically. So, if an attractant is occupied by the plasmodium of *Physarum polycephalum*, it cannot be occupied by the plasmodium of *Badhamia utricularis* and if it is occupied by the plasmodium of *Badhamia utricularis*, it cannot be occupied by the plasmodium of *Physarum polycephalum*. In this way we observe a competition between two plasmodia.

In order to implement the *performative syllogistic* in the concurrent Go games with *Physarum polycephalum* and *Badhamia utricularis* plasmodia, we will interpret data points denoted by appropriate syllogistic letters as black stones (attractants) if we assume that appropriate cells are occupied by the plasmodium of *Physarum polycephalum* and we will interpret data points denoted by appropriate syllogistic letters as white stones (attractants) if we assume that appropriate cells are occupied by the plasmodium of *Badhamia utricularis*. A data point S is considered empty for the Black player if and only if an appropriate attractant denoted by S is occupied by the white stone (plasmodium of *Badhamia utricularis*). A data point S is considered empty for the White player if and only if an appropriate attractant denoted by S is occupied by the black stone (plasmodium of *Physarum polycephalum*). Let us define syllogistic strings of the form SP with the following interpretation: (i) 'S is P', and with the following meaning: SP is true for the Black player if and only if S and P are reachable for each other by the plasmodium of *Physarum polycephalum* and both S and P are not empty for the Black player, otherwise SP is false; (ii) 'S is P', and with the following meaning: SP is true for the White player if and only if S and P are reachable for each other by the plasmodium of *Badhamia utricularis* and both S and P are not empty for the White player, otherwise SP is false. In other words, SP is true for the Black player (respectively, for the White player) if and only if S and P are not empty for the Black player (respectively, for the White player) and there is a line of non-empty cells for the Black player (respectively, for the White player) between points S and P, otherwise SP is false. Using this definition of syllogistic strings, we can define atomic syllogistic propositions as follows:

1. $Sa_{Go}^{non A}P$. *In the formal performative syllogistic*: there exists A such that A is S and for any A, AS is true and AP is true. *In the Go game model*: there is a black (white) stone in A connected by black (white) stones to S and connected by black (white) stones to P. *In the Physarum model*: a plasmodium of *Physarum polycephalum* (a plasmodium of *Badhamia utricularis*) in A occupies S and for any plasmodia A of *Physarum polycephalum* (for any plasmodia A of *Badhamia utricularis*) which is a neighbour for S and P, there are strings AS and AP. This means that we have a massive-parallel occupation of the region by plasmodia of *Physarum polycephalum* (plasmodia of *Badhamia utricularis*) where the cells S and P are located.
2. $Si_{Go}^{non A}P$. *In the formal performative syllogistic*: for any A, both AS is false and AP is false. *In the Go game model*: for any cell A there are no lines of black

(white) stones connecting A to S and A to P. *In the Physarum model*: for any plasmodium A of *Physarum polycephalum* (of *Badhamia utricularis*) which is a neighbour for S and P, there are no strings AS and AP. This means that the plasmodium of *Physarum polycephalum* (of *Badhamia utricularis*) cannot reach S from P or P from S immediately.

3. $Se_{Go}^{nonA}P$. *In the formal performative syllogistic*: there exists A such that if AS is false, then AP is true. *In the Go game model*: there exists a cell A with the black (white) stone which is a neighbour for cells S and P such that there is a string AS or there is a string AP. *In the Physarum model*: there exists the plasmodium of *Physarum polycephalum* (of *Badhamia utricularis*) A which is a neighbour for S and P such that there is a string AS or there is a string AP. This means that the plasmodium of *Physarum polycephalum* (of *Badhamia utricularis*) occupies S or P, but not the whole region where the cells S and P are located.
4. $So_{Go}^{nonA}P$. *In the formal performative syllogistic*: for any A, AS is false or there exists A such that AS is false or AP is false. *In the Go game model*: for any cell A with the black (white) stone which is a neighbour for S and P there is no string AS or there exists a black (white) stone in A which is a neighbour for S and P such that there is no string AS or there is no string AP. *In the Physarum model*: for any plasmodium of *Physarum polycephalum* (of *Badhamia utricularis*) A which is a neighbour for S and P there is no string AS or there exists A which is a neighbour for S and P such that there is no string AS or there is no string AP. This means that the plasmodium of *Physarum polycephalum* (of *Badhamia utricularis*) does not occupy S or there is a neighbouring cell which is not connected to S or P by a protoplasmic tube.

Notice that the same proposition $Sx_{Go}^{nonA}P$, where $x_{Go}^{nonA} \in \{a, e, i, o\}$, has a different meaning for the Black player and the White player:

Definition 8.1 There are the following semantic correlations between propositions in sense of the Black player and propositions in sense of the White player:

- $Sa_{Go}^{nonA}P$ is true for the Black player iff $Si_{Go}^{nonA}P$ is true for the White player with the same cells S and P;
- $So_{Go}^{nonA}P$ is true for the Black player iff $Se_{Go}^{nonA}P$ is true for the White player with the same cells S and P;
- $Sa_{Go}^{nonA}P$ is false for the Black player iff $Si_{Go}^{nonA}P$ is false for the White player with the same cells S and P;
- $So_{Go}^{nonA}P$ is false for the Black player iff $Se_{Go}^{nonA}P$ is false for the White player with the same cells S and P.

Composite propositions are defined in the standard way. The performative syllogistic as a Go game is an antagonistic game, where two players (Black and White) draw own conclusions without any coalition.

Let us also examine a game of Go at time step 10 to provide an example for performative syllogistic. Let this game be shown in Fig. 8.5. As usual, each Voronoi cell

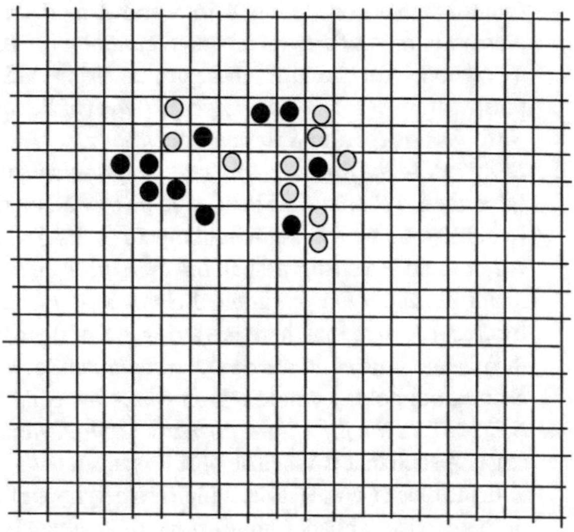

Fig. 8.5 The performative syllogistic concurrent Go game 1 at time step 10. The black stones are attractants occupied by the plasmodium of *Physarum polycephalum* and the white stones are attractants occupied by the plasmodium of *Badhamia utricularis*. So, we construct syllogisms from the point of view of the Black player, see [4]

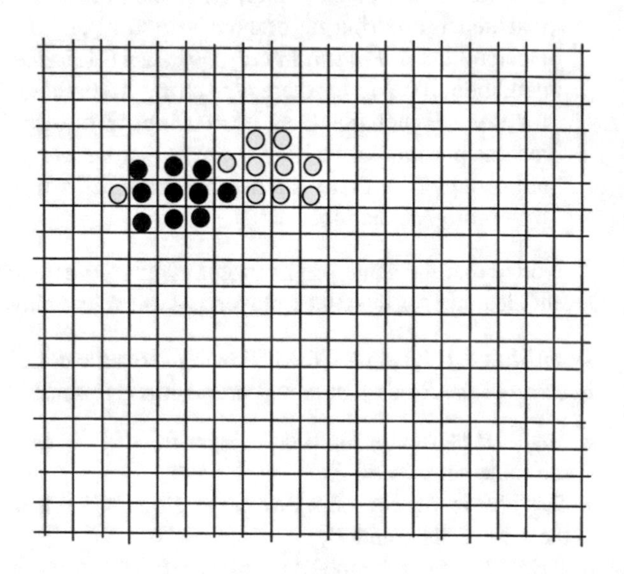

Fig. 8.6 The performative syllogistic concurrent Go game 2 at time step 10. The black stones are attractants occupied by the plasmodium of *Physarum polycephalum* and the white stones are attractants occupied by the plasmodium of *Badhamia utricularis*. We construct syllogisms from the point of view of the Black player, see [4]

is denoted from $S_{1,1}$ to $S_{18,18}$. In the universe of Fig. 8.5 there are no universal affirmative propositions and particular affirmative propositions. We face only universal negative propositions and particular negative propositions such as 'No $S'_{4,9}$ are $S'_{4,10}$', where $S'_{4,9}$ are neighbours for $S_{4,10}$ differing from $S_{4,9}$ and $S'_{4,10}$ are neighbours for $S_{4,9}$ differing from $S_{4,10}$, and 'Some $S'_{4,11}$ are not $S'_{5,11}$', where $S'_{4,11}$ are neighbours for $S_{5,11}$ differing from $S_{4,11}$ and $S'_{5,11}$ are neighbours for $S_{4,11}$ differing from $S_{5,11}$.

In the universe of Fig. 8.6 there is a universal affirmative proposition: 'All $S_{6,5}$ are $S_{6,6}$', and a particular affirmative proposition: 'Some $S_{5,8}$ are $S_{5,9}$'.

We have just shown that we can simulate the plasmodium motion as a concurrent Go game, where (i) black stones are interpreted as attractants and white stones as repellents in the Aristotelian version and (ii) black stones are interpreted as attractants occupied just by *Physarum polycephalum* and white stones as attractants occupied just by *Badhamia utricularis* in the non-Aristotelian version. We can consider configurations of stones as spatial reasoning. If we implement the Aristotelian syllogistic, we need a coalition of two players. If we implement the performative syllogistic [33], we deal with an antagonistic game. While Aristotelian syllogistic may describe concrete directions of *Physarum* spatial expansions, performative syllogistic may describe *Physarum* simultaneous propagations in all directions. Therefore, for the simulation of Aristotelian syllogistic we need repellents to avoid some possibilities in the *Physarum* propagations. Also, we can extend the performative syllogistic up to the case of p-adic valued fuzzy syllogistic, where quantifiers 'All', 'No', 'Some', 'Some... not' are interpreted in the massive-parallel way to describe the plasmodium propagation in different directions. Then we define p-adic probability (fuzzyness) to fix all properties (conditions) for any experiment with *Physarum polycephalum* or *Badhamia utricularis*.

Let us define concurrent games of swarms generally:

Definition 8.2 A (finite) *concurrent game of swarms* is a tuple $\mathscr{G} = \langle States, Agt, Act_n, Mov^n, Tab^n, (\preceq_A)_{A \in Agt}\rangle$, where

- $States$ is a (finite) set of *states* presented by attractants occupied by the swarm;
- $Agt = \{1, \ldots, k\}$ is a finite set of *players* presented by different active zones of swarm (compare to [34]);
- Act_n is a non-empty set of *actions* presented by logic gates or their inductive combinations with n inputs and one output, an element of Act_n^{Agt} is called a *move*;
- $Mov^n : States^n \times Act_n^{Agt} \to 2^{Act} \backslash \{\emptyset\}$ is a mapping indicating the *available* sets of actions to a given player in a given set of states, $n > 0$ is said to be a radius of swarm actions, a move $m^n_{Agt} = (m^n_A)_{A \in Agt}$ is legal at $\langle s_1, \ldots, s_n\rangle$ if $m^n_A \in Mov^n(\mathbf{s}, A)$ for all $A \in Agt$, where $\mathbf{s} = \langle s_1, \ldots, s_n\rangle$;
- $Tab^n : States^n \times Act_n^{Agt} \to States$ is the transition table which associates, with a given set of states and a given move of the players, the set of states resulting from that move;
- for each $A \in Agt$, $\preceq_A$ is a preorder (reflexive and transitive relation) over $States^\omega$, called the *preference relation* of player A, indicating the intensity of attractants; for each $\pi, \pi' \in States^\omega$, by $\pi \preceq_A \pi'$ we denote that π' is *at least as good as* π for A and when it is not $\pi \preceq_A \pi'$, we say that A *prefers* π over π'.

Thus, concurrent games are a natural form of bio-inspired games. They have been just exemplified by concurrent Go games (Aristotelian as well as non-Aristotelian ones).

8.2 Towards Theory of Context-Based Concurrency

Usually, concurrent processes are described in terms of labelled transition systems, in which we have a finite set of atomic actions and all processes are regarded as (co)inductive compositions of these atomic actions. Theoretically, the tools of labelled transition systems are defined as coalgebras [35]. Their entities may be presented as infinite streams [36, 37]. The main idea of labelled transition systems as coalgebras is that we can build up infinite trees of concurrent processes.

However, we can consider simple actions as not atomic, also. In this case, their compositions cannot be inductive and their universe does not satisfy the set-theoretic axiom of foundation [38]. Simple non-atomic actions are called *hybrid*. Their informal meaning is that in one simple action we can suppose the maximum of its modifications. Then we define k-agent transition systems, in which all actions are hybrid. We can define logical operations on hybrid actions and construct compositions of hybrid actions as well as compositions of hybrid actions with atomic ones. As a result, we can define some formal theories on these compositions: group theory (Proposition 8.1) and Boolean algebra (Proposition 8.2). Both formal theories are strong extensions of conventional group theory and Boolean algebra, respectively. They contain members which behaves unusually, such as hyperunit of group theory on hybrid actions that is a multiplicative zero for conventional members and a multiplicative unit for unconventional members (Proposition 8.1).

Hence, any action performed by a swarm can have an additional proto-symbolic value due to a lateral inhibition or lateral activation. But in fact there can be much more different modifications of the same actions. These modifications called hybrid actions are testimonies of *free will of swarms*. Bio-inspired games defined on these actions are said to be *context-based games*.

8.2.1 *Hybrid Simple Actions and Overlapping of Transition Systems*

Concurrency is observed almost at all composite processes. Usually, a concurrent behaviour is considered an interval of possible actions described in terms of a limited group of events. For instance, the concurrent activity of shop assistant and customer may be described by events marked by the members of the set $\{money, good\}$. This activity has the form of recursion:

$$money \rightarrow (good \rightarrow (money \rightarrow (good \rightarrow (money \rightarrow (good \rightarrow (\ldots)))))),$$

where $\rightarrow$ denotes a transition from one event to another. In the explicated form it is a *parallel recursion*:

$$\textbf{shop assistant} = (receiving\ money \rightarrow (giving\ a\ good \rightarrow \textbf{shop assistant}));$$

$$\textbf{customer} = (giving\ money \to (receiving\ a\ good \to \textbf{customer})).$$

In the given example we have two agents whose behaviour is absolutely expected and transparent for each other. In other words, their actions are automatic.

In order to analyze a concurrent behaviour from the point of view of computer science it is convenient to use a *labelled transition system*. Let $States$ be an area of values of this system, consisting of events, Act be a set of labels designating actions due to which we pass from one event to others, $\{\to_a: a \text{ belongs to } Act\}$ be a set of transitions which are defined on events. A transition is to be designated by $s_1 \to_a s_2$, where s_1 and s_2 are contained in $States$. This transition is a motion from a state (event) s_1 to a state (event) s_2 with a fulfilment of action a, states s_1 and s_2 are called a start point and an end point of this transition accordingly, and an action a is called a label of this transition.

For example, let us consider the process of making authorization on documents. A subordinate and manager participate in this process. Subordinate's action consists in the compilation of document (doc) and its preparation for authorization (aut). The manager's action consists in signing the document (sig) or its rejection (rej). The equations defining the alphabet and behaviour of three processes are defined as follows:

$$States = \{document, authorization, signed, rejected\}$$

$$Act = \{doc, aut, sig, rej\}$$

$$P = (document \to_{aut} a|document \to_{aut} b)$$

$$a = (authorization \to_{sig} signed \to_{doc} P|authorization \to_{rej} b)$$

$$b = (authorization \to_{rej} rejected \to_{doc} P|authorization \to_{sig} a)$$

Hence, it seems that any intelligent composite processes (business cycles, management systems, etc.) can be reduced to a labelled transition system. For instance, *Physarum polycephalum* is a medium of biological computations, because it is a kind of *natural labelled transition system*, $\mathscr{S} = \langle States, Acts, Edges\rangle$, where

- $States$ is a set of states presented by attractants occupying by the plasmodium step-by-step;
- $Acts$ is a set of stimuli for the plasmodium, consisting of neighbour attractants and neighbour repellents (different localizations of neighbour attractants and neighbour repellents cause different motions of the plasmodium);
- $Edges \subseteq States \times Act \times States$ is the set of transitions presenting the plasmodium propagation from one state to another in accordance with actions of Act.

And now, whether it is a canonical transition system with the finite set Act that covers all possible transitions, that is the question. If it is so, we can implement abstract automata (such as Kolmogorov-Uspensky automata [39, 40]) in the plasmodium transitions and observe plasmodium motions as spatial implementations of

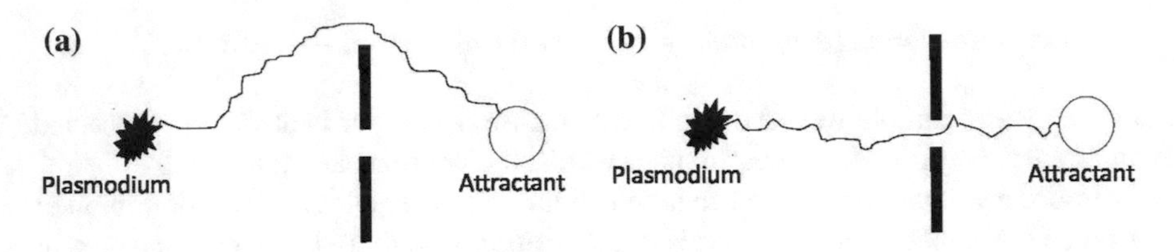

Fig. 8.7 The result of reaching the protoplasmic tubes at the attractant trough the barrier: **a** the plasmodium gets round the barrier; **b** the plasmodium goes through the slit of barrier. For more details on these experiments, see [41]

computable functions satisfying the Church-Turing thesis. If not, we can implement just some approximations of abstract automata.

Let us consider the following thought experiment as counterexample showing that the set *Act* for the plasmodium is infinite in principle. Assume that the transition system for the plasmodium consists just of one action presented by one neighbour attractant. The plasmodium is expected to propagate a protoplasmic tube towards this attractant. Now, let us place a barrier with one slit in front of the plasmodium. Because of this slit, the plasmodium can be propagated according to the shortest distance between two points and in this case the plasmodium does not pay attention on the barrier (Fig. 8.7b). However, sometimes the plasmodium can evaluate the same barrier as a repellent for any case and it gets round the barrier to reach the attractant according to the longest distance (Fig. 8.7a). So, even if the environment conditions change a little bit, the behaviour changes, too. The plasmodium is very sensible to the environment.

Thus, in the transition system with only one stimulus presented by one attractant, a passable barrier can be evaluated as a repellent 'for any case'. Therefore the transition system with only one stimulus and one passable barrier may have the following three simple actions: (i) pass trough (Fig. 8.7b), (ii) avoid from left (Fig. 8.7a), (iii) avoid from right. But in essence, we deal only with one stimulus and, therefore, with one action, although this action has the three modifications defined above.

Simple actions which have modifications depending on the environment are called *hybrid*. The problem is that the set of actions, *Act*, in any labelled transition systems must consist of the so-called *atomic* actions—simple actions that have no modifications. On atomic actions, we can define different operations to construct composite processes. Some examples of operations are as follows: (i) *choice* $(P_1 + P_2)$, when either the process P_1 holds, if an event x, such that $x \to_a P_1$, appears first, or the process P_2 holds, if an event y, such that $y \to_a P_2$, appears first, (ii) *concurrency* $(P_1|P_2)$, when x and y are different events, $x \to_a P_1$ and $y \to_a P_2$ do not have joint states and we obtain a new process $P_1|P_2$, such that in the same point of time actions can be fulfilled by the processes P_1, P_2, so that a new process $P_1|P_2$ can simultaneously commit the following couples of actions:

- for any transition $x_0 \to_a x_1$ from P_1 and for any state y from P_2 the process $P_1|P_2$ contains a transition $\langle x_0, y\rangle \to_a \langle x_1, y\rangle$;
- for any transition $y_0 \to_a y_1$ from P_2 and for any state x from P_1, the process $P_1|P_2$ contains a transition $\langle x, y_0\rangle \to_a \langle x, y_1\rangle$;

- for any couple of transitions $x_0 \rightarrow_a x_1$ from P_1 and $y_0 \rightarrow_{\overline{a}} y_1$ from P_2 with complementary actions a and $\overline{a}$ such that $a = \overline{\overline{a}}$, the process $P_1 | P_2$ contains a transition $\langle x_0, y_0 \rangle \rightarrow_\tau \langle x_1, y_1 \rangle$, where τ is an internal action that synchronizes complementary actions a and $\overline{a}$.

So, atomic and composite actions of any labelled transition system satisfy the set-theoretic axiom of foundation [38]. In other words, we have atomic actions and by induction we can obtain all composite actions. Nevertheless, hybrid actions do not satisfy the foundation axiom. The induction principle of obtaining entities is unapplied for hybrid actions. Their universe is non-well-founded [38].

Let us show that in the life-world we face hybrid actions of human beings almost all the time. First, imagine a vending machine which dispenses items such as snacks to customers automatically, after the customer inserts currency. This machine is a labelled transition system with a finite set *Act* of the following atomic actions: giving items, which were chosen by the customer, after they were paid and giving change in case the price is less, than the coins inserted. Now, assume that we contact not the vending machine, but a saleslady. Her hybrid actions in selling snacks and giving change can go with the following moods of modifications: smiling or not, talking or not, offending or not, being introduced or not, etc. The interval of possible modifications from typical actions to untypical is too large. So, she is not a labelled transition system in its conventional meaning as the vending machine. The main difference from conventional systems is that we have no atomic actions of the saleslady, therefore her set *Act* can be infinite. But it is impossible for conventional systems describing concurrency, *Act* must be always finite and well-founded.

Let us come back to our example of hybrid actions of *Physarum* plasmodia. In Fig. 8.7 we have one transition according to one stimulus. But at the same time, we observe a feature of another transition system consisting of one repellent that is avoided by the plasmodium. This phenomenon, when one transition system transmits some own features to another transition system, is said to be the *overlapping of two transition systems*. In this case we have a domination of one transition system, but with some features of another system. There may be an overlapping of many transition systems. These systems are called overlapped. In the *overlapped labelled transition systems*, actions are hybrid and sets of actions are non-well-founded and, hence, can always be infinite. Processes described in these systems are called *context-based concurrent processes*. Their system if we assume some players is called a *context-based game*. Notice that almost all natural (biological or social) processes are context-based concurrent such as swarm behaviours or human behaviours.

There are more actions in an overlapped transition system, more hybrid they are. All processes of the saleslady are context-based, because there can be many other transition systems such as joking, talking, being introduced, etc. which can be overlapped with selling. In human or swarm actions, we always can find an overlapping.

In the next section, we show, how we can define overlapped labelled transition systems and their hybrid actions formally.

8.2.2 Formal Logical Language for the Context-Based Concurrency

In order to know, how we can build up non-inductive compositions of hybrid actions, let us try first to define overlapped compositions of propositional formulas, i.e. let us try to present propositional formulas as hybrid entities. Any hybrid entity will be denoted by $()a, b, \ldots \{()$, where $a, b, \ldots$ are its modifications. Assume, a propositional formula is $()\varphi, \psi\{()$, where $\varphi = (\varphi \Rightarrow \psi)$ and $\psi = (\psi \Rightarrow \varphi)$ are defined recursively. In other words, we have an infinite formula $(((((\ldots \Rightarrow \psi) \Rightarrow \psi) \Rightarrow \psi) \Rightarrow \psi) \Rightarrow \psi)$ and $(((((\ldots \Rightarrow \varphi) \Rightarrow \varphi) \Rightarrow \varphi) \Rightarrow \varphi) \Rightarrow \varphi)$ and both are mutually defined. The composite formula $()\varphi, \psi\{()$ cannot be presented by linear words even if they are infinite. It is non-linear in principle. We do not know if it is either the first infinite implication $\varphi = (\varphi \Rightarrow \psi)$ or the second $\psi = (\psi \Rightarrow \varphi)$.

Now, let us consider the non-well-founded formula $()\varphi, \psi\{()$ as transition rule of a transition system. Suppose that agent A_1 moves from the state s_1^t to the state s_1^{t+1}, at time $t = 0, 1, \ldots$, and agent A_2 moves from the state s_2^t to the state s_2^{t+1}, at time $t = 0, 1, \ldots$, by the following transition rules: $s_1^{t+1} = (s_1^t \Rightarrow s_2^t)$ and $s_2^{t+1} = (s_2^t \Rightarrow s_1^t)$. Then we obtain the following infinite streams: $\langle s_1^0 \Rightarrow s_2^0, s_1^1 \Rightarrow s_2^1, s_1^2 \Rightarrow s_2^2, \ldots\rangle$ and $\langle s_2^0 \Rightarrow s_1^0, s_2^1 \Rightarrow s_1^1, s_2^2 \Rightarrow s_1^2, \ldots\rangle$. Let us pay attention that the stream $\langle s_1^0 \Rightarrow s_2^0, s_1^1 \Rightarrow s_2^1, s_1^2 \Rightarrow s_2^2, \ldots\rangle$ (resp. $\langle s_2^0 \Rightarrow s_1^0, s_2^1 \Rightarrow s_1^1, s_2^2 \Rightarrow s_1^2, \ldots\rangle$) may be understood as an infinite propositional formula $(((s_1^0 \Rightarrow s_2^0) \Rightarrow s_2^1) \Rightarrow s_2^2) \Rightarrow \ldots$ (resp. $(((s_2^0 \Rightarrow s_1^0) \Rightarrow s_1^1) \Rightarrow s_1^2) \Rightarrow \ldots$). Both formulas are mutually depended and they cannot be presented as linear sequence (inductive composition).

Assume, our hybrid formula is $()\varphi_1, \varphi_2, \ldots, \varphi_n\{()$. It means that formulas $\varphi_1, \varphi_2, \ldots, \varphi_n$ are infinite and mutually depended. Then we say that the hybrid formula is of radius $n > 1$. We suppose that in reality it can behave as one of its n modifications, but we do not know, how precisely. Let $(Act_{t,n})_{t\in\mathbf{N}}$ denote the set of all simple actions defined as hybrid propositional formulas of radius $n > 1$ and used in some transitions. We can define an overlapped transition system closed over these actions as the following extension of multi-agent transition systems (called also concurrent games [42–45]):

Definition 8.3 A (finite) *k-agent overlapped transition system* or a (finite) *k-agent context-based game* is a tuple $\mathcal{G} = \langle (States_t)_{t\in\mathbf{N}}, Agt, (Act_{t,n})_{t,n\in\mathbf{N}}, (Mov_t^n)_{t,n\in\mathbf{N}}, (Tab_t^n)_{t,n\in\mathbf{N}}\rangle$, where

- $States_t$ is a (finite) set of *states* at time $t = 0, 1, 2, \ldots$;
- $Agt = \{1, \ldots, k\}$ is a finite set of *players*;
- $Act_{t,n}$ is a non-empty set of *hybrid actions* with radius n at $t = 0, 1, 2, \ldots$, an element of $Act_{t,n}^{Agt}$ is called a *move* at time $t = 0, 1, 2, \ldots$;
- $Mov_t^n : States_t^n \times Act_{t,n}^{Agt} \to 2^{Act}\backslash\{\emptyset\}$ is a mapping indicating the *available* sets of actions to a given player in a given set of states, $n > 0$ is said to be a radius of hybrid actions, a move $m_{Agt}^n = (m_A^n)_{A\in Agt}$ is legal at $\langle s_1, \ldots, s_n\rangle$ if $m_A^n \in Mov^n(\mathbf{s}, A)$ for all $A \in Agt$, where $\mathbf{s} = \langle s_1, \ldots, s_n\rangle$;

- $Tab_t^n : States_t^n \times Act_{t,n}^{Agt} \to States_{t+1}^n$ is the transition table which associates, with a given set of states at t and a given move of the players at t, the set of states at $t+1$ resulting from that move.

The difference of $\mathcal{G}$ from concurrent games is that in concurrent games (i) we do not use time; (ii) the set of actions is always finite and consists of atomic actions.

Let us pay attention that actions of $(Act_{t,n})_{t\in\mathbf{N}}$ are simple, that is, not composite. However, they cannot be considered atomic actions in the way of conventional labelled transition systems.

For defining compositions of simple actions of $(Act_{t,n})_{t\in\mathbf{N}}$, let us construct the unconventional logic $\mathcal{L}$ for hybrid formulas. Let $\{p, q, r, \ldots\}$ be the set of all propositional variables. Then formulas are defined as follows: (1) if $\varphi, \psi \in \{p, q, r, \ldots\}$, then $\varphi\psi\star$, where $\star \in \{D, C, I\}$, i.e. $\star$ is disjunction, conjunction or implication respectively, is a formula; (2) by using D, C, I, we can construct finite or infinite logical combinations of formulas $\varphi_0\psi_0\star_0, \varphi_1\psi_1\star_1, \varphi_2\psi_2\star_2, \ldots$, where $\star_0, \star_1, \star_2 \ldots \in \{D, C, I\}$; (3) if φ is a finite formula and ψ is an infinite formula, then $\varphi\psi\star$, where $\star \in \{D, C, I\}$, is a formula. For instance, according to this definition $pqCrI$ is a well-formed formula that is $(p \wedge q) \Rightarrow r$ in the traditional notation. Also, this definition allows us to construct streams such as $p_0q_0Iq_1Iq_2Iq_3I\ldots$, i.e. infinite propositional formulas and logically combine different finite and infinite formulas. Let φ be a formula. Then $\varphi NN\ldots$, where N is negation which occurs finitely or infinitely many times, is a formula. Traditionally it is written thus: $\ldots\neg(\neg\varphi)$. The class of all finite or infinite strings is a maximal set of formulas closed under logical operations by two definitions above (for both D, C, I and N). That is, this set is defined non-inductively, although it contains all conventional propositional formulas that are defined inductively.

Hybrid formulas are sets of streams which have joint propositional variables. For example, the set consisting just of $p_0q_0Iq_1Iq_2Iq_3I\ldots$ and $q_0p_0Ip_1Ip_2Ip_3I\ldots$ is a hybrid formula. Let A_1, A_2 be hybrid propositional formulas and φ be a finite or infinite propositional formula. Then we can define their logical compositions as follows: (1) A_1A_2D is a union of A_1 and A_2; A_1A_2C is an intersection of A_1 and A_2 if they have joint streams and A_1A_2C is their union otherwise; A_1A_2I is a union of A_1N and A_2, where A_1N contains complements for each streams of A_1; (2) assume that φ is infinite, then φA_1D is a union of $\{\varphi\}$ and A_1; φA_1C is φ if φ belongs to A_1 and it is a union $\{\varphi\}$ and A_1 otherwise; $\varphi A_1I := \varphi NA_1D$; $A_1\varphi I := A_1N\varphi D$; (3) assume φ is finite and A_1 contains streams $\psi_1, \ldots, \psi_n$ $(n > 0)$ that begin with φ, then φA_1D is a union of $\{\varphi\psi_1D\}, \ldots, \{\varphi\psi_nD\}$, and A_1; φA_1C is a union of $\{\varphi\psi_1C\}, \ldots, \{\varphi\psi_nC\}$, and A_1; $\varphi A_1I := \varphi NA_1D$; $A_1\varphi I := A_1N\varphi D$.

In the logic $\mathcal{L}$, we can construct logical compositions of overlapped transition systems $\mathcal{G}_1, \mathcal{G}_2, \ldots, \mathcal{G}_i, \ldots$, where $\mathcal{G}_i = \langle (States_{t,i})_{t\in\mathbf{N}}, Agt_i, (Act_{t,n_i})_{t\in\mathbf{N}}, (Mov_t^{n_i})_{t\in\mathbf{N}}, (Tab_t^{n_i})_{t\in\mathbf{N}}\rangle$ for $i \geq 1$ as follows:

$$\mathcal{G}_1\mathcal{G}_2 \star \ldots \star \mathcal{G}_i \star \ldots,$$

where $\star \in \{D, C, I\}$, is an overlapped transition system $\mathcal{G} = \langle (States_t)_{t\in\mathbf{N}}, Agt, (Act_t)_{t\in\mathbf{N}}, (Mov_t)_{t\in\mathbf{N}}, (Tab_t)_{t\in\mathbf{N}} \rangle$, where

- $(States_t)_{t\in\mathbf{N}}$ is a union $\bigcup_i (States_{t,i})_{t\in\mathbf{N}}$;
- Agt is a union $\bigcup_i Agt_i$;
- $(Act_t)_{t\in\mathbf{N}}$ consists of actions $\alpha_1\alpha_2 \star \ldots \star \alpha_i \star \ldots$, where $\star \in \{D, C, I\}$, such that $\alpha_i \in (Act_{t,n_i})_{t\in\mathbf{N}}$ for $i \geq 1$;
- $(Mov_t)_{t\in\mathbf{N}}\colon (States_t)_{t\in\mathbf{N}} \times (Act_t)^{Agt}_{t\in\mathbf{N}} \to 2^{Act}\backslash\{\emptyset\}$;
- $(Tab_t)_{t\in\mathbf{N}}\colon (States_t)_{t\in\mathbf{N}} \times (Act_t)^{Agt}_{t\in\mathbf{N}} \to (States_t)_{t\in\mathbf{N}}$.

In this definition, we assume that the modifications of an action of $(Act_t)_{t\in\mathbf{N}}$ at time t would not possibly propagate to other atomic actions at time $t' > t$. In other words, we assume that at any time t, the radius n of propagations to other actions is constant, i.e. for any hybrid formula the radius n does not change for any $t = 0, 1, 2, \ldots$ It is an idealization to consider simple cases of overlapping.

Thus, using the language $\mathcal{L}$, we can define different formal theories on overlapped transitions systems. As an example, let us consider group theory extended by hybrid terms.

8.2.3 Context-Based Concurrent Group Theory

Assume $\{a, b, \ldots, c\}$ is a set of states of $States_t$. Let us define infinite terms as streams of states and hybrid terms as local transition functions. Let $\cdot$ be a two-place function. Terms are defined thus: (1) if $x, y \in \{a, b, \ldots, c\} \cup \{e\}$, then $x \cdot y$ is a term; (2) by using $\cdot$, we can construct finite or infinite combinations of terms $x_0 \cdot y_0$, $x_1 \cdot y_1, x_2 \cdot y_2, \ldots$; (3) if $x \in \{a, b, \ldots, c\} \cup \{e\}$ and y is a term, then $x \cdot y$ is a term. By this definition, each finite or infinite string of the form $x_0 \cdot x_1 \cdot x_2 \cdot \ldots$, where $x_i \in \{a, b, \ldots, c\}$, is a term and we can combine finite and infinite terms. Let $^{-1}$ be a one-place function. Then $(x_0 \cdot x_1 \cdot x_2 \cdot \ldots)^{-1}$, where $x_0 \cdot x_1 \cdot x_2 \cdot \ldots$ is a finite or infinite term, is a term, too, it is denoted by $x_0^{-1} \cdot x_1^{-1} \cdot x_2^{-1} \cdot \ldots$. Atomic formulas are defined thus: if x, y are terms, then $x = y$ is an atomic formula. Notice that this new language includes conventional language of group theory.

Now we can define hybrid terms. Let us consider an overlapped labelled transition system, where $n = 2$ is the radius. So, we have two infinite streams of states. Let a^0 be an initial state of the first stream and b^0 is an initial state of the second stream. Assume that two mutually dependent streams which we have are as follows: $\langle a^0 \cdot b^0, a^1 \cdot b^1, a^2 \cdot b^2, \ldots \rangle$, where $a^{t+1} = (a^t \cdot b^t)$; $\langle (a^{-1})^0 \cdot (b^{-1})^0, (a^{-1})^1 \cdot (b^{-1})^1, (a^{-1})^2 \cdot (b^{-1})^2, \ldots \rangle$, where $b^{t+1} = ((a^{-1})^t \cdot (b^{-1})^t)$. Notice that this local transition rule is a hybrid term consisting just of two streams $a^0 \cdot b^0 \cdot b^1 \cdot b^2 \cdot \ldots$ and $(a^{-1})^0 \cdot (b^{-1})^0 \cdot (b^{-1})^1 \cdot (b^{-1})^2 \cdot \ldots$ Both are infinite terms. If the given hybrid term contains all possible streams, i.e. streams for $n = \omega$, it is called *hyperunit* $\overline{e}$. It is a hybrid term that contains all streams which are dual to each other, i.e. for any stream $a \cdot b \cdot c \cdot \ldots$ it is true to say that $\overline{e}$ contains both $a \cdot b \cdot c \cdot \ldots$ and $a^{-1} \cdot b^{-1} \cdot c^{-1} \cdot \ldots$ It is an example of hybrid terms.

Let A_1, A_2 be hybrid terms and x be a finite or infinite term. Then we can define their multiplications as follows: (1) $A_1 \cdot A_2$ is a union of A_1 and A_2 if A_1 and A_2 do not include at least two joint streams and $A_1 \cdot A_2$ is an intersection of A_1 and A_2 if both include at least two joint streams; A_1^{-1} is a hybrid term which contains duals x^{-1} for each streams x of A_1; (2) assume that x is infinite, then $x \cdot A_1$ is a union of $\{x\}$ and A_1; (3) assume, x is finite and A_1 contains streams $y_1, \ldots, y_n$ $(n > 0)$ that begin with x or x^{-1}, then $x \cdot A_1$ is a union of $\{x \cdot y_1\}, \ldots, \{x \cdot y_n\}$, and A_1; assume, A_1 does not contain streams $y_1, \ldots, y_n$ $(n > 0)$ that begin with x or x^{-1}, then $x \cdot A_1$ is a union of $\{x \cdot x \cdot x \cdot x \cdot \ldots\}$ and A_1. This definition of complex terms allows hyperunit to annihilate any member a of conventional group: $a \cdot \overline{e} = \overline{e}$. Other properties: $\overline{e}^{-1} \cdot \overline{e} = \overline{e}$; $A \cdot \overline{e} = A$ for any hybrid term A; $y \cdot \overline{e} = \overline{e}$ for any infinite term y.

A hybrid term in group theory is denoted by $\overline{c}$. It may have the following notation:

$$\left(\right\}\begin{matrix} a^i \cdot b^j \cdot a^i \cdot b^j \cdot \ldots; & b^j \cdot a^i \cdot b^j \cdot a^i \cdot \ldots; & \ldots \\ a^{-i} \cdot b^{-j} \cdot a^{-i} \cdot b^{-j} \cdot \ldots; & b^{-j} \cdot a^{-i} \cdot b^{-j} \cdot a^{-i} \cdot \ldots; & \ldots \end{matrix}\left\{\right)$$

where $i, j = \pm 1$, $a^i \cdot b^j \cdot a^i \cdot b^j \cdot \ldots$ and $b^j \cdot a^i \cdot b^j \cdot a^i \cdot \ldots$, etc. are streams.

The theory closed over logical operations and group-theoretic operations defined above is called *context-based concurrent group theory*.

Proposition 8.1 *Let us assume that $\mathcal{G}$ is a group theory implemented in overlapped labelled transition systems and $a, b, c \in \mathcal{G}$. Then in an appropriate context-based concurrent group theory $\mathcal{G}'$ built as an extension of $\mathcal{G}$, the following properties hold true:*

1. *Let $\overline{b} = (\}b \cdot b \cdot a \cdot b \cdot b \cdot a \ldots; a \cdot a \cdot b \cdot a \cdot a \cdot b \ldots; \ldots\{)$. Then $\overline{b} \cdot a = a \cdot \overline{b} = (\}b \cdot b \cdot a \cdot b \cdot b \cdot a \ldots; a \cdot a \cdot a \cdot b \cdot a \cdot a \cdot b \ldots; \ldots\{)$. From this it follows that if $\overline{a} = (\}a \cdot a \cdot a \cdot a \cdot a \cdot a \ldots; \ldots\{)$ and $a \cdot a \cdot a \cdot a \cdot a \cdot a \ldots$ is the only stream of $\overline{a}$ that begins with a or a^{-1}, then*

$$\overline{a} \cdot a = a \cdot \overline{a} = \overline{a}.$$

2. *The rule of associativity holds true:*

$$(a \cdot b) \cdot c = a \cdot (b \cdot c),$$

where a, b, c are members of $\mathcal{G}$, and for its extension to the non-linear one:

$$(\overline{a} \cdot \overline{b}) \cdot \overline{c} = \overline{a} \cdot (\overline{b} \cdot \overline{c}),$$

where all $\overline{a}$, $\overline{b}$, $\overline{c}$ simultaneously have (do not have) at least two joint streams.

3. *In $\mathcal{G}$ there exists a unit e, i.e. the member satisfying the property:*

$$e \cdot a = a \cdot e = a \quad for \ any \ a \in \mathcal{G}.$$

We can obtain a hyperunit $\overline{e} \in \mathcal{G}'$ *defined thus:* $()a^{-1} \cdot a^{-1} \cdot a^{-1} \cdot a^{-1} \cdot \ldots; a \cdot a \cdot a \cdot a \cdot \ldots; \ldots()$, *where* $a^{-1} \cdot a^{-1} \cdot a^{-1} \cdot a^{-1} \cdot \ldots$ *and* $a \cdot a \cdot a \cdot a \cdot \ldots$ *are streams. The streams* $a^{-1} \cdot a^{-1} \cdot a^{-1} \cdot a^{-1} \cdot \ldots$ *and* $a \cdot a \cdot a \cdot a \cdot \ldots$ *are dual. The hyperunit satisfies two properties:*

$$\overline{e} \cdot \overline{a} = \overline{a} \cdot \overline{e} = \overline{a} \;\; for \;\; any \;\; \overline{a} \in \mathcal{G}',$$

$$\overline{e} \cdot a = a \cdot \overline{e} = \overline{e} \;\; for \;\; any \;\; a \in \mathcal{G}.$$

4. *For any* $a \in \mathcal{G}$ *there exists the inverse element* $a^{-1} \in \mathcal{G}$, *i.e. the member satisfying the property:*

$$a^{-1} \cdot a = a \cdot a^{-1} = e.$$

Also, for any $\overline{a} \in \mathcal{G}'$ *there exists a hybrid term:*

$$\overline{a^{-1}} = ()a^{-1} \cdot a^{-1} \cdot a^{-1} \cdot a^{-1} \cdot \ldots; \ldots(),$$

where $a^{-1} \cdot a^{-1} \cdot a^{-1} \cdot a^{-1} \cdot \ldots$ *is the only stream of* $\overline{a^{-1}}$ *that begins with* a *or* a^{-1}. *This hypermember satisfies the following two properties:*

$$\overline{a^{-1}} \cdot a = a \cdot \overline{a^{-1}} = \overline{a^{-1}}, \qquad \overline{a^{-1}} \cdot \overline{a} = \overline{a} \cdot \overline{a^{-1}} \subseteq \overline{e},$$

where $\overline{a}$ *contains just streams that are dual for streams of* $\overline{a^{-1}}$, *e.g.* $\overline{a} = ()a \cdot a \cdot a \cdot a \cdot \ldots; \ldots()$ *and* $a \cdot a \cdot a \cdot a \cdot \ldots$ *is a stream.* □

To sum up, the context-based concurrent group theory $\mathcal{G}'$ is a strong extension of conventional group theory $\mathcal{G}$. $\mathcal{G}'$ contains some members with unusual properties such as hyperunit $\overline{e}$ that behaves as zero in multiplicative groups for finite or infinite terms and behaves as unit in multiplicative groups for hybrid terms (Proposition 8.1(3)). Infinite logical formulas of $\mathcal{G}'$ specify teleological goals for each hybrid action based on the group operations $\cdot$, $^{-1}$. Therewith $\overline{e}$ is the greatest hybrid action based on the group operations $\cdot$, $^{-1}$, i.e. it contains the largest number of modifications.

The group theory $\mathcal{G}'$ can be implemented into the swarm (e.g. slime mould) behaviour as follows. Each atomic $a \in \mathcal{G}$ is a transition of swarm to the north. Each atomic $a^{-1} \in \mathcal{G}$ is a transition of swarm to the south. Each $a \cdot b$ is a composition of transition a and then transition b. For example, the unit $e = a \cdot a^{-1}$ means that there are no transitions of swarm. Hybrid actions of radius $n > 1$ mean that there are n linear directions of swarm transitions which can be mixed at different time $t = 0, 1, 2, \ldots$ For instance, if $n = 2$ we have the following two directions: (i) either to the north or to the south; (ii) either to the east or to the west. The hyperunit $\overline{e}$ of radius 2 means the total transitions in any directions at any time.

8.2.4 *Context-Based Concurrent Boolean Algebra*

Let us assume again that $\{a, b, \ldots, c\}$ is a set of states of $States_t$. Let $\cap$, $\cup$ be two-place functions and $^{-1}$ be one-place function of a Boolean algebra $\mathscr{A}$, which are defined on $States_t$. They can be extended to hybrid terms in the way of the previous section. Denote a *context-based concurrent Boolean algebra* that we are going to define by $\mathscr{A}'$. Assume, in $\mathscr{A}'$ the following operations are defined for any $a, b \in \mathscr{A}$:

$$\overline{a} \cap b = \overline{c},\ \overline{a} \cup b = \overline{c},$$
$$a \cap \overline{b} = \overline{c},\ a \cup \overline{b} = \overline{c},$$
$$\overline{a} \cap \overline{b} = \overline{c},\ \overline{a} \cup \overline{b} = \overline{c},$$

where $\overline{a}, \overline{b}, \overline{c}$ designate hybrid terms such that the hybrid term $\overline{a}$ contains a stream that begins with a, the hybrid term $\overline{b}$ contains a stream that begins with b, the hybrid term $\overline{c}$ contains a stream that begins with c. These hybrid terms are called *hypermembers*. Sometimes, the hypermember $\overline{c}$ may be denoted by $\overline{a \cap b}$ or $\overline{a \cup b}$. It may have the following meaning:

$$\left(\left\}\begin{matrix} a^i \cap b^j \cap a^i \cap b^j \cap \ldots; & b^j \cap a^i \cap b^j \cap a^i \cap \ldots; & \ldots \\ a^{-i} \cap b^{-j} \cap a^{-i} \cap b^{-j} \cap \ldots; & b^{-j} \cap a^{-i} \cap b^{-j} \cap a^{-i} \cap \ldots; & \ldots \end{matrix}\right\{\right)$$

$$or \left(\left\}\begin{matrix} a^i \cup b^j \cup a^i \cup b^j \cup \ldots; & b^j \cup a^i \cup b^j \cup a^i \cup \ldots; & \ldots \\ a^{-i} \cup b^{-j} \cup a^{-i} \cup b^{-j} \cup \ldots; & b^{-j} \cup a^{-i} \cup b^{-j} \cup a^{-i} \cup \ldots; & \ldots \end{matrix}\right\{\right)$$

where $i, j = \pm 1$, $a^i \cap b^j \cap a^i \cap b^j \cap \ldots$ and $b^j \cap a^i \cap b^j \cap a^i \cap \ldots$, etc. are streams. Suppose that all streams belonging to hybrid terms are cyclic, i.e. they are of the form either infinite conjunction or infinite disjunction. The extensions of both linear operations ($\cap$, $\cup$) are defined as follows:

- formulas $\overline{b} \cap a = \overline{c}$, $a \cap \overline{b} = \overline{c}$ (respectively, $\overline{b} \cup a = \overline{c}$, $a \cup \overline{b} = \overline{c}$) mean that for any $a \in \mathscr{A}$ and a hybrid term $\overline{b}$, the member a interacts with all streams of $\overline{b}$ which (i) begin with a or a^{-1} and (ii) are of the form of infinite conjunction (respectively, infinite disjunction), otherwise we add a new stream $a \cap a \cap a \cap a \cap a \ldots$ (respectively, $a \cup a \cup a \cup a \cup a \cup a \ldots$) into the hybrid term $\overline{b}$. For instance, let $\overline{b} = (\}(b \cap b) \cup a \cup (b \cap b) \cup a \cup \ldots; (a \cup a \cup b) \cap (a \cup a \cup b) \cap \ldots; \ldots\{)$. Then $\overline{b} \cap a = a \cap \overline{b} = (\}(b \cap b) \cup a \cup (b \cap b) \cup a \cup \ldots; a \cap (a \cup a \cup b) \cap (a \cup a \cup b) \cap \ldots; \ldots\{)$. From this it follows that if $\overline{a} = (\}a \cap a \cap a \cap a \cap a \cap \ldots; \ldots\{)$ and $a \cap a \cap a \cap a \cap a \cap a \cap \ldots$ is the only stream of $\overline{a}$ that begins with a or a^{-1}, then
$$\overline{a} \cap a = a \cap \overline{a} = \overline{a}.$$
- formula $\overline{a} \cup \overline{b} = \overline{c}$ means that $\overline{c}$ contains all streams belonging to both $\overline{a}$ and $\overline{b}$.
- formula $\overline{a} \cap \overline{b} = \overline{c}$ means that (1) if $\overline{a}$ and $\overline{b}$ have at least two joint streams, then $\overline{c} = \overline{a} \cap \overline{b}$, i.e. it contains these joint streams; (2) if $\overline{a}$ and $\overline{b}$ do not have at least two joint streams, then $\overline{c} = \overline{a} \cup \overline{b}$.

For any conventional terms $a, b \in \mathscr{A}$, we define: $a \leq b$ iff $a \cap b = a$ or equivalently $a \leq b$ iff $a \cup b = b$. We can extend this ordering relation to hybrid terms of $\mathscr{A}'$ by the following condition: $\overline{a} \leq \overline{b}$ iff $\overline{a} \cap \overline{b} = \overline{a}$ or equivalently $\overline{a} \leq \overline{b}$ iff $\overline{a} \cup \overline{b} = \overline{b}$, $\overline{a} \leq b$ iff $\overline{a} \cap b = \overline{a}$, $a \leq \overline{b}$ iff $a \cup \overline{b} = \overline{b}$.

Proposition 8.2 *Let $\mathscr{A}'$ be a context-based concurrent Boolean algebra extended from a conventional Boolean algebra $\mathscr{A}$. Then the following properties hold true:*

1. *Associativity:*

$$(a \cap b) \cap c = a \cap (b \cap c); \quad (a \cup b) \cup c = a \cup (b \cup c),$$

where a, b, c are conventional terms of $\mathscr{A}$;

$$(\overline{a} \cap \overline{b}) \cap \overline{c} = \overline{a} \cap (\overline{b} \cap \overline{c}),$$

where all $\overline{a}$, $\overline{b}$, $\overline{c}$ simultaneously have (do not have) at least two joint streams, and

$$(\overline{a} \cup \overline{b}) \cup \overline{c} = \overline{a} \cup (\overline{b} \cup \overline{c})$$

for any hybrid terms $\overline{a}$, $\overline{b}$, $\overline{c}$.

2. *Commutativity:*

$$(a \cap b) = (b \cap a); \ (\overline{a} \cap b) = (b \cap \overline{a}); \ (\overline{a} \cap \overline{b}) = (\overline{b} \cap \overline{a});$$
$$(a \cup b) = (b \cup a); \ (\overline{a} \cup b) = (b \cup \overline{a}); \ (\overline{a} \cup \overline{b}) = (\overline{b} \cup \overline{a}).$$

3. *Idempotence:*

$$(a \cap a) = a; \quad (\overline{a} \cap \overline{a}) = \overline{a}; \quad (a \cup a) = a; \quad (\overline{a} \cup \overline{a}) = \overline{a}.$$

4. *Absorption:*

$$\begin{array}{ll} (a \cap b) \cup a = a; & (\overline{a} \cap \overline{b}) \cup \overline{a} \geq \overline{a}; \\ (a \cup b) \cap a = a; & (\overline{a} \cup \overline{b}) \cap \overline{a} = \overline{a}; \\ (a \cap \overline{b}) \cup a \neq a; & (a \cup \overline{b}) \cap a \neq a; \\ (\overline{a} \cap b) \cup \overline{a} \geq \overline{a}; & (\overline{a} \cup b) \cap \overline{a} = \overline{c} \leq \overline{a}, \end{array}$$

where $\overline{c}$ consists of all streams of $\overline{a}$ without streams beginning with b.

5. *Distributivity:*

$$\begin{array}{ll} a \cup (b \cap c) = (a \cup b) \cap (a \cup c); & a \cap (b \cup c) = (a \cap b) \cup (a \cap c); \\ a \cup (\overline{b} \cap c) = (a \cup \overline{b}) \cap c; & a \cap (\overline{b} \cup c) = (a \cap \overline{b}) \cup c; \\ a \cap (b \cup \overline{c}) = (a \cap \overline{c}) \cup b; & a \cup (b \cap \overline{c}) = (a \cup \overline{c}) \cap b; \\ a \cap (\overline{b} \cup \overline{c}) \leq (a \cap \overline{b}) \cup (a \cap \overline{c}); & a \cup (\overline{b} \cap \overline{c}) \geq (a \cup \overline{b}) \cap (a \cup \overline{c}); \\ \overline{a} \cap (\overline{b} \cup c) \leq (\overline{a} \cap \overline{b}) \cup (\overline{a} \cap c); & \overline{a} \cup (\overline{b} \cap c) \geq (\overline{a} \cup \overline{b}) \cap (\overline{a} \cup c); \\ \overline{a} \cap (\overline{b} \cup \overline{c}) \leq (\overline{a} \cap \overline{b}) \cup (\overline{a} \cap \overline{c}); & \overline{a} \cup (\overline{b} \cap \overline{c}) \geq (\overline{a} \cup \overline{b}) \cap (\overline{a} \cup \overline{c}). \end{array}$$

6. *There is a conventional term that behaves as unit e, i.e. the member satisfying the properties: (i) identity:* $e \cap a = a$ *for any conventional term a; (ii) annihilator:* $e \cup a = e$ *for any conventional term a.*
7. *There is a hybrid term that behaves as hyperunit* $\overline{e}$*:*

$$\left(\left\{ \begin{matrix} a^{-1} \cup a^{-1} \cup a^{-1} \cup a^{-1} \cup \ldots; \; a \cup a \cup a \cup a \cup \ldots; \; \ldots \\ a^{-1} \cap a^{-1} \cap a^{-1} \cap a^{-1} \cap \ldots; \; a \cap a \cap a \cap a \cap \ldots; \; \ldots \end{matrix} \right\}\right)$$

where $a^{-1} \cup a^{-1} \cup a^{-1} \cup a^{-1} \cup \ldots$ *and* $a \cup a \cup a \cup a \cup \ldots$*, etc. are streams and the radius of* $\overline{e}$ *is equal* $n = \omega$*. The streams* $a^{-1} \cap a^{-1} \cap a^{-1} \cap a^{-1} \cap \ldots$ *and* $a \cup a \cup a \cup a \cup \ldots$ *are called* $\cup$*-dual, the streams* $a^{-1} \cup a^{-1} \cup a^{-1} \cup a^{-1} \cup \ldots$ *and* $a \cap a \cap a \cap a \cap \ldots$ *are called* $\cap$*-dual. The hyperunit satisfies properties: (i) identity:* $\overline{e} \cap \overline{a} = \overline{a}$ *for any hybrid term* $\overline{a}$*; (ii) annihilator:* $\overline{e} \cap a \leq \overline{e}$ *and* $\overline{e} \cup a \leq \overline{e}$ *for any conventional term a, and* $\overline{e} \cup \overline{a} = \overline{e}$ *for any hybrid term* $\overline{a}$*.*
8. *There is a conventional term that behaves as zero o, i.e. the member satisfying the properties: (i) identity:* $o \cup a = a$ *for any conventional term a; (ii) annihilator:* $o \cap a = o$ *for any conventional term a.*
9. *For any conventional term a, there exists the complement* a^{-1}*, i.e. the member satisfying properties for conventional terms* a, b*:*

   - *complementation:* $a \cup a^{-1} = e$*;* $a \cap a^{-1} = o$*;*
   - *double negation:* $(a^{-1})^{-1} = a$*;*
   - *de Morgan:* $(a \cup b)^{-1} = (a^{-1} \cap b^{-1})$*;* $(a \cap b)^{-1} = (a^{-1} \cup b^{-1})$*.*

10. *For any hybrid term* $\overline{a}$ *with the radius* $n = \omega$*, there exists the complement* $\overline{a^{-1}}$*, satisfying properties:*

$$\overline{a} \cup \overline{a^{-1}} = \overline{e}; \qquad \overline{a} \cap \overline{a^{-1}} = \overline{e};$$

for any hybrid term $\overline{a}$ *with the radius* $n < \omega$*, there exists the complement* $\overline{a^{-1}}$*, satisfying properties:*

$$\overline{a} \cup \overline{a^{-1}} \leq \overline{e}; \qquad \overline{a} \cap \overline{a^{-1}} \leq \overline{e};$$

$$(\overline{a}^{-1})^{-1} = (\overline{a^{-1}})^{-1} = \overline{(a^{-1})^{-1}} = \overline{a},$$

where $\overline{a}$ *contains just streams that are* $\cap$*- and* $\cup$*-dual for streams of* $\overline{a^{-1}}$*;*

$$(a \cup \overline{b})^{-1} = (a^{-1} \cap \overline{b^{-1}}); \qquad (a \cap \overline{b})^{-1} = (a^{-1} \cup \overline{b^{-1}});$$

$$(\overline{a} \cup \overline{b})^{-1} \geq (\overline{a^{-1}} \cap \overline{b^{-1}}); \qquad (\overline{a} \cap \overline{b})^{-1} \leq (\overline{a^{-1}} \cup \overline{b^{-1}}).$$

□

As we see, the context-based concurrent Boolean algebra $\mathscr{A}'$ possesses many unusual properties caused by containing hybrid terms.

The informal meaning of hybrid actions (e.g. hybrid terms or hybrid formulas) is that any hybrid action is defined just on streams and we cannot say in accordance with which stream the hybrid action will be embodied in the given environment. It can behave like any stream it contains but there is an uncertainty how exactly. The hybrid action looks like the quantum bit (or *qubit* for short). Let us remember that a classical bit can be 0 or 1. The qubit exists in various superpositions of 0 and 1 at the same time. There is an uncertainty of either 0 or 1, but qubit behaves like both 0 and 1. In other words, each quantum state is described in terms of classical states, associating two numbers with each: an absolute amplitude (between 0 and 1) which when squared gives you the probability of obtaining that classical state when you measure the system, and a complex phase which governs interference effects. The classical states 0 and 1 are the top point and bottom point of the sphere, while the other states are in superposition. The points around the equator represent states where there is an equal probability of measuring 0 or 1. As a result, we can represent the state of a qubit as simply the point on the surface of a sphere, known as a *Bloch sphere*, see [46, 47]. As a result, no matter what two points you pick, it is impossible to determine either of them by making a measurement on such a system. So if the user sends such randomized states to the server, together with measurement angles, there is nothing to be learned by any measurement of the system. Therefore everything is perfectly hidden from a malicious server. In the same way any hybrid action is a computation which is hidden from outside observers. The hybrid action includes as many versions of behaviour of the whole system as possible.

A saleslady selling some items, plasmodium behaviours (Fig. 8.7), and the Bloch sphere are examples of hybrid actions which do not satisfy the foundation axiom [38] and cannot be reduced to atomic actions. But in our opinion, using hybrid actions, we can construct some interesting extensions of formal theories such as context-based concurrent group theory (Proposition 8.1) and context-based concurrent Boolean algebra (Proposition 8.2). Both theories are defined on some concurrent processes, which cannot be obtained as inductive compositions of atomic actions.

Hence, *the bio-inspired game theory is based on context-based (or hybrid) actions assuming an unlimited variety of possible modifications*. The two formal theories for these actions are given in Propositions 8.1 and 8.2. The same idea of non-well-foundedness of behaviour can be reformulated as rough transitions. About rough set models of *Physarum polycephalum* transitions in games, please see [1, 6, 48, 49].[1]

References

1. Pancerz, K., Schumann, A.: Rough set description of strategy games on Physarum machines. In: Adamatzky, A. (ed.) Advances in Unconventional Computing, Volume 2: Prototypes, Models and Algorithms, Emergence, Complexity and Computation, vol. 23, pp. 615–636. Springer International Publishing (2017)

[1] About the notion of rough set, please see [50–52].

2. Schumann, A.: Go games on plasmodia of Physarum polycephalum. In: 2015 Federated Conference on Computer Science and Information Systems, FedCSIS 2015, Lódz, Poland, 13–16 Sept 2015, pp. 615–626 (2015). https://doi.org/10.15439/2015F236
3. Schumann, A.: Towards context-based concurrent formal theories. Parallel Proc. Lett. **25**, 1540,008 (2015)
4. Schumann, A.: Syllogistic versions of go games on Physarum. In: Adamatzky, A. (ed.) Advances in Physarum Machines: Sensing and Computing with Slime Mould, pp. 651–685. Springer International Publishing, Cham (2016)
5. Schumann, A., Pancerz, K.: Interfaces in a game-theoretic setting for controlling the plasmodium motions. In: Proceedings of the 8th International Conference on Bio-inspired Systems and Signal Processing (BIOSIGNALS'2015), pp. 338–343, Lisbon, Portugal (2015)
6. Schumann, A., Pancerz, K.: A rough set version of the go game on Physarum machines. In: Suzuki, J., Nakano, T., Hess, H. (eds.) Proceedings of the 9th International Conference on Bio-inspired Information and Communications Technologies (BICT'2015), pp. 446–452. New York City, New York, USA (2015)
7. Schumann, A., Pancerz, K., Adamatzky, A., Grube, M.: Bio-inspired game theory: the case of Physarum polycephalum. In: Suzuki, J., Nakano, T. (eds.) Proceedings of the 8th International Conference on Bio-inspired Information and Communications Technologies (BICT'2014), pp. 9–16, Boston, Massachusetts, USA (2014)
8. Schumann, A., Pancerz, K., Adamatzky, A., Grube, M.: Context-based games and Physarum polycephalum as simulation model. In: Proceedings of the Workshop on Unconventional Computation in Europe, London, UK (2014)
9. Schumann, A., Pancerz, K., Szelc, A.: The swarm computing approach to business intelligence. Studia Humana **4**(3), 41–50 (2015). https://doi.org/10.1515/sh-2015-0019
10. Schumann, A., Szelc, A.: Towards new probabilistic assumptions in business intelligence. Studia Humana **3**(4), 11–21 (2014). https://doi.org/10.1515/sh-2015-0003
11. Arellano, M.: Panel Data Econometrics. Oxford University Press (2003)
12. Davidson, R., MacKinnon, J.G.: Econometric Theory and Methods. Oxford University Press (2004)
13. McPherson, G.: Statistics in Scientific Investigation: Its Basis. Springer, Application and Interpretation (1990)
14. Wooldridge, J.M.: Econometric Analysis of Cross Section and Panel Data. MIT Press, Cambridge MA (2002)
15. Bourdieu, P.: Acts of Resistance: Against the Tyranny of the Market. New Press (1999)
16. Bourdieu, P., Wacquant, L.J.D.: An Invitation to Reflexive Sociology. University of Chicago Press, Chicago and London (1992)
17. Althusser, L., Balibar, E.: Reading Capital. Verso, London (2009)
18. Marx, K., Engels, F.: Collected Works of Karl Marx and Frederick Engels. International Publishers, New York (1988)
19. Mead, G.H.: Mind, Self, and Society. University of Chicago Press, Chicago (1934)
20. Mead, G.H.: Movements of Thought in the Nineteenth Century. University of Chicago Press, Chicago (1936)
21. Mead, G.H.: The Philosophy of the Act. University of Chicago Press, Chicago (1938)
22. Blümer, H.G.: George Herbert Mead and Human Conduct (2004)
23. Blümer, H.G.: Symbolic Interactionism: Perspective and Method (1969)
24. Crawford, V.P., Iriberri, N.: Level-k auctions: can a nonequilibrium model of strategic thinking explain the winner's curse and overbidding in private-value auctions? Econometrica **75**(6), 1721–1770 (2007)
25. Winrich, J.S.: Self-reference and the incomplete structure of neoclassical economics. J. Econ. Issues **18**(4), 987–1005 (1984)
26. Bourdieu, P.: Language and Symbolic Power. Harvard University Press (1991)
27. Schumann, A.: Reflexive games and non-archimedean probabilities. P-Adic Numbers Ultrametr. Anal. Appl. **6**(1), 66–79 (2014)

28. Pancerz, K., Schumann, A.: Some issues on an object-oriented programming language for Physarum machines. In: Bris, R., Majernik, J., Pancerz, K., Zaitseva, E. (eds.) Applications of Computational Intelligence in Biomedical Technology, Studies in Computational Intelligence, vol. 606, pp. 185–199. Springer International Publishing, Switzerland (2016)
29. Schumann, A., Pancerz, K.: Towards an object-oriented programming language for Physarum polycephalum computing. In: Szczuka, M., Czaja, L., Kacprzak, M. (eds.) Proceedings of the Workshop on Concurrency, Specification and Programming (CS&P'2013), pp. 389–397, Warsaw, Poland (2013)
30. Pancerz, K., Schumann, A.: Principles of an object-oriented programming language for Physarum polycephalum computing. In: Proceedings of the 10th International Conference on Digital Technologies (DT'2014), pp. 273–280, Zilina, Slovak Republic (2014)
31. Schumann, A., Pancerz, K.: Towards an object-oriented programming language for Physarum polycephalum computing: A Petri net model approach. Fundam. Informaticae **133**(2–3), 271–285 (2014)
32. Kim, J., Jeong, S.: Learn to Play Go. Good Move Press, New York (1994)
33. Schumann, A.: On two squares of opposition: the lesniewski's style formalization of synthetic propositions. Acta Anal. **28**, 71–93 (2013)
34. Jones, J., Mayne, R., Adamatzky, A.: Representation of shape mediated by environmental stimuli in Physarum polycephalum and a multi-agent model. JPEDS **32**(2), 166–184 (2017)
35. Jacobs, B., Rutten, J.: A tutorial on (co)algebras and (co)induction. EATCS Bull. **62**, 222–259 (1997)
36. Pavlović, D., Escardó, M.H.: Calculus in coinductive form. In: Proceedings of the 13th Annual IEEE Symposium on Logic in Computer Science, pp. 408–417 (1998)
37. Rutten, J.J.M.M.: A coinductive calculus of streams. Math. Struct. Comput. Sci. **15**(1), 93–147 (2005)
38. Aczel, A.: Non-Well-Founded Sets. Stanford University Press (1988)
39. Adamatzky, A.: Physarum machine: implementation of a Kolmogorov-Uspensky machine on a biological substrate. Parallel Proc. Lett. **17**(4), 455–467 (2007)
40. Uspensky, V.U.: Kolmogorov and mathematical logic. J. Symb. Logic **57**, 385–412 (1992)
41. Schumann, A., Adamatzky, A.: The double-slit experiment with Physarum polycephalum and p-adic valued probabilities and fuzziness. Int. J. Gen. Syst. **44**(3), 392–408 (2015)
42. Abramsky, S., Mellies, P.A.: Concurrent games and full completeness. In: Proceedings of the 14th Symposium on Logic in Computer Science, pp. 431–442 (1999)
43. Brenguier, R.: Praline: a tool for computing Nash equilibria in concurrent games. In: CAV, pp. 890–895 (2013)
44. Bouyer, P., Brenguier, R., Markey, N., Ummels, M.: Concurrent games with ordered objectives. In: FoSSaCS, pp. 301–315 (2012)
45. Bouyer, P., Brenguier, R., Markey, N., Ummels, M.: Nash equilibria in concurrent games with büchi objectives. In: FSTTCS, pp. 375–386 (2011)
46. Feynman, R.P.: Quantum mechanical computers. Found. Phys. **16**(6), 507–531
47. Sleator, T., Weinfurter, H.: Realizable universal quantum logic gates. Phys. Rev. Lett. **74**, 4087–4090 (1995)
48. Pancerz, K., Schumann, A.: Rough set models of Physarum machines. Int. J. Gen. Syst. **44**(3), 314–325 (2015)
49. Schumann, A., Pancerz: Roughness in timed transition systems modeling propagation of plasmodium. In: Ciucci, D., Wang, G., Mitra, S., Wu, W.Z. (eds.) Rough Sets and Knowledge Technology, Lecture Notes in Artificial Intelligence, vol. 9436. Springer International Publishing (2015)
50. Dubois, D., Prade, H.: Rough fuzzy sets and fuzzy rough sets. Int. J. Gen. Syst. **17**(2–3), 191–209 (1990)
51. Petri, C.A.: Kommunikation mit automaten. Schriften des IIM nr. 2, Institut für Instrumentelle Mathematik, Bonn (1962)
52. Ziarko, W.: Variable precision rough set model. J. Comput. Syst. Sci. **46**(1), 39–59 (1993)

Chapter 9
Logics for Preference Relations in Swarm Behaviours

Each action of swarm can have many possible modifications. It means that preference relations of swarm are being modified, too. As far as we know, there are at least the following two ways of modifications in sensing and motoring: (i) the lateral inhibition with increasing the proto-symbolic value of item and, as a result, with concentrating on the chosen item; and (ii) the lateral activation with decreasing the proto-symbolic value of concrete items which causes a consideration of many items simultaneously. In this chapter I am going to examine influences of lateral inhibition and lateral activation on preference relations of swarms. See [1–5].

9.1 Preference Relation

A *preference relation* $R_i(a, b)$ that is read as follows: "agent i prefers alternative a to alternative b", is one of the basic notions in decision making and games. Due to preference relations any choice is possible. There are two main possibilities in defining preferences: (i) by evaluating each alternative individually; (ii) by evaluating each pair of alternatives. The second approach was proposed first as the Thurstone's law of comparative judgment [6]. Then it was developed as the *pairwise comparisons method* [7, 8] that is applied in the analytic hierarchy process [9] and the complex decision process [10, 11].

The pairwise comparisons method allows us to express preference relations in a qualitative way. So, given a set of n objects and a set of binary comparisons between pairs of objects, we face $n \cdot (n-1)/2$ possible comparisons. These comparisons can be measured either actively, or recursively. As a result, we find all binary comparisons $c_{i,j}$ for any pairs (i, j), $i, j = 1, 2, \ldots, n$, of n objects such that $c_{i,j} = 1 \Leftrightarrow \Pi(i) < \Pi(j)$, where $\Pi : \{1, \ldots, n\} \mapsto \mathbf{R}^+$ is a scoring function. Due to the scoring function, a decision maker can claim how many times the j-th object is preferred than the i-th one (i.e. $a_{ij} \cdot \Pi(i) = \Pi(j)$). This result of comparisons is denoted by a_{ij}. Usually, it is assumed that the scores have the following properties: (i) $a_{ij} > 0$; (ii) $a_{ii} = 1$;

A. Schumann, *Behaviourism in Studying Swarms: Logical Models of Sensing and Motoring*, Emergence, Complexity and Computation 33,
https://doi.org/10.1007/978-3-319-91542-5_9

(iii) $a_{ij} = 1/a_{ji}$. The scores are involved in matrices $A = [a_{ij}]$ which are called *pairwise comparison matrices* or *positive reciprocal matrices*. See [12–17].

Comparisons can be formulated in many forms: “player j won against player i” or “the boss accepted report j instead of i”, etc. In any case it is supposed that the decision maker can compare any two items of n objects. This assumption is called *comparability*. It is very strong, because comparisons can ever give not only a preference relation, but also an indifference relation or an incomparability relation.

The comparison is one of the most fundamental cognitive activities in decision making from swarms to human beings. In this chapter, I propose two logical models of comparisons with four relations (‘S is at least as good as P’, ‘S is not at least as good as P’, ‘S is at least as bad as P’, and ‘S is not at least as bad as P’) on the basis of two models of preferences on the level of neuron networks: with lateral inhibition and lateral activation. Also, I show that the logical model that satisfies the lateral inhibition in preferring some items can be formalized as an Aristotelian system, while the logical model that satisfies lateral activation effects can be formalized as a non-Aristotelian system. I demonstrate that in the Aristotelian system the preference relation is quantitative (numerically expressible), while in the non-Aristotelian system the preference relation is qualitative (without numerical shaping). To sum up, we show that there are cases of comparisons when the scoring function is just subjective and has no real meaning. Hence, we can build effective pairwise comparisons without the comparability assumption.

9.1.1 Preferences with Lateral Inhibition and Lateral Activation

Let us start with considering preference relations on the level of neurons. It is known from neurobiology that there are two different synaptic effects: (i) excitatory effect (depolarization) that increases the membrane potential to make neuron more negative and to decrease the likelihood of an action potential and (ii) inhibitory effect (hyperpolarization) that decreases the membrane potential to make neuron more positive and to increase the likelihood of an action potential. So, *lateral activation* is the structuring of a neural network so that neurons activate their neighbours to decrease their own responses and *lateral inhibition* is the structuring of a neural network so that neurons inhibit their neighbours in proportion to their own excitation. In other words, the more neighbouring neurons stimulated, the less strongly a neuron responds and the fewer neighbouring neurons stimulated, the more strongly a neuron responds. For example, the lateral activation decreases the contrast and sharpness in visual response to describe more explicitly all the edges and regions in the image. In this case we deal with the so-called *low-level vision*. The lateral inhibition increases the contrast and sharpness in visual response to perform an overall action-oriented interpretation of the scene. It is the so-called *high-level vision*.

Thus, in the lateral inhibition and lateral activation modes we prefer items in a different manner: in the lateral inhibition we compare a few items with their possible (numerical) comparability to formulate explicit preference relations, while in the lateral activation we compare many items and put forward general estimations without details and without scoring functions. As we see, only the lateral inhibition model corresponds to the classical pairwise comparisons method [7, 8].

There are populations which behave as a distributed network, capable of responding to a wide range of spatially represented stimuli, for example, colonies of ants or fungi have such a behaviour. In their behaviours we can observe effects of neural networks with lateral activation and lateral inhibition mechanisms. It was shown experimentally that effects of lateral activation and lateral inhibition are detected in the plasmodium of *Physarum polycephalum*, the supergroup *Amoebozoa*, phylum *Mycetozoa*, class *Myxogastria* (see [18]). This means that we can perform experiments how the plasmodium prefers items in the two different modes: within lateral inhibition and lateral activation. Let us recall that the plasmodium is an active feeding stage of *Physarum polycephalum* that moves by protoplasmic streaming and can switch its direction or even multiply in accordance with appropriate attractants (chemical signals which attract the organism) and repellents (chemical signals which repel it). This behaviour is intelligent and can be controlled by different locations of chemical signals attracting and repelling the plasmodium.

The true slime mould (plasmodium) of *Physarum polycephalum* has the following two distinct stages in responding to signals: first, the sensory stage (perceiving signals) and, second, the motor stage (action as responding). The effect of lateral activation in the plasmodium is to decrease contrast between spatial environmental stimuli at the sensory stage and to split protoplasmic tubes towards two or more attractants at the motor stage (Fig. 9.1a). The effect of lateral inhibition is to increase contrast between spatial environmental stimuli at the sensory stage and to fuse protoplasmic tubes towards one attractant at the motor stage (Fig. 9.1b). Hence, in the lateral activation the plasmodium prefers items by splitting tubes and in lateral inhibition the plasmodium prefers items by fusing the same tubes.

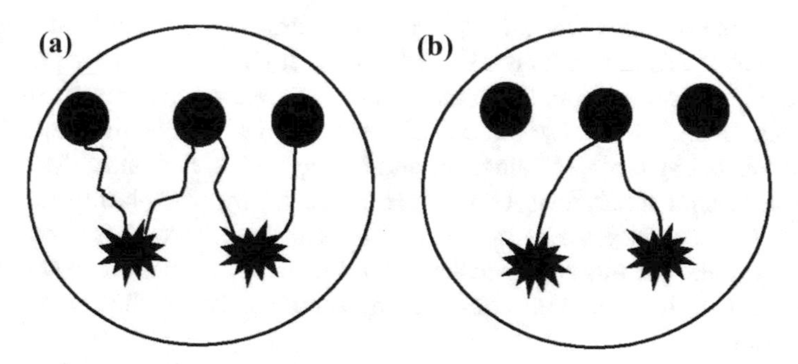

Fig. 9.1 The two plasmodia propagate protoplasmic tubes towards three attractants denoted by black circles: **a** the splitting of protoplasmic tubes of each plasmodium. **b** The fusion of two plasmodia by the fusion of their protoplasmic tubes, see [5]

Thus, in the plasmodium approximation of the neural response, the lateral activation is represented by chemoattraction with splitting of plasmodium and lateral inhibition is represented by chemoattraction with fusion of plasmodia. Both are the two different ways of distribution density of protoplasmic network, i.e. the two different ways of the plasmodium concentration in its networking. So, the phenomenon of lateral activation on plasmodium can be explained by means of splitting in plasmodium motions towards attractants and the phenomenon of lateral inhibition on plasmodium can be explained by means of fusion in plasmodium motions towards attractants. Fusion and splitting are key motions in the plasmodium propagation as well as lateral inhibition and lateral activation are key reactions of neural nets to stimuli. Notably, the fusion and splitting of plasmodium can be interpreted syllogistically. The spatial deducing according to fusion is formalized as a *spatial version of Aristotelian syllogistic* (see Sect. 7.2.2) and the spatial deducing according to splitting is formalized as a *spatial version of performative syllogistic* (see Sect. 7.2.3) defined in [2] (the word 'performative' refers to modes of symbolic behaviour: 'think', 'like', 'order', 'ask', etc.).

9.1.2 Aristotelian System for Preference Relations

Let us first consider the syllogistic of plasmodium fusion as a cognition model by lateral inhibition. In this syllogistic, all data points are denoted by appropriate syllogistic letters as attractants. These attractants are scattered at different places and the swarm (the plasmodium or ant nest) tries to occupy them. A data point S is considered empty if and only if an appropriate attractant denoted by S is not occupied by the swarm (the plasmodium or ant nest). We have syllogistic strings of the form SP with the following interpretation: 'S and P are comparable positively', and with the following meaning: SP is true if and only if S and P are neighbours and both S and P are not empty, otherwise SP is false. By this definition of syllogistic strings, we can define atomic syllogistic propositions as follows:

$Sa_{LI}P$ *In the formal syllogistic*: "for agent i, alternative P is at least as good as alternative S", formally: there exists A such that AS and for any A, if AS, then AP. *In the Physarum model*: "for the plasmodium, alternative P is at least as good as alternative S", formally: there is a plasmodium at A such that cells A and S are connected by protoplasmic tubes and for any cell A, if A and S are connected by protoplasmic tubes, then A and P are connected by protoplasmic tubes. *In the ant model*: "for the ant nest, alternative P is at least as good as alternative S", formally: there are ants at A such that cells A and S are connected by ant trails and for any cell A, if A and S are connected by ant trails, then A and P are connected by ant trails.

$Si_{LI}P$ *In the formal syllogistic*: "for agent i, alternative P is not at least as bad as alternative S", formally: there exists A such that both AS is true and AP is true. *In the Physarum model*: "for the plasmodium, alternative P is not at least as bad as alternative S", formally: there exists a plasmodium at A such that A and S are connected by protoplasmic tubes and A and P are connected by protoplasmic tubes. *In the ant model*: "for the ant nest, alternative P is not at least as bad as alternative S", formally: there are ants at A such that A and S are connected by ant trails and A and P are connected by ant trails.

$Se_{LI}P$ *In the formal syllogistic*: "for agent i, alternative P is at least as bad as alternative S", formally: for all A, AS is false or AP is false. *In the Physarum model*: "for the plasmodium, alternative P is at least as bad as alternative S", formally: for all cells A, A and S are not connected by protoplasmic tubes or A and P are not connected by protoplasmic tubes. *In the ant model*: "for the ant nest, alternative P is at least as bad as alternative S", formally: for all cells A, A and S are not connected by ant trails or A and P are not connected by ant trails.

$So_{LI}P$ *In the formal syllogistic*: "for agent i, alternative P is not at least as good as alternative S", formally: for any A, AS is false or there exists A such that AS is true and AP is false. *In the Physarum model*: "for the plasmodium, alternative P is not at least as good as alternative S", formally: for any cell A, A and S are not connected by protoplasmic tubes or there exists cell A such that A and S are connected by protoplasmic tubes and A and P are not connected by protoplasmic tubes. *In the ant model*: "for the ant nest, alternative P is not at least as good as alternative S", formally: for any cell A, A and S are not connected by ant trails or there exists cell A such that A and S are connected by ant trails and A and P are not connected by ant trails.

According to these definitions, it is easy to see that we deal there with the Aristotelian syllogistic formalized within Leśniewski's ontology [2] (see Definition 7.3 and Proposition 7.1). The semantics for this system is represented by swarm motions with emphasis on fusions, e.g. on a fusion of plasmodia or a fusion of ant trails. Let M be a set of attractants. Take a subset $|X| \subseteq M$ of attractants occupied by the swarm as a meaning for each syllogistic variable X. Let an ordering relation $\subseteq$ be defined on subsets $|S|, |P| \subseteq M$ thus: $|S| \subseteq |P|$ iff all attractants from $|S|$ are connected by swarm transport nets with some attractants from $|P|$. Notice that this means that at the points of $|P|$ we observe fusions of some swarm transport roads started from the points of $|S|$. Hence, $|S| \cap |P| \neq \emptyset$ means that some attractants from $|S|$ are connected by the swarm road with the attractants from $|P|$. In other words, at the data points of $|P|$ we detect fusions of transport roads started from the points of $|S|$ more rarely than in case $|S| \subseteq |P|$. Finally, $|S| \cap |P| = \emptyset$ means that no attractants from $|P|$ are connected by roads with the attractants from $|S|$. In other words, at the points of $|P|$ we cannot see any fusion of transport roads started from the points of $|S|$. This gives rise to models $\mathscr{M} = \langle M, |\cdot|\rangle$ such that

$\mathscr{M} \models Sa_{LI}P$ iff $|S| \subseteq |P|$;
$\mathscr{M} \models Si_{LI}P$ iff $|S| \cap |P| \neq \emptyset$;
$\mathscr{M} \models Se_{LI}P$ iff $|S| \cap |P| = \emptyset$;
$\mathscr{M} \models p \wedge q$ iff $\mathscr{M} \models p$ and $\mathscr{M} \models q$;
$\mathscr{M} \models p \vee q$ iff $\mathscr{M} \models p$ or $\mathscr{M} \models q$;
$\mathscr{M} \models \neg p$ iff it is false that $\mathscr{M} \models p$.

In these models, the inclusion relation $\subseteq$ has the meaning of fusions. We have:

Proposition 9.1 *The Aristotelian syllogistic is sound and complete relatively to $\mathscr{M}$ if we understand $\subseteq$ formally as an inclusion relation and semantically as fusions of swarm roads.* □

9.1.3 Performative Syllogistic System for Preference Relations

The fusion is a motor stage of swarm that corresponds to the lateral inhibition in responding to stimuli. However, the swarm (such as the plasmodium or the ant nest) more often prefers to be propagated in all possible directions, i.e. more often it performs the splitting of transport roads towards the attractants. Now, let us construct the syllogistic of swarm splitting as a cognition model by lateral activation, the so-called *performative syllogistic verified on swarms* (see Propositions 7.2 and 7.3 and Axioms (7.21)–(7.27)). In the same way as in the case of Aristotelian syllogistic implementation, we interpret all data points denoted by appropriate syllogistic letters as attractants. A data point S is considered empty if and only if an appropriate attractant denoted by S is not occupied by the swarm. Let us define syllogistic strings of the form SP with the following interpretation: 'S and P are comparable positively', and with the following meaning: SP is true if and only if S and P are reachable for each other by the swarm and both S and P are not empty, otherwise SP is false. Using this definition of syllogistic strings, we can define atomic syllogistic propositions as follows:

$Sa_{LA}P$ *In the formal performative syllogistic*: "for agent i, alternative P is at least as good as alternative S", formally: there exists A such that AS is true and for any A, AS is true and AP is true. *In the Physarum model*: "for the plasmodium, alternative P is at least as good as alternative S", formally: there is a string AS and for any cell A which is a neighbour for S and P, there are strings AS and AP. This means that we have a massive-parallel occupation of the region where the cells S and P are located. *In the ant model*: "for the ant nest, alternative P is at least as good as alternative S", formally: there is a string AS and for any cell A which is a neighbour for S and P, there are strings AS and AP.

$Si_{LA}P$ *In the formal performative syllogistic*: "for agent i, alternative P is not at least as bad as alternative S", formally: for any A, both AS is false and AP is false. *In the Physarum model*: "for the plasmodium, alternative P is not at least as bad as alternative S", formally: for any cell A which is a neighbour for S and P, there are no strings AS and AP. This means that the plasmodium cannot reach S from P or P from S immediately. *In the ant model*: "for the ant nest, alternative P is not at least as bad as alternative S", formally: for any cell A which is a neighbour for S and P, there are no strings AS and AP.

$Se_{LA}P$ *In the formal performative syllogistic*: "for agent i, alternative P is at least as bad as alternative S", formally: there exists A such that if AS is false, then AP is true. *In the Physarum model*: "for the plasmodium, alternative P is at least as bad as alternative S”, formally: there exists cell A which is a neighbour for S and P such that there is a string AS or there is a string AP. This means that the plasmodium occupies S or P, but not the whole region where the cells S and P are located. *In the ant model*: “for the ant nest, alternative P is at least as bad as alternative S”, formally: there exists cell A which is a neighbour for S and P such that there is a string AS or there is a string AP.

$So_{LA}P$ *In the formal performative syllogistic*: “for agent i, alternative P is not at least as good as alternative S”, formally: for any A, AS is false or there exists A such that AS is false or AP is false. *In the Physarum model*: “for the plasmodium, alternative P is not at least as good as alternative S”, formally: for any cell A which is a neighbour for S and P there is no string AS or there exists cell A which is a neighbour for S and P such that there is no string AS or there is no string AP. This means that the plasmodium does not occupy S or there is a neighbouring cell which is not connected to S or P by a protoplasmic tube. *In the ant model*: “for the ant nest, alternative P is not at least as good as alternative S”, formally: for any cell A which is a neighbour for S and P there is no string AS or there exists cell A which is a neighbour for S and P such that there is no string AS or there is no string AP.

In the syllogistic for the lateral activation in the swarm behaviour we have the following Axioms:

$$Sa_{LA}P := (\exists A(A \,\text{is}\, S) \wedge (\forall A(A \,\text{is}\, S \wedge A \,\text{is}\, P))); \tag{9.1}$$

$$\begin{aligned} So_{LA}P &:= \neg(\exists A(A \,\text{is}\, S) \vee (\forall A(A \,\text{is}\, P \wedge A \,\text{is}\, S))), \text{ i.e.} \\ &(\forall A \neg(A \,\text{is}\, S) \wedge \exists A(\neg(A \,\text{is}\, P) \vee \neg(A \,\text{is}\, S))); \end{aligned} \tag{9.2}$$

$$\begin{aligned} Se_{LA}P &:= \neg\forall A(\neg(A \,\text{is}\, S) \wedge \neg(A \,\text{is}\, P)), \text{ i.e.} \\ &\exists A(A \,\text{is}\, S \vee A \,\text{is}\, P). \end{aligned} \tag{9.3}$$

$$Si_{LA}P := \forall A(\neg(A \,\text{is}\, S) \wedge \neg(A \,\text{is}\, P)) \tag{9.4}$$

$$Sa_{LA}P \Rightarrow Se_{LA}P; \tag{9.5}$$

$$Sa_{LA}P \Rightarrow Pa_{LA}S; \tag{9.6}$$

$$Si_{LA}P \Rightarrow Pi_{LA}S; \tag{9.7}$$

$$Sa_{LA}M \Rightarrow Se_{LA}P; \tag{9.8}$$

$$Ma_{LA}P \Rightarrow Se_{LA}P; \tag{9.9}$$

$$(Ma_{LA}P \wedge Sa_{LA}M) \Rightarrow Sa_{LA}P; \tag{9.10}$$

$$(Mi_{LA}P \wedge Si_{LA}M) \Rightarrow Si_{LA}P. \tag{9.11}$$

The formal properties of this axiomatic system are considered in [2]. Within this system we can study how the swarm realizes the lateral activation at the motor stage in occupying all possible attractants in any direction if it can reach them. A model $\mathcal{M}' = \langle M', |\cdot|_x \rangle$ for this syllogistic, where M' is the set of attractants and $|X|_x \subseteq M'$ is a meaning of syllogistic letter X which is understood as all attractants reachable for the swarm from the point x, is defined as follows:

$\mathcal{M}' \models$ 'alternative y is at least as good as alternative x' iff $|X|_x \neq \emptyset$, $|X|_y \neq \emptyset$, and $|X|_x \cap |X|_y \neq \emptyset$, i.e. the swarm can move from neighbours of y to x and it can move from neighbours of x to y;

$\mathcal{M}' \models$ 'alternative y is not at least as bad as alternative x' iff $y \notin |X|_x$ and $x \notin |X|_y$, i.e. the swarm cannot move from neighbours of y to x and it cannot move from neighbours of x to y;

$\mathcal{M}' \models$ 'alternative y is at least as bad as alternative x' iff $y \in |X|_x$ or $x \in |X|_y$, i.e. the swarm can move from neighbours of y to x or it can move from neighbours of x to y;

$\mathcal{M}' \models$ 'alternative y is not at least as good as alternative x' iff $y \notin |X|_x$ or $x \notin |X|_y$, i.e. the swarm cannot move from neighbours of y to x or it cannot move from neighbours of x to y;

$\mathcal{M}' \models p \wedge q$ iff $\mathcal{M}' \models p$ and $\mathcal{M}' \models q$;

$\mathcal{M}' \models p \vee q$ iff $\mathcal{M}' \models p$ or $\mathcal{M}' \models q$;

$\mathcal{M}' \models \neg p$ iff it is false that $\mathcal{M}' \models p$.

In these models, the non-empty intersection $\cap$ has the meaning of splitting:

Proposition 9.2 *The performative syllogistic is sound and complete relatively to* $\mathcal{M}'$ *if we understand* $\cap$ *formally as a non-empty intersection and semantically as splitting of swarm roads.* □

Hence, lateral inhibition effects in the networking behaviour can be formalized within the Aristotelian syllogistic (Proposition 9.1) and lateral activation effects can be formalized within the performative syllogistic (Proposition 9.2). Both models provide an account of how we can carry out comparisons to build up preference relations. So, preference relations can be realized in the following two basic ways: two items can be compared either with lateral inhibition effects within the Aristotelian syllogistic, when we increase "contrast" of events, or with lateral activation effects within the performative syllogistic, when we decrease "contrast" of events to observe all observable phenomena as one picture.

9.2 Measurable and Non-measurable Pairwise Comparisons

Notably, the Aristotelian syllogistic (Proposition 9.1) and the performative syllogistic (Proposition 9.2) have so different sets of theorems (true propositions). In particular, the *conventional logical square of opposition* (CLSO) holds true in the Aristotelian syllogistic and it is false in the performative syllogistic [2].

9.2.1 Conventional Logical Square of Opposition for Preference Relations

CLSO is a propositional function $\mathbf{SQUARE}_{CLSO}(\alpha, \beta, \gamma, \delta)$, where α, β, γ, δ are well-formed formulas of Aristotelian syllogistic such that (i) $\vdash (\alpha \Rightarrow \gamma)$, (ii) $\vdash \neg(\alpha \Leftrightarrow \delta)$, (iii) $\vdash (\beta \Rightarrow \delta)$, and (iv) $\vdash \neg(\beta \Leftrightarrow \gamma)$, where α is '$Sa_{LI}P$', β is '$Se_{LI}P$', γ is '$Si_{LI}P$', and δ is '$So_{LI}P$'.[1]

Formally speaking, CLSO is defined by the following provable formulas (we neglect possible simplifications):

$$\vdash \neg(\alpha \wedge \beta) \quad (\alpha \;\; and \;\; \beta \;\; are \;\; contrary);$$

$$\vdash (\alpha \Rightarrow \gamma) \quad (\alpha \;\; entails \;\; \gamma; \;\; \gamma \;\; is \;\; subordinated \;\; to \;\; \alpha);$$

$$\vdash (\beta \Rightarrow \delta) \quad (\beta \;\; entails \;\; \delta; \;\; \delta \;\; is \;\; subordinated \;\; to \;\; \beta);$$

$$\vdash \neg(\alpha \Leftrightarrow \delta) \quad (\alpha \;\; and \;\; \delta \;\; are \;\; contradictory);$$

$$\vdash \neg(\beta \Leftrightarrow \gamma) \quad (\beta \;\; and \;\; \gamma \;\; are \;\; contradictory);$$

$$\vdash (\gamma \vee \delta) \quad (\gamma \;\; and \;\; \delta \;\; are \;\; complementary).$$

It is readily seen that CLSO also displays the dependencies between alethic modal statements, like 'it is necessary that A' $(=\alpha)$, 'it is impossible that A' $(=\beta)$, 'it is possible that A' $(=\gamma)$ and 'it is possible that not A' $(=\delta)$. Also, CLSO can have many other logical interpretations, such as deontic, temporal, epistemic, argumentative, etc. They constitute the extended variety of modalities, that is, not restricted to its alethic species. About interpretations and theorems of modal logic see for example [19].

Assume that in pairwise comparisons analysis we deal with two items α and γ with the same propositional content A, such that α is modally stronger and γ is modally weaker. It means that $\alpha \Rightarrow \gamma$. For example, let α mean 'ordering to do A' and γ mean 'asking to do A' or let α mean 'it is a reason for concluding A' and γ mean 'it is a reason supporting A', etc. Let us introduce two modalities $\bullet$ and $\circ$ with the following meaning: $\bullet A \Rightarrow \circ A$, i.e. $\bullet A$ is modally stronger and $\circ A$ is modally weaker, e.g. $\bullet A$ means 'believing that A' and $\circ A$ means 'not believing that not-A' (i.e. 'assuming that A').

[1] In the previous section we have proposed the following interpretations of atomic syllogistic propositions: $Sa_{LI}P$ is 'alternative P is at least as good as alternative S', $Se_{LI}P$ is 'alternative P is at least as bad as alternative S', $Si_{LI}P$ is 'alternative P is not at least as bad as alternative S', and $So_{LI}P$ is 'alternative P is not at least as good as alternative S'.

We can construct now many preference relations of the form:

Weak agent 'agent i prefers $\circ A$ instead of $\bullet A$' iff $\bullet A \Rightarrow \circ A$;
Strong agent 'agent i prefers $\bullet A$ instead of $\circ A$' iff $\bullet A \Rightarrow \circ A$,

where $\bullet$ and $\circ$ are performative verbs: 'think', 'like', 'order', 'ask', etc.

An example of the weak agent: (s)he prefers not to believe that not-A instead of that to believe that A.[2] An example of the strong agent: (s)he prefers to insist that A instead of that to ask that A.[3]

Between both items $\bullet A$ and $\circ A$ the relationship can be measurable by numbers, that is, there is a function f (in fact, it is a scale) running over modal formulas with values in $\mathbf{R}^+$, such that $f(\bullet A) \leq f(\circ A)$ for weak agents and $f(\bullet A) \geq f(\circ A)$ for strong agents. This function f can partially order all modal formulas, where $\bullet$ would mean any performative verb (expressing illocutionary acts, like commanding, believing, asking, suggesting, etc.; note that in the case of humans performative verbs mostly refer to actions with explicit mental parameters) that is stronger than a performative verb denoted by $\circ$ (represented by $\neg \bullet \neg$).

Due to f, performative verbs with the same propositional content satisfy the relationships of

$$\mathbf{SQUARE}_{CLSO}(\alpha, \beta, \gamma, \delta),$$

where α is interpreted as $\bullet A$, β as $\bullet\neg A$, γ as $\circ A$ and δ as $\circ\neg A$, where A stands for arbitrary well-formed formula of classical propositional logic. So, we have:

$$\mathbf{SQUARE}_{CLSO}(\alpha, \beta, \gamma, \delta) := (a) \vdash (\bullet A \Rightarrow \circ A);\ (b) \vdash \neg(\bullet A \Leftrightarrow \circ\neg A);$$

$$(c) \vdash \neg(\circ A \Leftrightarrow \bullet\neg A).$$

Then we have the following facts:

$$\vdash (\bullet A \Leftrightarrow \neg \circ \neg A) \quad (\bullet \ is\ definable\ as\ \neg \circ \neg);$$

$$\vdash (\circ A \Leftrightarrow \neg \bullet \neg A) \quad (\circ \ is\ definable\ as\ \neg \bullet \neg),$$

which additionally illuminates logical behaviour of modalities (of all kinds) modulo their mutual inter-definability.

Returning to lateral inhibition as described in the previous section we deal with pairwise comparisons satisfying $\mathbf{SQUARE}_{CLSO}(\alpha, \beta, \gamma, \delta)$. In this case pairs α and γ (respectively, β and δ) can be measurable by numbers, provided that a function f is available. This scheme works for any kind of scaling, that is, purely comparative, interval or quotient. Note that we do not decide which kind of scaling is proper and deserve to be termed as reasonably mathematical.

[2]In this case $\bullet A$ means 'believe that A' and $\circ A$ means 'not believe that not-A'.

[3]In this case $\bullet A$ means 'insist that A' and $\circ A$ means 'ask that A'.

So, lateral inhibition effects in preference relations with performative verbs consist in that we increase a "contrast" of one performative verb denoted by weak modalities $\circ A$ to obtain another performative verb denoted by strong modalities $\bullet A$. As a result, all possible preference relations with performative verbs running the same propositional content are described by $\mathbf{SQUARE}_{CLSO}(\alpha, \beta, \gamma, \delta)$ and we can deal with a numerical scale of comparisons. In the meanwhile, we can deal there with two different agents: weak and strong. A weak agent prefers weak modalities (weak performative verbs). A strong agent prefers strong modalities (strong performative verbs).

9.2.2 *Unconventional Logical Square of Opposition for Preference Relations*

Notably, CLSO is not valid in performative syllogistic. Instead of CLSO, the so-called *unconventional logical square of opposition* (ULSO) holds [2]. It is a propositional function $\mathbf{SQUARE}_{ULSO}(\alpha, \beta, \gamma, \delta)$, where α, β, γ, δ are well-formed formulas of performative syllogistic such that (i) $\vdash (\alpha \Rightarrow \gamma)$, (ii) $\vdash \neg(\alpha \Leftrightarrow \delta)$, (iii) $\vdash (\beta \Rightarrow \delta)$, and (iv) $\vdash \neg(\beta \Leftrightarrow \gamma)$, where α is '$Sa_{LA}P$', β is '$Si_{LA}P$', γ is '$Se_{LA}P$', and δ is '$So_{LA}P$'. Now let us show how $\mathbf{SQUARE}_{ULSO}(\alpha, \beta, \gamma, \delta)$ is verified on preference relations with performative verbs. Let $\bullet A$ be interpreted as a modally stronger performative verb and $\circ A$ as a modally weaker performative verb with the same propositional content A. Let us consider pairs $\bullet A$ and $\bullet' \neg A$, where different performative verbs $\bullet$ and $\bullet'$ occur and these verbs belong to different groups of illocutions, i.e. both cannot be simultaneously representatives, directives, declaratives, expressive, or comissives. For instance, 'believing' and 'knowing' are both representatives and 'ordering' and 'insisting' are both directives. Let 'believing' be denoted by $\bullet$ and 'ordering' by $\bullet'$. Notice that 'assuming' is modally weaker than 'believing', and 'asking' is modally weaker than 'commanding'. So, 'assuming' can be denoted by $\circ$, and 'asking' can be denoted by $\circ'$, such that $\bullet A \Rightarrow \circ A$ and $\bullet' A \Rightarrow \circ' A$ and both pairs are measurable by numbers. Nevertheless, pairs $\bullet A$ and $\bullet' \neg A$ (respectively, $\circ' A$ and $\circ \neg A$) cannot be measurable by numbers in any way, although we can deduce $\bullet A \Rightarrow \bullet' \neg A$ (respectively, $\circ' A \Rightarrow \circ \neg A$) if we assume that the same propositional content cannot be used by different illocutions in the same speech situation. For example, if I believe that A, then I order to do not A, and if I ask to do A, then I assume that not A. In other words, in the same speech situation only one illocution concerning A holds, all other illocutions can concern just not-A, where 'not' is understood in the way of performative syllogistic: something from the complement of A. The construction $\bullet A \Rightarrow \bullet' \neg A$ (respectively, $\circ' A \Rightarrow \circ \neg A$) fits the situation that a belief that A, a question that A, etc. are ever stronger than some other illocutions (belonging to other illocution groups) related to not-A.

Now, we can offer many new preference relations of the form:

Meditative agent (i) 'agent i prefers $\bullet'\neg A$ instead of $\bullet A$' iff $\bullet A \Rightarrow \bullet'\neg A$; and (ii) 'agent i prefers $\circ\neg A$ instead of $\circ' A$' iff $\circ' A \Rightarrow \circ\neg A$;

Active agent (i) 'agent i prefers $\bullet A$ instead of $\bullet'\neg A$' iff $\bullet A \Rightarrow \bullet'\neg A$; (ii) and 'agent i prefers $\circ' A$ instead of $\circ\neg A$' iff $\circ' A \Rightarrow \circ\neg A$.

where $\bullet$, $\bullet'$, $\circ$, and $\circ'$ are performative verbs: 'think', 'like', 'order', 'ask', etc.

An example of the meditative agent: (s)he prefers to believe that not-A instead of that to order that A.[4] An example of the active agent: (s)he prefers to insist that A instead of that to believe that not-A.[5] Let us notice that the lateral inhibition assumes that agents are either weak or strong and the lateral activation assumes that agents are either meditative or active.

ULSO is defined as follows [2]:

$$\mathbf{SQUARE}_{ULSO}(\alpha, \beta, \gamma, \delta) := (a) \quad \vdash (\bullet A \Rightarrow \bullet'\neg A); \quad (b) \quad \vdash (\circ' A \Rightarrow \circ\neg A);$$

$$(c) \quad \vdash \neg(\bullet A \Leftrightarrow \circ'\neg A); \quad (d) \quad \vdash \neg(\circ A \Leftrightarrow \bullet'\neg A).$$

Between pairs $\bullet A$ and $\bullet'\neg A$ (respectively, $\circ' A$ and $\circ\neg A$) there is, on the one hand, the entailment relation, but, on the other hand, there are no numerical functions (scales) such that $f(\bullet A) \leq f(\bullet'\neg A)$ or $f(\circ' A) \leq f(\circ\neg A)$.

The lateral activation effects in comparisons of performative verbs consist in that we decrease "contrast" of modalities to accept different performative verbs at once, i.e. $\bullet A$ and $\bullet'\neg A$ (respectively, $\circ' A$ and $\circ\neg A$). As a consequence, our preference relations with performative verbs running the same propositional content satisfy $\mathbf{SQUARE}_{ULSO}(\alpha, \beta, \gamma, \delta)$.

Hence, this analysis is related to the lateral activation as dealing with pairwise comparisons satisfying $\mathbf{SQUARE}_{ULSO}(\alpha, \beta, \gamma, \delta)$—they are concluded from (9.1)–(9.11). In this case pairs α and γ (respectively, β and δ) cannot be measurable by numbers, because they are not mutually related by the quantitative ordering $\leq$. In fact, it is not meaningful to say that our asking is stronger than our assuming that not, although, on the other hand, we claim to ask A, therefore we assume not-A. Putting this in other words, we only compare asking and assuming that not, but not asking and not-asking. Now, the difference between the conventional square and unconventional ones as devices for accounting claims for pairwise comparisons consists in the following fact. If we take $\mathbf{SQUARE}_{CLSO}(\alpha, \beta, \gamma, \delta)$ as our general model we can appeal to quantitative comparisons. If we take $\mathbf{SQUARE}_{ULSO}(\alpha, \beta, \gamma, \delta)$ as our general model we can appeal only to qualitative comparisons. So, $\mathbf{SQUARE}_{CLSO}(\alpha, \beta, \gamma, \delta)$ can require scales, but $\mathbf{SQUARE}_{ULSO}(\alpha, \beta, \gamma, \delta)$ cannot.

Thus, from the point of view of symbolic logic, our preference relations with numerical scales can be formalized within the Aristotelian syllogistic with the basic

[4] In this case $\bullet'\neg A$ means 'believe that not-A' and $\bullet A$ means 'order that A'.

[5] In this case $\bullet A$ means 'insist that A' and $\bullet'\neg A$ means 'believe that not-A'.

propositions $\mathbf{SQUARE}_{CLSO}(\alpha, \beta, \gamma, \delta)$ and our preference relations without numerical scales can be formalized within the performative syllogistic with the basic propositions $\mathbf{SQUARE}_{ULSO}(\alpha, \beta, \gamma, \delta)$. The propositions $\mathbf{SQUARE}_{CLSO}(\alpha, \beta, \gamma, \delta)$ satisfy lateral inhibition in our cognitions and the propositions $\mathbf{SQUARE}_{ULSO}(\alpha, \beta, \gamma, \delta)$ satisfy the lateral activation in our cognitions. Weak and strong agents are focused on lateral inhibition effects, while meditative and active agents are focused on lateral activation effects. As a result, we deal with two different ways of pairwise comparisons and preference relations with two different approaches to the duality between concepts.

9.3 Preference Relations for Swarm Intelligence

9.3.1 Social Behaviour and Swarm Behaviour

Usually, a *social behaviour* is understood as a synonymous to a collective animal behaviour. It is claimed that there are many forms of this behaviour from bacteria and insects to mammals including humans. So, bacteria and insects performing a collective behaviour are called *social*.

For example, a *prokaryote*, a one-cell organism that lacks a membrane-bound nucleus (karyon), can build colonies in a way of growing slime. These colonies are called 'biofilms'. Cells in biofilms are organized in dynamic networks and can transmit signals (the so-called quorum sensing) [20]. As a result, these bacteria are considered social. Social insects may be presented by ants—insects of the family *Formicidae*. Due to a division of labour, they construct a real society of their nest even with a pattern to make slaves. Also, *Synalpheus regalis* sp., a species of snapping shrimp that commonly live in the coral reefs, demonstrates a collective behaviour like ants. Among shrimps of the same colony there is one breeding female, as well, and a labour division of other members [21]. The ant-like organization of colony is observed among some mammals, too, e.g. among naked mole-rats (*Heterocephalus glaber* sp.). In one colony they have only one queen and one to three males to reproduce, while other members of the colony are just workers [22]. The same collective behaviour is typical for Damaraland blesmols (*Fukomys damarensis* sp.), another mammal species—they have one queen and many workers [23, 24]. There are swarms which even make slaves from other groups, see the case of ants making slaves in [25].

All these patterns of ant-like collective behaviour (a brood care and a division of labour into reproductive and non-reproductive groups) are evaluated as a form of *eusociality*, the so-called highest level of organization of animal sociality [26]. Nevertheless, it is quite controversial if we can regard the ant-like collective behaviour as a social behaviour, indeed. We can do it if and only if we concentrate, first, on outer stimuli controlling individuals and, second, on 'social roles' ('worker', 'queen', etc.) of individuals as functions with some utilities for the group as such, i.e. if and only if we follow, first, *behaviourism* which represents any collective behaviour as a complex system that is managed by stimulating individuals (in particular by their reinforcement and punishment) [27] and, second, if and only if we share the ideas of *structural functionalism* which considers the whole society as a system of functions ('roles') of its constituent elements [28]. In case we accept both behaviourism and structural functionalism, we can state that a collective animal behaviour has the same basic patterns from 'social' bacteria and 'social' insects to humans whose sociality is evident for ourselves.

However, there are different approaches to sociality. One of the approaches, alternative to behaviourism and structural functionalism, is represented by *symbolic interactionism*. In this approach, a collective behaviour is social if in the process of interaction it involves a thought with a symbolic meaning that arises out of the interaction of agents [29]. In other words, social behaviour is impossible without *material culture*, e.g. without using some tools which always have symbolic meanings. Obviously, in this sense the collective behaviour of ants cannot be regarded as social. There are no tools and no symbolic meanings for the ants.

But not only humans perform social behaviour in the meaning of symbolic interactionism. It is known that wild bottlenose dolphins (*Tursiops* sp.) "apparently use marine sponges as foraging tools" [30] and this behaviour of them cannot be explained genetically or ecologically. This means that "sponging" is an example of an existing material culture in a marine mammal species and this culture is transmitted, presumably by mothers teaching the skills to their sons and daughters [30].

Also, chimpanzees involve tools in their behaviours: large and small sticks as well as large and small stones. In [31], the authors discover 39 different behaviour patterns of chimpanzees, including tool usage, grooming and courtship behaviours. It is a very interesting fact that some patterns of chimpanzees are habitual in some communities but are absent in others because of different traditions of chimpanzee material cultures [31]. Hence, we see that the collective behaviour of chimpanzees can be evaluated as social, as well.

So, within symbolic interactionism we cannot consider any complex collective behaviour, like the ant nest, as a social behaviour. The rest of complex behaviours can be called a *swarm behaviour*. Its examples are as follows: swarming of insects, flocking of birds, herding of quadrupeds, schooling of fish. In swarms, animals behave collectively, e.g. in schools or flocks each animal moves in the same direction as its neighbour, it remains close to its neighbours, it avoids collisions with its neighbours [32].

A group of people, such as *pedestrians*, can also exhibit a swarm behaviour like a flocking or schooling: humans prefer to avoid a person conditionally designated by them as a possible predator and if a substantial part of the group (not less than 5%) changes the direction, then the rest follows the new direction [33]. An ant-based algorithm can explain aircraft boarding behaviour [34]. Under the conditions of *escape panic* the majority of people perform a swarm behaviour, too [35]. The point is that a risk of predation is the main feature of swarming at all [36, 37] and under these risk conditions (like a terrorist act) symbolic meanings for possible human interactions are promptly reduced. As a consequence, the social behaviour transforms into a swarm behaviour.

Thus, *we distinguish the swarm behaviour from the social one*. The first is fulfilled without symbolic interactions, but it is complex, as well, and has an appearance from a collective decision making. In the swarm behaviour, there are only two ways to ascribe a proto-symbolic meanings to things: lateral inhibition and lateral activation. In this chapter, I am going to show that an *addictive behaviour of humans* can be considered a kind of swarm behaviour, also. The risk of predation is a main reason of reducing symbolic interactions in human collective behaviours, but there are possible other reasons like addiction. An addiction increases roles of addictive stimuli (e.g. alcohol, morphine, cocaine, sexual intercourse, gambling, etc.) by their reinforcing and intrinsically rewarding.

9.3.2 Common Patterns of Swarm Behaviour. The Müller-Lyer Illusion

There are many optical illusions which show that there are different modalities in perceiving signals which change our picture of reality. Let us consider the *Müller-Lyer illusion*, see Fig. 9.2. Traditionally, this illusion was explained as a combination of two opposing factors: (i) *lateral inhibition* increasing contrast when two points are seen closer than the objective display would justify (Fig. 9.2a), and (ii) *lateral activation* decreasing contrast when two points are seen too far apart to be considered currently (Fig. 9.2b). The matter is that simple cells in primary visual cortex have small receptive fields and respond preferentially to oriented bars. Then neurons increase or decrease a receptive field of visual cortex as a whole according to concentrations of stimuli, see [38]. In case of increasing we see Fig. 9.2a, in case of decreasing we see Fig. 9.2b.

Hence, perceiving the same line lengths is strongly biased by the neural-computation process in accordance with the stimulus distributions causing lateral inhibition or lateral activation. It is possible to show that the same patterns of perceiving signals can be detected at different levels of behaviours: from behaviours of unicellular organisms even to swarming, such as to group behaviours of ant nests. So, in [39] it was shown that the Müller-Lyer illusion holds for *foraging ants*, as well, see Fig. 9.3. It means that their swarm behaviour embodies lateral activation and lateral inhibition in the group perceptions of signals. The authors of [39] explain this phenomenon by that each swarm of ants has the following two main logistic tasks: (i) to build a global route system connecting the nest with food sources to monopolize all reachable food sources (it corresponds to the lateral activation, that is, to the colony's ability to discover new food sources through exploration); (ii) to exploit effectively and efficiently each found food source (it corresponds to the lateral inhibition, i.e. to the colony's ability to concentrate on some food sources). And there is an economic balance which is analogous to the neurophysiological balance that generates the Müller-Lyer illusion [39]. In this analogy, each ant in a colony corresponds to a neuron

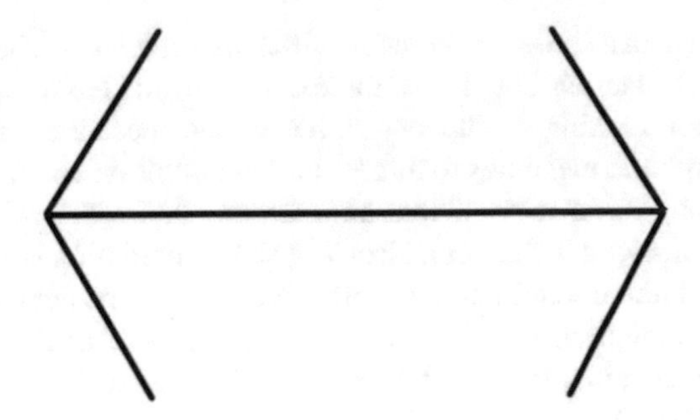

(a) Line with inward wings.

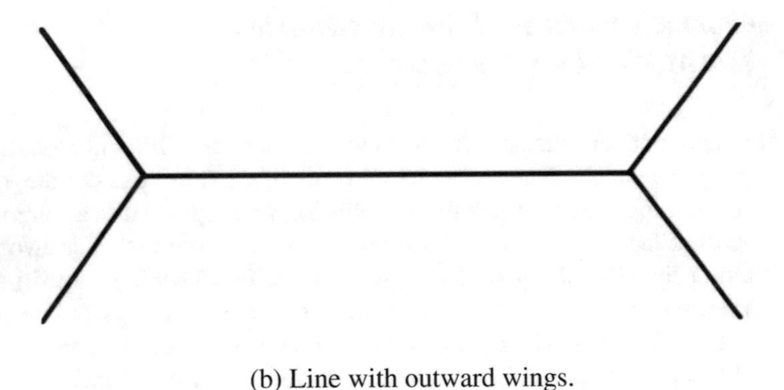

(b) Line with outward wings.

(c) Line without wings.

Fig. 9.2 The Müller-Lyer illusion. The lines of (**a**, **b**, **c**) are of the same length. Nevertheless, it seems to us that the line with the inward wings (i.e. (**a**)) is shorter than line (**b**), and the line with the outward wings (i.e. (**b**)) is longer than line (**a**)

or retinal cell and the behaviour of a swarm of ants corresponds to the behaviour of a neurological field.

The patterns of lateral activation and lateral inhibition in sensing and motoring can be detected even at the level of unicellular organisms, such as the *slime mould*, also called *Physarum polycephalum*. So, it is shown in [40] that Kanizsa illusory contours appear in the plasmodium pattern of *Physarum polycephalum*. We can show that the Müller-Lyer illusion holds for the plasmodium too, see Fig. 9.4. Indeed, we can observe (i) a lateral inhibition when there are two points of higher concentration of plasmodium with a distance that is shorter than the distance between two extremal points of the straight line (Fig. 9.4a), and (ii) a lateral activation when there are two

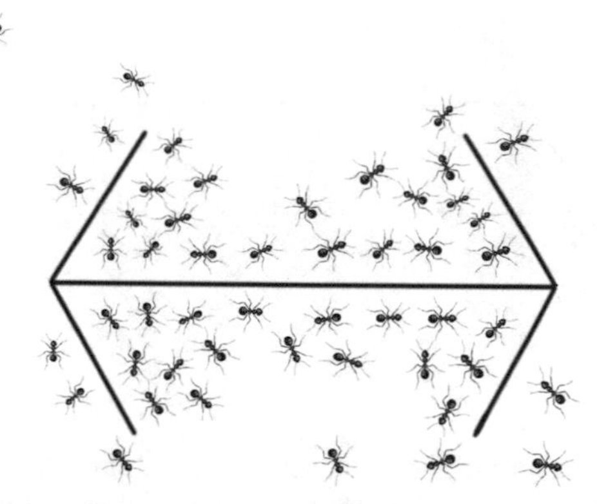

(a) Two points of highest concentrations of ants have a smaller distance than the distance between extremal points of the straight line.

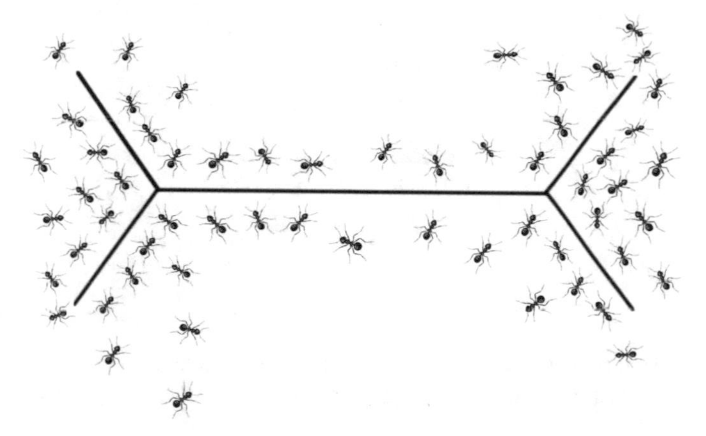

(b) Two points of highest concentrations of ants have a longer distance than the distance between extremal points of the straight line.

Fig. 9.3 An example from experiments performed in [39] to show that the Müller-Lyer illusion is detected in the swarm behaviour of ants. The authors of [39] located an attractant (honeydew solution of 50% w/w) on cardboard in the shape of the Müller-Lyer figure. The figure consisted of a 7.5-cm central shaft with two 3-cm wings pointing either inward or outward. As a result of this experiment proved statistically, the distribution of ants repeats the Müller-Lyer illusion. It means that the group sensing of ants repeats two patterns of perceiving performed by our neurons: **i** the first pattern is under condition of lateral inhibition (**a**); and **ii** the second pattern is under conditions of lateral activation (**b**)

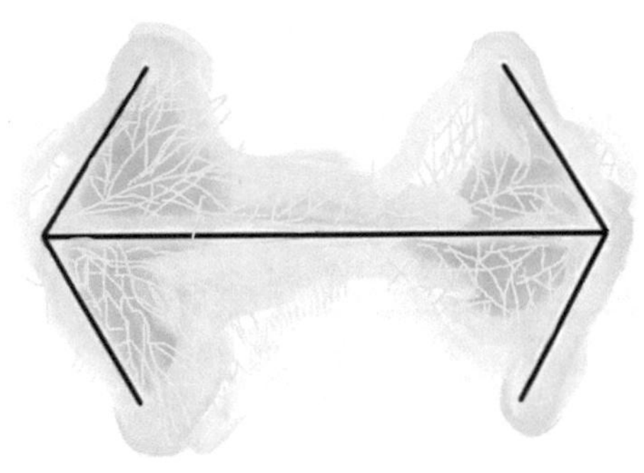

(a) Two points of highest concentrations of plasmodium have a smaller distance than the distance between extremal points of the straight line.

(b) Two points of highest concentrations of plasmodium have a longer distance than the distance between extremal points of the straight line.

Fig. 9.4 A possible experiment analogous to the experiment carried out in [39] to show that the Müller-Lyer illusion is detected in the swarm behaviour of *Physarum polycephalum*. We can locate an attractant (high nutrient concentration) in the shape of the Müller-Lyer figure. As a result, we can expect that the concentration of plasmodium repeats the Müller-Lyer illusion

points of higher concentration of plasmodium with a distance that is longer than the distance between two extremal points of the straight line (Fig. 9.4b).

Thus, some optical illusions, such as the Müller-Lyer illusion, are connected to combining lateral inhibition and lateral activation effects. Therefore, these illusions are repeated in the swarm behaviour of *Physarum polycephalum* and foraging ants.

9.3.3 *Lateral Inhibition and Lateral Activation in the Swarm Behaviour*

Each animal behaviour can be stimulated by attractants ("pull") and repellents ("push"). For instance, in [41] there was proposed a combination of "pull" and "push" methods for managing populations of *German cockroach* (Dictyoptera: Blattellidae).

Different patterns in reacting to attractants and repellents are well visible on bacteria. In the absence of an attractant or repellent a bacterium such as *Salmonella typhimurium* has the following stages in its motions: (i) the stage of run (it swims in a smooth, straight line for several seconds); (ii) the stage of tumble (it thrashes around for a fraction of a second); (ii) the stage of twiddle (it changes its direction); (iv) the new stage of run (it again swims in a straight line, but in a new, randomly chosen direction), see [42]. Under conditions of sensing attractants or repellents, a bacterium has the same stages, but they lasted longer or shorter.

If there are different gradients of attractant, we face the following changes in the bacterium stages:

- if a concentration of attractant increases, cells tumble less frequently.
- if a concentration of attractant decreases, bacteria tumble more frequently.

Under the conditions of different gradients of repellent, we see the following opposite reactions:

- if a concentration of repellent increases, bacteria tumble more frequently.
- if a concentration of repellent decreases, they tumble less frequently.

Some bacteria can be grouped into swarms. For example, *Myxococcus xanthus* is "a predatory surface-associated bacterium that moves in large multicellular groups and secretes digestive enzymes to destroy and consume other bacteria in the environment" [43]. Sometimes these multicellular groups form *biofilms* in which sessile bacteria secrete an extracellular matrix. Now, it is on open question how to define all the details for attracting and repelling bacterial swarms [43]. Nevertheless, there are some unicellular organisms, such as *Physarum polycephalum*, which have all the basic swarm patterns in their behaviour, but for them the mechanism of attracting and repelling is studied well.

Each bacterium can be attracted or repelled directly. But there are no bacteria which can occupy several attractants simultaneously or which can avoid several repellents at once. Only swarms can do it. And *Physarum polycephalum* can do it, too. Let us consider an example of Fig. 9.5. On the one hand, the slime mould can react to both repellents (Fig. 9.5a) to run away from them. On the other hand, the slime mould can react to both attractants (Fig. 9.5b) to occupy them. As a consequence, we see a behavioural pattern of the Müller-Lyer illusion: (i) the distance between two extremal points of plasmodium distribution became shorter under the condition of lateral inhibition in Fig. 9.5a, and (ii) the distance between two extremal points of plasmodium distribution became longer under the condition of lateral activation in Fig. 9.5b. To compare, in the case of Fig. 9.4a the plasmodium behaves under the

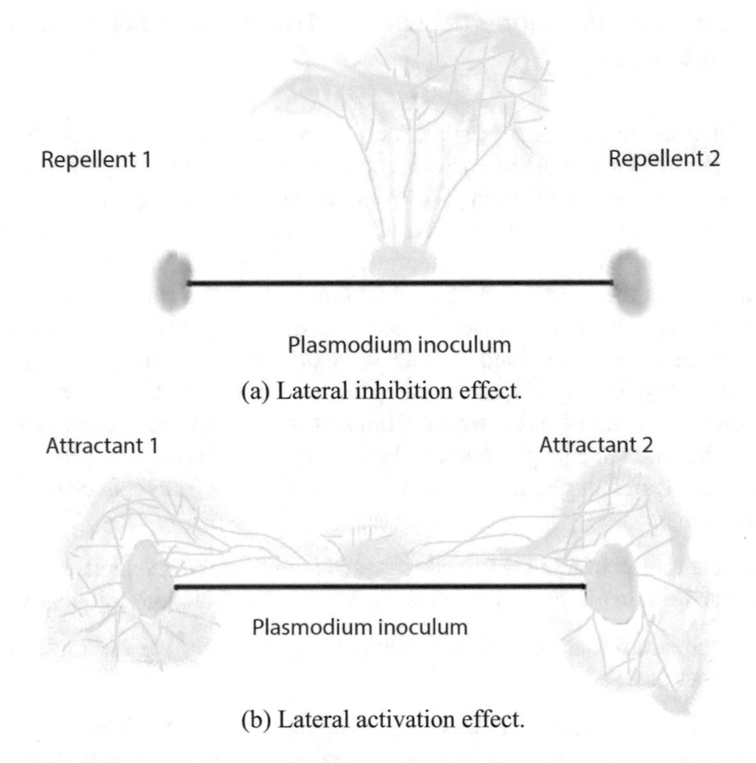

Fig. 9.5 The Müller-Lyer illusion can be detected in the following behaviour of plasmodium of *Physarum polycephalum*. Let an inoculum of plasmodium be located at the centre of a short line. Then the Müller-Lyer illusion holds if (i) at first we locate two repellents at two ends of the straight line respectively, (ii) then second we locate two attractants at both ends respectively. In the first case we observe a lateral inhibition. In the second case we observe a lateral activation. Hence, the distance between active zones is different for (**a**) and for (**b**), although the straight line is of the same length

conditions of informational noise, when there were too much attractants at a closer distance, to perform a lateral inhibition.

Hence, we can assume that lateral inhibition and lateral activation are important fundamental mechanisms which are combined for sensing and motoring in reactions of different swarms. So, swarms can have different biochemistries (different attractants and repellents), but algorithmically they realize the same patterns such as the Müller-Lyer illusion. In the next section, the alcoholic behaviour will be regarded as a new example of swarming with the same lateral inhibition and lateral activation in sensing and motoring of groups.

9.4 Swarm Behaviour of Alcoholics

9.4.1 Statistical Data on Group Behaviours of Alcoholics

Vadim Fris and me have carried out a statistical research of group behaviour of 107 people addicted to alcohol who have been actually treated in the rehabilitation centre "Iscelenie," Minsk, Belarus. Vadim Fris is the Chief Psychiatrist of this centre. Some main characteristics of the sample for our survey are collected in Table 9.1. First, the majority of interviewees were men (82.2%). Second, most respondents were older than 41 (52.3%). Third, most respondents have a job at the moment (71%).

Most interviewees (68.2%) claim that their need for alcohol is not satisfied properly, see Table 9.2. It means that logistics optimizing the drink behaviour is an actual trouble for them still.

Furthermore, most respondents (77.6%) buy alcohol at different places, see Table 9.3. It is correlated to their group behaviour. Joining different small groups for drinking or meeting the same small group at different places, they buy alcohol not at the same place. Some respondents live in the country, where there is only one store. This fact simplifies logistics for them.

Some respondents (19.6%) prefer to drink alone, but sometimes they drink jointly. Others drink only in small groups (19.6%). So, the majority (60.8%) prefer to drink jointly, but sometimes they drink alone. See Table 9.4.

Table 9.1 Some general characteristics of people questioned by us

	Quantity	Working	Jobless	Age less than 30	Age from 30 to 40	Age from 41 to 60
Women	19	15	4	1	8	10
Men	88	61	27	13	29	46

Table 9.2 The question whether the need for alcohol is satisfied

	There are ever possibilities to drink as much as you want	There are no possibilities to drink as much as you want
Women	6	13
Men	28	60

Table 9.3 The question where they usually buy alcohol

	There is the same place where you buy alcohol	There are different places where you buy alcohol
Women	4	15
Men	20	68

Table 9.4 The question whether they prefer to drink alone or jointly

	You prefer to drink alone	You prefer to drink with somebody	You drink only with companions
Women	5	11	3
Men	16	54	18

Table 9.5 The question whether the group for drinking jointly consists of men or women

	The community of your comrades who are your companions for drinking consists only of women	The community of your comrades who are your companions for drinking consists only of men	The community of your comrades who are your companions for drinking consists only of both men and women
Women	8	0	11
Men	0	39	49

Table 9.6 The question whether the companions buy drinks for you sometimes

	Your companions buy drinks sometimes to help you if you need	Your companions never buy drinks to help you if you need
Women	13	6
Men	76	12

Table 9.7 The question whether you buy drinks for your companions sometimes

	You help your companions and sometimes you buy drinks for them if they need	You never help your companions to buy drinks if they need
Women	14	5
Men	81	7

Hence, we have detected that drinking is a form of group behaviour. The gender characteristics of these groups are collected in Table 9.5. Some women (42.1% of all women) drink in small groups consisting only of women and some men (44.3% of all men) drink in small groups consisting only of men. So, 56% of the respondents drink in mixed-gender groups. In the meanwhile, a sex/gender behaviour is mainly reduced in these groups.

Joining small groups for drinking jointly supposes a kind of solidarity from participants. So, the companions can buy drinks if the interviewee has not enough money, see Table 9.6. For 68.4% of women and for 86.4% of men it is a common practice.

On the other hand, joining small groups for drinking jointly has a requirement to help others; and 73.7% of women and 92% of men buy alcohol for others sometimes if their companions have not enough money currently, see Table 9.7.

According to our survey, all the respondents have affirmed that sometimes or always they drink in small groups from 3 to 7 people, but the same respondents can join different small groups in due course. The number of stable friends to drink commonly is from 2 to 5. The alcohol-addicted people distinguish their groups from relatives or colleagues and 63% of the respondents think that their family and job hinder them to drink safely.

Thus, these small groups from 3 to 7 people can be regarded as human swarms which help their members to drink safely and to logistically optimize the task to drink. 83.1% of all the respondents have responded that members of the group can pay for drinks if the respondent does not have money (Table 9.6); and 88.8% of all the respondents have claimed that they can buy alcohol for somebody from the group who does not have money (Table 9.7). So, we deal with a form of solidarity in helping to drink.

In the case of involving new members into groups the main reasons are as follows: they are neighbours or colleagues and they can help (it means, pay). Entering new groups is possible if a friend/acquaintance has invited to join them because it is more safe and interesting for the respondent to join the new group where there is his or her friend. Without an invitation it is impossible to enter the group.

Groups are very friendly and the only reason to expel somebody from the group is that (s)he quarrels (in particular, (s)he does not want to pay). 32% of respondents have noticed that it would be better to expel one member in their groups.

Only 28% of respondents have stated that in their groups there are leaders. They are men or women more than 40 years old. The leadership consists in a support of the group to drink together.

We have discovered that alcoholics form a *network consisting of several small groups*. And the task of optimizing common drinks is solved not by a small group, but by the whole network, i.e. by several groups whose members are interconnected. The point is that each small group of alcoholics appears and disappears under different conditions, but the network, these alcoholics belong to, is almost the same. We have studied that small groups of alcohol-addicted people are not stable and, by exchanging their members, *they can fuse or split in the optimization of drinking*. The same behavioural patterns are observed in the slime mould: fusing and splitting in front of attractants to optimize their occupation. Outer stimuli (attractants) for the slime mould are pieces of nutrients scattered before this organism. *Attractants for alcoholics are represented by places where they can drink in small groups safely*: flat or outside. 38% of the respondents prefer to drink at the same place and 62% at different places. The arguments in choosing the places are as follows: the short distance from the home, low price, quality of drinks.

To sum up, the alcohol-dependent people realize a version of swarm intelligence to optimize drinking in the way of fusing or splitting the groups under different conditions. In case the groups are splitting, we face a *lateral activation effect*; and in case the groups are fusing, we deal with a *lateral inhibition effect* of alcoholic networks. The lateral activation is observed in the following cases: (i) the behaviour is carried out under conditions of attractants rather than repellents; (ii) the behaviour is performed in relation to many well visible attractants simultaneously (there are

Table 9.8 The question why you choose this place to drink (jointly)

Why do you choose this place to drink?	Pleasant company	Comfortable atmosphere	Closer to home	No one interferes	Easier to drink quickly
Women	5	2	5	5	2
Men	29	10	13	22	14

many places where it is possible to drink safely). The lateral inhibition is examined in the following cases: (i) the behaviour is conducted under pressure or stress (there are more repellents than attractants); (ii) there are a few attractants for a choice or there is an information noise in choosing the attractants. For instance, some respondents have performed the swarm behaviour to drink jointly under stress recently (it means, they have fulfilled a lateral inhibition). Some other interviewees have performed the same behaviour under favourable conditions (it means, they have fulfilled a lateral activation). And both effects can be detected by answering the question by them why they choose this place to drink, see Table 9.8. We have proposed the following variants: (i) "pleasant company" (the most favourable condition); (ii) "comfortable atmosphere" (the less favourable condition); (iii) "closer to home" (the neutral condition); (iv) "no one interferes" (the less stress condition); (v) "easier to drink quickly" (the most stress condition). We do not know circumstances of drinking for the respondents currently, but according to the data, 43% of them have fulfilled a lateral activation recently, 40,1% of them fulfilled a lateral inhibition, see Table 9.8.

Thus, small groups of alcoholics are considered by us as a kind of *human swarms*. These swarms build a network and within the same network alcoholics can freely move from one swarm to another. As a consequence, the swarms fuse or split.

9.4.2 Simulating the Lateral Activation and Lateral Inhibition Performed by Alcoholics

So, the alcohol-addicted people prefer to drink in small groups from 3 to 7 persons. These groups are said to be *agents* of swarm intelligence. Within this intelligence an appropriate network of alcoholics solves optimization tasks to drink. These swarm solutions are unconscious, but very effective. Each agent follows an unconscious collective decision-making mechanism that is decentralized and distributed among all members of the group. The same situation of distribution of intelligence is observed in any swarm.

The agents (swarms) are denoted by small letters $i, j, \ldots$ As well as all swarms, these agents can fuse and split to optimize a group occupation of attractants. Usually, there are many agents who communicate among themselves by exchanging people (their members), e.g. someone can be a member of agent i today and later become a member of agent j.

The places where agents $i, j, \ldots$ (appropriate small groups of humans addicted to alcohol) can drink safely are called *attractants* for swarm intelligence. The attractants are denoted by S, P, ...

There are two different ways in occupying attractants by swarm agents: (i) with high concentration of people (lateral inhibition effect) at places of meeting and (ii) with low concentration of people (lateral activation effect) at places of meeting [18]. In the first case much less attractants are occupied. In the second case much more attractants are occupied. For instance, in snow winter there are less attractants (places to drink jointly and safely) and this causes a lateral inhibition effect in alcoholic swarming. In sunny summer there are more attractants (places to drink in a group) and this implies a lateral activation effect in alcoholic networking.

Lateral inhibition and lateral activation can be detected in any forms of swarm networking. For example, this mechanism is observed also in the true slime mould (plasmodium) of *Physarum polycephalum*. The plasmodium has the two distinct stages in responding to signals: (i) the sensory stage (perceiving signals) and (ii) the motor stage (action as responding). The effect of lateral activation in the plasmodium is to decrease contrast between attractants at the sensory stage and to split protoplasmic tubes towards two or more attractants at the motor stage (Fig. 9.5b). The effect of lateral inhibition is to increase contrast between attractants at the sensory stage and to fuse protoplasmic tubes towards one attractant at the motor stage or to increase contrast in front of repellents at the sensory stage and to fuse protoplasmic tubes just in one safe direction at the motor stage (Fig. 9.5a).

In human groups there are (i) the one sensory stage consisting in perceiving signals (as well as for the plasmodium) and the following two motor stages consisting in actions as responding: (ii) illocutionary stage and (iii) perlucotionary stage.

For the first time the well-known 20th-century philosopher John L. Austin has investigated speech acts as a way of coordination for human behaviour by a verbal communication as well as by a non-verbal communication (e.g. by gestures or mimics). His main philosophical claim that was accepted then by almost all later language philosophers has based on the idea that we coordinate our joint behaviour by *illocutionary acts*—some utterances which express our intentions and expectations to produce joint symbolic meanings for symbolic interactions: "I hereby declare," "I sentence you to ten years' imprisonment", "I promise to pay you back," "I pray to God", etc. These utterances can produce an effect on the hearer that is called a *perlocutionary act*. Hence, according to Austin, in order to commit a group behaviour, the humans should start with illocutionary acts (uttering illocutions) to coordinate their common symbolic meanings. As a result, their group behaviour appears as a kind of perlocutionary act grounded on previous illocutionary utterances.

Thus, the motor stage for the plasmodium is just a direct behaviour, while the motor stage for the humans starts from illocutionary acts to produce symbolic meanings for performing an interaction and then this stage is continued in perlocutionary acts (a direct coordinated behaviour of a human group).

Attractants S, P, ... are detected by alcoholics at the sensory stage. Then alcoholics perform illocutionary acts to share preference relations on detected attractants. Later they commit perlocutionary acts to occupy some detected attractants. A data point S is considered empty if and only if an appropriate attractant (the place denoted by S where it is possible to drink jointly) is not occupied by the group of alcohol-dependent people. Otherwise, it is not-empty. Let us define syllogistic strings of the form SP with the following interpretation: 'S and P are comparable positively', and with the following meaning: SP is true if and only if S and P are reachable for each other by members of the group i and both S and P are not empty, otherwise SP is false. Let $\mathcal{S}$ be a set of all true syllogistic strings.

Now we can construct an *illocutionary logic of alcohol-dependent people*. In this logic we deal with preference relations about detected attractants from $\mathcal{S}$.

9.4.3 Agents in the Case of Lateral Inhibition

Let us construct an extension of modal logic **K**, please see [19] about **K** (also [44]), for preference relations of agents in case of lateral inhibition. Let 'A' and 'B' be metavariables ranging over syllogistic letters $S, P, \ldots$ or over standard propositional compositions of syllogistic letters by means of conjunctions, disjunction, implication, negation. Let us introduce two modalities $\bullet$ and $\circ$ with the following meaning:

$$\bullet A \Rightarrow \circ A,$$

i.e. $\bullet A$ is modally stronger and $\circ A$ is modally weaker, e.g. $\bullet A$ means 'I like A' (or 'I desire A') and $\circ A$ means 'maybe A' (or 'it can be A'). So, the performative verb of $\bullet$ is stronger and the performative verb of $\circ$ is weaker with the same type of performativity (modality) to prefer A.

In our logic **K** for preference relations we have also only two axioms as in the standard **K** (the inference rules are the same also):

Necessitation Rule If A is a theorem of **K**, then so is $\bullet A$.
Distribution Axiom $\bullet(A \Rightarrow B) \Rightarrow (\bullet A \Rightarrow \bullet B)$.

The operator $\circ$ can be defined from $\bullet$ as follows:

$$\circ A := \bullet A,$$

where $\bullet$ are any performative verbs for expressing a preference relation with a strong modality: 'like', 'want', 'desire', etc.

Now let us add countable many new one-place sentential connectives $\bullet_{ki}$ to the language of **K**:

if A is a formula, then $\bullet_{ki} A$ is a formula, too.

These $\bullet_{ki} A$ are read as follows: "the k-th utterance of preference relation uttered by agent i to fulfil an illocutionary act". The weaker modality $\circ_{ki}$ is defined thus:

$$\circ_{ki} A := \neg \bullet_{ki} \neg A.$$

We assume that $\bullet_{ki}$ and $\circ_{ki}$ satisfy the necessity rule and distribution axiom as well.

Let us denote the new extension by $\mathbf{K}_i$.

Now let us define in $\mathbf{K}_i$ the four basic preference relations as atomic syllogistic propositions: $k_i(S \preceq_{LI}^{good} P), k_i(S \npreceq_{LI}^{bad} P), k_i(S \preceq_{LI}^{bad} P), k_i(S \npreceq_{LI}^{good})$. They are defined as follows.

$$k_i(S \preceq_{LI}^{good} P) := \bullet_{ki}(S \Rightarrow P) \tag{9.12}$$

The atomic proposition $k_i(S \preceq_{LI}^{good} P)$ means: "for agent i, alternative P is at least as good as alternative S by the k-th utterance" and it is defined under conditions of lateral inhibition. *In the model of alcohol-addicted swarms* it means: "for the grouping of alcohol-dependent people i, alternative P is at least as good as alternative S at the k-th utterance under conditions of lateral inhibition".

Let us define a model $\mathscr{M}$.

Semantic meaning of $k_i(S \preceq_{LI}^{good} P)$:

$\mathscr{M} \models k_i(S \preceq_{LI}^{good} P) :=$ at the utterance k uttered by i, there exists a data point $A \in \mathscr{M}$ such that $AS \in \mathscr{S}$ and for any $A \in \mathscr{M}$, if $AS \in \mathscr{S}$, then $AP \in \mathscr{S}$ and it is defined under conditions of lateral inhibition.

Semantic meaning of $k_i(S \preceq_{LI}^{good} P)$ *in alcohol-addicted swarms*: there is a group of alcoholics i at a place A such that places A and S are connected by exchanging of some members of i and for any place A, if A and S are connected by exchanging of some members of i, then A and P are connected by exchanging of some members of i.

$$k_i(S \npreceq_{LI}^{bad} P) := \circ_{ki}(S \wedge P) \tag{9.13}$$

The atomic proposition $k_i(S \npreceq_{LI}^{bad} P)$ means: "for agent i, alternative P is not at least as bad as alternative S by the k-th utterance" and it is defined under conditions of lateral inhibition. *In the model of alcohol-addicted swarms* it means: "for the grouping of alcohol-dependent people i, alternative P is not at least as bad as alternative S at the k-th utterance under conditions of lateral inhibition".

Semantic meaning of $k_i(S \npreceq_{LI}^{bad} P)$:

$\mathscr{M} \models k_i(S \npreceq_{LI}^{bad} P) :=$ at the utterance k uttered by i, there exists a data point $A \in \mathscr{M}$ such that both $AS \in \mathscr{S}$ and $AP \in \mathscr{S}$ and it is defined under conditions of lateral inhibition.

Semantic meaning of $k_i(S \npreceq_{LI}^{bad} P)$ *in alcohol-addicted swarms*: there exists a group of alcoholics i at A such that A and S are connected by exchanging of some members of i and A and P are connected by exchanging of some members of i.

$$k_i(S \preceq_{LI}^{bad} P) := \bullet_{ki}(S \Rightarrow \neg P) \tag{9.14}$$

The atomic proposition $k_i(S \preceq_{LI}^{bad} P)$ means: "for agent i, alternative P is at least as bad as alternative S by the k-th utterance" and it is defined under conditions of lateral inhibition. *In the model of alcohol-addicted swarms*: "for the grouping of alcohol-dependent people i, alternative P is at least as bad as alternative S by the k-th utterance under conditions of lateral inhibition".

Semantic meaning of $k_i(S \preceq_{LI}^{bad} P)$:

$\mathscr{M} \models k_i(S \preceq_{LI}^{bad} P) :=$ at the utterance k uttered by i, for all data points $A \in \mathscr{M}$, AS is false or AP is false and it is defined under conditions of lateral inhibition.

Semantic meaning of $k_i(S \preceq_{LI}^{bad} P)$ *in alcohol-addicted swarms*: for all groups of alcoholics i at places A, A and S are not connected by exchanging of some members of i or A and P are not connected by exchanging of some members of i.

$$k_i(S \not\preceq_{LI}^{good} P) := \circ_{ki}(S \wedge \neg P) \tag{9.15}$$

The atomic proposition $k_i(S \not\preceq_{LI}^{good} P)$ means: "for agent i, alternative P is not at least as good as alternative S by the k-th utterance" and it is defined under conditions of lateral inhibition. *In the alcohol-addicted swarms*: "for the grouping of alcoholics i, alternative P is not at least as good as alternative S by the k-th utterance under conditions of lateral inhibition".

Semantic meaning of $k_i(S \not\preceq_{LI}^{good} P)$:

$\mathscr{M} \models k_i(S \not\preceq_{LI}^{good} P) :=$ at the utterance k uttered by i, for any data points $A \in \mathscr{M}$, AS is false or there exists $A \in \mathscr{M}$ such that $AS \in \mathscr{S}$ and AP is false and it is defined under conditions of lateral inhibition.

Semantic meaning of $k_i(S \not\preceq_{LI}^{good} P)$ *in alcohol-addicted swarms*: for all groups of alcoholics i at places A, A and S are not connected by exchanging of some members of i or there exists place A such that A and S are connected by exchanging of some members of i and A and P are not connected by exchanging of some members of i.

We can distinguish different swarms according to the acceptance of stronger or weaker modality:

Weak agent Agent i prefers $\circ_{ki} A$ instead of $\bullet_{ki} A$ iff $\bullet_{ki} A \Rightarrow \circ_{ki} A$.
Strong agent Agent i prefers $\bullet_{ki} A$ instead of $\circ_{ki} A$ iff $\bullet_{ki} A \Rightarrow \circ_{ki} A$.

An example of the weak agent: (s)he prefers not to like not-A instead of that to like A. An example of the strong agent: (s)he prefers to desire A instead of that to accept A.

Hence, in logic $\mathbf{K}_i$ we have the four kinds of atomic syllogistic propositions: $k_i(S \preceq_{LI}^{good} P)$, $k_i(S \not\preceq_{LI}^{bad} P)$, $k_i(S \preceq_{LI}^{bad} P)$, $k_i(S \not\preceq_{LI}^{good} P)$ for different k, i, S, and P. All other propositions of $\mathbf{K}_i$ are derivable by Boolean combinations of atomic propositions. Models for these combinations are defined conventionally:

$$\mathscr{M} \models \neg A \text{ iff } A \text{ is false in } \mathscr{M};$$

$$\mathcal{M} \models A \vee B \text{ iff } \mathcal{M} \models A \text{ or } \mathcal{M} \models B;$$

$$\mathcal{M} \models A \wedge B \text{ iff } \mathcal{M} \models A \text{ and } \mathcal{M} \models B;$$

$$\mathcal{M} \models A \Rightarrow B \text{ iff if } \mathcal{M} \models A, \text{ then } \mathcal{M} \models B.$$

Proposition 9.3 *Logic* $\mathbf{K}_i$ *is a conservative extension of* $\mathbf{K}$. □

Proposition 9.4 *In* $\mathbf{K}_i$, *the conventional square of opposition holds, i.e. there are the following tautologies:*

$$k_i(S \preceq_{LI}^{good} P) \Rightarrow k_i(S \npreceq_{LI}^{bad} P);$$

$$k_i(S \preceq_{LI}^{bad} P) \Rightarrow k_i(S \npreceq_{LI}^{good} P);$$

$$\neg(k_i(S \preceq_{LI}^{good} P) \wedge k_i(S \preceq_{LI}^{bad} P));$$

$$k_i(S \npreceq_{LI}^{bad} P) \vee k_i(S \npreceq_{LI}^{good} P);$$

$$k_i(S \preceq_{LI}^{good} P) \vee k_i(S \npreceq_{LI}^{good} P);$$

$$\neg(k_i(S \preceq_{LI}^{good} P) \wedge k_i(S \npreceq_{LI}^{good} P));$$

$$k_i(S \preceq_{LI}^{bad} P) \vee k_i(S \npreceq_{LI}^{bad} P);$$

$$\neg(k_i(S \preceq_{LI}^{bad} P) \wedge k_i(S \npreceq_{LI}^{bad} P)).$$

Proof It follows from (9.12)–(9.15). □

The fusion of two swarms i and j for syllogistic propositions is defined in $\mathbf{K}_i$ in the way:

$$\frac{k_i(S_1 \preceq_{LI}^{good} P); \quad m_j(S_2 \preceq_{LI}^{good} P)}{(k \cup m)_{i \cup j}((S_1 \vee S_2) \preceq_{LI}^{good} P)}.$$

The splitting of one swarm $i \cup j$ is defined in $\mathbf{K}_i$ thus:

$$\frac{(k \cup m)_{i \cup j}(S \preceq_{LI}^{good} (P_1 \wedge P_2))}{k_i(S \preceq_{LI}^{good} P_1); \quad m_j(S \preceq_{LI}^{good} P_2)}.$$

Hence, the illocutionary logic $\mathbf{K}_i$ describes the preference relations of alcoholics towards attractants under the conditions of lateral inhibition.

9.4.4 Agents in the Case of Lateral Activation

When the concentration of attractants (different places of grouping for common drinks) is high (i.e. we deal with the lateral activation), the logic **K** for preference relations is unacceptable. Instead of **K** we will use its modification **K**$'$ (with the same inference rules):

Necessitation Rule If A is a theorem of $\mathbf{K}'$, then so is $\bullet A$.
Distribution Weak Axiom $\bullet(A \wedge B) \Rightarrow (\bullet A \wedge \bullet B)$.

Now let us construct $\mathbf{K}'_i$ by adding countable one-place sentential connectives $\bullet_{ki}$ and $\circ_{ki}$ to the language of $\mathbf{K}'_i$ and then define the four basic preference relations $k_i(S \preceq^{good}_{LA} P), k_i(S \npreceq^{bad}_{LA} P), k_i(S \preceq^{bad}_{LA} P), k_i(S \npreceq^{good}_{LA} P)$ in the following manner:

$$k_i(S \preceq^{good}_{LA} P) := \bullet_{ki}(S \wedge P). \tag{9.16}$$

The atomic proposition $k_i(S \preceq^{good}_{LA} P)$ means: "for agent i, alternative P is at least as good as alternative S by the k-th utterance" and it is defined under conditions of lateral activation. *In the model of alcohol-addicted swarms*: "for the group of alcoholics i, alternative P is at least as good as alternative S by the k-th utterance under conditions of lateral activation".

Let us define a model $\mathcal{M}'$.
Semantic meaning of $k_i(S \preceq^{good}_{LA} P)$:

$\mathcal{M}' \models k_i(S \preceq^{good}_{LA} P) :=$ there exists a data point $A \in \mathcal{M}'$ such that $AS \in \mathcal{S}$ and for any $A \in \mathcal{M}'$, $AS \in \mathcal{S}$ and $AP \in \mathcal{S}$ and it is defined under conditions of lateral activation.

Semantic meaning of $k_i(S \preceq^{good}_{LA} P)$ *in alcohol-addicted swarms*: there is a string AS and for any place A which is reachable for S and P by exchanging of members of i, there are strings AS and AP. This means that we have an occupation of the whole region where the places S and P are located.

$$k_i(S \npreceq^{bad}_{LA} P) := \bullet_{ki}(\neg S \wedge \neg P). \tag{9.17}$$

The atomic proposition $k_i(S \npreceq^{bad}_{LA} P)$ means: "for agent i, alternative P is not at least as bad as alternative S by the k-th utterance" and it is defined under conditions of lateral activation. *In the model of alcohol-addicted swarms*: "for the group of alcoholics i, alternative P is not at least as bad as alternative S by the k-th utterance under conditions of lateral activation".
Semantic meaning of $k_i(S \npreceq^{bad}_{LA} P)$:

$\mathcal{M}' \models k_i(S \npreceq^{bad}_{LA} P) :=$ for any data point $A \in \mathcal{M}'$, both AS is false and AP is false and it is defined under conditions of lateral activation.

Semantic meaning of $k_i(S \npreceq_{LA}^{bad} P)$ *in alcohol-addicted swarms*: for any place A which is reachable for S and P by exchanging of members of i, there are no strings AS and AP. This means that the group of alcoholics cannot reach S from P or P from S immediately.

$$k_i(S \preceq_{LA}^{bad} P) := \circ_{ki}(S \vee P). \tag{9.18}$$

The atomic proposition $k_i(S \preceq_{LA}^{bad} P)$ means: "for agent i, alternative P is at least as bad as alternative S by the k-th utterance" and it is defined under conditions of lateral activation. *In the model of alcohol-addicted swarms*: "for the group of alcoholics i, alternative P is at least as bad as alternative S by the k-th utterance under conditions of lateral activation".

Semantic meaning of $k_i(S \preceq_{LA}^{bad} P)$:

$\mathscr{M}' \models k_i(S \preceq_{LA}^{bad} P) :=$ there exists a data point $A \in \mathscr{M}'$ such that if AS is false, then $AP \in \mathscr{S}$ and it is defined under conditions of lateral activation.

Semantic meaning of $k_i(S \preceq_{LA}^{bad} P)$ *in alcohol-addicted swarms*: there exists a place A which is reachable for S and P by exchanging of members of i such that there is a string AS or there is a string AP. This means that the group of alcoholics i occupies S or P, but not the whole region where the places S and P are located.

$$k_i(S \npreceq_{LA}^{good} P) := \circ_{ki}(\neg S \vee \neg P). \tag{9.19}$$

The atomic proposition $k_i(S \npreceq_{LA}^{good} P)$ means: "for agent i, alternative P is not at least as good as alternative S by the k-th utterance" and it is defined under conditions of lateral activation. *In the model of alcohol-addicted swarms*: "for the group of alcoholics i, alternative P is not at least as good as alternative S by the k-th utterance under conditions of lateral activation".

Semantic meaning of $k_i(S \npreceq_{LA}^{good} P)$:

$\mathscr{M}' \models k_i(S \npreceq_{LA}^{good} P) :=$ for any data point $A \in \mathscr{M}'$, AS is false or there exists a data point $A \in \mathscr{M}'$ such that AS is false or AP is false and it is defined under conditions of lateral activation.

Semantic meaning of $k_i(S \npreceq_{LA}^{good} P)$ *in alcohol-addicted swarms*: for any place A which is reachable for S and P by exchanging of members of i there is no string AS or there exists a place A which is reachable for S and P by exchanging of members of i such that there is no string AS or there is no string AP. This means that the group of alcoholics i does not occupy S or there is a place which is not connected to S or P by exchanging of members of i.

Models for the Boolean combinations of atomic proposition of $\mathbf{K}_i'$ are defined thus:

$$\mathscr{M}' \models \neg A \text{ iff } A \text{ is false in } \mathscr{M}';$$

$$\mathscr{M}' \models A \vee B \text{ iff } \mathscr{M}' \models A \text{ or } \mathscr{M}' \models B;$$

$$\mathcal{M}' \models A \wedge B \text{ iff } \mathcal{M}' \models A \text{ and } \mathcal{M}' \models B;$$

$$\mathcal{M}' \models A \Rightarrow B \text{ iff if } \mathcal{M}' \models A, \text{ then } \mathcal{M}' \models B.$$

Proposition 9.5 *Logic* $\mathbf{K}'_i$ *is a conservative extension of* $\mathbf{K}'$. □

Proposition 9.6 *In* $\mathbf{K}'_i$, *the unconventional square of opposition holds, i.e. there are the following tautologies:*

$$k_i(S \preceq^{good}_{LA} P) \Rightarrow k_i(S \preceq^{bad}_{LA} P);$$

$$k_i(S \npreceq^{bad}_{LA} P) \Rightarrow k_i(S \npreceq^{good}_{LA} P);$$

$$\neg(k_i(S \preceq^{good}_{LA} P) \wedge k_i(S \npreceq^{bad}_{LA} P));$$

$$k_i(S \preceq^{bad}_{LA} P) \vee k_i(S \npreceq^{good}_{LA} P);$$

$$k_i(S \preceq^{good}_{LA} P) \vee k_i(S \npreceq^{good}_{LA} P);$$

$$\neg(k_i(S \preceq^{good}_{LA} P) \wedge k_i(S \npreceq^{good}_{LA} P);$$

$$k_i(S \preceq^{bad}_{LA} P) \vee k_i(S \npreceq^{bad}_{LA} P);$$

$$\neg(k_i(S \preceq^{bad}_{LA} P) \wedge k_i(S \npreceq^{bad}_{LA} P)).$$

Proof It follows from (9.16)–(9.19). □

Now, let us consider pairs $\bullet_{ki} A$ and $\bullet_{mi} A$, where different performative verbs $\bullet_{ki}$ and $\bullet_{mi}$ occur and these verbs belong to different groups of illocutions in expressing a preference relation, i.e. both cannot be simultaneously representatives, directives, declaratives, expressive, or comissives. For instance, 'believing' and 'knowing' are both representatives and 'ordering' and 'insisting' are both directives. Assume, 'believing' be denoted by $\bullet_{ki}$ and 'advising' by $\bullet_{mi}$. Notice that 'assuming' is modally weaker than 'believing', and 'advising' is modally weaker than 'insisting'. So, 'assuming' can be denoted by $\circ_{ki}$, and 'advising' can be denoted by $\circ_{mi}$, such that $\bullet_{ki} A \Rightarrow \circ_{ki} A$ and $\bullet_{mi} A \Rightarrow \circ_{mi} A$. The construction $\bullet_{ki} A \Rightarrow \bullet_{mi} A$ (respectively, $\circ_{mi} A \Rightarrow \circ_{ki} A$) fits the situation that a belief that A is ever stronger than some other illocutions (belonging to other illocution groups) related to not-A.

Let us distinguish different swarms according to the acceptance of stronger or weaker modality:

Meditative agent (i) agent i prefers $\bullet_{mi} A$ instead of $\bullet_{ki} A$ iff $\bullet_{ki} A \Rightarrow \bullet_{mi} A$; and (ii) agent i prefers $\circ_{ki} A$ instead of $\circ_{mi} A$ iff $\circ_{mi} A \Rightarrow \circ_{ki} A$.

Active agent (i) agent i prefers $\bullet_{ki} A$ instead of $\bullet_{mi} A$ iff $\bullet_{ki} A \Rightarrow \bullet_{mi} A$; (ii) and agent i prefers $\circ_{mi} A$ instead of $\circ_{ki} A$ iff $\circ_{mi} A \Rightarrow \circ_{ki} A$.

An example of the meditative agent: (s)he prefers to believe that not-A instead of that to order that A. An example of the active agent: (s)he prefers to insist that A instead of that to believe that not-A.

The fusion of two swarms i and j universal affirmative syllogistic propositions is defined in $\mathbf{K}_i'$ as follows:

$$\frac{k_i(S_1 \preceq_{LA}^{good} P); \quad m_j(S_2 \preceq_{LA}^{good} P)}{(k \cup m_{i \cup j}((S_1 \wedge S_2) \preceq_{LA}^{good} P)}.$$

The splitting of one swarm $i \cup j$ is defined in $\mathbf{K}_i'$:

$$\frac{(k \cup m)_{i \cup j}(S \preceq_{LA}^{good} (P_1 \wedge P_2))}{k_i(S \preceq_{LA}^{good} P_1); \quad m_j(S \preceq_{LA}^{good} P_2)}.$$

The illocutionary logic $\mathbf{K}_i'$ is to express the preference relations of alcoholics towards attractants under the conditions of lateral activation.

We have shown that the lateral inhibition and lateral activation are two fundamental patterns in sensing and motoring of swarms. The point is that both patterns allow swarms to occupy several attractants and to avoid several repellents at once. Then we have shown that a habit of joint drinking of alcohol-addicted people in small groups can be considered a swarm behaviour controlled by outer stimuli (places to drink jointly). And this behaviour follows the lateral inhibition and lateral activation, also. The swarms of alcoholics can be managed by localization of places for meeting to drink. Generally, the logic of propagation of groups of alcoholics has the same axioms as the logic of parasite propagation for *Schistosomatidae* sp. as well as the same axioms as the logic of slime mould expansion. The difference is that instead of syllogistics for *Schistosomatidae* sp. and for slime mould, where preference relations are simple and express only attractions by food, we involve many performative actions (verbs), which express a desire to drink together, within modal logics $\mathbf{K}_i$ and $\mathbf{K}_i'$. The logic $\mathbf{K}_i$ is used to formalize lateral inhibition in distributing people to drink jointly and the logic $\mathbf{K}_i'$ is used to formalize lateral activation in distributing people to drink jointly.

The main outcome of our research is to show that some forms of human group behaviour are not social in fact. A kind of unsocial group behaviours is designated by us as swarm behaviour. Many forms of human swarming have recently been studied – from crowds of people in escape panic [35] to aircraft boarding [34]. However, some stable patterns of interconnected people have never been analyzed as a swarm. We have proposed to consider a network of coordinated alcoholics as human swarming. The reasons are as follows: (i) their behaviour is controlled by replacing stimuli: attractants (places where they can drink jointly and safely) and repellents (some interruptions which can appear for drinking); this control is executed by the same algorithms as for standard swarms from social bacteria to eusocial mammals; (ii) the behaviour of alcoholics is collective and even cooperative, but it is subordinated to the only one uncontrolled intention, namely, how to drink; so, this motivation bears no symbolic meanings in the terms of symbolic interactionism [45] and, then, it cannot be evaluated as social.

Each alcoholic realizes a group adaptation and belongs to a network of people with the same addiction. This network allows its members to optimize the task to drink. Therefore, it is a substitute of social groups (from family to other institutions) and it is a displacement of standard social behaviour. One of the effective means to recover alcoholism is a back replacement of ways of group optimization how to drink by that how not to drink. It is possible within a network of the so-called Alcoholics Anonymous where alcoholics can help each other to stay sober.

To sum up, the preference relations for swarms are being modified during their behaviour in accordance with a lateral inhibition or lateral activation.

References

1. Schumann, A.: Two squares of opposition: for analytic and synthetic propositions. Bulletin Sect. Logic **40**(3–4), 165–178 (2011)
2. Schumann, A.: On two squares of opposition: the Lesniewski's style formalization of synthetic propositions. Acta Analytica **28**, 71–93 (2013)
3. Schumann, A., Fris, V.: Swarm intelligence among humans—the case of alcoholics. In: Proceedings of the 10th International Joint Conference on Biomedical Engineering Systems and Technologies—Volume 4: BIOSIGNALS, (BIOSTEC 2017), pp. 17–25. ScitePress (2017)
4. Schumann, A., Woleński, J.: Decisions involving databases, fuzzy databases and codatabases. Op. Res. Decis. **25**(3), 59–72 (2015). https://doi.org/10.5277/ord150304
5. Schumann, A., Woleński, J.: Two squares of oppositions and their applications in pairwise comparisons analysis. Fundamenta Informaticae **144**(3–4), 241–254 (2016)
6. Thurstone, L.L.: A law of comparative judgment, reprint of an original work published in 1927. Psychol. Rev. **101**(2), 266–270 (1994)
7. Kułakowski, K.: A heuristic rating estimation algorithm for the pairwise comparisons method. CEJOR **23**(1), 187–203 (2015)
8. Kułakowski, K.: On the properties of the priority deriving procedure in the pairwise comparisons method. Fundamentae Informaticae **139**(4), 403–419 (2015)
9. Saaty, T.L.: Relative measurement and its generalization in decision making. Why pairwise comparisons are central in mathematics for the measurement of intangible factors. The analytic hierarchy/network process. Stat. Op. Res. (RACSAM) **102**, 251–318 (2008)
10. Chu, A.T.W., Kalaba, R.E., Spingarn, K.: A comparison of two methods for determining the weight belonging to fuzzy sets. J. Opt. Theory Appl. **4**, 531–538 (1979)
11. Koczkodaj, W.W., Herman, M.W., Orlowski, M.: Using consistency-driven pairwise comparisons in knowledge-based systems. In: A. Press (ed.) Proceedings of the Sixth International Conference on Information and Knowledge Management, pp. 91–96 (1997)
12. Ailon, N.: An active learning algorithm for ranking from pairwise preferences with an almost optimal query complexity. J. Machine Learn. Res. **13**, 137–164 (2012)
13. David, H.A.: The Method of Paired Comparisons. Oxford University Press, New York (1988)
14. Gass, S.I., Standard, S.M.: Characteristics of positive reciprocal matrices in the analytic hierarchy process. J. Op. Res. Soc. **53**, 1385–1389 (2002)
15. Jamieson, K.G., Nowak, R.: Active ranking using pairwise comparisons. In: J. Shawe-Taylor, R.S. Zemel, P. Bartlett, F.C.N. Pereira, K.Q. Weinberger (eds.) Advances in Neural Information Processing Systems 24 (NIPS), pp. 2240–2248. MIT Press (2011)
16. Negahban, S., Oh, S., Shah, D.: Iterative ranking from pairwise comparisons. In: P. Bartlett, F. Pereira, C. Burges, L. Bottou, K.Q. Weinberger (eds.) Advances in Neural Information Processing Systems 25, pp. 2483–2491. MIT Press (2012)

17. Peláez, J.I., Lamata, M.T.: A new measure of consistency for positive reciprocal matrices. Comput. Math. Appl. **46**, 1839–1845 (2003)
18. Jones, J.D.: Towards lateral inhibition and collective perception in unorganised non-neural systems. In: Pancerz, K., Zaitseva, E. (eds.) Computational Intelligence. Selected Links. Springer, Medicine and Biology (2015)
19. Bull, R., Segerberg, K.: Basic Modal Logic, pp. 1–88. Springer Netherlands, Dordrecht (1984)
20. Costerton, J.W., Lewandowski, Z., Caldwell, D.E., Korber, D.R., Lappin-Scott, H.: Microbial biofilms. Annu. Rev. Microbiol. **49**, 711–745 (1995). https://doi.org/10.1146/annurev.mi.49.100195.003431
21. Duffy, J.E.: The ecology and evolution of eusociality in sponge-dwelling shrimp. In: Kikuchi, T. (ed.) Genes, Behavior, and Evolution in Social Insects, pp. 1–38. University of Hokkaido Press, Sapporo, Japan (2002)
22. Jarvis, J.: Eusociality in a mammal: Cooperative breeding in naked mole-rat colonies. Science **212**(4494), 571–573 (1981). https://doi.org/10.1126/science.7209555
23. Jacobs, D.S., Bennett, N.C., Jarvis, J.U.M., Crowe, T.M.: The colony structure and dominance hierarchy of the damaraland mole-rat, cryptomys damarensis (rodentia: Bathyergidae), from namibia. J. Zool. **224**(4), 553–576 (1991). https://doi.org/10.1111/j.1469-7998.1991.tb03785.x
24. Jarvis, J.U.M., Bennett, N.C.: Eusociality has evolved independently in two genera of bathyergid mole-rats but occurs in no other subterranean mammal. Behav. Ecol. Sociobiol. **33**(4), 253–360 (1993). https://doi.org/10.1007/BF02027122
25. D'Ettorre, P., Heinze, J.: Sociobiology of slave-making ants. Acta Ethologica **3**(2), 67–82 (2001). https://doi.org/10.1007/s102110100038
26. Michener, C.: Comparative social behavior of bees. Annu. Rev. Entomol. **14**, 299–342 (1969). https://doi.org/10.1146/annurev.en.14.010169.001503
27. Skinner, B.F.: About Behaviorism. Random House Inc, New York (1976)
28. Parsons, T.: Social Systems and The Evolution of Action Theory. The Free Press, New York (1975)
29. Beni, G., Wang, J.: Swarm intelligence in cellular robotic systems. In: Proceeding of NATO Advanced Workshop on Robots and Biological Systems. Tuscany, Italy, pp. 26–30 (1989)
30. Krützen, M., Mann, J., Heithaus, M.R., Connor, R.C., Bejder, L., Sherwin, W.B.: Cultural transmission of tool use in bottlenose dolphins. PNAS **102**(25), 8939–8943 (2005)
31. Whiten, A., Goodall, J., McGrew, W.C., Nishida, T., Reynolds, V., Sugiyama, Y., Tutin, C.E., Wrangham, R.W., Boesch, C.: Cultures in chimpanzees. Nature **399**(6737), 682–685 (1999)
32. Viscido, S., Parrish, J., Grunbaum, D.: Individual behavior and emergent properties of fish schools: a comparison of observation and theory. Marine Ecol. Progress Series **273**, 239–249 (2004)
33. Helbing, D., Keltsch, J., Molnar, P.: Modelling the evolution of human trail systems. Nature **388**, 47–50 (1997)
34. John, A., Schadschneider, A., Chowdhury, D., Nishinari, K.: Characteristics of ant-inspired traffic flow. Swarm Intel. **2**(1), 25–41 (2008). https://doi.org/10.1007/s11721-008-0010-8
35. Helbing, D., Farkas, I., Vicsek, T.: Simulating dynamical features of escape panic. Nature **407**(6803), 487–490 (2000). https://doi.org/10.1038/35035023
36. Abrahams, M., Colgan, P.: Risk of predation, hydrodynamic efficiency, and their influence on school structure. Environ. Biol. Fishes **13**(3), 195–202 (1985)
37. Olson, R.S., Hintze, A., Dyer, F.C., Knoester, D.B., Adami, C.: Predator confusion is sufficient to evolve swarming behaviour. J. R. Soc. Interface **10**(85), 20130305 (2013). https://doi.org/10.1098/rsif.2013.0305
38. Riesenhuber, M., Poggio, T.: Neural mechanisms of object recognition. Curr Opin Neurobiol. **12**(2), 162–168 (2002)
39. Sakiyama, T.: Gunji, Y.P.: The Müller-Lyer illusion in ant foraging. PLOS ONE **12**(8), 1–12 (2013)
40. Tani, I., Yamachiyo, M., Shirakawa, T., Gunji, Y.P.: Kanizsa illusory contours appearing in the plasmodium pattern of *Physarum polycephalum*. Front. Cell. Infect. Microbiol. **4** (2014)

41. Nalyanya, G., Moore, C.B., Schal, C.: Integration of repellents attractants, and insecticides in a "push-pull" strategy for managing German cockroach (dictyoptera: Blattellidae) populations. J. Med. Entomol. **37**(3), 427–434 (2000)
42. Tsang, N., Macnab, R., Koshland, D.E.: Common mechanism for repellents and attractants in bacterial chemotaxis. Science **181**(4094), 60–63 (1973)
43. Kearns, D.B.: A field guide to bacterial swarming motility. Nature Rev. Microbiol. **8**(9), 634–644 (2010)
44. Kripke, S.A.: A completeness theorem in modal logic. J. Symbol. Logic **24**(1), 1–14 (1959)
45. Blumer, H.: Symbolic Interactionism. Perspective and Method. Prentice-Hall, Englewood Cliffs, NJ (1969)

Chapter 10
Non-Archimedean Probabilities and Reflexive Games

Any swarm can change its preference relations in accordance with some conditions outside and inside the intercommunication of swarm members. These conditions can make the same stimuli more or less proto-symbolic (that is, laterally inhibited or laterally activated, respectively). So, some primitive forms of symbolic interactionism can be observed even at the level of swarming. Each player of swarms can change a decision during the game. This circumstance contradicts to the Aumann's agreement theorem, according to that each rational player makes the same decision, basing on the same facts and appealing to Bayesian probabilities. Nevertheless, even a swarm does not follow this theorem. So what does that mean for human beings who interact first of all symbolically? The theorem becomes absurd.

In this chapter, I am going to introduce reflexive games as a generalization of bio-inspired games. For these games the Aumann's agreement theorem does not hold. The results of this chapter are published in [1–8].

10.1 Reflexive Games

10.1.1 Perlocutionary Effects and Reflexivity

Any everyday dialogue can be considered a reflexive game. Each person, speaking those or other things, tries to obtain something from us. We always try to understand the motives (s)he has before talking to us. Are the motives only to learn something from us or to influence us? How exclusive is the message which (s)he utters? Will we begin to know more on the topic after the talk? Is (s)he sincere? How sincerely does (s)he express for us his/her strategy of reasoning?

A. Schumann, *Behaviourism in Studying Swarms: Logical Models of Sensing and Motoring*, Emergence, Complexity and Computation 33,
https://doi.org/10.1007/978-3-319-91542-5_10

In reflexive games both conflicting parties simulate reasoning as well as emotional reactions of each other and aspire to foresee them. This reasoning and these reactions, which we expect from our interlocutor/opponent, are called *perlocutionary effects* of our *dialogue*. Let us recall that any dialogue has three levels: *locutionary* (information expressed in verbal or non-verbal exposition of states of affairs), *illocutionary* (cognitive and emotional estimations of considered information), *perlocutionary* (effects of cognitive and emotional estimations of interlocutors/opponents). What is the perlocutionary effects of our utterance, we do not know, but we try to foresee them. The situation where my cognitive estimations and feelings are not transparent for the interlocutor, while his/her cognitive estimations and feelings are transparent for me, shows that my level of reflexion is higher than my interlocutor's level.

Emotions, which are expressed in illocutions, are one of the main forms of reflexion. The interchanging of emotions is always a reflexive game, a method of manipulation of others. The character played by Sharon Stone in *Basic Instinct* (the 1992 movie) shows reflexive abilities in emotional management. How transparent are her emotions? Are we capable of winning emotionally in games with her or at least of reaching an emotional consensus? Her emotions are not at all transparent for us as are the emotions of coaching trainers who better know strategies of management struggle and overcome us in any reflexive game.

In this chapter, I will show that we can appeal to non-Archimedean probabilities in formalizing reflexive games. The main features of these probabilities are that Aumann's agreement theorem [9, 10] does not hold there and we can define reflexive games for finite as well as infinite levels of reflexion. The basic idea to deny Aumann's theorem in some cases is as follows. He used inductive sets for defining knowledge operators. However, on non-inductive sets, when we apply non-Archimedean probabilities, this theorem cannot be proved. But these sets allow us to define sophisticated reflexive games.

10.1.2 Knowledge Operators and Aumann's Agreement Theorem

Let Ω be a finite set of possible states of the world. It is finite, since any physically realistic agent cannot have more than finitely many possible experiences. Subsets of Ω will be called propositions. Suppose we have a set of N agents, call them $i = 1, \ldots, N$. Agent i's *knowledge structure* is a function $\mathbf{P}_i$ which assigns to each $\omega \in \Omega$ a non-empty subset of Ω. $\mathbf{P}_i$ is a partition of Ω: each world ω belongs to exactly one element of each $\mathbf{P}_i$, i.e. Ω is a set of mutually disjoint subsets $\mathbf{P}_i$ whose union is Ω. Then $\mathbf{P}_i(\omega)$ is called i's *knowledge state at* ω. This means that if the true state is ω, the individual only knows that the true state is in $\mathbf{P}_i(\omega)$. This is important because it restricts the assignment of probabilities about the true state ω. The elements of $\mathbf{P}_i(\omega)$ are those states of the world that are compatible with everything that i knows at ω. In other words, we can interpret $\mathbf{P}_i(\omega)$ probabilistically as follows: $\mathbf{P}_i(\omega) =$

$\{\omega' : P_i(\omega'|\omega) > 0\}$. Since i cannot distinguish among states in the cell $\mathbf{P}_i(\omega)$ of his knowledge partition $\mathbf{P}_i$, his subjective prior must satisfy $P_i(\omega'', \omega) = P_i(\omega'', \omega')$ for all $\omega'' \in \Omega$ and all $\omega' \in \mathbf{P}_i(\omega)$. Also, we assume a player believes the actual state is possible, so $P_i(\omega|\omega) > 0$ for all $\omega \in \Omega$. From these both assumptions it follows that the possibility operator $\mathbf{P}_i$ has the following two properties: for all $\omega', \omega \in \Omega$:

$$\omega \in \mathbf{P}_i(\omega) \tag{10.1}$$

$$\omega' \in \mathbf{P}_i(\omega) \Rightarrow \mathbf{P}_i(\omega') = \mathbf{P}_i(\omega) \tag{10.2}$$

All propositions in Ω including those in any of the N partitions form a σ-field $\mathscr{A}$. $\mathbf{P}_i(\omega) \subseteq A$ is interpreted as meaning that at ω agent i knows that A, i.e. $\omega' \in A$ for all states ω' that i considers possible at ω.

For each i, the expression below defines a *knowledge operator* K_i which, applied to any set $A \in \mathscr{A}$, yields the set $K_i A \in \mathscr{A}$ of worlds in which i knows A:

$$K_i A = \{\omega : \mathbf{P}_i(\omega) \subseteq A\}.$$

If $\mathbf{P}_i(\omega) \subset A$, an individual i who observes ω, will know that a state of the event A has occurred. The most important property of the knowledge operator is $K_i A \subseteq A$; i.e. if an agent knows an event A in state ω (i.e. $\omega \in K_i A$), then A is true in state ω (i.e. $\omega \in A$) as it follows from (10.1).

We can prove the following statements:

$$K_i \Omega = \Omega; \tag{10.3}$$

$$K_i(A \cap B) = K_i A \cap K_i B; \tag{10.4}$$

$$A \subseteq B \Rightarrow K_i A \subseteq K_i B; \tag{10.5}$$

$$K_i A \subseteq A; \tag{10.6}$$

$$K_i K_i A = K_i A; \tag{10.7}$$

$$\neg K_i \neg K_i A \subseteq K_i A. \tag{10.8}$$

Now let us try to define "*common knowledge*" in a state ω. First notice that "mutual knowledge" appears if and only if that state ω is an element of the knowledge function $K_i A$ for all individuals i. "Common knowledge" between two individuals assumes as well an infinite mutual reflexion: both have mutual knowledge of A, but also both know that both know A, both know that both know that both know A etc. ad infinitum. The *common knowledge operator* KA is then defined by

$$KA = K_1 A \cap K_2 A \cap K_1 K_2 A \cap K_2 K_1 A \cap K_1 K_2 K_1 A \cap \ldots$$

More formally, for each natural number n we can define an operator M_n expressing "n-th degree mutual reflexion:"

$$M_0 A = A; \quad M_{n+1} A = \bigcap_{i=1}^{N} K_i M_n A$$

Finally, the next expression defines *common knowledge*, κ, as mutual knowledge and mutual reflexion of all finite degrees:

$$\kappa A = \bigcap_{n=0}^{+\infty} M_n A.$$

Lemma 10.1 *If $\omega \in \kappa A$, then for each i, $\mathbf{P}_i(\omega) \subseteq \kappa A$.*

Proof If $\omega \in \kappa A$ then $\omega \in K_i M_n A$ for all agents i and degrees n of mutual reflexion. Therefore $\mathbf{P}_i(\omega) \subseteq M_n A$ for all n, and thus $\mathbf{P}_i(\omega) \subseteq \kappa A$. □

Theorem 10.1 (Aumann's Agreement Theorem) *Let us consider a hypothesis $H \in \mathscr{A}$ for which the various agents' probabilities are $q_1, \ldots, q_N$ after they condition P on priors. The proposition $C \in \mathscr{A}$ identifies these probabilities:*

$$C = \bigcap_{i=1}^{N} \{\omega : P(H|\mathbf{P}_i(\omega)) = q_i\}.$$

If (i) $\langle \Omega, \mathscr{A}, P \rangle$ is a probability space, where $\mathscr{A}$ includes each of the partitions $\Omega_1, \ldots, \Omega_N$ of Ω, and each of those partitions is countable ($\mathscr{A}$ is then closed under all of the operators K_i and M_n, and under κ) and P is the old probability measure that is common to all the agents; (ii) the possibility of C's becoming common knowledge is not equal to zero: $P(\kappa C) \neq 0$, then

$$P(H|\kappa C) = q_1 = \ldots = q_N.$$

Proof By the lemma $\kappa C = \bigcup_j D_{ij}$, where D_{ij} are distinct members of $\mathbf{P}_i$. Then $P(H|\kappa C) = \frac{P(H \cap \bigcup_j D_{ij})}{P(\bigcup_j D_{ij})} = \frac{\sum_j P(H|D_{ij})P(D_{ij})}{\sum_j P(D_{ij})} = \frac{\sum_j q_i P(D_{ij})}{\sum_j P(D_{ij})} = q_i.$ □

As we see, the condition $K_i A \subseteq A$ is basic for Aumann's theorem. It allows him to appeal to inductive sets in defining knowledge operators. Nevertheless, we can change this condition by another: $A \subseteq K_i' A$ to appeal to non-inductive sets. The condition $K_i A \subseteq A$ is understood as the least fixed point, but we can propose to use $A \subseteq K_i' A$ as the greatest fixed point. From $A \subseteq K_i' A$ it follows that $K_i'^n A \subseteq K_i'^{n+1} A$.

10.1.3 Why Can We Reject Aumann's Agreement Theorem?

Aumann's agreement theorem actually says that two agents acting rationally (according to Bayesian formulas) and with common knowledge of each other's beliefs cannot agree to disagree. More specifically, if two people share common priors, and have common knowledge of each other's current probability assignments (their posteriors for a given event A are common knowledge), then they must have equal probability assignments (these posteriors must be equal). It is one of the most important statement of game theory, epistemic logic and so on. For example, according to this statement, each rational player has to behave in the same manner under the same circumstances. Rational players have always a common knowledge, they know all parameters of the game and have to be sure that their opponents know that they know parameters of the game, that they know that they know and so on ad infinitum.

To prove his theorem, Aumann appeals to representing the possibility operator $\mathbf{P}_i(\omega)$ and the common knowledge operator K_i as the least fixed points, i.e. as *inductive sets*. Nevertheless, we can define the possibility operator $\mathbf{P}_i(\omega)$ and the common knowledge operator K_i as the greatest fixed points as well, i.e. as *coinductive sets*. In this way we cannot prove Aumann's agreement theorem. Instead of the latter statement we then prove the *reflexion disagreement theorem* as an appropriate negation of Aumann's theorem. While for Aumann's theorem we need the property $\mathbf{P}_i(\omega) = \bigcap\{A : \omega \in K_i A\}$, for its negation we need the property $\mathbf{P}_i(\omega) = \bigcup\{A : \omega \in K_i A\}$. In other words, this new statement can be proven if we change some standard philosophical presuppositions in game theory by the following new assumptions: each rational agent can cheat (disagree in his heart with) other rational agents, each player cannot know everything prior the game, each agent can try to foresee knowledge (beliefs) of his/her opponents and manipulate them, therefore the common knowledge does not mean that an agent cannot disagree and will be completely predictable for all others.

These philosophical presuppositions contradicting to Aumann's ideas were first formulated by Vladimir Lefebvre in his notion '*reflexive games*' in 1965 [11–13]. A game is called reflexive if to choose the action the agent has to model (predict) actions of his/her opponents, e.g. (s)he can try to manipulate them or cheat them. The Early Levebvre formulated reflexive games assuming many reflexion levels. At the zero level I ignore beliefs of opponents, at the first level I take into account their beliefs, at the second level I take into account that they try to predict my beliefs, at the third level I foresee their beliefs in which my beliefs are foreseen by them, etc. The game-theoretic mathematics for ideas of the Early Levebvre has been developed by Novikov and Chkhirtishvili [14]. The reflexion disagreement theorem that will be proved in the next section holds true for their approach, namely if we suppose a possibility of reflexive games on a reflexion level of any natural number. I was inspirited and suggested by the ideas of the Early Levebvre in the same measure as them.

The Later Lefebvre tried to simulate decision making in reflexive games by means of Boolean functions [12]. The main disadvantages of approaches encouraged by the

Later Lefebvre consist in that reflexive levels are ignored and agents are presented as automata. However, the self-estimation is a variable. Reflexion varies depending on characteristic moods (illocutionary acts) as well the persuasiveness and emotionality of our interlocutors (perlocutionary effects). In this sense, the dynamism of reflexion quite corresponds to the well-known *paradox of* Belzung and Chevalley [15] which is formulated as such: an emotional response of the same person in the same situation can be various in different points of time. Thus, the simulation of decision making in reflexive games by means of Boolean functions is too speculative and cannot help in analyzing everyday situations. Within this approach we assume that reactions and evaluations of the same agent remain the same forever. Nevertheless, it is false. This approach can be useful only in explicating some basic features of reflexive management that take place in the given situation (such as the case of Soviet and American ethical patterns [12]).

The ideas of the Early Lefebvre which I try to develop in this chapter are very close to ideas of metagame which were proposed by Howard [16]. According to him, for any game G and any player i there can be a *metagame* iG, in which player i chooses in knowledge of the choices of all the others. More formally, let $G = \langle S_1, S_2; M_1, M_2\rangle$ be the normal form of the game, where S_1 (resp. S_2) is the set of strategies for player 1 (resp. 2) and M_1 (resp. M_2) is his/her preference function. The set of outcomes is $S = S_1 \times S_2$, i.e. an outcome is an ordered pair $s = \langle s_1, s_2\rangle$. $M_i(s) = \{s' : \text{is not preferred to } s \text{ by player } i\}, i = 1, 2$. Let $B(S_1)$ (resp. $B(S_2)$) be the set of non-null subsets of S_1 (resp. of S_2) and $K_1 \subseteq B(S_1)$ (resp. $K_2 \subseteq B(S_2)$). Then the *first level metagame* KG is defined as the normal form $KG = \langle X_1, X_2; M'_1, M'_2\rangle$, where $X_1 = \{x_1 : x_1 = \langle f_1, c_1\rangle; c_1 \in K_1; f_1 : K_2 \to c_1\}$, $X_2 = \{x_2 : x_2 = \langle f_2, c_2\rangle; c_2 \in K_2; f_2 : K_1 \to c_2\}$, and for $i = 1, 2$, M'_i satisfies the following property:

$$x' \in M'_i(x) \text{ iff } \beta x' \in M'_i(\beta x),$$

where $\beta(f_1, c_1; f_2, c_2) = (f_1(c_2), f_2(c_1))$.

By induction, we can obtain an n-th-level metagame $K_n K_{(n-1)} \ldots K_1 G$. The set of all metagames $K_n K_{(n-1)} \ldots K_1 G$ for any natural number n is called the *infinite metagame* based on G. This metagame corresponds to Lefebvre's *reflexive game of infinite level.*

Nigel Howard proposed to use metagames to have possibilities to require "more of rationality than that each player should optimize given its beliefs about the others' choices" [16]. Now (s)he should be able "to know the others' choices, and know how the other would choose to react to such knowledge, and know each others' reactions to such reactions, and so on" [16]. These ideas are almost the same as ideas introduced in reflexive games. The difference consists in other ways of defining preference functions.

The reflexion disagreement theorem holds true for the infinite metagame in Howard's meaning as well as for the reflexive game of infinite level in Lefebvre's meaning. This theorem shows *limits in infinite mutual predictions of others' knowledge*.

The mathematical meaning of the reflexion disagreement theorem is that we cannot prove the agreement theorem using probabilities running over streams (e.g. using probabilities with values on hypernumbers or p-adic numbers) in any way. In non-standard fields Aumann's theorem is false, because the powerset of any infinite set of streams is not a Boolean algebra and the Bayesian theorem does not hold in general case for streams (see Theorem 5.4). Notice that we cannot avoid streams in the case of infinite metagame or reflexive game of infinite level, because we face there an infinite data structure consisting of streams. The fuzzy and probability logic with values on streams was obtained in [3]. This logic can be used for developing a probability theory and epistemic logic for the infinite metagame and reflexive game of infinite level.

Mathematically, the infinite metagame is a *coalgebra* [17]. Graphically, coalgebras (e.g. processes or games) can be represented as infinite trees. The reflexion disagreement theorem is valid for games presented in the form of coalgebra. Recently many researchers [18–21] have focused on the idea that in economics, in particular in decision theory, we cannot avoid coalgebraic notions such as process dynamics, behavioural instability, self-reference, or circularity. There exist many more cases of non-equilibriums in economics, because we engage coinductive databases more often as a matter of fact [22]. For example, repeated games may be defined only coalgebraically [23] and as well it is better to define epistemic games and belief functors as coalgebras [18].

Thus, the reflexion disagreement theorem can be proved if (i) we assume that rational agents can become unpredictable for each other and try to manipulate; (ii) we define probabilities on streams (e.g. on hypernumbers or p-adic numbers); (iii) games are presented as coalgebras. As we see, this new theorem is a very important statement within the new mathematics (coalgebras, transition systems, process calculi, etc.) which has been involved into game theory recently. Sets of streams which have been modelled coalgebraically cannot generate inductive sets [24], therefore Aumann's agreement theorem is meaningless on these sets, but we face just the sets of streams in many kinds of games (e.g. if we deal with repeated games, games with infinite states, concurrent games, infinite metagame, reflexive game of infinite level, etc.). Instead of the agreement theorem, the reflexion disagreement theorem is valid if we cannot obtain inductive sets, e.g. in the case of sets of streams. Notice that according to Aczel [25], the universum of coinductive sets is much larger than the universe of inductive sets.

10.2 Reflexive Disagreement Theorem

10.2.1 Streams and p-*Adic Numbers*

One of the most useful non-well-founded mathematical object is a *stream*—a recursive data-type of the form $s = \langle a, s' \rangle$, where s' is another stream. The notion of *stream*

Table 10.1 Coinductive definitions of sum, product and inverse

Differential equation	Initial value	Name
$(\sigma+\tau)' = \sigma' + \tau'$	$(\sigma+\tau)(0) = \sigma(0) + \tau(0)$	Sum
$(\sigma \times \tau)' =$ $(\|\sigma(0)\| \times \tau') + (\sigma' \times \tau)$	$(\sigma \times \tau)(0) = \sigma(0) \times \tau(0)$	Product
$(\sigma^{-1})' =$ $\|-1\| \times \|\sigma(0)^{-1}\| \times \sigma' \times \sigma^{-1}$	$(\sigma^{-1})(0) = \sigma(0)^{-1}$	Inverse

calculus was introduced by Pavlović and Escardó [26] as a means to do symbolic computation using the coinduction principle instead of the induction one. Let A be any set. We define the set A^{ω} of all streams over A as $A^{\omega} = \{\sigma : \{0, 1, 2, \ldots\} \to A\}$. For a stream σ, we call $\sigma(0)$ the initial value of σ. We define the *derivative* $\sigma(0)$ of a stream σ, for all $n \geq 0$, by $\sigma'(n) = \sigma(n+1)$. For any $n \geq 0$, $\sigma(n)$ is called the n-th element of σ. It can also be expressed in terms of higher-order stream derivatives, defined, for all $k \geq 0$, by $\sigma^{(0)} = \sigma$; $\sigma^{(k+1)} = (\sigma^{(k)})'$. In this case the n-th element of a stream σ is given by $\sigma(n) = \sigma^{(n)}(0)$. Also, the stream is understood as an infinite sequence of derivatives. It will be denoted by an infinite sequence of values or by an infinite tuple: $\sigma = \sigma(0) :: \sigma(1) :: \sigma(2) :: \cdots :: \sigma(n-1) :: \sigma^{(n)}$, $\sigma = \langle \sigma(0), \sigma(1), \sigma(2), \ldots \rangle$.

Streams are defined by coinduction: two streams σ and τ in A^{ω} are equal if they are *bisimilar*: (i) $\sigma(0) = \tau(0)$ (they have the same *initial value*) and (ii) $\sigma' = \tau'$ (they have the same *differential equation*). To set addition and multiplication by coinduction, we should use the following facts about differentiation of sums and products by applying the basic operations: $(\sigma + \tau)' = \sigma' + \tau'$, $(\sigma \times \tau)' = (|\sigma(0)| \times \tau') + (\sigma' \times \tau)$, where $|\sigma(0)| = \langle \sigma(0), 0, 0, 0, \ldots \rangle$. Now we can define them as well as another stream operation as follows (Table 10.1).

We can embed the real numbers into the streams by defining the following constant stream. Let $r \in \mathbf{R}$. Then $|r| = \langle r, 0, 0, 0, \ldots \rangle$ is defined so: its differential equation is $|r|' = [0]$, its initial value is $|r|(0) = r$. We are to rely on our intuitions that it would be natural to define the positive real numbers to be less than the positive streams.

Consider the set of streams $[0, 1]^{\omega}$ and extend the standard order structure on $[0, 1]$ to a partial order structure on $[0, 1]^{\omega}$. Further define this order as follows:

$\mathscr{O}_{[0,1]^{\omega}}$ (1) For any streams $\sigma, \tau \in [0, 1]^{\omega}$, we set $\sigma \leq \tau$ if $\sigma(n) \leq \tau(n)$ for every $n \in \mathbf{N}$. For any streams $\sigma, \tau \in [0, 1]^{\omega}$, we set $\sigma = \tau$ if σ, τ are bisimilar. For any streams $\sigma, \tau \in [0, 1]^{\omega}$, we set $\sigma < \tau$ if $\sigma(n) \leq \tau(n)$ for every $n \in \mathbf{N}$ and there exists n_0 such that $\sigma(n_0) \neq \tau(n_0)$. (2) Each stream of the form $|r| \in [0, 1]^{\omega}$ (i.e. constant stream) is less than an inconstant stream σ.

This ordering relation is not linear, but partial, because there exist streams $\sigma, \tau \in [0, 1]^{\omega}$, which are incompatible.

Introduce two operations sup, inf in the partial order structure $\mathscr{O}_{[0,1]^{\omega}}$. Assume that $\sigma, \tau \in [0, 1]^{\omega}$ are either both constant streams or both inconstant streams.

Then their supremum and infimum are defined by coinduction: the differential equation of supremum is $(\sup(\sigma, \tau))' = \sup(\sigma', \tau')$ and its initial value is $(\sup(\sigma, \tau))(0) = \sup(\sigma(0), \tau(0))$, the differential equation of infimum is $(\inf(\sigma, \tau))' = \inf(\sigma', \tau')$ and its initial value is $(\inf(\sigma, \tau))(0) = \inf(\sigma(0), \tau(0))$. Suppose now that one and only one of $\sigma, \tau \in [0, 1]^\omega$ is constant, then an inconstant stream is greater than a constant one, therefore their supremum gives an inconstant stream, but their infimum gives a constant stream.

According to $\mathcal{O}_{[0,1]^\omega}$, there exist the maximal stream $[1] \in [0, 1]^\omega$ and the minimal stream $[0] = |0| \in [0, 1]^\omega$.

Now introduce the following new operations defined for all $\sigma, \tau \in [0, 1]^\omega$ in the partial order structure $\mathcal{O}_{[0,1]^\omega}$:

- $\sigma \to \tau = [1] + |-1| \times \sup(\sigma, \tau) + \tau$,
- $\neg\sigma = [1] + |-1| \times \sigma$, i.e. $\sigma \to [0]$,
- $\sigma \& \tau = \sup(\sigma, [1] + |-1| \times \tau) + \tau[-1]$, i.e. $\sigma \& \tau = \neg(\sigma \to \neg\tau)$,
- $\sigma \oplus \tau := \neg\sigma \to \tau$,
- $\sigma \ominus \tau := \sigma \& \neg\tau$,
- $\sigma \wedge \tau = \inf(\sigma, \tau)$, i.e. $\sigma \wedge \tau = \sigma \& (\sigma \to \tau)$,
- $\sigma \vee \tau = \sup(\sigma, \tau)$, i.e. $\sigma \vee \tau = (\sigma \to \tau) \to \tau$,
- $\sigma \leftrightarrow \tau := (\sigma \to \tau) \wedge (\tau \to \sigma)$,

Obviously, all above operations are defined by coinduction.

In 1897 the German mathematician Kurt Hensel presented an idea how to use an analogy of Taylor and Laurent series to study algebraic numbers by expressing them as an expansion in terms of powers of a prime number. He was mainly inspired by the work of Kummer. This approach by Hensel led him to introduce the p-adic numbers. There are many books which give a good introduction to the p-adic theory, see for instance Koblitz [27].

It can be easily shown that p-adic numbers may be represented as potentially infinite data structures such as streams. Each stream of the form $\sigma = \sigma(0) :: \sigma(1) :: \sigma(2) :: \cdots :: \sigma(n-1) :: \sigma^{(n)}$, where $\sigma(n) \in \{0, 1, \ldots, p-1\}$ for every $n \in \mathbf{N}$, may be converted into a *p-adic integer* by the following rule:

$$\forall n \in \mathbf{N}, \sigma(n) = \sum_{k=0}^{n} \sigma(k) \cdot p^k \wedge \sigma(n) = \sigma(0) :: \sigma(1) :: \cdots :: \sigma(n). \tag{10.9}$$

And vice versa, each p-dic integer may be converted into a stream taking rule (10.9). Such a stream is called *p-adic*.

Extend rule (10.9) as follows. Suppose that we have a stream of the form $\sigma = \sigma(0) :: \sigma(1) :: \sigma(2) :: \cdots :: \sigma(n-1) :: \sigma^{(n)}$, where $\sigma(n) \geq 0$ for every $n \in \mathbf{N}$. Then its p-adic representation is

$$\forall n \in \mathbf{N}, \sigma(n) = \sum_{k=0}^{m} \tau(k) \cdot p^k \wedge \sigma(n) = \tau(0) :: \tau(1) :: \cdots :: \tau(m), \quad (10.10)$$

where $\tau(i) \in \{0, 1, \ldots, p-1\}$ for every $i = \overline{1, m}$ and $\sum_{k=0}^{m} \tau(k) \cdot p^k = \sum_{k=0}^{n} \sigma(k) \cdot p^k$. (In the case $\sigma(i) \in \{0, 1, \ldots, p-1\}$ for every $i = \overline{1, n}$, we have $n = m$ and then $\sigma(i) = \tau(i)$ for every $i = \overline{1, n}$.) Such a stream is called *p-adic* too. Its *canonical form* is $\tau(0) :: \tau(1) :: \cdots :: \tau(m) :: \tau^{(m+1)}$, where $\tau(n) \in \{0, 1, \ldots, p-1\}$ for every $n \in \mathbf{N}$.

Using (10.9), (10.10), we can show that sum, product and inverse have the same differential equations and initial values as in stream calculus. This proves that *p-adic numbers are one of the natural interpretations of streams.*

It is easily shown that the set A^{ω} of all p-adic streams includes the set of natural numbers. Let n be a natural number. It has a finite p-adic expansion $n = \sum_{k=0}^{m} \alpha_k \cdot p^k$. Then we can identify n with a p-adic stream $\sigma = \sigma(0) :: \sigma(1) :: \cdots :: \sigma(m) :: \sigma^{(m+1)}$, where $\sigma(i) = \alpha_i$ for $i = \overline{0, m}$ and $\sigma^{(m+1)} = [0]$.

Extend the standard order structure on $\mathbf{N}$ to a partial order structure on p-adic streams (i.e. on $\mathbf{Z}_p$).

1. for any p-adic streams $\sigma, \tau \in \mathbf{N}$ we have $\sigma \leq \tau$ in $\mathbf{N}$ iff $\sigma \leq \tau$ in $\mathbf{Z}_p$,
2. each p-adic stream $\sigma = \sigma(0) :: \sigma(1) :: \cdots :: \sigma(m) :: \sigma^{(m+1)}$, where $\sigma^{(m+1)} = [0]$ (i.e. each finite natural number), is less than any infinite number τ, i.e. $\sigma < \tau$ for any $\sigma \in \mathbf{N}$ and $\tau \in \mathbf{Z}_p \backslash \mathbf{N}$.

Define this partial order structure on $\mathbf{Z}_p$ as follows:

$\mathcal{O}_{\mathbf{Z}_p}$ Let $\sigma = \sigma(0) :: \sigma(1) :: \cdots :: \sigma(n-1) :: \sigma^{(n)}$ and $\tau = \tau(0) :: \tau(1) :: \cdots :: \tau(n-1) :: \tau^{(n)}$ be p-adic streams. (1) We set $\sigma < \tau$ if the following three conditions hold: (i) there exists n such that $\sigma(n) < \tau(n)$; (ii) $\sigma(k) \leq \tau(k)$ for all $k > n$; (iii) σ is a finite integer, i.e. there exists m such that $\sigma^{(m)} = [0]$. (2) We set $\sigma = \tau$ if σ and τ are bisimilar. (3) Suppose that σ, τ are infinite integers. We set $\sigma \leq \tau$ by coinduction: $\sigma \leq \tau$ iff $\sigma(n) \leq \tau(n)$ for every $n \in \mathbf{N}$. We set $\sigma < \tau$ if we have $\sigma \leq \tau$ and there exists $n_0 \in \mathbf{N}$ such that $\sigma(n_0) < \tau(n_0)$.

Now introduce two operations sup, inf in the partial order structure on $\mathbf{Z}_p$. Suppose that p-adic streams σ, τ represent infinite p-adic integers. Then their sup and inf may be defined by coinduction as follows: the differential equation of supremum is $(\sup(\sigma, \tau))' = \sup(\sigma', \tau')$ and its initial value is $(\sup(\sigma, \tau))(0) = \sup(\sigma(0), \tau(0))$, the differential equation of infimum is $(\inf(\sigma, \tau))' = \inf(\sigma', \tau')$ and its initial value is $(\inf(\sigma, \tau))(0) = \inf(\sigma(0), \tau(0))$. Now suppose that at most one of two streams σ, τ represents a finite p-adic integer. In this case $\sup(\sigma, \tau) = \tau$ if and only if $\sigma \leq \tau$ under condition $\mathcal{O}_{\mathbf{Z}_p}$ and $\inf(\sigma, \tau) = \sigma$ if and only if $\sigma \leq \tau$ under condition $\mathcal{O}_{\mathbf{Z}_p}$.

It is important to remark that there exists the maximal p-adic stream $N_{max} \in \mathbf{Z}_p$ under condition $\mathcal{O}_{\mathbf{Z}_p}$. It is easy to see: $N_{max} = [p-1] = -1 = (p-1) + (p-1) \cdot p + \cdots + (p-1) \cdot p^k + \cdots$.

Further consider the following new operations defined for all p-adic streams σ, τ in the partial order structure $\mathcal{O}_{\mathbf{Z}_p}$:

- $\sigma \to \tau = N_{max} + |-1| \times \sup(\sigma, \tau) + \tau$,
- $\neg\sigma = N_{max} + |-1| \times \sigma$, i.e. $\sigma \to [0]$,
- $\sigma \& \tau = \sup(\sigma, N_{max} + |-1| \times \tau) + \tau + |-1| \times N_{max}$, i.e. $\sigma \& \tau = \neg(\sigma \to \neg\tau)$,
- $\sigma \oplus \tau := \neg\sigma \to \tau$,
- $\sigma \ominus \tau := \sigma \& \neg\tau$,
- $\sigma \wedge \tau = \inf(\sigma, \tau)$, i.e. $\sigma \wedge \tau = \sigma \& (\sigma \to \tau)$,
- $\sigma \vee \tau = \sup(\sigma, \tau)$, i.e. $\sigma \vee \tau = (\sigma \to \tau) \to \tau$,
- $\sigma \leftrightarrow \tau := (\sigma \to \tau) \wedge (\tau \to \sigma)$,

It is evident that all these operations are defined by coinduction.

10.2.2 Non-Well-Founded Probabilities

There is a problem how it is possible to define probabilities on stream structures if we have no opportunity to put them on an algebra of subsets, taking into account the following result:

Proposition 10.1 *Define union, intersection and complement in the standard way. The powerset $\mathscr{P}(A^\omega)$, where A^ω is the set of all streams over A, is not a Boolean algebra.*

Proof Consider a counterexample on 7-adic streams. Let $A_1 = \{x\colon 0 \leq x \leq \ldots 11234321\}$ and $A_2 = \{x\colon \ldots 66532345 \leq x \leq \ldots 66666\}$ be subsets of $\mathbf{Z}_7$. It is readily seen that $\neg(A_1 \cap A_2) = \mathbf{Z}_7$, but $(\neg A_1 \cup \neg A_2) \subset \mathbf{Z}_7$, because $\neg A_1 = \{x\colon \ldots 11234321 < x \leq \ldots 66666\}$ and $\neg A_2 = \{x\colon 0 \leq x < \ldots 66532345\}$, therefore $\mathbf{Z}_7 \backslash (\neg A_1 \cup \neg A_2) = A_3 = \{x\colon x = \ldots y_5 y_4 y_3 y_2 y_1 y_0$, where $y_i \in \{0, 1, \ldots, 6\}$ for each $i \in \mathbf{N}\backslash\{3\}\}$. It is obvious that the set A_3 is infinite. As a result, we obtain that $\neg(A_1 \cap A_2) \neq \neg A_1 \cup \neg A_2$ in the general case. □

This proposition is a particular case of the following provable statement: *if A is a non-well-founded set, then its powerset will not be a Boolean algebra in the general case.*

In stream calculus and p-adic calculus we have, evidently, a different partial ordering relation and obtain different powersets $\mathscr{P}^{[0,1]^\omega}(A^\omega)$, $\mathscr{P}^{\mathbf{Z}_p}(A^\omega)$, but in any case there is no Boolean algebra, because *the complement in them is not Boolean.* The powersets $\mathscr{P}^{[0,1]^\omega}(A^\omega)$, $\mathscr{P}^{\mathbf{Z}_p}(A^\omega)$ should be interpreted as a corresponding class $\mathscr{F}^V(A^\omega)$ of fuzzy subsets $Y \subset A^\omega$, where V is equal one of sets $[0, 1]^\omega$, $\mathbf{Z}_p$.

We can try to get non-well-founded probabilities on a non-well-founded algebra $\mathscr{F}^V(A^\omega)$ of fuzzy subsets $Y \subset A^\omega$ that consists of the following: (1) union, intersection, and difference of two *non-well-founded fuzzy* subsets of A^ω; (2) $\emptyset$ and A^ω. In this case a *finitely additive non-well-founded probability measure* is a nonnegative set function $P(\cdot)$ defined for sets $Y \in \mathscr{F}^V(A^\omega)$ that runs the set V and satisfies the following properties: (1) $P(A) \geq [0]$ for all $A \in \mathscr{F}^V(A^\omega)$,

(2) $P(A^{\omega}) = |1|$ and $P(\emptyset) = [0]$, (3) if $A \in \mathscr{F}^V(A^{\omega})$ and $B \in \mathscr{F}^V(A^{\omega})$ are disjoint, then $P(A \cup B) = P(A) + P(B)$, (4) $P(\neg A) = |1| + |-1| \times P(A)$ for all $A \in \mathscr{F}^V(A^{\omega})$.

This probability measure is called *non-well-founded probability*. Their main originality is that conditions 3, 4 are independent. As a result, in a probability space $\langle X, \mathscr{F}^V(X), P\rangle$ some Bayes' formulas do not hold in the general case (see Theorem 5.4).

As an example of *trivial non-well-founded probability* we can introduce the following function defined on streams by coinduction: (1) $P(\sigma) = \inf(\sigma, [1]) \times [1]^{-1}$ for every $\sigma \in [0, 1]^{\omega}$, (2) $P(\sigma) = \inf(\sigma, N_{max}) \times N_{max}^{-1}$ for every $\sigma \in \mathbf{Z}_p$.

Consider a random experiment $\mathscr{S}$ and by $L = \{s_1, \ldots, s_m\}$ denote the set of all possible results of this experiment. The set $\mathscr{S}$ is called the label set, or the set of attributes. Suppose there are N realizations of $\mathscr{S}$ and write a result x_j after each realization. Then we obtain the finite sample: $x = (x_1, \ldots, x_N)$, $x_j \in L$. A collective is an infinite idealization of this finite sample: $x = (x_1, \ldots, x_N, \ldots)$, $x_j \in L$. Let us compute frequencies $\nu_N(\alpha; x) = n_N(\alpha; x)/N$, where $n_N(\alpha; x)$ is the number of realizations of the attribute α in the first N tests.

There exists the statistical stabilization of relative frequencies: the frequency $\nu_N(\alpha; x)$ approaches a limit as N approaches infinity for every label $\alpha \in L$. This limit $P(\alpha) = \lim \nu_N(\alpha; x)$ is said to be the probability of the label α in the frequency theory of probability. Sometimes this probability is denoted by $P_x(\alpha)$ to show a dependence on the collective x. Notice that the limits of relative frequencies have to be stable with respect to a place selection (a choice of a subsequence) in the collective.

The statistical stabilization of relative frequencies $\nu_N(\alpha; x)$ can be considered not only in the real topology on the field of rational numbers $\mathbf{Q}$ but also in any other topology on $\mathbf{Q}$. For instance, it is possible to construct the frequency theory in which probabilities were defined as limits of relative frequencies $\nu_N(\alpha; x)$ in the p-adic topology. The frequency theory of p-adic probability was proposed in [28]. It is a kind of non-well-founded probability.

Since stream calculus and as well as p-adic calculus contain infinitely large numbers, they give the possibility to consider statistical ensembles with an infinite number of elements.

Define a *non-well-founded operation of cardinality* $/\cdot/$ as follows: suppose $X \subseteq A^{\omega}$ and $K(\cdot)$ is the conventional operation of cardinality. Represent X by a Cartesian product $\prod_{j=0}^{\infty} X_j$, where X_0 is the set of all values of the form $\sigma(0)$ belonging to all streams of X, X_1 is the set of all values of the form $\sigma(1)$ belonging to all streams of X, etc. Then $/X/$ is defined by coinduction: its initial value is $K(X_0)$, its differential equation is $(/X/)' = /X'/$. The informal meaning of non-well-founded operation of cardinality is that we obtain an infinite sequence of conventional cardinalities $K(X_0)$, $K(X_1), \ldots, K(X_m), \ldots$ that coinductively calculates not the number of streams from X but the number of their possible values at every step. It is evident, therefore, that the values of $/\cdot/$ are streams.

We study now some ensembles $S = S_N$, which have a non-well-founded volume N, i.e. $/S/ = N$, where N is the stream of $[0, 1]^{\omega}$ or $\mathbf{Z}_p$. Consider a sequence of ensembles S_j having volumes $K(S_j)$, $j = 0, 1, \ldots$ Get $S = \prod_{j=0}^{\infty} S_j$. Then the cardinality $/S/ = N$. We may imagine an ensemble S as being the population of a tower $T = T_S$, which has an infinite number of floors with the following distribution of population through floors: population of j-th floor is S_j. Set $T_k = \prod_{j=0}^{k} S_j \times \prod_{m=k+1}^{\infty} \emptyset_m$. This is a population of the first $k+1$ floors. Let $A \subset S$ and let there exists: $n(A) = \lim_{k \to \infty} n_k(A)$, where $n_k(A) = /A \cap T_k/$. The quantity $n(A)$ is said to be a *non-well-founded volume of the set* A.

We define the probability of A by the standard proportional relation:

$$P(A) := P_S(A) = n(A) \times N^{-1},$$

where $/S/ = N, n(A) = /A \cap S/$.

We denote the family of all $A \subset S$, for which $P(A)$ exists, by $\mathcal{G}_S$. The sets $A \in \mathcal{G}_S$ are said to be events. The ordered system $\langle S, \mathcal{G}_S, P_S \rangle$ is called a *non-well-founded ensemble probability space for the ensemble* S.

Proposition 10.2 *Let $\mathcal{F}$ be the non-well-founded algebra of fuzzy subsets. Then $\mathcal{F} \subseteq \mathcal{G}_S$.*

Proof Let A be a set of streams. Then $n(A) = /A/$ and the probability of A has the form: $P(A) = /A/ \times /S/^{-1}$.

For instance, let $B = \neg A$. Then $/B \cap T_k/ = /T_k/ + |-1| \times /A \cap T_k/$. Hence there exists $\lim_{k \to \infty} /B \cap T_k/ = N + |-1| \times /A/$. This equality implies the standard formula: $P(\neg A) = |1| + |-1| \times P(A)$.

In particular, we have: $P(S) = |1|$. □

Proposition 10.3 *Let $A_1, A_2 \in \mathcal{G}_S$ and $A_1 \cap A_2 = \emptyset$. Then $A_1 \cup A_2 \in \mathcal{G}_S$ and $P(A_1 \cup A_2) = P(A_1) + P(A_2)$.* □

Proposition 10.4 *Let $A \in \mathcal{G}_S$, $P_S(A) \neq 0$ and $B \in \mathcal{G}_A$. Then $B \in \mathcal{G}_S$ and the following Bayes formula holds:*

$$P_A(B) = P_S(B/A) = P_S(B) \times P_S(A)^{-1}.$$

□

Proposition 10.5 *Let $N \in \mathbf{Z}_p$, $N \neq 0$ and let the ensemble S_{-1} have the p-adic volume $-1 = N_{max}$ (it is the largest ensemble, because N_{max} is the largest p-adic integer in accordance with $\mathcal{O}_{\mathbf{Z}_p}$).*

1. *Then $S_N \in \mathcal{G}_{S_{-1}}$ and $P_{S_{-1}}(S_N) = /S_N/ \times /S_{-1}/^{-1} = -N$.*
2. *Then $\mathcal{G}_{S_N} \subset \mathcal{G}_{S_{-1}}$ and probabilities $P_{S_N}(A)$ are calculated as conditional probabilities with respect to the subensemble S_N of ensemble S_{-1}: $P_{S_N}(A) = P_{S_{-1}}(\frac{A}{S_N}) = P_{S_{-1}}(A) \times P_{S_{-1}}(S_N)^{-1}, A \in \mathcal{G}_{S_N}$.*

□

If we take the p-adic case of non-well-founded probability theory, then we observe essentially new properties of relative frequencies that do not appear on real numbers. For example, consider two attributes α_1 and α_2. Suppose that in the first $N := N_k = (\sum_{j=0}^{k} 2^j)^2$ tests the label α_1 has $n_N(\alpha_1; x) = 2^k$ realizations, α_2 has $n_N(\alpha_2; x) = \sum_{j=0}^{k} 2^j$ realizations. According to our intuition, their probabilities should be different, but in real probability theory we obtain: $P_x(\alpha_1) = \lim n_N(\alpha_1; x)/N = P_x(\alpha_2) = \lim n_N(\alpha_2; x)/N = 0$. In 2-adic probability theory we have $P_x(\alpha_1) = 0 \neq P_x(\alpha_2) = -1$, because in $\mathbf{Q}_2$, $2^k \to 0$, $k \to 0$, and $-1 = 1 + 2 + 2^2 + \cdots + 2^n + \ldots$.

This example shows that in p-adic probability theory there are statistical phenomena for that relative sequences of observed events have non-zero probabilities in the p-adic metric, but do not have positive probabilities in the standard real metric.

10.2.3 Frequencies and Non-Well-Founded Probabilities in Logic

Suppose that our logical language $\mathscr{L}$ is associated with p-valued semantics (resp. infinite-valued semantics) and the set of truth values for well-formed formulas is $V = \{0, 1, \ldots, p-1\} \subseteq \mathbf{Z}$ (resp. $V = [0, 1]$).

Consider a random experiment $\mathscr{S}$ and by $L = \{\varphi_1, \ldots, \varphi_m\}$ denote the set of all possible p-valued (resp. infinite-valued) sentences that describe results of this experiment. Suppose there are N (that approaches infinity) realizations of $\mathscr{S}$ and write a sentence φ_j if it holds with the truth degree $v(\varphi_j) \in \{0, 1, \ldots, p-1\}$ (resp. $v(\varphi_j) \in [0, 1]$). Then we obtain the infinite collective of sentences: $\varphi = \langle \varphi_1, \ldots, \varphi_N, \ldots \rangle$, $\varphi_j \in L$. We can define the truth valuation v of this collective as follows: $v(\varphi) = \langle v(\varphi_1), \ldots, v(\varphi_N), \ldots \rangle$, $v(\varphi_j) \in \{0, 1, \ldots, p-1\}$ (resp. $v(\varphi_j) \in [0, 1]$). Evidently, v takes each collective of sentences to a p-adic integer (resp. real-valued stream).

Assume that $\top$ is a tautology of p-valued logic, i.e. $\top$ has the only truth value $p-1$ (resp. 1), and $\bot$ is a contradiction of p-valued logic (resp. infinite-valued logic) (it has the only truth value 0). Then we have two ideal experiments: $\mathscr{S}_1$ (when only $\top$ can describe the results of this experiment) and $\mathscr{S}_2$ (when only $\bot$ can describe the results of this experiment). In the case of $\mathscr{S}_1$ we have the infinite collective of sentences that has the only truth valuation $N_{max} = \sum_{k=0}^{\infty}(p-1) \cdot p^k \in \mathbf{Z}_p$ (resp. $[1] \in [0, 1]^{\omega}$), in the case of $\mathscr{S}_2$ the infinite collective of sentences has the only truth valuation $0 \in \mathbf{Z}_p$ (resp. $[0] \in [0, 1]^{\omega}$).

We shall identify every experiment with an appropriate infinite collective of sentences: $\psi, \varphi, \ldots$ Now we can set a *support function* $\mathbf{P}$ that is defined on the sets of all experiments (more precisely, on the set of all infinite collectives of sentences) as follows.

$$\mathbf{P}(\varphi/\psi) = \frac{\mathbf{P}(\varphi \wedge \psi)}{\mathbf{P}(\psi)} =$$

$$\frac{v(\varphi \wedge \psi)}{v(\psi)} = \frac{\inf(v(\varphi), v(\psi))}{v(\psi)},$$

when ψ is not logically false, where $v(\varphi)$ is a truth valuation of an infinite collective of sentences φ and $v(\psi)$ is a truth valuation of an infinite collective of sentences ψ.

The support function shows a logical connection between different experiments. Thus we can build a special probability logic that computes probabilities on the class of all infinite collectives of p-valued sentences, i.e. a common probability logic of different experiments. It is one of natural interpretations of p-adic valued probability logic.

10.2.4 *Reflexion Disagreement Theorem Proved on Streams and* p*-Adic Integers*

Now, using the given notion of streams, let us prove the reflexion disagreement theorem. For Ω, the finite set of possible states of the world, and N, the set of agents, we can unconventionally define agent i's accepted performances as a function $\mathbf{Q}_i$ which assigns to each $o \in \Omega$ a non-empty subset of Ω, so that each world o belongs to one or more elements of each $\mathbf{Q}_i$, i.e. Ω is contained in a union of $\mathbf{Q}_i$, but $\mathbf{Q}_i$ are not mutually disjoint. Then $\mathbf{Q}_i(o)$ is called i's *accepted performative state* at o. If the successful performance is o, the individual knows (accepts) that the performative state is in $\mathbf{Q}_i(o)$. The elements of $\mathbf{Q}_i(o)$ are those states of the world that are considered as types of situations for performative states making the latter successful at o.

We can propose a stream interpretation of $\mathbf{Q}_i(o)$ and construct Ω^ω. We know that the set Ω^ω is much larger than Ω. According to the orders $\mathcal{O}_{[0,1]^\omega}$ and $\mathcal{O}_{\mathbf{Z}_p}$, we can identify all members of Ω with some streams of Ω^ω. Let the set of these streams be denoted by ${}^\sigma\Omega$. Evidently, ${}^\sigma\Omega \subset \Omega^\omega$. Assume that Ω^ω is a union of $\mathbf{Q}_i$. Therefore $\mathbf{Q}_i$ contains ${}^\sigma\Omega$ (it means that by our assumption Ω).

Now let us define probabilities on streams as follows: a *finitely additive probability measure* is a nonnegative set function $P(.)$ defined for sets $A \subseteq \Omega^\omega$, running the set $[0, 1]^\omega$, and satisfying the following properties:

1. $P(\emptyset) \geq |0|$ for all $A \subseteq \Omega^\omega$.
2. $P(\Omega^\omega) = [1]$ and $P(\emptyset) = |0|$.
3. If $A \subseteq \Omega^\omega$ and $B \subseteq \Omega^\omega$ are disjoint, i.e. $\inf(P(A), P(B)) = |0|$, then $P(A \cup B) = P(A) + P(B)$. Otherwise, $P(A \cup B) = P(A) + P(B) - \inf(P(A), P(B)) = \sup(P(A), P(B))$.
4. $P(A) = [1] - P(A)$ for all $A \subseteq \Omega^\omega$, where $A = \Omega^\omega \setminus A$.
5. Relative probability functions $P(A|B) \in [0, 1]^\omega$ are characterized by the following constraint:

$$P(A|B) = \frac{P(A \cap B)}{P(B)},$$

where $P(B) \neq |0|$ and $P(A \cap B) = \inf(P(A), P(B))$. Notice that since there are no partitions of sets of streams in general case, there are some problems in defining the conditional relation $P(A|B)$ between events also. There are much more dependent events than in the usual σ-field. For example, any real number of [0, 1] is less than any inconstant stream of $[0, 1]^\omega$. Let $P(B) = a$, $P(A) = b$, where a is a number of [0, 1] and b is an inconstant stream of $[0, 1]^\omega$. Then according to $\mathcal{O}_{[0,1]^\omega}$, $P(A|B) = 1$. However, this case cannot be defined within the traditional independence condition $P(B) = P(B|A)$. Instead of that we use the following condition: $P(A) \cdot P(B) = \inf(P(A), P(B))$.

The main originality of those probabilities is that conditions (3), (4) are independent. As a result, in a probability space $\langle \Omega^\omega, P \rangle$ some Bayes' formulas do not hold in the general case (see Theorem 5.4).

A particular case of stream probabilities is presented by p-adic probabilities. Let us define them on any subsets of Ω^ω as follows: a *finitely additive probability measure* is a set function $P(.)$ defined for sets $A \subseteq \Omega^\omega$, running over the set $\mathbf{Z}_p$ and satisfying the following properties:

1. $P(\Omega^\omega) = -1$ and $P(\emptyset) = 0$.
2. If $A \subseteq \Omega^\omega$ and $B \subseteq \Omega^\omega$ are disjoint, i.e. $\inf(P(A), P(B)) = 0$, then $P(A \cup B) = P(A) + P(B)$. Otherwise, $P(A \cup B) = P(A) + P(B) - \inf(P(A), P(B)) = \sup(P(A), P(B))$. Let us exemplify this property by 7-adic probabilities. Let $P(A) = \ldots 323241$ and $P(B) = \ldots 354322$ in 7-adic metrics. Then $P(A) + P(B) = \ldots 010563$; $\inf(P(A), P(B)) = \ldots 323221$; $P(A) + P(B) - \inf(P(A), P(B)) = \sup(P(A), P(B)) = \ldots 354342$.
3. $P(A) = -1 - P(A)$ for all $A \subseteq \Omega^\omega$, where $A = \Omega^\omega \setminus A$.
4. Relative probability functions $P(A|B) \in \mathbf{Q}_p$ are characterized by the following constraint:

$$P(A|B) = \frac{P(A \cap B)}{P(B)},$$

where $P(B) \neq 0$ and $P(A \cap B) = \inf(P(A), P(B))$.

Now we can interpret $\mathbf{Q}_i(|o|)$, where $|o| \in {}^\sigma\Omega$, probabilistically as follows: $\mathbf{Q}_i(|o|) = \{\tau \colon P_i(\tau \mid |o|) > [0]\}$. These relative probabilities cannot set a partition of Ω^ω. In other words, using them we cannot define an equivalence relation corresponding to a partition. Instead of that the following properties hold, as we can prove on the basis of the orders $\mathcal{O}_{[0,1]^\omega}$ and $\mathcal{O}_{\mathbf{Z}_p}$:

- If $P_i(\tau|\pi) > [0]$, then $P_i(\rho|\pi) > [0]$ and $P_i(\tau|\rho) > [0]$ for some ρ. This property takes place instead of the usual transitivity in real probability logic: if $P_i(\rho|\pi) > 0$ and $P_i(\tau|\rho) > 0$, then $P_i(\tau|\pi) > 0$.
- $P_i(\tau||\omega|) = |1|$, where $|\omega| \in {}^\sigma\Omega$ and $\tau \in \Omega^\omega \setminus {}^\sigma\Omega$.
- $P_i(\tau|\tau) > [0]$.

Thus, the possibility operator $\mathbf{Q}_i$ has the following properties: for all $\tau, \pi \in \Omega^\omega$:

$$\tau \in \mathbf{Q}_i(\tau) \tag{10.11}$$

$$\pi \in \mathbf{Q}_i(\tau) \Rightarrow \mathbf{Q}_i(\pi) = \mathbf{Q}_i(\tau) \tag{10.12}$$

Now we consider the relation $A \subseteq \mathbf{Q}_i(o)$, where $A \subseteq {}^\sigma\Omega$, as the statement that at o agent i accepts the performance A, i.e. $|o| \in A$ for all states $|\upsilon|$ that i considers possible at $|o|$:

$$K_i A = \{|o| \colon A \subseteq \mathbf{Q}_i(|o|)\}. \tag{10.13}$$

This set is another interpretation of knowledge operator which is coinductive now. If $A \subseteq \mathbf{Q}_i(|o|)$, an individual i who observes $|o|$, will accept a state of the performance A. The most important property of the knowledge operator is $A \subseteq K_i A$; i.e. if A is successful in state $|o|$ (i.e. $|o| \in A$), then an agent accepts a performance A in state $|o|$ (i.e. $|o| \in K_i A$).

The following statements can be proved in relation to *coinductive knowledge operator* defined in (10.13):

$${}^\sigma\Omega \subseteq K_i^\sigma \Omega \subseteq \Omega^\omega; \tag{10.14}$$

$$(K_i A \cap K_i B) \Rightarrow K_i(A \cap B); \tag{10.15}$$

$$K_i(A \cup B) \Rightarrow (K_i A \cup K_i B); \tag{10.16}$$

$$K_i(A \cup B) = (K_i A \cap K_i B); \tag{10.17}$$

$$A \subseteq B \Rightarrow K_i B \subseteq K_i A; \tag{10.18}$$

$$A \subseteq K_i A; \tag{10.19}$$

$$K_i K_i A = K_i A. \tag{10.20}$$

We can compare Aumann's statements (10.3)–(10.8) with statements (10.14)–(10.20) to notice that the latter assume a new epistemic logic with a stream interpretation. For example, it is possible to build up a kind of multy-valued *illocutionary logic*, where streams are values for perlocutionary effects. So, the coinductive knowledge operator K_i of (10.14)–(10.20) designates *perlocutionary effects of illocutionary acts*, i.e. it takes into account just successful performative propositions (it defines what influence was made on hearer's behaviour). Let $K_j A$ (respectively $K_i A$) means agent j's (respectively i's) performative (cognitive or emotional) estimations of states of affairs A with an expected perlocutionary effect of these estimations on agent j (respectively i). So, $K_j A$ means 'j + performative verb + that A' (e.g. 'j thinks that A', 'j likes A', 'j hates A', etc.) and agent j follows this statement in his(her) behaviour.

On the basis of the standard propositional language $\mathscr{L}$ with the set of values $[0, 1]$ (or $\{0, 1, \ldots, p-1\}$) we can construct an extension $\mathscr{L}''$ containing new modal operators $E, E_1, E_2, E_3, \ldots$ said to be *perlocutionary effects*. The semantics of $\mathscr{L}''$ is defined in the following way. Assume that V is a valuation of well-formed formulas of $\mathscr{L}$ and it takes values in $[0, 1]$ (or $\{0, 1, \ldots, p-1\}$). Let us extend V to V_e as follows:

(a) If for $\varphi \in \mathscr{L}$, $V(\varphi) = r$, then $V_e(E_i(\varphi)) = \langle \sigma(0) = r, \sigma(1), \sigma(2), \ldots \rangle$, i.e. $V_e(E_i(\varphi))$ is a mapping from $V(\varphi)$ to an inconstant stream σ starting with $V(\varphi)$.

(b) If for $\varphi \in \mathscr{L}$, $V(\varphi) = r$, then $V_e(\varphi) = |r|$.

(c) For all $\varphi, \psi \in \mathscr{L}$, $V_e(E_i(\varphi \vee \psi)) = V_e(E_i(\varphi) \wedge V_e(E_i(\psi))$.

In this semantics the following propositions will be perlocutionary tautologies, i.e. they will be true:

$$\varphi \Rightarrow E(\varphi), \tag{10.21}$$

$$\neg\varphi \Rightarrow E(\neg\varphi), \tag{10.22}$$

$$E(\neg\varphi) \Rightarrow \neg E(\varphi), \tag{10.23}$$

$$(E(\varphi) \wedge E(\psi)) \Rightarrow E(\varphi \wedge \psi), \tag{10.24}$$

$$E(\varphi \vee \psi) \Rightarrow (E(\varphi) \vee E(\psi)), \tag{10.25}$$

$$(E(\varphi \Rightarrow \psi)) \Rightarrow (E(\varphi) \Rightarrow E(\psi)), \tag{10.26}$$

$$(E(\varphi \vee \psi)) \Leftrightarrow E(\varphi) \wedge E(\psi). \tag{10.27}$$

The epistemic logic closed under tautologies (10.21)–(10.27) is a kind of many-valued illocutionary logic with values on streams or non-Archimedean numbers.

Let us notice that the non-well-founded knowledge operator designates *perlocutionary effects of illocutionary acts*, i.e. it takes into account only successful performative propositions (it defines what influence was made on the listener's behaviour).

Aumann's understanding of common knowledge satisfies the classical intuition of well-foundedness of all logical entities, i.e. the presupposition that we can appeal only to inductive sets in our reasoning. For example, we can always find an infinite intersection κA according to knowledge operators of different people. However, this intuition contradicts to the possibility of reflexive games where I can cheat or make false public announcements and should detect if I am cheated by other people.

Under conditions of reflexive games I cannot define common perlocutionary effects as the infinite intersection κA. An infinite mutual reflexion between two individuals assumes an infinite union: both have mutual knowledge of A or both know that both know A or both know that both know that both know A etc. ad infinitum. In other words, the *common perlocutionary operator* $\overline{KA}$ is defined as follows:

$$\overline{KA} = K_1 A \cup K_2 A \cup K_1 K_2 A \cup K_2 K_1 A \cup K_1 K_2 K_1 A \cup \ldots$$

For each natural number n an operator $\overline{M_n}$ expressing "n-th degree mutual reflexion for perlocutionary effects" is defined so:

$$\overline{M_0 A} = A; \quad \overline{M_{n+1} A} = \bigcup_{i=1}^{N} K_i \overline{M_n A}$$

Common perlocutionary effect, $\overline{\kappa}$, is understood as a mutual reflexion for perlocutionary effects of all finite degrees:

$$\overline{\kappa A} = \bigcup_{n=0}^{+\infty} \overline{M_n A}.$$

Also, let us define for each natural number n an operator M_n expressing "n-th degree mutual reflexion":

$$M_0 A = A; \quad M_{n+1} A = \bigcap_{i=1}^{N} K_i M_n A$$

and common knowledge, κ, as mutual reflexion for common knowledge of all finite degrees:

$$\kappa A = \bigcap_{n=0}^{+\infty} M_n A.$$

Lemma 10.2 *If $|o| \in \kappa A$, then for any i, $\kappa A \subseteq \mathbf{Q}_i(|o|)$. And if $|o| \in \overline{\kappa A}$, then for some i, $\overline{\kappa A} \subseteq \mathbf{Q}_i(|o|)$.*

Proof If $|o| \in \kappa A$, then $|o| \in K_i M_n A$ for all agents i and degrees n of mutual reflexion for common knowledge. Therefore $\kappa A \subseteq \mathbf{Q}_i(|o|)$ for any i. If $|o| \in \overline{\kappa A}$, then $|o| \in K_i \overline{M_n A}$ for some agents i and degrees n of mutual reflexion for perlocutionary effects. Therefore $\overline{M_n A} \subseteq \mathbf{Q}_i(|o|)$ for some n, and thus $\overline{\kappa A} \subseteq \mathbf{Q}_i(|o|)$ for some i. □

Theorem 10.2 (Reflexion Disagreement Theorem) *Let us consider a hypothesis* H *of coinductive/stream probability logic for which the various agents' coinductive/stream probabilities are* $q_1, \ldots, q_N$ *after they condition* $P(\cdot)$ *on priors. The propositions* C *and* $\overline{C}$ *of coinductive/stream probability logic identify these probabilities:*

$$C = \bigcap_{i=1}^{N} \{|o| : P(H|\mathbf{Q}_i(|o|)) = q_i\}.$$

$$\overline{C} = \bigcup_{i=1}^{N} \{|o| : P(H|\mathbf{Q}_i(\omega)) = q_i\}.$$

Let the coinductive/stream probability space $\langle \Omega^{\omega}, P \rangle$ *be closed under all of the operators* K_i, M_n, κ, $\overline{K_i}$, $\overline{M_n}$, *and* $\overline{\kappa}$ *and let* P *be the old probability measure that is common to all the agents. Assume that the possibility of* C*'s and* $\overline{C}$*'s becoming common knowledge or common perlocution is not equal to zero:* $P(\kappa C) \neq [0]$ *and* $P(\overline{\kappa C}) \neq [0]$, *then*

$$P(H|\kappa C) \neq q_i \text{ for some } i.$$

$$P(H|\overline{\kappa C}) \neq q_i \text{ for some } i.$$

Proof By Lemma 10.2, $\kappa C = \bigcup_j D_{ij}$, where $\bigcup_j D_{ij}$ covers $\mathbf{Q}_i$, but it is not a partition of $\mathbf{Q}_i$ because of basic properties of coinductive probabilities. Then $P(H|\kappa C) = \frac{P(H \cap \bigcup_j D_{ij})}{P(\bigcup_j D_{ij})} = \frac{\inf(P(H), \sup_j P(D_{ij}))}{\sup_j P(D_{ij})} \neq \frac{\sum_j P(H|D_{ij})P(D_{ij})}{\sum_j P(D_{ij})} = \frac{\sum_j q_i P(D_{ij})}{\sum_j P(D_{ij})} = q_i$. Thus, $P(H|\kappa C) \neq q_i$ in general case. In the same way we can show that $P(H|\overline{\kappa C}) \neq q_i$ in general case. □

10.2.5 *Reflexion Disagreement Theorem Proved on Hypernumbers*

We can propose a non-standard interpretation of $\mathbf{Q}_i(\omega)$ and construct the ultrapower Ω^I/U, where U is an ultrafilter that contains the Frechet filter on I (the set that contains all complements for finite subsets of I). Ω^I/U is said to be a *proper non-standard extension* of Ω and it is denoted by ${}^*\Omega$. There exist two groups of members of ${}^*\Omega$: (i) functions that are constant, e.g. $f(\alpha) = \omega \in \Omega$ for an infinite index subset $\{\alpha \in I\}$; a constant function $[f = \omega]$ is denoted by ${}^*\omega$; (ii) functions that aren't constant. The set of all constant functions of ${}^*\Omega$ is called *standard set* and it is denoted by ${}^\sigma\Omega$. The members of ${}^\sigma\Omega$ are called *standard*. It is readily seen that ${}^\sigma\Omega$ and Ω are isomorphic: ${}^\sigma\Omega \cong \Omega$. If Ω was a number system, then members of ${}^*\Omega$ are called *hypernumbers*.

Let us identify Ω with ${}^\sigma\Omega$ and assume that ${}^*\Omega$ is a union of $\mathbf{Q}_i$. Therefore $\mathbf{Q}_i$ contains ${}^\sigma\Omega$ (by our assumption Ω).

Assume that ${}^*Q_{[0,1]} = Q^{\mathbf{N}}_{[0,1]}/\mathrm{U}$ is a nonstandard extension of the subset $Q_{[0,1]} = \mathbf{Q} \cap [0, 1]$ of rational numbers and ${}^\sigma Q_{[0,1]} \subset {}^*Q_{[0,1]}$ is the subset of standard members. Let us try our best to define probabilities on that set. We can extend the usual order structure on $Q_{[0,1]}$ to a partial order structure on ${}^*Q_{[0,1]}$: (i) for rational numbers $x, y \in Q_{[0,1]}$ we have $x \leq y$ in $Q_{[0,1]}$ iff $[f] \leq [g]$ in ${}^*Q_{[0,1]}$, where $\{\alpha \in \mathbf{N} : f(\alpha) = x\} \in \mathrm{U}$ and $\{\alpha \in \mathbf{N} : g(\alpha) = y\} \in \mathrm{U}$, i.e. f and g are constant functions such that $[f] = {}^*x$ and $[g] = {}^*y$, (ii) each positive rational number ${}^*x \in {}^\sigma Q_{[0,1]}$ is greater than any number $[f] \in {}^*Q_{[0,1]} \backslash {}^\sigma Q_{[0,1]}$, i.e. ${}^*x > [f]$ for any positive $x \in Q_{[0,1]}$ and $[f] \in {}^*Q_{[0,1]}$, where $[f]$ isn't constant function.

These conditions have the following informal sense: (i) The sets ${}^\sigma Q_{[0,1]}$ and $Q_{[0,1]}$ have isomorphic order structure. (ii) The set ${}^*Q_{[0,1]}$ contains actual infinities that are less than any positive rational number of ${}^\sigma Q_{[0,1]}$. These members are called *infinitesimals*.

Define this partial order structure on ${}^*Q_{[0,1]}$ as follows:

$\mathscr{O}_{*Q}$ For any hyperrational numbers $[f], [g] \in {}^*Q_{[0,1]}$, we set $[f] \leq [g]$ if $\{\alpha \in \mathbf{N} : f(\alpha) \leq g(\alpha)\} \in \mathrm{U}$. For any hyperrational numbers $[f], [g] \in {}^*Q_{[0,1]}$, we set $[f] < [g]$ if $\{\alpha \in \mathbf{N} : f(\alpha) \leq g(\alpha)\} \in \mathrm{U}$ and $[f] \neq [g]$, i.e. $\{\alpha \in \mathbf{N} : f(\alpha) \neq g(\alpha)\} \in \mathrm{U}$. For any hyperrational numbers $[f], [g] \in {}^*Q_{[0,1]}$, we set $[f] = [g]$ if $f \in [g]$.

This ordering relation is not linear, but partial, because there exist elements $[f], [g] \in {}^*Q_{[0,1]}$, which are incompatible.

Introduce two operations sup, inf in the partial order structure $\mathscr{O}_{*Q}$: (i) $\inf([f], [g]) = [h]$ iff there exists $[h] \in {}^*Q_{[0,1]}$ such that $\{\alpha \in \mathbf{N} : \min(f(\alpha), g(\alpha)) = h(\alpha)\} \in \mathrm{U}$; (ii) $\sup([f], [g]) = [h]$ iff there exists $[h] \in {}^*Q_{[0,1]}$ such that $\{\alpha \in \mathbf{N} : \max(f(\alpha), g(\alpha)) = h(\alpha)\} \in \mathrm{U}$.

Note there exist the maximal number ${}^*1 \in {}^*Q_{[0,1]}$ and the minimal number ${}^*0 \in {}^*Q_{[0,1]}$ under condition $\mathscr{O}_{*Q}$.

Recall that for each $i \in [0, 1]$, ${}^*i = [f = i]$, i.e. it is a constant function. Every element of ${}^*Q_{[0,1]}$ has the form of infinite tuple (i.e. a stream) $[f] = \langle y_0, y_1, \ldots \rangle$, where $y_i \in Q_{[0,1]}$ for each $i = 0, 1, 2, \ldots$

Now let us define non-Archimedean probabilities on any subsets of ${}^*\Omega$ as follows: a *finitely additive probability measure* is a nonnegative set function $P_{{}^*Q_{[0,1]}}(\cdot)$ defined for sets $A \subseteq {}^*\Omega$, it runs over the set ${}^*Q_{[0,1]}$ and satisfies the following properties:

1. $P_{{}^*Q_{[0,1]}}(A) \geq {}^*0$ for all $A \subseteq {}^*\Omega$,
2. $P_{{}^*Q_{[0,1]}}({}^*\Omega) = {}^*1$ and $P_{{}^*Q_{[0,1]}}(\emptyset) = {}^*0$,
3. if $A \subseteq {}^*\Omega$ and $B \subseteq {}^*\Omega$ are disjoint, i.e. $\inf(P_{{}^*Q_{[0,1]}}(A), P_{{}^*Q_{[0,1]}}(B)) = {}^*0$, then $P_{{}^*Q_{[0,1]}}(A \cup B) = P_{{}^*Q_{[0,1]}}(A) + P_{{}^*Q_{[0,1]}}(B)$.
   Otherwise, $P_{{}^*Q_{[0,1]}}(A \cup B) = P_{{}^*Q_{[0,1]}}(A) + P_{{}^*Q_{[0,1]}}(B)) - \inf(P_{{}^*Q_{[0,1]}}(A), P_{{}^*Q_{[0,1]}}(B)) = \sup(P_{{}^*Q_{[0,1]}}(A), P_{{}^*Q_{[0,1]}}(B))$.
4. $P_{{}^*Q_{[0,1]}}(\neg A) = {}^*1 - P_{{}^*Q_{[0,1]}}(A)$ for all $A \subseteq {}^*\Omega$, where $\neg A = {}^*\Omega \backslash A$.

5. relative probability functions $P_{^*Q_{[0,1]}}(A|B) \in {}^*Q_{[0,1]}$ are characterized by the following constraint:
$$P_{^*Q_{[0,1]}}(A|B) = \frac{P_{^*Q_{[0,1]}}(A \cap B)}{P_{^*Q_{[0,1]}}(B)}, \tag{10.28}$$
where $P_{^*Q_{[0,1]}}(B) \neq 0$ and $P_{^*Q_{[0,1]}}(A \cap B) = \inf(P_{^*Q_{[0,1]}}(A), P_{^*Q_{[0,1]}}(B))$.

Let $\mathbf{Q}_i(^*\omega)$, where $^*\omega \in {}^\sigma\Omega$, be defined thus: $\mathbf{Q}_i(^*\omega) = \{[f] : P^i_{^*Q_{[0,1]}}([f]|\, ^*\omega) > {}^*0\}$. Its properties are as follows:

1. If $P^i_{^*Q_{[0,1]}}([f]|[f']) > {}^*0$, then $P^i_{^*Q_{[0,1]}}([f'']|[f']) > {}^*0$ and $P^i_{^*Q_{[0,1]}}([f]|[f'']) > {}^*0$ for some $[f'']$.
2. $P^i_{^*Q_{[0,1]}}(^*\omega|[f]) = {}^*1$, where $^*\omega$ is constant and $[f]$ is not a constant function of $^*\Omega$.
3. $P^i_{^*Q_{[0,1]}}([f]|[f]) > {}^*0$.

From these properties it follows that if $P^i_{^*Q_{[0,1]}}([f]|[f']) > {}^*0$, then $P^i_{^*Q_{[0,1]}}([f]|[f]) > {}^*0$ and $P^i_{^*Q_{[0,1]}}([f']|[f]) > {}^*0$. Thus, the possibility operator $\mathbf{Q}_i$ has the following properties: for all $^*\omega', {}^*\omega \in {}^*\Omega$:

$$^*\omega \in \mathbf{Q}_i(^*\omega) \tag{10.29}$$

$$^*\omega' \in \mathbf{Q}_i(^*\omega) \Rightarrow \mathbf{Q}_i(^*\omega') = \mathbf{Q}_i(^*\omega) \tag{10.30}$$

The relation $A \subseteq \mathbf{Q}_i(^*\omega)$, where $A \subseteq {}^\sigma\Omega$, is understood as the statement that at $^*\omega$ agent i accepts the performance A:

$$K_i A = \{^*\omega : A \subseteq \mathbf{Q}_i(^*\omega)\}. \tag{10.31}$$

From this definition we can obtain statements (10.14)–(10.20) and then we can prove Lemma 10.2 on hypernumbers.

Theorem 10.3 (Reflexion Disagreement Theorem) *Let us consider a hypothesis H of non-Archimedean probability logic for which the various agents' non-Archimedean probabilities are $q_1, \ldots, q_N$ after they condition $P_{^*Q_{[0,1]}}$ on priors. The propositions C and $\overline{C}$ of non-Archimedean probability logic identify these probabilities:*

$$C = \bigcap_{i=1}^{N} \{^*\omega : P_{^*Q_{[0,1]}}(H|\mathbf{Q}_i(^*\omega)) = q_i\}.$$

$$\overline{C} = \bigcup_{i=1}^{N} \{^*\omega : P_{^*Q_{[0,1]}}(H|\mathbf{Q}_i(^*\omega)) = q_i\}.$$

Let the non-Archimedean probability space $\langle {}^*\Omega, P_{^*Q_{[0,1]}}\rangle$ be closed under all of the operators K_i, M_n, κ, $\overline{K_i}$, $\overline{M_n}$, and $\overline{\kappa}$ and let $P_{^*Q_{[0,1]}}$ be the old probability measure *that is common to all the agents. Assume that the possibility of C's and $\overline{C}$'s becoming common knowledge or common perlocution is not equal to zero:* $P_{^*Q_{[0,1]}}(\kappa C) \neq^* 0$ *and* $P_{^*Q_{[0,1]}}(\overline{\kappa C}) \neq^* 0$, *then*

$$P_{^*Q_{[0,1]}}(H|\kappa C) \neq q_i \text{ for some } i.$$

$$P_{^*Q_{[0,1]}}(H|\overline{\kappa C}) \neq q_i \text{ for some } i.$$

Proof Let us start with the same reasoning as in the Auman's agreement theorem. By Lemma 10.2, $\kappa C = \bigcup_j D_{ij}$, where $\bigcup_j D_{ij}$ covers $\mathbf{Q}_i$, but it is not a partition of $\mathbf{Q}_i$ because of basic properties of non-Archimedean probabilities. Then

$$P_{^*Q_{[0,1]}}(H|\kappa C) = \frac{P_{^*Q_{[0,1]}}(H \cap \bigcup_j D_{ij})}{P_{^*Q_{[0,1]}}(\bigcup_j D_{ij})} = \frac{\inf(P_{^*Q_{[0,1]}}(H), \sup_j P_{^*Q_{[0,1]}}(D_{ij}))}{\sup_j P_{^*Q_{[0,1]}}(D_{ij})} \neq$$

$$\neq \frac{\sum_j P_{^*Q_{[0,1]}}(H|D_{ij}) P_{^*Q_{[0,1]}}(D_{ij})}{\sum_j P_{^*Q_{[0,1]}}(D_{ij})} = \frac{\sum_j q_i P_{^*Q_{[0,1]}}(D_{ij})}{\sum_j P_{^*Q_{[0,1]}}(D_{ij})} = q_i$$

Thus, $P_{^*Q_{[0,1]}}(H|\kappa C) \neq q_i$ in the general case. In the same way we can show that $P_{^*Q_{[0,1]}}(H|\overline{\kappa C}) \neq q_i$ in the general case. □

This theorem shows us that there is no common knowledge in Aumann's meaning if we appeal to common performative effects with interpretation on non-Archimedean probabilities. It is more natural and allows us to use reflexive games with different mental accounts.

10.3 Hierarchies of Reflexion

10.3.1 Orders of Reflexion

In reflexive games we deal with an unlimited hierarchy of cognitive pictures. Let us consider a bimatrix game with agents i and j. Each of them can have their own picture about a state of affairs A. Denote these pictures by $K_i A$ and $K_j A$ respectively. *The first-order reflexion* (thoughts about pictures of the opponent) is expressed by means of *pictures of the second order* which are designated by $K_j K_i A$ and $K_i K_j A$ where $K_j K_i$ are pictures of agent j about pictures of agent i, $K_i K_j A$ are pictures of agent i about pictures of agent j. *The reflexion of the second order* defines which pictures of one opponent are related to pictures of another opponent. At this level of reflexion pictures of *the third order* $K_i K_j K_i A$ and $K_j K_i K_j A$ are generated. And so ad infinitum. The collection of all pictures $K_i A$, $K_j A$, $K_j K_i A$, $K_i K_j A$, $K_i K_j K_i A$, $K_j K_i K_j A$ etc. makes an infinite hierarchy.

Definition 10.1 The reflexion of agent i at the n-th level in bimatrix games is expressed by $(n+1)$-order knowledge operators $K_i^{n+1}A = K_i K_j K_i \dots A$, where on the right side there are $n+1$ K_m-operators ($m = i, j$).

Let us consider two agents i and j and suppose that the reflexive game takes place on level n. This means that we have $K_i^{n+1}A$ and/or $K_j^{n+1}A$ which are understood as perlocutionary effects of illocutionary acts and satisfy requirements (10.14)–(10.20) (as well as (10.21)–(10.27)). We know that $A \subseteq \dots \subseteq K_j^n A \subseteq K_i^{n+1}A$ and $A \subseteq \dots \subseteq K_i^n A \subseteq K_j^{n+1}A$. Therefore $K_i^{n+1}A \cap K_j^{n+1}A \neq \emptyset$.

Decision rules of each player depend on their knowledge. Let $\mathbf{B}_i^{n+1}$ be a n-th-order decision rule of agent i that corresponds to knowledge $K_i^{n+1}A$ and $\mathbf{B}_j^{n+1}$ be an n-th-order decision rule of agent j that corresponds to his knowledge $K_j^{n+1}A$. We suppose that i knows j's $(n-1)$-th-order decision rule and j knows i's $(n-1)$-th-order decision rule.

Definition 10.2 A payoff of reflexive game at the n-th level in accordance with $\mathbf{B}_i^{n+1}$ or $\mathbf{B}_j^{n+1}$ is called performative equilibrium of this game.

We have the following possibilities:

1. both $K_i^{n+1}A$ and $K_j^{n+1}A$ are used in a performative equilibrium—this means that agents i and j are at the n-th level of reflexion, simultaneously;
2. only $K_i^{n+1}A$ is used in a performative equilibrium (then we can take $K_j^{n+1}A = K_j^n A$)—this means that agent i stays at the n-th level of reflexion, but agent j stays at the $(n-1)$-th level of reflexion;
3. only $K_j^{n+1}A$ is used in a performative equilibrium (then we can take $K_i^{n+1}A = K_i^n A$)—this means that agent j stays at the n-th level of reflexion, but agent i stays at the $(n-1)$-th level of reflexion.

For any reflexive game at the n-th level of reflexion we can build up a tree of graphs. Vertices of the tree correspond to real or phantom agents, participating in reflexive game. Branches of the tree simulate a mutual knowledge of agents at the reflexion of level n: if from (real or phantom) agent i there exists a path to agent j, then agent j is correctly informed about agent i. In this case $K_j^{n+1}A$ is a performative equilibrium. If both $K_i^{n+1}A$ and $K_j^{n+1}A$ are a performative equilibrium of the same game, then an appropriate tree has a loop.

In a reflexive game at level n it is important for agent j that $K_i^n A \subseteq K_j^{n+1}A$ holds, because it means that agent j has really corresponded to level n. Correctly defining the n-th level of reflexion implies a victory in a game. Let us consider the game of two brokers to show how it is sophisticated sometimes to define n. Two brokers at a stock exchange have appropriate expert systems which have been used for the support of decision making. The network administrator illegally copied both expert systems and sold each broker an expert system of his opponent. Then he tries to sell each of them the following information: "Your opponent has your expert system." Then the administrator tries to sell the information: "Your opponent knows that you have his expert system," etc. How should brokers use the information received from the administrator and also what information on what iteration is essential? Theoretically, reflexive level n can be any natural number.

10.3.2 Non-Archimedean Probabilities for Discrete and Continuous Reflexive Levels

Let us designate reflexive players by i and j. Let $K_j A$ (accordingly, $K_i A$) mean agent j's (accordingly, agent i's) performative (cognitive or emotional) estimations of states of affairs A with a subsequent perlocutionary effect of these estimations on agent j (accordingly, i). So, $K_j A$ means 'j + performative verb + that A' (e.g. 'j thinks that A,' 'j likes A,' 'j hates A,' etc.) and agent j follows this statement in behaviours. Further, let $K_j K_i A$ (accordingly, $K_i K_j A$) mean agent j's performative estimation of perlocutionary effect $K_i A$ (accordingly, agent i's estimation of perlocutionary effect $K_j A$) and $K_j K_i A$ (accordingly, $K_i K_j A$) is a performative equilibrium for both agents i, j. For instance, $K_j K_i A$ means 'j knows that i hates A' and this statement determines the behavioural strategies of both agents: i hates A and j takes into account this perlocution of i. Thus, state of affairs A can be interpreted differently by agents i, j and k: $K_i A$ ('A from i's viewpoint'), $K_j A$ ('A from j's viewpoint'), $K_k A$ ('A from k's viewpoint'). But images which are generated by one of the players can be foreseen by others, too. As a result, there are performative effects $K_j K_i A$ ('$K_i A$ from j's viewpoint'), $K_i K_j A$ ('$K_j A$ from i's viewpoint'), $K_k K_i A$ ('$K_i A$ from k's viewpoint'), etc.

Let us generalise it up to the case of $K_i^{n+1} A = K_i K_j K_i \ldots A$. Let $K_i^{n+1} A$ mean agent i's performative estimation of perlocutionary effect $K_j^n A$ of level n and performative equilibrium at level $n + 1$. In this general case, i can simulate j's performative evaluations up to level $n + 1$. An example of $K_j K_i K_j A$: in a chess game chess player j lays a trap for chess player i that is grounded on the fact that j knows how i imagines j's thought course.

Now let us define $K_i^{n+1} A$ on p-adic probabilities. For the first time Andrei Khrennikov proposed to apply p-adic probabilities in modelling cognitive processes [29]. Assume we have $p \in \mathbf{N}$ reflexive players $i, j, \ldots$ Then all possible combinations $\ldots K_{\alpha_n} \ldots K_{\alpha_2} K_{\alpha_1} A$, where $\alpha_k \in \{i, j, \ldots\}$, can be presented by finite p-adic integers $\ldots 00\beta_n \ldots \beta_2\beta_1\beta_0 = \sum_{k=0}^{n} \beta_k p^k$, where $\beta_k \in \{0, \ldots, p-1\}$ for each $k = 0, \ldots, n$ and there is a bijection between the sets $\{0, \ldots, p-1\}$ and $\{i, j, \ldots\}$.

Let Ω be a finite set of possible states of the world and $A \subseteq \Omega$. Then a *finite p-adic probability measure* P_i^{n+1} is defined on sets $A, B \subseteq \Omega$ as follows:

$$P_i^{n+1}(\emptyset) = 0 \quad and \quad P_i^{n+1}(\Omega) = 1\,;$$

$$if \quad P_j^n(A) > 0, \quad then \quad P_i^{n+1}(A) > 0\,;$$

$$P_i^n(A) > 0 \quad iff \quad P_i^{n+1}(A) > 0;$$

$$if \quad P_j^{n+1}(A) = 1, \quad then \quad P_i^n(A) = 1;$$

$$P_i^{n+1}(A) = \frac{\sum_{k=0}^{n} \alpha_k p^k}{\sum_{k=0}^{n} (p-1)p^k} \quad and \quad P_i^{n+1}(B) = \frac{\sum_{k=0}^{n} \beta_k p^k}{\sum_{k=0}^{n} (p-1)p^k},$$

where $\alpha_k, \beta_k \in \{0, \ldots, p-1\}$ for each $k = 0, \ldots, n$;

$$P_i^{n+1}(A \cup B) = P_i^{n+1}(A) + P_i^{n+1}(B) \; if \; A \cap B = \emptyset; \; P_i^{n+1}(\neg A) = 1 - P_i^{n+1}(A);$$

$$P_i^{n+1}(A \mid B) = \frac{\inf(\sum_{k=0}^{n} \alpha_k p^k, \sum_{k=0}^{n} \beta_k p^k)}{\sum_{k=0}^{n} \beta_k p^k},$$

where inf is defined digit by digit. For instance, if we have just two agents, then at the zero level of reflexion we have only two probability values: either 0 or 1 (meaning e.g. that an agent either does not follow the content $A \subseteq \Omega$ or does). At the first level of reflexion we have already the following four probability values: 0, 1/3, 2/3, 1 (meaning e.g. that both agents do not follow the content $A \subseteq \Omega$, one of them does not follow, another does, and both of them follow), etc.

Now we can define $K_i^{n+1} A$ in the following way:

$$K_i^{n+1} A = \{\omega : A \subseteq \mathbf{Q}_i(\omega) = \{a : P_i^{n+1}(a|\omega) > 0\}\}.$$

Notice that according to this definition, taking into account our assumption that if $P_j^n(a|\omega) > 0$, then $P_i^{n+1}(a|\omega) > 0$, we have $K_j^n A \subseteq K_i^{n+1} A$ for each agent j participating in the reflexive game.

Let us suppose that there are just three reflexive players k, l, m at the reflexion level of $n = 2$. Then $P_i^2(A \subseteq \Omega) \in \{0, 1/8, 2/8, 3/8, 4/8, 5/8, 6/8, 7/8, 1\}$ for each $i \in \{k, l, m\}$. At the infinite level of reflexion we have the following p-adic probabilities:

$$P_i^{\infty}(A) = \lim_{n \to \infty} P_i^n(A).$$

The knowledge operators $K_i^{n+1} A$ satisfy the following relations:

$$(K_i^{n+1} A \cap K_i^{n+1} B) \Rightarrow K_i^{n+1}(A \cap B);$$

$$K_i^{n+1}(A \cup B) \Rightarrow (K_i^{n+1} A \cup K_i^{n+1} B);$$

$$K_i^{n+1}(A \cup B) = (K_i^{n+1} A \cap K_i^{n+1} B);$$

$$A \subseteq B \Rightarrow K_i^{n+1} A \subseteq K_i^{n+1} B;$$

$$A \subseteq K_i^{n+1} A;$$

$$K_i^{n+1} K_i^n A = K_i^n A.$$

Using (finite) p-adic probabilities, we understand reflexion levels discretely. For any finite number of agents we can always define a reflexive level n such that probabilities are distributed on an appropriate finite set of p-adic numbers. The larger the n, the more finite p-adic probabilities. Thereby between n and $n + 1$ there are no other reflexive levels.

Nevertheless, reflexive levels can be treated continuously as well, when between reflexive levels n and $n + 1$ there are always other reflexive levels. For instance, let the level of reflexion $n = 1/c$, where $c \in (0, 1]$. Let $^*[0, 1] = [0, 1]^{\mathbf{N}}/\mathrm{U}$. In this case $K_i^n A$ is defined in non-Archimedean probabilities as follows:

$$K_i^n A = \{\omega : A \subseteq \mathbf{Q}_i(\omega) = \{a : {}^*P_i^n(a|\omega) > {}^*0\}\}.$$

The *non-Archimedean probability measure* $^*P_i^n$ is defined in sets $A, B \subseteq \Omega$ thus:

$${}^*P_i^n(\emptyset) = {}^*0 \quad and \quad {}^*P_i^n(\Omega) = {}^*1\,;$$

$$if \quad {}^*P_j^n(A) > {}^*0, \quad then \quad {}^*P_i^k(A) > {}^*0\,, where\ k > n;$$

$${}^*P_i^n(A) > {}^*0 \quad iff \quad {}^*P_i^k(A) > {}^*0\,, where\ k > n;$$

$$if \quad {}^*P_j^k(A) = {}^*1, where\ k > n, \quad then \quad {}^*P_i^n(A) = {}^*1\,;$$

$${}^*P_i^n(A) = a \in {}^*[0, 1];$$

$${}^*P_i^n(A \cup B) = {}^*P_i^n(A) + {}^*P_i^n(B)\ if\ A \cap B = \emptyset;\ {}^*P_i^n(\neg A) = {}^*1 - {}^*P_i^n(A);$$

$${}^*P_i^n(A\,|B\,) = \frac{\inf(a,b)}{b}\,,$$

where $^*P_i^n(A) = a \in {}^*[0, 1]$, $^*P_i^n(B) = b \in {}^*[0, 1]$ and inf is defined digit by digit.

Any reflexive game is carried out in a synchronous response and synchronous reading of thoughts. Each illocutionary force has an appropriate perlocutionary effect. The task of each player is to foresee the perlocutionary effect of dialogue on their interlocutor's behaviour.

Let us consider the set $\mathbf{K}_i^n$ of different perlocutionary effects $K_{1,i}^n A$, $K_{2,i}^n A, \ldots,$ $K_{m,i}^n A, \ldots$ for each agent i and each reflexion level n. Hence, we have infinitely many sets $\mathbf{K}_i^1, \mathbf{K}_i^2, \ldots, \mathbf{K}_i^n, \ldots, \mathbf{K}_j^1, \mathbf{K}_j^2, \ldots, \mathbf{K}_j^n, \ldots$ of perlocutionary effects. We can combine different effects in the same evaluation, for example: $K_{l,j} K_{k,i} K_{k,j} A =$ $K_{l,j} K_{k,i}^2 A$.

The reflexive game is being described as a process of the form:

$$X_0 A_0.X_1 A_1.X_2 A_2.\cdots.X_n A_n.\cdots,$$

where $A_0, A_1, A_2, \ldots, A_n, \ldots$ are different states of affair uttered at different time $t = 0, 1, \ldots, n \ldots$ and $X_0, X_1, X_2, \ldots, X_n, \ldots$ are different combinations of perlocutionary effects for appropriate states of affair, e.g. $X_n = K_{l,j} K^2_{k,i}$.

Assume, we have just two reflexive players i, j and an appropriate reflexive game is presented by the process:

$$K_{l,j} K^2_{k,i} A_0 . K^3_{m,j} A_1 . \cdots . K_{a,j} K_{b,i} A_n$$

At each step, a perlocutionary effect, belonging to i, is simulated in a perlocutionary effect belonging to player j. Player j has a model of his opponent, such that any thought or emotion of i is simulated by j who on the basis of the outcome of the simulation makes a decision. Here there is prevalence over the opponent. In particular, $K_{a,j} K_{b,i} A_n$ says that in i's subjective world a real state of affairs A_n is simulated for decision-making and i deals only with it. Player j possesses a more complex picture: it includes a real state of affairs A_n together with i's mapping $K_{b,i} A_n$ from this state (it can include a simulation of i's purpose or i's doctrines due to which i solves problems). In this case j makes decision on the basis of reality picture $K_{b,i} A_n$, which allows j to set the task of control in i's decision-making process.

To sum up, the non-Archimedean probabilities can be used for explicating knowledge operators engaged in reflexive games. It helps us to formally describe the notion of reflexive game. Nevertheless, the meaning of reflexive game depends on contexts of human interactions. So, a reflexive game is probable only in a case where agents can reach *performative equilibrium*—they can act concordantly at reflexion levels $n > 0$. This condition is fulfilled in the case where there are mechanisms of intercommunication broadly agreed upon among people. These mechanisms have been preserved within an appropriate discourse community (*Kommunikationsgemeinschaft*) shared by members with a suitable degree of discourse expertise (i.e. members possess one or more genres in the communicative furtherance of its aims and know a specific lexis) and with a degree of relevant content to provide information and feedback.

It is worth noting that *our reflexive games grow up just from bio-inspired games*. Swarms are capable to play in some primitive proto-reflexive games. For instance, there are two species of the slime mould: *Physarum polycephalum* and *Badhamia utricularis*. And both can compete with each other, see Sect. 8.1.5. The point is that the swarm behaviour is based on some proto-symbolic values within a lateral inhibition and lateral activation. The human behaviour is just more symbolic. Therefore, while the swarms can reach a small level of reflexion in their games, the human beings can reach an unlimited level of reflexion. A kind of free will (a possibility to modify own actions) is detected already at the level of swarms that allows them to interact quite proto-symbolically, too, see Sect. 8.2.1.

References

1. Khrennikov, A., Schumann, A.: p-adic physics, non-well-founded reality and unconventional computing. p-Adic Num. Ultramet. Anal. Appl. **1**(4), 297–306 (2009)
2. Schumann, A.: Non-well-founded probabilities and coinductive probability logic. In: I.C.S. Press (ed.) Eighth International Symposium on Symbolic and Numeric Algorithms for Scientific Computing (SYNASC'08), pp. 54–57 (2008)
3. Schumann, A.: Non-well-founded probabilities on streams. In: Dubois, D., Lubiano, M.A., Prade, H., Gil, M.Á., Grzegorzewski, P., Hryniewicz, O. (eds.) Soft Methods for Handling Variability and Imprecision, pp. 59–65. Springer, Berlin, Heidelberg (2008)
4. Schumann, A.: Non-well-founded probabilities within unconventional computing. In: U.K. Durham (ed.) 6th International Symposium on Imprecise Probability: Theories and Applications (2009)
5. Schumann, A.: Probabilities on streams and reflexive games. Op. Res. Decis.**24**(1), 71–96 (2014). https://doi.org/10.5277/ord140105
6. Schumann, A.: Reflexive games and non-archimedean probabilities. p-adic Num. Ultrametr. Anal. Appl. **6**(1), 66–79 (2014)
7. Schumann, A.: Reflexive games in e-university. In: Proceedings of the Science and Information Conference (SAI), Science and Information Organization, London, England, July 28–30, 2015, Sponsor(s): Nvidia; IEEE; The Future and Emerging Technol FET at the European Comm; EUREKA; Cambridge Wireless; British Comp Soc; Digital Catapult; Springer; Media Partner for this conference, 2015 Science and Information Conference (SAI), pp. 770–775 (2015)
8. Schumann, A., Khrennikov, A.: Physics beyond the set-theoretic axiom of foundation. In: Khrennikov (ed.) AIP Conference Proceedings, Foundations of Probability and Physics-5, vol. 1101, pp. 374–380. AIP (2009)
9. Aumann, R.J.: Agreeing to disagree. Annals. Stat. **4**(6), 1236–1239 (1976)
10. Aumann, R.J.: Notes on Interactive Epistemology. Mimeo. Hebrew University of Jerusalem, Jerusalem (1989)
11. Lefebvre, V.A.: The basic ideas of reflexive game's logic. In: Problems of Systems and Structures Research (in Russian) pp. 73–79 (1965)
12. Lefebvre, V.A.: Algebra of Conscience. D. Reidel (1982)
13. Lefebvre, V.A.: Lect. Reflex. Game Theory. Leaf & Oaks, Los Angeles (2010)
14. Novikov, D.A., Chkhartishvili, A.G.: Stability of information equilibrium in reflexive games. Auto. Remote Control **66**(3), 441–448 (2005)
15. Belzung, C., Chevalley, C.: Emotional behaviour as the result of stochastic interactions. A process crucial for cognition. Behav. Process. **60**(2), 115–132 (2002). https://doi.org/10.1016/S0376-6357(02)00079-7
16. Howard, N.: General metagames: an extension of the metagame concept. In: Game Theory as a Theory of Conflict Resolution. Reidel, Dordrecht (1974)
17. Rutten, J.J.M.M.: Universal coalgebra: a theory of systems. Theor. Comput. Sci. **249**(1), 3–80 (2000)
18. Brandenburger, A., Friedenberg, A., Keisler, H.J.: Fixed points in epistemic game theory, Mathematical. Foundations of Information Flow, American Mathematical Society (2012)
19. Brandenburger, A., Keisler, H.J.: An impossibility theorem on beliefs in games. Studia Logica **84**(2), 211–240 (2006)
20. Knudsen, C.: Equilibrium, perfect rationality and the problem of self-reference in economics, pp. 204–208. Routledge (1993)
21. Winrich, J.S.: Self-reference and the incomplete structure of neoclassical economics. J. Econ. Issues **18**(4), 987–1005 (1984)
22. Berger, S.: The Foundations of Non-Equilibrium Economics: The Principle of Circular Cumulative Causation. Chapman & Hall, Routledge (2009)
23. Mailath, G.J., Samuelson, L.: Repeated Games and Reputations: Long-Run Relationships. Oxford University Press (2006)

24. S. Abramsky and P.-A. Mellies: Concurrent games and full completeness. In: Proceedings of the 14th Symposium on Logic in Computer Science, pp. 431–442 (1999)
25. Aczel, A.: Non-Well-Founded Sets. Stanford University Press (1988)
26. Pavlović, D., Escardó, M.H.: Calculus in coinductive form. In: Proceedings of the 13th Annual IEEE Symposium on Logic in Computer Science, pp. 408–417 (1998)
27. Koblitz, N.: p-adic numbers, p-adic analysis and zeta functions, 2nd edn. Springer (1984)
28. Khrennikov, A.Y.: p-adic quantum mechanics with p-adic valued functions. J. Math. Phys. **32**(4), 932–937 (1991)
29. Khrennikov, A.: p-adic discrete dynamical systems and collective behaviour of information states in cognitive models. Discrete Dyn. Nature Soc. **5**, 59–69 (2000)

Chapter 11
Payoff Cellular Automata and Reflexive Games

Reflexive games are just an abstract continuation of bio-inspired games. A high reflexivity of game is explained by a high level of modifications of involved actions and strategies at each game move. In this chapter, I will show that bio-inspired and reflexive games can be represented in the form of cellular automata. The more modified a cellular-automatic transition rule is at each step, the more reflexive our game is. For example, the Belousov-Zhabotinsky reaction can be described as a game, for which the transition rule does not change. It is a game of zero reflexion. If a transition rule is being modified at each time step, we deal with a reflexion of infinite level. We assume that only human beings are capable to change transition rules in their behaviour often. See [1–4]. Some relevant papers on cellular automata: [5–7].

11.1 Cellular-Automatic Approach in Logical Simulations of Swarm Behaviour

In this chapter I propose to use cellular automata as kind of coalgebras for formalizing reflexive games. This approach corresponds to Vernon L. Smith's idea of *ecological rationality* in behavioural finance [8] and my idea of aposterioristic symbolic logics [3]. Within this approach it is possible to define classical decision rules such as looking for saddle points, minimax and maximin, the Nash equilibrium, PROMETHEE methods, etc., but these rules become context-based and behaviourally instable. Nevertheless, we can define much more sophisticated decision rules as well, when players follow different cognitive pictures of reality and try to influence on decisions of each others. These situations are called reflexive games. In reflexive games players try to logically combine and continuously change different decision rules.

A. Schumann, *Behaviourism in Studying Swarms: Logical Models of Sensing and Motoring*, Emergence, Complexity and Computation 33,
https://doi.org/10.1007/978-3-319-91542-5_11

I define games as interactions among rational players, where decisions impact the payoffs of others, but players are limited by contexts that permanently change. A game is described by (i) its players who are presented in appropriate *transition rules of cellular automata*; (ii) players' passible strategies which are supposed known before the game and a combination of all possible payoffs from each strategy outcome gives the resulting payoffs which are collected as a *set of states of cellular automata*; (iii) a *neighbourhood of cellular automata* that makes some strategies actual and others non-actual (i.e. accepts the most important strategies in the given context at time t) and also changes or correct strategies. So, in this form of game description, players analyze strategies not purely logically, but contextually. Therefore players take decisions not only in an environment given by the payoff that corresponds to each possible outcome, but also in an environment of different other circumstances, e.g. by defining: which strategies can be accepted in this context, how they can be changed by the given context, how past contexts have influenced present contexts, whether some public announcements are false in fact, etc. Therefore I agree that the so-called payoff matrix is fixed throughout the game, but in these tools it is not possible to take contexts into account. In payoff cellular automata we expand payoff matrices by a dynamic massive-parallel logic.

Payoff matrices define the logic of the game, when all players are rational and each of them follows a strategy after considering all possible strategies of all players. In this chapter such a logic is not denied, but it is extended a lot by the assumption that in reality we deal with bounded rationality that is limited by contexts. The conventional logic given in payoff matrices becomes unconventional, e.g. dynamic, temporal, and massive-parallel. Usually, it is claimed that the logic of the game can be extended not as a more general unconventional logical system, but by adding heuristics, i.e. studying how different rules of behaviour are performed in different game situations. For instance, it is assumed that for reflexive games, where I try to predict emotions and cognitions of the other player, there is no complex logic, but there is the conventional logic given in payoff matrices and extended by a special kind of heuristics presented by introspections, methods of internal thinking in the player's mind, etc. However, in my opinion we can state that there is a logical system with special inference rules and that system is more complex than the payoff matrix logic. This complex logic is described by payoff cellular automata proposed in this chapter.

In conventional game theory an equilibrium of the game is nothing else that a prediction of expected outcome of the game. In payoff cellular automata the players make decision in accordance with their predictions and the past decisions imply future decisions. Therefore equilibria can change at different time. The difference between payoff matrices and payoff cellular automata are the same as the difference between conventional computing with reducing all computations to Turing machines and unconventional computing where heuristics is considered logically. So, payoff cellular automata is just a way to extend conventional approaches by the logical formalization of infinitely many contexts influencing computations. Unconventional means in game theory were first formulated by Albin [9]. Since his pioneering work, the cellular-automatic approach has been used in multi-agent games. Peter S. Albin has analyzed economic systems by means of generative linguistics and cellular automata to explore complex dynamic interactions of human beings such as business cycles, decentralized market trading, etc. He has showed that economic systems have the complexity level at least of a Turing machine, therefore in macroeconomic predictions we face problems which are undecidable within conventional computing. My work is a continuation of his ideas and it is an attempt to involve cellular automata in epistemic game theory. Whithin my approach it is possible to show that Aumann's agreement theorem [10] is invalid, see Theorems 10.2 and 10.3 [11]. This result is more natural for our intuition and it shows that the new game logic given by payoff cellular automata can describe much more real games logically—from bio-inspired games and up to reflexive games.

11.1.1 Ecological Rationality and Cellular Automata

In the new branch of social psychology dealing with economic behaviour of different agents such as customers there are collected many examples contradicting the rationality assumptions of the standard sociocconomic model. In this model called the *constructivist* one by Vernon L. Smith it is assumed that decision making is algorithmically transparent and may be presented as an automaton (black box) where inputs are algorithmically transformed into outputs. For instance, we can ever find the Nash competitive equilibrium if agents are rational, i.e. they possess both complete information on basic logical rules of non-cooperative decision making, and common knowledge in the meaning of Aumann's agreement theorem (all know that all know that they have this information). Rational agents have common expectations of equilibria and therefore it holds in their behaviour. In this way an equilibrium is a certain algorithm that transforms inputs (agents' needs) into outputs (agents' decisions). So, we have the same situations as in a black box, because equilibria describe outcomes and say nothing about the way players make decisions. As we see, decision making becomes automatic. It is just a constructivist approach, according to which every agent is rational in exactly the same sense as theorists of game theory. Nevertheless, experimental economists have reported that people usually do not follow constructivist rationality and can be better or worse in achieving aims than it is predicted in the way of this classical rationality. The situation is even much worse, because the swarms are not subordinated to the pure behaviourism, also—they can modify own actions and even have a proto-symbolic values. Hence, the classical rationality does not cover the swarm behaviour, too.

Instead of constructivist rationality Vernon L. Smith proposed the so-called *ecological rationality* that emphasizes contexts and frames of culturally-based action and norms for the case of the human beings and context-based modifications of actions for the case of the swarms:

> Ecological rationality uses reason—rational reconstruction—to examine the behaviour of individuals based on their experience and folk knowledge, who are "naive" in their ability to apply constructivist tools to the decisions they make; to understand the emergent order

in human cultures; to discover the possible intelligence embodied in the rules, norms, and institutions of our cultural and biological heritage that are created from human interactions but not by deliberate human design [8].

In ecological rationality we deal not with a black box of automatic decisions, but with kinds of *habitus in decision behaviour*, i.e. with decision lifestyles embodied through the activities and experiences of everyday life. The black box model of decisions, in particular the idea of looking for the Nash equilibrium, assumes that there are decisions a priori (e.g. the Nash equilibrium) and just these decisions are considered rational. For ecological rationality there are no decisions a priori (contexts should always be taken into account). Whether this means that our decision or the swarm decisions cannot be rational?

Let us examine the following simplified abstract model of market. There is a huge number of agents. Each of them has a different sum of money and would like to invest. These agents can (1) either loan money at 10% for one year (in this case my payoff consists of +10% for one year), (2) or borrow money at 10% for one year for investing them somewhere directly (in this case if I do not invest money, my payoff is −10%), (3) or invest money in a joint business venture with expecting economic profit consisting of 20% of the investment for one year (i.e. my payoff is +20%). We assume in this model that I always can earn money if I have people I know and they agree to loan or invest in a joint business. Each agent loses 10% of his/her money for one year. It means that if money were loaned, my payoff is 0%. If money were borrowed and not invested, my payoff is −20%. If money were invested, my payoff consists of +10%. Each agent becomes inactive (cannot invest or borrow) if (s)he does not have money. Every agent has a possibility to behave wrongly one time, when (1) either (s)he spends all outside investments aimed to be used in a joint venture; (2) or (s)he does not give back borrowed money and does not pay an interest. If an agent behaves wrongly second time, after that (s)he becomes inactive. Let us suppose, each agent can interact only with agents whom (s)he knows. Let us call them his/her neighbours. For two different agents, the sets of neighbours cannot be the same, but their intersection is either empty, or non-empty. Each state of agent is defined by the sum of money which (s)he possesses including outside investments and can invest further. The dynamics of states is fixed by discrete time $t = 0, 1, 2, \ldots$, where each step t is one year. Each agent should fulfil only one investment at t and no more than one borrowing/loaning at this t. A state of agent at t depends on states of neighbours, whether they are active and ready to invest in a joint venture at t.

Is it possible to find a priori decisions for this abstract model of market? We know that at $t > 0$, the state of agent A_1 depends on a state of neighbour A_2, the state of A_2 depends on a state of neighbour A_3, the state of A_3 depends on a state of neighbour A_4, and so on k times, where k is a number of all market agents. To sum up, the larger t, the more difficult is to predict my state. For predicting at $t > 0$, I should know the behaviour of all agents of market. Nevertheless, it is impossible; I know the behaviour just of my neighbours.

This abstract model of market shows that in everyday life we can resort to a black box model of automatic (a priori) decisions only at $t = 0$. If $t > 0$, I am dealing with

$t=0$

6	8	1	−3	2
−2	1	−3	−1	3
−2	−1	6	−1	8
1	0	8	6	−2
8	0	−3	4	1

$t=1$

8	8	8	8	8
8	8	8	8	8
8	8	8	8	8
8	8	8	8	8
8	8	8	8	8

Fig. 11.1 Let $S = \{-3, -2, \ldots, 7, 8\}$ be the set of states. Let p be a state of cell z, q a state of its right neighbour $z_0 \in \{z_0\} = N(z)$. The operations $\neg$, $\wedge$, and $\Rightarrow$ are understood as follows: $\neg p = N_{max} - p$ if $p \geq 0$ and $\neg p = N_{max} + p$ if $p < 0$; $p \wedge q = \max(p, q)$; $p \Rightarrow q = N_{max} - \max(p, q) + q$, where N_{max} is a maximal payoff, i.e. it is equal 8. Then an initial configuration at $t = 0$ is transformed into a configuration at $t = 1$ by the following local transition function: $p^{t+1}(z) = (p^t(z) \wedge \neg q^t(z_0)) \Rightarrow p^t(z)$. This cellular automaton cannot vary at $t > 1$

ecological rationality, when there cannot be a priori decisions even if all agents are rational in the classical meaning (e.g. they aim to remain active and to maximize their profits).

Thus, in ecological rationality we assume multi-agentness and massive parallelism of decision making, which exclude the possibility of a priori decisions. To formally simulate this kind of rationality we can appeal to a cellular automata formalism. Recall that any cellular automaton consists of cells belonging to the set $\mathbf{Z}^d$, thereby each of cell takes its value in S, a finite or infinite set of elements called the states of an automaton. Usually, cells are considered unchangeable, but their states change permanently. This dynamics depends on local transition rule $\delta : S^{n+1} \to S$ that transforms states of cells taking into account states of n neighbour cells. The ordered set of n elements, N, is said to be a neighbourhood. Each step of dynamics is fixed by discrete time $t = 0, 1, 2, \ldots$.

At the moment t, the configuration of the whole system (or the global state) is given by the mapping x^t from $\mathbf{Z}^d$ into S, and the evolution is the sequence $x^0 x^1 x^2 x^3 \ldots$ defined as follows: $x^{t+1}(z) = \delta(x^t(z), x^t(z + \alpha_1), x^t(z + \alpha_2), \ldots, x^t(z + \alpha_n))$, where $\langle \alpha_1, \alpha_2, \ldots, \alpha_n \rangle \in N(z)$, i.e. they are neighbours of z. Here is the initial configuration, and it fully determines the future behaviour of the automaton.

Local transition functions may present different logical notions: syntactic, semantic, proof-theoretic, etc. For instance, let us consider $(p \wedge \neg q) \Rightarrow p$, a complex Boolean formula, as local transition function in a cellular automaton of Fig. 11.1, where $S = \{-3, -2, \ldots, 7, 8\}$. All states are considered different payoffs. This automaton will simulate an infinite process of truth valuation of $(p \wedge \neg q) \Rightarrow p$. At $t > 0$ the automaton of Fig. 11.1 is reversible.

This example of cellular-automatic consideration of $(p \wedge \neg q) \Rightarrow p$ shows how deep the massive-parallel understanding of logical notions differs from the conventional one. First, in standard sequential thinking I analyzes sequences and in massively parallel thinking the dynamics of the whole system. Second, in conventional logic all sequential possibilities are considered a priori, e.g. in semantic valuation of a well-formed formula I regard all cases in what way different values of atomic propositions influence on the value of the whole proposition. Due to this fact we could claim that all the basic notions of conventional logic are a priori, they do not depend on our experience and cover all possible cases. There is no initial configuration of the whole that should be taken into account. It would be so aposterioristic. On the contrary, in unconventional logic we focus on initial configuration of the whole system, because it causes all further dynamics. Hereby each step of dynamics is unique and evolution may be infinite. Under these circumstances all logical notions in unconventional logic could be said to be a posteriori, they depend on initial configuration, i.e. on our experience.

An unconventional logic whose notions are presented in some cellular automata is called *aposterioristic*, see [12]. About cellular automaton approach, see [13].

11.1.2 Proof-Theoretic Cellular Automata

Non-well-founded proofs including cyclic proofs have been actively studying recently (see [14–17]). Their features consist in that in the classical theory of deduction, derivation trees, on the one hand, are finite and, on the other hand, they are without cycles, while in the non-well-founded approach they can be infinite and, at the same time, circles occur in them. Non-well-founded proofs have different applications in computer science. In the chapter I am proposing a more radical approach than other non-well-founded approaches to deduction by defining massive-parallel proofs and rejecting axioms in proof theory. This novel approach is characterized as follows:

- Deduction is considered as a transition in cellular automata, where states of cells are regarded as well-formed formulas of a logical language.
- We build up derivations without using axioms, therefore there is no sense in distinguishing logic and theory (i.e. logical and nonlogical axioms), derivable and provable formulas, etc.
- In deduction we do not obtain derivation trees and instead of the latter we find out derivation traces, i.e. a linear evolution of each singular premise.
- Some derivation traces are circular, i.e. some premises are derivable from themselves.
- Some derivation traces are infinite.

For any logical language $\mathscr{L}$ we can construct a *proof-theoretic cellular automaton* (instead of conventional deductive systems) simulating massive-parallel proofs.

Definition 11.1 A proof-theoretic cellular automaton is a 4-tuple $\mathscr{A} = \langle \mathbf{Z}^d, S, N, \delta \rangle$, where

- $d \in \mathbf{N}$ is a number of dimensions and the members of $\mathbf{Z}^d$ are referred as cells,
- S is a finite or infinite set of elements called the states of an automaton $\mathscr{A}$, the members of $\mathbf{Z}^d$ take their values in S, the set S is collected from well-formed formulas of a language $\mathscr{L}$.
- $N \subset \mathbf{Z}^d \setminus \{0\}^d$ is a finite ordered set of n elements, N is said to be a neighbourhood,
- $\delta \colon S^{n+1} \to S$ that is δ is the inference rule of a language $\mathscr{L}$, it plays the role of local transition function of an automaton $\mathscr{A}$.

As we see, the automaton is considered on the endless d-dimensional space of integers, i.e. on $\mathbf{Z}^d$. Discrete time is introduced for $t = 0, 1, 2, \ldots$ fixing each step of inferring.

For any given $z \in \mathbf{Z}^d$, its neighbourhood is determined by $z + N = \{z + \alpha : \alpha \in N\}$. There are two often-used neighbourhoods:

- Von Neumann neighbourhood $N_{VN} = \{z \in \mathbf{Z}^d : \sum_{k=1}^{d} |z_k| = 1\}$
- Moore neighbourhood $N_M = \{z \in \mathbf{Z}^d : \max\limits_{k=\overline{1,d}} |z_k| = 1\} = \{-1, 0, 1\}^d \setminus \{0\}^d$

For example, if $d = 2, N_{VN} = \{(-1, 0), (1, 0), (0, -1), (0, 1)\}; N_M = \{(-1, -1), (-1, 0), (-1, 1), (0, -1), (0, 1), (1, -1), (1, 0), (1, 1)\}$.

In the case $d = 1$, von Neumann and Moore neighbourhoods coincide. It is easily seen that $|N_{VN}| = 2d$, $|N_M| = 3^d - 1$.

At the moment t, the *configuration* of the whole system (or the *global state*) is given by the mapping $x^t : \mathbf{Z}^d \to S$, and the *evolution* is the sequence $x^0, x^1, x^2, \ldots$ defined as follows: $x^{t+1}(z) = \delta(x^t(z), x^t(z + \alpha_1), \ldots, x^t(z + \alpha_n))$, where $\langle \alpha_1, \ldots, \alpha_n \rangle \in N$. Here x^0 is the initial configuration, and it fully determines the future behaviour of the automaton. It is the set of all premises (not axioms).

We assume that δ is an inference rule, i.e. a mapping from the set of premises (their number cannot exceed $n = |N|$) to a conclusion. For any $z \in \mathbf{Z}^d$ the sequence $x^0(z)$, $x^1(z), \ldots, x^t(z), \ldots$ is called a *derivation trace from a state* $x^0(z)$. If there exists t such that $x^t(z) = x^l(z)$ for all $l > t$, then a derivation trace is *finite*. It is *circular/cyclic* if there exists l such that $x^t(z) = x^{t+l}(z)$ for all t.

Definition 11.2 In case all derivation traces of a proof-theoretic cellular automaton $\mathscr{A}$ are circular, this automaton $\mathscr{A}$ is said to be reversible, see [18].

Notice that x^{t+1} depends only upon x^t, i.e. the previous configuration. It enables us to build the function $G_{\mathscr{A}} : C_{\mathscr{A}} \to C_{\mathscr{A}}$, where $C_{\mathscr{A}}$ is the set of all possible configurations of the cellular automaton $\mathscr{A}$ (it is the set of all mappings $\mathbf{Z}^d \to S$, because we can take each element of this set as the initial configuration x^0, though not every element can arise in the evolution of some other configuration). $G_{\mathscr{A}}$ is called the *global function* of the automaton.

Example 11.1 (Modus ponens) Consider a propositional language $\mathscr{L}$ that is built in the standard way with the only binary operation of implication $\supset$. Let us suppose that well-formed formulas of that language are used as the set of states for a proof-theoretic cellular automaton $\mathscr{A}$. Further, assume that *modus ponens* is a transition rule of this automaton $\mathscr{A}$ and it is formulated for any $\varphi, \psi \in \mathscr{L}$ as follows:

$$x^{t+1}(z) = \begin{cases} \psi, & \text{if } x^t(z) = \varphi \supset \psi \text{ and } \varphi \in (z + N); \\ x^t(z), & \text{otherwise.} \end{cases}$$

The further dynamics will depend on the neighbourhood. If we assume the Moor neighbourhood in the 2-dimensional space, this dynamics will be exemplified by the evolution of cell states in Figs. 11.2, 11.3 and 11.4.

$(p \supset q) \supset r$	$p \supset (p \supset q)$	$p \supset q$	$(p \supset q) \supset (p \supset q)$	$(r \supset p) \supset r$
$(p \supset r) \supset (q \supset r)$	$p \supset q$	p	$p \supset (p \supset q)$	$r \supset p$
$p \supset r$	p	$p \supset (q \supset (p \supset q))$	p	r
$p \supset (q \supset r)$	$p \supset p$	$p \supset q$	$(p \supset r) \supset (q \supset p)$	$p \supset r$
$p \supset q$	$p \supset (q \supset p)$	q	$p \supset r$	p

Fig. 11.2 An initial configuration of a proof-theoretic cellular automaton $\mathscr{A}$ with the Moor neighbourhood in the 2-dimensional space, its states run over formulas set up in a propositional language $\mathscr{L}$ with the only binary operation $\supset$, $t = 0$. Notice that p, q, r are propositional variables

Fig. 11.3 An evolution of $\mathscr{A}$ described in Fig. 11.2 at the time step $t = 1$

r	$p \supset q$	q	$p \supset q$	r
$q \supset r$	q	p	$p \supset q$	p
r	p	$q \supset (p \supset q)$	p	r
$q \supset r$	p	q	$q \supset p$	r
$p \supset q$	$p \supset (q \supset p)$	q	r	p

Fig. 11.4 An evolution of $\mathscr{A}$ described in Fig. 11.2 at the time step $t = 3$. Its configuration cannot vary further

r	q	q	q	r
r	q	p	q	p
r	p	q	p	r
r	p	q	p	r
q	p	q	r	p

This example shows that first we completely avoid axioms and secondly we take premisses from the cell states of the neighbourhood according to a transition function. As a result, we do not come across proof trees in our novel approach to deduction taking into account that a cell state has just a linear dynamics (the number of cells and their location do not change). This allows us evidently to simplify deductive systems.

Now we are trying to consider a cellular-automaton presentation of two basic deductive approaches: Hilbert's type and sequent ones.

Example 11.2 (Hilbert's inference rules) Suppose a propositional language $\mathscr{L}$ contains two basic propositional operations: negation and disjunction. As usual, the set of all formulas of $\mathscr{L}$ is regarded as the set of states of an appropriate proof-theoretic cellular automata. In that we will use the exclusive disjunction of the following five inference rules converted from Joseph R. Shoenfield's deductive system:

$$x^{t+1}(z) = \begin{cases} \psi \vee \varphi, & \text{if } x^t(z) = \varphi; \\ \varphi, & \text{if } x^t(z) = \varphi \vee \varphi; \\ (\chi \vee \psi) \vee \varphi, & \text{if } x^t(z) = \chi \vee (\psi \vee \varphi); \\ \chi \vee \psi, & \text{if } x^t(z) = \varphi \vee \chi \text{ and } (\neg\varphi \vee \psi) \in (z + N); \\ \chi \vee \psi, & \text{if } x^t(z) = \neg\varphi \vee \psi \text{ and } (\varphi \vee \chi) \in (z + N). \end{cases}$$

Example 11.3 (Sequent inference rules) Let us take a sequent propositional language $\mathscr{L}$, in which the classical propositional language with negation, conjunction, disjunction and implication is extended by adding the sequent relation $\hookrightarrow$. Recall that a *sequent* is an expression of the form $\Gamma_1 \hookrightarrow \Gamma_2$, where $\Gamma_1 = \{\varphi_1, \ldots, \varphi_j\}$, $\Gamma_2 = \{\psi_1, \ldots, \psi_i\}$ are finite sets of well-formed formulas of the standard propositional language, that has the following interpretation: $\Gamma_1 \hookrightarrow \Gamma_2$ is logically valid iff

$$\bigwedge_j \varphi_j \supset \bigvee_i \psi_i$$

is logically valid. Let S denote the set of all sequents of $\mathscr{L}$, furthermore let us assume that this family S is regarded as the set of states for a proof-theoretic cellular automaton $\mathscr{A}$. The transition rule of $\mathscr{A}$ is an exclusive disjunction of the 14 singular rules (6 structural rules and 8 logical rules):

$$x^{t+1}(z) = \Gamma_1 \hookrightarrow \Gamma_2, \begin{cases} \text{if } \Gamma_1 \hookrightarrow \Gamma_2 \text{ is a result of applying to } x^t(z) \\ \text{eather one of structural rules} \\ \text{or the left (right) introduction of negation} \\ \text{or the left introduction of conjunction} \\ \text{or the right introduction of disjunction} \\ \text{or the right introduction of implication.} \end{cases}$$

$$x^{t+1}(z) = \begin{cases} \Gamma \hookrightarrow \Gamma', \psi \wedge \chi, & \text{if } x^t(z) = \Gamma \hookrightarrow \Gamma', \psi \text{ and} \\ & (\Gamma \hookrightarrow \Gamma', \chi) \in (z+N); \\ \Gamma, \psi \vee \chi \hookrightarrow \Gamma', & \text{if } x^t(z) = \Gamma, \psi \hookrightarrow \Gamma' \text{ and} \\ & (\Gamma, \chi \hookrightarrow \Gamma') \in (z+N); \\ \psi \supset \chi, \Gamma, \Delta \hookrightarrow \Gamma', \Delta', & \text{if } x^t(z) = \Gamma \hookrightarrow \Gamma', \psi \text{ and} \\ & (\chi, \Delta \hookrightarrow \Delta') \in (z+N). \end{cases}$$

Example 11.4 (Brotherston's cyclic proofs) The sequent language used in the previous example we extend by adding predicates N, E, O and appropriate inference rules of Fig. 11.5 for them. Further, let us extend also the automaton of Example 11.3 in the same way by representing inference rules of Fig. 11.5 in the cellular-automatic form.

Now we assume that a cell has an initial state $[\Gamma, N(z) \hookrightarrow \Delta, O(z), E(z)]$ and its neighbour cell an initial state $[\Gamma, z = 0 \hookrightarrow \Delta]$ that is equal to $[\Gamma, z = 0 \hookrightarrow \Delta, O(z), E(z)]$ for any $t = 4, 14, 24, \ldots$ and to $[\Gamma, z = 0 \hookrightarrow \Delta, E(z), O(z)]$ for any $t = 9, 19, 29, \ldots$. Then we will have the following infinite cycle:

$$[\Gamma, N(z) \hookrightarrow \Delta, O(z), E(z)] \longrightarrow^{(substitution)} [\Gamma, N(y) \hookrightarrow \Delta, O(y), E(y)] \longrightarrow$$
$$[\Gamma, N(y) \hookrightarrow \Delta, O(y), O(y+1)] \longrightarrow [\Gamma, N(y) \hookrightarrow \Delta, E(y+1), O(y+1)] \longrightarrow$$
$$[\Gamma, z = (y+1), N(y) \hookrightarrow \Delta, O(z), E(z)] \longrightarrow^{(case\,N)} [\Gamma, N(z) \hookrightarrow \Delta, E(z), O(z)] \longrightarrow \ldots$$

Another instance of cyclic proof is given in the next subsection. As we see, the possibility of cyclic derivation traces depends on configuration of cell states.

$$\frac{\Gamma\hookrightarrow N(x)}{\Gamma\hookrightarrow N(x+1)}, \quad \frac{\Gamma\hookrightarrow\Delta}{\Gamma\hookrightarrow\Delta,N(0)}, \quad \frac{\Gamma\hookrightarrow E(x)}{\Gamma\hookrightarrow O(x+1)}, \quad \frac{\Gamma\hookrightarrow O(x)}{\Gamma\hookrightarrow E(x+1)}, \quad \frac{\Gamma\hookrightarrow\Delta}{\Gamma\hookrightarrow\Delta,E(0)},$$

$$\frac{N(x)\hookrightarrow\Delta}{N(x+1)\hookrightarrow\Delta}, \quad \frac{\Gamma\hookrightarrow\Delta}{\Gamma,N(0)\hookrightarrow\Delta}, \quad \frac{E(x)\hookrightarrow\Delta}{O(x+1)\hookrightarrow\Delta}, \quad \frac{O(x)\hookrightarrow\Delta}{E(x+1)\hookrightarrow\Delta}, \quad \frac{\Gamma\hookrightarrow\Delta}{\Gamma,E(0)\hookrightarrow\Delta},$$

$$\frac{\Gamma,t=0\hookrightarrow\Delta \qquad \Gamma,t=x+1,N(x)\hookrightarrow\Delta}{\Gamma,N(t)\hookrightarrow\Delta}\ (Case\ N), \text{where } x\notin FV(\Gamma\cup\Delta\cup\{N(t)\}),$$

$$\frac{\Gamma\hookrightarrow\Delta}{\Gamma[x]\hookrightarrow\Delta[x]}\ (Substitution).$$

Fig. 11.5 Inference rules for predicates N ('being a natural number'), E ('being an even number'), O ('being an odd number'), see [14]

Traditional tasks concerning proof theory like completeness and independence of axioms lose their sense in massive-parallel proof theory, although it can be readily shown that we can speak about consistency:

Proposition 11.1 *Proof theories given in Examples 11.2 and 11.3 are consistent, i.e. we cannot deduce a contradiction within them.* □

11.1.3 Proof-Theoretic Simulation of Belousov-Zhabotinsky Reaction

In this subsection I am analyzing simulating the *Belousov-Zhabotinsky reaction* within the framework of the theory of massive-parallel proofs.

Let us consider a proof-theoretic cellular automaton with circular proofs for the Belousov-Zhabotinsky reaction containing feedback relations. The mechanism of this reaction (namely cerium(III) $\longleftrightarrow$ cerium(IV) catalyzed reaction) is very complicated: its recent model contains 80 elementary steps and 26 variable species concentrations. Let us consider a simplification of Belousov-Zhabotinsky reaction assuming that the set of states consists just of the following reactants: Ce^{3+}, $HBrO_2$, BrO_3^-, H^+, Ce^{4+}, H_2O, $BrCH\ (COOH)_2$, Br^-, $HCOOH$, CO_2, $HOBr$, Br_2, $CH_2(COOH)_2$ which interact according to inference rules (reactions) (11.1)–(11.7). In this reaction we observe sudden oscillations in color from yellow to colorless, allowing the oscillations to be observed visually. In spatially nonhomogeneous systems (such as a simple petri dish), the oscillations propagate as spiral wave fronts. The oscillations last about one minute and are repeated over a long period of time. The color changes are caused by alternating oxidation-reductions in which cerium changes its oxidation state from cerium(III) to cerium(IV) and vice versa: $Ce^{3+} \longrightarrow Ce^{4+} \longrightarrow Ce^{3+} \longrightarrow \ldots$.

When Br^- has been significantly lowered, the reaction pictured by inference rule (11.1) causes an exponential increase in bromous acid ($HBrO_2$) and the oxidized form of the metal ion catalyst and indicator, cerium(IV). Bromous acid is subsequently converted to bromate (BrO_3^-) and $HOBr$ [the step (11.3)]. Meanwhile, the step (11.2) reduces the cerium(IV) to cerium(III) and simultaneously increase bromide (Br^-) concentration. Once the bromide concentration is high enough, it reacts with bromate (BrO_3^-) and $HOBr$ in (11.4) and (11.6) to form Br_2, further Br_2 reacts with $CH_2(COOH)_2$ to form $BrCH(COOH)_2$ and the process begins again. Thus, parallel processes in (11.1)–(11.7) have several cycles which are performed synchronously.

The proof-theoretic simulation of Belousov-Zhabotinsky reaction can be defined as follows:

Definition 11.3 Consider a propositional language $\mathscr{L}$ with the only binary operation $\oplus$, it is built in the standard way over the set of variables $S = \{Ce^{3+}, HBrO_2, BrO_3^-, H^+, Ce^{4+}, H_2O, BrCH(COOH)_2, Br^-, HCOOH, CO_2, HOBr, Br_2, CH_2(COOH)_2\}$. Let S be the set of states of proof-theoretic cellular automaton $\mathscr{A}$. The inference rule of the automaton is presented by the conjunction of singular inference rules (11.1)–(11.7):

$$(11.1) \wedge (11.2) \wedge (11.3) \wedge (11.4) \wedge (11.5) \wedge (11.6) \wedge (11.7).$$

The operation $\oplus$ has the following meaning: $A \oplus B$ defines a probability distribution of events A and B in neighbour cells participated in a reaction caused the appearance of $A \oplus B$. Then $\mathscr{A}$ simulates the Belousov-Zhabotinsky reaction.

Definition 11.4 Let $p, s_i, s_{i+1} \in \{Ce^{3+}, HBrO_2, BrO_3^-, H^+, Ce^{4+}, H_2O, BrCH(COOH)_2, Br^-, HCOOH, CO_2, HOBr, Br_2, CH_2(COOH)_2\}$. A state p is called a premise for deducing s_{i+1} from s_i by the inference rule $(11.1) \wedge (11.2) \wedge \ldots \wedge (11.7)$ iff

- p is s_i or
- in a neighbour cell we find out an expression of the form $p \oplus A, B \oplus C$, where A, B, C are propositional metavariables, i.e. they run over either the empty set or the set of states closed under the operation $\oplus$. Thus, we assume that each premise should occur in a separate cell. This means that if we find out an expression $p_i \oplus p_j, B \oplus C$ or $p_i \oplus A, p_j \oplus C$ in a neighbour cell and both p_i and p_j are needed for deducing, whereas p_i, p_j do not occur in other neighbour cells, then p_i, p_j could not be considered as premises.

$$x^{t+1}(z) = \begin{cases} (1)\ Ce^{4+} \oplus HBrO_2 \oplus H_2O, & \text{if } x^t(z) \in \{Ce^{3+}\} \\ \text{and premises } HBrO_2, BrO_3^-, H^+ \in (z+N); & \\ (2)\ x^t(z), & \text{otherwise.} \end{cases} \tag{11.1}$$

$$x^{t+1}(z) = \begin{cases} (1)\ Br^- \oplus Ce^{3+} \oplus HCOOH \oplus CO_2 \oplus H^+, & \text{if } x^t(z) \in \{Ce^{4+}\} \\ \text{and premises } BrCH(COOH)_2, H_2O \in (z+N); & \\ (2)\ x^t(z), & \text{otherwise.} \end{cases} \tag{11.2}$$

$$x^{t+1}(z) = \begin{cases} (1)\ HOBr \oplus BrO_3^- \oplus H^+, & \text{if } x^t(z) \in \{HBrO_2\}; \\ (2)\ x^t(z), & \text{otherwise.} \end{cases} \quad (11.3)$$

$$x^{t+1}(z) = \begin{cases} (1)\ HOBr \oplus HBrO_2, & \text{if } x^t(z) \in \{BrO_3^-\} \text{ and} \\ & \text{premises } Br^-, H^+ \in (z+N); \\ (2)\ x^t(z), & \text{otherwise.} \end{cases} \quad (11.4)$$

$$x^{t+1}(z) = \begin{cases} (1)\ HOBr, & \text{if } x^t(z) \in \{Br^-\} \text{ and} \\ & \text{premises } HBrO_2, H^+ \in (z+N); \\ (2)\ x^t(z), & \text{otherwise.} \end{cases} \quad (11.5)$$

$$x^{t+1}(z) = \begin{cases} (1)\ Br_2 \oplus H_2O, & \text{if } x^t(z) \in \{HOBr\} \text{ and} \\ & \text{premises } Br^-, H^+ \in (z+N); \\ (2)\ x^t(z), & \text{otherwise.} \end{cases} \quad (11.6)$$

$$x^{t+1}(z) = \begin{cases} (1)\ Br^- \oplus H^+ \oplus BrCH(COOH)_2, \ \text{if } x^t(z) \in \{Br_2\} \\ \text{and premises } CH_2(COOH)_2 \in (z+N); \\ (2)\ x^t(z), \ \text{otherwise.} \end{cases} \quad (11.7)$$

Example 11.5 (Belousov-Zhabotinsky's cyclic proofs) We can simplify the automaton defined above assuming that $\oplus$ is a metatheoretic operation with the following operational semantics:

$$\frac{A \oplus B}{A} \quad \frac{A \oplus B}{B},$$

where A and B are metavariables defined on S. The informal meaning of that operation is that we can ignore one of both variables coupled by $\oplus$. In the cellular automaton $\mathscr{A}$ this metaoperation will be used as follows:

$$x^{t+1}(z) = \begin{cases} X, Y, & \text{if } x^t(z) = A \oplus B \text{ and according to rules (11.1)–(11.7),} \\ & X \text{ changes from } A \text{ and } Y \text{ changes from } B; \\ X, & \text{if } x^t(z) = A \oplus B \text{ and according to rules (11.1)–(11.7),} \\ & X \text{ changes from } A \text{ and } B \text{ does not change;} \\ Y, & \text{if } x^t(z) = A \oplus B \text{ and according to rules (11.1)–(11.7),} \\ & Y \text{ changes from } B \text{ and } A \text{ does not change;} \\ A \oplus B, & \text{if } x^t(z) = A \oplus B \text{ and rules (11.1)–(11.7)} \\ & \text{cannot be applied to } A \text{ or } B. \end{cases} \quad (11.8)$$

Let us suppose now that X, Y run over the set of states closed under the operation $\oplus$.

$$x^{t+1}(z) = \begin{cases} X, Y, & \text{if (i)} \, x^t(z) = X, Y \text{ and (ii) both } X, \text{ and } Y \\ & \text{are simultaneously usable (not usable)} \\ & \text{as premises in at least two different rules} \\ & \text{of (11.1)–(11.7) (see Definition 11.4);} \\ X, & \text{if (i)} \, x^t(z) = X, Y \text{ and (ii) only } X \text{ is usable} \\ & \text{as a premise in at least one rule of (11.1)–(11.7);} \\ Y, & \text{if (i)} \, x^t(z) = X, Y \text{ and (ii) only } Y \text{ is usable} \\ & \text{as a premise in at least one rule of (11.1)–(11.7).} \end{cases} \quad (11.9)$$

$$\text{idempotency:} A := A, A. \quad (11.10)$$

$$\text{commutativity:} A, B := B, A. \quad (11.11)$$

Hence, we cannot ignore one of both variables coupled by $\oplus$ and should accept both them if in the neighbourhood there are reactants that catenate both variables and change them. This rule is the simplest interpretation of $A \oplus B$. We have three cases: (i) both variables are catenated with reactants from the neighbourhood, in this case we mean that the probability distribution of events A and B is the same and equal to 0.5 and, as a result, we cannot choose one of them and accept both; (ii) only A is catenated with reactants from the neighbourhood, then the probability distribution of event A is equal to 1.0 and that of B to 0.0; (iii) only B is catenated with reactants from the neighbourhood, then the probability distribution of event B is equal to 1.0 and that of A to 0.0. Thus, $A \oplus B$ is a function that associates either exactly one value with its arguments (i.e. either A or B) or simultaneously both values (i.e. A and B).

This simplified version of the automaton $\mathscr{A}$ is exemplified in Fig. 11.6.

Evidently, reducing the complicated dynamics of Belousov-Zhabotinsky reaction to conventional logical proofs is a task that cannot be solved so easily I have just done.

The Belousov-Zhabotinsky reaction is one of the best examples of self-organization in nature. Hence, an opportunity to have a cycle in proof-theoretic automata testifies to a self-organization.

11.1.4 Proof-Theoretic Simulation of Slime Mould Dynamics

The dynamics of plasmodium of *Physarum polycephalum* could be regarded as another simple example of the natural proof-theoretic automata. The point is that when the plasmodium is cultivated on a nutrient-rich substrate (agar gel containing crushed oat flakes) it exhibits uniform circular growth similar to the excitation waves in the excitable Belousov-Zhabotinsky medium. If the growth substrate lacks nutrients, e.g. the plasmodium is cultivated on a non-nutrient and repellent containing gel,

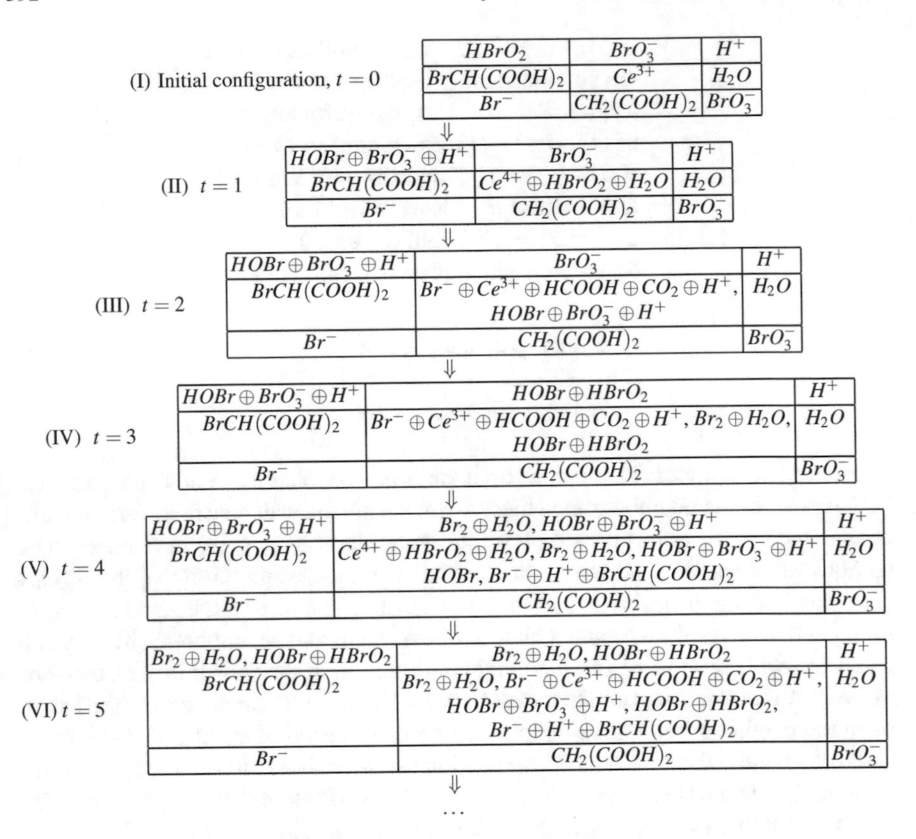

Fig. 11.6 The evolution of a reversible proof-theoretic cellular automaton $\mathscr{A}$ with the Moor neighbourhood in the 2-dimensional space for the Belousov-Zhabotinsky reaction. This automaton simulates the circular feedback $Ce^{3+} \longrightarrow Ce^{4+} \longrightarrow Ce^{3+} \longrightarrow \ldots$ (more precisely temporal oscillations in a well-stirred solution): Ce^{3+} is colorless and Ce^{4+} is yellow. The initial configuration of $\mathscr{A}$-cells described in (I) occurs in the same form at the further steps and the cycle repeats several times. For entailing (I) $\longrightarrow$ (II) we have just used inference rule (11.1) (row 2, column 2) and inference rule (11.3) (row 1, column 1), for entailing (II) $\longrightarrow$ (III) inference rules (11.2), (11.3) and (11.8) (row 2, column 2), for entailing (III) $\longrightarrow$ (IV) inference rule (11.4) (row 1, column 2) and inference rules (11.4), (11.6), (11.8) and (11.9) (row 2, column 2), for entailing (IV) $\longrightarrow$ (V) inference rules (11.3) and (11.6) (row 1, column 2) inference rules (11.1), (11.3), (11.5), (11.6), (11.7), (11.8) and (11.9) (row 2, column 2), for entailing (V) $\longrightarrow$ (VI) inference rules (11.4), (11.6) (row 1, column 1), inference rules (11.4), (11.6), (11.10) (row 1, column 2), inference rules (11.2), (11.3), (11.4), (11.6), (11.7), (11.9), (11.10), (11.11) (row 2, column 2)

a wet filter paper or even glass surface localizations emerge and branching patterns become clearly visible.

The plasmodium continues its spreading, reconfiguration and development as long as there are enough nutrients. When the supply of nutrients is over, the plasmodium either switches to a fructification state (if the level of illumination is high enough), when sporangia are produced, or forms sclerotium (if there is darkness).

The pseudopodium propagates in a manner analogous to the formation of wave-fragments in sub-excitable Belousov-Zhabotinsky systems. Starting in the initial conditions, the plasmodium exhibits a foraging behaviour, searching for sources of nutrients. When such sources are located and taken over, the plasmodium forms characteristic veins of protoplasm. The Belousov-Zhabotinsky reaction and the plasmodium are light-sensitive, which gives us the means to program them. The plasmodium of *Physarum polycephalum* exhibits an articulated negative phototaxis, the Belousov-Zhabotinsky reaction is inhibited by light. Therefore by using masks of illumination one can control the dynamics of localizations in these media. The light-sensitivity of plasmodium has been already explored in the design of robotics controllers [19].

Experiments with *Physarum polycephalum* were carried out by Andrew Adamatzky as follows. The plasmodia of *Physarum polycephalum* were cultured on wet paper towels, fed with oat flakes, and moistened regularly. He subcultured the plasmodium every 5–7 days. The repellents were implemented with illumination domains using blue electroluminescent sheets, see details in [19]. Masks were prepared from black plastic, namely the triangle was cut in the plastic, when this mask was placed on top of the electro-luminescent sheet, the light was passing only through the cuts [19].

Results of experiments may be described in the terms of proof-theoretic cellular automata. Let us assume that its set of states consists of the entities from the following sets.

1. The set of *neutral zones*, $\{Z_1, Z_2, \ldots\}$, where nothing goes.
2. The set of *growing pseudopodia*, $\{P_1, P_2, \ldots\}$, localized in *active zones*. On a nutrient-rich substrate, the plasmodium propagates as a typical circular, target wave, while on the nutrient-poor substrates localized wave-fragments are formed.
3. The set of *attractants* $\{A_1, A_2, \ldots\}$, they are sources of nutrients, on which the plasmodium feeds. Based on previous experiments by Prof. Adamatzky, we can assume that if the whole experimental area is about 8–10 cm in diameter, then the plasmodium can locate and colonize nearby sources of nutrients.
4. The set of *repellents* $\{R_1, R_2, \ldots\}$. The plasmodium of *Physarum polycephalum* avoids light. Thus, domains of high illumination are repellents such that each repellent R is characterized by its position and intensity of illumination, or the force of repelling.
5. The set of *protoplasmic tubes* $\{C_1, C_2, \ldots\}$. Typically plasmodium spans sources of nutrients with protoplasmic tubes/veins. The plasmodium builds a planar graph, where nodes are sources of nutrients, e.g. oat flakes, and edges are protoplasmic tubes.

Hence, the set of states in the proof-theoretic cellular automaton for dynamics of plasmodium of *Physarum polycephalum* is equal to $\{Z_1, Z_2, \ldots\} \cup \{P_1, P_2, \ldots\} \cup \{A_1, A_2, \ldots\} \cup \{R_1, R_2, \ldots\} \cup \{C_1, C_2, \ldots\}$.

The proof-theoretic simulation of *Physarum polycephalum* is defined as follows:

Definition 11.5 Consider a propositional language $\mathscr{L}$ with the only binary operation $\oplus$, it is built in the standard way over the set of variables $S = \{Z_1, Z_2, \ldots, P_1, P_2, \ldots, A_1, A_2, \ldots, R_1, R_2, \ldots, C_1, C_2, \ldots\}$. Let S be the set of states of proof-theoretic cellular automaton $\mathscr{A}$. The inference rule of the automaton is presented by the conjunction of singular inference rules (11.12)–(11.18):

$$(11.12) \wedge (11.13) \wedge (11.14) \wedge (11.15) \wedge (11.16) \wedge (11.17) \wedge (11.18).$$

The operation $\oplus$ has the following meaning: $A \oplus B$ defines a probability distribution of events A and B in cells. The operation $\oplus$ is idempotent, commutative and associative. Then $\mathscr{A}$ simulates the dynamics of plasmodium of *Physarum polycephalum*.

Definition 11.6 Let $p, s_i, s_{i+1} \in \{Z_1, Z_2, \ldots, P_1, P_2, \ldots, A_1, A_2, \ldots, R_1, R_2, \ldots, C_1, C_2, \ldots\}$. A state p is called a premise for deducing s_{i+1} from s_i by inference rules (11.12)–(11.18) iff

- p is s_i or
- in a neighbour cell we find out an expression of the form $p \oplus X$, where X is a propositional metavariable, i.e. it runs over either the empty set or the set of states closed under the operation $\oplus$. Thus, we assume that each premise should occur in a separate cell. This means that if we find out an expression $p_i \oplus p_j \oplus X$ in a neighbour cell and both p_i and p_j are needed for deducing, whereas p_i, p_j do not occur in other neighbour cells, then p_i, p_j could not be considered as premises. This restriction is just for rule (11.18).

$$x^{t+1}(z) = \begin{cases} (1)\ X \ \text{ if } x^t(z) = X \oplus A_i \text{ and and } X = Y \oplus P_j \\ (2)\ x^t(z), \ \text{ otherwise.} \end{cases} \tag{11.12}$$

$$x^{t+1}(z) = \begin{cases} (1)\ X \oplus P_j \ \text{ if } x^t(z) = X \oplus C_n \oplus A_i \\ (2)\ x^t(z), \ \text{ otherwise.} \end{cases} \tag{11.13}$$

$$x^{t+1}(z) = \begin{cases} (1)\ X \oplus P_j \ \text{ if } x^t(z) = X \oplus P_j \oplus A_i \\ (2)\ x^t(z), \ \text{ otherwise.} \end{cases} \tag{11.14}$$

$$x^{t+1}(z) = \begin{cases} (1)\ X \oplus P_i \vee X \oplus P_j \ \text{ if } x^t(z) = X \oplus P_i \oplus P_j \\ (2)\ x^t(z), \ \text{ otherwise.} \end{cases} \tag{11.15}$$

$$x^{t+1}(z) = \begin{cases} (1)\ X \oplus C_i \vee X \oplus C_j \ \text{ if } x^t(z) = X \oplus C_i \oplus C_j \\ (2)\ x^t(z), \ \text{ otherwise.} \end{cases} \tag{11.16}$$

$$x^{t+1}(z) = \begin{cases} (1)\ X \oplus C_i, & \text{if}\, x^t(z) = X \text{ and} \\ & \text{premises}\, R_i \notin (z+N), \\ & \text{premises}\, C_j, A_i \in (z+N); \\ (2)\ x^t(z), & \text{otherwise.} \end{cases} \tag{11.17}$$

$$x^{t+1}(z) = \begin{cases} (1)\ X \oplus C_i, & \text{if}\, x^t(z) = X \text{ and} \\ & \text{premises}\, R_i \notin (z+N), \\ & \text{premises}\, P_j, A_i \in (z+N); \\ (2)\ x^t(z), & \text{otherwise.} \end{cases} \tag{11.18}$$

This automaton can simulate the plasmodium behaviour completely only in case the plasmodium faces not more than one attractant and one repellent at each time step. The matter is that in this situation, the swarm behaviour of plasmodium is reduced to primitive reactions of bacteria such as *Salmonella typhimurium*—either to react to one attractant or to react to one repellent. In this way, we can implement the pure behaviourism (a complete outer control of all actions) in the strict sense. In the case of several neighbour attractants and repellents there can be lateral inhibition or lateral activation effects that would change a cellular-automatic transition rule at some time steps. Thus, it gives the main difference of the slime mould from the Belousov-Zhabotinsky reaction. The slime mould can play in proto-reflexive games—it can change the own decisions. About other cellular-automatic formalizations of slime mould, see [20, 21] (Fig. 11.7).

(I) Initial configuration, $t = 0$

A_{11}	Z_{12}	A_{31}
R_{12}	Z_{22}	Z_{32}
R_{13}	Z_{23}	P_{33}

$\Downarrow$

(II) $t = 1$

A_{11}	Z_{12}	A_{31}
R_{12}	Z_{22}	$Z_{32} \oplus C_{32}$
R_{13}	Z_{23}	P_{33}

$\Downarrow$

(III) $t = 2$

A_{11}	Z_{12}	$A_{31} \oplus C_{31}$
R_{12}	Z_{22}	$Z_{32} \oplus C_{32}$
R_{13}	Z_{23}	P_{33}

$\Downarrow$

(IV) $t = 3$

A_{11}	Z_{12}	P_{31}
R_{12}	Z_{22}	$Z_{32} \oplus C_{32}$
R_{13}	Z_{23}	P_{33}

Fig. 11.7 The evolution of a proof-theoretic cellular automaton $\mathscr{A}$ with the Moor neighbourhood in the 2-dimensional space for *Physarum polycephalum*

11.2 Payoff Cellular Automata

The swarms such as foraging ants and plasmodia can make context-based decisions and change them. Usually, for representing databases of games, payoff matrices are involved (see Fig. 11.8). However, in the case of accepting the ecological rationality (an opportunity of changing past decisions) we cannot appeal to payoff matrices. For example, we cannot appeal to them if we deal with games limited by some contexts or with infinite games. Some kinds of databases for making decisions within aposterioristic logics could be presented by *payoff cellular automata*. These automata are constructed as follows. Cells of automata belong to the set $\mathbf{Z}^d$ and they take its value in S. The set S of states consists of payoff for all n players. The cardinality $|S|$ is equal $i_1 \cdot i_2 \cdot \ldots \cdot i_n$, where i_j is the number of all strategies for the j-th player, $j = 1, \ldots, n$. Each state has the form of n-tuple $\langle a_{ij...k}, b_{ij...k}, \ldots, c_{ij...k} \rangle$, where

(1) $a_{ij...k}$ is the payoff to player 1 when (1) he plays a_i; (2) player 2 plays b_j; ..., (n) player n plays c_k;
(2) $b_{ij...k}$ is the payoff to player 2 when (1) player 1 plays a_i; (2) player 2 plays b_j; ..., (n) player n plays c_k;

... ...

(n) $c_{ij...k}$ is the payoff to player n when (1) player 1 plays a_i; (2) player 2 plays b_j; ..., (n) player n plays c_k.

The local transition function, δ_j, for the player j, where $j = 1, \ldots, n$, is presented by a decision rule based on the past payoff for all players. The rule δ_j can be the same for all players or different. The initial configuration of payoff cellular automaton is the set of all premises which fully determines the future behaviour of the automaton. They may be understood as the expected payoff for different contexts before the game. The *game context* is defined by the neighbourhood $N(z)$ for the cell z. The number of premises (the past payoff that we can take into account) cannot exceed the number $n = |N(z) \cup z|$. The decision rule δ_j is a mapping from the set of premises $N(z) \cup z$ to a conclusion. This rule generates the sequence $\mathbf{a}^0(z), \mathbf{a}^1(z), \mathbf{a}^2(z), \ldots, \mathbf{a}^t(z), \ldots$ for any $z \in \mathbf{Z}^d$, where $\mathbf{a} = \langle a_{ij...k}, b_{ij...k}, \ldots, c_{ij...k} \rangle$ and $\mathbf{a}^i(z)$ means the state of z at the i-th step of application of δ_j to $\mathbf{a}^0(z)$, the state of z at step 0. That sequence is called a *derivation trace from an initial state* $\mathbf{a}^0(z)$.

		Player1	
		a_1	a_2
Player2	b_1	$\langle a_{11}, b_{11} \rangle$	$\langle a_{21}, b_{21} \rangle$
	b_2	$\langle a_{12}, b_{12} \rangle$	$\langle a_{22}, b_{22} \rangle$

Fig. 11.8 An example of a payoff matrix showing the possible strategies available to player 1 (namely a_1 and a_2) and player 2 (namely b_1 and b_2) and the payoff that each player receives for his choice, depending on what the other players do. The payoff is in the form $\langle a_{ij}, b_{ij} \rangle$, where a_{ij} is the payoff to player 1 when he plays a_i and player 2 plays b_j and b_{ij} is the payoff to player 2 when he plays b_j and player 2 plays a_i

11.2.1 Saddle Point

Let us consider a simple payoff cellular automaton for the game with two players 1 and 2. Let a_{ij} be a payoff for the i-th strategy of player 1 and j-th strategy of player 2. If $\max_i \min_j a_{ij} = \min_j \max_i a_{ij}$ for $z \cup N(z)$ at step t, then a_{ij} is called a *saddle point* for z at time t. Thus, a saddle point is the element of the payoff cellular automaton at time t which is both a maximum of the minimums of each row within the neighbourhood $N(z)$ and a minimum of the maximums of each column within the same neighbourhood. The cells $z \cup N(z)$ may have no saddle points, one saddle point, or multiple saddle points. Let the payoff states pictured in Fig. 11.9 be an initial configuration of the automaton. The set S of states consists of the integers −5, −4, −3, ..., 7, 8.

The local transition function is defined as follows:

$$a^{t+1}(z) = \begin{cases} a_{ij}, & \text{if } a_{ij} = \max(a_{kl}, a_{mn}), \\ & \text{where } a_{kl}, a_{mn} \text{ are saddle points of } N(z) \cup z; \\ a^t(z), & \text{otherwise.} \end{cases}$$

At time $t = 1$, the configuration of Fig. 11.9 has the form of Fig. 11.10.

11.2.2 Minimax and Maximin

Assume that a_{ij} (respectively b_{ij}) is the payoff to player 1 (respectively to player 2) when he uses the i-th strategy of his 3 total strategies and player 2 uses the j-th strategy of his 4 total strategies. In the neighbourhood for each cell, a different number of strategies can be engaged (see Figs. 11.11, 11.12). For example, in the cell of Fig. 11.11 whose state is marked by bold, the neighbourhood contains only

8	2	3	2	3
−3	0	2	−5	−4
−2	−1	6	−1	8
4	1	9	2	4
5	−2	3	0	2

Fig. 11.9 An initial configuration of a payoff cellular automaton with the neighbourhood consisting of 8 members in a 2-dimensional space and with players 1 and 2

Fig. 11.10 Values of given in Fig. 11.9 at time $t = 1$

2	2	2	2	2
2	2	2	2	2
1	1	1	2	2
1	1	1	2	2
1	1	1	2	2

$\langle a_{21}, b_{21}\rangle$	$\langle a_{31}, b_{31}\rangle$	$\langle a_{22}, b_{22}\rangle$	$\langle a_{11}, b_{11}\rangle$	$\langle a_{33}, b_{33}\rangle$
$\langle a_{13}, b_{13}\rangle$	$\langle a_{23}, b_{23}\rangle$	$\langle a_{11}, b_{11}\rangle$	$\langle a_{13}, b_{13}\rangle$	$\langle a_{14}, b_{14}\rangle$
$\langle a_{34}, b_{34}\rangle$	$\langle a_{12}, b_{12}\rangle$	$\langle \mathbf{a_{33}}, \mathbf{b_{33}}\rangle$	$\langle a_{23}, b_{23}\rangle$	$\langle a_{11}, b_{11}\rangle$
$\langle a_{11}, b_{11}\rangle$	$\langle a_{11}, b_{11}\rangle$	$\langle a_{14}, b_{14}\rangle$	$\langle a_{11}, b_{11}\rangle$	$\langle a_{12}, b_{12}\rangle$
$\langle a_{22}, b_{22}\rangle$	$\langle a_{33}, b_{33}\rangle$	$\langle a_{24}, b_{24}\rangle$	$\langle a_{22}, b_{22}\rangle$	$\langle a_{34}, b_{34}\rangle$

Fig. 11.11 An initial configuration of a payoff cellular automaton with the neighbourhood consisting of 8 members in a 2-dimensional space and with players 1 and 2. In this automata we involve an appropriate payoff matrix. The given infinite table will be interpreted as infinitely many payoff matrices, where payoff oucomes are mixed. This mix for any neighbourhood may be interpreted as a correction of initial strategies or false public announcement. It is a result of that an appropriate context holds

$\langle 10,3\rangle$	$\langle 9,8\rangle$	$\langle 4,3\rangle$	$\langle 5,7\rangle$	$\langle 10,10\rangle$
$\langle 6,7\rangle$	$\langle 3,5\rangle$	$\langle 5,7\rangle$	$\langle 6,7\rangle$	$\langle 1,1\rangle$
$\langle 2,1\rangle$	$\langle 2,5\rangle$	$\langle 10,10\rangle$	$\langle 3,5\rangle$	$\langle 5,7\rangle$
$\langle 5,7\rangle$	$\langle 5,7\rangle$	$\langle 1,1\rangle$	$\langle 5,7\rangle$	$\langle 2,5\rangle$
$\langle 4,3\rangle$	$\langle 10,10\rangle$	$\langle 8,7\rangle$	$\langle 4,3\rangle$	$\langle 2,1\rangle$

Fig. 11.12 Values of given in Fig. 11.11 at time $t = 0$

5 payoff combinations of 12 possible ones. Let us assume that all payoff values of Fig. 11.11 are given in Fig. 11.12.

Let us suppose that player 1 knows that player 2 prefers to choose $j(i)$ for the i-th strategy of player 1 for all cells $z \cup N(z)$ at step t such that $a_{ij}(i) \leq a_{ij}$ for any $j = 1, \ldots, n$ occurring in $z \cup N(z)$ at step t. Let us denote $a_i = a_{ij}(i) = \min_{1 \leq j \leq n} a_{ij}$, where $i = 1, \ldots, m$. Then the best strategy for player 1 is a choice of i_0 such that

$$\alpha = \max_i a_i = \max_{1 \leq i \leq m} \min_{1 \leq j \leq n} a_{ij} = \alpha_{i_0}$$

This decision rule in choosing α is called a *maximin strategy*. By this strategy player 1 avoids the worst possible performance, maximizing the minimal performing criterion of player 2.

Let player 2 assume that for any j player 1 chooses $i(j)$ for cells $z \cup N(z)$ at step t such that $b_{i(j)j} \geq b_{ij}$ for any i, i.e. $\beta_j = \max_{1 \leq i \leq m} b_{ij}$. Then player 2 chooses j with minimal β_j for cells $z \cup N(z)$ at step t:

$$\beta = \min_{1 \leq j \leq m} \max_{1 \leq i \leq n} b_{ij} = \beta_{j_0}$$

This decision rule in choosing β is called a *minimax strategy*. This strategy of player 2 minimizes the expected payoff under the assumption that player 1 aims to maximize this payoff.

The dynamics of the given cellular automaton is pictured in Figs. 11.13, 11.14 and 11.15.

Fig. 11.13 Values of given in Fig. 11.12 at time $t = 1$

⟨6,7⟩	⟨6,7⟩	⟨5,7⟩	⟨5,7⟩	⟨5,7⟩
⟨2,5⟩	⟨4,7⟩	⟨4,7⟩	⟨4,7⟩	⟨3,7⟩
⟨2,5⟩	⟨2,7⟩	⟨3,7⟩	⟨3,7⟩	⟨3,7⟩
⟨2,5⟩	⟨2,7⟩	⟨3,7⟩	⟨3,7⟩	⟨3,3⟩
⟨5,7⟩	⟨5,7⟩	⟨5,7⟩	⟨4,7⟩	⟨4,3⟩

Fig. 11.14 Values of given in Fig. 11.12 at time $t = 2$. Notice that b_{ij} cannot vary at any time $t > 2$

⟨4,7⟩	⟨4,7⟩	⟨4,7⟩	⟨4,7⟩	⟨4,7⟩
⟨2,7⟩	⟨3,7⟩	⟨3,7⟩	⟨3,7⟩	⟨3,7⟩
⟨2,7⟩	⟨3,7⟩	⟨3,7⟩	⟨3,7⟩	⟨3,7⟩
⟨2,7⟩	⟨3,7⟩	⟨3,7⟩	⟨3,7⟩	⟨3,7⟩
⟨2,7⟩	⟨3,7⟩	⟨3,7⟩	⟨3,7⟩	⟨3,7⟩

Fig. 11.15 Values of given in Fig. 11.12 at time $t = 3$. Notice that a_{ij} cannot vary at any time $t > 3$

⟨3,7⟩	⟨3,7⟩	⟨3,7⟩	⟨3,7⟩	⟨3,7⟩
⟨3,7⟩	⟨3,7⟩	⟨3,7⟩	⟨3,7⟩	⟨3,7⟩
⟨3,7⟩	⟨3,7⟩	⟨3,7⟩	⟨3,7⟩	⟨3,7⟩
⟨3,7⟩	⟨3,7⟩	⟨3,7⟩	⟨3,7⟩	⟨3,7⟩
⟨3,7⟩	⟨3,7⟩	⟨3,7⟩	⟨3,7⟩	⟨3,7⟩

Thus, player 1 has followed the maximin strategy, while player 2 has followed the minimax strategy. As a result, at any time $t > 3$ the payoff cellular automaton has become reversible.

In payoff cellular automata their local transition functions can be set by logical (e.g. Boolean) superpositions of decision rules. As an example of such automata, let us consider a context-based infinite game with the following decision rule: 'if in the game context there is an equilibrium in dominant pure strategies, then the current state holds.'

11.2.3 Equilibrium in Dominant Strategies as Conditional

Recall that an *equilibrium in dominant pure strategies* in two-player games is presented by a pair of strategies (a, b) if and only if (i) while player 1 chooses any of his possible strategies, strategy b is the only strategy that will produce the highest possible payoff for player 2; (ii) while player 2 chooses any of his possible strategies, strategy a is the only strategy that will produce the highest possible payoff for player 1. Let us suppose that an initial configuration of cellular automaton is given in Fig. 11.16 and the set S of states consists of all pairs $\langle a_{ij}, b_{ij}\rangle$, where a_{ij}, b_{ij} are integers of $[-15, 13]$.

Now we can define implication for all pairs $\langle a_{ij}, b_{ij}\rangle, \langle a_{kl}, b_{kl}\rangle \in S$:

$$\langle a_{ij}, b_{ij}\rangle \Rightarrow \langle a_{kl}, b_{kl}\rangle = \langle N_{max} - \sup(a_{ij}, a_{kl}) + a_{kl}, N_{max} - \sup(b_{ij}, b_{kl}) + b_{kl}\rangle,$$

where $N_{max} = 13$.

$\langle 3,3\rangle$	$\langle 12,-12\rangle$	$\langle 13,-15\rangle$
$\langle -12,12\rangle$	$\langle -5,-5\rangle$	$\langle -2,-2\rangle$
$\langle -1,1\rangle$	$\langle 0,-6\rangle$	$\langle 2,-3\rangle$

Fig. 11.16 An initial configuration of a payoff cellular automaton with the neighbourhood consisting of 8 members in a 2-dimensional space and with players 1 and 2

$\langle -5,-5\rangle \Rightarrow \langle 3,3\rangle$	$\langle -2,-2\rangle \Rightarrow \langle 12,-12\rangle$	$\langle -2,-2\rangle \Rightarrow \langle 13,-15\rangle$
$\langle -5,-5\rangle \Rightarrow \langle -12,12\rangle$	$\langle -2,-2\rangle \Rightarrow \langle -5,-5\rangle$	$\langle -2,-2\rangle \Rightarrow \langle -2,-2\rangle$
$\langle -5,-5\rangle \Rightarrow \langle -1,1\rangle$	$\langle -2,-2\rangle \Rightarrow \langle 0,-6\rangle$	$\langle -2,-2\rangle \Rightarrow \langle 2,-3\rangle$

Fig. 11.17 Values of given in Fig. 11.16 at time $t = 1$

$\langle 13-\sup(-5,3)+3,$ $13-\sup(-5,3)+3\rangle$	$\langle 13-\sup(-2,12)+12,$ $13-\sup(-2,-12)-12\rangle$	$\langle 13-\sup(-2,13)+13,$ $13-\sup(-2,-15)-15\rangle$
$\langle 13-\sup(-5,-12)-12,$ $13-\sup(-5,12)+12\rangle$	$\langle 13-\sup(-2,-5)-5,$ $13-\sup(-2,-5)-5\rangle$	$\langle 13-\sup(-2,-2)-2,$ $13-\sup(-2,-2)-2\rangle$
$\langle 13-\sup(-5,-1)-1,$ $13-\sup(-5,1)+1\rangle$	$\langle 13-\sup(-2,0)+0,$ $13-\sup(-2,-6)-6\rangle$	$\langle 13-\sup(-2,2)+2,$ $13-\sup(-2,-3)-3\rangle$

Fig. 11.18 Values of given in Fig. 11.16 at time $t = 1$

Fig. 11.19 Values of given in Fig. 11.16 at time $t = 1$

$\langle 13,13\rangle$	$\langle 13,3\rangle$	$\langle 13,0\rangle$
$\langle 6,13\rangle$	$\langle 10,10\rangle$	$\langle 13,13\rangle$
$\langle 13,13\rangle$	$\langle 13,9\rangle$	$\langle 13,12\rangle$

Then the local transition function is defined thus:

$$a^{t+1}(z) = \begin{cases} \langle a_{ij}, b_{ij}\rangle \Rightarrow \langle a_{kl}, b_{kl}\rangle, & \text{if } \langle a_{ij}, b_{ij}\rangle \text{ is a single equilibrium} \\ & \text{in dominated strategies} \\ & \text{of } N(z) \cup z \text{ and } \langle a_{kl}, b_{kl}\rangle = a^t(z); \\ a^t(z), & \text{otherwise.} \end{cases}$$

After applying this function to the initial configuration given in Fig. 11.16, we obtain the data of Figs. 11.17, 11.18 and 11.19.

To sum up notice that equilibria in dominant strategies could be different for different game contexts (neighbourhoods for different cells) and our reasoning 'if in the game context there is an equilibrium in dominant strategies, then the current state holds' is a kind of reflexion formalized in payoff cellular automaton of Fig. 11.16. There can be very complex forms of reflexion in game decisions. They may be formulated only within aposterioristic logics, because in different contexts our reasoning will have a different meaning.

Now let us analyze a simple case of game reflexion where players follow different reasoning.

11.2.4 PROMETHEE Method

Let us consider a payoff cellular automaton when a decision is being made from alternative solutions under multiple criteria of choice for each player. Assume that the set S of states consists of payoff for all n players and S is equal $k = i_1 \cdot i_2 \cdot \ldots \cdot i_n$, where i_j is the number of all strategies for the j-th player, $j = 1, \ldots, n$. Thus, there are k alternatives $\mathbf{a}_1, \ldots, \mathbf{a}_k$, where

$$\mathbf{a} = \left\langle \underbrace{a_{ij\ldots k}, b_{ij\ldots k}, \ldots, c_{ij\ldots k}}_{n} \right\rangle.$$

They are being evaluated by m criteria $g_1(\cdot), \ldots, g_m(\cdot)$. Let a *preference relation* $P_i(\mathbf{a}_r, \mathbf{a}_q)$ be associated to each criterion g_i. It represents the degree of the preference of alternative $\mathbf{a}_r$ over $\mathbf{a}_q$ for criterion g_i and satisfies the condition $0 \leq P_i(\mathbf{a}_r, \mathbf{a}_q) \leq 1$. The equality $P_i(\mathbf{a}_r, \mathbf{a}_q) = 0$ means an *indifference*, $P_i(\mathbf{a}_r, \mathbf{a}_q) = 1$ means a *strict preference*, $P_i(\mathbf{a}_r, \mathbf{a}_q) \approx 0$ means a *weak preference*, and $P_i(\mathbf{a}_r, \mathbf{a}_q) \approx 1$ means a *strong preference*. Let us take weights w_i of relative importance of the different criteria g_i such that $\sum_{i=1}^{m} w_i = 1$. Assume that the higher the weight, the more important the criterion. Now we can define a multicriteria preference index $\pi(\mathbf{a}_r, \mathbf{a}_q)$ of $\mathbf{a}_r$ over $\mathbf{a}_q$:

$$\pi(\mathbf{a}_r, \mathbf{a}_q) = \sum_{i=1}^{m} w_i \cdot P_i(\mathbf{a}_r, \mathbf{a}_q).$$

This index takes values between 0 and 1 as well. It represents the global intensity of preference among alternatives. The method for obtaining $\pi(\mathbf{a}_r, \mathbf{a}_q)$ is called the *PROMETHEE method*. If $\pi(\mathbf{a}_r, \mathbf{a}_q) > \pi(\mathbf{a}_q, \mathbf{a}_r)$, we say that $\mathbf{a}_r$ is preferred over $\mathbf{a}_q$ and if $\pi(\mathbf{a}_r, \mathbf{a}_q) = \pi(\mathbf{a}_q, \mathbf{a}_r)$, we say that $\mathbf{a}_r$ and $\mathbf{a}_q$ are indifferent. This allows us to set a partial ordering of relations $\geq_\pi$ on all alternatives $\mathbf{a}_r, \mathbf{a}_q \in S$: $\mathbf{a}_r \geq_\pi \mathbf{a}_q$ if and only if $\pi(\mathbf{a}_r, \mathbf{a}_q) > \pi(\mathbf{a}_q, \mathbf{a}_r)$ or $\pi(\mathbf{a}_r, \mathbf{a}_q) = \pi(\mathbf{a}_q, \mathbf{a}_r)$. In that ordered structure the following logical operations are defined:

$$\max(\mathbf{a}_r, \mathbf{a}_q) = \mathbf{a}_r \text{ if and only if } \mathbf{a}_r \geq_\pi \mathbf{a}_q;$$

$$\min(\mathbf{a}_r, \mathbf{a}_q) = \mathbf{a}_r \text{ if and only if } \mathbf{a}_q \geq_\pi \mathbf{a}_r;$$

$$\mathbf{a}_r \vee \mathbf{a}_q := \max(\mathbf{a}_r, \mathbf{a}_q);$$

$$\mathbf{a}_r \wedge \mathbf{a}_q := \min(\mathbf{a}_r, \mathbf{a}_q);$$

$$\neg\mathbf{a}_r = \max_q \mathbf{a}_q \text{ for all } \mathbf{a}_q \text{ such that } \pi(\mathbf{a}_r, \mathbf{a}_q) = 0 \text{ or } \pi(\mathbf{a}_q, \mathbf{a}_r) = 0;$$

$$\mathbf{a}_r \Rightarrow \mathbf{a}_q := \neg\mathbf{a}_r \vee \mathbf{a}_q.$$

Let $\mathbf{B}_\pi$ denote a superposition of the operations above. Suppose that for the j-th player ($j = 1, \ldots, n$) the sign $\mathbf{B}_\pi^j$ denotes a logical superposition defined only on all j-th projections of all tuples occurring in $\mathbf{B}_\pi$, while its $\geq_\pi$ concerns the whole tuples. Then we can define the local transition function:

$$\mathbf{a}^{t+1}(z) = \langle \mathbf{B}_\pi^1, \mathbf{B}_\pi^2, \ldots, \mathbf{B}_\pi^n \rangle,$$

where $\mathbf{B}_\pi^1$, $\mathbf{B}_\pi^2$, ..., $\mathbf{B}_\pi^n$ are different logical superpositions defined on alternatives of cells $z \cup N(z)$ at step t.

Hence, each j-th player possesses reflexion $\mathbf{B}_\pi^j$ that gives different results in different game contexts (different neighbourhoods) and reasoning of all players at time t influence their reasoning at time $t + 1$.

11.3 Reflexive Games

In *reflexive games* we assume an unlimited hierarchy of cognitive pictures of reality. In the bimatrix game with agents i and j, there are two pictures of the same A: K_iA (belonging to i) and K_jA (belonging to j). Then there are K_jK_iA and K_iK_jA, where K_jK_iA are pictures of agent j about pictures of agent i, K_iK_jA are pictures of agent i about pictures of agent j. Then there are $K_iK_jK_iA$ and $K_jK_iK_jA$, etc. Please see Definitions 10.1 and 10.2.

Let us define reflexive games within payoff cellular automata. Designate reflexive players by 1 and 2. Let A, a state of affairs, be identified with a set of payoffs within a game context (i.e. within a neighbourhood). In other words, let $A_{\langle a_{ij}, b_{ij}\rangle}$ be a set of payoffs at the point $z \in \mathbf{Z}^d$ consisting of all payoffs of $N(z) \cup z$, where z has a state $\langle a_{ij}, b_{ij}\rangle$, see Fig. 11.16, where the initial configuration of the payoff cellular automaton presents 9 states of affairs: $A_{\langle 3,3\rangle} = \{\langle 3, 3\rangle, \langle 12, -12\rangle, \langle -12, 12\rangle, \langle -5, -5\rangle\}$; $A_{\langle 12,-12\rangle} = \{\langle 3, 3\rangle, \langle 12, -12\rangle, \langle 13, -15\rangle, \langle -12, 12\rangle, \langle -5, -5\rangle, \langle -2, -2\rangle\}$, $A_{\langle 13,-15\rangle} = \{\langle 12, -12\rangle, \langle 13, -15\rangle, \langle -5, -5\rangle, \langle -2, -2\rangle\}$; $A_{\langle -12,12\rangle} = \{\langle 3, 3\rangle, \langle 12, -12\rangle, \langle -12, 12\rangle, \langle -5, -5\rangle, \langle -1, 1\rangle, \langle 0, -6\rangle\}$, ...

Let $\mathbf{B}_1^1 A_{\langle a_{ij}, b_{ij}\rangle}$ (accordingly, $\mathbf{B}_1^2 A_{\langle a_{ij}, b_{ij}\rangle}$) mean agent 1's (accordingly, agent 2's) Boolean superpositions of 1's payoffs of $A_{\langle a_{ij}, b_{ij}\rangle}$ (accordingly, 2's payoffs) for each first (accordingly, second) projection of all points of $A_{\langle a_{ij}, b_{ij}\rangle}$. Then $K_1 A_{\langle a_{ij}, b_{ij}\rangle} = A_{\langle a_{ij}, b_{ij}\rangle} \cup \mathbf{B}_1^1 A_{\langle a_{ij}, b_{ij}\rangle}$ and $K_2 A_{\langle a_{ij}, b_{ij}\rangle} = A_{\langle a_{ij}, b_{ij}\rangle} \cup \mathbf{B}_1^2 A_{\langle a_{ij}, b_{ij}\rangle}$. Let $\mathbf{B}_2^1 A_{\langle a_{ij}, b_{ij}\rangle}$ (accordingly, $\mathbf{B}_2^2 A_{\langle a_{ij}, b_{ij}\rangle}$) mean agent 1's (accordingly, agent 2's) Boolean superpositions of $\mathbf{B}_1^1 A_{\langle a_{ij}, b_{ij}\rangle}$ and $\mathbf{B}_1^2 A_{\langle a_{ij}, b_{ij}\rangle}$ for each first (accordingly, second) projection of all points of $A_{\langle a_{ij}, b_{ij}\rangle}$. Then $K_1 K_2 A_{\langle a_{ij}, b_{ij}\rangle} = A_{\langle a_{ij}, b_{ij}\rangle} \cup \mathbf{B}_1^2 A_{\langle a_{ij}, b_{ij}\rangle} \cup \mathbf{B}_2^1 A_{\langle a_{ij}, b_{ij}\rangle}$ and $K_2 K_1 A_{\langle a_{ij}, b_{ij}\rangle} = A_{\langle a_{ij}, b_{ij}\rangle} \cup \mathbf{B}_1^1 A_{\langle a_{ij}, b_{ij}\rangle} \cup \mathbf{B}_2^2 A_{\langle a_{ij}, b_{ij}\rangle}$. Let $\mathbf{B}_3^1 A_{\langle a_{ij}, b_{ij}\rangle}$ (accordingly, $\mathbf{B}_3^2 A_{\langle a_{ij}, b_{ij}\rangle}$) mean agent 1's (accordingly, agent 2's) Boolean superpositions of $\mathbf{B}_2^1 A_{\langle a_{ij}, b_{ij}\rangle}$ and $\mathbf{B}_2^2 A_{\langle a_{ij}, b_{ij}\rangle}$ for each first (accordingly, second) projection of all points of $A_{\langle a_{ij}, b_{ij}\rangle}$. Then $K_2 K_1 K_2 A_{\langle a_{ij}, b_{ij}\rangle} = A_{\langle a_{ij}, b_{ij}\rangle} \cup \mathbf{B}_1^2 A_{\langle a_{ij}, b_{ij}\rangle} \cup \mathbf{B}_2^1 A_{\langle a_{ij}, b_{ij}\rangle} \cup \mathbf{B}_3^2 A_{\langle a_{ij}, b_{ij}\rangle}$ and $K_1 K_2 K_1$ $A_{\langle a_{ij}, b_{ij}\rangle} = A_{\langle a_{ij}, b_{ij}\rangle} \cup \mathbf{B}_1^1 A_{\langle a_{ij}, b_{ij}\rangle} \cup \mathbf{B}_2^2 A_{\langle a_{ij}, b_{ij}\rangle} \cup \mathbf{B}_3^1 A_{\langle a_{ij}, b_{ij}\rangle}$. And so on.

$\langle 3,13\rangle$	$\langle -12,13\rangle$	$\langle -15,13\rangle$
$\langle -12,13\rangle$	$\langle -5,13\rangle$	$\langle -2,13\rangle$
$\langle -1,1\rangle$	$\langle -6,3\rangle$	$\langle -3,13\rangle$

Fig. 11.20 The configuration of the payoff cellular automaton of Fig. 11.16 at time $t = 1$

Let $\mathbf{B}_0^k A_{\langle a_{ij}, b_{ij}\rangle} = A_{\langle a_{ij}, b_{ij}\rangle}$, $k = 1, 2$. Then, according to Definitions 10.1 and 10.2, a reflexive game at the n-th level ($n \geq 0$) can be reformulated thus: if player 1 follows the decision rule $\mathbf{B}_{n+1}^1 A_{\langle a_{ij}, b_{ij}\rangle}$, then player 2 follows the decision rule $\mathbf{B}_{n+1}^2 A_{\langle a_{ij}, b_{ij}\rangle}$ or $\mathbf{B}_n^2 A_{\langle a_{ij}, b_{ij}\rangle}$; if player 2 follows the decision rule $\mathbf{B}_{n+1}^2 A_{\langle a_{ij}, b_{ij}\rangle}$, then player 1 follows the decision rule $\mathbf{B}_{n+1}^1 A_{\langle a_{ij}, b_{ij}\rangle}$ or $\mathbf{B}_n^1 A_{\langle a_{ij}, b_{ij}\rangle}$.

11.3.1 Reflexive Game of the Second Level

Let us consider a payoff cellular automaton of Fig. 11.16 where the set S of states consists of all pairs $\langle a_{ij}^t, b_{ij}^t\rangle$, where a_{ij}^t, b_{ij}^t at time $t = 0, 1, 2, \ldots$ are integers of $[-15, 13]$ and the local transition function is as follows: $a^{t+1}(z) = \langle a_{ij}^{t+1}, b_{ij}^{t+1}\rangle$, where $a_{ij}^{t+1} = ((\bigvee_m b_m^t \Rightarrow \bigvee_k a_k^t) \wedge (a_{ij}^t \wedge b_{ij}^t))$ and $b_{ij}^{t+1} = (\bigvee_k b_k^t \Rightarrow \bigvee_m a_m^t)$ and $\bigvee_k a_k^t$, $\bigvee_m b_m^t$ are maximal payoffs of player 1 and player 2 respectively at cells $N(z) \cup z$ at time t, the logical operations are understood thus: $a \Rightarrow b := 13 - \max(a, b) + b$; $a \vee b := \max(a, b)$.

This automaton simulates the reflexive game, where player 1 has the second level of reflexion, while player 2 has the first level of reflexion. Its evolution at time $t = 1$ is pictured in Fig. 11.20.

Let us extend the formulas $\mathbf{B}_n^1 A_{\langle a_{ij}, b_{ij}\rangle}$ and $\mathbf{B}_n^2 A_{\langle a_{ij}, b_{ij}\rangle}$ up to the case $\mathbf{B}_n^1 A$ and $\mathbf{B}_n^2 A$, where A is a union of sets $A_{\langle a_{ij}, b_{ij}\rangle}$, $A_{\langle a_{i_1 j_1}, b_{i_1 j_1}\rangle}$, $A_{\langle a_{i_2 j_2}, b_{i_2 j_2}\rangle}$, ..., $A_{\langle a_{i_n j_n}, b_{i_n j_n}\rangle}$ such that $\langle a_{ij}, b_{ij}\rangle$ belongs to $A_{\langle a_{i_1 j_1}, b_{i_1 j_1}\rangle}$, $\langle a_{i_1 j_1}, b_{i_1 j_1}\rangle$ belongs to $A_{\langle a_{i_2 j_2}, b_{i_2 j_2}\rangle}$, ..., $\langle a_{i_{n-1} j_{n-1}}, b_{i_{n-1} j_{n-1}}\rangle$ belongs to $A_{\langle a_{i_n j_n}, b_{i_n j_n}\rangle}$.

Hence, in payoff cellular automata at the n-th level of reflexion, we apply an infinite sequence of Boolean operations at the point $z \in \mathbf{Z}^d$:

$$\mathbf{B}_{l,m}^{1,2} := \langle \mathbf{B}_l^1 A, \mathbf{B}_m^2 A\rangle \text{ at } t = 0, \langle \mathbf{B}_l^1 A, \mathbf{B}_m^2 A\rangle \text{ at } t = 1, \langle \mathbf{B}_l^1 A, \mathbf{B}_m^2 A\rangle \text{ at } t = 2, \text{ etc.},$$

where if $l = n + 1$, then $m = n + 1$ or n; if $m = n + 1$, then $l = n + 1$ or n.

We can suppose that players change their reflexion foundations at different t on the basis of past payoffs. For instance, they can change their reflexive levels or decision rules of the same levels:

$\mathbf{B}^{1,2}_{il_i,im_i} := \langle \mathbf{B}^1_{0l_0}A, \mathbf{B}^2_{0m_0}A\rangle$ at $t = 0$, $\langle \mathbf{B}^1_{1l_1}A, \mathbf{B}^2_{1m_1}A\rangle$ at $t = 1$, $\langle \mathbf{B}^1_{2l_2}A, \mathbf{B}^2_{2m_2}A\rangle$ at $t = 2$, etc.,

where if $l_k = n_k + 1$, then $m_k = n_k + 1$ or n_k; if $m_k = n_k + 1$, then $l_k = n_k + 1$ or n_k for any $k = 0, 1, 2\ldots$ Notice that n_k is a reflexive level that can be different for different k.

Notice that the formulas $\mathbf{B}^{1,2}_{l,m}$ and $\mathbf{B}^{1,2}_{il_i,im_i}$ are unconventional. On the one hand, they present an infinite sequence of states for each cell of $A = A_{\langle a_{ij},b_{ij}\rangle} \cup A_{\langle a_{i_1j_1},b_{i_1j_1}\rangle} \cup A_{\langle a_{i_2j_2},b_{i_2j_2}\rangle} \cup \ldots \cup A_{\langle a_{i_nj_n},b_{i_nj_n}\rangle}$. So, this sequence is not linear and consists of parallel mutually dependent infinite streams $\mathbf{a}^0(z)$, $\mathbf{a}^1(z)$, $\mathbf{a}^2(z)$, ..., $\mathbf{a}^t(z)$, ... for any z of A. On the other hand, $\mathbf{B}^{1,2}_{l,m}$ and $\mathbf{B}^{1,2}_{il_i,im_i}$ are self-applicable. They are continuously being applied to results of their past applications ad infinitum. About formal definitions of these formulas, please see Sect. 8.2.2.

The formulas $\mathbf{B}^{1,2}_{l,m}$ and $\mathbf{B}^{1,2}_{il_i,im_i}$ are called *wave sets*. The feature of wave sets is to contain parallel mutually dependent infinite streams of states (values) generated by self-applicable logical formula(s). The number $2 \cdot n$, where $|A| = n$, is a *radius* of an appropriate wave set. In $\mathbf{B}^{1,2}_{l,m}$ and $\mathbf{B}^{1,2}_{il_i,im_i}$ we have just $2 \cdot n$ parallel mutually dependent infinite streams. Notice that n can be equal $|\mathbf{Z}^d|$, the maximal possible case. The minimal case takes place if $n = 1$, then we have only two streams. If formulas $\mathbf{B}^{1,\ldots,k}_{l,m}$ and $\mathbf{B}^{1,\ldots,k}_{il_i,im_i}$ hold true, then the number $k \cdot n$, where $|A| = n$, is a *radius* of an appropriate wave set.

Let us consider an example of $\mathbf{B}^{1,2}_{l,m}$. Let $|A| = 1$. Assume that two streams which we have are as follows: $\langle a^0 \Rightarrow b^0, a^1 \Rightarrow b^1, a^2 \Rightarrow b^2, \ldots\rangle$, where $a^{t+1} = (a^t \Rightarrow b^t)$; $\langle b^0 \Rightarrow a^0, b^1 \Rightarrow a^1, b^2 \Rightarrow a^2, \ldots\rangle$, where $b^{t+1} = (b^t \Rightarrow a^t)$. Let us pay attention that the stream $\langle a^0 \Rightarrow b^0, a^1 \Rightarrow b^1, a^2 \Rightarrow b^2, \ldots\rangle$ (resp. $\langle b^0 \Rightarrow a^0, b^1 \Rightarrow a^1, b^2 \Rightarrow a^2, \ldots\rangle$) may be understood as an infinite propositional formula $(((a^0 \Rightarrow b^0) \Rightarrow b^1) \Rightarrow b^2) \Rightarrow \ldots$ (resp. $(((b^0 \Rightarrow a^0) \Rightarrow a^1) \Rightarrow a^2) \Rightarrow \ldots$). Hence, we have a wave set of radius 2 presented by two infinite mutually dependent propositional formulas. Also, we can show that any wave set of radius n can be formulated as n infinite mutually dependent propositional formulas.

Thus, in payoff cellular automata we deal with cumulative expected payoffs for any $t = 0, 1, 2, \ldots$, where strategies can change in an appropriate neighbourhood in accordance with beliefs and real actions of mine and the other player. It means, in payoff cellular automata we take into account not only payoffs, which have changed for any new context, but also transition rules over payoffs. We assume that players' decision rules at the zero level of reflexion are presented by knowledge functions, which include the set of payoffs with their Boolean superpositions and the choices (past and/or expected). At the first level of reflexion players' decision rules have the form of knowledge functions which include a Boolean superposition of other agents' payoffs of the zero level and the suitable choices, etc.

11.3.2 Logic for Reflexive Decision Making

Now let us construct the unconventional logic for reflexive games formally. Let $\{p, q, r, \ldots\}$ be the set of all propositional variables. Then (1) each variable is a formula; (2) we can construct finite or infinite Boolean combinations of formulas $\varphi_0\psi_0\star_0$, $\varphi_1\psi_1\star_1$, $\varphi_2\psi_2\star_2$, ..., where $\star_0, \star_1, \star_2 \ldots \in \{D, C, I\}$; (3) if φ is a finite formula and ψ is an infinite formula, then $\varphi\psi\star$, where $\star \in \{D, C, I\}$, is a formula. The class of all finite or infinite strings is a maximal set of formulas closed under logical operations.

Wave sets are sets of streams which have joint propositional variables. For example, the set consisting just of $p_0q_0Iq_1Iq_2Iq_3I\ldots$ and $q_0p_0Ip_1Ip_2Ip_3I\ldots$ is a wave set. Let A_1, A_2 be wave sets of propositional formulas and φ be a finite or infinite propositional formula. Then their logical compositions are as follows: (1) A_1A_2D is a union of A_1 and A_2; A_1A_2C is an intersection of A_1 and A_2 if they have joint streams and A_1A_2C is their union otherwise; A_1A_2I is a union of A_1N and A_2, where A_1N contains complements for each streams of A_1; (2) let φ be infinite, then φA_1D is a union of $\{\varphi\}$ and A_1; φA_1C is φ if φ belongs to A_1 and it is a union $\{\varphi\}$ and A_1 otherwise; $\varphi A_1I := \varphi NA_1D$; $A_1\varphi I := A_1N\varphi D$; (3) let φ be finite and A_1 contain streams ψ_1, ..., ψ_n $(n > 0)$ that begin with φ, then φA_1D is a union of $\{\varphi\psi_1D\}$, ..., $\{\varphi\psi_nD\}$, and A_1; φA_1C is a union of $\{\varphi\psi_1C\}$, ..., $\{\varphi\psi_nC\}$, and A_1; $\varphi A_1I := \varphi NA_1D$; $A_1\varphi I := A_1N\varphi D$. For more details please see Sect. 8.2.2.

Hence, each infinite formula or wave set can be interpreted in payoff cellular automata. Let us consider some simple cases assuming that each propositional sign p_i^t corresponds to a state at z_i at time t.

Definition 11.7 In the payoff cellular automaton $\mathscr{A}$, where $S = \{1, 0\}$ and $|N(z)| = n - 1$, let us consider the truth valuation of infinite conjunction $\bigwedge_{i=0}^{n} p_i^0 \wedge \bigwedge_{j=0}^{n-1} p_j^1 \wedge \bigwedge_{j=0}^{n-1} p_j^2 \wedge \ldots$, where $\bigwedge_{i=0}^{n} p_i^0$ is a minimal payoff of $z \cup N(z)$ at time $t = 0$, $\bigwedge_{j=0}^{n-1} p_j^1$ is a minimal payoff of $N(z)$ at time $t = 1$, $\bigwedge_{j=0}^{n-1} p_j^2$ is a minimal payoff of $N(z)$ at time $t = 2$, etc. Its local transition rule is formulated as follows:

$$x^{t+1}(z) = \begin{cases} 1, \textit{ if } x^t(z) = 1 \textit{ and } 0 \notin N(z); \\ 0, \textit{ otherwise}. \end{cases}$$

Definition 11.8 In the payoff cellular automaton $\mathscr{A}$, where $S = \{1, 0\}$ and $|N(z)| = n - 1$, let us consider the truth valuation of infinite disjunction $\bigvee_{i=0}^{n} p_i^0 \vee \bigvee_{j=0}^{n-1} p_j^1 \vee \bigvee_{j=0}^{n-1} p_j^2 \vee \ldots$, where $\bigvee_{i=0}^{n} p_i^0$ is a maximal payoff of $z \cup N(z)$ at time $t = 0$, $\bigvee_{j=0}^{n-1} p_j^1$ is a maximal payoff of $N(z)$ at time $t = 1$, $\bigvee_{j=0}^{n-1} p_j^2$ is a maximal payoff of $N(z)$ at time $t = 2$, etc. Its local transition rule is formulated as follows:

$$x^{t+1}(z) = \begin{cases} 1, \textit{ if } x^t(z) = 1 \textit{ or } 1 \in N(z); \\ 0, \textit{ otherwise}. \end{cases}$$

Definition 11.9 In the payoff cellular automaton $\mathscr{A}$, where $S = \{1, 0\}$ and $|N(z)| = n - 1$, let us consider the truth valuation of infinite implication $\ldots \Rightarrow (\bigwedge_{i=0}^{n-1} p_i^2 \Rightarrow (\bigwedge_{i=0}^{n-1} p_i^1 \Rightarrow (\bigwedge_{i=0}^{n-1} p_i^0 \Rightarrow p_j^0)))$, where $\bigwedge_{i=0}^{n-1} p_i^0$ is a minimal payoff of $N(z)$ at time $t = 0$, $\bigwedge_{j=0}^{n-1} p_j^1$ is a minimal payoff of $N(z)$ at time $t = 1$, $\bigwedge_{j=0}^{n-1} p_j^2$ is a minimal payoff of $N(z)$ at time $t = 2$, etc. Its local transition rule is formulated as follows:

$$x^{t+1}(z) = \begin{cases} 0, \ \textit{if} \ x^t(z) = 0 \ \textit{and} \ 0 \notin N(z); \\ \\ 1, \ \textit{otherwise}. \end{cases}$$

Definition 11.10 In the payoff cellular automaton $\mathscr{A}$, where $S = \{1, 0\}$ and $|N(z)| = n - 1$, let us consider the truth valuation of infinite negation $\ldots \neg(\neg(\neg p_i^0))$, where $i = 1, \ldots, n$. Its local transition rule is formulated as follows:

$$x^{t+1}(z) = \begin{cases} 0, \ \textit{if} \ x^t(z) = 1; \\ \\ 1, \ \textit{otherwise}. \end{cases}$$

Truth valuation of any propositional stream:
Let φ be a propositional stream $\ldots \star_3 (\bigodot_{i=0}^{\prime\prime n-1} p_i^2 \star_2 (\bigodot_{i=0}^{\prime n-1} p_i^1 \star_1 (\bigodot_{i=0}^{n-1} p_i^0 \star_0 p_j^0)))$, where $\star_m \in \{\vee, \wedge, \Rightarrow\}$ for $m = 0, 1, 2, \ldots$ and $\bigodot, \bigodot', \bigodot'', \ldots \in \{\bigvee, \bigwedge\}$. Its initial truth value is an initial configuration of a cellular automaton $\mathscr{A}$ with the set of states $\{1, 0\}$. Then we start the evaluation of φ with the very first connective $\star_0$ that connects states at time $t = 0$. By using one of Definitions 11.7–11.10 that corresponds to $\star_0$, we transform the initial configuration of $\mathscr{A}$ at time $t = 0$ to a configuration at time $t = 1$. Further, we move to evaluate the second connective $\star_1$ that connects states at time $t = 1$. By using one of Definitions 11.7–11.10 that corresponds to $\star_1$, we transform the configuration of $\mathscr{A}$ at time $t = 1$ to a configuration at time $t = 2$. Then we go to evaluate the third connective $\star_2$ that connects states at time $t = 2$. By using one of Definitions 11.7–11.10 that corresponds to $\star_2$, we transform the configuration of $\mathscr{A}$ at time $t = 2$ to a configuration at time $t = 3$, etc.

In order to provide an example, let us evaluate the propositional stream $\ldots \vee (\bigwedge_{i=0}^{n-1} p_i^2 \wedge (\bigwedge_{i=0}^{n-1} p_i^1 \wedge (\bigwedge_{i=0}^{n-1} p_i^0 \Rightarrow p_j^0)))$ in a cellular automaton $\mathscr{A}$ with the Moor neighbourhood in the 2-dimensional space:

(I) Initial configuration, $t = 0$

1	0	0
0	1	1
0	1	0

We begin our evaluation with the connective $\Rightarrow$ as the first. Thereby we are using Definition 11.9:

(II) $t = 1$

1	1	1
1	1	1
1	1	0

Then we move to the second connective $\wedge$ and by 11.7 we obtain the following data over (II):

(III) $t = 2$

1	1	1
1	0	0
1	0	0

(IV) $t = 3$

0	0	0
0	0	0
0	0	0

In aposterioristic logics of these cellular automata we can define the context-based concurrent group theory, see Proposition 8.1, and the context-based concurrent Boolean algebra, see Proposition 8.2. In these logics we cannot use conventional variables, because each propositional variable (individual variable) should designate some concrete state(s) at cell(s) z and at time t.

Definition 11.11 Let $c_m, c_n, \ldots,$ be states of payoff cellular automata. A *logical language for reflexive decisions* $\mathscr{L}_M$ in these automata consists of the following symbols: (i) a non-empty set of constant symbols $c_m, c_n, \ldots$; (ii) a non-empty set of function symbols of arity n: $f_0^n, f_1^n, f_2^n, \ldots$; (iii) a non-empty set of predicate symbols of arity n: $P_0^n, P_1^n, P_2^n, \ldots$; (iv) the truth constants 0, 1; (v) propositional connectives of arity n_j: $\Box_0^{n_0}, \Box_1^{n_1}, \ldots, \Box_r^{n_r}$, which are built by superposition of D, C, I, N; (vi) first-order quantifiers $\forall, \exists$; (vii) auxiliary symbols (,).

Terms and well-formed atomic formulas of $\mathscr{L}_M$ are defined non-inductively:

Definition 11.12 Terms are defined as follows: (1) if $t_1, \ldots, t_n \in \{c_m, c_n, \ldots\}$, then $f_i^n(t_1, \ldots, t_n)$ is a term; (2) by using f_i^n, we can construct finite or infinite combinations of terms $t_1, t_2, t_3, \ldots$ which are terms; (3) if $t_1 \in \{c_m, c_n, \ldots\}$ and $t_2, \ldots, t_n$ are terms, then $f_i^n(t_1, \ldots, t_n)$ is a term.

Well-formed formulas of $\mathscr{L}_M$ have a non-inductive definition, too.

Definition 11.13 Formulas are defined as follows: (i) if $t_1, \ldots, t_n$ are terms, then $P_i^n(t_1, \ldots, t_n)$ is an atomic formula; (ii) if $x_1, \ldots, x_n$ are atomic, then $\Box_i^{n_i}(x_1, \ldots, x_n)$ is a formula; by using $\Box_i^{n_i}$, we can construct finite or infinite combinations of formulas $x_1, x_2, x_3 \ldots$, which are formulas; if x_1 is atomic and $x_2, \ldots, x_n$ are formulas, then $\Box_i^{n_i}(x_1, \ldots, x_n)$ is a formula.

The language $\mathscr{L}_M$ defined in these definitions allows us to logically combine different payoff cellular automata with the same set of states (possible payoffs) or their local parts with the same set of states.

11.3.3 Biological Implementations of Reflexive Games

Reflexive games in payoff cellular automata can describe any intelligent behaviour of swarms where different decisions in respect to possible payoffs are being made at different time $t = 0, 1, 2, \ldots$

As we know, in swarm intelligence, we can distinguish the following stimuli constituting swarm data nodes:

- *Attractants*, sources of food, which a swarm feeds. Each attractant A is characterized by its position and intensity.
- *Repellents*, dangerous places avoided by a swarm. Each repellent R is characterized by its position and intensity.

Attractants may be regarded as payoffs for a swarm. Swarms occupy attractants step by step. By different localizations of attractants we can affect on swarm motions differently. These localizations combined with a different intensity of attractants are stimuli for swarm motions. We can interpret these stimuli as Boolean functions on payoffs. For example, we can define the following simplest logical gates (simplest Boolean functions):

- The *AND gate* (*swam conjunction*), the serial connection of contacts. The swarm P follows conjunction $A_1 \wedge A_2$ if and only if it achieves both attractants A_1 and A_2, i.e. the swarm is attracted to A_1, when it is placed in its region of influence, then it is is attracted by A_2 in the region of its influence. If we have deactivated at least one of the attractants A_1, A_2, the swarm cannot achieve conjunction $A_1 \wedge A_2$, i.e. the latter is false.
- The *OR gate* (*swarm disjunction*), the parallel connection of contacts. The swarm P follows disjunction $A_1 \vee A_2$ if and only if one of the attractants A_1 or A_2 or both are occupied by the swarm. In case of deactivation of both attractant A_1 and attractant A_2 the swarm cannot start from the initial position. This means that $A_1 \vee A_2$ is false.
- The *NOT gate* (*swarm negation*). The swarm P follows negation $\neg A_1$ if and only if its behaviour is simulated by the repellent R before the attractant A_1, which avoids the swarm to be attracted by A_1.

By using these simplest functions we can design more complex Boolean functions on attractants (payoffs). In this way we can implement payoff cellular automata in the swarm behaviour. First, let us design a tunnel consisting of attractants and repellents combined as logical gates. Swarms propagate towards the tunnel and face different lines. The first line of logical gates in the tunnel which is occupied by swarms implements Boolean functions at time $t = 0$. The second line of logical gates implements Boolean functions at time $t = 1$. The third line of logical gates implements Boolean functions at time $t = 2$, etc. Second, the number of swarm groups/members before the tunnel corresponds to the number of propositional streams we are going to implement. Third, each swarm is active within a neighbourhood consisting of an appropriate set of attractants. Thus, the number of swarm groups/members before the tunnel corresponds to the number of neighbourhoods as well.

Now on the basis of swarm implementations of payoff cellular automata we can propose swarm implementations of reflexive games of different level of reflexion for two players:

- *The zero level of reflexion.* We design the one tunnel T_1 for player 1 and the second tunnel T_2 for player 2. If the players follow different decision rules presented by two wave sets of formulas, φ_1 and φ_2, then T_1 implements φ_1 and T_2 implements φ_2. If both players follow the same decision rule presented by a wave set of formulas, φ, then both T_1 and T_2 implement φ.
- *The first level of reflexion.* Let player 1 be at the first reflexive level and follow a decision rule φ. Then T_1 implements φ and all logical gates of T_2 at the zero level of reflexion of player 2 occur in T_1.
- *The second level of reflexion.* Let player 1 be at the second reflexive level and follow a decision rule φ. Then T_1 implements φ and all logical gates of T_2 at the first level of reflexion of player 2 occur in T_1, etc.

Thus, swarm implementations of reflexive games allow us to claim that payoff cellular automata can be considered fundamental forms of context-based decisions, indeed.

References

1. Schumann, A.: Towards theory of massive-parallel proofs. cellular automata approach. Bull. Sect. Log. **39**(3/4), 133–145 (2010)
2. Schumann, A.: Qal wa-homer and theory of massive-parallel proofs. History Philos. Log. **32**(1), 71–83 (2011)
3. Schumann, A.: Proof-theoretic cellular automata as logic of unconventional computing. Int. J. Unconv. Comput. **8**(3), 263–280 (2012)
4. Schumann, A.: Payoff cellular automata and reflexive games. J. Cell. Autom. **9**(4), 287–313 (2014)
5. Alonso-Sanz, R.: On a three-parameter quantum battle of the sexes cellular automaton. Quantum Inf. Process 1–16 (2013)
6. Alonso-Sanz, R.: A quantum prisoner's dilemma cellular automaton. In: Proceedings of the Royal Society of London, vol. 470, p. 20130793. The Royal Society (2014)

7. Alonso-Sanz, R., Martín, M.C., Martin, M.: The effect of memory in the spatial continuous-valued prisoner's dilemma. Int. J. Bifurc. Chaos **11**(08), 2061–2083 (2001)
8. Smith, V.L.: Constructivist and ecological rationality in economics. Am. Econ. Rev. **93**(3), 465–508 (2003)
9. Albin, P.S.: Essays on economic complexity and dynamics in interactive systems. In: Barriers and Bound to Rationality. Princeton University Press (1998)
10. Aumann, R.J.: Agreeing to disagree. The Ann. Stat. **4**(6), 1236–1239 (1976)
11. Schumann, A.: Reflexive games and non-archimedean probabilities. P-Adic Numbers Ultrametr. Anal. Appl. **6**(1), 66–79 (2014)
12. Schumann, A.: Anti-platonic logic and mathematics. Mult. Valued Log. Soft Comput. **21**(1–2), 53–88 (2013)
13. Chopard, B., Droz, M.: Cellular Automata Modeling of Physical Systems. Cambridge University Press (2005)
14. Brotherston, J.: Cyclic proofs for first-order logic with inductive definitions. In: B. Beckert (ed.) TABLEAUX 2005, volume 3702 of LNAI, pp. 78–92. Springer (2005)
15. Brotherston, J.: Sequent calculus proof systems for inductive definitions. Ph.D. thesis, University of Edinburgh (2006)
16. Brotherston, J., Simpson, A.: Complete sequent calculi for induction and infinite descent. In: 22nd Annual IEEE Symposium on Logic in Computer Science (LICS 2007), pp. 51–62 (2007)
17. Santocanale, L.: A calculus of circular proofs and its categorical semantics. In: Proceedings of FoSSaCS 2002, LNCS 2303, pp. 357–371. Springer, Grenoble (2002)
18. Morita, K., Imai, K.: Logical Universality and Self-Reproduction in Reversible Cellular Automata. ICES (1996)
19. Adamatzky, A.: Physarum machines: computers from slime mould. In: World Scientific, Series on Nonlinear Science. Series A (2010)
20. Shirakawa, T., Sato, H., Ishiguro, S.: Constrcution of living cellular automata using the Physarum polycephalum. Int. J. General Syst. **44**, 292–304 (2015)
21. Tsompanas, M.I., Adamatzky, A., Ieropoulos, I., Phillips, N., Sirakoulis, G.C., Greenman, J.: Cellular non-linear network model of microbial fuel cell. Biosystems **156**, 53–62 (2017)

Chapter 12
Foundations of Mathematics Within Lateral Inhibition and Lateral Activation

In the previous chapters, I have shown that conventional logics can be suitable for modelling the swarm behaviour under the condition of lateral inhibition. For example, for his purpose we can use the Aristotelian syllogistic (see Sects. 7.1.3, 7.2.2, 8.1.4, 9.1.2 and 9.2.1) and modal logic **K** (see Sect. 9.4.3). Nevertheless, these logics are inappropriate to simulate the behaviour under the condition of lateral activation. It means that the swarm behaviour is irreducible to conventional logics. Furthermore, the swarm behaviour is context-based and it can be described rather my means of aposterioristic logics (see Chaps. 8 and 11). In this chapter, I am going to demonstrate that mathematics is irreducible to conventional logics, too, since mathematical cognitions are grounded on both lateral inhibition and lateral activation. It shows that the lateral inhibition and lateral activation as basic forms of swarm sensing and motoring are fundamental for all cognitions, including the mathematical one. See [1–5].

12.1 Lateral Inhibition and Lateral Activation in Mathematical Cognitions

12.1.1 *What Are Foundations of Mathematics?*

The *Principia Mathematica*, a three-volume work written jointly by Alfred North Whitehead and Bertrand Russell and published in 1910, 1912, and 1913, was one of the first books devoted to 'foundations of mathematics' in the strict sense, i.e. in fact it was one of the first attempts to make explicitly mathematics from the point of view of symbolic logic, that is an attempt to consider mathematical theorems as logical statements which are automatically inferred from axioms by logical inference rules.

A. Schumann, *Behaviourism in Studying Swarms: Logical Models of Sensing and Motoring*, Emergence, Complexity and Computation 33,
https://doi.org/10.1007/978-3-319-91542-5_12

Their work was written by a direct influence of Gottlob Frege's ideas presented in his fundamental book *Die Grundlagen der Arithmetik* (the *Foundations of Arithmetic*) published in 1884. To continue and enhance the approach established by Whitehead and Russell, David Hilbert, the German mathematician (1862–1943), put forward a new proposal for the foundation of mathematics called the *Finitist Program* (or Hilbert's Program). In this proposal all of mathematics should have been formalized in axiomatic form, together with a proof by 'finitary' methods proposed by Hilbert that this axiomatization is consistent. One of the attempts to axiomatize all the mathematics was made by 'Nicolas Bourbaki'—the group of 20th-century mathematicians written jointly the many-volume work entitled *Éléments de mathématique* (the *Elements of Mathematics*).

Now there are some basic formal theories which are regarded as start points in the foundations of mathematics. This means that these theories, in the way how it seems to mathematicians, can cover big fragments of mathematics by their extensions. For instance, it is assumed that in the foundations for number theory we should start from the five Peano's axioms, introduced by Giuseppe Peano in 1889 and now called the *Peano arithmetic* **PA**. Also, it is supposed that any set-theoretic reasoning in mathematics (like reasoning in topology) can be reduced to statements formalized in the *Zermelo-Fraenkel set theory*, constructed by mathematicians Ernst Zermelo and Abraham Fraenkel and denoted by the abbreviation **ZFC**, where **C** means axiom of choice.

To sum up, mathematicians believe still that the foundations of mathematics in the meaning of *Principia Mathematica* are possible and any correct well-done mathematical reasoning can be rewritten in a logical theory such as **PA** or **ZFC**. So, they believe that all the mathematics can be reduced to a logic. Is it true indeed? Is it possible?

In this chapter, we assume that the mathematicians in proving new significant theorem, such as Fermat's Last Theorem, stay beyond any conventional foundations of mathematics and rather deal with combining proof trees on tree forests by using analogy as inference metarule. In other words, the real mathematical proofs of complicated new theorems cannot be completely formalized as discrete sequences, because they are concurrent and can be better formalized as analogue processes within a space with some topological properties. For the first time, inference metarules in a metric space were proposed in the Talmud within a general Judaic approach to concurrent or even massive-parallel conclusions. The mathematicians do not think sequentially like a logical automaton, but concurrently, also. Hence, we suppose that the proof technique of real mathematics cannot be formalized only by discrete methods. It is just a hypothesis of the foundations of mathematics that we can use discrete tools and it is enough, since mathematics is totally reducible to logic. Nevertheless, we show in the paper that the mathematical proof can be formalized also by analogue computations, not only by discrete ones.

Thus, there are two extreme positions in respect to proving new mathematical theorems: (i) *discrete foundations of mathematics*, the traditional belief that we can make all proof steps visible and transparent, i.e. that an absolute detailing (digitalization) in proving is ever possible; (ii) *Talmudic* or *analogue foundations of mathematics*, the opposite belief that at least for some new significant theorems a digitalization can be impossible. According to the first approach, the mathematicians are concentrated on details in proving. According to the second approach, the mathematicians try to be captured by the whole picture of many possible theorems. We know from cognitive neuroscience that a concentration on details is caused by a *lateral inhibition* in our cognitions and an attempt to cover all themes at once is caused by a *lateral activation*.

The lateral inhibition is responsible for a cognitive blindness in seeing the whole picture and the lateral activation is responsible for a cognitive blindness in seeing the details. So, there are detected many cognitive illusions (first of all optical illusions) that are explained by different contexts actualizing the lateral inhibition or lateral activation for different purposes. So, we assume that the mathematicians can actualize both mechanisms also in their proof cognitions. As a consequence, on the one hand, they can focus on logical details and follow the discrete foundations of mathematics (lateral inhibition) and, on the other hand, they can keep in mind the whole picture with thousand concurrent theorems and, thereby, follow the analogue foundations of mathematics (lateral activation).

It is worth noting that I have no ambitions to propose something like the *Principia Mathematica*, but I am going only to focus on mathematical cognitions beyond any discrete (standard) foundations. So, recently there have been proposed by Vladimir Voevodsky (1966–2017) real new foundations called the *univalent foundations of mathematics and homotopy type theory* as a good alternative to set theory, such as **ZFC**. These ideas have been well expressed by the 21st-century group of mathematicians in their legendary book [6] published in 2013. In my terminology, these foundations are discrete also, because they ignore the process of mathematical thinking under the condition of lateral activation.

12.1.2 Cognitive Blindness and Mechanisms of Lateral Activation and Lateral Inhibition

One of the best examples of inattentional cognitive blindness is called the *invisible gorilla*. It is a well-known experiment [7], whose authors ask the viewers to count the number of times a basketball passes between members of one of two distinct teams. If the viewers count them carefully, i.e. if they are too focused on the assigned task, they miss a "gorilla" walking through the scene. The point is that the task to count the passes under the conditions of the information noise brings out the lateral inhibition—we should be completely concentrated on details of passes and, therefore, we should ignore the whole picture. When the authors ask the viewers whether they have seen the gorilla and show them the video again, the viewers can see the gorilla immediately, but they cannot count all the passes. In this case, the viewers look for something new and, as a result, they involve the lateral activation to see the whole picture and ignore small details.

Fig. 12.1 The screenshot from the video at https://www.youtube.com/watch?v=Ahg6qcgoay4. It is a modification of the invisible gorilla. During this awareness test, we should count how many passes the team in white makes. If we count them carefully, we miss the "moonwalking bear"

There are many modifications of the invisible gorilla, e.g. see Fig. 12.1. In the most experiments naïve observers engage in the task to count. Nevertheless, in [8] it was shown that expert searchers who have spent years honing their ability to detect small abnormalities in specific types of image miss the gorilla as well. So, the authors of [8] have asked 24 radiologists to perform a familiar lung nodule detection task. A gorilla, 48 times larger than the average nodule, was inserted into the video. And 83% of radiologists did not see the gorilla. It means that even expert searchers, operating in their domain of expertise, have the same inattentional cognitive blindness because of the lateral inhibition. In the same way, the mathematicians can have this blindness in proof cognitions caused by the lateral inhibition or lateral activation.

Hence, cognitive illusions, such as the invisible gorilla, cannot be interpreted as failures of our perception or dysfunctions indicating the typical limits of our perceptual or cognitive system, but they are rather an effect of economic balance between analyzing some details and the whole picture [9]. The lateral inhibition is a mechanism in our sensing that allows us to be more concentrated on the needed details and the lateral activation allows us to focus on the whole picture.

A type of cognitive illusions caused by the lateral activation and lateral inhibition is detected even at the level of amoeboid organisms, such as *Amoeba proteus* and *Physarum polycephalum*. For example, the amoeba of *Physarum polycephalum* moves differently under both conditions: (i) in the situation of nutrient-poor substrate (corresponding to the condition of lateral inhibition) the amoeba builds a well-distinguished rooted tree (a kind of directed acyclic graphs, see Fig. 12.2(1)); (ii) in the situation of nutrient-rich substrate (corresponding to the condition of lateral

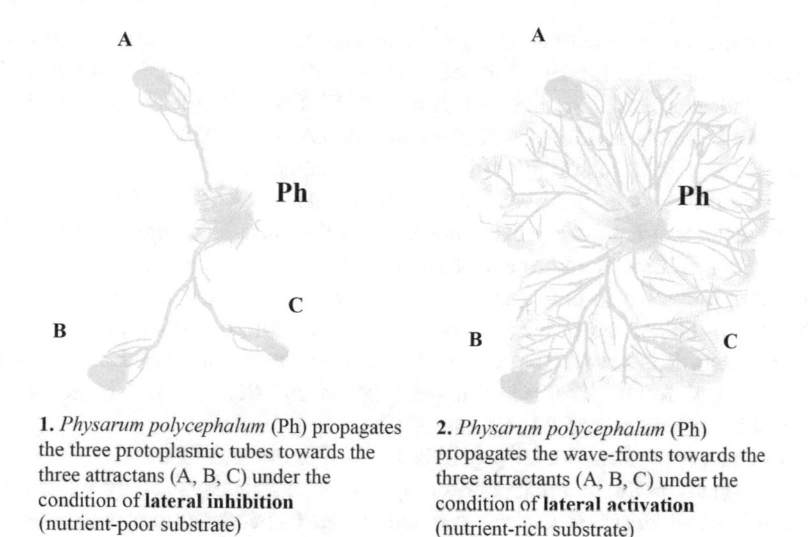

1. *Physarum polycephalum* (Ph) propagates the three protoplasmic tubes towards the three attractans (A, B, C) under the condition of **lateral inhibition** (nutrient-poor substrate)

2. *Physarum polycephalum* (Ph) propagates the wave-fronts towards the three atrractants (A, B, C) under the condition of **lateral activation** (nutrient-rich substrate)

Fig. 12.2 The motions of *Physarum polycephalum* amoeba under the following two conditions: (1) the lateral inhibition and (2) the lateral activation. The first condition is to construct a rooted tree. It is analogous to a proof tree in mathematics for the foundations of mathematics under lateral inhibition. The second condition is to construct a network of concurrent trees with cycles. It is analogous to concurrent cyclic proof trees in mathematics for the foundations of mathematics under lateral activation

activation) the amoeba builds wave-fronts with complex networks of many and many trees (see Fig. 12.2(2)).

My inspiration from neuroscience is that a mathematical cognition proceeds under both conditions, also. As a consequence, we can build a rooted proof tree proving a theorem (similarly as an amoeba does in the nutrient-poor substrate, see Fig. 12.2(1)) or a metric space for combining many and many proof trees simultaneously for many and many theorems at once (similarly as an amoeba is propagated in the nutrient-rich substrate in many directions simultaneously, see Fig. 12.2(2)).

12.1.3 Proof Trees and Proof Forests

In mathematics there are really non-trivial theorems which are so deep that they cannot be inferred without introducing absolutely new mathematical constructions. For example, *Fermat's Last Theorem* (**FLT**) is well formulated in **PA**. Therefore this statement seems to be so simple. For the first time, **FLT** was put forward by Pierre de Fermat in 1637 in the margin of a copy of *Arithmetica*. However, this statement was proven formally only after 358 years of effort by mathematicians, namely by Andrew Wiles in 1995 [10]. The most dramatic problem of **FLT** recently is that this theorem is proved mathematically and this proof was accepted by mathematical communities, but this statement was not checked by logicians at all. It is unknown still whether there is a logical proof of **FLT**. In other words, **FLT** is not covered by any foundations of mathematics still. It means, **FLT** was obtained due to the lateral activation in proof cognitions, but the small details of **FLT** are invisible yet. So, nobody has applied a lateral inhibition for this statement.

As we said, **FLT** is well written in a first-order sentence of **PA**. However, it does not mean that it can be proved in **PA**. After the Paris-Harrington theorem [11], it is well known that there are ever first-order arithmetic statements written in **PA** which cannot be proved in **PA**, such as the *Strong Ramsey Theorem* that can easily be proved in the second-order arithmetic from the infinite version of the standard theorem. Also, it is known that there are many other combinatorial problems that are beyond **PA**. In [12], Colin McLarty supposed that **FLT** can be proved in some higher-order extensions of **PA**, but nobody has checked it still.

Another hypothesis of McLarty [12] is that **FLT** is beyond **ZFC**. It is quite evident taking into account the fact that cohomological number theory used by Wiles [10] is based on *Grothendieck's universes* which model **ZFC**, but the existence of a universe is not provable in **ZFC**. Grothendieck's own axiom of universes, which was added to **ZFC**, affirms that every set is contained in some universe (there is an uncountable strongly inaccessible cardinal for sets) [13]. Hence, in cohomology we deal with **ZFC+U** consisting of **ZFC** with the assumption **U** of a universe. So, **FLT** can be proved at least in **ZFC+U** or even in higher extensions, and evidently not in **ZFC**. Nevertheless, there is no formal proof still what is set theory looks like for **FLT**.

Perhaps, Vladimir Voevodsky's homotopy type theory [6] is a good candidate for a formal proof of **FLT**. But for us the most significant think is that some theorems like **FLT** are being proved at first beyond any formal (logical) system.

Thus, there are ever serious mathematical theorems such as **FLT** which are beyond the recent foundations of mathematics (for instance, beyond **PA** or **ZFC**). However, mathematicians and logicians unconsciously obey the quite religious faith and follow the deep-inner intuition that any mathematical theorem can be reduced to a theorem of existed symbolic logic. In symbolic logic we appeal to a *formal theory* T_i that possesses logical axioms/theorems $(a_{i1}^L, ..., a_{in}^L)$ within a logical system L like the classical propositional logic and non-logical axioms/theorems $(a_{i1}^T, \dots, a_{im}^T)$ for defining properties of predicates and functions introduced in T_i. Then by using inference rules of L we can infer in T_i all possible provable sentences from $(a_{i1}^L, \dots, a_{in}^L)$ + $(a_{i1}^T, \dots, a_{im}^T)$. Surely, it does work in symbolic logic indeed, but it is unknown still whether it gives something to real mathematics. In real mathematics, i.e. in a proof of deep sentences such as **FLT**, we use some axioms/presuppositions/ sentences $(a_{i1}^M, \dots, a_{ik}^M)$ and the main task of the foundations of mathematics is to set up a formal theory T_i to find out an injective mapping from $(a_{i1}^M, \dots, a_{ik}^M)$ into $(a_{i1}^T, \dots, a_{im}^T)$. The problem is that, on the one hand, some sentences from $(a_{i1}^M, \dots, a_{ik}^M)$ in **FLT** can be just intuitive and not well-formulated (or even not conscious). On the other hand, their true symbolic-logical analogues in T_i can be absent from the list $(a_{i1}^T, \dots, a_{im}^T)$ and not sketched still.

Let us assume that $(a^M_{i1}, \ldots, a^M_{ik})$ for serious theorems like **FLT** are not non-logical axioms / theorems which can be represented as $(a^T_{i1}, \ldots, a^T_{im})$. So, we assume that higher mathematics cannot be reduced to symbolic logic.

Let us suppose now that $(a^M_{i1}, \ldots, a^M_{ik}) = (a^M_{i1}, \ldots, a^M_{ik}) + (a^L_{i1}, \ldots, a^L_{in})$ and $(a^T_{i1}, \ldots, a^T_{im}) = (a^T_{i1}, \ldots, a^T_{im}) + (a^L_{i1}, \ldots, a^L_{in})$, i.e. they are closed under the inference rules of L. Each proof in $T_i = (a^T_{i1}, \ldots, a^T_{im})$ is a *tree* t. Let t_{e_j} be an inference rule of T_i, i.e. it is an elementary tree—it has one parent (the inferred statement) and several its direct children (axioms as premisses). Then t consists of $t_{e_1}, \ldots, t_{e_j}$ as its subtrees. Let {} be an empty tree. Then each *edge/branch* of t can be labelled by a sentence $\alpha \in (a^T_{i1}, \ldots, a^T_{im})$ as follows: $\alpha[t']$, where t' is an elementary subtree (inference rule) of t that was used to obtain α. Then a tree can be denoted by all its branches. For instance, the notation $t = \{c[\{a[\{\}], b[\{\}]\}]\}$ means that we have a tree consisting only of two edges/premisses $a[\{\}]$ and $b[\{\}]$ and one conclusion from them $c[\{a[\{\}], b[\{\}]\}]$:

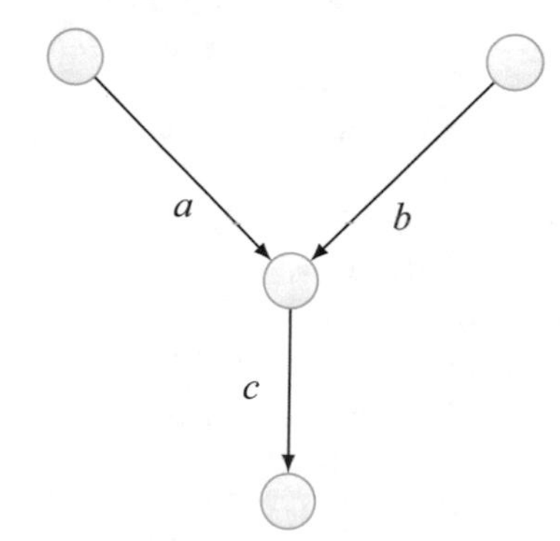

The notation $t' = \{c[\{a[\{\}], b[\{\}]\}], a[\{\}]\}$ designates the following tree:

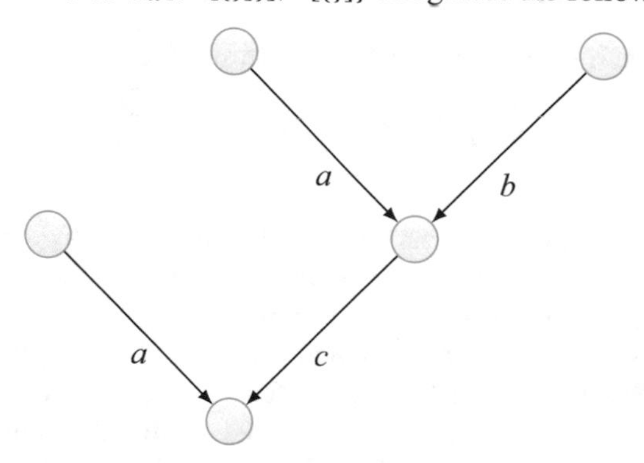

The formula $t'' = \{d[\{c[\{a[\{\}], b[\{\}]\}], a[\{\}]\}], f[\{\}], e[\{\}]\}$ satisfies the following tree:

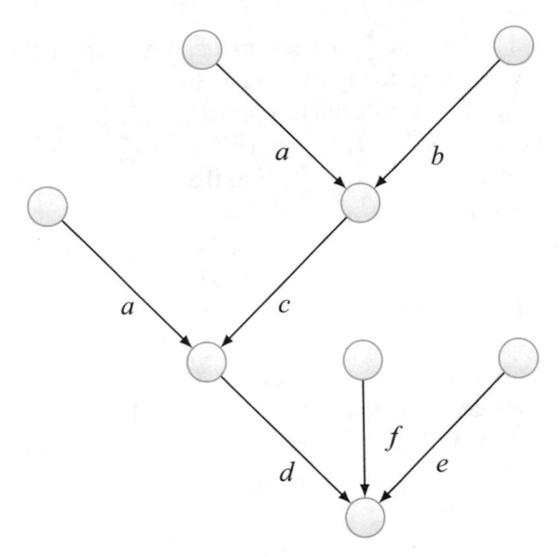

Hence, in T_i we deal just with finite trees labelled in $(a^T_{i1}, \ldots, a^T_{im})$ in the way defined above.

Let an ordered sequence of trees be called a *forest*. We say a forest $s_i = \langle t_i, t'_i, t''_i, \ldots \rangle$ is a *piece* of a forest $s_j = \langle t_j, t'_j, t''_j, \ldots \rangle$, written $s_i \preceq s_j$ if s_i can be obtained from s_j by removing nodes. In other words, there is an injective mapping from nodes of s_i to nodes of s_j that preserves the lexicographic and descendant ordering. Then a reduction of $(a^M_{i1}, \ldots, a^M_{im})$ to $(a^T_{i1}, \ldots, a^T_{in})$ would mean that $(a^M_{i1}, \ldots, a^M_{im})$ is a piece of $(a^T_{i1}, \ldots, a^T_{in})$, i.e. $(a^M_{i1}, \ldots, a^M_{im}) \preceq (a^T_{i1}, \ldots, a^T_{in})$.

In deep theorems like **FLT** we have ever a forest s_j, which is a substantial part of $(a^M_{i1}, \ldots, a^M_{im})$, such that s_j is not a piece for any forest of $(a^T_{i1}, \ldots, a^T_{in})$. So, we can distinguish two kinds of forests from $(a^M_{i1}, \ldots, a^M_{im})$:

1. a forest s_j which is a piece of some forest from $(a^T_{i1}, \ldots, a^T_{in})$, these pieces are called *logical fragments*;
2. a forest s_j which is not a piece of any forest from $(a^T_{i1}, \ldots, a^T_{in})$, this s_j is called a *new construction*.

We assume that logical fragments are well studied in the foundations of mathematics, but new constructions are not yet. The matter is that the conventional foundations of mathematics was concentrated only on some details under the condition of lateral inhibition. In this way, new constructions, such as **FLT**, remain invisible for mathematicians like the gorilla is invisible for the observers counting the passes. The true mathematical task is to give non-trivial essential theorems containing new constructions. For this purpose, the lateral activation holds, but the latter is out of precise details. The ambitious mathematicians start their work with logical fragments to obtain new constructions later.

Thus, the higher mathematics is eternally to extend forests of $(a^M_{i1}, \ldots, a^M_{im})$ by proposing new constructions. Therefore, the ambitious mathematicians deal not only with proofs/trees (what can be reduced completely to pieces from $(a^T_{i1}, \ldots, a^T_{in})$), but more often with forests which are irreducible to pieces of $(a^T_{i1}, \ldots, a^T_{in})$. The mathematicians know that sometimes reasoning by analogy allows them to propose really something new. They take a piece from one theory $\mathcal{T}_i = (a^M_{i1}, \ldots, a^M_{im})$ to combine it with forests of another theory $\mathcal{T}_j = (a^M_{j1}, \ldots, a^M_{jk})$ and then to obtain the next theory $\mathcal{T}_f = (a^M_{f1}, \ldots, a^M_{fl})$.

In the Talmud for the first time there was proposed a metareasoning for extending the dataset (forests) of the Torah, $(a^M_{i1}, \ldots, a^M_{ik})$, by using analogies. It is a way how we can add new axioms to $(a^M_{i1}, \ldots, a^M_{ik})$ just looking at existing forests. Then we will show how we can use these methods in the recent foundations of mathematics. Hence, the Talmudic metareasoning in the foundations of mathematics are called by us '*Talmudic foundations of mathematics*', see [1, 2, 4]. We assume that in a real mathematical practice mathematicians use a similar metareasoning to build up new forests by looking at existing forests (let us notice that this metareasoning is considered often as a mathematical intuition).

12.2 Foundations of Mathematics Within the Lateral Activation

12.2.1 What Is Talmudic Logic?

Usually, the term 'Talmudic logic' means the Judaic hermeneutic rules (Hebrew: *middot*, מידות) [1], first formulated as a special hermeneutics by Hillel in the 1st century B.C. (he proposed the 7 rules), by Rabbi Ishmael in the 2nd century A.D. (he established the 13 rules for inferring the law, *halakhah*), and by Rabbi Eliezer ben Jose HaGelili in the same 2nd century A.D. (he examined the 32 rules for inferring the holy stories, *haggadah*). As a result, in the Talmud we face a logical point of view in respect to the Torah: all the Biblical statements are considered 'particular' (פרט) or 'general' (כלל) implying some conclusions by using hermeneutic rules. A Biblical statement, φ^i_{part}, is regarded as *particular*, if its verification $|\varphi^i_{part}|$ (i.e. all denotates of φ^i_{part}) is a subset of a verification $|\varphi^j_{gen}|$ for another statement, φ^j_{gen}, namely: $|\varphi^i_{part}| \subseteq |\varphi^j_{gen}|$. In this case the implication $\varphi^i_{part} \to \varphi^j_{gen}$ is true. Hence, we can draw appropriate proof trees by using *modus ponens* and other classical inference rules, if we define all the last particulars of the Torah as axioms of the Talmud: $(a^{part}_{i1}, \ldots, a^{part}_{in})$.

Now, let us consider an example from *Exodus* 22 and let us try to build up a proof tree for the notion 'responsibility for the property of his neighbour', just basing on the text of the Torah. We have the following statements mentioned in this chapter of the Holy Book:

G "He is responsible for the property of his neighbour."
G1 "The neighbour's property is given for safekeeping for free of charge."
G2 "The neighbour's property is given for safekeeping for money."
G3 "The neighbour's property is borrowed."
G1P1 "The safekeeping for free of charge is stolen. He should swear that he did not lay his hand upon his neighbour's property."

> If a man shall deliver unto his neighbour money or stuff to keep, and it be stolen out of the man's house; if the thief be found, let him pay double. If the thief be not found, then the master of the house shall be brought unto the judges, to see whether he have put his hand unto his neighbour's goods (KJV, *Exodus*, 22:7-8).

G2P1 "The safekeeping for money is stolen. He should pay the loss."
G2P2 "The safekeeping for money is destroyed for natural reasons. He should swear that he did not lay his hand upon his neighbour's property."

> If a man deliver unto his neighbour an ass, or an ox, or a sheep, or any beast, to keep; and it die, or be hurt, or driven away, no man seeing it: Then shall an oath of the Lord be between them both, that he hath not put his hand unto his neighbour's goods; and the owner of it shall accept thereof, and he shall not make it good. And if it be stolen from him, he shall make restitution unto the owner thereof (KJV, *Exodus*, 22:10-12).

G3P2 "The borrowed is destroyed for natural reasons. He should pay the loss."

> And if a man borrow ought of his neighbour, and it be hurt, or die, the owner thereof being not with it, he shall surely make it good. But if the owner thereof be with it, he shall not make it good: if it be an hired thing, it came for his hire (KJV, *Exodus*, 22:14-15).

As we see, we deal here with the four last particulars (i.e. Talmudic axioms): (**G1P1**, **G2P1**, **G2P2**, **G3P2**). However, the data set for 'responsibility for the property of his neighbour' is not complete—we know nothing about the following two particulars (axioms) which are supposed also:

G1P2 "The safekeeping for free of charge is destroyed for natural reasons. What should he do?"
G3P1 "The borrowed is stolen. What should he do?"

In the picture form, the complete data set must be seen as follows:

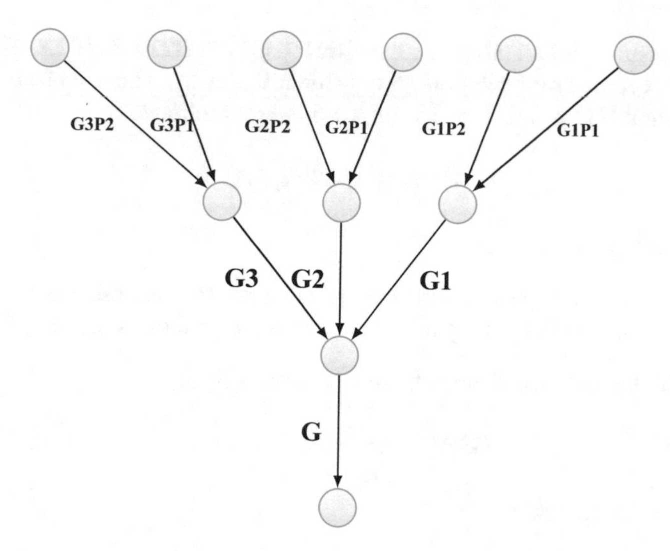

Nevertheless, in the Torah an appropriate data set is sketched in the following manner, i.e. it is absolutely incomplete for inferring 'responsibility for the property of his neighbour':

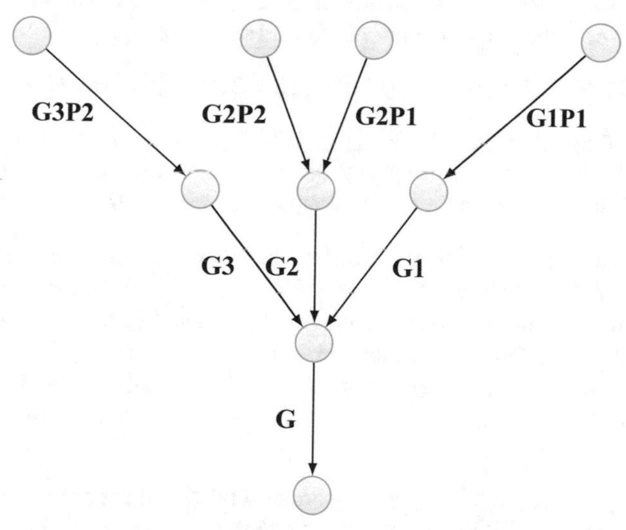

Thus, we need to find out possible ways for defining the two new axioms, **G1P2** and **G3P1**. But how? Do not worry. In the *Bava Metzia*, the second of the first three tractates of the Babylonian Talmud in the order of *Nezikin* ("damages"), Chap. 8, there is a metareasonig for this purpose—the inference metarule for defining new axioms, called *qal wa-ḥomer* (קל וחמר). This rule occurs among the 7 rules of Hillel, as well as among the 13 rules of Rabbi Ishmael and the 32 rules of Rabbi Eliezer ben Jose HaGelili. Let us take all the particulars for **G2**, because they are complete. Between **G2P1** and **G2P2** there is a strong ordering relation: **G2P2** $\precsim$ **G2P1**. Indeed, **G2P2** means that he is free of paying and **G2P1** means that he should pay the loss.

Hence, we assume that all last particulars (**G1P1**, **G1P2**, **G2P1**, **G2P2**, **G3P1**, **G3P2**) are partially ordered by a relation '$\precsim$ or =' denoted by $\precapprox$, where = means the same payment. Then **G1P2** is defined by *qal wa-ḥomer* as follows:

$$\mathbf{G1P2} = \min(\mathbf{G2P2}, \mathbf{G1P1}).$$

From this it follows that

G1P2 "The safekeeping for free of charge is destroyed for natural reasons. He should swear that he did not lay his hand upon his neighbour's property."

Analogically, **G3P1** is defined by *qal wa-ḥomer* thus:

$$\mathbf{G3P1} = \min(\mathbf{G2P1}, \mathbf{G3P2}).$$

Then it is inferred that

G3P1 "The borrowed is stolen. He should pay the loss."

So, the main goal of *qal wa-ḥomer* is to add new axioms for the Torah data sets to make the proof trees more symmetrical: in the same tree t with only one root all subtrees must bear the same number of edges. For instance, in the tree

t = {**G**[{**G1**[{**G1P1**[{}]}], **G2**[{**G2P1**[{}], **G2P2**[{}]}], **G3**[{**G3P2**[{}]}]}]}

we had the subtree t' = {**G2**[{**G2P1**[{}], **G2P2**[{}]}]} with the number of edges 2 that exceeds the numbers of edges of all other subtrees. Then we must add new branches to each subtree where the number of edges is less than 2.

Definition 12.1 (Inference Metarule *qal wa-ḥomer*) Let a tree t have i subtrees: t_1, ..., t_i and let N_j be a number of axioms $(a_{j1}^{part}, \ldots, a_{jN}^{part})$ for each $j = \overline{1, i}$ and let all axioms for each subtree be lexicographically ordered in the same way. Assume that $\max_{j=1}^{i} N_j = n$. This means that there exists l such that $1 \leq l \leq i$ and $N_l = n$. Let us suppose now that there exists $k \in \{1, \ldots, i\}$ such that $k \neq l$ and $N_k = m < n$. Then we can add new axioms by *qal wa-ḥomer* to complete the axioms of t_k up to the set $(a_{k1}^{part}, \ldots, a_{kn}^{part})$:

1. the set $(a_{l1}^{part}, \ldots, a_{ln}^{part})$ is partially ordered by $\precapprox$ and the order $\precapprox$ of $(a_{o1}^{part}$, ..., $a_{on}^{part})$ is the same for any subtree t_o of t such that $N_o = n$;
2. the set $(a_{k1}^{part}, \ldots, a_{km}^{part})$ is partially ordered by the same $\precapprox$, but the order can be different;
3. the item $a_{k(m+1)}^{part} = \min(a_{km}^{part}, a_{l(m+1)}^{part})$;
4. the item $a_{k(m+p)}^{part} = \min(a_{k(m+p-1)}^{part}, a_{l(m+p)}^{part})$ for each $p = \overline{1, n-m}$.

According to this metarule, we can have the same number of axioms for each subtree of the same tree t. In other words, for all subtrees t_1, ..., t_i of t we have the axioms $(a_{11}^{part}, \ldots, a_{1n}^{part}, a_{21}^{part}, \ldots, a_{2n}^{part}, \ldots, a_{i1}^{part}, \ldots, a_{in}^{part})$ due to *qal wa-ḥomer*.

In the previous section we have said that the task of every ambitious mathematician is to extend a set of mathematical axioms. In the statements like **FLT** we exceed the set of existing axioms (i.e. we put proofs outside of the foundation of mathematics). And we assume that the ambitious mathematicians appeal to some inference metarule to obtain new axioms for proving their non-trivial sentences. In other words, they deal not with mechanical proofs from existing axioms within the foundations of mathematics (i.e. they do not follow the lateral inhibition), but they combine different trees to expand the set of possible axioms beyond any foundations of mathematics (i.e. they follow the lateral activation to see the "invisible gorilla" in math). This way of proving is called by us *Talmudic* because of the priority of the Talmudic logic in proposing some inference metarules for defining axioms. So, let us generalize Definition 12.1 as follows.

Definition 12.2 (Inference Metarule $\mathscr{I}$) Let a tree t have i subtrees: t_1, ..., t_i and let N_j be a number of axioms $(a^M_{j1}, \ldots, a^M_{jN})$ for each $j = \overline{1, i}$. Assume that $\max^i_{j=1} N_j = n$. This means that there exists l such that $1 \leq l \leq i$ and $N_l = n$. Let us suppose now that there exists $k \in \{1, \ldots, i\}$ such that $k \neq l$ and $N_k = m < n$. Then we can add new axioms by the inference metarule $\mathscr{I}$ to complete the axioms of t_k up to the set $(a^M_{k1}, \ldots, a^M_{kn})$:

1. the sets $(a^M_{l1}, \ldots, a^M_{ln})$, ..., $(a^M_{o1}, \ldots, a^M_{on})$ are partially ordered by $\precapprox$ for all subtrees t_l, ..., t_o with the number of axioms $N_l = \cdots = N_o = n$;
2. the set $(a^M_{k1}, \ldots, a^M_{km})$ is partially ordered by the same $\precapprox$, but the order can be different;
3. the item $a^M_{k(m+p)} = \Box_{\mathscr{I}}(a^M_{k1}, \ldots, a^M_{k(m+p-1)}, a^M_{l1}, \ldots, a^M_{l(m+p)}, \ldots, a^M_{o1}, \ldots, a^M_{o(m+p)})$ for each $p = \overline{1, n-m}$, where $\Box_{\mathscr{I}}$ is a Boolean function.

12.2.2 *Talmudic Reasoning for Simulating Swarms*

The rule *qal wa-ḥomer* can be represented as a form of spatial reasoning. In this form the *qal wa-ḥomer* can simulate swarm behaviours. To show this, let us consider a new Talmudic example. In the *Baba Qama* (one of the Talmudic books), different kinds of damages (*nezeqin*) are analyzed, among which three genera are examined: foot action (*regel*), tooth action (*šen*) and horn action (*qeren*). These three are damages that could be caused by an ox (he can trample (foot), eat (tooth) and gore (horn)). Due to the Torah it is known that tooth damage (as well as foot damage) by an ox at a public place needs to pay zero compensation. Horn damage at a public place entails payment of 50% the damage costs as compensation. In a private area foot/tooth damage must be paid in full. What can we say now about payments for horn actions of private places?

Damages (nezeqin)	Public place	Private place
Horn action (*qeren*)	50%	?
Foot action (*regel*)	0	100%
Tooth action (*šen*)	0	100%

In order to draw up a conclusion by *qal wa-ḥomer*, we should define a two-dimensional ordering relation on the set of data: (i) on the one hand, according to the *dayo* principle, we know that payment for horn action in a private area cannot be greater than the same in a public area, (ii) on the other hand, payment for horn action at a private place cannot be greater than foot/tooth action at the same place. Hence, we infer that payment of compensation for horn action at a private place is equal to 50% of the damage costs.

The table above shows us that *qal wa-ḥomer* can be interpreted spatially, too. Already Yisrael Ury [14] proposed to use the Carroll's bilateral diagrams for modelling conclusions by *qal wa-ḥomer*. Let us take now the diagram

xy'	xy
$x'y'$	$x'y$

that plays the role of 'universe of discourse' for Talmudic reasoning over adjuncts x, y, non-x in the neighbourhood (that is denoted by x'), non-y in the neighbourhood (that is denoted by y'). Assume that we have only black counters and if a black counter is placed within a cell, this means that "this cell is occupied" (i.e. "there is at least one thing in it", "an appropriate Talmudic rule should be obeyed"). Thus, the cell that does not contain a black counter indicates a situation in which the obligation is not fulfilled, whereas the cell containing a black counter indicates a situation in which the obligation is fulfilled.

Hence, if we have two rows and two columns, there are sixteen possible ways to cover such a diagram by means of black counters, in connect to Yisrael Ury who accepts only six of them (Fig. 12.3).

Let x mean proposition 1 and y mean proposition 2. Then the above mentioned accepted diagrams (Fig. 12.3) have the following sense: (a) It is necessary and sufficient to obey x; (b) It is necessary and sufficient to obey y; (c) It is sufficient to obey either x or y; (d) It is not sufficient to obey x and/or y; (e) It is necessary and sufficient to obey both x and y; (f) It is not necessary to obey either x or y.

Let us return to our example. Let x_3 be a 'foot action', x_2 a 'tooth action', x_1 a 'horn action', y' 'at public place', y 'at private place'. So, we obtain the following diagram:

x_1y'	x_1y
x_2y'	x_2y
x_3y'	x_3y

Assume that a black counter means an obligation to pay 100% of the damage costs as compensation and a grey counter means an obligation to pay any 50%. We can cover this diagram by counters as shown in Fig. 12.4.

In so doing, we have supposed that there is a different power of intensity in obligation. In this case our rule for inferring by *qal wa-ḥomer* is formulated thus:

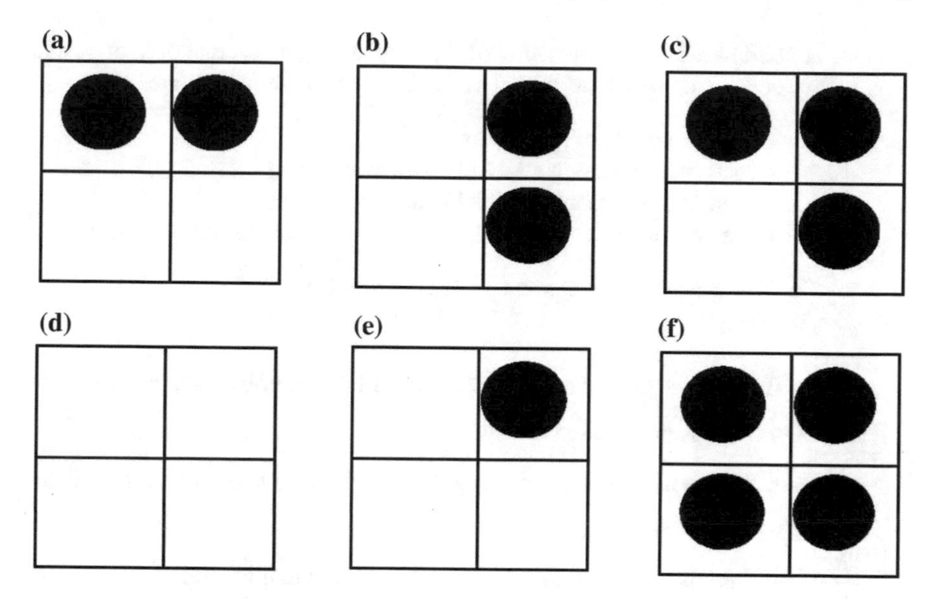

Fig. 12.3 Ury's diagrams for conclusions by *qal wa-ḥomer*, see [15]

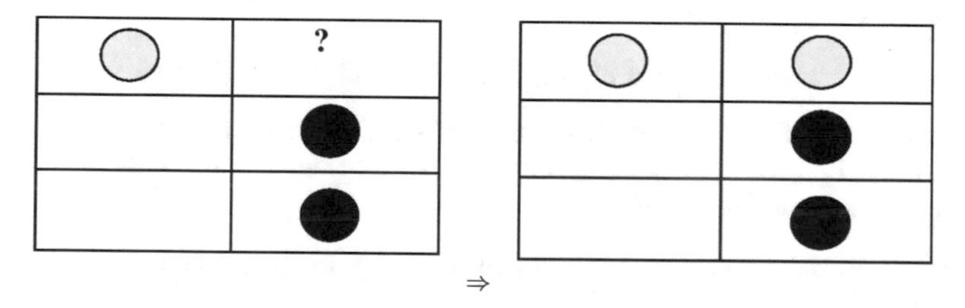

Fig. 12.4 Ury's diagrams for inferring whether we should pay a horn action at a private place, see [15]

Definition 12.3 (Inference Metarule *qal wa-ḥomer*) If a cell contains a black or grey counter, all cells above it and to its right also contain a black or grey counter and the color of that counter has the minimal hardness of black and grey in counters of the neighbour cells; if a cell does not contain any counter, all cells below it and to its left are also without counters.

Talmudic diagrams defined just above are close to Carroll's diagrams and can simulate the swarm behaviour under the condition of lateral activation.

Let us consider a nest of ants. Each cell x, y, x', y' of the universe

xy'	xy
$x'y'$	$x'y$

means an attractant located at this cell which attract the ants. In this way we obtain the following syllogistic strings: xy, yx, $x'y$, yx', xy', $y'x$, $x'y'$, $y'x'$, where x and y in xy are interpreted as two neighbour attractants connected by a network of ants,

x' is understood as all attractants which differ from x, but are neighbours of y, and y' is understood as all attractants which differ from y and are neighbours of x:

xy there is a road of ants between points x and y;
$x'y$ there is a road of ants among all neighbours of x and y;
xy' there is a road of ants among all neighbours of y and x;
$x'y'$ there is a road of ants among all neighbours of x and all neighbours of y;

The rule of *qal wa-ḥomer* defined in Fig. 12.3 and in Definition 12.3 has the following meaning:

xy'	xy

—if there is a road of ants among x and all the neighbours of y, then there is a road of ants between x and y;

	xy
	$x'y$

—if there is a road of ants among y and all the neighbours of x, then there is a road of ants between x and y;

xy'	xy
	$x'y$

—if there is a road of ants among x and all the neighbours of y and there is a road of ants among y and all the neighbours of x, then there is a road of ants between x and y;

—if there is no road of ants between x and y, then (i) there is no road of ants among x and all the neighbours of y and (ii) there is no road of ants among y and all the neighbours of x and (iii) there is no road of ants among all the neighbours of x and all the neighbours of y;

	xy

—if there is a road of ants between x and y, then we do not know anything more;

xy'	xy
$x'y'$	$x'y$

—if there is a road of ants among all neighbours of x and all the neighbours of y, then (i) there is a road of ants among x and all the neighbours of y and (ii) there is a road of ants among y and all the neighbours of x and (iii) there is a road of ants between x and y.

We can propose also the following Talmudic diagrams for simulating the swarm sensing and motoring:

x	y
y'	x'

, where x' is a non-empty class of neighbour attractants for x reachable from y and y' is a non-empty class of neighbour attractants for y reachable from x. Then the *qal wa-ḥomer* tells us whether a multiplication took place during the propagation of ants at points x and/or y. In Fig. 12.5, all the possible conclusions inferred by *qal wa-ḥomer* in relation to x and y are considered, and they are defined if we have a multiplication at those points.

x	y
y'	

—if there is a road between y' and x, then there is a road between x and y (see Fig. 12.5a);

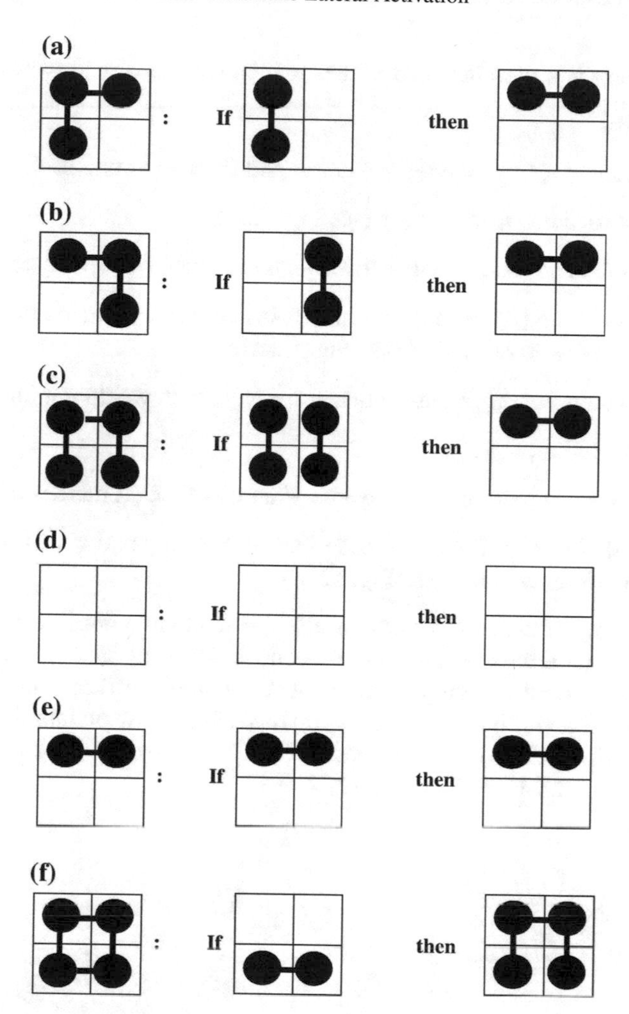

Fig. 12.5 The *diagrams for qal wa-homer* simulating the ant networking: **a** If the string xy' is verified, then the string xy is verified, too (i.e. if xy' is verified, then x has a multiplication of ant roads). **b** If the string yx' is verified, then the string xy is verified, too (i.e. if yx' is verified, then y has a multiplication of ant roads). **c** If the strings xy' and yx' are verified, then the string xy is verified, too (i.e. if both xy' and yx' are verified, then both x and y have multiplications of ant roads). **d** If no string is verified, then there is no multiplication of ant roads. **e** If the string xy is verified, then the there is no multiplication of ant roads. **f** If the strings $x'y'$ is verified, then the strings xy', $x'y$, xy are verified, too (i.e. if $x'y'$ is verified, then both x and y have multiplications of ant roads), see [15]

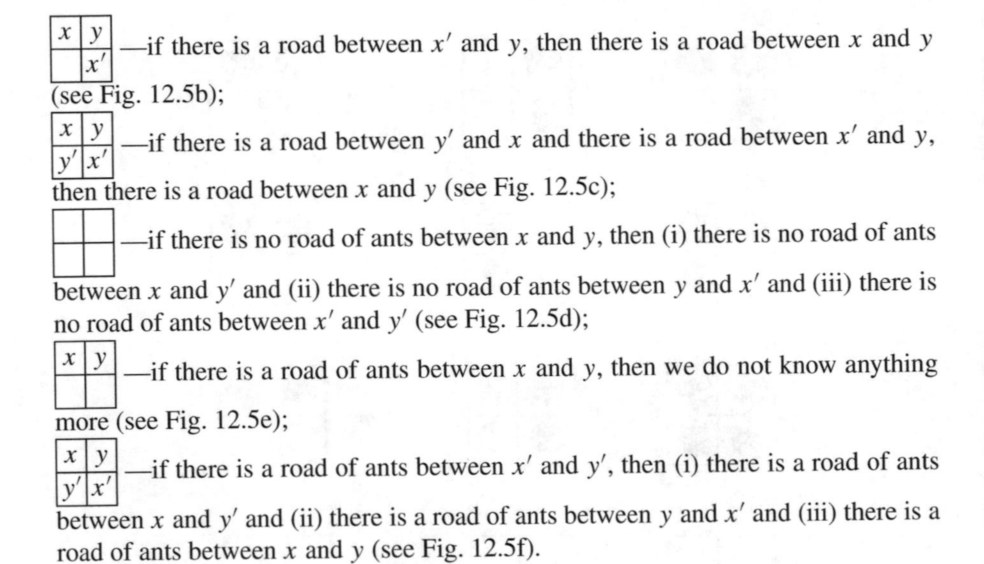

x	y
	x'

—if there is a road between x' and y, then there is a road between x and y (see Fig. 12.5b);

x	y
y'	x'

—if there is a road between y' and x and there is a road between x' and y, then there is a road between x and y (see Fig. 12.5c);

—if there is no road of ants between x and y, then (i) there is no road of ants between x and y' and (ii) there is no road of ants between y and x' and (iii) there is no road of ants between x' and y' (see Fig. 12.5d);

x	y

—if there is a road of ants between x and y, then we do not know anything more (see Fig. 12.5e);

x	y
y'	x'

—if there is a road of ants between x' and y', then (i) there is a road of ants between x and y' and (ii) there is a road of ants between y and x' and (iii) there is a road of ants between x and y (see Fig. 12.5f).

Hence, the main difference between the Aristotelian syllogistic and Talmudic reasoning is that, on the one hand, we are concentrating on fusions of swarm roads in the case of the Aristotelian syllogistic (lateral inhibition effects) and, on the other hand, we deal with multiplications of swarm roads in the case of Talmudic reasoning (lateral activation effects). Let us exemplify Fig. 12.5 by different ant roads, see Figs. 12.6, 12.7, 12.8 12.9 12.10 and 12.11.

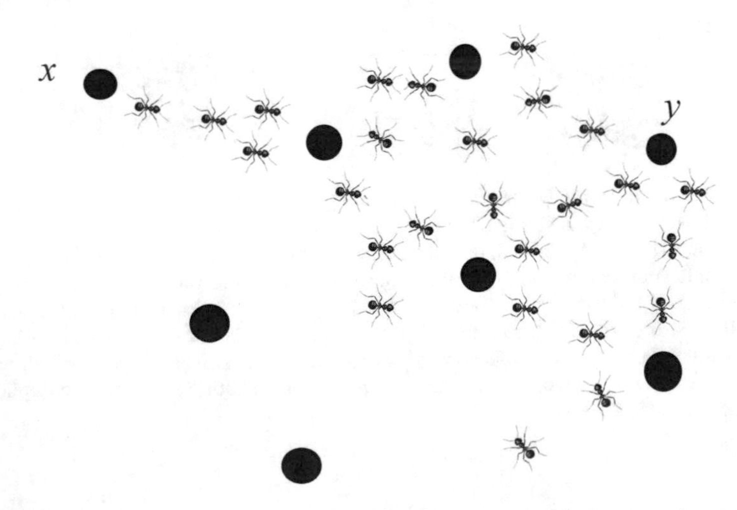

Fig. 12.6 The *qal wa-ḥomer* for the ant networking. There are several attractants denoted by black circles. We designate two attractants as x and y. The picture shows that if there is a road between y' and x, then there is a road between x and y (see Fig. 12.5a)

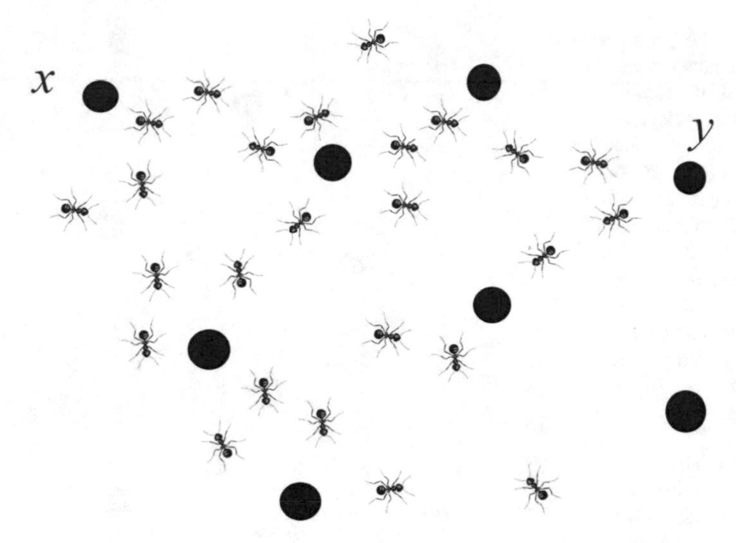

Fig. 12.7 The *qal wa-ḥomer* for the ant networking. There are several attractants denoted by black circles. We designate two attractants as x and y. The picture shows that if there is a road between x' and y, then there is a road between x and y (see Fig. 12.5b)

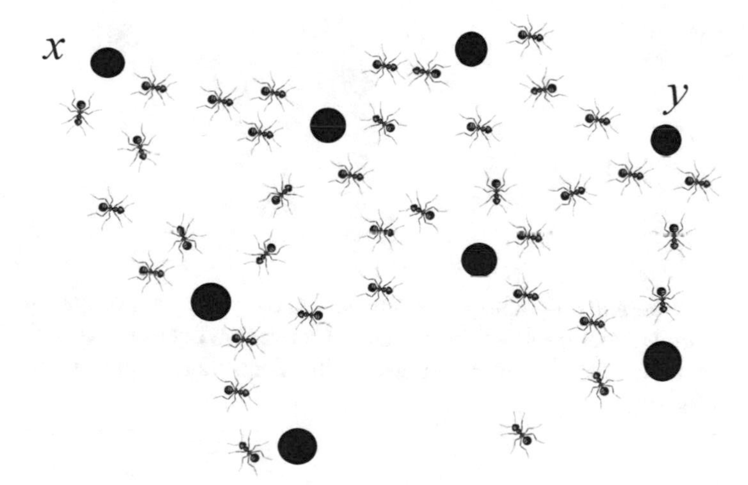

Fig. 12.8 The *qal wa-ḥomer* for the ant networking. There are several attractants denoted by black circles. We designate two attractants as x and y. If there is a road between y' and x and there is a road between x' and y, then there is a road between x and y (see Fig. 12.5c)

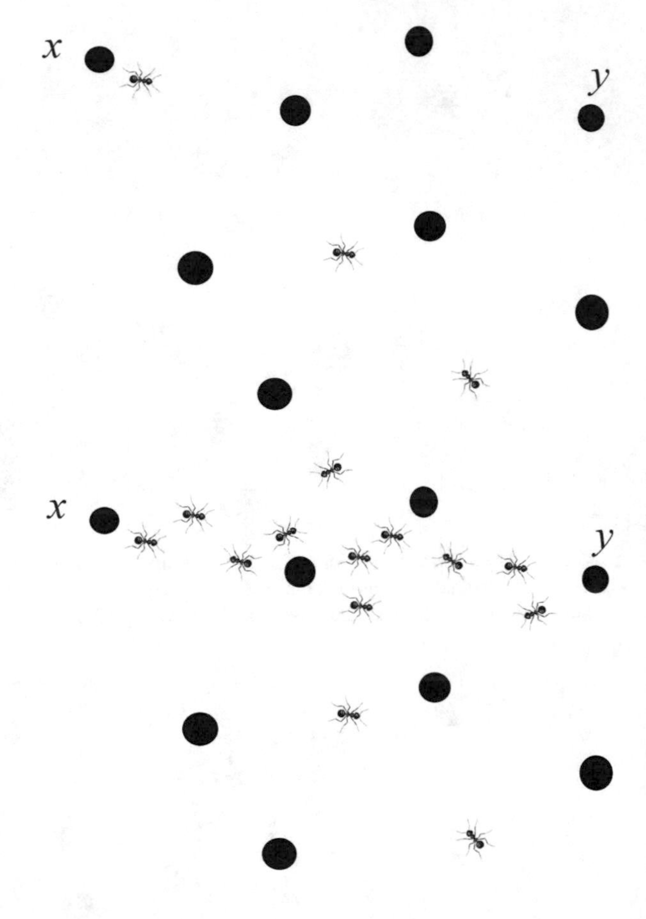

Fig. 12.9 The *qal wa-ḥomer* for the ant networking. There are several attractants denoted by black circles. We designate two attractants as x and y. If there is no road of ants between x and y, then (i) there is no road of ants between x and y' and (ii) there is no road of ants between y and x' and (iii) there is no road of ants between x' and y' (see Fig. 12.5d)

Fig. 12.10 The *qal wa-ḥomer* for the ant networking. There are several attractants denoted by black circles. We designate two attractants as x and y. If there is a road of ants between x and y, then we do not know anything more (see Fig. 12.5e)

As we see, the *qal wa-ḥomer* can simulate some forms of ant propagations, i.e. the ant behaviour under the conditions of lateral activation. Let us return to the lateral activation in the mathematical cognitions. In the same way we can simulate plant roots, see [16–19].

12.2.3 Metareasoning in Mathematical Analysis

Thus, in the *Talmudic foundations of mathematics* we transform one metric space of mathematical proof trees into another space by some inference metarules. We make it by analogy—how in the Talmud we make a transformation from one space of Biblical particular-and-general relationships to another one. The proof trees supposed in the Torah are not complete and by the Talmudic inference metarules such as *qal wa-ḥomer* we can make trees more symmetric so that their subtrees must have the same number of branches at the end.

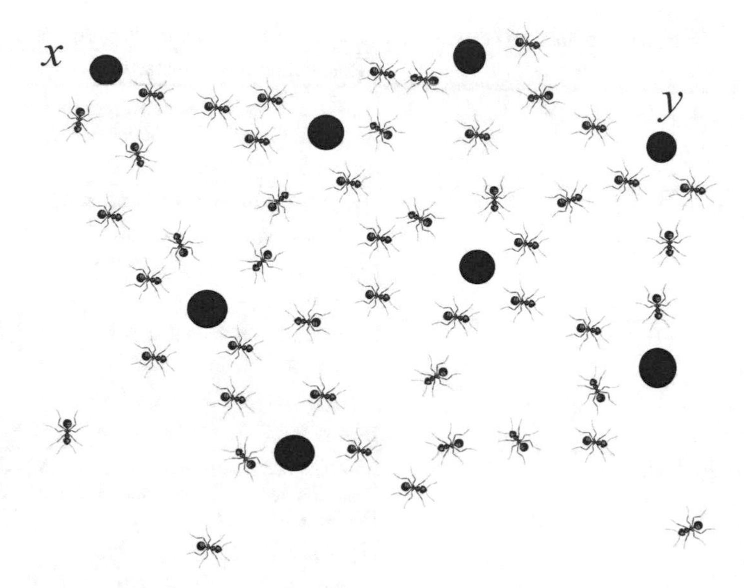

Fig. 12.11 The *qal wa-ḥomer* for the ant networking. There are several attractants denoted by black circles. We designate two attractants as x and y. If there is a road of ants between x' and y', then (i) there is a road of ants between x and y' and (ii) there is a road of ants between y and x' and (iii) there is a road of ants between x and y (see Fig. 12.5f)

Let us consider an example from mathematical analysis to show that a mathematician, extending a mathematical horizon, operates with forests and appeals to metarules for combining different trees, indeed. Great mathematicians do not prove theorems in the way like a logical automaton does it. For example, Augustin-Louis Cauchy, the French mathematician of the 19th century who became a founder of modern mathematical analysis, put forward an axiom now called Dedekind Completeness for the expansion of mathematics. Due to this axiom, some properties of real numbers (non-logical axioms describing real numbers) can have transmitted to complex numbers, vectors, and even infinite sequences by a mathematical analogy of the Talmudic *qal wa-ḥomer* rule.

Let us denote an infinite sequence (i.e. a countable set) of objects $x_1, x_2, \ldots, x_k, \ldots$ by $\{x_n\}$. Then some basic definitions of the Cauchy approach are given in Table 12.1, for more details see [4].

Among the definitions of Table 12.1 there are relationships of logical derivability. For example, the theorem proving $\mathbf{CCP}(\{\alpha_n\}; \mathbf{R}, |\cdot|)$ can be represented as the following tree:

$$t_1 = \{\mathbf{CSP}[\{\mathbf{ECP}[\{\mathbf{Ax3}[\{\{a_n\}[\{\}], \{b_n\}[\{\}], \mathbf{SES}[\{\}]\}]\}],$$

$$b_n - a_n \overset{\mathbf{R}}{\to} 0[\{\}]\}]\},$$

the same is in the graph form:

Table 12.1 Some basic definitions of mathematical analysis formulated in the 19th century

Abbreviations	Basic definitions of mathematical analysis
$\mathbf{Conv}(\{x_n\}; \mathbf{R}, \lvert\cdot\rvert)$	$x_n \overset{\mathbf{R}}{\to} x \overset{\text{def}}{=} \{x_n\} \subset \mathbf{R}$ converges to $x \in \mathbf{R}$ what is defined thus: $\{x_n\} \subset \mathbf{R} \wedge \exists(x \in \mathbf{R})\forall(\varepsilon > 0)$ $\exists(N_\varepsilon)\ \forall(n \geq N_\varepsilon)[\lvert x_n - x\rvert < \varepsilon]$
$\mathbf{CC}(\{x_n\}; \mathbf{R}, \lvert\cdot\rvert)$	$\{x_n\} \subset \mathbf{R}$ satisfies the Cauchy condition iff $\{x_n\} \subset \mathbf{R} \wedge$ $\forall(\varepsilon > 0)\exists(N_\varepsilon \in \mathbf{N})\forall(m, n \in \mathbf{N} \mid n \geq N_\varepsilon \wedge$ $m \geq N_\varepsilon)[\lvert x_n - x_m\rvert < \varepsilon]$
$\mathbf{Ax3}(A, B, a, b; \mathbf{R}, \lvert\cdot\rvert)$	Dedekind completeness: $\forall(A \subseteq \mathbf{R} \mid A \neq \emptyset)$ $\forall(B \subseteq \mathbf{R} \mid B \neq \emptyset)$ $[\forall(a \in A)\forall(b \in B)[a \leqslant b] \to$ $\exists(\xi \in \mathbf{R})\forall(a \in A)\forall(b \in B)[a \leqslant \xi \leqslant b]]$
$\mathbf{SES}(\{[a_n, b_n]\}; \mathbf{R}, \lvert\cdot\rvert)$	A sequence $[a_n, b_n]$ of closed intervals of $\mathbf{R}$ is called a nested closed intervals system iff $\forall(i \in \mathbf{N})[a_i \in \mathbf{R} \wedge$ $b_i \in \mathbf{R} \wedge [a_i, b_i] \supset [a_{i+1}, b_{i+1}]]$
$\mathbf{ECP}(\{[a_n, b_n]\}; \mathbf{R}, \lvert\cdot\rvert)$	$\forall(\{[a_n, b_n]\} \subset 2^{\mathbf{R}})[\mathbf{SES}(\{[a_n, b_n]\}; \mathbf{R}, \lvert\cdot\rvert) \to$ $\exists(\xi \in \mathbf{R})\forall(i \in \mathbf{N})[\xi \in [a_i, b_i]]]$
$\mathbf{SCP}(\{[a_n, b_n]\}; \mathbf{R}, \lvert\cdot\rvert)$	$\mathbf{ECP}(\{[a_n, b_n]\}; \mathbf{R}, \lvert\cdot\rvert) \wedge b_n - a_n \overset{\mathbf{R}}{\to} 0 \to$ $\exists!(\xi \in \mathbf{R})\forall(i \in \mathbf{N})[\xi \in [a_i, b_i]]]$
$\mathbf{CCP}(\{\alpha_n\}; \mathbf{R}, \lvert\cdot\rvert)$ where $\alpha_n = [a_n, b_n]$ is a closed interval	Cauchy-Cantor's intersection theorem: $\mathbf{ECP}(\{[a_n, b_n]\}; \mathbf{R}, \lvert\cdot\rvert) \wedge$ $\mathbf{SCP}(\{[a_n, b_n]\}; \mathbf{R}, \lvert\cdot\rvert)$
$\mathbf{B}(\{x_n\}; \mathbf{R}, \lvert\cdot\rvert)$	A sequence $\{x_n\}$ is called bounded iff $\exists(C)\forall(n \in \mathbf{N})[\lvert x_n\rvert \leq C]$
$\mathbf{BWT}(\{x_n\}; \mathbf{R}, \lvert\cdot\rvert)$	Bolzano-Weierstrass' theorem: Each bounded sequence $\{x_n\} \subseteq \mathbf{R}$ has a convergent subsequence $\{x_{n_k}\} \subseteq \mathbf{R}$, $\exists(x \in \mathbf{R})[x_{n_k} \overset{\mathbf{R}}{\to} x]$
$\mathbf{Tr}(x, y; \mathbf{R}, \lvert\cdot\rvert)$	Triangle inequality: $\forall(x, y \in \mathbf{R})[\lvert x + y\rvert \leq \lvert x\rvert + \lvert y\rvert]$
$\mathbf{CCrit}(\{x_n\}; \mathbf{R}, \lvert\cdot\rvert)$	Cauchy criterion: $(\{x_n\} \subset \mathbf{R} \wedge x_n \overset{\mathbf{R}}{\to} x) \leftrightarrow$ $\mathbf{CC}(\{x_n\}; \mathbf{R}, \lvert\cdot\rvert)$

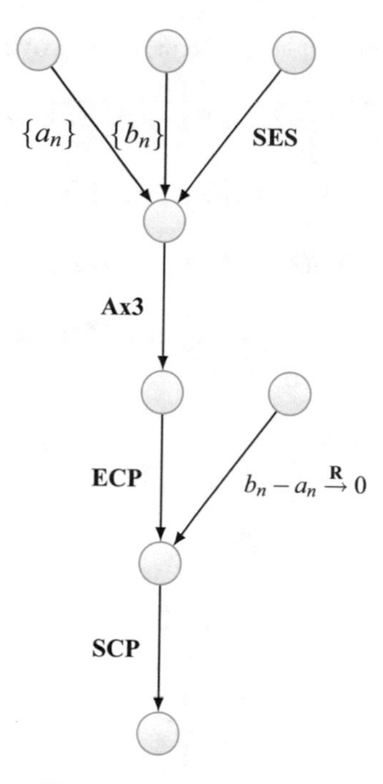

The proof tree for theorem $\mathbf{BWT}(\{x_n\}; \mathbf{R}, |\cdot|)$ is built up as follows:

$$t_2 = \{\mathbf{SES}[\{\{x_n\} \text{ is bounded } [\{\}]\}],$$

$$\exists!(\xi \in \mathbf{R})\forall(m \in \mathbf{N})[\xi \in [a_m, b_m]][\{\mathbf{CCP}[t_1]\}]\},$$

which is pictured in the following manner:

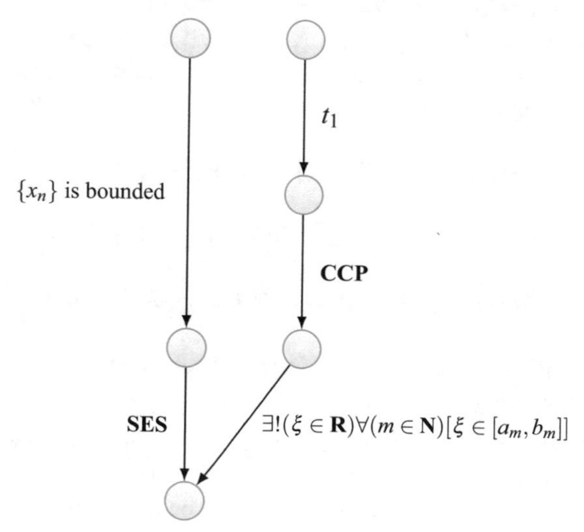

Let us define now the proof tree for the theorem proving $\mathbf{CCrit}(\{x_n\}; \mathbf{R}, |\cdot|)$:

$$t_3 = \{\mathbf{CC}[\{\forall(\varepsilon > 0)\exists(N_\varepsilon \in \mathbf{N})\forall(n, m > N_\varepsilon)[|x_n - x| < \varepsilon \wedge$$

$$|x_m - x| < \varepsilon][\{\mathbf{Conv}[\{\}]\}], \mathbf{Tr}[\{\}]\}],$$

$$\forall(\varepsilon > 0)\exists(N_\varepsilon \in \mathbf{N})\forall(m, k \in \mathbf{N}|n \geq N_\varepsilon \wedge$$

$$m \geq N_\varepsilon)[|x_{n_k} - x_m| < \varepsilon][\{\mathbf{B}[\{\mathbf{CC}[\{\}]\}]\mathbf{BWT}[t_2]\}]\},$$

with the following graph:

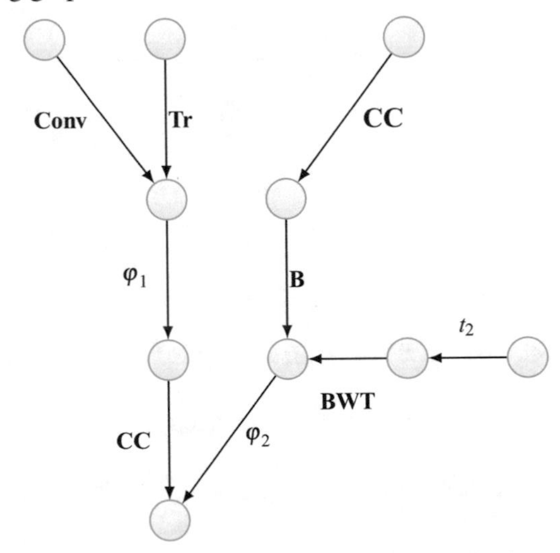

where $\varphi_1 = \forall(\varepsilon > 0)\exists(N_\varepsilon \in \mathbf{N})\forall(n, m > N_\varepsilon)[|x_n - x| < \varepsilon \wedge |x_m - x| < \varepsilon]$; $\varphi_2 = \forall(\varepsilon > 0)\exists(N_\varepsilon \in \mathbf{N})\forall(m, k \in \mathbf{N}|n \geq N_\varepsilon \wedge m \geq N_\varepsilon)[|x_{n_k} - x_m| < \varepsilon]$.

Let us define an ordering relation among non-logical axioms, $\prec$, as follows: $A \prec B$ iff an axiom A is contained in the proof tree of B. Hence, for t_3 we obtain:

$$[\mathbf{Ax3}(A, B, a, b; \mathbf{R}, |\cdot|) \prec \mathbf{CCP}(\{\alpha_n\}; \mathbf{R}, |\cdot|) \prec$$

$$\mathbf{BWT}(\{x_n\}; \mathbf{R}, |\cdot|) \prec \mathbf{CCrit}(\{x_n\}; \mathbf{R}, |\cdot|)] \wedge$$

$$[\mathbf{Tr}(x, y; \mathbf{R}, |\cdot|) \prec \mathbf{CCrit}(\{x_n\}; \mathbf{R}, |\cdot|)].$$

Theorem t_3 holds for real numbers. Nevertheless, this theorem can be extended to some other numbers by a *qal wa-ḥomer*:

1. A sequence of complex numbers $\{x_n + iy_n\} \subset \mathbf{C}$ can be defined in the way of $\mathbf{CCrit}(\{x_n + iy_n\}; \mathbf{C}, |\cdot|_{\mathbf{C}})$ where $|x + iy|_{\mathbf{C}} = \sqrt{x^2 + y^2}$, i.e. it is sufficient for us to replace all occurrences of $\mathbf{R}$ to $\mathbf{C}$ and all occurrences of the metrics $|\cdot|$ to the metrics $|\cdot|_{\mathbf{C}}$ in the proof statements of the theorem.

2. Also, sequences of vectors from $\mathbf{R}^s$ can be formulated as $\mathbf{CCrit}(\{x_n\}; \mathbf{R}^s, \|\cdot\|_{\mathbf{R}^s})$ where

$$\|(v^1, \ldots, v^s)\|_{\mathbf{R}^s} = \left(\sum_{i=1}^{s}(v^i)^2\right)^{1/2}.$$

So, we can generalize theorem t_3 (the Cauchy criterion) due to the fact that there is an analogue for the Dedekind completeness axiom (**Ax3**) (defined in real numbers) that holds for aforesaid types of numbers which can be represented as some tuples of real numbers. In this case some new inequalities $a_i \leqslant \xi_i \leqslant b_i$, $i = 1, \ldots, n$ for $a = (a_1, \ldots, a_n)$, $b = (b_1, \ldots, b_n)$, $\xi = (\xi_1 \ldots, \xi_n)$ take place instead of the one inequality of **Ax3**.

We can formulate the same generalization by using Definition 12.2. Assume that we have some axioms $(a_1^{\mathbf{R}}, \ldots, a_N^{\mathbf{R}})$ for real numbers, some axioms $(a_1^{\mathbf{C}}, \ldots, a_M^{\mathbf{C}})$ for complex numbers, and some axioms $(a_1^{\mathbf{V}}, \ldots, a_L^{\mathbf{V}})$ for vectors. We know that $N > M$ (respectively, $N > L$), because in the beginning, **R** was studied better than **C** (respectively, better than **V**). Let us take a subset of axioms $(a_1^{\mathbf{R}}, \ldots, a_n^{\mathbf{R}})$ for real numbers, which express the real metrics. So, $a_n^{\mathbf{R}}$ is **Ax3**. Now let us define a partial ordering relation $\precapprox$ on $(a_1^{\mathbf{R}}, \ldots, a_n^{\mathbf{R}})$. Let $\precapprox$ be a standard implication. For instance, $\mathbf{Conv}(\{x_n\}; \mathbf{R}, |\cdot|) \rightarrow \mathbf{Ax3}$ means that $\mathbf{Conv}(\{x_n\}; \mathbf{R}, |\cdot|) \precapprox \mathbf{Ax3}$. Suppose that $\precapprox$ is extended to an ordering relation on the set of axioms $(a_1^{\mathbf{C}}, \ldots, a_m^{\mathbf{C}})$, where $m < n$, from which we are going to infer the complex metrics. By Definition 12.2 we can add new axioms for **C**, for example: $a_n^{\mathbf{C}} = [[\mathbf{Conv}(\{x_n\}; \mathbf{R}, |\cdot|) \wedge [\mathbf{CC}(\{x_n\}; \mathbf{R}, |\cdot|)] \rightarrow \mathbf{Ax3}] \rightarrow [[\mathbf{Conv}(\{x_n\}; \mathbf{C}, |\cdot|_{\mathbf{C}}) \wedge \mathbf{CC}(\{x_n\}; \mathbf{C}, |\cdot|_{\mathbf{C}})] \rightarrow \mathbf{Ax3}]$. We have just added this axiom by *qal wa-ḥomer* (see Definition 12.2).

As we see, the new axioms in **C** and **V** are obtained due to the extrapolation from the axioms of **R** to move the proof forest $\langle t_1, t_2, t_3\rangle$ from real numbers to complex numbers and vectors. It is an effect of lateral activation in math that can be analyzed within the Talmudic or analogue foundations of mathematics.

By the end of the 19th century, the mathematical objects for which applying **Ax3** seems quite natural have been exhausted. The worlds of real numbers, complex numbers, vectors, and even infinite sequences have already been studied enough and mathematicians wanted something more, especially because there were problems of variation calculus that works with maps of maps. It has been assumed for a long period that the spaces of maps usually are infinite and so different from the well-studied finite-dimensional spaces in their basic properties. Further, in the minds of founders of the functional analysis there were born metric spaces X, where you can select a distance (metric) $\rho : X^2 \rightarrow \mathbf{R}$ among elements, but it is impossible to determine a module (norm) of elements in any reasonable way.

Immediately, a question arose: what properties should X and ρ have in order to satisfy

$$\mathbf{CCrit}(\{x_n\}; X, \rho) = \text{true?}$$

It is clear that in this case we have to replace all expressions of the form $|x - y|$ by $\rho(x, y)$ at all occurrences in

$$\mathbf{CCrit}(\{x_n\}; \mathbf{R}, |\cdot|)$$

Obviously, the field of real numbers is a special case of metric space with the following metric: $\rho_{\mathbf{R}}(x, y) = |x - y|$. At the same time, the Cauchy criterion holds not in every metric space, for example, it does not hold in the space of continuous functions $C[0, 1]$ with the metric $\rho_{L_1}(x, y) = \int_0^1 |x(t) - y(t)|dt$. Cauchy himself, of course, knew nothing about the metric spaces and could hardly suspect that there will be the following definition:

Definition 12.4 The metric space (X, ρ) where

$$\forall(\{x_n\} \subset X)[\mathbf{CCrit}(\{x_n\}; X, \rho) \leftrightarrow \exists(x \in X)[x_n \stackrel{X}{\to} x]],$$

is called complete.

The idea was to define the property $\mathbf{Tr}(x, y; X, \rho)$, natural for $(\mathbf{R}, |\cdot|)$, on the elements of (X, ρ), by transforming this property from the derived theorem to the preexisted axiom, also by transforming $\mathbf{CCrit}(\{x_n\}; X, \rho)$ from the theorem to the axiom and by expecting that the final object will behave like numbers with almost the same structural theorems. Surprisingly, it became true! For example, for **R** there exists the compactness of the bounded and closed set and for the complete metric space of functions, continuous on $[0, 1]$, with the norm $\|x\|_C = \max_{t\in[0,1]} |x(t)|$, also there exists the compactness of the bounded set (with an additional condition of equicontinuity)—it is the claim of the Arzelà-Ascoli theorem, denoted by **AAT**, the analogue of **BWT**. The Cauchy-Cantor intersection theorem, i.e. **CCP**, gets its counterpart in the form of the topological Cauchy-Cantor intersection theorem, denoted by **TCCP**, for the closed non-empty nested subsets of X.

Hence, as Alexander V. Kuznetsov noticed, the mathematicians of the end of the 19th century proposed to replace the axiom **Ax3** by the theorem $\mathbf{CCrit}(\{x_n\}; X, \rho)$ in the proof trees and, as a result, they changed all proof sequences invented by the mathematicians of the early 19th century. They did it to extend the mathematical limits in building new proof trees. For instance, if we define an ordering relation among non-logical axioms, $\prec$, as follows: $A \prec B$ iff an axiom A is contained in the proof tree of B, then the reasoning proposed by the mathematicians of the late 19th century is as follows:

$$[\mathbf{CCrit}(\{x_n\}; X, \rho) \prec \mathbf{AAT}(\{x_n\}; X, \rho)] \wedge$$

$$[\mathbf{Tr}(x, y; X, \rho) \prec \mathbf{AAT}(\{x_n\}; X, \rho)];$$

$$[\mathbf{CCrit}(\{x_n\}; X, \rho) \prec \mathbf{TCCP}(X, \rho)] \wedge$$

$$[\mathbf{Tr}(x, y; X, \rho) \prec \mathbf{TCCP}(X, \rho)].$$

To sum up, it is important to point out that if the mathematicians followed the standard foundations of mathematics in fact, they would not change the proof sequence.

But they did it indeed, because it is significant for drawing new trees by inference rules defined in Definition 12.2. So, the new theorems **AAT**$(\{x_n\}; X, \rho)$ and **TCCP**(X, ρ) allow the mathematicians to apply Definition 12.2 in their reasoning more often and in more cases than it can be done by using **Ax3**.

Hence, the axiom **CCrit**$(\{x_n\}; X, \rho)$ was taken by the mathematicians instead of **Ax3** to move more proof forests from metric spaces to new mathematical constructions. This movement of proof forests is out of discrete foundations of mathematics (out of the lateral inhibition in proof cognitions). It gives a lateral activation effect in math.

12.2.4 Metric Space for Proof Forests

Let us start with defining a dissimilarity metric for proof forests on the basis of generalization of the graph edit distance (see, e.g. [20]). This metric allows us to compare weighted graphs with vertices and weights in arbitrary metric spaces. Let $\langle V, \rho\rangle$, $\langle W, \xi\rangle$ be different metric spaces. We define $\mathscr{G}$ as the set of forests $\Gamma = \langle V', E', W', \varphi, \eta\rangle$ with weighted edges presenting deduction steps, $V' \subseteq V$, $W' \subseteq W$ are finite sets, $\varphi : (V')^2 \to \{\text{True}, \text{False}\}$ is an incidence function, $\eta : (V')^2 \to W'$ is the weight function, and

$$E' = \{(v_1, v_2) | v_1, v_2 \in V' \wedge v_1 \varphi v_2\}.$$

Let $\mathscr{G}^*$ be a set of bijections $\chi : V \to V$ and $\mathscr{G}^*_{adm} \subseteq \mathscr{G}^*$.

Let us choose $\chi \in \mathscr{G}^*_{adm}$ and $\Gamma_i \in \mathscr{G}$, $i = \overline{0, 1}$. The map χ can be an isomorphism between some subforests of Γ_0 and Γ_1, preserving their orders. Let $\Gamma_i' \subset \Gamma_i$ are the maximal subforests that $\chi : \Gamma_0' \to \Gamma_1'$ is the graph's isomorphism, $\Gamma_i' = \langle V_i'', E_i'', W_i'' \varphi_i, \eta_i\rangle$. The "poorness" of the map χ as an isomorphism of Γ_i can be formally expressed as

$$q(\chi; \Gamma_0, \Gamma_1) = F(b(\chi; \Gamma_0', \Gamma_1'), d_0(\chi; \Gamma_0', \Gamma_1'), d_1(\chi; \Gamma_0', \Gamma_1')),$$

where F is the non-negative, monotonically increasing in each variable function, $F(x, y, z) = 0$ iff $x = y = z = 0$, $d_i = |V_i' \setminus V_i''|$, b is the dissimilarity between Γ_0' and Γ_1'. Let also β be the non-negative, monotonically increasing in each variable function, $W(x, y) = 0$ iff $x = y = 0$, and

$$\mathbf{v} = \{\rho(v, \chi(v)) \colon v \in V_0''\},$$

$$\mathbf{e} = \{\xi(\eta(v_1, v_2), \eta(\chi(v_1), \chi(v_2))) \colon v_1, v_2 \in V_0'', v_1 \varphi_0 v_2\},$$

$$\mathbf{v} = \{0\}, V_0'' = \emptyset, \quad \mathbf{e} = \{0\}, E_0'' = \emptyset.$$

In this case, we can define

$$b(\chi; \Gamma_0', \Gamma_1') = \beta(\mathbf{v}, \mathbf{e}).$$

The following definition was proposed by Alexander V. Kuznetsov:

Definition 12.5 Define the dissimilarity distance between two proof forests Γ_0 and Γ_1 as follows:

$$diss(\Gamma_0, \Gamma_1) = \min_{\chi \in \mathcal{G}^*_{adm}} q(\chi; \Gamma_0, \Gamma_1).$$

The set of forests $\mathcal{G}$ can be ordered by a partial order relation in different ways. At first, if Γ_0 is a subforest of Γ_1, while $\Gamma_0, \Gamma_1 \in \mathcal{G}$, we can put that $\Gamma_0 \leq \Gamma_1$.

Let us define a new operation on forests $\Gamma_0, \Gamma_1 \in \mathcal{G}$: $\Gamma_1 \backslash \Gamma_0$ for $\Gamma_0 \leq \Gamma_1$. Let 1_n be $\bigcup_{n=0}^{\infty} \{\Gamma_{n+1} \colon \Gamma_n \leq \Gamma_{n+1}, n \in \mathbf{N}\}$. Now, we can define the Boolean operations on forests $\Gamma_i, \Gamma_j \in \mathcal{G}$:

$$\neg_n \Gamma_i := 1_n \backslash \Gamma_i \quad \text{(negation)}.$$

Notice that we have the following two cases: (i) $\Gamma_i \subseteq 1_n$, then $1_n \backslash \Gamma_i$ is not trivial, and (ii) $\Gamma_i \nsubseteq 1_n$, then we assume, $1_n \backslash \Gamma_i$ is trivial and gives 1_n.

$$\Gamma_i \wedge_n \Gamma_j := \Gamma_i \backslash (1_n \backslash \Gamma_j) \quad \text{(conjunction)}.$$

There are four cases: (i) $\Gamma_i, \Gamma_j \subseteq 1_n$, then $\Gamma_i \wedge_n \Gamma_j$ gives a new forest consisting of joint proof trees of Γ_i and Γ_j; (ii) $\Gamma_i \subseteq 1_n$ and $\Gamma_j \nsubseteq 1_n$, then $\Gamma_i \wedge \Gamma_j = \emptyset$; (iii) $\Gamma_i \nsubseteq 1_n$ and $\Gamma_j \subseteq 1_n$, then we assume $\Gamma_i \wedge_n \Gamma_j$ gives $\emptyset$; (iv) $\Gamma_i \nsubseteq 1_n$ and $\Gamma_j \nsubseteq 1_n$, then we assume $\Gamma_i \wedge_n \Gamma_j$ gives $\emptyset$.

$$\Gamma_i \vee_n \Gamma_j := 1_n \backslash ((1_n \backslash \Gamma_i) \backslash \Gamma_j) \quad \text{(disjunction)}.$$

There are four cases: (i) $\Gamma_i, \Gamma_j \subseteq 1_n$, then $\Gamma_i \vee_n \Gamma_j$ gives a new forest consisting of all proof trees of Γ_i and Γ_j; (ii) $\Gamma_i \subseteq 1_n$ and $\Gamma_j \nsubseteq 1_n$, then we assume $\Gamma_i \vee_n \Gamma_j = 1_n$; (iii) $\Gamma_i \nsubseteq 1_n$ and $\Gamma_j \subseteq 1_n$, then we assume $\Gamma_i \vee_n \Gamma_j$ gives Γ_j; (iv) $\Gamma_i \nsubseteq 1_n$ and $\Gamma_j \nsubseteq 1_n$, then we assume $\Gamma_i \vee_n \Gamma_j = \emptyset$.

$$\Gamma_i \supset_n \Gamma_j := 1_n \backslash (\Gamma_i \backslash \Gamma_j) \quad \text{(implication)}.$$

There are four cases: (i) $\Gamma_i, \Gamma_j \subseteq 1_n$, then $\Gamma_i \supset_n \Gamma_j$ gives a new forest consisting of all proof trees of $1_n \backslash \Gamma_i$ and Γ_j; (ii) $\Gamma_i \subseteq 1_n$ and $\Gamma_j \nsubseteq 1_n$, then we assume $\Gamma_i \supset_n \Gamma_j = 1_n$; (iii) $\Gamma_i \nsubseteq 1_n$ and $\Gamma_j \subseteq 1_n$, then we assume $\Gamma_i \supset_n \Gamma_j$ gives 1_n; (iv) $\Gamma_i \nsubseteq 1_n$ and $\Gamma_j \nsubseteq 1_n$, then we assume $\Gamma_i \supset_n \Gamma_j = 1_n$.

Definition 12.6 Let 1_n denote a n-th maximal forest of $\mathcal{G}$. Let us take maximal forests $1_i, \ldots, 1_j$, for them we can build the system $\langle \mathcal{G}, 1_i, \ldots, 1_j, \neg_i, \ldots, \neg_j, \wedge_i, \ldots, \wedge_j, \vee_i, \ldots, \vee_j, \supset_i, \ldots, \supset_j \rangle$ that is called a difference analogue logic on proof forests.

This system gives a kind of lattice, but it cannot allow us to change proof forests how it was possible due to *qal wa-ḥomer*. Let us introduce a new partial order relation $\preccurlyeq$:

Definition 12.7 We will say that $\Gamma_1 \preccurlyeq \Gamma_0$, if there exists an isomorphism $\chi \in \mathcal{G}^*_{adm}$ such that $\chi(\Gamma_1) \leq \Gamma_0$.

If we provide $\mathcal{G}$ or some of its subsets with the selected forest Γ_0, then we can improve the order relation as follows.

Definition 12.8 If $\Gamma_1 \preccurlyeq \Gamma_0$ and $\Gamma_2 \preccurlyeq \Gamma_0$, $\Gamma_i \in \mathcal{G}(\Gamma_0) \subseteq \mathcal{G}$, $i = \overline{1,2}$, there *does not* exist $\chi_{12} \in \mathcal{G}^*_{adm}$ such that $\chi_{12}(\Gamma_1) \leq \Gamma_2 \vee \chi_{12}(\Gamma_2) \leq \Gamma_1$, and

$$diss(\Gamma_1, \Gamma_0) \leq diss(\Gamma_2, \Gamma_0),$$

then define $\Gamma_2 \preccurlyeq \Gamma_1$.

Now, let us introduce logical operations on forests $\Gamma_i, \Gamma_j \in \mathcal{G}$:

$$\neg\Gamma_i \quad \text{(negation)};$$

$$\Gamma_i \bigvee \Gamma_j \quad \text{(disjunction)};$$

$$\Gamma_i \bigwedge \Gamma_j \quad \text{(conjunction)};$$

$$\Gamma_i \Rightarrow \Gamma_j \quad \text{(implication)}.$$

We can represent $\mathcal{G}_0 \subseteq \mathcal{G}$ as $\mathcal{G}_0 = \bigcup_i \mathcal{G}(\Gamma_i)$, where $\mathcal{G}(\Gamma_i) \subseteq \mathcal{G}$ is the set of all forests Γ such that there exists $\chi \in \mathcal{G}^*_{adm}$, $\chi(\Gamma) \leq \Gamma_i$ or $\Gamma_i \leq \chi(\Gamma)$. All sets $\mathcal{G}(\Gamma_i)$ may be ordered in the aforementioned way except $\mathcal{G}(\Gamma_i) \cap \mathcal{G}(\Gamma_j)$. Then

Definition 12.9 We will say that $\mathcal{G}' = \{\Gamma_i \in \mathcal{G} : i = \overline{1,n}\}$ is a model for $\Gamma \in \mathcal{G}$ and write $\mathcal{G}' \models \Gamma$, if

$$\exists(\chi_i \in \mathcal{G}^*_{adm}) \bigcup_{i=1}^{n} \chi_i(\Gamma_i) = \Gamma' \preccurlyeq \Gamma.$$

Let us define $\neg, \bigvee, \bigwedge, \Rightarrow$ on models:

$$\mathcal{G}' \models \neg\Gamma_i \text{ if and only if } \mathcal{G}' \not\models \Gamma_i.$$

$$\mathcal{G}' \models \Gamma_i \bigvee \Gamma_j \text{ if and only if } \mathcal{G}' \models \Gamma_i \text{ or } \mathcal{G}' \models \Gamma_j.$$

$$\mathcal{G}' \models \Gamma_i \bigwedge \Gamma_j \text{ if and only if } \mathcal{G}' \models \Gamma_i \text{ and } \mathcal{G}' \models \Gamma_j.$$

$$\mathcal{G}' \models \Gamma_i \Rightarrow \Gamma_j \text{ if and only if } \mathcal{G}' \not\models \Gamma_i \text{ or } \mathcal{G}' \models \Gamma_j.$$

In this way we obtain a new analogue logic on proof forests:

Definition 12.10 The system $\langle \mathcal{G}, \neg, \bigwedge, \bigvee, \Rightarrow \rangle$ is called a *diss* analogue logic on proof forests.

This logic allows us to make dissimilation of forests shorter due to logical conclusions. Hence, we suppose that a process of constructing a new theory can be modeled with this logic as follows:

0. Let $\mathcal{R} : \mathcal{G} \to \mathbf{R}$ is the graph's fitness functional. If $\Gamma \in \mathcal{G}, \mathcal{R}(\Gamma) > 0$ then we will think that forest Γ fits.
1. We choose the base theories set $\mathcal{G}'$.
2. We generate forests Γ'' which are homotopical equivalent to all $\Gamma' \in \mathcal{G}'$ until

$$\mathcal{R}(\Gamma'') > 0 \wedge \forall(\Gamma' \in \mathcal{G}') \quad \Gamma' \preccurlyeq \Gamma''.$$

3. We add Γ'' to $\mathcal{G}'$ and go to the step 1.

In practice, it seems that the aforesaid construction process can be represented as a problem of an integer programming in most cases and it is possible to use existing solvers like SCIP [21] for the theories generation.

Thus, Definitions 12.6 and 12.10 introduce for us the analogue logics that can formalize the math reasoning performed under the condition of lateral activation in the proof observation.

We have just tried to show that the mathematicians can deal not only with a logical way of automatic proving from some axioms (the lateral inhibition in math), but also with combining proof trees on tree forests by using the analogies as inference metarules (the lateral activation in math). For the first time, such metarules were proposed in the Talmud within a general Judaic approach to concurrent or even massive-parallel conclusions, see [1, 2, 4]. The mathematician can think not only sequentially like a logical automaton, but concurrently, also.

In the logical foundations of mathematics there are two approaches in drawing computer-assisted proofs: (i) automated theorem proving (i.e. proving mathematical theorems by computer programs) and (ii) automated proof checking (i.e. using computer programs for checking proofs for correctness). There are many objections for these approaches. For instance, for (i) one the main objections is that these methods do not give new and useful concepts in mathematics in fact, but they present just a long gloomy calculation. For (ii) one of the main objections is that these methods can check just very simple theorems. There are no even insights how to check **FLT** by computer programs.

In our opinion, the most significant problem of existing logical foundations of mathematics is that a mathematical proof is considered a discrete process that can be formalized by discrete methods, i.e. the proof is regarded only under the condition of lateral inhibition. However, it is only a hypothesis that mathematics can be reduced to logic and the mathematical thinking is discrete. We can assume that it is not only so and a mathematical proof can be an operation in a metric space of proof trees with some topological properties, i.e. we can start to analyze the lateral activation in proof cognitions. As a result, the mathematical proof can be formalized as well by analogue computations, not only discrete ones. The meaning of mathematical proofs in the analogue logic we have just introduced is to transform some forests with a bigger dissimilarity to forests with a smaller dissimilarity. For example, in the Talmud this transformation means that each branch in proof trees with one root should have the same number of subbranches and inferring allows us to construct additional subbranches to make the trees more symmetric. In mathematics the goal of proofs is quite similar and it is to extend the mathematical limits to make proof trees in forests more symmetric, too.

Hence, we suppose that computer-assisted proofs can be based on some analogue computations involving topological properties of proof trees. This idea was inspirited by swarm intelligence.

References

1. Schumann, A.: Preface. Hist. Philos. Log. **32**(1), 1–8 (2011)
2. Schumann, A.: Qal wa-homer and theory of massive-parallel proofs. Hist. Philos. Log. **32**(1), 71–83 (2011)
3. Schumann, A.: Physarum polycephalum syllogistic L-systems and Judaic roots of unconventional computing. Stud. Log. Gramm. Rhetor. **44**(1), 181–201 (2016). https://doi.org/10.1515/slgr-2016-0011
4. Schumann, A., Kuznetsov, A.V.: Talmudic foundations of mathematics. In: 10th EAI International Conference on Bio-inspired Information and Communications Technologies (formerly BIONETICS), pp. 67–74 (2017)
5. Schumann, A., Kuznetsov, A.V.: Foundations of mathematics under neuroscience conditions of lateral inhibition and lateral activation. Int. J. Parallel Emerg. Distrib. Syst. **33**(3), 237–256 (2018)
6. Univalent Foundations Program, T.: Homotopy Type Theory: Univalent Foundations of Mathematics, Institute for Advanced Study (2013). https://homotopytypetheory.org/book
7. Chabris, C., Crown, D.S.: The Invisible Gorilla and Other Ways Our Intuitions Deceive Us. New York (2010)
8. Drew, T., Vo, M.L.H., Wolfe, J.M.: The invisible gorilla strikes again: Sustained inattentional blindness in expert observers. Psychol. Sci. **24**(9), 1848–1853 (2013)
9. Carbon, C.C.: Understanding human perception by human-made illusions. Front. Human Neurosci. **8**(566) (2014). https://doi.org/10.3389/fnhum.2014.00566
10. Wiles, A.: Modular elliptic curves and Fermat's last theorem. Ann. Math. **141**, 443–551 (1995)
11. Paris, J., Harrington, L.A.: Mathematical incompleteness in Peano arithmetic. In: Barwise, J. (ed.) Handbook for Mathematical Logic. North-Holland, Amsterdam, Netherlands (1977)
12. McLarty, C.: What does it take to prove Fermat's last theorem? Grothendieck and the logic of number theory. Bullet. Symbol. Log. **16**(3), 359–377 (2010)
13. Artin, M., Grothendieck, A., l. Verdier, J.: Theorie des topos et cohomologie et ale des schemas, Seminaire de Geometrie Algebrique du Bois-Marie, vol. 4. Springer (1972)
14. Ury, Y.: Charting the Sea of Talmud: A Visual Method for Understanding the Talmud (2012)
15. Schumann, A., Adamatzky, A.: Physarum polycephalum diagrams for syllogistic systems. J. Log. Appl. **2**(1), 35–68 (2015)
16. Adamatzky, A., Sirakoulis, G.C., Martínez, G.J., Baluska, F., Mancuso, S.: On plant roots logical gates. Biosystems **156**, 40–45 (2017)
17. Lindenmayer, A.: Mathematical models for cellular interaction in development. Parts i and ii. J. Theor. Biol. **18**(280–299), 300–315 (1968)

18. Niklas, K.: Computer simulated plant evolution. Sci. Am. (1985)
19. Prusinkiewicz, P., Lindenmayer, A.: The Algorithmic Beauty of Plants. Springer (1990)
20. Gao, X., Xiao, B., Tao, D., Li, X.: A survey of graph edit distance. Patt. Anal. Appl. **13**, 113–129 (2010)
21. Maher, S.J., Fischer, T., Gally, T., Gamrath, G., Gleixner, A., Gottwald, R.L., Hendel, G., Koch, T., Lübbecke, M.E., Miltenberger, M., Müller, B., Pfetsch, M.E., Puchert, C., Rehfeldt, D., Schenker, S., Schwarz, R., Serrano, F., Shinano, Y., Weninger, D., Witt, J.T., Witzig, J.: The SCIP optimization suite 4.0. Technical Report 17–12, ZIB, Takustr. 7, 14195, Berlin (2017)

Chapter 13
Conclusion and Future Work

13.1 Main Results

In accordance with behaviourism, any animal behaviour based on unconditioned and conditioning reflexes can be controlled or even managed by stimuli in the environment: attractants (motivational reinforcement) and repellents (motivational punishment). In the meanwhile, there are the following two main stages in reactions to stimuli: *sensing* (perceiving signals) and *motoring* (appropriate direct reactions to signals). In this book, the strict limits of behaviourism have been studied from the point of view of symbolic logic and algebraic mathematics: How far can animal behaviours be controlled by the topology of stimuli? In other words, how far can we design unconventional computers on the basis of animal reactions to stimuli?

On the one hand, we can try to design reversible logic gates in which the number of inputs is the same as the number of outputs. In our case, the behaviouristic stimuli for swarms are inputs and appropriate reactions of swarms at the motoring stage are outputs. It means that behaviourism can hold true, indeed. Nevertheless, on the other hand, at the sensing stage the same signal can be perceived so differently. The problem is that at the sensing stage even each unicellular organism can be regarded as a logic gate in which the number of outputs (means of perceiving signals) greatly exceeds the number of inputs (signals). It is connected to actin filament networks and other subcellular protein mechanisms perceiving the stimuli to react to them in various ways in accordance with the needs and tasks of the cell at the current moment. As a consequence, even one cell can resist outside influences. Hence, we face some strict biological limits in applying behaviourism. From the standpoint of symbolic logic and algebraic mathematics, this means that we cannot examine animal behaviours as conventional spatial algorithms, such as Kolmogorov-Uspensky machines. The mathematics of animal behaviours is much more complicated. The matter is that we should know how logical-mathematically we can design logic gates in which the number of inputs is far exceeded by the number of outputs. The universe of such incorrect mathematical "functions" is known for mathematicians well

A. Schumann, *Behaviourism in Studying Swarms: Logical Models of Sensing and Motoring*, Emergence, Complexity and Computation 33,
https://doi.org/10.1007/978-3-319-91542-5_13

and called by them *non-well-founded*. In this book, some new mathematical tools for studying the non-well-founded universe of animal behaviours were proposed: p-adic valued logic, non-Archimedean probability logic, non-Aristotelian syllogistic, context-based games, reflexive games.

The main results obtained in this way are as follows:

13.1.1 Morphological Processors

My ideas represented in this book have been in line with *morphological processors* within the project *Physarum Chip: Growing Computers from Slime Mould* [1]. These processors compute by configurations of protoplasmic networks of *Physarum polycephalum*. But the same processors can be designed on a medium of different swarms—the logic will be the same, only the chemistry will vary. In the morphological processors, data are represented as a spatial configuration of attractants and repellents. Halting of computation is detected by stable state aided by compressibility changes as proposed and experimentally demonstrated in [2]. In the project *Physarum Chip: Growing Computers from Slime Mould* the results of computation were programmed by morphology of the slime mould. The morphological processors can implement p-adic arithmetics if the swarm can see not more than $p - 1$ attractants at each step of its propagation at all places of its directions [3]. In case the swarm can be well controlled in its motions (by using repellents or by distributing attractants on a nutrient-poor substrate in the case of *Physarum polycephalum*), functions on finite p-adic integers can be implemented, i.e. we deal with standard arithmetics [4]. In case the swarm moves massive-parallelly (e.g. *Physarum polycephalum* moves on a nutrient-rich substrate or many attractants are localized near each other) functions on infinite p-adic integers can be implemented, i.e. we deal with non-Archimedean metrics [4]. The morphological processors solve a substantial range of problems from graph optimisation (e.g. shortest path, spanning trees, transport networks, space exploration, risk averse behaviour), combinatorial optimisation (Travelling Salesman Problem), and computational geometry (e.g. tessellations, triangulations, hulls) and collision-based computing (e.g. ballistic logical gates).

The morphological processors are parallel: the data space is explored by numerous growth cones simultaneously. These processors can implement concurrent and context-based games [5, 6]. So, they can be used as universal behavioural models for simulating group or swarm behaviour [4]. The time complexity of the computation is determined by a diameter of the physical data set. Thus, in the case of *Physarum polycephalum* a computation of one problem can proceed for days if a size of the data set is tens of centimetres. The processors can be scaled down only to a one element of data per square centimetre. The morphological processors are not reusable, but can be sequentially arranged for more complex applications.

In this book, I have proposed different ways of designing reversible logic gates for morphological processors using controlling stimuli such as attractants and repellents. Repellents are needed because of uncertainty in the direction of plasmodium propagation to eliminate some directions as unimportant. In this way, we can construct conventional reversible logic gates: the CNOT gate, the FREDKIN gate, the TOFFOLI gate, etc. [7, 8]. Combinations of reversible logic gates are regarded as matrix multiplications. Nevertheless, the plasmodium in its networking as well as any other swarm can permanently change its decisions and without repellents we have an extension of matrix multiplication group theory [9]. Within this extension we can design unconventional reversible logic gates, where the number of inputs and outputs is uncertain. For designing logic gates Krzysztof Pancerz and me have proposed to use Petri net models [10, 11] that can be treated as a high level description. Petri net models enable us to reflect propagation of protoplasmic veins of the plasmodium in consecutive time instants (step by step).

13.1.2 Actin Filament Networks

Morphological processors can be designed on the slime mould due to actin filament networks—networks of proteins in the plasmodium. I have proposed artificial actin filament networks where inputs are different stresses and outputs are formations and destructions of filaments, on the one hand, and as assemblies and disassemblies of actin filament networks, on the other hand. Hence, under different external conditions we observe dynamic changes in the length of actin filaments and in the outlook of filament networks [12–14]. As we see, the main difference of actin filament networks from others including neural networks is that the topology of actin filament networks changes in responses to dynamics of external stimuli.

Artificial actin filament networks employ transmitting information by means of building and rebuilding actin filament networks in responses to dynamics of intra-cellular and extra-cellular stimuli. Some new processors (filaments) can appear in one conditions and they can disappear in other conditions. This system is much more complex, than artificial neural networks, where we have a fixed number of processors (neurons) [12–14].

13.1.3 p-*Adic Valued Logic and Arithmetics*

Each swarm can be considered a natural fuzzy processor with fuzzy values on the set of p-adic integers [3, 15, 16]. The point is that in any experiment with the swarm we deal with attractants which can be placed differently to obtain different topologies and to induce different transitions of the swarm. If the set A of attractants, involved into the experiment, has the cardinality number $p - 1$, then any subset of A can be regarded as a condition for the experiment such as 'Attractants occupied by the swarm'. These conditions change during the time, $t = 0, 1, 2, \ldots$, and for the infinite time, we obtain p-adic integers as values of fuzzy (probability) measures defined on conditions (properties) of the experiment. This space is a semantics for p-adic valued fuzzy syllogistic I have constructed for describing the propagation of swarms [15, 17]. This syllogistic can be extended to a p-adic valued logic and p-adic valued arithmetics. Within this logic I have developed a context-based game theory [18]. All these logical tools can be implemented on plasmodia of *Physarum polycephalum* by conventional and unconventional reversible logic gates [8].

I have proposed to use p-adic valued fuzziness and probabilities for measuring behaviours which cannot be measured additively. Then I have constructed a natural deductive system for describing all possible experiments with swarms [19]. This system is p-adic many valued. I have considered possibilities for applying p-adic valued logic $BL\forall$ (see [20, 21]) to the task of designing the morphological processors on the slime mould. If it is assumed that at any time step t of propagation the slime mould can discover and reach not more than $p - 1$ attractants, then this behaviour can be coded in terms of p-adic numbers. As a result, this behaviour implements some p-adic valued arithmetic circuits and can verify p-adic valued logical propositions. I have offered two unconventional arithmetic circuits: adder and subtractor defined on finite p-adic integers [4]. Adder and subtractor are designed by means of spatial configurations of several attractants and repellents which are stimuli for the swarm behaviour [22]. For instance, the plasmodium could form a network of protoplasmic veins connecting attractants and original points of the plasmodium. Occupying new attractants is considered in the way of adders and leaving some attractants because of repelling is considered in the way of subtracters. On the basis of p-adic adders and subtractors we can design complex p-adic valued arithmetic circuits within a p-adic valued logic proposed by me. So, p-adic valued logic and p-adic valued arithmetic are implementable on swarms (e.g. on plasmodia). In the meanwhile, the *Physarum polycephalum* computing on nutrient-poor substrate is expressible by finite p-adic integers and the *Physarum polycephalum* computing on nutrient-rich substrate is expressible by infinite p-adic integers.

13.1.4 Process-Algebraic Formalization of Swarm Behaviour and Hybrid Actions

In my research I have been based on process-algebraic formalizations of swarm behaviour [23–25]. So, I have considered some instructions to program swarms in terms of process algebra like: add node, remove node, add edge, remove edge [26]. Adding and removing nodes can be implemented through activation and deactivation of attractants, respectively. Adding and removing edges can be implemented by means of repellents put in proper places in the space. An activated repellent can avoid a swarm transition between attractants. Adding and removing edges can change dynamically over time. To model such a behaviour, I have proposed a high-level model, based on timed-transition systems. In this model I have defined the following four basic forms of swarm transitions (motions): *direct* (direction: a movement from one point, where the swarm is located, towards another point, where there is a neighbouring attractant), *fuse* (fusion of two swarms at the point, where they meet the same attractant), *split* (splitting the swarm from one active point into two active points, where two neighbouring attractants with a similar power of intensity are located), and *repel* (repelling of swarm or inaction). In swarm motions, we can perceive some ambiguity influencing on exact anticipation of states of swarms in time. In the case of splitting, there is some uncertainty in determining next active points (attractants occupied by the swarm), if a given active point is known. This uncertainty does not occur in the case of direction, where the next active point is uniquely determined. To model ambiguity in anticipation of states of swarms, I propose to use some tools of non-well-founded mathematics and their extensions: rough set theory (together with [27]) and some extensions of concurrent calculi [18].

I have defined an extension of process algebra [18], where simple actions of labelled transition systems cannot be atomic; consequently, their compositions cannot be inductive. Their informal meaning is that in one simple action we can suppose the maximum of its modifications. Such actions are called hybrid. Then I have proposed two formal theories on hybrid actions (the hybrid actions are defined there as non-well-founded terms and non-well-founded formulas): group theory and Boolean algebra. The group theory proposed by me can be used as the new design method to construct reversible logic gates on plasmodia. In this way, we should appeal to the so-called nonlinear permutation groups. These groups contain non-well-founded objects such as infinite streams and their families. The theory of non-linear permutation groups can be used for designing reversible logic gates on any behavioural systems [9]. The simple versions of these gates are represented by logic circuits constructed on the basis of the performative syllogistic [17, 28–30]. It seems to be natural for behavioural systems and these circuits have a very high accuracy in implementing. My general motivation in designing logic circuits in behavioural systems without repellents is as follows: in this way, we can present behavioural systems as a calculation process more naturally; we can design devices, where there are much more outputs than inputs, for performing massive-parallel computations in the bio-inspired way; we can obtain unconventional (co)algorithms by programming behavioural systems. For instance, computations on protoplasmic trees are understood as a kind of extension of concurrent processes defined in concurrent games [5, 6]. This extension is called context-based processes and they are defined in the theory of context-based games proposed by me [18].

13.1.5 Computations on Trees

Computations on trees are usually represented by spatial algorithms like Kolmogorov-Uspensky machines [31]. Theoretically, Turing machines, Kolmogorov-Uspensky machines, Schönhage's storage modification machines, and random-access machines have the same expressibility power. Unfortunately, the computational power of their implementations on swarms is too low [26]. To make computations on trees more expressive I have proposed the performative syllogistic—a syllogistic system of propagation [17, 28–30]. This system can logically simulate a massive-parallel behaviour in the propagation of collectives of *Trematode larvae* (miracidia and cercariae) and other swarms. While the Aristotelian syllogistic may describe concrete directions of swarm spatial expansions, the performative syllogistic proposed by me may describe swarm simultaneous propagations in all directions. Therefore, while for the implementation of Aristotelian syllogistic we need repellents to avoid some possibilities in the swarm propagations, for the implementation of performative syllogistic we do not need them. The performative syllogistic has a p-adic valued semantics and satisfies p-adic valued probabilities. In the performative syllogistic we can build non-well-founded trees for which there cannot be logical atoms. This theory is much more expressive than standard spatial algorithms in simulating the swarm motions. The syllogistic can be extended to a more general theory of context-based games. Within this theory we can define algorithms for computing on trees. Computations on trees are treated as an extension of concurrent processes defined in concurrent games proposed by Samson Abramsky [32]. This extension is called context-based processes [18].

13.1.6 Neural Properties of Swarm Behavioural Patterns

There are two main properties of neural networks: lateral activation and lateral inhibition. The same properties are observed in swarm networks [33, 34]. I have generalized my studies of lateral inhibition effects in the swarm behaviour in the way of constructing new syllogistics and modal logics. So, I have shown that there are two main possibilities of pairwise comparisons analysis in computer science: first, pairwise comparisons within a lattice, in this case these comparisons can be measurable by numbers (this one corresponds to lateral inhibition); second, comparisons beyond any lattice, in this case these comparisons cannot be measurable in principle (this one corresponds to lateral activation of neural networks). I have shown that the first approach to pairwise comparisons analysis is based on the conventional square of opposition and its generalization, but the second approach is based on unconventional squares of opposition [34]. Furthermore, the first approach corresponds to lateral inhibition in transmission signals and the second approach corresponds to lateral activation in transmission signals. For combining lateral inhibition and lateral activation in the same behaviour I have introduced the notion of the so-called context-based games to describe rationality of swarms. In these games we assume that, first, strategies can change permanently, second, players cannot be defined as individuals performing just one action at each time step. They can perform many actions simultaneously. Under favourable conditions the swarm has lateral activation effects and under stress conditions the swarm has lateral inhibition effects. I have shown that modal logic **K** corresponds to the lateral inhibition property, but, as I have demonstrated, we can construct new modal logics, dual to **K**, for embodying the lateral activation property [33].

13.1.7 Bio-inspired Games

The universe, where each swarm lives, consists of cells possessing different topological properties according to the intensity of chemo-attractants and chemo-repellents. The intensity entails the natural or geographical neighbourhood of the set's elements in accordance with the spreading of attractants or repellents. As a result, we obtain Voronoi cells. In this structure we can implement cellular automata. Taking into account the fact that each swarm is very sensitive to the environment and can change its strategies we can extent the standard notion of cellular automata and assume that transition rules can change in an appropriate neighbourhood in accordance with some outer conditions at time step $t = 0, 1, 2, \ldots$ The swarm implements cellular automata with changing transition rules. Within these automata we can define context-based games.

I have proposed a bio-inspired game theory on plasmodia, i.e. an experimental game theory, where all basic definitions are verified in the experiments with swarms, such as *Physarum polycephalum* and *Badhamia utricularis* [5, 6, 35]. Any swarm can serve as a model for concurrent games and context-based games. In context-based games, players can move concurrently as well as in concurrent games, but the set of actions is ever infinite. In our experiments, we follow the following interpretations of basic entities: (i) attractants as payoffs; (ii) attractants occupied by the swarm as states of the game; (iii) active zones of swarm as players; (iv) logic gates for behaviours as moves (available actions) for the players; (v) a propagation of swarm as the transition table which associates, with a given set of states and a given move of the players, the set of states resulting from that move. In games of *Physarum polycephalum* and *Badhamia utricularis* we can demonstrate creativity of primitive biological substrates of plasmodia. The point is that plasmodia do not strictly follow spatial algorithms like Kolmogorov-Uspensky machines, but perform many additional actions. So, the plasmodium behaviour can be formalized within strong extensions of spatial algorithms, e.g. within concurrent games or context-based games.

13.1.8 Chemical Interface

In moving, the plasmodium switches its direction or even multiplies in accordance with different bio-signals attracting or repelling its motions, e.g. in accordance with pheromones of bacterial food, which attract the plasmodium, and high-salt concentrations, which repel it. So, the plasmodium motions can be controlled by different topologies of attractants and repellents so that the plasmodium can be considered a programmable biological device in the form of a timed-transition system, where attractants and repellents determine the set of all plasmodium transitions. Furthermore, we can define p-adic probabilities on these transitions and, using them, we can define a knowledge state of plasmodium and its game strategy in occupying attractants as payoffs for the plasmodium. We can regard the task of controlling the plasmodium motions as a game and we can design different interfaces in a game-theoretic setting for the controllers of plasmodium transitions by chemical signals [36, 37].

In the universe of 5-adic integers, I have simulated the motions of *Physarum polycephalum* and *Badhamia utricularis* by the game of Go [17, 38, 39]. I have considered two syllogistic systems implemented as Go games: the Aristotelian syllogistic and performative syllogistic. In the Aristotelian syllogistic, the locations of black and white stones are understood as locations of attractants and repellents, respectively. In the performative syllogistic, we consider the locations of black stones as locations of attractants occupied by plasmodia of *Physarum polycephalum* and the locations of white stones as locations of attractants occupied by plasmodia of *Badhamia utricularis*. The Aristotelian syllogistic version of Go game is a coalition game. The performative syllogistic version of Go game is an antagonistic game.

13.1.9 Reflexive Games

Within payoff cellular automata I have formalized *reflexive games* as an extension of bio-inspired games. These games can explain the decision-making behaviour in cases of hidden processes such as log-rolling, side-payments and package-deals. In reflexive games I should see the maximum of different paths in decision making including straight and non-straight paths. Meanwhile I should know which paths are chosen by my opponents and take them into account, but my paths should be hidden and unseen by opponents [40–43]. Therefore straight paths are always an epic failure in reflexive games. Thus, the logic of reflexive decision-making should be constructed on the greatest fixed points, i.e. on non-well-founded (coinductive) data. This logic is built on non-Archimedean lattices [41, 42]. In reflexive games, we appeal to the following game-theoretic assumptions:

- Each game can become infinite, because its rules can change.
- Players can possess different levels of reflexion: one player can know all information about another, but the second may know only false announcements from the first.
- Some utilities can have symbolic meanings. These meanings are results of players accepting symbolic values. The higher the symbolism of payoffs, the higher the level of reflexion of appropriate players. At the zero level of reflexion, the payoffs do not have symbolic meanings at all. For consensus the players are looking for joint symbolic meanings.
- Resistance points for players are reduced to the payoffs at the zero level of reflexion.
- The joint symbolic meanings can change throughout the game if a player increases his/her level of reflexion.
- For any game there is performative efficiency, where all symbolic meanings of one player are shared by other players.

We can use probability measures defined on streams (e.g. on hypernumbers and p-adic numbers) for simulating reflexive games. In particular, it can be proved that Aumann's agreement theorem does not hold for these probabilities. Instead of this theorem, there is a statement called the *reflexion disagreement theorem* [41, 42]. Based on this theorem, probabilistic and knowledge conditions can be defined for reflexive games at various reflexion levels up to the infinite level. This new statement can be proven if we change some of the standard philosophical presuppositions in game theory for the following new assumptions: each rational agent can cheat (disagree in his heart with) other rational agents, no player can know everything prior to the game, each agent can try to foresee the knowledge (beliefs) of his/her opponents and manipulate them, therefore a common knowledge does not mean that an agent will agree with or be completely predictable to all others. The reflection disagreement theorem is valid for the infinite metagame in Howard's meaning [44] and for the reflexive game of infinite level in Lefebvre's meaning [45].

An opponent's decision-making can be managed by means of suggestions to him/her of some foundations from which (s)he could logically infer decisions favourable to us. Such a process of suggesting foundations for an opponent's decision-making is called *reflexive management*. Reflexive management can be performed by means of stating false information about a state of affairs (creation of false objects), by means of suggesting an opponent's purposes (provocations and intrigues, acts of terrorism and ideological diversions), or by means of suggesting decisions (false advice). A reflexive game is probable only in a case where agents can reach *performative equilibrium*—they can act concordantly at reflexion levels $n > 0$. This condition is fulfilled in the case where there are mechanisms of intercommunication broadly agreed upon among people. These mechanisms have been preserved within an appropriate *discourse community* (*Kommunikationsgemeinschaft*), in the meaning of Karl-Otto Apel (1922–2017), shared by members with a suitable degree of discursive expertise (i.e. members possess one or more genres in the communicative furtherance of its aims and know a specific lexis) and with a degree of relevant content to provide information and feedback.

Any discourse community represents a group of people who are in permanent interactions with each other and exchange performative propositions. This community is self-organized. Due to common discourse it can reach an informational equilibrium, and also a parity of creative reasoning as well as emotional consensus in interchanging performative propositions. Members of a discourse community have a common *speech competence* (*Sprachkompetenz*), in the meaning of Apel, sufficient for interactions. Let us recall that speech competence as such is comprehended neither by members of a discourse community, nor by outside agents, but its possession is a necessary condition for entry into an appropriate discourse community. Speech competence is understood as the knowledge and ability to use language in accordance with different contexts. Thus, modeling by speech competence is a key notion for managing a discourse community. In reflexive games both conflicting parties simulate reasoning as well as emotional reactions of each other and aspire to foresee them. This reasoning and these reactions, which we expect from our interlocutor/opponent, are called perlocutionary effects of our dialogue. Let us recall that any dialogue has three levels: locutionary (information expressed in verbal or non-verbal exposition of states of affairs), illocutionary (cognitive and emotional estimations of the considered information), perlocutionary (effects of cognitive and emotional estimations of interlocutors/opponents) [46]. What will be the perlocutionary effects of our utterance, we do not know, but we try to foresee them. The situation where my cognitive estimations and feelings are not transparent for my interlocutor, while his/her cognitive estimations and feelings are transparent for me, shows that my level of reflexion is higher than my interlocutor's. Thus, reflexive management can be formally described within *illocutionary logic*, the logic studying perlocutionary effects as a part of illocutionary acts [47]. Whenever a speaker utters a sentence in an appropriate context with certain intentions, he performs one or more illocutionary acts. The simple illocutionary act is denoted by $E(\varphi)$. This denotation means that each simple illocutionary act can be regarded as consisting of an illocutionary force E and a propositional content φ. For example, the utterance 'I promise you (E) to come (φ)' has such a structure. Composite illocutionary acts are formed from simple illocutionary acts non-inductively [42].

The formal system of illocutionary logic proposed in my work is a modal logic with an infinitely large number of modal operators of the form E. This logic has non-well-founded syntax and semantics. The point is that its formulas and their meanings are presented by infinite streams. This logic is an extension of non-Archimedean many-valued fuzzy and probability logic [42]. Illocutionary logic, as it is defined in my work, deals with self-referent notions such as reflexion, mental accounting, cognitive biases and it focuses on non-equilibrium states. This logic puts emphasis on social relations among people, where valuations and illocutionary forces are more important in human interactions than facts. Due to the digitization of our life-world (e.g. the Internet and new media), we have many more opportunities to share our interests, activities, and backgrounds. For example, online social networking services such as Facebook consisting of representations of users by their profiles extend real-life connections among people significantly. As a consequence, Russellian semantics accepting only factual propositions is losing its function in our everyday life. It is assumed in my work that subjective meanings, illocutionary forces and perlocutionary effects are more important now in perceiving any kind of information than pure facts. For example, an event can be perceived as more important and more 'real' if it gets more 'like' on Facebook. Propositions about this event will more likely be accepted as successful utterances.

Illocutionary logic as a theory of performative propositions, self-reference, and reflexing in games and decisions describes the illocutionary reality in which the focus lies exclusively on the logical structure of human activities with reflexion.

13.2 Designing a Protein Monster as Future Work

13.2.1 Main Objectives

Hence, we assume that it is possible to design the so-called *protein robot* (*biological computer*) as an artificial protein broth (metacell) that can effectively solve different logistic tasks (orientation, transition, food transporting, etc.), please compare to [48, 49]. This robot will be designed on the basis of actin filament networks [50–53]. The point is that in accordance with our preliminary results the swarm intelligence which is responsible for effective solutions of logistic tasks for different swarms from social bacteria to social insects and even mammals is based on some properties of actin filaments which are involved into bio-chemical mechanisms of attracting and repelling the organism movements. Attractants and repellents stimulate the directed movement of swarms towards and away from the stimulus, respectively. Various compounds can act as attractants and repellents and they are different for different swarms. For the social bacteria, nutrients (including sugars, such as maltose, ribose, galactose, and amino acids, such as L-aspartate and L-serine) are attractants and some chemical conditions damaging to bacteria (including the high concentration of salts or weak acids) are their repellents. For different insects, there are many different sex attractants produced by males or females as well as feeding stimulants and attractants produced by plants and there are many different repellents produced by arthopods or repellents found in plants. So, on the one hand, attractants and repellents for different swarms are different, but, on the other hand, their functions in stimulating animal behaviours are the same. In the meanwhile, we know that attractants and repellents are external stimuli for assemblies and disassemblies of actin filament waves. Hence, at the level of one cell we deal with the same attracting and repelling due to actin filaments.

Thus, if a protein robot is possible indeed, it will be built up on using *actin filament networks*. We are going to develop an *artificial actin filament networks* (AAFN) with finding computational tasks for AAFN, such as information processing, controlling processes of Big Data, navigation, vision recognition, etc. The AAFN can find a huge number of different commercial and societal applications. The matter is that AAFN is a high generalization of artificial neural networks and it can be used more effectively than the latter. We are going to represent AAFN as a new game theory that can be applied, for instance, in formalization of court processes within e-justice.

13.2.2 Previous Project and Preliminary Results

In the project *Physarum Chip: Growing Computers from Slime Mould* [1] (for short: *PhyChip project*), we have studied *Physarum polycephalum* known as the slime mould. *Physarum polycephalum* reacts to attractants by the plasmodium growing towards nutrients and building the network structure for transporting the food pieces.

The network architecture of *Physarum polycephalum* is highly dynamic, but it can be managed by localizations of nutrients [54–56]. In our project, we have built some biomorphic computing devices operated by plasmodia of *Physarum polycephalum*. In these devices we have combined the self-assembled and fault-tolerant networks of plasmodia with conventional electronic components into a hybrid chip, called the *Physarum* Chip. This chip can solve the following tasks: (i) plane tessellation, concave and convex hull in computational geometry; (ii) contouring, filtering, erosion, dilation in image processing; (iii) proximity graphs, spanning trees, triangulations in graph theory; (iv) p-adic arithmetic circuits in formal arithmetics; (v) universal logical circuits as well as storage modification machines (Kolmogorov and Schoenhage machines) in logics. The *Physarum* Chip is a network of processing elements consisting of protoplasmic tubes (i) coated with conductive substances as fast information channels and (ii) acting as a decentralized non-linear transducer of information. It has different parallel inputs (optical, chemo- and electro-based) and different outputs (reactions). It is worth noting that *Physarum* Chip is a first step in bio-physical modeling future nano-chips based on biomorphic mineralisation.

So, in the *PhyChip project* we have obtained a mathematics that is suitable for programming the behaviour of plasmodia (my part of responsibility in the project). Now we can extend this mathematics to a general theory of AAFN as a theory of massive-parallel behaviour of appearing and disappearing agents with a permanent changing of topology of their links. The innovation idea proposed in the new project will concern programming Big Data structures and representing them as a calculation process. Actin filaments will be a bio-inspired model of Big Data structure. Our results will be summarized as a new game theory suitable for modelling communication networks of millions mobile agents. These results can be used in market analysis, in Web mining and so on. Usually, if someone deals with Big Data, (s)he appeals to some statistic tools such as random graphs. We are going to propose some logical-algebraic tools for analysis Big Data within the AAFN. The main advantage of such tools is that they can be applied in programming Big Data.

13.2.3 Concreteness and Pertinence of the Objective

In the *PhyChip project* we have found out that the plasmodium behaviour has the same algorithms of attracting and repelling as any swarm behaviour. For example, a prokaryote, a one-cell organism that lacks a membrane-bound nucleus (karyon), can build colonies and transmit attracting and repelling signals within the *quorum sensing* to react collectively. Human beings, such as pedestrians, can also exhibit a swarm behaviour like a flocking or schooling: we prefer to avoid a person conditionally designated by us as a possible predator and if a substantial part of the group (not less than 5%) changes the direction, then the rest follows the new direction.

It is known that any swarm can be controlled by replacing stimuli: attractants and repellents what we have studied in the *PhyChip project*, therefore we can design logic circuits based on topology of stimuli. In one of our preliminary researches we have found out that alcohol-dependent humans embody a version of swarm intelligence, also, to optimize the alcohol-drinking behaviour (these results were obtained jointly with the psychotherapist Vadim Fris on the basis of the Health Centre "Iscelenie" in Minsk, we have discovered 107 addicted people).

Thus, we have seen that the model of intelligent behaviour in reacting to external stimuli (attractants and repellents) and embodied in any swarm can be simulated by the plasmodia of *Physarum polycephalum*, the unicellular organism. It means that the subcellular mechanism of assembling and disassembling actin filament waves in reacting to attractants and repellents is enough for studying the intelligent reactions to external stimuli mathematically and logically. This mechanism is repeated then at the level of biofilms, many-cellular organisms, ant nests, swarms of naked mole-rats, swarms of humans under the conditions of escape panic, etc. For different instances of swarm behaviour, we detect different attractants and repellents chemically and biologically, but their mathematical and logical mechanism remains unchanged. So, it is enough for us to construct the AAFN to know how to design protein robots.

Furthermore, at the level of actin filaments we deal with the following two main mechanisms of swarm intelligence in rout optimizations: (i) lateral activation (LA) as structuring a network so that some units activate their neighbours to decrease their own responses and (ii) lateral inhibition (LI) as structuring a network so that some units inhibit their neighbours in proportion to their own excitation. Thus, the bio-chemical model of actin filament networks is fundamental for all swarms: from bacteria to humans, because LA and LI in responses to stimuli (it is the main way to optimize logistics) are incorporated at the subcellular level of actin filaments.

In machine learning there have been used some biologically inspired networks such as artificial neural networks, where we have a fixed number of processors involved into computations. So, in neural networks we deal with a system $\langle N, V, w \rangle$, where N is the set of processors called "neurons", V is a set of tuples $\{(i, j) : i, j \in N\}$ whose elements are connections between neuron i and neuron j, and w is a function from V to $\mathbf{R}$ such that $w((i, j))$, for short $w_{i,j}$, is called the weight of the connection between neuron i and neuron j. Nevertheless, elementary computational units (processors) understood in real biological networks cannot be fixed. They are being built and rebuilt continuously in responses to external stimuli. One of the examples of these units is given by actin filaments used in remodelling cell configurations and in the cell motility.

In the research we are going to propose a mathematical theory of AAFN. This theory can be applied everywhere where we deal with swarm intelligence. In a sense, it will be a new approach to supercomputing. Traditional ways of communication networks do not concern a possibility of interchanging of communication facility among mobile agents, with a necessity of urgent reorganization of topology. In our research we are going to propose a general mathematical tool for exposition of communication networks organized with orientation to hierarchy and with application of such a tool to concrete self-organizing systems in the Web.

First of all, we are planning to formalize the processes of polymerization and depolymerisation of actin filaments to control these reactions in building and rebuilding the AAFN. Then we are going to design the AAFN as a mathematical theory with some applications in designing logic circuits on the basis of proteins. Then we are going to consider the AAFN as an extension of random graphs theory for modelling Web data to obtain new tools of Web mining for the Internet monitoring with Big Data. And at the end, we are going to apply the AAFN in designing an e-lawyer with formalization of concurrent and massive-parallel court processes for supporting court decisions. All these results can become impacts in robotics and e-justice.

To sum up, we are going to consider all the possible states in the cell deformation as a form of signal transmission by filaments. Each deformation state is caused by an appropriate attractant or repellent. In turn, the mechanical signal generated within the AAFN is transmitted in different ways according to a kind of the cell deformation. So, the cell can be regarded as a reversible logic gate which has the same number of inputs (mechanical stresses implied by attracting or repelling stimuli) and outputs (stress transmissions by deformation). The form of the cell deformation is a form of signal transmission. And all these transitions can be coded by p-adic arithmetic functions.

13.2.4 Expected Impacts

In the *PhyChip project* for the first time the entire biosystem has been integrated into the electronics and information treating systems. The organism of *Physarum polycephalum* has been converted into an architecture called *Physarum* chip and consisting of bio-electronic compounds. In continuation of this project we are going to consider the treatment of information according to a combination of external and internal conditions at the subcellular level, namely we are going to regard actin filaments as an intercellular mechanism of calculation. We expect a realization of calculations, based on the polymerization and depolymerisation of actin filaments. We suppose that it is possible to create an artificial protein broth (i.e. a protein robot) which may possess an intelligence ability in the meaning of solving the complex of various tasks, such as learning, orientation in space, decision making, optimization in transporting of foods, etc. This artificial protein broth will consist of actin filaments controlled by us. In the project we propose AAFN which are grounded on a logical-mathematical formalization of chemical reactions, responsible for controlling polymerization and depolymerisation of actin filaments. This research will be a theoretical part in designing a prototype of protein robot in future.

We are going to apply the AAFN to the information processing under conditions of Big Data. The AAFN we are going to obtain can be used, first, for exposition of the communication networks organized among mobile agents, communicated in large groups which can joint higher-order groups, second, for designing such communication networks, with emphasis on ever-ending modifications of links among driving agents and under conditions of strong noise. Hence, this theory can be used in signal processing, controlling processes of Big Data, navigation, vision recognition, time series analysis, stock market prediction, etc. In this respect, this will be a first-time realization of bio-based communication networks for Big Data. The

AAFN are a massive parallel communicative structure where old agents disappear and new agents appear. Growing networks of actin filaments will be used for information transmission in terms of potential impact of neuro-morphic processors. Such processors will have a look of artificial protein broth. This will change of interface to optical, chemical, and mechanical at once.

The AAFN will be a mathematical model which can be used more widely than artificial neural networks and more effectively. We are going to present the AAFN as a new game theory—the so-called massive-parallel game theory, where there are involved millions players with hidden strategies. This game theory we are planning to build will be used by us for new tools of Web mining—monitoring Web sources to protect before information warfare in public sphere, and for designing a model of e-justice. Thus, the realization of AAFN provides the society with new horizons in robotics (possibility to design robots as protein broths) and in e-democracy. The *PhyChip project* has allowed us to design communication networks of plasmodia as a calculation system (the so-called *Physarum* Chip). Now it is possible to improve this result and to obtain the AAFN—a general mathematical theory of biological reactions to external and internal stimuli at the subcellular and intercellular level. This theory will be a game theory of massive-parallel behaviours of millions mobile agents.

First of all, it will be possible to design robots consisting of artificial proteins. Then it will be possible to use bio-inspired mathematical tools in designing massive-parallel communication networks of millions mobile agents. As a societal application of AAFN, we are going to propose a model of e-justice and a model of Web monitoring. In e-justice we are going to propose a system of electronic trial processes, where there is a game-theoretic formalization of court processes that allows to program decision making in courts (it is really important for the further development and realization of e-government—digital interacting between government and citizens, government and firms, government and employees, and also between government and governments). Notice that e-justice is an integral part of e-democracy. E-justice is the higher form of justice, because it protects a transparency and logicality of court judgments. Within the AAFN we are going to design a version of e-lawyer with a possibility to make court processes electronically with a mechanization of decisions on the basis of documents, data, and existing laws. Also, we are going to apply our game-theoretic model for monitoring Web sources to protect before information warfare in public sphere. Reflexive management tools which can be proposed by us can support a more efficient and effective government management within the so-called e-government, to transform relations with citizens, businesses, and other arms of government, aiming at less corruption, increased transparency, greater convenience, and cost reductions. Reflexive management in public communication can allow us to find new forms of e-democracy. In case a reflexive game is carried out when we have disclosed all the hidden agents (corporations, companies, institutes), success in a reflexive game can be reached by a purposeful modification of some components of a controlled system.

The impact of the project results on the development of the research field and scientific discipline is divided into the following groups: (i) new methods of Web data mining and Big Data analysis for automatic monitoring of public sources concerning national interests; (ii) new methods of reflexive management in information warfare in public sphere, (iii) new methods of reflexive management in e-democracy. The main outcomes of our project: a real-time process control for analytics in the public sphere, scaling up analytics to gain insights from Big Data, it is possible to design a platform for the automatic analysis of the incoming data in real time to make correlations for grouping data and to produce insights and conclusions.

References

1. Adamatzky, A., Erokhin, V., Grube, M., Schubert, T., Schumann, A.: Physarum chip project: growing computers from slime mould. Int. J. Unconv. Comput. **8**(4), 319–323 (2012)
2. Adamatzky, A., Jones, J.: On using compressibility to detect when slime mould completed computation. Complexity **21**(5), 162–175 (2016)
3. Schumann, A.: p-Adic valued models of swarm behaviour. AIP Conf. Proc. **1863**(1), 360,009 (2017). https://doi.org/10.1063/1.4992538
4. Schumann, A.: Group theory and p-adic valued models of swarm behaviour. Math. Method Appl. Sci. https://doi.org/10.1002/mma.4540
5. Schumann, A., Pancerz, K., Adamatzky, A., Grube, M.: Bio-inspired game theory: the case of Physarum polycephalum. In: Suzuki, J., Nakano (eds.) Proceedings of the 8th International Conference on Bio-inspired Information and Communications Technologies (BICT'2014), pp. 9–16. Boston, Massachusetts, USA (2014)
6. Schumann, A., Pancerz, K., Adamatzky, A., Grube, M.: Context-based games and Physarum polycephalum as simulation model. In: Proceedings of the Workshop on Unconventional Computation in Europe. London, UK (2014)
7. Schumann, A.: Reversible logic gates on Physarum polycephalum. AIP Conf. Proc. **1648**(1), 580,011 (2015). https://doi.org/10.1063/1.4912819
8. Schumann, A.: Conventional and unconventional reversible logic gates on Physarum polycephalum. Int. J. Parallel Emerg. Distrib. Syst. **32**(2), 218–231 (2017). https://doi.org/10.1080/17445760.2015.1068775
9. Schumann, A.: Non-linear permutation groups on Physarum polycephalum. In: The 2014 2nd International Conference on Systems and Informatics (ICSAI 2014), pp. 246–251. Shanghai (2014)
10. Schumann, A., Pancerz, K.: Towards an object-oriented programming language for Physarum polycephalum computing: a Petri net model approach. Fundam. Inform. **133**(2–3), 271–285 (2014)
11. Schumann, A., Pancerz, K.: Petri net models of simple rule-based systems for programming Physarum machines: extended abstract. In: Z. Suraj, L. Czaja (eds.) Proceedings of the 24th International Workshop on Concurrency, Specification and Programming (CS&P'2015), vol. 2, pp. 155–160. Rzeszow, Poland (2015)
12. Schumann, A.: Toward a computational model of actin filament networks. In: Proceedings of the 9th International Joint Conference on Biomedical Engineering Systems and Technologies (BIOSTEC 2016)—Volume 4: BIOSIGNALS, Rome, Italy, 21–23 February, 2016, pp. 290–297 (2016). https://doi.org/10.5220/0005828902900297
13. Schumann, A.: Decidable and undecidable arithmetic functions in actin filament networks. J. Phys. D Appl. Phys. (2017). https://doi.org/10.1088/1361-6463/aa9d7b
14. Schumann, A.: On arithmetic functions in actin filament networks. In: 10th EAI International Conference on Bio-inspired Information and Communications Technologies (formerly BIONETICS). ACM (2017). https://doi.org/10.4108/eai.22-3-2017.152402

15. Schumann, A.: p-Adic valued fuzzyness and experiments with Physarum polycephalum. In: 11th International Conference on Fuzzy Systems and Knowledge Discovery (FSKD), pp. 466–472. IEEE Xplore (2014)
16. Schumann, A., Adamatzky, A.: The double-slit experiment with Physarum polycephalum and p-adic valued probabilities and fuzziness. Int. J. General Syst. **44**(3), 392–408 (2015)
17. Schumann, A.: Syllogistic versions of go games on Physarum. In: Adamatzky, A. (ed.) Advances in Physarum Machines: Sensing and Computing with Slime Mould, pp. 651–685. Springer International Publishing, Cham (2016)
18. Schumann, A.: Towards context-based concurrent formal theories. Parallel Process. Lett. **25**, 1540,008 (2015)
19. Schumann, A.: p-Adic valued logical calculi in simulations of the slime mould behaviour. J. Appl. Non-Class. Log. **25**(2), 125–139 (2015). https://doi.org/10.1080/11663081.2015.1049099
20. Schumann, A.: Non-archimedean fuzzy and probability logic. J. Appl. Non-Class. Log. **18**(1), 29–48 (2008)
21. Schumann, A.: Non-archimedean valued extension of logic $l\pi$ and p-adic valued extension of logic bl. J. Uncertain Syst. **4**(2), 99–115 (2010)
22. Schumann, A., Pancerz, K.: p-Adic computation with Physarum. In: A. Adamatzky (ed.) Advances in Physarum Machines: Sensing and Computing with Slime Mould. Emergence, Complexity and Computation, vol. 21, pp. 619–649. Springer International Publishing (2016)
23. Schumann, A., Adamatzky, A.: Physarum spatial logic. New Math. Nat. Comput. **7**(3), 483–498 (2011)
24. Schumann, A., Akimova, L.: Simulating of schistosomatidae (trematoda: Digenea) behavior by Physarum spatial logic. In: Annals of Computer Science and Information Systems, Volume 1. Proceedings of the 2013 Federated Conference on Computer Science and Information Systems, pp. 225–230. IEEE Xplore (2013)
25. Schumann, A., Akimova, L.: Process calculus and illocutionary logic for analyzing the behavior of schistosomatidae (trematoda: Digenea). In: Pancerz, K., Zaitseva, E. (eds.) Computational Intelligence, Medicine and Biology: Selected Links, pp. 81–101. Springer International Publishing, Cham (2015)
26. Schumann, A.: Towards slime mould based computer. New Math. Nat. Comput. **12**(2), 97–111 (2016). https://doi.org/10.1142/S1793005716500083
27. Pancerz, K., Schumann, A.: Rough set models of Physarum machines. Int. J. General Syst. **44**(3), 314–325 (2015)
28. Schumann, A.: Physarum syllogistic L-systems. In: FUTURE COMPUTING 2014, The Sixth International Conference on Future Computational Technologies and Applications. ThinkMind (2014)
29. Schumann, A., Adamatzky, A.: Physarum polycephalum diagrams for syllogistic systems. IfCoLog J. Log. Appl. **2**(1), 35–68 (2015)
30. Schumann, A., Akimova, L.: Syllogistic system for the propagation of parasites. The case of schistosomatidae (trematoda: Digenea). Stud. Log. Gramm. Rhetor. **40**(53), 303–319 (2015)
31. Adamatzky, A.: Physarum machine: implementation of a Kolmogorov-Uspensky machine on a biological substrate. Parallel Process. Lett. **17**(4), 455–467 (2007)
32. Abramsky, S., Mellies, P.A.: Concurrent games and full completeness. In: Proceedings of the 14th Symposium on Logic in Computer Science, pp. 431–442 (1999)
33. Schumann, A., Fris, V.: Swarm intelligence among humans—the case of alcoholics. In: Proceedings of the 10th International Joint Conference on Biomedical Engineering Systems and Technologies—Volume 4: BIOSIGNALS, (BIOSTEC 2017), pp. 17–25. ScitePress (2017)
34. Schumann, A., Woleński, J.: Two squares of oppositions and their applications in pairwise comparisons analysis. Fundam. Inform. **144**(3–4), 241–254 (2016)
35. Pancerz, K., Schumann, A.: Rough set description of strategy games on Physarum machines. In: A. Adamatzky (ed.) Advances in Unconventional Computing, Volume 2: Prototypes, Models and Algorithms. Emergence, Complexity and Computation, vol. 23, pp. 615–636. Springer International Publishing (2017)

36. Schumann, A., Pancerz: Physarumsoft—a software tool for programming Physarum machines and simulating Physarum games. In: M. Ganzha, L. Maciaszek, M. Paprzycki (eds.) Proceedings of the 2015 Federated Conference on Computer Science and Information Systems (FedCSIS'2015), pp. 607–614. Lodz, Poland (2015)
37. Schumann, A., Pancerz, K.: Interfaces in a game-theoretic setting for controlling the plasmodium motions. In: Proceedings of the 8th International Conference on Bio-inspired Systems and Signal Processing (BIOSIGNALS'2015), pp. 338–343. Lisbon, Portugal (2015)
38. Schumann, A.: Go games on plasmodia of Physarum polycephalum. In: 2015 Federated Conference on Computer Science and Information Systems, FedCSIS 2015, Lódz, Poland, 13–16 Sept, 2015, pp. 615–626 (2015). https://doi.org/10.15439/2015F236
39. Schumann, A., Pancerz, K.: A rough set version of the go game on Physarum machines. In: J. Suzuki, T. Nakano, H. Hess (eds.) Proceedings of the 9th International Conference on Bio-inspired Information and Communications Technologies (BICT'2015), pp. 446–452. New York City, New York, USA (2015)
40. Schumann, A.: Payoff cellular automata and reflexive games. J. Cell. Autom. **9**(4), 287–313 (2014)
41. Schumann, A.: Probabilities on streams and reflexive games. Oper. Res. Decis. **24**(1), 71–96 (2014). https://doi.org/10.5277/ord140105
42. Schumann, A.: Reflexive games and non-archimedean probabilities. P-Adic Number Ultrametr. Anal. Appl. **6**(1), 66–79 (2014)
43. Schumann, A.: Reflexive games in e-university. In: Proceedings of the Science and Information Conference (SAI), Science and Information Organization, London, England, 28–30 July, 2015, Sponsor(s): Nvidia; IEEE; The Future & Emerging Technol FET at the European Comm; EUREKA; Cambridge Wireless; British Comp Soc; Digital Catapult; Springer; Media Partner for this conference, 2015 Science and Information Conference (SAI), pp. 770–775 (2015)
44. Howard, N.: General metagames: an extension of the metagame concept. In: Game Theory as a Theory of Conflict Resolution. Reidel, Dordrecht (1974)
45. Lefebvre, V.A.: Lectures on Reflexive Game Theory. Leaf & Oaks, Los Angeles (2010)
46. Searle, J.R.: Speech Acts; An Essay in the Philosophy of Language. Cambridge University Press, Cambridge (1969)
47. Searle, J.R., Vanderveken, D.: Foundations of Illocutionary Logic. Cambridge University Press, Cambridge (1984)
48. Fechner, G.T.: Elements of psychophysics, vol. 1. Holt, Rinehart and Winston, New York (1966)
49. Semertzidou, C., Dourvas, N.I., Tsompanas, M.I., Adamatzky, A., Sirakoulis, G.C.: Introducing chemotaxis to a mobile robot. In: Proceedings of 12th IFIP WG 12.5 International Conference and Workshops on Artificial Intelligence Applications and Innovations, AIAI 2016, Thessaloniki, Greece, 16–18 Sept, 2016, pp. 396–404 (2016)
50. Alonso-Sanz, R., Adamatzky, A.: Actin automata with memory. Int. J. Bifurc. Chaos **26**(1) (2016)
51. Siccardi, S., Adamatzky, A.: Actin quantum automata: communication and computation in molecular networks. Nano Commun. Netw. **6**(1), 15–27 (2015)
52. Siccardi, S., Adamatzky, A.: Logical gates implemented by solitons at the junctions between one-dimensional lattices. Int. J. Bifurc. Chaos **26**(6), 1650,107 (2016)
53. Siccardi, S., Tuszynski, J.A., Adamatzky, A.: Boolean gates on actin filaments. Phys. Lett. A **380**(1), 88–97 (2016)
54. Mayne, R., Adamatzky, A.: Slime mould foraging behaviour as optically coupled logical operations. Int. J. General Syst. **44**(3), 305–313 (2015). https://doi.org/10.1080/03081079.2014.997528
55. Shirakawa, T., Sato, H., Ishiguro, S.: Constrcution of living cellular automata using the Physarum polycephalum. Int. J. General Syst. **44**, 292–304 (2015)
56. Shirakawa, T., Yokoyama, K., Yamachiyo, M., Y-p, G., Miyake, Y.: Multi-scaled adaptability in motility and pattern formation of the Physarum plasmodium **4**, 131–138 (2012)

Index

A

A. Schumann, *Behaviourism in Studying Swarms: Logical Models of Sensing and Motoring*, Emergence, Complexity and Computation 33,
https://doi.org/10.1007/978-3-319-91542-5

Q

Printed in the United States
By Bookmasters